The World in Eleven Dimensions: Supergravity, Supermembranes and M-theory

The World in Eleven Dimensions
Supergravity, supermembranes and M-theory

M J Duff, Texas A & M University, USA

A unified theory embracing all physical phenomena is a major goal of theoretical physics. In the early 1980s, many physicists looked to eleven-dimensional supergravity in the hope that it might provide that elusive superunified theory. In 1984 supergravity was knocked off its pedestal by ten-dimensional superstrings, one-dimensional objects whose vibrational modes represent the elementary particles. Superstrings provided a perturbative finite theory of gravity which, after compactification to four spacetime dimensions, seemed in principle capable of explaining the Standard Model. Despite these major successes, however, nagging doubts persisted about superstrings.

Then in 1987 and 1992 respectively the elementary supermembrane and its dual partner, the solitonic superfivebrane were discovered. These are supersymmetric extended objects with respectively two and five dimensions moving in an eleven-dimensional spacetime.

Over the period since 1996, perturbative superstrings have been superseded by a new non-perturbative called M-theory which describes, amongst other things, supermembranes and superfivebranes, which subsumes string theories, and which has as its low-energy limit, eleven-dimensional supergravity! M-theory represents the most exciting development in the subject since 1984 when the superstring revolution first burst on the scene.

This book brings together seminal papers that have shaped our current understanding of this eleven-dimensional world: from supergravity through supermembranes to M-theory. Included at the beginnings of the six chapters are commentaries intended to explain the importance of these papers and to place them in a wider perspective. Each chapter also has an extensive bibliography.

This is the first book devoted to M-theory, and will be of great interest to researchers and postgraduate students in particle physics, mathematical physics, gravitation and cosmology.

November 1999 c500 pages illustrated
hardcover 0 7503 0671 8 **c£80.00/cUS$150.00**
paperback 0 7503 0672 6 **c£30.00/cUS$55.00**

Studies in High Energy Physics, Cosmology and Gravitation

Other books in the series

Electron–Positron Physics at the Z
M G Green, S L Lloyd, P N Ratoff and D R Ward

Non-accelerator Particle Physics
Paperback edition
H V Klapdor-Kleingrothaus and A Staudt

Ideas and Methods of Supersymmetry and Supergravity
***or* A Walk Through Superspace**
Revised edition
I L Buchbinder and S M Kuzenko

Pulsars as Astrophysical Laboratories for Nuclear and Particle Physics
F Weber

Classical and Quantum Black Holes
Edited by P Fré, V Gorini, G Magli and U Moschella

The World in Eleven Dimensions: Supergravity, Supermembranes and M-theory

Edited by

M J Duff

University of Michigan

'Nature requires five,
Custom allows seven,
Idleness takes nine,
And wickedness eleven.'

Mother Goose

Institute of Physics Publishing
Bristol and Philadelphia

British Library Cataloguing-in-Publication Data

A catalogue record for this book is available from the British Library.

ISBN 0 7503 0671 8 (hbk)
0 7503 0672 6 (pbk)

Library of Congress Cataloging-in-Publication Data are available

Commissioning Editor: Jim Revill
Editorial Assistant: Victoria Le Billon
Production Editor: Simon Laurenson
Production Control: Sarah Plenty and Jenny Troyano
Cover Design: Jeremy Stephens
Marketing Executive: Colin Fenton

Published by Institute of Physics Publishing, wholly owned by The Institute of Physics, London

Institute of Physics Publishing, Dirac House, Temple Back, Bristol BS1 6BE, UK

US Office: Institute of Physics Publishing, The Public Ledger Building, Suite 1035, 150 South Independence Mall West, Philadelphia, PA 19106, USA

Typeset in LaTeX using the IOP Bookmaker Macros
Printed in the UK by J W Arrowsmith Ltd, Bristol

Contents

Preface **ix**

1 **Eleven-dimensional supergravity** **1**

Introduction 1

W Nahm 1978 Supersymmetries and their representations *Nucl. Phys.* B **135** 149–66 7

E Cremmer, B Julia and J Scherk 1978 Supergravity theory in eleven-dimensions *Phys. Lett.* B **76** 409–12 25

E Witten 1981 Search for a realistic Kaluza–Klein theory *Nucl. Phys.* B **186** 412–28 29

M A Awada, M J Duff and C N Pope 1983 $N = 8$ supergravity breaks down to $N = 1$ *Phys. Rev. Lett.* **50** 294–7 46

M J Duff, B E W Nilsson and C N Pope 1983 Compactification of $D = 11$ supergravity on $K3 \times T^3$ *Phys. Lett.* B **129** 39–42 50

C M Hull 1984 Exact *pp*-wave solutions of 11-dimensional supergravity *Phys. Lett.* B **139** 39–41 54

S K Han and I G Koh 1985 $N = 4$ remaining supersymmetry in Kaluza–Klein monopole background in $D = 11$ supergravity *Phys. Rev.* D **31** 2503–6 57

2 **The eleven-dimensional supermembrane** **61**

Introduction 61

E Bergshoeff, E Sezgin and P K Townsend 1987 Supermembranes and eleven-dimensional supergravity *Phys. Lett.* B **189** 75–8 69

B de Wit, J Hoppe and H Nicolai 1988 On the quantum mechanics of supermembranes *Nucl. Phys.* B **305** 545–81 73

M J Duff and K Stelle 1991 Multimembrane solutions of $d = 11$ supergravity *Phys. Lett.* B **253** 113-18 110

A Strominger 1996 Open *p*-branes *Phys. Lett.* B **383** 44–7, hep-th/9512059 116

P K Townsend 1996 *D*-branes from *M*-branes *Phys. Lett.* B **373** 68–75, hep-th/9512062 120

3 **The eleven-dimensional superfivebrane** **129**

Introduction 129

Preface

A vital ingredient in the quest for a unified theory embracing all physical phenomena is *supersymmetry*, a symmetry which (a) unites bosons and fermions, (b) requires the existence of gravity and (c) places an upper limit of *eleven* on the dimension of spacetime. For these reasons, in the early 1980's, many physicists looked to eleven-dimensional supergravity in the hope that it might provide that elusive super-unified theory. Then, in 1984, superunification underwent a major paradigm shift: eleven-dimensional supergravity was knocked off its pedestal by ten-dimensional superstrings, one-dimensional objects whose vibrational modes represent the elementary particles. Unlike eleven-dimensional supergravity, superstrings provided a perturbatively finite theory of gravity which, after compactification to four spacetime dimensions, seemed in principle capable of explaining the Standard Model of the strong, weak and electromagnetic forces including the required *chiral* representations of quarks and leptons.

Despite these major successes, however, nagging doubts persisted about superstrings. First, many of the most important questions in string theory—How do strings break supersymmetry? How do they choose the right vacuum state? How do they explain the smallness of the cosmological constant? How do they resolve the apparent paradoxes of quantum black holes?—seemed incapable of being answered within the framework of a weak-coupling perturbation expansion. They seemed to call for some new, *non-perturbative*, physics. Second, why did there appear to be *five* different mathematically consistent superstring theories: the $E_8 \times E_8$ heterotic string, the $SO(32)$ heterotic string, the $SO(32)$ Type I string, the Type IIA and Type IIB strings? If one is looking for a unique *Theory of Everything*, this seems like an embarrassment of riches! Third, if supersymmetry permits eleven dimensions, why do superstrings stop at ten? This question became more acute with the discoveries of the elementary *supermembrane* in 1987 and its dual partner, the solitonic *superfivebrane*, in 1992. These are supersymmetric extended objects with respectively two and five dimensions moving in an eleven-dimensional spacetime. Finally, therefore, if we are going to generalize zero-dimensional point particles to one-dimensional strings, why stop there? Why not two-dimensional membranes or more generally p-dimensional objects (inevitably dubbed *p-branes*)? Although this latter possibility was actively pursued by a small but dedicated group of theorists, starting around 1986, it was largely ignored by the mainstream physics community.

The year 1995 witnessed a new paradigm shift: perturbative ten-dimensional superstrings were in their turn superseded by a new *non-perturbative* theory called *M-theory* which describes, amongst other things, supermembranes and superfivebranes, which subsumes the above five consistent string theories, and which has, as its low-energy limit, eleven-dimensional supergravity! According to Fields Medalist Edward Witten, 'M stands for magical, mystery or membrane, according to taste'. New evidence in favour of this theory is appearing daily on the Internet and represents the most exciting development in the subject since 1984 when the superstring revolution first burst on the scene. These new insights hold promise of a deeper understanding of the Standard Model of particle physics, of the unification of the four fundamental forces, of the quantum theory of gravity, of the mysteries of black holes, of big-bang cosmology and, ultimately, of their complete synthesis in a final theory of physics.

The first purpose of this volume is to bring together the seminal papers that have shaped our current understanding of this eleven-dimensional world: from supergravity through supermembranes to M-theory. Second, I have included at the beginning of each of the six chapters a commentary intended to explain the importance of these papers and to place them in a wider perspective. Each chapter also has an extensive bibliography. For reasons of space, I have limited to 33 this selection of important papers on eleven dimensions: a daunting task. This has meant omitting long review articles, and also significant papers on string theory dualities, membrane theory, D-branes and F-theory which, though important for our present state of knowledge, did not have eleven dimensions as their primary theme. I have tried to combine originality and topicality by including not only the well-cited classic papers but also some very recent works which, in the editor's judgement, will prove to be influential with the passage of time.

M-theory has sometimes been called the *Second Superstring Revolution*, but I feel this is really a misnomer. It certainly involves new ideas every bit as significant as those of the 1984 string revolution, but its reliance on supermembranes and eleven dimensions renders it sufficiently different from traditional string theory to warrant its own name. One cannot deny the tremendous historical influence of the last decade of superstrings on our current perspectives. Indeed, it is the pillar upon which our belief in a quantum-consistent M-theory rests. In the editor's opinion, however, the focus on the perturbative aspects of one-dimensional objects moving in a ten-dimensional spacetime that prevailed during this period will ultimately be seen to be a small corner of M-theory. Whatever the fate of the world in eleven dimensions, I hope this volume will help chart its course.

In making my, sometimes treacherous, way through the world in eleven dimensions over the last two decades I have been guided by many colleagues. I owe a particular debt to Eric Bergshoeff, Leonardo Castellani, Stanley Deser, Ricardo D'Auria, Sergio Ferrara, Pietro Fré, Gary Gibbons, Chris Hull, Paul Howe, Takeo Inami, Ramzi Khuri, Jim Liu, Hong Lu, Jianxin Lu, Ruben Minasian, Bengt Nilsson, Chris Pope, Joachim Rahmfeld, the late Abdus Salam, Ergin Sezgin, Kelly Stelle, Paul Townsend, Peter van Nieuwenhuizen, Steven Weinberg, Nick Warner, Peter West and Edward Witten. I am also grateful to all the authors who kindly gave their permission to reproduce the papers. Special thanks are due to Hisham

Sati for help in preparing the manuscript. Finally, I would like to acknowledge my editor Jim Revill of Institute of Physics Publishing for his enthusiasm and advice.

Michael Duff
College Station, Texas, 1998

Chapter 1

Eleven-dimensional supergravity

Eleven is the maximum spacetime dimension in which one can formulate a consistent supersymmetric theory, as was first recognized by Nahm [1] in his classification of supersymmetry algebras. The easiest way to see this is to start in four dimensions and note that one supersymmetry relates states differing by one half unit of helicity. If we now make the reasonable assumption that there be no massless particles with spins greater than two, then we can allow up to a maximum of $N = 8$ supersymmetries taking us from helicity -2 through to helicity $+2$. Since the minimal supersymmetry generator is a Majorana spinor with four off-shell components, this means a total of 32 spinor components. Now in a spacetime with D dimensions and signature $(1, D-1)$, the maximum value of D admitting a 32 component spinor is $D = 11$. (Going to $D = 12$, for example, would require 64 components.) See table 1.1. Furthermore, as we shall see in chapter 2, $D = 11$ emerges naturally as the maximum dimension admitting supersymmetric extended objects, without the need for any assumptions about higher spin. Not long after Nahm's paper, Cremmer, Julia and Scherk [2] realized that supergravity not only permits up to seven extra dimensions but in fact takes its simplest and most elegant form when written in its full eleven-dimensional glory. The unique $D = 11, N = 1$ supermultiplet is comprised of a graviton g_{MN}, a gravitino ψ_M and 3-form gauge field C_{MNP} with 44, 128 and 84 physical degrees of freedom, respectively. The theory may also be formulated in superspace [3, 4]. Ironically, however, these extra dimensions were not at first taken seriously but rather regarded merely as a useful device for deriving supergravities in four dimensions. Indeed $D = 4, N = 8$ supergravity was first obtained by Cremmer and Julia [5] via the process of *dimensional reduction*, i.e. by requiring that all the fields of $D = 11, N = 1$ supergravity be independent of the extra seven coordinates.

In the early 1920's, in their attempts to unify Einstein's gravity and Maxwell's electromagnetism, Theodore Kaluza and Oskar Klein suggested that spacetime may have a hidden fifth dimension. This idea was quite successful: Einstein's equations in five dimensions not only yield the right equations for gravity in four dimensions but Maxwell's equations come for free. Conservation of electric charge is just conservation of momentum in the fifth direction. By taking this fifth dimension to have the topology of a circle, moreover, the quantization of electric charge would then

Dimension (D or d)	**Minimal spinor** (M or m)	**Supersymmetry** (N or n)
11	32	1
10	16	2, 1
9	16	2, 1
8	16	2, 1
7	16	2, 1
6	8	4, 3, 2, 1
5	8	4, 3, 2, 1
4	4	8, ..., 1
3	2	16, ..., 1
2	1	32, ..., 1

Table 1.1. Minimal spinor components and supersymmetries. Upper and lower case refer to spacetime and worldvolume quantities, respectively

be automatic: the gauge group is $U(1)$. To get the right value for the charge on the electron, however, the radius of the circle would have to be tiny, $R \sim 10^{-35}$ metres, which satisfactorily explains why we are unaware of its existence in our everyday lives[1]. For many years this Kaluza–Klein idea of taking extra dimensions seriously was largely forgotten, but the arrival of eleven-dimensional supergravity provided the missing impetus. The kind of four-dimensional world we end up with depends on how we *compactify* these extra dimensions: maybe seven of them would allow us to give a gravitational origin, *à la* Kaluza–Klein, to the strong and weak forces as well as the electromagnetic. In a very influential paper, Witten [15] drew attention to the fact that in such a scheme the four-dimensional gauge group is determined by the *isometry* group of the compact manifold $\mathcal{K}$. Moreover, he proved (what to this day seems to be merely a gigantic coincidence) that seven is not only the maximum dimension of $\mathcal{K}$ permitted by supersymmetry but the minimum needed for the isometry group to coincide with the standard model gauge group $SU(3) \times SU(2) \times U(1)$.

Round about this time there was great interest in N-extended supergravities for which the global $SO(N)$ is promoted to a gauge symmetry, in particular the maximal $N = 8$, $SO(8)$ theory of De Wit and Nicolai [16]. In these theories the underlying symmetry is described by the $D = 4$ anti-de Sitter (AdS_4) supersymmetry algebra, and the Lagrangian has a non-vanishing cosmological constant proportional to the square of the gauge coupling constant. This suggested that these theories might admit a Kaluza–Klein interpretation, and indeed this maximal theory was seen to correspond to the massless sector of $D = 11$ supergravity compactified on an S^7 whose metric admits an $SO(8)$ isometry [17]. An important ingredient in these developments that had been insufficiently emphasized in earlier work on Kaluza–

[1] A variation on the Kaluza–Klein theme is that our universe is a 3-brane embedded in a higher dimensional spacetime [6, 7]. This is particularly compelling in the context of the Type IIB threebrane [8] since the worldvolume fields necessarily include gauge fields [9]. Thus the strong, weak and electromagnetic forces might be confined to the worldvolume of the brane while gravity propagates in the bulk. It has recently been suggested that, in such schemes, the extra dimensions might be much larger than 10^{-35} metres [10, 11] and may even be a large as a millimetre [12–14].

Klein theory was that the $AdS_4 \times S^7$ geometry was not fed in by hand but resulted from a *spontaneous compactification*, i.e. the vacuum state was obtained by finding a stable solution of the higher-dimensional field equations [18]. The mechanism of spontaneous compactification appropriate to the $AdS_4 \times S^7$ solution of eleven-dimensional supergravity was provided by the Freund–Rubin mechanism [19] in which the 4-form field strength in spacetime $G_{\mu\nu\rho\sigma}$ ($\mu = 0, 1, 2, 3$) is proportional to the alternating symbol $\epsilon_{\mu\nu\rho\sigma}$ [20].

By deforming this geometry while keeping the same S^7 topology, one could find a new stable vacuum solution, the *squashed seven-sphere* with only $N = 1$ supersymmetry and $SO(5) \times SU(2)$ gauge symmetry [21]. Moreover, this admitted the four-dimensional interpretation of a Higgs mechanism in which some of the scalars acquired non-vanishing vacuum expectation values [22]. More general solutions were also found for which the internal components of the 3-form C_{mnp} ($m = 1, 2, ...7$) are non-vanishing and correspond to the parallelizing torsion on S^7 [23]. These also admit a $D = 4$ Higgs interpretation. Of course, there was still the problem of the huge cosmological constant of AdS_4 unless one could arrange to cancel it via fermion condensates [24]. A summary of this S^7 and other X^7 compactifications of $D = 11$ supergravity down to AdS_4 may be found in [25, 26]. By applying a similar mechanism to the 7-form dual of this field strength one could also find compactifications on $\mathrm{AdS}_7 \times S^4$ [28] whose massless sector describes gauged maximal $N = 4$, $SO(5)$ supergravity in $D = 7$ [29, 30]. Type IIB supergravity in $D = 10$, with its self-dual 5-form field strength, also admits a Freund–Rubin compactification on $\mathrm{AdS}_5 \times S^5$ [31–33] whose massless sector describes gauged maximal $N = 8$ supergravity in $D = 5$ [34, 35].

Compactification	**Supergroup**	**Bosonic subgroup**
$AdS_4 \times S^7$	$OSp(4\|8)$	$SO(3,2) \times SO(8)$
$AdS_5 \times S^5$	$SU(2,2\|4)$	$SO(4,2) \times SO(6)$
$AdS_7 \times S^4$	$OSp(6,2\|4)$	$SO(6,2) \times SO(5)$

Table 1.2. Compactifications and their symmetries.

In the three cases given above, the symmetry of the vacuum is described by the supergroups $OSp(4|8)$, $SU(2,2|4)$ and $OSp(6,2|4)$ for the S^7, S^5 and S^4 compactifications respectively, as shown in table 1.2. As discussed in chapters 4 and 6, these compactifications were later to prove crucial in the so-called AdS/CFT correspondence which relates supergravity theories in the bulk of AdS to conformal field theories on its boundary.

That the four-dimensional manifold $K3$ plays a ubiquitous role in much of present day M theory is also discussed in chapters 4 and 6. It was first introduced as a compactifying manifold in 1983 [27] when it was realized that the number of unbroken supersymmetries surviving compactification in a Kaluza–Klein theory depends on the *holonomy group* of the extra dimensions [21]. By virtue of its $SU(2)$ holonomy, $K3$ preserves precisely half of the supersymmetry. This means, in particular, that an $N = 2$ theory on $K3$ has the same number of supersymmetries as an $N = 1$ theory on T^4, a result which was subsequently to prove of vital importance for string/string duality, as we shall see in chapter 4. $K3$ also provided another novel phenomenon in Kaluza–Klein theory; the appearance of massless particles as a consequence of the *topology*, as opposed to the *geometry* of the compactifying manifold, determined by Betti numbers and index theorems [27]. It was thus the forerunner of the very influential *Calabi–Yau* compactifications of ten-dimensional supergravity and string theory [36].

The Kaluza–Klein approach to $D = 11$ supergravity eventually fell out of favour for two reasons. First, as emphasized by Witten [37], it is impossible to derive by the conventional Kaluza–Klein technique of compactifying on a manifold a *chiral theory* in four spacetime dimensions starting from a non-chiral theory such as eleven-dimensional supergravity. (Ironically, as discussed in chapter 6, Horava and Witten were to solve this problem years later by compactifying on something that is not a manifold!). Secondly, in spite of its maximal supersymmetry and other intriguing features, eleven dimensional supergravity was, after all, still a *field theory* of gravity with all the attendant problems of non-renormalizability. (For a recent proof of this, see [43].) The solution to this problem also had to await the dawn of M-theory.

Finally, we have included a paper by Hull [38] which displays a plane wave solution of $D = 11$ supergravity and one by Han and Koh [39] which displays a Kaluza–Klein monopole [40, 41, 42] solution of $D = 11$ supergravity. Both solutions are special because, in common with the the supermembrane of chapter 2 and the superfivebrane of chapter 3, they preserve half of the supersymmetry. As discussed in chapter 5, the eleven-dimensional plane-wave, supermembrane, superfivebrane and Kaluza–Klein monopole are the progenitors of the lower dimensional supersymmetric objects of M-theory.

References

[1] Nahm W 1978 Supersymmetries and their representations *Nucl. Phys.* B **135** 149

[2] Cremmer E, Julia B and Scherk J 1978 Supergravity theory in eleven-dimensions *Phys. Lett.* B **76** 409

[3] Brink L and Howe P S 1980 Eleven-dimensional supergravity on the mass-shell in superspace *Phys. Lett.* B **91** 384

[4] Cremmer E and Ferrara S 1980 Formulation of 11-dimensional supergravity in superspace *Phys. Lett.* B **91** 61

[5] Cremmer E and Julia B 1979 The $SO(8)$ supergravity *Nucl. Phys.* B **159** 141

[6] Rubakov V A and Shaposhnikov M E 1983 Do we live inside a domain wall? *Phys. Lett.* B **125** 136

[7] Gibbons G W and Wiltshire D L 1987 Space-time as a membrane in higher dimensions *Nucl. Phys.* B **287** 717

[8] Horowitz G T and Strominger A 1991 Black strings and p-branes *Nucl. Phys.* B **360** 197

[9] Duff M J and Lu J X 1991 The self-dual Type IIB superthreebrane *Phys. Lett.* B **273** 409

[10] Antoniadis I 1990 A possible new dimension at a few TeV *Phys. Lett.* B **246** 377

[11] Witten E 1996 Strong coupling expansion of Calabi–Yau compactification *Nucl.Phys.* B **471** 135
[12] Lykken J 1996 Weak scale superstrings *Phys. Rev.* D **54** 3693
[13] Arkani-Hamed N, Dimopoulos S and Dvali G 1998 The hierarchy problem and new dimensions at a millimeter *Phys. Lett.* B **429**
[14] Dienes K R, Dudas E and Gherghetta T 1998 Extra space-time dimensions and unification *Phys. Lett.* B **436** 55
[15] Witten E 1981 Search for a realistic Kaluza–Klein theory *Nucl. Phys.* B **186** 412–28
[16] De Wit B and Nicolai H 1982 $N = 8$ supergravity *Nucl. Phys.* B **208** 323
[17] Duff M J and Pope C N 1983 Kaluza–Klein supergravity and the seven sphere *Supersymmetry and Supergravity 82, Proc. Trieste Conf.* ed S Ferrara, J G Taylor and P van Nieuwenhuizen (Singapore: World Scientific) pp 183–228
[18] Cremmer E and Scherk J 1977 Spontaneous compactification of extra space dimensions *Nucl. Phys.* B **118** 61
[19] Freund P G O and Rubin M A 1980 Dynamics of dimensional reduction *Phys. Lett.* B **97** 233
[20] Duff M J and van Nieuwenhuizen P 1980 Quantum inequivalence of different field representations *Phys. Lett.* B **94** 179
[21] Awada M A, Duff M J and Pope C N 1983 $N = 8$ supergravity breaks down to $N = 1$ *Phys. Rev. Lett.* **50** 294
[22] Duff M J, Nilsson B E W and Pope C N 1983 Spontaneous supersymmetry breaking by the squashed seven-sphere *Phys. Rev. Lett.* **50** 2043
[23] Englert F 1982 Spontaneous compactification of eleven dimensional supergravity *Phys. Lett.* B **119** 339
[24] Duff M J and Orzalesi C 1983 The cosmological constant in spontaneously compactified $D = 11$ supergravity *Phys. Lett.* B **122** 37
[25] Duff M J, Nilsson B E W and Pope C N 1986 Kaluza–Klein supergravity *Phys. Rep.* **130** 1
[26] Castellani L, D'Auria R and Fre P 1991 *Supergravity and Superstrings: A Geometric Perspective* in 3 volumes (Singapore: World Scientific)
[27] Duff M J, Nilsson B E W and Pope C N 1983 Compactification of $D = 11$ supergravity on $K3 \times T^3$ *Phys. Lett.* B **129** 39
[28] Pilch K, Townsend P K and van Nieuwenhuizen P 1984 Compactification of $d = 11$ supergravity on S^4 (or $11 = 7 + 4$, too) *Nucl. Phys.* B **242** 377
[29] Pernici M, Pilch K and van Nieuwenhuizen P 1984 Gauged maximally extended supergravity in seven dimensions *Phys. Lett.* B **143** 103
[30] Townsend P K and van Nieuwenhuizen P 1983 Gauged seven-dimensional supergravity *Phys. Lett.* B **125** 41–6
[31] Schwarz J H 1983 Covariant field equations of chiral $N = 2$ $D = 10$ supergravity *Nucl. Phys.* B **226** 269
[32] Gunaydin M and Marcus N 1985 The spectrum of the S^5 compactification of the $N = 2$, $D = 10$ supergravity and the unitary supermultiplet *Class. Quantum Grav.* **2** L11
[33] Kim H J, Romans L J and van Nieuwenhuizen P 1985 Mass spectrum of chiral ten-dimensional $N = 2$ supergravity on S^5 *Phys. Rev.* D **32** 389
[34] Pernici M, Pilch K and van Nieuwenhuizen P 1985 Gauged $N = 8$, $d = 5$ supergravity *Nucl. Phys.* B **259** 460
[35] Gunaydin M, Romans L J and Warner N P 1985 Gauged $N = 8$ supergravity in five dimensions *Phys. Lett.* B **154** 268

[36] Candelas P, Horowitz G, Strominger A and Witten E 1985 Vacuum configurations for superstrings *Nucl. Phys.* B **258** 46

[37] Witten E 1985 Fermion quantum numbers in Kaluza–Klein theory *Shelter Island II, Proc. 1983 Shelter Island Conf. on Quantum Field Theory and the Fundamental Problems of Physics* ed R Jackiw, N Khuri, S Weinberg and E Witten (Cambridge, MA: MIT Press) p 227

[38] Hull C M 1984 Exact *pp*-wave solutions of 11-dimensional supergravity *Phys. Lett.* B **139** 39

[39] Han S K and Koh I G 1985 $N = 4$ supersymmetry remaining in Kaluza–Klein monopole background in $D = 11$ supergravity *Phys. Rev.* D **31** 2503

[40] Pollard D 1983 Antigravity and classical solutions of five dimensional Kaluza–Klein theory *J. Phys. A: Math. Gen.* **16** 569

[41] Sorkin R 1983 Kaluza–Klein monopole *Phys. Rev. Lett.* **51** 87

[42] Gross D and Perry M 1983 Magnetic monopoles in Kaluza–Klein theories *Nucl. Phys.* B **226** 29

[43] Deser S and Seminara D 1999 Counterterms/M-theory corrections to $D = 11$ supergravity *Phys. Rev. Lett.* **82** 2435

Reprinted from Nucl. Phys. B **135** (1978) 149–66

SUPERSYMMETRIES AND THEIR REPRESENTATIONS

W. NAHM
CERN, Geneva, Switzerland

Received 5 September 1977

We determine all manifest supersymmetries in more than 1 + 1 dimensions, including those with conformal or de Sitter space-time symmetry. For the supersymmetries in flat space we determine the structure of all representations and give formulae for an effective computation. In particular we show that at least for masses $m^2 = 0, 1, 2$ the states of the spinning string form supersymmetry multiplets.

1. Introduction

All supersymmetries of the S-matrix in 3 + 1 dimensions are known [1]. However, there are further interesting possibilities, e.g., supersymmetries in de Sitter space-time or in higher dimensions. Particularly important is the conjecture that a suitably restricted version of the Neveu-Schwarz-Ramond string yields a renormalizable sypersymmetric Yang-Mills and gravity theory in 9 + 1 dimensions [2]. Such theories may be reduced to 3 + 1 dimensions by compactifying some directions [3].

In sect. 2 we shall classify all manifest supersymmetries in more than 1 + 1 dimensions. For a flat space-time we determine the structure of the corresponding little groups in sect. 3. In sect. 4 we determine their representations and derive formulae to calculate them explicitly. In sect. 5 we consider as examples the theories which admit multiplets with spins at most 1. In particular we shall see that the lowest mass levels of the spinning string indeed can be regarded as supersymmetry representations, thus confirming the conjecture of ref. [2].

The notations are those of ref. [4]. In particular, the bracket $\langle l, l' \rangle$ will denote the anticommutator, if both l, l' are odd, and the commutator, if at least one of them is even. We shall always work with the supersymmetry algebra, not with groups.

2. Classification of supersymmetries

Let $L = G \oplus U$ be a finite dimensional supersymmetry algebra, where G, U denote the even and odd subspace respectively. We assume that the generators exhibit the usual relation between spin and statistics (in fact it is sufficient to assume that U

contains no Lorentz scalars). Furthermore, L must admit an adjoint operation +. This is true, if L commutes with some unitary S-matrix, but we shall also consider theories with massless particles, or in de Sitter space, where the usual S-matrix formalism runs into difficulties. However, we restrict ourselves to manifest supersymmetries, acting on some Hilbert space of particle states.

Taking the subspace of L generated by the elements which obey

$$g^+ = -g \qquad \text{for} \quad g \in G\,,$$

$$u^+ = iu \qquad \text{for} \quad u \in U\,, \tag{1}$$

we obtain a real form of L.

For the even part we write

$$G = S \oplus J\,, \tag{2}$$

where S is the space-time symmetry, and J a compact internal symmetry of the form

$$J = T \oplus A\,, \tag{3}$$

with T semi-simple and A Abelian.

Consider first the case where S is simple, i.e., a conformal or a de Sitter algebra. Let $C(X)$ denote the centre of X.

Proposition 2.1.

$L/C(L)$ is the direct sum of an internal symmetry J_c of type (3) plus a supersymmetry which is simple up to a possible extension by an algebra of outer automorphisms (of course J_c may be zero).

Proof:

$C(L) \subset G$, as

$$\langle uu^+\rangle > 0 \tag{4}$$

for all non-zero $u \in U$. Let $V \subset U$ be invariant under $(S, +)$, i.e., under S and the adjoint operation. Then $\langle S\,\langle VV\rangle\rangle$ is an S-invariant subspace of S, thus either equal to S or zero. In the latter case, $\langle VV^+\rangle$ defines a positive-definite S-invariant Hermitian form on V, thus that part of S which is faithfully represented on V is contained in some compact unitary algebra. As S is simple and non-compact, $\langle SV\rangle$ has to vanish. But because of the spin-statistics relation, U contains no scalars. Thus for any $V \neq 0$ we have

$$\langle S\langle VV\rangle\rangle = S\,. \tag{5}$$

Therefore no ideal of L which contains an odd element can be soluble. Now let C be the maximal soluble ideal of L. Because of $C \subset G$ one has

$$\langle CU\rangle = 0\,. \tag{6}$$

In addition because of eqs. (2) and (3)

$$C \subset A \,. \tag{7}$$

Thus

$$C = C(L)\,, \tag{8}$$

and $L/C(L)$ is semi-simple. All semi-simple graded Lie algebras are described in ref. [5]. Because G is of type (2) with simple S, and U contains no scalars, $L/C(L)$ has to be a direct sum. Its summands must be simple modulo extensions by outer automorphisms. Because of eq. (5) all odd elements of $L/C(L)$ belong to that direct summand which contains S. Apart from the outer automorphisms the direct summand which contains U can be written as

$$U \oplus \langle UU \rangle$$

and this supersymmetry algebra is simple modulo "central charges" as defined in ref. [1]. Here

$$\langle UU \rangle = S \oplus J\,, \tag{10}$$

where J' is a direct summand of J, thus again of type (3).

All real simple graded Lie algebras have been classified [5,6]. Thus we just have to select the algebras which are compatible with our assumptions. We use the notation (G, U as representation space of G).

Proposition 2.2.
The simple supersymmetry algebras are:

$$(\mathrm{o}(2,1) \oplus \mathrm{u}(N), (2,N) + (2,\bar{N})), \qquad N \neq 2\,, \tag{I}$$

$$(\mathrm{o}(2,1) \oplus \mathrm{su}(2), (2,2) + (2,2)), \tag{I_1}$$

$$(\mathrm{o}(2,1) \oplus \mathrm{o}(N), (2,N)), \qquad N = 1, 2, \ldots\,, \tag{II}$$

$$(\mathrm{o}(2,1) \oplus \mathrm{o}(4), (2,4))_\alpha\,, \tag{II_α}$$

$$(\mathrm{o}(2,1) \oplus \mathrm{o}(3) \oplus \mathrm{su}(N,\mathrm{H}), (2,2,2N)), \qquad N = 1, 2, \ldots\,, \tag{III}$$

$$(\mathrm{o}(2,1) \oplus \mathrm{o}(7), (2,8)), \tag{IV}$$

$$(\mathrm{o}(2,1) \oplus g_2, (2,7)), \tag{V}$$

$$(\mathrm{o}(3,1), (2,1) + (1,2)), \tag{VI}$$

$$(\mathrm{o}(3,2) \oplus \mathrm{o}(N), (4,N)), \qquad N = 1, 2, \ldots\,, \tag{VII}$$

$$(o(4,1) \oplus u(1), 4 + \bar{4}), \qquad (VII_1)$$

$$(o(4,2) \oplus u(N), (4,N) + (\bar{4},\bar{N})), \qquad N \neq 4, \qquad (VIII)$$

$$(o(4,2) \oplus su(4), (4,4) + (\bar{4},\bar{4})), \qquad (VIII_1)$$

$$(o(6,1) \oplus su(2), (8,2)), \qquad (IX_1)$$

$$(o(5,2) \oplus su(2), (8,2)), \qquad (IX_2)$$

$$(o(6,2) \oplus su(N, H), (8, 2N)), N = 1, 2, \ldots . \qquad (X)$$

The lie algebras are denoted by lower case latters, capitals are reserved for the groups. SU(N, H) is the group of unitary quaternionic $N \times N$ matrices; it is the compact real form of Sp($2N$, C). In particular su(1, H) $\sim$ su(2), su(2, H) $\sim$ o(5). In (II_α), α is a real constant which enters only into the structure constants for $\langle UU \rangle$. The algebras involving o(4, 2) have been classified in ref. [1].

Recently, Euclidean supersymmetries have been studied [7]. The algebras o(R + 1, 1) may be interpreted as conformal algebras of an R-dimensional Euclidean space. For compact de Sitter spaces we obtain the additional possibilities

$$(o(3) \oplus u(N), (2,N) + (2,\bar{N})), \qquad N \neq 2, \qquad (I')$$

$$(o(3) \oplus su(2), (2,2) + (2,2)), \qquad (I'_1)$$

$$(o(5) \oplus u(1), 4 + \bar{4}), \qquad (VII'_1)$$

$$(o(6) \oplus u(N), (4,N) + (\bar{4},\bar{N})), \qquad N \neq 4, \qquad (VIII')$$

$$(o(6) \oplus su(4), (4,4) + (\bar{4},\bar{4})). \qquad (VIII'_1)$$

Note that most of the orthosymplectic algebras are unsuitable, as the sp($2N$) are non-compact. Thus we had to use the isomorphisms [8]

o(2, 1) $\sim$ su(1, 1),	o(4, 1) $\sim$ su(1, 1, H),
o(3) $\sim$ su(2),	o(5) $\sim$ su(2, H),
o(2, 1) $\oplus$ o(3) $\sim$ s α u(2, H),	o(4, 2) $\sim$ su(2, 2),
o(3, 1) $\sim$ sp(2, C),	o(6) $\sim$ su(4),
o(3, 2) $\sim$ sp(4),	o(6, 2) $\sim$ s α u(4, H).

SαU(N, H) denotes the group of anti-unitary quaternionic $N \times N$ matrices.

The representations of S in U are always spinor representations.

To find central charges, one has to look at the decomposition of the symmetric part of the tensor product of U with itself. For algebras involving a u(1), any G scalar can be absorbed into it. Most other algebras yield no G scalars, with the exception of (I_1), (I'_1), (VIII_1) and ($VIII'_1$). These algebras admit one central charge. In addition, only they admit outer automorphisms U(1), and (I_1) even SU(2). This extension of (I_1) by SU(2) can be obtained from (II_α) in the limit $\alpha = 0$. In this latter case no central charge is allowed.

The algebras for de Sitter spaces may be contracted to algebras of flat spaces. Here any direct summand of J may either be left unchanged or contracted to a vector space of central charges.

For supersymmetries where S is a conformal algebra we have a natural grading over the integers

$$L = L^{(-2)} \oplus L^{(-1)} \oplus L^{(0)} \oplus L^{(1)} \oplus L^{(2)}, \tag{11}$$

$$\langle L^{(m)} L^{(n)} \rangle \subset L^{(m+n)}, \tag{12}$$

defined by

$$L^{(n)} = \{ l \in L \mid \langle dl \rangle = nl \}, \tag{13}$$

with a suitably normalized dilation generator d. Here

$$L^{(-2)} = P \tag{14}$$

is the subspace of translation generators, and

$$L^{(-1)} \oplus L^{(1)} = U. \tag{15}$$

Let us now consider the supersymmetries with

$$S = \mathrm{io}(R, 1), \tag{16}$$

i.e., the Poincaré algebra in $R + 1$ dimensions. As we have seen, these supersymmetries cannot be simple. A typical example is the subalgebra

$$L^{(-2)} \oplus L^{(-1)} \oplus L_d^{(0)}, \tag{17}$$

of (11), where

$$L_d^{(0)} \oplus \text{dilations} = L^{(0)}. \tag{18}$$

In fact, we shall show that such a grading by dimension can always be constructed. Provisionally we define recursively a filtration

$$P \oplus C(L) = \tilde{L}^{(-2)} \subset \tilde{L}^{(0)} \subset \ldots \subset \tilde{L}^{(2k)} = L \tag{19}$$

by

$$\tilde{L}^{(m)} = \{ l \in L \mid \langle Pl \rangle \subset \tilde{L}^{(m-2)} \} \qquad \text{for } m \geqslant 0. \tag{20}$$

The proof of ref. [9] that k is finite applies also to supersymmetries. Obviously

$$\langle \widetilde{L}^{(m)} \widetilde{L}^{(n)} \rangle \subset \widetilde{L}^{(m+n)} , \tag{21}$$

$$\widetilde{L}^{(m)+} = \widetilde{L}^{(m)} , \tag{22}$$

$$G \subset \widetilde{L}^{(0)} . \tag{23}$$

Proposition 2.3.
In (19), $L = \widetilde{L}^{(0)}$.
Proof: Put

$$U^{(i)} = U \cap \widetilde{L}^{(i+1)} , \qquad i = -1, 1, \ldots . \tag{24}$$

We have

$$\langle P U^{(-1)} \rangle = 0 , \tag{25}$$

and therefore

$$\langle P \langle U^{(-1)} U^{(-1)} \rangle \rangle = 0 . \tag{26}$$

Thus

$$\langle U^{(-1)} U^{(-1)} \rangle \subset P \oplus J . \tag{27}$$

From

$$\langle P \langle P U^{(1)} \rangle \rangle = 0 \tag{28}$$

we obtain for any $u \in U^{(1)}$

$$\langle \langle Pu \rangle \langle Pu \rangle \rangle = \langle P \langle \langle Pu \rangle u \rangle \rangle$$
$$= \langle P \langle P \langle uu \rangle \rangle \rangle \subset \langle P \langle PG \rangle \rangle = 0 . \tag{29}$$

For $u \in U^{(1)}$ with $u^+ = u$ this means

$$\langle Pu \rangle = 0 . \tag{30}$$

But these elements span $U^{(1)}$. Thus

$$U^{(1)} = U^{(-1)} = U . \tag{31}$$

Proposition 2.4.
U consists of spinor representations of o(R, 1).
Proof: Consider a Cartan subalgebra of the Lorentz algebra spanned by the Hermitian generators $M_{01}, M_{23}, \ldots$. Decompose U into eigenspaces of this Cartan algebra. For any element u of one of these spaces

$$\langle M_{2i,\, 2i+1}\, u \rangle = \alpha_i(u)\, u . \tag{32}$$

Because of the compactness properties of o(R, 1), $\alpha_0(u)$ must be purely imaginary, and all the other $\alpha_i(u)$ real. This yields

$$\langle M_{2i,2i+1}\,\langle u\,u^+\rangle\rangle = 2\alpha_0(u)\,\delta_{0i}\,\langle u\,u^+\rangle\,. \tag{33}$$

But there is no element with this property in G, unless

$$\alpha_0(u) = \pm\tfrac{1}{2} \text{ or } 0\,. \tag{34}$$

This must be true for all eigenspaces of the Cartan algebra. As scalars have been excluded, U must consist of spinors.

From (4) and eq. (33) one obtains that the coefficient of P^0 in $\langle uv^+\rangle$ defines a positive-definite $o(R) \oplus J$ invariant Hermitian form on U. We write

$$\langle uv^+\rangle = (uv)\,P^0 + \dots\,. \tag{35}$$

From the existence of this form it follows that even the representation of A in U is completely reducible.

Proposition 2.5.

For $R > 2$ the filtration (19) can be refined to a grading

$$L = L^{(-2)} \oplus L^{(-1)} \oplus L^{(0)}\,, \tag{36}$$

where

$$L^{(-2)} = P \oplus C(L)\,, \tag{37}$$

$$L^{(-2)} = U\,, \tag{38}$$

$$L^{(0)} = o(R,\,1) \oplus T \oplus A_c\,, \tag{39}$$

with

$$A_c \oplus C(L) = A\,. \tag{40}$$

Proof: Because of (4), $C(L)$ contains only even elements. Taking into account eqs. (25), (27) and (31), we only have to prove that $\langle UU\rangle \cap J$ lies in $C(L)$. Put

$$M = U \oplus \langle UU\rangle\,, \tag{41}$$

$$M_c = M/(L^{(-2)} \cap \langle UU\rangle) = V \oplus \langle VV\rangle\,. \tag{42}$$

We have to prove that M_c contains only odd elements. Let $W \oplus B$ be an Abelian ideal of M_c with B even, W odd. As the representation of B in V is completely reducible,

$$\langle BV\rangle = \langle B\langle BV\rangle\rangle \subset \langle BW\rangle = 0\,. \tag{43}$$

Thus

$$B \subset \langle VV\rangle \cap C(M_c) = 0\,, \tag{44}$$

$$\langle VW\rangle \subset B = 0\,. \tag{45}$$

This yields

$$W \oplus B = W \subset C(M_c) \subset V\,. \tag{46}$$

Moreover, $C(M_c)$ is even a direct summand of M_c, as the complete reducibility of the representation of $\langle VV\rangle$ in V yields

$$\langle\langle VV\rangle V\rangle \cap C(M_c) \subset \langle\langle VV\rangle\, C(M_c)\rangle = 0\,. \tag{47}$$

Thus

$$M_c = M_s \oplus C(M_c) \tag{48}$$

with semi-simple or vanishing M_s.

But M_s admits o(R, 1) as outer automorphism. Thus for $R > 2$ it has to vanish.

For $R = 2$, M_s has to be a direct sum whose summands are all of the form (I'_1), i.e., (su(2) $\oplus$ su(2), (2, 2) + (2, 2)). This algebra has the outer automorphism algebra o(2, 1). As M_s admits no further outer automorphisms, all direct summands of J which act non-trivially on M_s are contained in $\langle UU\rangle$. Furthermore, M_s admits no central charges. Thus $(U \oplus P \oplus J)/P$ is a direct sum of an algebra isomorphic to M_s and one of type (36). This yields all L with $S = i$o(2, 1).

In the simplest case, where M_s is just (I'_1), L can be obtained from (II_α) in the limit $\alpha = -1$. In this limit, o(2, 1) may become the outer automorphism, or it may be scaled down to a three-dimensional centre. If one does both, the centre transforms as the adjoint, i.e., the vector representation of the outer automorphism o(2, 1). This doubling of o(2, 1) thus yields the Poincaré algebra io(2,1).

The supersymmetry algebra just described apparently has not been discussed before.

For $R = 1$, no new possibilities for M_s appear, as (I'_1) is the only real form of a simple graded Lie algebra which admits o(1, 1) as outer automorphism. For example, (su(N) $\oplus$ su(N), (N, N) + ($\bar{N}$, $\bar{N}$)) with $N > 2$ has the outer automorphism algebra o(2), which prevents the introduction of momenta.

However, the extension of (I'_1) by o(1, 1) admits now one central charge which may be obtained from one momentum component in 2 + 1 dimensions by reduction to 1 + 1 dimensions.

3. Little groups

We shall only determine the representations of the supersymmetries graded according to eqs. (36)–(39). For the conformal supersymmetries this means that we represent only a subalgebra. For the de Sitter case we are anyhow only interested in representations which have a limit for the contraction to flat space-time.

The representations can be induced in the usual way from the representations of the little group. Thus we fix some subspace H of the Hilbert space on which P is constant and which is irreducible with respect to

$$L' = G' \oplus U\,, \tag{49}$$

where

$$G' = S' \oplus J \tag{50}$$

and

$$S' = \mathrm{o}(R) \qquad \text{for the massive case}\,,$$

$$S' = \mathrm{io}(R-1) \text{ for the massive case}\,. \tag{51}$$

In the latter case the "Galilei-transformations" of $\mathrm{io}(R-1)$ have to be represented by zero, as otherwise the representations become infinite dimensional. Thus S' can be restricted to $\mathrm{o}(R-1)$.

On H, A is represented by constants. Thus $U|_H$ (U restricted to H) yields a Clifford algebra with bilinear form $\langle UU\rangle|_H$. This bilinear form is not necessarily positive definite, though by (4) it is non-negative definite.

Proposition 3.1.

U can be decomposed into $(G', +)$ invariant subspaces

$$U = U^0 \oplus U'\,, \tag{52}$$

such that

$$U^0/_H = 0 \tag{53}$$

and $\langle U' U'\rangle|_H$ is positive-definite.

Proof: We may choose H such that $p^0 = p^R$ in the massless case and $p = (m, 0)$ for massive particles. Write

$$U = U_+ \oplus U_-\,, \tag{54}$$

where

$$\langle M_{0R} u\rangle = \pm\tfrac{1}{2}u \qquad \text{for } u \in U_\pm\,. \tag{55}$$

Eq. (33) yields

$$\langle u^+ u\rangle = c(P^0 \pm P^R) \qquad \text{for } u \in U_\pm\,, \tag{56}$$

with $c > 0$ for any non-trivial u. In the massless case we obtain

$$U^0 = U_-\,,$$

$$U' = U_+\,. \tag{57}$$

In the massive case we use the positive-definite Hermitian form on U defined by eq. (35). Note that eq. (33) yields

$$(U_+ U_-) = 0\,. \tag{58}$$

In general, let U^0 be the subspace of U which annihilates H and take its orthogonal complement U' with respect to this form. As the form is $(G', +)$ invariant, this is also true for the decomposition.

Note that in the massless case all central charges have to vanish, as

$$C(L)|_H \subset \langle U_+ U_- \rangle|_H = 0 \,. \tag{59}$$

In the massive case without central charges one obtains

$$\langle U_+ U_- \rangle|_H = 0 \,, \tag{60}$$

such that $\langle UU \rangle|_H$ is positive definite and U^0 vanishes. Even with central charges according to eq. (56)

$$\dim U' \geqslant \dim U_+ = \tfrac{1}{2} \dim U \,. \tag{61}$$

However, there may be linear relations between $U_+|_H$ and $U_-|_H$

Proposition 3.2.

$C(L)|_H$ forms a compact, convex set. At its boundary and only at its boundary $U_0 \neq 0$.

Proof: Choose a basis u^k of $U|_H$, c^i of $C(L)|_H$. We have

$$\langle \lambda_j u^i, (\lambda'_k u^k)^+ \rangle = m a_0(\lambda\lambda') + c^i a_i(\lambda\lambda') \,, \tag{62}$$

where the a_0, a_i are Hermitian forms. We have seen that a_0 has to be positive-definite. In contrast, no linear combination of the a_i can be positive- or negative-definite, as the non-compact algebra G has no invariant finite-dimensional positive-definite Hermitian forms.

The allowed values for $C(L)|_H$ are those for which

$$\mathcal{P}(\lambda) = m a_0(\lambda\lambda) + c^i a_i(\lambda\lambda) \geqslant 0 \qquad \text{for all } \lambda \,. \tag{63}$$

U^0 is non-vanishing, if in addition

$$\mathcal{P}(\lambda) = 0 \tag{64}$$

for some non-zero λ. We may restrict λ to the compact space

$$\Lambda = \{\lambda | \textstyle\sum |\lambda_i|^2 = 1\} \,. \tag{65}$$

Then

$$\min_{\Lambda} (\mathcal{P}(\lambda)/a_0(\lambda\lambda))$$

is a continuous function of the c^i. Along any ray in $C(L)|_H$ from zero to infinity it is a linear function which will take at first positive, then negative values, with one zero in between. The convexity follows from

$$\min_{\Lambda} ((c^i + d^i)\, a_i(\lambda\lambda)) \geqslant \min_{\Lambda} (c^i a_i(\lambda\lambda)) + \min_{\Lambda} (d^i a_i(\lambda\lambda)) \,. \tag{66}$$

An important special case arises, if $S = \mathrm{io}(R, 1)$ is reduced to $\mathrm{io}(R', 1)$ with $R' < R$. Here the superfluous momentum components become central charges, and the boundary of $C(L)|_H$ corresponds to zero mass in $R + 1$ dimensions.

Now we shall show that as far as the representations of L' are concerned, the central charges enter only *via* the determination of U^0.

Proposition 3.3.

Let the representation $U'|_H$ of $(G', +)$ be given. Then $\langle U' U' \rangle|_H$ is fixed up to an isomorphism.

Proof: Let

$$U' = U^{(1)} \oplus U^{(2)} \oplus \dots \tag{67}$$

be the decomposition of U' into inequivalent representations of $(G', +)$. As $\langle U' U' \rangle|_H$ contains only G' scalars,

$$\langle U^{(m)} U^{(n)} \rangle|_H = 0 \qquad \text{for } m \neq n . \tag{68}$$

Let u_{ai} be a basis of some $U^{(m)}$, where $(G', +)$ acts irreducibly on the first index, whereas i counts the multiplicity of the representation. We may write

$$\langle u_{ai} u^{+}_{bj} \rangle = K_{ab} X_{ij} , \tag{69}$$

where K is the uniquely defined positive-definite Hermitian invariant form of the corresponding representation of $(G', +)$. $K \otimes X$ has to be positive-definite Hermitian, thus also X. In particular, we may choose a basis such that

$$K_{ab} = \delta_{ab} , \qquad X_{ij} = \delta_{ij} . \tag{70}$$

Now let us classify the supersymmetries with regard to the representations. As we have seen, this requires the classification of all possible U'.

Proposition 3.4.

Let $G' = S' \oplus J$ be an algebra of type (3), (51) and U' a spinorial representation of $(G', +)$. Then one can always find a supersymmetry L which yields $G' \oplus U'$ as algebra of the little group, both in the massless and in the massive case.

Proof: It is sufficient to consider an irreducible U', otherwise one just takes a direct sum with components orthogonal under the Lie bracket. For any irreducible spinorial representation of some $(G, +)$ with $G = S \oplus J$ one can define a supersymmetry by

$$\langle Q^{+}_{\alpha i} Q_{\beta j} \rangle = (\gamma^0 \gamma_\mu)_{\alpha\beta} P^\mu X_{ij} . \tag{71}$$

Here S acts on the first and J on the second index of Q. X_{ij} is the positive-definite Hermitian form on the representation of J.

Let

$$S' = \mathrm{o}(r) , \tag{72}$$

where $r = R$ for the massive and $r = R - 1$ for the massless case. Take

$$G = \mathrm{io}(r + 1, 1) \oplus J \tag{73}$$

and choose a U which transforms under J according to the given representation. Furthermore let its transformation properties under $o(r+1, 1)$ be given by the embedding of the representation of $o(r)$ into the spinorial representation of $o(r+1, 1)$ of twice its dimension. As the real, quaternionic, or non-self-conjugated nature of the spinorial representations of $o(r+n, n)$ is independent of n [8], the representation of $(G', +)$ is embedded into a representation of $(G, +)$ of twice its dimension. Now consider the corresponding algebra (71). For an H with $p^0 = p^{r+1}$ it yields the wanted algebra of the little group. In the massless case, we have finished. In the massive case we just have to take the subalgebra $io(R, 1)$ of $io(R+1, 1)$, and to interpret p^{R+1} as central charge.

4. Representations

In this section we always take the restriction to H, without noting it explicitly.

It remains to determine the representations of $(L', +)$. As the undecomposable representations of Clifford algebras with non-degenerate bilinear form are fixed up to isomorphisms, this problem is completely solved by

Proposition 4.1.

The universal associative enveloping algebra $U(L')$ of L' decomposes as

$$U(L') = \bar{U}(G') \otimes U(U'), \tag{74}$$

where $\bar{U}(G')$ is isomorphic to $U(G')$.

Proof: Take a basis g^i of G', Q_α of U', such that

$$\langle Q_\alpha Q_\beta^+ \rangle = \delta_{\alpha\beta}, \tag{75}$$

$$\langle g^i Q_\alpha \rangle = Q_\beta \sigma^i_{\beta\alpha}. \tag{76}$$

Let

$$\langle Q_\alpha Q_\gamma \rangle = T_{\alpha\gamma} = T_{\gamma\alpha}. \tag{77}$$

Then

$$Q_\alpha = T_{\alpha\beta} Q_\beta^+. \tag{78}$$

The Jacobi identity yields

$$T_{\beta\gamma}\sigma^i_{\beta\alpha} + T_{\alpha\beta}\sigma^i_{\beta\gamma} = 0, \tag{79}$$

$$\langle g^i - \tfrac{1}{2} Q_\beta \sigma^i_{\beta\gamma} Q_\gamma^+, Q_\alpha \rangle = 0 \qquad \text{for all } \alpha, i. \tag{80}$$

Thus we obtain a set of elements

$$\bar{g}^i = g^i - \tfrac{1}{2} Q_\beta \sigma^i_{\beta\gamma} Q_\gamma^+ \tag{81}$$

of $U(L')$ which commute with all elements of $U(U')$ and form a Lie algebra isomorphic to G'. The enveloping algebra $\bar{U}(G')$ of this Lie algebra fulfils eq. (74).

Thus all representations of L' are products of a representation of G' with the irreducible representation F of U'. Taking the trivial representation of $\bar{U}(G')$ we obtain the fundamental representation $1 \otimes F$ of L', for which the generators g^i are represented according to

$$g^i = \tfrac{1}{2}Q_\beta \sigma^i_{\beta\gamma} Q^+_\gamma \,. \tag{82}$$

Its dimension is

$$\dim F = 2^{\dim U'/2} \,. \tag{83}$$

Proposition 4.2.

The representations of L' contain the same number of fermion as of boson states.

Proof: Let f be the fermion number. Because of

$$(-)^f Q = -Q(-)^f \qquad Q \in U' \,, \tag{84}$$

one has

$$\mathrm{Tr}((-)^f \langle QQ' \rangle) = 0 \qquad \text{for } Q,\, Q' \in U' \,. \tag{85}$$

The number $\langle QQ' \rangle$ is in general not zero.

To tackle the calculation of the representations it is convenient to use characters, i.e., the traces of elements of the group generated by the g^i. We may restrict ourselves to a maximal Abelian subgroup, because this determines already all the weights. Let $\bar{A}$ be any representation of $\bar{U}(G')$. It corresponds, *via* the isomorphism to $U(G')$, to a representation A of G' and *vice versa*. Thus we obtain

$$\begin{aligned} \chi_{\bar{A}\,\otimes\, F}(\exp(\zeta_i g^i)) &= \chi_{\bar{A}\,\otimes\, F}(\exp(\zeta_i \bar{g}^i)\exp(\tfrac{1}{2}\zeta_i Q_\alpha \sigma^i_{\alpha\beta} Q^+_\beta)) \\ &= \chi_{\bar{A}}(\exp(\zeta_i \bar{g}^i))\, \chi_F(\tfrac{1}{2}Q_\alpha \sigma^i_{\alpha\beta} Q^+_\beta) \\ &= \chi_A(\exp(\zeta_i g^i))\, \chi_{1\,\otimes\, F}(\exp(\zeta_i g^i)) \,. \end{aligned} \tag{86}$$

Furthermore let us use eq. (67). Let $U^{(m)}$ contain c_m irreducible representations of type m, which by themselves yield fundamental representations $1 \otimes Fm$. Then

$$\chi_{1\,\otimes\, F}(g) = \prod_m \chi_{1\,\otimes\, Fm}(g)^{c_m} \,. \tag{87}$$

Thus we can restrict ourselves to irreducible U'. We can even reduce the calculation of $\chi_{1\,\otimes\, Fm}$, using the same formula, to the corresponding one for irreducible representations of a maximum Abelian subgroup of G'.

Now take an r-dimensional Abelian group with generators $g^1, \ldots, g^r$ and a 2^r dimensional real representation U' for which no g^i is represented trivially. We may assume that the eigenvalues of all g^i are $\pm\frac{1}{2}i$. For convenience we define

$$\chi_r(\zeta_1, \ldots, \zeta_r) = \chi_{1\,\otimes\, F}(\exp(\zeta_i g^i)) \,. \tag{88}$$

A change of base from g^1, g^2 to $\frac{1}{2}(g^1 \pm g^2)$ yields the recursion relation

$$\chi_r(\zeta_1, \zeta_2, \zeta_3, ...) = \chi_{r-1}(\zeta_1 + \zeta_2, \zeta_3, ...)\, \chi_{r-1}(\zeta_1 - \zeta_2, \zeta_3, ...)\,. \tag{89}$$

χ_1 can easily be calculated directly. We take a base $Q_\pm$ of U', where Q_- is the adjoint of Q_+. The σ of eq. (76) has the form

$$\sigma^1 = \tfrac{1}{2} i \begin{pmatrix} 1 & 0 \\ 0 & -1 \end{pmatrix}. \tag{90}$$

Thus

$$g^1 = \tfrac{1}{4}(Q_+Q_- - Q_-Q_+)\,. \tag{91}$$

Then $Q_\pm$ may be represented by the Pauli matrices $\frac{1}{2}(\sigma_x \pm i\sigma_y)$. Thus

$$\chi_1(\zeta) = 2 \cos \tfrac{1}{4}\zeta\,. \tag{92}$$

Eq. (89) then yields

$$\chi_2(\zeta_1, \zeta_2) = 2 \cos \tfrac{1}{2}\zeta_1 + 2 \cos \tfrac{1}{2}\zeta_2\,, \tag{93}$$

$$\chi_3(\zeta_1, \zeta_2, \zeta_3) = 2 + 2 \sum_{i=1}^{3} \cos \zeta_i + \sum_{\pm} \exp(\tfrac{1}{2}(\pm\zeta_1 \pm \zeta_2 \pm \zeta_3)i)\,. \tag{94}$$

Eq. (92) yields in general

$$\chi_{1 \otimes F}(\exp(i\zeta M_{12})) = (2 \cos \tfrac{1}{4}\zeta)^{\dim U'/2}\,. \tag{95}$$

Eq. (93) shows that the state of $1 \otimes F$ which yields the highest eigenvalue of M_{12} has zero eigenvalue for all generators commuting with M_{12}. Thus for any $g \in J$

$$\chi_{1 \otimes F}(\exp(i \sum_i \zeta_i M_{2i-1,2i} + g)) = \sum_i 2(\cos \tfrac{1}{8}\zeta_i \dim U') + \dots\,. \tag{96}$$

One sees that for $\dim U' \equiv 0 \pmod 8$, the fundamental representation contains a totally symmetric $\frac{1}{8}d$ tensor. For $\dim U' = 4$, one has a spinor, more generally for $\dim U' \equiv 4 \pmod 8$ some spinor-tensor. For $\dim U' \equiv 2 \pmod 4$, which may happen for massless particles in 3 + 1 dimensions, $\chi_A = 1$ is obviously impossible. $\chi_A(\exp(i\zeta M_{12}))$ has to be a sum of terms of the form $2\cos(\frac{1}{4}(2n+1)\zeta)$.

For o(10, 1), the spin representation has dimension 32, such that for massless particles $\dim U'$ is at least 16. Thus any representation in more than 9 + 1 dimensions contains at least a symmetric tensor field. For o(11, 1), no Majorana-Weyl-spinor exists [8], thus $\dim U'$ is at least 32 and higher spins have to occur. Consequently, supergravity theories are impossible in more than 10 + 1 dimensions, supersymmetric Yang-Mills theories in more than 9 + 1 dimensions.

Note that the minimal value of $\dim U'$ grows exponentially with the dimension. According to eq. (83), $\dim F$ grows like an iterated exponential.

5. Examples

At first we shall list the fundamental representations of the supersymmetries which allow multiplets with highest spin one. This requires dim $U' \leqslant 8$.

(*A*) $S' = o(2)$.

For $J = 0$ we obtain the character of eq. (92). With

$$\chi_A(\exp(i\zeta M_{12})) = 2\cos(\tfrac{1}{4}(2n+1)\zeta), \qquad n = 0, 1, \ldots, \tag{97}$$

we obtain

$$\chi_{\bar{A}} \otimes_F(\exp(i\zeta M_{12})) = 2\cos(\tfrac{1}{2}(n+1)\zeta) + 2\cos(\tfrac{1}{2}n\zeta). \tag{98}$$

These are the well-known massless multiplets of the standard supersymmetry in 3 + 1 dimensions.

For $J = su(2)$ and isospin $\frac{1}{2}$ we may embed the representation U' into the corresponding one for $S' = o(3)$ without changing dim U'. Therefore we need not treat this case separately. From now on we omit most representations for which such an embedding is possible.

For $J = su(3)$ and representation $3 + \bar{3}$ a slightly more complicated embedding into the vector representation of o(6) is possible. As fundamental multiplets, one obtains su(3) singlets for "spin" $\pm\frac{3}{4}$ and triplet, antitriplet for "spin" $\pm\frac{1}{4}$ respectively. Multiplying by the χ_A of eq. (97) with $n = 0$ one obtains singlets with spins $\pm 1, \pm\frac{1}{2}$, triplet, antitriplet for spin $\pm\frac{1}{2}$ respectively, and both triplet and antitriplet for spin 0. Multiplying by the octet of su(3) one obtains the particles of a possible supersymmetric Yang-Mills theory.

However, here a general difficulty of those theories becomes apparent. In eq. (96) we have seen that for the fundamental representation the particles with highest spin are J singlets. Thus either one has to except multiplets with spin larger than one, or one has to multiply by the adjoint representation of some gauge group. But this procedure yields unreasonably high representations of the gauge group for the fermions. If one takes supercharges which commute with the gauge group, one obtains only adjoint representations of this group, otherwise higher representations have to occur. But, of course, one has to keep in mind that our investigation concerns only manifest symmetries.

For $J = su(4)$ and representation $4 + \bar{4}$ compare the case $S' = o(6), J = o(2)$, which may be embedded into the case $G' = o(8)$ discussed below. The fundamental representation has a singlet for spin ± 1, quartet, antiquartet for spin $\pm\frac{1}{2}$ respectively, and an antisymmetric tensor for spin zero. Note that the simplest multiplet for $J = su(3)$ discussed above admits the larger symmetry $J = su(4)$.

(*B*) $S' = o(3)$

Even for $J = 0$ the invariance under the adjoint operation requires that U' contains an even number of spinors. As smallest multiplet one obtains

$$\chi_1 \otimes_F(\zeta) = \chi_1(\zeta)^2 = 2\cos\tfrac{1}{2}\zeta + 2. \tag{99}$$

Taking the spinor representation for $\bar{A}$ one obtains in $\bar{A} \otimes F$ a vector, a scalar and two Majorana spinors. These multiplets are well known [10].

Here an embedding into the (2,2) representation of o(3) ⊕ o(3) is possible. The character of the fundamental representation has already been given in eq. (93).

If one chooses the isospin-$\frac{3}{2}$ representation of J = su(2), the representation of the maximum Abelian subgroup in U' is reducible. One obtains

$$\chi_{1 \otimes F}(\zeta_1, \zeta_2) = \chi_2(\zeta_1, \zeta_2)\, \chi_2(\zeta_1, 3\zeta_2)\,, \tag{100}$$

i.e., (3, 1) + (1, 5) + (2, 4).

o(4) = o(3) ⊕ o(3) need not be considered separately, as for its spinors one o(3) is represented trivially.

From S' = o(5) on, representations with dim $U' < 8$ no longer occur and those with dim $U' = 8$ can be embedded into a Majorana-Weyl spinor of o(8). This representation occurs for the massless particles of the supersymmetric spinning string. Let us consider this system in detail.

For all supersymmetries in more than 5 + 1 dimensions, dim U is at least 16. All supersymmetries with $R > 5$, dim $U = 16$ can be considered as subsymmetries of the supersymmetry $L = G \oplus U$ with G = io(9, 1), U = Majorana-Weyl spinor. Central charges can be interpreted as components of the momentum in 9 + 1 dimensions. There are at most $9 - R$ of them. The maximal internal symmetry for S = io(R, 1) is just o($9 - R$). This can easily be checked case by case.

For the massless multiplet in 9 + 1 dimensions, the fundamental representation is essentially determined by dim $F = 16$ and proposition 4.2. Alternatively it can be read of from eq. (94). According to the chirality of U one finds

$$\chi_\pm(\zeta_1, \zeta_2, \zeta_3, \zeta_4) = \sum_\epsilon \exp\Big(i \sum_{i=1}^{4} \tfrac{1}{2}\epsilon_i \zeta_i\Big) + \sum_{i=1}^{4} 2\cos\zeta_i\,, \tag{101}$$

where all $\epsilon_i \in \{1, -1\}$ and the sum goes over all quadruples $(\epsilon_1, \epsilon_2, \epsilon_3, \epsilon_4)$ with

$$\prod_i \epsilon_i = \pm 1 \qquad \text{for } \chi_\pm \text{ respectively}\,. \tag{102}$$

The r.h.s. of eq. (101) represents a Majorana-Weyl spinor plus a vector. These are just the $m = 0$ states of the spinning string of ref. [2]. The supersymmetry admits no central charges. Thus according to eq. (60), one finds, for the fundamental representation of the massive case,

$$\chi_4(\zeta_1, \zeta_2, \zeta_3, \zeta_4) = \chi_+(\zeta_1, \zeta_2, \zeta_3, \zeta_4)\, \chi_-(\zeta_1, \zeta_2, \zeta_3, \zeta_4)\,. \tag{103}$$

As can easily be read off from the helicity partition function [11] this yields exactly the multiplet which occurs at the $m^2 = 1$ level of the Neveu-Schwarz-Ramond string as considered in ref. [2]. For $m^2 = 2$ one finds the $\bar{V} \otimes F$ representation, where V is the o(9) vector with

$$\chi_V(\zeta_1, \zeta_2, \zeta_3, \zeta_4) = 2 \sum_{i=1}^{4} \cos\zeta_i + 1\,. \tag{104}$$

This is a strong confirmation for the conjecture that this model is supersymmetric.

For the closed string the representations considered in ref. [2] are just the tensor product of the open-string representations with the corresponding representation of the boson sector alone. Thus for a supersymmetric open string, the closed string has to be supersymmetric too. This yields one possible supergravity theory in 9 + 1 dimensions, which by reduction yields the o(4) supergravity [12] plus an additional sector considered below. The representation is given by

$$\chi(\zeta_1, \zeta_2, \zeta_3, \zeta_4) = \chi_+(\zeta_1, \zeta_2, \zeta_3, \zeta_4)\, 2 \sum_{i=1}^{4} \cos \zeta_i \,. \tag{105}$$

To restrict the representations of io(9, 1) to those of io(3, 1) + su(4) one just has to interpret M_{45}, M_{67} and M_{89} in the character formulae as generators of SU(4). For the fundamental massless representation, eq. (101) yields a su(4) singlet with spin ±1, a su(4) quartet, antiquartet with spin $\pm\frac{1}{2}$ respectively, and a su(4) sextet with spin 0.

Multiplication with $2 \cos \zeta_1$ yields the o(4) supergravity. The remaining part of the character of the o(8) vector in eq. (105) yields a vector of o(6) ~ su(4). This is the adjoint representation of o(4) ⊂ su(4). Thus one obtains the multiplet of the o(4) Yang-Mills theory.

Supergravity theories are possible in at most 10 + 1 dimensions, as we have seen. For G = io(10, 1), U = Majorana spinor, one obtains the fundamental representation (103).

If seven dimensions are compactified, one finds G = io(3, 1) ⊕ o(7), while U transforms as Majorana spinor both under io(3, 1) and o(7). Now, one can enlarge o(7) to o(8) without changing the representation space U. As the Majorana-Weyl spinor and the vector representations of o(8) are connected by outer automorphisms of o(8), the embedding of o(7) into o(8) may be done in such a way that U transforms as a vector under o(8). Thus one should obtain the o(8) supergravity by dimensional reduction, if the supergravity in 10 + 1 dimensions can be constructed.

In 9 + 1 dimensions, there is one further supergravity, which arises, if one takes all tensor products of the open string with itself including the fermion-fermion sector. This yields an internal symmetry J = o(2). Taking into account only the space-time symmetry, one obtains for the fundamental representation

$$\chi(\zeta_1, \zeta_2, \zeta_3, \zeta_4) = \chi_+(\zeta_1, \zeta_2, \zeta_3, \zeta_4)^2 \,. \tag{106}$$

Scherk has discovered that dimensional reduction of this theory probably yields the o(8) supergravity [13]. Indeed, as far as the little group G' = o(8) ⊕ o(2) is concerned, an exchange of S' = o(8) and J = o(2) would yield the representations of this supergravity. This exchange may arise automatically by dimensional reduction of G' to o(2) ⊕ o(6) ⊕ o(2). Now, the o(6) counts as part of the internal symmetry J = o(6) ⊕ o(2). As discussed for o(7) above, the representation of J in U admits an extension to the vector representation of o(8).

Thus in 9 + 1 dimensions three supergravity theories may exist, with multiplets given by the eqs. (105), (103) and (106), respectively. Dimensional reduction of the first should yield the o(4) supergravity, whereas from the other two one might obtain the o(8) supergravity.

I wish to thank D. Olive, B. Zumino and J. Scherk for stimulating discussions.

References

[1] R. Haag, J.T. Łopuszanski and M. Sohnius, Nucl. Phys. B88 (1975) 257.
[2] F. Gliozzi, J. Scherk and D. Olive, Nucl. Phys. B122 (1977) 253.
[3] E. Cremer and J. Scherk, Nucl. Phys. B108 (1976) 409.
[4] W. Nahm, V. Rittenberg and M. Scheunert, Phys. Lett. 61B (1976) 383.
[5] G.I. Kac, Comm. Math. Phys. 54 (1977) 31.
[6] M. Scheunert, W. Nahm and V. Rittenberg, J. Math. Phys. 18 (1977) 146.
[7] B. Zumino, Phys. Lett. 69B (1977) 369.
[8] J. Tits, Tabellen zu den einfachen Lie Gruppen und ihren Darstellungen, Lecture notes in mathematics 40 (Springer, 1967).
[9] S. Coleman and J. Mandula, Phys. Rev. 159 (1967) 1251.
[10] S. Ferrara and B. Zumino, Nucl. Phys. B79 (1974) 125.
[11] W. Nahm, Nucl. Phys. B120 (1977) 125.
[12] L. Brink, J.H. Schwarz and J. Scherk, Nucl. Phys. B121 (1977) 77.
[13] J. Scherk, Private communication.

SUPERGRAVITY THEORY IN 11 DIMENSIONS

E. CREMMER, B. JULIA and J. SCHERK
Laboratoire de Physique Théoriquè de l'Ecole Normale Supérieure [1], *Paris, France*

Received 4 April 1978

We present the action and transformation laws of supergravity in 11 dimensions which is expected to be closely related to the O(8) theory in 4 dimensions after dimensional reduction.

Extended $O(N)$ ($N = 1, \ldots 8$) supergravity theories [1–5] are notoriously difficult to construct beyond $N = 3$. The difficulty lies partly in the large number of fields involved (for $N = 8$, which is the largest theory that can be constructed in this frame-work, one has 1 graviton, 8 spin 3/2 gravitinos, 28 vectors, 56 spinors, 35 scalar and 35 pseudoscalar particles) but mostly in the fact that the spin 0 fields appear in a non-polynomial way, thus forbidding a step-by-step construction of the action and transformation laws. So far, only the $N = 4$ theory has been constructed in a closed form [3], the simplest form of it exhibiting a manifest SU(4) invariance [5], while the $N = 8$ theory [4] has been constructed only to order K^2. Further, geometrical methods [6] do not seem readily applicable to these theories beyond $N = 2$, due to the presence of fields not bearing a vector index, which makes their interpretation as gauge fields difficult.

On the other hand, an elegant method has been found to be very useful to circumvent the similar problem of constructing in a simple way the $O(N)$ ($N = 1, \ldots 4$) supersymmetric Yang–Mills theories in 4 dimensions. One first establishes the existence of a supersymmetric Yang–Mills theory in 10 dimensions [7], which was suggested by the study of the dual spinor model. Then one reduces the theory to four dimensions by assuming all fields to be independent of the extra 6 spacial coordinates. In this way, the $N = 4$ theory, which has a vanishing β-function at the first two non-trivial orders [8] was found, and also a systematic search of all supersymmetric Yang–Mills in less than 10 dimensions was conducted [9].

As shown by Nahm [10], $D = 10$ is the highest number of dimensions in which supersymmetry representations with $J \leqslant 1$ can exist, while supergravity theories ($J \leqslant 2$) can exist up to $D = 11$. The interest in constructing the 11 dimensional theory lies in the fact that its reduction to four dimensions is automatically guaranteed to yield an O(7) invariant supergravity theory which has exactly the same field content as the O(8) theory, both theories being presumably equivalent just as the O(4) and SU(4) supergravity theories have been shown to be.

The field content of the $D = 11$ theory is remarkably simple. It consists of the vierbein V^a_μ, a Majorana spin 3/2 ψ_μ, and of a completely antisymmetric gauge tensor with 3 indices $A_{\mu\nu\rho}$. To arrive at this set of fields a simple argument is to count the number of physical states. In $D = 11$, the Dirac matrices are 32×32 and a Majorana spin 3/2 field ψ_μ represents $1/2 \cdot 32\,(D-3) = 128$ degrees of freedom. The vierbein field represents $\{(D-1)(D-2)\}/2 - 1 = 44$ degrees of freedom. The mismatch is $84 = \binom{9}{3}$, which is just the number of components of a transverse, antisymmetric gauge field with 3 indices in 11 dimensions. Transversality amounts to requiring that beside coordinate invariance, local Lorentz invariance and local supersymmetry, the action is also invariant under the Abelian gauge transformation:

$$\delta A_{\mu\nu\rho} = \partial_\mu \zeta_{\nu\rho} + \partial_\nu \zeta_{\rho\mu} + \partial_\rho \zeta_{\mu\nu} \,,$$

[1] Laboratoire Propre du C.N.R.S., associé à l'Ecole Normale Supérieure et à l'Université de Paris-Sud. Postal address: 24 rue Lhomond, 75231 Paris Cedex 05, France.

where $\zeta_{\mu\nu} = -\zeta_{\nu\mu}$.

This gauge invariance, together with the requirement of absence of terms with more than two derivatives, implies that the action is polynomial in the $A_{\mu\nu\rho}$ field. This is a considerable simplification compared to the non-polynomial reduced forms.

Another way to arrive at this simplest set of fields is to start from the spinor dual model of closed strings. This model is obtained by doubling the Neveu–Schwarz (Bose states, NS)–Ramond (Fermi states, R) model [11] and a priori contains four possible sectors: NS ⊗ NS, NS ⊗ R, R ⊗ NS and R ⊗ R. In ref. [7] it was shown that the first sector, at zero mass, contains a graviton V_μ^a an antisymmetric tensor $A_{\mu\nu}$, and a scalar ϕ, while the second sector contains Majorana–Weyl spin 3/2 ψ_μ and spin 1/2 χ fields forming an irreducible representation of $D = 10$ supergravity, and the two other sectors were neglected. However, if we include them we get in addition a new spin 3/2 ψ'_μ and a spin 1/2 χ' field of opposite helicities and a bispinor field representing 64 degrees of freedom, equivalent to an antisymmetric gauge field with 3 indices $A_{\mu\nu\rho}$ (56 degrees of freedom) and a gauge vector field A_ρ (8 degrees of freedom). This set of fields forms an irreducible representation of extended supergravity in $D = 10$ dimensions, a point which was also realized by Schwarz [12]. It is easy to see how this set of fields arises from reducing the $D = 11$ theory to $D = 10$. The V_M^A decomposes into V_μ^a, A_ρ, ϕ fields (taking into account the fact that only the symmetric part propagates). The A_{MNP} decomposes into $A_{\mu\nu\rho}$, $A_{\mu\nu}$. Finally the Majorana ψ_M field decomposes into the Majorana–Weyl fields ψ_μ, ψ'_μ, χ, χ'.

It is also interesting to show that reducing the theory down to 4 dimensions we obtain the same counting of fields as in the O(8) theory. The Majorana ψ_M field decomposes into 8 spin 3/2 fields ψ_μ^i ($i = 1, \ldots 8$), and 56 spin 1/2 fields ψ_a^i ($i = 1, \ldots 8; a = 1, \ldots 7$). The graviton field decomposes into 1 graviton $g_{\mu\nu}$, 7 vectors $g_{\mu a}$ and 28 scalars ($g_{ab} = g_{ba}$). The antisymmetric tensor decomposes into an $A_{\mu\nu\rho}$ field equivalent, in 4 dimensions, to an auxiliary scalar field carrying no degrees of freedom, 7 antisymmetric $A_{\mu\nu a}$ gauge fields equivalent in 4 dimensions to 7 scalar fields ϕ_a, 21 vector fields $A_{\mu ab}$, and 35 pseudoscalar fields A_{abc}. The total content of fields is thus the same as in the O(8) theory, although dimensional reduction will only make an O(7) invariance manifest.

Let us now present the action and transformation laws of the $D = 11$ supergravity theory. Our metric is $(+ - - \ldots -)$; Greek indices are world indices while Latin indices refer to the tangent space. The Γ^a matrices are in the Majorana representations and form a purely imaginary representation of the Clifford algebra in 11 dimensions. $\Gamma^{a_1 \cdots a_N}$ represents the product of $N\,\Gamma$ matrices completely antisymmetrized, i.e. for unequal indices

$$\Gamma^{a_1 \ldots a_N} = \Gamma^{a_1} \cdots \Gamma^{a_N} .$$

The lagrangian we find is the following:

$$\begin{aligned}\mathcal{L} = &-\frac{V}{4K^2} R(\omega) \\ &-\frac{\mathrm{i}V}{2} \bar{\psi}_\mu \Gamma^{\mu\nu\rho} D_\nu \left(\frac{\omega + \hat{\omega}}{2}\right) \psi_\rho - \frac{V}{48} F_{\mu\nu\rho\sigma} F^{\mu\nu\rho\sigma} \\ &+ \frac{KV}{192} (\bar{\psi}_\mu \Gamma^{\mu\nu\alpha\beta\gamma\delta} \psi_\nu + 12\, \bar{\psi}^\alpha \Gamma^{\gamma\delta} \psi^\beta)(F_{\alpha\beta\gamma\delta} + \hat{F}_{\alpha\beta\gamma\delta}) \\ &+ \frac{2K}{(144)^2} \epsilon^{\alpha_1\alpha_2\alpha_3\alpha_4\beta_1\beta_2\beta_3\beta_4\mu\nu\rho} F_{\alpha_1\alpha_2\alpha_3\alpha_4} F_{\beta_1\beta_2\beta_3\beta_4} A_{\mu\nu\rho} .\end{aligned}$$

The quartic terms in this Lagrangian are absorbed in the supercovariant fields (i.e. fields which under supersymmetry transform without derivatives of ϵ) $\hat{\omega}$ and $\hat{F}$ which will be defined below.

$F_{\mu\nu\rho\sigma}$ is the field strength associated with the gauge field $A_{\mu\nu\rho}$:

$$F_{\mu\nu\rho\sigma} = 4\, \partial_{[\mu} A_{\nu\rho\sigma]} ,$$

where the brackets represent the antisymmetrized sum over all permutations, divided by their number. The covariant derivative of ψ_μ is given by:

$$D_\nu(\omega)\psi_\mu = \partial_\nu \psi_\mu + \tfrac{1}{4} \omega_{\nu ab} \Gamma^{ab} \psi_\mu ,$$

and the convention for R is the same as in ref. [13].

The Lorentz connection coefficients $\omega_{\mu ab}$ are given by:

$$\omega_{\mu ab} = \omega^0_{\mu ab}(V) + K_{\mu ab} ,$$

where the contorsion tensor is

$$\begin{aligned}K_{\mu ab} = (\mathrm{i}K^2/4) [&-\bar{\psi}_\alpha \Gamma_{\mu ab}{}^{\alpha\beta} \psi_\beta \\ &+ 2(\bar{\psi}_\mu \Gamma_b \psi_a - \bar{\psi}_\mu \Gamma_a \psi_b + \bar{\psi}_b \Gamma_\mu \psi_a)] .\end{aligned}$$

The torsion tensor is given by:

$$C_{\mu\nu}{}^{a} = K_{\nu\mu}{}^{a} - K_{\mu\nu}{}^{a}$$
$$= (\mathrm{i}K^2/2)\,[\bar{\psi}_\alpha \Gamma_{\mu\nu}{}^{a\alpha\beta}\psi_\beta - 2\bar{\psi}_\mu \Gamma^a \psi_\nu]\,.$$

The transformation laws are given by:

$$\delta V^a_\mu = -\mathrm{i}K\bar{\epsilon}\Gamma^a\psi_\mu\,,$$

$$\delta\psi_\mu = \frac{1}{K}D_\mu(\hat{\omega})\epsilon + \frac{\mathrm{i}}{144}(\Gamma^{\alpha\beta\gamma\delta}{}_\mu$$

$$-8\Gamma^{\beta\gamma\delta}\delta^\alpha_\mu)\epsilon\hat{F}_{\alpha\beta\gamma\delta} \equiv \frac{1}{K}\hat{D}_\mu\epsilon\,,$$

$$\delta A_{\mu\nu\rho} = \tfrac{3}{2}\bar{\epsilon}\Gamma_{[\mu\nu}\psi_{\rho]}\,,$$

$$\hat{F}_{\mu\nu\rho\sigma} = F_{\mu\nu\rho\sigma} - 3K\bar{\psi}_{[\mu}\Gamma_{\nu\rho}\psi_{\sigma]}\,.$$

In order to obtain $\omega_{\mu ab}$ from its equation of motion $\hat{\omega}_{\mu ab}$ should be put equal to:

$$\hat{\omega}_{\mu ab} = \omega_{\mu ab} + (\mathrm{i}K^2/4)\bar{\psi}_\alpha\Gamma_{\mu ab}{}^{\alpha\beta}\psi_\beta\,.$$

As one can easily verify, $\hat{\omega}_{\mu ab}$ is a supercovariant tensor.

To obtain this action and transformation laws, we have proceeded in the following deductive fashion: Firstly, in the 1.5 order formalism and taking the linear part in δV and $\delta\psi_\mu$, the terms of the form $\bar{\epsilon}\psi$ in δS, the variation of the action vanishes as for $D = 4$ supergravity. Then, as we have a kinetic term for $A_{\mu\nu\rho}$ (whose scale has been fixed by the conventional factor 1/48), to cancel the $\bar{\epsilon}\psi F^2$, we need a $\bar{\psi}X\psi F$ coupling and a $Z\epsilon F$ term in $\delta\psi$, where X and Z are unknown tensors made of Γ matrices.

To determine these terms, it is shortest to use the requirement that the equations of motion of ψ_μ must be supercovariant. That is defining $\hat{D}_\mu$ by $\delta\psi_\mu = (1/K)\hat{D}_\mu\epsilon$ and the supercovariant field strength by $\psi_{\nu\rho} = \hat{D}_\nu\psi_\rho - \hat{D}_\rho\psi_\nu$, the fermion field equation should read:

$$\Gamma^{\mu\nu\rho}\hat{D}_\nu\psi_\rho = 0\,.$$

Looking at the terms of the form $Z\psi F$ in this equation and comparing them with what is obtained from the $\bar{\psi}X\psi F$ terms in the action fixes the form of the X and Z tensors and relate their coefficients.

Then we consider the terms of the form $\bar{\epsilon}\psi F^2$, in the variation of the action. They all indeed disappear if the coefficient of Z is fixed properly, except for one term involving a product of 9 Γ matrices which can be cancelled by adding to S the gauge invariant expression:

$$a\int \mathrm{d}x\,\epsilon^{\alpha\beta\gamma\delta\,\mu\nu\rho\sigma ijk}F_{\alpha\beta\gamma\delta}F_{\mu\nu\rho\sigma}A_{ijk}\,,$$

and only if we have: $\delta A_{\mu\nu\rho} = b\bar{\epsilon}\Gamma_{[\mu\nu}\psi_{\rho]}$.

This fixes the product ab. Then looking at the terms of the type $\bar{\epsilon}\partial\psi F$ and $\bar{\epsilon}\psi\partial F$ in the variation, we determine b so that all terms are fixed, up to trilinear terms in $\delta\psi$ and up to quartic terms in the fermion fields in the Lagrangian.

In order to fix these, we require that $\psi_{\nu\rho}$ be indeed supercovariant. This imposes that we replace F by $\hat{F}$ in $\delta\psi$, but also ω by $\hat{\omega}$ since ω is not supercovariant. Now the transformation laws are fully fixed and a crucial test transformation laws are fully fixed and a crucial test is to see whether the supersymmetry algebra closes at least on the Bose fields. This is indeed the case, confirming the need to replace F and ω by their supercovariant versions and the correctness of the transformation laws.

Finally the quartic terms in the action are fixed so as to reproduce the supercovariant fermion field equation. The fact that this is possible is a test of consistency and requires the following identity derived by a Pauli–Fierz transformation:

$$\tfrac{1}{8}\Gamma^{\mu\nu\alpha\beta\gamma\delta}\psi_\nu\bar{\psi}_\alpha\Gamma_{\beta\gamma} - \tfrac{1}{8}\Gamma_{\beta\gamma}\psi_\nu\bar{\psi}_\alpha\Gamma^{\mu\nu\alpha\beta\gamma\delta}$$
$$-\tfrac{1}{4}\Gamma^{\mu\nu\alpha\beta\delta}\psi_\nu\bar{\psi}_\alpha\Gamma_\beta + \tfrac{1}{4}\Gamma_\beta\psi_\nu\bar{\psi}_\alpha\Gamma^{\mu\nu\alpha\beta\delta}$$
$$-2g^{\beta[\alpha}\Gamma^{\delta\mu\nu]}\psi_\nu\bar{\psi}_\alpha\Gamma_\beta - 2\Gamma_\beta\psi_\nu\bar{\psi}_\alpha g^{\beta[\alpha}\Gamma^{\delta\mu\nu]}$$
$$+2g^{\beta[\alpha}\Gamma^{\delta\mu\nu]}\bar{\psi}_\alpha\Gamma_\beta\psi_\nu = 0\,.$$

As usual the supersymmetry algebra closes on shell. Explicitly we find:

$$[\delta_2,\delta_1]V^a_\mu = \partial_\mu\xi^\nu V^a_\nu + \xi^\nu\partial_\nu V^a_\mu$$
$$-\mathrm{i}\kappa\bar{\epsilon}'\Gamma^a\psi_\mu + \Omega^a{}_b V_\mu{}^b\,,$$

where

$$\xi^\nu = -\mathrm{i}\bar{\epsilon}_1\Gamma^\nu\epsilon_2\,,\qquad \epsilon' = -K\xi^\nu\psi_\nu\,,$$

and

$$\Omega^a{}_b = (K/72)\,\bar{\epsilon}_1(\Gamma^{ab\alpha\beta\gamma\delta}$$
$$-24V_b{}^\alpha V^{a\beta}\Gamma^{\gamma\delta})\epsilon_2\hat{F}_{\alpha\beta\gamma\delta} + \xi^\nu\hat{\omega}_\nu{}^a{}_b\,,$$

$$[\delta_2,\delta_1]A_{\mu\nu\rho} = 3(\partial_{[\mu}\xi^\sigma)A_{\sigma\nu\rho]} + \xi^\sigma\partial_\sigma A_{\mu\nu\rho}$$
$$+\tfrac{3}{2}\bar{\epsilon}'\Gamma_{[\mu\nu}\psi_{\rho]} + \partial_{[\mu}\Lambda_{\nu\rho]}\,,$$

where

$$\Lambda_{\nu\rho} = (3/2K)\bar{\epsilon}_1 \Gamma_{\nu\rho}\epsilon_2 - 3\xi^\sigma A_{\sigma\nu\rho} ,$$

$$[\delta_2, \delta_1]\,\psi_\mu = \partial_\mu \xi^\nu \psi_\nu + \xi^\nu \partial_\nu \psi_\mu$$
$$+ (1/K)\hat{D}_\mu \epsilon' + \tfrac{1}{4}\Omega_{ab}\Gamma^{ab}\psi_\mu + \mathcal{R}_\mu ,$$

where $\mathcal{R}_\mu$ is proportional to the fermion equation of motion.

To check the full invariance of the action, we have successively verified the vanishing of terms of type: $\epsilon\psi^3(F + \hat{F}/2)$ and $\epsilon\psi^2 D(\hat{\omega})\psi$ to minimize the number of terms of the type $\epsilon\psi^5$ which have been finally shown analytically to cancel.

In conclusion, two things remain to be done with our theory. First, we are studying the reduction to four dimensions and the connection with the O(8) theory and also the reduction to 10 dimensions to get the zero slope limit of the closed string dual model of ref. [7]. The second is to find geometrical interpretations analogous to the one obtained for $D = 4$ in the case of O(1) and O(2) supergravity theories. The natural candidate for the graded Lie algebra is OSp(32, 1) which contains an internal O(8) subalgebra. Works along these lines are presently in progress.

References

[1] D.Z. Freedman, P. van Nieuwenhuizen and S. Ferrara, Phys. Rev. D13 (1976) 3214;
D.Z. Freedman and P. van Nieuwenhuizen, Phys. Rev. D14 (1976) 912;
S. Deser and B. Zumino, Phys. Lett. 62B (1976) 335;
S. Ferrara and P. van Nieuwenhuizen, Phys. Rev. Lett. 37 (1976) 1669.

[2] D.Z. Freedman, Phys. Rev. Lett. 38 (1976) 105;
S. Ferrara, J. Scherk and B. Zumino, Phys. Lett. 66B (19 35 and Nucl. Phys. B121 (1977) 393.

[3] A. Das, Phys. Rev. D15 (1977) 2805;
E. Cremmer, J. Scherk and S. Ferrara, Phys. Lett. 68B (1977) 234;
E. Cremmer and J. Scherk, Nucl. Phys. B127 (1977) 259.

[4] B. deWit and D.Z. Freedman, Nucl. Phys. B130 (1977) 105.

[5] E. Cremmer, J. Scherk and S. Ferrara, LPTENS 77/19 preprint, to be published in Physics Letters B.

[6] S.W. MacDowell and F. Mansouri, Phys. Rev. Lett. 38 (1977) 739;
A. Chamseddine and P.C. West, Nucl. Phys. B129 (1977) 34;
P.K. Townsend and P. van Nieuwenhuizen, Phys. Lett. 67B (1977) 439;
Y. Ne'eman and T. Regge, Univ. of Texas preprint (Dec. 1977).

[7] F. Gliozzi, J. Scherk and D. Olive, Nucl. Phys. B122 (197 253.

[8] D.R.T. Jones, Phys. Lett. 72B (1977) 199;
E.C. Poggio and H.N. Pendleton, Phys. Lett. 72B (1977) 200.

[9] L. Brink, J.H. Schwarz and J. Scherk, Nucl. Phys. B121 (1977) 77.

[10] W. Nahm, CERN TH 2341 preprint (1977).

[11] A. Neveu and J.H. Schwarz, Nucl. Phys. B31 (1971) 86, Phys. Rev. D4 (1971) 1109;
P. Ramond, Phys. Rev. D3 (1971) 2415.

[12] J.H. Schwarz, private communication. He also computed some of the couplings and transformation laws of the $D = 10$ supergravity theory.

[13] D.Z. Freedman and P. van Nieuwenhuizen, Phys. Rev. D14 (1976) 912.

Reprinted from Nucl. Phys. B 186 (1981) 412–28

SEARCH FOR A REALISTIC KALUZA–KLEIN THEORY*

Edward WITTEN

Joseph Henry Laboratories, Princeton University, Princeton, New Jersey 08544, USA

Received 12 January 1981

An attempt is made to construct a realistic model of particle physics based on eleven-dimensional supergravity with seven dimensions compactified. It is possible to obtain an $SU(3)\times SU(2)\times U(1)$ gauge group, but the proper fermion quantum numbers are difficult to achieve.

In 1921 Kaluza suggested [1] that gravitation and electromagnetism could be unified in a theory of five-dimensional riemannian geometry. The idea was further developed by Klein [2] and was the subject of considerable interest during the classical period of work on unified field theories [3]. Readable expositions of some of the classical work have been given in text books by Bergmann and by Lichnerowicz [4]; more recent discussions have been given by Rayski and by Thirring [5].

While the Kaluza–Klein approach has always been one of the most intriguing ideas concerning unification of gauge fields with general relativity, it has languished because of the absence of a realistic model with distinctive and testable predictions. Yet the urgency of the unification of gauge fields with general relativity has surely greatly increased with the growing importance of gauge fields in physics. Moreover, the Kaluza–Klein theory has generalizations to non-abelian gauge fields which actually were first proposed [6] well before real applications were known for Yang–Mills fields in physics.

In the last few years this approach has been revived by Scherk and Schwarz and by Cremmer and Scherk, originally in connection with dual models [7]. These authors introduced many new ideas as well as new focus. In contrast to much of the classical literature, they advocated that the extra dimensions should be regarded as true, physical dimensions, on a par with the four observed dimensions. Cremmer and Scherk suggested that the obvious differences between the four observed dimensions and the extra microscopic ones could arise from a spontaneous breakdown of the vacuum symmetry, or, as they called it, from a process of "spontaneous compactification" of the extra dimensions.

These ideas have motivated much recent work. The idea of spontaneous compactification has been developed in more detail by Luciani [8]. An interesting idea by Palla [9] about massless fermions in theories with extra compact dimensions

* Research partially supported by NSF grant PHY78-01221.

will figure in some of the discussion below. Manton [10] has discussed some questions that arise in trying to generate Higgs fields as components of the gauge field in extra dimensions. The idea of extra hidden dimensions has stimulated much work in supersymmetry theory, including the successful construction of the $N = 8$ supergravity theory by Cremmer, Julia and Scherk and by Cremmer and Julia [11]. This work has been generalized to give models with broken sypersymmetry [12].

In many respects, of course, the modern approaches to this subject tend to differ from the classical point of view. In view of the proliferation of new particles in the last thirty years, one may be more willing today than in the past to postulate the infinite number of new degrees of freedom that must exist if extra dimensions really exist. Much of the classical literature focussed on the need to eliminate a massless spin-zero particle that naturally exists in the original Kaluza–Klein theory; the question seems less urgent today because the obvious answer is that quantum mechanical mass renormalization could easily account for the failure to observe this particle (a mass of 10^{-4} eV would make it undetectable). Some of the early work was motivated by the hope that the fifth dimension could provide the hidden variables that would eliminate indeterminacy from quantum mechanics. Despite the many generalizations and changes in emphasis that have occurred, I will refer generically to theories in which gauge fields are unified with gravitation by means of extra, compact dimensions as Kaluza–Klein theories.

It has often been suggested that spontaneous compactification and supergravity could be usefully combined together. The $N = 8$ supergravity theory was constructed by "dimensional reduction" starting from an eleven-dimensional theory. In this context, "dimensional reduction" just means that the fields are taken to be independent of seven of the original eleven coordinates, to which physical reality need not be attributed. However, Cremmer and Julia [11] suggested that one might wish to consider seriously the eleven dimensions and interpret seven of them as compact dimensions in the spirit of Kaluza and Klein. This idea has been raised, on occasion, by various other theorists. In this paper, I will describe an attempt – not completely successful, but not completely unsuccessful either – to construct a realistic theory of Kaluza–Klein type, based on eleven-dimensional supergravity.

As discussed by some of the authors mentioned above, from a modern point of view the Kaluza–Klein unified theory of gravitation and electromagnetism is probably best understood as a theory of spontaneous symmetry breaking in which the group of general coordinate transformations in five dimensions is spontaneously broken to the product of the four-dimensional general coordinate transformation group and a local U(1) gauge group.

Let us review how this arises. One considers standard general relativity in five dimensions with the standard Einstein–Hilbert action

$$A = \int d^5x \sqrt{g} R. \tag{1}$$

Instead of assuming that the ground state of this system is five-dimensional Minkowski space, which we will denote as M^5, one takes the ground state to be the product $M^4 \times S^1$ of four-dimensional Minkowski space M^4 with the circle S^1. The space $M^4 \times S^1$ is, like M^5, a solution of the five-dimensional Einstein equations. Classically it is difficult to decide which of the spaces M^5 and $M^4 \times S^1$ is a more appropriate choice as the ground state, since they both have zero energy, insofar as energy can be defined in general relativity*. Conventionally, one might assume that the ground state is M^5. In the Kaluza–Klein approach one assumes, instead, that the ground state is $M^4 \times S^1$, and the physical spectrum is determined by studying small oscillations around this ground state. One assumes that the radius of the circle S^1 is microscopically small, perhaps of order of the Planck length, and this accounts for why the existence of this fifth dimension is not noted in everyday experience.

The symmetries of the Kaluza–Klein ground state $M^4 \times S^1$ are the four-dimensional Poincaré symmetries, acting on M^4, and a U(1) group of rotations of the circle S^1. These symmetries would be observed as local or gauge symmetries in the apparent four-dimensional world because the whole theory started with the Einstein action (1) which is generally covariant. In fact, if one considers small oscillations around the "ground state" $M^4 \times S^1$, one finds an infinite number of massive excitations, the masses being of order the inverse of the circumference of S^1. One finds also a finite number of massless modes, which presumably would constitute the low-energy physics. The massless modes turn out to be a spin-two graviton and a spin-one photon, which are gauge particles of the symmetries of $M^4 \times S^1$, and a Brans–Dicke scalar.

The ansatz which exhibits the massless modes is the following. The metric tensor of this theory is a five by five matrix $g_{AB}(x^\mu, \phi)$ which in general may depend on the four coordinates x^μ, $\mu = 1 \cdots 4$, of M^4, and on the angular coordinate ϕ of S^1. The massless modes are those for which g_{AB} is a function of x^μ only. One can then write g_{AB} in block form

$$g_{AB}(x^\mu, \phi) = \left(\begin{array}{c|c} g_{\mu\nu}(x) & A_\mu(x) \\ \hline A_\mu(x) & \sigma(x) \end{array}\right), \tag{2}$$

where $g_{\mu\nu}$ is a four by four matrix (the first four rows and columns of g_{AB}), $A_\mu = g_{\mu 5}$, and $\sigma = g_{55}$. Then $g_{\mu\nu}$ is the ordinary metric tensor of the apparent four-dimensional world, and describes a massless spin-two particle; A_μ is the gauge field of the U(1) symmetry, and σ is the Brans-Dicke scalar.

In the classical work on the Kaluza–Klein theory, it is shown that the five-dimensional Einstein action (1), when expanded in terms of $g_{\mu\nu}$, A_μ, and σ (and the other modes, which decouple from these at low energies) contains a four-dimensional Einstein action $\sqrt{g}R^{(4)}$ for $g_{\mu\nu}$, a Maxwell action $F^2_{\mu\nu}$ for A_μ, and the usual

* The definition of energy in general relativity depends on the boundary conditions, so while both M^5 and $M^4 \times S^1$ have zero energy, a comparison between them is meaningless, like comparing zero apples to zero oranges.

kinetic energy for σ. Also, one can readily check that A_μ transforms as a gauge field $A_\mu \to A_\mu + \partial_\mu \varepsilon$ under coordinate transformations of the special type $(x^i, \phi) \to (x^i, \phi + \varepsilon(x^i))$ if the metric g_{AB} is transformed by the standard rule

$$g_{AB} \to g_{A'B'} \frac{\partial x'^{A'}}{\partial x^A} \frac{\partial x'^{B'}}{\partial x^B}.$$

The Kaluza–Klein theory thus unifies the metric tensor $g_{\mu\nu}$ and a gauge field A_μ into the unified structure of five-dimensional general relativity. This theory is surely one of the most remarkable ideas ever advanced for unification of electromagnetism and gravitation.

The Kaluza–Klein theory, as noted above, also has a non-abelian generalization, which has been extensively discussed over the years. In this generalization, one starts with general relativity in $4+n$ dimensions, possibly with additional matter fields or with a cosmological constant. Instead of assuming the ground state to be M^{4+n}, Minkowski space of $4+n$ dimensions, one assumes the ground state to be a product space $M^4 \times B$, where B is a compact space of dimension n. $M^4 \times B$ should be a solution of the classical equations of motion, or possibly, as will be discussed later, a minimum of some effective potential.

As in the previous discussion, symmetries of B will be observed as gauge symmetries in the effective four dimensional world. With a suitable choice of B, one may unify an arbitrary gauge group, abelian or non-abelian, with ordinary general relativity, in a $4+n$ dimensional theory.

The ansatz which generalizes (2) is the following. Let ϕ_i, $i = 1 \cdots n$, be coordinates for the internal space B. Let T^a, $a = 1 \cdots N$, be the generators of the symmetry group G of B. Let the action of the symmetry generator T^a on the ϕ_i be $\phi_i \to \phi_i + K_i^a(\phi)$, where $K_i^a(\phi)$ is the "Killing vector" associated with the symmetry T^a. Then the massless excitations of the candidate "ground state" $M^4 \times B$ correspond to an ansatz of the following form:

$$g_{AB}(x^\alpha, \phi^k) = \left(\frac{g_{\mu\nu}(x^\alpha)}{\sum_a A_\mu^a(x^\alpha) K_i^a(\phi^k)} \middle| \frac{\sum_a A_\mu^a(x^\alpha) K_i^a(\phi^k)}{\gamma_{ij}(\phi^k)} \right), \tag{3}$$

where γ_{ij} is the metric tensor of the internal space B. The fields $A_\mu^a(x^\alpha)$ are massless gauge fields of the group G. In this way one may obtain the gauge fields of an arbitrary abelian or non-abelian gauge group as components of the gravitational field in $4+n$ dimensions.

One may verify that the $4+n$ dimensional gravitational action really contains the proper kinetic energy term $\sum_a (F_{\mu\nu}^a)^2$. It is also straightforward to check that under infinitesimal coordinate transformations of the special form $(x^\alpha, \phi_i) \to (x^\alpha, \phi_i + \sum_a \varepsilon^a(x^\alpha) K_i^a(\phi))$, which is an x-dependent symmetry transformation of the internal space B, the field $A_\mu^a(x)$ transforms in the expected fashion, $A_\mu^a(x) \to A_\mu^a(x) + D_\mu \varepsilon^a(x)$. Thus, A_μ^a really has the properties expected of an ordinary four-dimensional gauge field. This gauge field is a remnant of the original coordinate invariance group in $4+n$ dimensions, which has been spontaneously broken down to the symmetries of $M \times B$.

As has been noted before, there is a fairly extensive literature on this construction. The case which has been discussed most widely is the case in which B is itself the manifold of some group H. It should be noted that, if H is a non-abelian group, the symmetry group G of the group manifold is not H but $H \times H$, since the group manifold can be transformed by either left or right multiplication. If one starts with general relativity in $4+n$ dimensions, the ansatz (3) will automatically give massless gauge mesons of the full symmetry group $H \times H$.

What problems arise if we try to construct a realistic theory along these lines? Known particle interactions can be described by the gauge group $SU(3) \times SU(2) \times U(1)$. So the symmetry group G of the compact space B must at least contain this as a subgroup,

$$SU(3) \times SU(2) \times U(1) \subset G . \tag{4}$$

So B must at least have $SU(3) \times SU(2) \times U(1)$ as a symmetry group.

To be as economical as possible, we may wish to choose B to be a manifold of minimum dimension with an $SU(3) \times SU(2) \times U(1)$ symmetry. What is the minimum dimension of a manifold which can have $SU(3) \times SU(2) \times U(1)$ symmetry?

$U(1)$ is the symmetry group of the circle S^1, which has dimension one. The lowest dimension space with symmetry $SU(2)$ is the ordinary two-dimensional sphere S^2. The space of lowest dimension with symmetry group $SU(3)$ is the complex projective space CP^2, which has real dimension four. (CP^2 is the space of three complex variables (Z^1, Z^2, Z^3), not all zero, with the identification $(Z^1, Z^2, Z^3) \simeq (\lambda Z^1, \lambda Z^2, \lambda Z^3)$ for any non-zero complex number λ. CP^2 can also be defined as the homogeneous space $SU(3)/U(2)$.) Therefore, the space $CP^2 \times S^2 \times S^1$ has $SU(3) \times SU(2) \times U(1)$ symmetry, and it has $4+2+1=7$ dimensions.

As we will see below, seven dimensions is in fact the minimum dimensionality of a manifold with $SU(3) \times SU(2) \times U(1)$ symmetry, although $CP^2 \times S^2 \times S^1$ is not the only seven-dimensional manifold with this symmetry. If, therefore, we wish to construct a theory in which $SU(3) \times SU(2) \times U(1)$ gauge fields arise as components of the gravitational field in more than four dimensions, we must have at least seven extra dimensions. With also four non-compact "space-time"dimensions, the total dimensionality of our world must be at least $4+7=11$.

This last number is most remarkable, because eleven dimensions is probably the maximum for supergravity. Eleven-dimensional supergravity has been explicitly constructed, and it is strongly believed that supergravity theories do not exist in dimensions greater than eleven. (The reason for this belief is that, on purely algebraic grounds [13], a supergravity theory in $d>11$ would have to contain massless particles of spin greater than two. But there are excellent reasons, both *S*-matrix theoretic [14] and field theoretic [15], to believe that consistent field theories with gravity coupled to massless particles of spin greater than two do not exist.) It is consequently just barely possible to obtain $SU(3) \times SU(2) \times U(1)$ gauge fields as part of the gravitational field in a supergravity theory, if we use the unique, maximal, eleven-dimensional supergravity theory.

It is certainly a very intriguing numerical coincidence that eleven dimensions, which is the maximum number for supergravity, is the minimum number in which one can obtain SU(3) × SU(2) × U(1) gauge fields by the Kaluza–Klein procedure. This coincidence suggests that the approach is worth serious consideration.

Let us now discuss in more detail the question of why seven dimensions is the minimum number of dimensions for a space with SU(3) × SU(2) × U(1) symmetry – and the related matter of determining all seven-dimensional manifolds with this symmetry.

The space of lowest dimension with any symmetry group G is always a homogeneous space G/H, where H is a maximal subgroup of G. (The space G/H is defined as the set of all elements g of G, with two elements g and g' regarded as equivalent, $g \simeq g'$, if they differ by right multiplication by an element of H, that is, if $g = g'$ with $h \in$ H.) The dimension of G/H is always equal to the dimension of G minus the dimension of H.

In the case G = SU(3) × SU(2) × U(1), the largest dimension subgroup that is suitable is SU(2) × U(1) × U(1). Any larger subgroup of G would contain as a subgroup one of the three factors SU(3), SU(2), or U(1) of G, and this factor would then not have any non-trivial action on G/H – it would not really be a symmetry group of G/H. Since the dimension of SU(3) × SU(2) × U(1) is 8 + 3 + 1 = 12 and the dimension of SU(2) × U(1) × U(1) is 3 + 1 + 1 = 5, the dimension of (SU(3) × SU(2) × U(1))/(SU(2) × U(1) × U(1)) is 12 − 5 = 7. It is for this reason that a space with SU(3) × SU(2) × U(1) symmetry must have at least seven dimensions. However, there are many ways to embed SU(2) × U(1) × U(1) in SU(3) × SU(2) × U(1), and as a result there are many seven-dimensional manifolds with SU(3) × SU(2) × U(1) symmetry.

To embed SU(2) × U(1) × U(1) in SU(3) × SU(2) × U(1) we first embed SU(2). SU(2) can be embedded in SU(3) × SU(2) × U(1) in a variety of ways. The only embedding that turns out to be relevant is for SU(2) to be embedded in SU(3) as an "isospin" subgroup, so that the fundamental triplet of SU(3) transforms as 2 + 1 under SU(2). [Other embeddings of SU(2) lead to spaces G/H on which some of the SU(3) × SU(2) × U(1) symmetries act trivially, as discussed in the previous paragraph.] We still must embed U(1) × U(1) in SU(3) × SU(2) × U(1).

SU(3) × SU(2) × U(1) has three commuting U(1) generators which commute with the SU(2) subgroup of SU(3) that we have just chosen. There is a "hypercharge" generator of SU(3), which we may call λ_8, which commutes with the "isospin" subgroup. Also, we have the U(1) factor of SU(3) × SU(2) × U(1), which will be called Y, and we may choose an arbitrary U(1) generator of the SU(2) factor, which will be called T_3.

So SU(3) × SU(2) × U(1) contains an essentially unique subgroup SU(2) × U(1) × U(1) × U(1), where the three U(1) factors are λ_8, T_3, and Y. We do not want to divide

SU(3) × SU(2) × U(1) by the full SU(2) × U(1) × U(1) × U(1) subgroup because this would yield a space ($CP^2 \times S^2$, to be precise) on which the U(1) of SU(3) × SU(2) × U(1) would act trivially and would not really be a symmetry. So we delete one of the three U(1) factors, and divide only by SU(2) × U(1) × U(1).

The U(1) factor that is deleted may be an arbitrary linear combination $p\lambda_8 + qT_3 + rY$ of λ_8, T_3, and Y where p, q, and r are any three integers which have no common divisor*. So we define H as SU(2) × U(1) × U(1), where the SU(2) is our "isospin" subgroup of SU(3), and the two U(1)'s are the two linear combinations of λ_8, T_3, and Y which are orthogonal to $p\lambda_8 + qT_3 + rY$. The space G/H is then a seven-dimensional space with SU(3) × SU(2) × U(1) symmetry, which we may call M^{pqr}.

In a few cases the M^{pqr} are familiar spaces. M^{001} is our previous example $CP^2 \times S^2 \times S^1$. But in most cases the M^{pqr} are not familiar spaces, and are not products.

In a few cases the M^{pqr} have greater symmetry than SU(3) × SU(2) × U(1). M^{101} is $S^5 \times S^2$, which has the symmetry O(6) × SU(2). M^{011} is $CP^2 \times S^3$, whose full symmetry is SU(3) × SU(2) × SU(2). Except for these two cases, one cannot obtain from seven extra dimensions a symmetry "larger" than SU(3) × SU(2) × U(1). Therefore, the observed gauge group in nature is practically the "largest" group one could obtain from a Kaluza–Klein theory with seven extra dimensions.

Although the M^{pqr} for general values of p, q, and r are not familiar spaces, it is possible to give a rather explicit description of them. Consider first the eight dimensional space $S^5 \times S^3$ [S^n is the n-dimensional sphere, with symmetry group $O(n+1)$]. The symmetry group of $S^5 \times S^3$ is O(6) × O(4). Let us introduce a particular generator of O(6),

$$K = \begin{pmatrix} 0 & 1 & 0 & 0 & 0 & 0 \\ -1 & 0 & 0 & 0 & 0 & 0 \\ 0 & 0 & 0 & 1 & 0 & 0 \\ 0 & 0 & -1 & 0 & 0 & 0 \\ 0 & 0 & 0 & 0 & 0 & 1 \\ 0 & 0 & 0 & 0 & -1 & 0 \end{pmatrix}, \tag{5}$$

and a particular generator of O(4),

$$L = \begin{pmatrix} 0 & 1 & 0 & 0 \\ -1 & 0 & 0 & 0 \\ 0 & 0 & 0 & 1 \\ 0 & 0 & -1 & 0 \end{pmatrix}. \tag{6}$$

Then the subgroup of O(6) that commutes with K is SU(3) × U(1) [the U(1) being

* And r should be non-zero to avoid obtaining a space on which U(1) is realized as the identity.

generated by K itself] and the subgroup of O(4) that commutes with L is $SU(2)\times U(1)$ [the U(1) being generated by L].

For any non-zero p and q, we now define $N = -qK + pL$. Then N generates a U(1) subgroup of $O(6)\times O(4)$, consisting of elements of the form $\exp tN$, $0 \leq t \leq 2\pi$. We may now form from $S^5\times S^3$ a seven-dimensional space $M^{pq} = (S^5\times S^3)/U(1)$, where two points in $S^5\times S^3$ are considered to be identical if they are mapped into each other by the action of the U(1) subgroup generated by N.

This space M^{pq} is equal to the $r = 1$ case of what we have previously called M^{pqr}. The M^{pq} are actually the most general simply connected seven-dimensional manifolds with $SU(3)\times SU(2)\times U(1)$ symmetry. To obtain M^{pqr} for $r \neq 1$ one must factor out from $S^5\times S^3$ an additional discrete subgroup consisting of elements of the form $\exp(2\pi qK/r)$ $(q = 0, 1, 2, \ldots, r-1)$. We define $M^{pqr} = M^{pq}/Z^r = (S^5\times S^3)/(U(1)\times Z^r)$.

To verify that the construction of the M^{pqr} just presented is equivalent to the previous definition as $(SU(3)\times SU(2)\times U(1))/(SU(2)\times U(1)\times U(1))$, one uses the fact that SU(3)/SU(2) is S^5, while SU(2) is S^3, so $(SU(3)\times SU(2)\times U(1))/(SU(2)\times U(1)\times U(1))$ is $(S^5\times S^3\times U(1))/(U(1)\times U(1))$. Dividing out the two U(1) factors, one arrives at the above definition of M^{pqr} as $(S^5\times S^3)/(U(1)\times Z^r)$.

The M^{pqr} are not quite the most general seven-dimensional manifold with $SU(3)\times SU(2)\times U(1)$ symmetry, because for special values of p, q, and r it is possible to supplement $SU(2)\times U(1)\times U(1)$ with an additional twofold discrete symmetry. One obtains in this way some non-orientable manifolds with one of the M^{pqr} as a double covering space. These spaces are the following. Dividing M^{001} by a discrete symmetry one can get $CP^2\times P^2\times S^1$ (P^k is real projective space of dimension k), or $CP^2\times(S^2\times S^1)/Z_2$, where Z_2 is a simultaneous inversion of S^2 and S^1. From M^{101} one gets $S^5\times P^2$ and $(S^5\times S^2)/Z_2$, where the Z_2 is a simultaneous inversion of S^5 and S^2. Likewise, by dividing M^{10r} by an additional two-fold symmetry one can make $S^5/Z^r\times P^2$ and $(S^5/Z^r\times S^2)/Z_2$. These spaces are non-orientable. This completes the list of seven-dimensional manifolds with $SU(3)\times SU(2)\times U(1)$ symmetry.

If one is willing to suppose that the ground state of eleven-dimensional supergravity is a product of four-dimensional Minkowski space with one of the M^{pqr}, one can obtain an $SU(3)\times SU(2)\times U(1)$ gauge group, the gauge fields being components of the gravitational field, according to the ansatz of eq. (3). Of course, to describe nature, it is not sufficient to have the gauge group. It is also necessary to have quarks and leptons of essentially zero mass [very light compared to the energy scale of gravitation; massless in any approximation in which $SU(3)\times SU(2)\times U(1)$ is not spontaneously broken] which should be in the appropriate representation of the gauge group. And it is necessary to find Higgs bosons whose vacuum expectation value could ultimately trigger $SU(2)\times U(1)$ breaking.

How can one obtain massless quarks and leptons in the Kaluza–Klein framework? To understand the basic idea*, suppose that in a $4+n$ dimensional theory we have a

* See also a discussion by Palla [9].

massless spin one half fermion. It satisfies the $4+n$ dimensional Dirac equation,

$$\not{D}\psi = 0\,, \tag{7}$$

or explicitly

$$\sum_{i=1}^{4+n} \gamma^i D_i \psi = 0\,. \tag{8}$$

This Dirac operator can be written in the form

$$\not{D}^{(4)}\psi + \not{D}^{(\mathrm{int})}\psi = 0, \tag{9}$$

where $\not{D}^{(4)} = \sum_{i=1}^{4} \gamma^i D_i$ is the ordinary four-dimensional Dirac operator, and $\not{D}^{(\mathrm{int})} = \sum_{i=5}^{4+n} \gamma^i D_i$ is the Dirac operator in the internal space of n compact dimensions.

The expression (9) immediately shows that the eigenvalue of $\not{D}^{(\mathrm{int})}$ will be observed in practice as the four-dimensional mass. If $\not{D}^{(\mathrm{int})}\psi = \lambda\psi$, then ψ will be observed by four-dimensional observers who are unaware of the existence of the extra microscopic dimensions as a fermion of mass $|\lambda|$.

The operator $\not{D}^{(\mathrm{int})}$ acts on a compact space, so its spectrum is discrete. Its eigenvalues either are zero or are of order $1/R$, R being the radius of the extra dimensions. Since $1/R$ is, in the Kaluza–Klein approach, presumably of order the Planck mass, the non-zero eigenvalues of $\not{D}^{(\mathrm{int})}$ correspond to extremely massive fermions which would not have been observed. The observed quarks and leptons must correspond to the zero modes of $\not{D}^{(\mathrm{int})}$.

If, in eleven-dimensional supergravity, the ground state is a product of four-dimensional Minkowski space with one of the M^{pqr}, then the zero modes of the Dirac operator in the internal space will, if there are any zero modes at all, automatically form multiplets of $SU(3)\times SU(2)\times U(1)$, since this is the symmetry of the internal space. It therefore is reasonable to wonder whether for an appropriate choice of p, q, and r, zero modes could exist and form the appropriate representation of the symmetry group, so as to reproduce the observed spectrum of quarks and leptons.

Of course, to reproduce what is observed in nature, we would need quite a few zero modes of the internal space Dirac operator. If the top quark exists, there are in nature at least 45 fermion degrees of freedom of given helicity, counting all colors and flavors of quarks and leptons. We would therefore need at least 45 Dirac zero modes. However, when a Dirac operator has zero modes, the number usually depends on topological invariants. Perhaps by choosing suitable values of p, q, and r we could suitably "twist" the topology and obtain the required 45 zero modes lying in the appropriate representation of $SU(3)\times SU(2)\times U(1)$.

Actually, if one has in mind eleven-dimensional supergravity, one must modify this program slightly. In eleven-dimensional supergravity, there is no fundamental spin one half field. The only fundamental Fermi field in that theory is the Rarita–Schwinger field $\psi_{\mu\alpha}$, of spin $\frac{3}{2}$ (μ is a vector index, α a spinor index).

Although this field has spin $\frac{3}{2}$ from the point of view of eleven dimensions, the components of ψ_μ with $5\leq\mu\leq 11$ are spin one half fields from the point of view of ordinary four-dimensional physics. For $\mu \geqslant 5$, μ would be observed as an internal symmetry index, not a space-time index; it carries spin zero. Although the components ψ_μ with $\mu = 1 \cdots 4$ are spin-$\frac{3}{2}$ fields in the four-dimensional sense, the components with $\mu = 5 \cdots 11$ are spin one half fields. So zero-mode solutions of the spin-$\frac{3}{2}$ wave equation in the extra dimensions would be observed as massless spin-$\frac{1}{2}$ fermions in four dimensions. These would be the ordinary light fermions of the spontaneously compactified eleven-dimensional theory.

In one sense, it is an advantage to have to consider the Rarita–Schwinger operator rather than the Dirac operator. The Rarita–Schwinger operator can have zero modes more easily and in more abundance than the Dirac operator, because the Dirac operator has positivity properties which tend to suppress the number of zero modes. For instance, with four extra dimensions, it is known [16] that there is only a single non-flat compact solution of Einstein's equations on which the Dirac operator has zero modes. This is the Kahler manifold K3 (which has no Killing vectors). On this space there are two zero modes of the Dirac operator – but 42 zero modes of the Rarita–Schwinger operator. The large discrepancy is caused, in this case, by a much larger coefficient of the axial anomaly for Rarita–Schwinger fields. This example shows, incidentally, that the rather large number of zero modes that would be required to describe what is observed in physics is not necessarily out of reach.

In the approach considered here, the solution of the problem of flavor – the problem of the existence of several "generations" of fermions with the same quantum numbers – would be that the extra dimensions have a sufficiently complex topology that there are several zero modes with the same $SU(3)\times SU(2)\times U(1)$ quantum numbers. When an operator has several zero modes, they are not necessarily related by any symmetry. For instance, the isospinor Dirac operator in a Yang–Mills instanton of topological number K has K modes; these modes are not related by any symmetry. This is fortunate, because the various generations of fermions have very different masses and are not obviously related by any symmetry.

Unfortunately, there is a basic reason that this idea does not work, at least not in the form described above. The reason for this is related to one of the most basic facts about the observed quarks and leptons: the fermions of given helicity transform in a complex representation of the gauge group, or, to put it differently, right-handed fermions do not transform the same way that the left-handed fermions transform. For instance, left-handed color triplets (quarks) are SU(2) doublets, but right-handed color triplets are SU(2) singlets. This is the reason that quarks and leptons do not have bare masses but receive their mass from the Higgs mechanism – from $SU(2)\times U(1)$ symmetry breaking. This is a very important fact theoretically, because it is the basis for our theoretical understanding of why the quarks and leptons are very light compared to the mass scale of grand unification or the Planck mass. If left- and right-handed fermions transformed the same way under the the gauge group, bare masses would have been possible and could have been arbitrarily large.

In the framework that has been described above, right- and left-handed fermions would inevitably transform the same way under $SU(3)\times SU(2)\times U(1)$. The reason for this is that low mass fermions are supposed to arise as zero modes of the Rarita–Schwinger operator in the extra dimensions. But the Rarita–Schwinger operator in the seven extra dimensions does not "know" whether a spinor field is left- or right-handed with respect to four-dimensional Lorentz transformations. It treats four-dimensional left- and right-handed fermions in the same way. One therefore could not get the observed $SU(3)\times SU(2)\times U(1)$ representation. One would inevitably get vector-like rather than V–A weak interactions, with bare masses being possible for all fermions. (Indeed, precisely because bare masses would be possible for all fermions, it is not natural to get any massless fermions at all.)

There is an intriguing mechanism by which, at first sight, it seems that the internal space Rarita–Schwinger equation could treat left and right fermions differently. Eleven-dimensional spinors are constructed with eleven gamma matrices γ_i, $i = 1 \cdots 11$. Let us define an operator $\Gamma_{11} = i\gamma_1 \cdots \gamma_{11}$ which is a sort of eleven-dimensional helicity operator. Let us also define an operator $\Gamma_4 = i\gamma_1\gamma_2\gamma_3\gamma_4$ which measures the ordinary four-dimensional helicity, and an operator $\Gamma_7 = \gamma_5 \cdots \gamma_{11}$ which one might think of as "helicity" in the internal eleven-dimensional space. Then $\Gamma_{11}^2 = \Gamma_4^2 = \Gamma_7^2 = 1$ and $\Gamma_{11} = \Gamma_4\Gamma_7$.

The Rarita–Schwinger field ψ of eleven-dimensional supergravity satisfies a Weyl condition $\psi = \Gamma_{11}\psi$. (This condition must be imposed; otherwise there would be more Fermi than Bose degrees of freedom and supersymmetry would not be possible.) This identity may equivalently be written $\Gamma_4\psi = \Gamma_7\psi$.

The latter equation shows that in eleven-dimensional supergravity the four-dimensional helicity of fermions is correlated with the seven-dimensional "helicity". Components with $\Gamma_4 = +1$ (or -1) have $\Gamma_7 = +1$ (or -1). If the quantum numbers of zero modes of the seven-dimensional Rarita–Schwinger equation depended on Γ_7, as one might intuitively expect, they would also depend on Γ_4.

Unfortunately, the spectrum of the seven-dimensional Rarita–Schwinger operator does not depend on Γ_7. The reason for this is very simple (and depends only on the fact that the number of extra dimensions is odd). In defining how spinors transform under coordinate transformations in riemannian geometry one needs the matrices $\sigma_{ij} = [\gamma_i, \gamma_j]$. One does not (on an orientable manifold) need the γ_i themselves. The transformation $\gamma_i \leftrightarrow -\gamma_i$ does not change the σ_{ij}, so it does not affect the definition of spinors. It does, however, change the sign of $\Gamma_7 = \gamma_1\gamma_2 \cdots \gamma_7$. Consequently, spinors with opposite values of Γ_7 transform the same way under coordinate transformations. Since, in the approach discussed here, $SU(3)\times SU(2)\times U(1)$ transformations are coordinate transformations, spinors with opposite values of Γ_7 have the same $SU(3)\times SU(2)\times U(1)$ quantum numbers.

One could try to avoid this conclusion by taking the extra seven dimensions to be a non-orientable manifold. In a non-orientable manifold, the definition of spinors is subtle and involves the γ_i as well as σ_{ij}. However, seven-dimensional non-orientable manifolds with SU(3) × SU(2) × U(1) symmetry are not abundant (they have all been listed above), and it is not difficult to show that none of them are suitable.

One might also try to avoid the above stated conclusion by going beyond riemannian geometry to include some variant of torsion. What possibilities this would offer is not very clear; the matter will be discussed at the end of this paper.

Obtaining the right quantum numbers for quarks and leptons is, of course, not the only problem that must be faced in order to obtain a realistic theory, although it may be the most difficult problem. We must also worry about spontaneous breaking of supersymmetry, spontaneous breaking of *CP*, spontaneous breaking of SU(2) × U(1) gauge symmetry, and obtaining the proper values of the low-energy parameters (coupling constants, masses, and mixing angles); and we must worry about what the true ground state of the theory really is. These questions will now be briefly discussed in turn.

For spontaneous breaking of supersymmetry the prospects are very bright; in fact, supersymmetry almost inevitably is spontaneously broken as part of any scheme in which there are compact dimensions with a non-abelian symmetry.

The reason for this is the following. Unbroken supersymmetry means that under a supersymmetry transformation the vacuum expectation values of the fields do not change. The vacuum expectation values of the Bose fields automatically are invariant under supersymmetry, since their supersymmetric variation would be proportional to the (vanishing) vacuum expectation values of the Fermi fields. The delicate question is whether the vacuum expectation values of the fermi fields change under supersymmetry.

To illustrate the point, let us ignore the possible presence in the theory of Bose fields other than the gravitational field. Then the transformation law for the Rarita–Schwinger field is $\delta\psi_\mu = D_\mu\varepsilon$, ε being the gauge parameter. An unbroken supersymmetry – a symmetry of the vacuum – must have $\delta\psi_\mu = 0$, so unbroken supersymmetry transformations correspond to solutions of $D_\mu\varepsilon = 0$.

On a curved manifold, this equation will almost certainly not have solutions, since $D_\mu\varepsilon = 0$ implies the integrability condition $[D_\mu, D_\nu]\varepsilon = 0$ or $R_{\mu\nu\alpha\beta}[\gamma^\alpha, \gamma^\beta]\varepsilon = 0$, which on most curved manifolds is not satisfied by any non-zero ε. For instance, on none of the M^{pqr} does a solution exist. (The properties of seven-dimensional manifolds admitting solutions of $D_\mu\varepsilon = 0$ have been discussed in the mathematical literature [17], but non-trivial examples do not seem to be known.) So in theories with curved extra dimensions, there will generally not be any unbroken supersymmetries.

The picture does not change greatly when one includes Bose fields other than the gravitational field. We now have $\delta\psi_\mu = \bar{D}_\mu\varepsilon$, where $\bar{D}_\mu = D_\mu$ plus non-minimal terms involving the vacuum expectation values of other Bose fields (and possibly involving the expectation values of fermion bilinears, as discussed below). Unbroken supersymmetries are now solutions of $\bar{D}_\mu\varepsilon = 0$, but solutions will still typically not exist because the integrability condition $[\bar{D}_\mu, \bar{D}_\nu]\varepsilon = 0$ will still not have solutions.

Although solutions will generally not exist, the extra dimensions and the vacuum expectation values of the fields may be just such that one or more solutions of $\bar{D}_\mu \varepsilon = 0$ would exist. Each solution of $\bar{D}_\mu \varepsilon = 0$ in the internal space would correspond to an unbroken supersymmetry charge in four dimensions. If there is precisely one such solution, and so only one unbroken supersymmetry generator, this corresponds to a theory in which $N = 8$ supersymmetry has been spontaneously broken down to $N = 1$ supersymmetry. If there are K solutions, there is an unbroken $N = K$ supersymmetry.

A particularly attractive possibility would be a theory in which the equation $\bar{D}_\mu \varepsilon = 0$ has precisely one solution in the extra dimensions, corresponding to unbroken $N = 1$ supersymmetry. With $N = 1$ supersymmetry it is possible to construct more or less realistic models of observed particle physics. With $N \geqslant 2$ it is not possible to make a realistic model, because the supersymmetry algebra for $N \geqslant 2$ forces left- and right-handed fermions to transform in the same way under the gauge group, in contrast with what is observed. It is attractive to believe that $N = 1$ supersymmetry might survive after compactification of seven dimensions because this would severely constrain the theory, would make many predictions that might be testable in accelerators, and [19] might shed light on SU(2) × U(1) breaking and the gauge hierarchy problem. Of course, we would then have to explain how $N = 1$ supersymmetry is eventually spontaneously broken at low energies.

In addition to supersymmetry breaking, we must also explain P and CP breaking in order to construct a realistic theory. The eleven-dimensional supergravity langrangian is invariant under inversions of space (or time) combined with a change of sign of the antisymmetric tensor gauge field that exists in this theory. After compactification of seven dimensions, the eleven-dimensional symmetry could be manifested as both P (inversion of space) and C (inversion of the compact dimensions). These potential invariances must be spontaneously broken.

A natural mechanism for spontaneous breaking of P, C, and CP involves the antisymmetry tensor gauge field of the eleven-dimensional supergravity theory. The curl $F_{\alpha\beta\gamma\delta}$ of this field may have a vacuum expectation value without breaking Lorentz invariance or SU(3) × SU(2) × U(1). In fact, as discussed recently by several authors [20], a vacuum expectation value of F_{1234} is Lorentz invariant. It would violate P and CP but conserve C. The components F_{ijkm} for $i \cdots m \geqslant 5$ may also have expectation values, which would spontaneously break C and CP but conserve P. It is not difficult to see (by considering the little group of a point on M^{pqr}) that on any of the M^{pqr}, the most general SU(3) × SU(2) × U(1) invariant vacuum expectation value of F_{ijkl} depends on two real parameters.

Although the eleven-dimensional theory can have spontaneous breaking of C, P, and CP, the strong interaction angle θ will inevitably vanish at the tree level. The reason for this is that in the eleven-dimensional theory, there is no operator which might be added to the lagrangian which reduces in four dimensions to $\theta \int d^4x F_{\mu\nu} \tilde{F}_{\mu\nu}$. There simply does not exist in eleven dimensions any topological invariant that can be written as the integral of a lagrangian density. Of course, the question of how large a vacuum angle might be generated by quantum corrections must wait until we understand how to do calculations in this (presumably) non-renormalizable theory.

It is also necessary, of course, to obtain $SU(2)\times U(1)$ symmetry breaking; this presumably means that we must find, at the tree level, a massless Higgs doublet which could later obtain a very tiny negative mass squared.

There are various ways that, in a Kaluza–Klein theory, one might obtain massless charged scalars. In the original Kaluza–Klein theory, with a single compact dimension (a circle) there is a massless scalar (at least at the tree level) because the classical field equations do not determine the radius of the circle. Space-time dependent fluctuations of this radius would be observed as a massless scalar degree of freedom.

If the equations that determine our hypothetical ground state $M^4\times M^{pqr}$ admit not a unique solution for the metric of M^{pqr} but a whole family of solutions, then oscillations within this family would be observed as massless scalars. Some of these oscillations might involve departures from $SU(2)\times U(1)$ symmetry and could be the desired Higgs bosons.

One might also obtain massless scalars as components of the antisymmetric tensor gauge field. In fact, massless scalars can be obtained in this way, but tend to be neutral under the gauge group.

Regardless of where the scalars come from, why would they be massless? The most plausible explanation would be an unbroken supersymmetry relating the massless bosons to massless fermions. This could involve the possibility discussed above that the equation $\tilde{D}_\mu\varepsilon=0$ has a unique non-trivial solution, leaving $N=1$ supersymmetry unbroken. In this case, of course, we must hope to find a non-perturbative mechanism spontaneously breaking the supersymmetry and giving a small vacuum expectation value to the scalar bosons. (Some relevant issues will be discussed in a future paper [21].)

Without understanding the Higgs bosons and the low-energy symmetry breaking, it is of course not possible to predict the quark and lepton masses and mixing angles. If we understood the dynamics that determines the metric of M^{pqr} (assuming that the ground state really is $M^4\times M^{pqr}$), we could predict the strong, weak, and electromagnetic coupling constants, since the gauge fields all arise, by the ansatz of eq. (3), as part of the metric tensor in eleven dimensions, and the gauge field kinetic energy is part of the Einstein action. [The most general $SU(3)\times SU(2)\times U(1)$ invariant metric on M^{pqr} depends on three arbitrary parameters. If we understood the dynamics and could calculate the three parameters, we could predict the SU(3), SU(2), and U(1) coupling constants.] Even though we do not understand this dynamics (see below), it is possible to make a useful comment.

In a theory of this kind, the gauge coupling constants, which are determined by integrating the action over the compact dimensions, would scale as a rather high power of $1/(M_pR)$, where M_p is the Planck mass and R is the radius of the extra dimensions. The fact that the observed gauge coupling constants in nature differ from one by only one or two orders of magnitude shows that R cannot be too much greater than $1/M_p$; the extra dimensions really have a radius not too different from 10^{-33} cm.

The eleven-dimensional supergravity theory has no global symmetry that could be interpreted as baryon number, so in this theory nucleons are almost surely unstable. The mass scale in nucleon decay, however, would probably be $1/R$, which is the mass scale of the heavy quanta in this theory. Since, as just noted, $1/R$ cannot be much less than M_p, the nucleon lifetime will probably be very long, perhaps 10^{45} years, which is far too long for nucleon decay to be observable. If the present nucleon decay experiments give a positive result, the approach described in this paper would become significantly less attractive.

It is now time to finally discuss the question of whether one can really sensibly expect $M^4 \times M^{pqr}$ to be the ground state of this theory.

The most attractive possibility would be that $M^4 \times M^{pqr}$ might be a solution of the classical equations of motion, possibly with a suitable vacuum expectation assumed for $F_{\mu\nu\alpha\beta}$. Unfortunately, a straightforward calculation shows that this is not true (regardless of what vacuum expectation value one assumes). If one arbitrarily adds to the lagrangian a cosmological constant (with a sign corresponding to a positive energy density) then $M^4 \times M^{pqr}$ can be a solution. However, local supersymmetry does not permit a cosmological constant in the eleven-dimensional lagrangian.

This problem is not necessarily fatal, since one can always hope that $M^4 \times M^{pqr}$, although not a solution of the classical equations of motion, is the minimum of the appropriate effective potential. In eleven-dimensional supergravity, there is no small dimensionless parameter whose smallness could justify the use of the classical field equations as an approximation. So the fact that $M^4 \times M^{pqr}$ does not satisfy the classical equations, while not encouraging, is not necessarily critical.

In any case, there is absolutely no obvious reason that $M^4 \times M^{pqr}$, rather than the more obvious possibility of eleven dimensional Minkowski space, should be the ground state of this theory.

It will be shown in a separate paper that even when Kaluza–Klein vacuum states are stable classically, they can be destabilized by quantum mechanical tunneling [22]. However, unbroken supersymmetry (plus a technical requirement that the extra dimensions be simply connected; this is not satisfied in the original Kaluza–Klein theory) seems to be a sufficient condition for stability. This is another reason that theories in which $\bar{D}_\mu \varepsilon = 0$ has a solution and there is an unbroken supersymmetry at the energies of compactification would be attractive.

As has been pointed out above, the most serious obstacle to a realistic model of the type considered in this paper is that the fermion quantum numbers do not turn out right. It is conceivable that this problem could be overcome if instead of riemannian geometry one considered geometry with torsion or some generalization of torsion; in such a theory the fermion transformation laws might be different.

How can one obtain torsion in eleven-dimensional supergravity? As has been noted [11], the theory formally contains torsion in the sense that certain fermion bilinears enter, formally, in the way that torsion would appear. Of course, a "torsion" that is bilinear in Fermi fields does not have a classical limit. However, by analogy with QCD, in which $\bar{q}q$ has a vacuum expectation value, one may be willing in supergravity to assume a vacuum expectation value for the "torsion field" $K \sim \bar{\psi}\psi$ (or perhaps for some other bilinears). Perhaps in this way the predictions for fermion quantum numbers can be modified. This possibility is under study.

I wish to acknowledge discussions with V. Bargmann and J. Wolf.

Note added in proof

For a recent discussion of Dirac zero modes in Kaluza–Klein theories, see ref. [23].

References

[1] Th. Kaluza, Sitzungsber. Preuss. Akad. Wiss. Berlin, Math. Phys. K1 (1921) 966
[2] O. Klein, Z. Phys. 37 (1926) 895; Arkiv. Mat. Astron. Fys. B 34A (1946); Contribution to 1955 Berne Conf., Helv. Phys. Acta Suppl. IV (1956) 58
[3] A. Einstein and W. Mayer, Preuss. Akad. (1931) p. 541, (1932) p. 130;
A. Einstein and P. Bergmann, Ann. Math. 39 (1938) 683;
A. Einstein, V. Bargmann, and P.G. Bergmann, Theodore von Kármán Anniversary Volume (Pasadena, 1941) p. 212;
O. Veblen, Projektive Relativitats Theorie (Springer, Berlin, 1933);
W. Pauli, Ann. der Phys. 18 (1933) 305; 337;
P. Jordan, Ann. der Physik (1947) 219;
Y. Thirz, C. R. Acad. Sci. 226 (1948) 216
[4] P.G. Bergmann, Introduction to the theory of relativity (Prentice-Hall, New York, 1942);
A. Lichnerowicz, Theories relativistes de la gravitation et de l'electromagnetisme (Masson, Paris, 1955)
[5] J. Rayski, Acta Phys. Pol. 27 (1965) 89;
W. Thirring, *in* 11th Schladming Conf., ed. P. Urban (Springer-Verlag, Wien, New York, 1972)
[6] B. DeWitt, *in* Lectures at 1963 Les Houches School, Relativity, groups, and topology, ed. B. DeWitt and C. DeWitt (New York, Gordon and Breach, 1964), published separately under the title Dynamical theory of groups and fields (New York, Gordon and Breach, 1965);
R. Kerner, Ann. Inst. H. Poincaré 9 (1968) 143;
A. Trautman, Rep. Math. Phys. 1 (1970) 29;
Y.M. Cho, J. Math. Phys. 16 (1975) 2029;
Y.M. Cho and P.G.O. Freund, Phys. Rev. D12 (1975) 1711;
Y.M. Cho and P.S. Jang, Phys. Rev. D12 (1975) 3789;
L.N. Chang, K.I. Macrae, and F. Mansouri, Phys. Rev. D13 (1976) 235
[7] J. Scherk and J.H. Schwarz, Phys. Lett. 57B (1975) 463;
E. Cremmer and J. Scherk, Nucl. Phys. B103 (1976) 393; B108 (1976) 409
[8] J.-F. Luciani, Nucl. Phys. B135 (1978) 111
[9] L. Palla, Proc. 1978 Tokyo Conf. on High-energy physics, p. 629
[10] N. Manton, Nucl. Phys. B158 (1979) 141
[11] E. Cremmer, B. Julia and J. Scherk, Phys. Lett. 76B (1978) 409;
E. Cremmer and B. Julia, Phys. Lett. 80B (1978) 48; Nucl. Phys. B159 (1979) 141
[12] J. Scherk and J.H. Schwarz, Phys. Lett. 82B (1979) 60; Nucl. Phys. B153 (1979) 61;
E. Cremmer, J. Scherk, and J.H. Schwarz, Phys. Lett. 84B (1979) 83

[13] W. Nahm, Nucl. Phys. B135 (1978) 149
[14] M.T. Grisaru, H.N. Pendleton and P. van Nieuwenhuysen, Phys. Rev. D15 (1977) 996
[15] F.A. Berends, J.W. van Holten, B. deWit and P. van Nieuwenhuizen, Nucl. Phys. B154 (1979) 261; Phys. Lett. 83B (1979) 188; J. Phys. A13 (1980) 1643;
C. Aragone and S. Deser, Phys. Lett. 85B (1979) 161; Phys. Rev. D21 (1980) 352;
K. Johnson and E.C.G. Sudarshan, Ann. of Phys. 13 (1961) 126;
G. Velo and D. Zwanziger, Phys. Rev. 186 (1967) 1337
[16] N. Hitchin, J. Diff. Geom. 9 (1974) 435;
S.T. Yau, Proc. Nat. Acad. Sci. US 74 (1977) 1798;
S. Hawking, Nucl. Phys. B146 (1978) 381
[17] R. Brown and A. Gray, Trans. Amer. Math. Soc. 141 (1969) 465
[18] P. Fayet, Phys. Lett. 69B (1977) 489; 84B (1979) 421;
G.R. Farrar and P. Fayet, Phys. Lett. 76B (1978) 575; 79B (1978) 442; 89B (1980) 191
[19] S. Weinberg, Phys. Lett. 62B (1976) 111
[20] A. Aurilia, H. Nicolai and P.K. Townsend, preprint, CERN Th. 2884 (1980);
P. Freund and M.A. Rubin, Phys. Lett. 97B (1980) 233
[21] E. Witten, preprint, Princeton Univ. (April, 1981)
[22] E. Witten, in preparation
[23] W. Mecklenburg, preprint, ICTP (March, 1981)

Reprinted from Phys. Rev. Lett. **50** (1983) 294–7

$N = 8$ Supergravity Breaks Down to $N = 1$

M. A. Awada, M. J. Duff,[a] and C. N. Pope
Blackett Laboratory, Imperial College, London SW7 2BZ, England
(Received 9 December 1982)

The field equations of $N = 1$ supergravity in $d = 11$ dimensions admit a spontaneous compactification on the seven-sphere to an $N = 8$ theory in $d = 4$ with local SO(8) invariance. A spontaneously broken version of this theory is provided by another solution of the field equations for which the metric on S^7 is distorted. The isometry group of this new solution is $SO(5) \otimes SU(2)$, the relevant holonomy group is G_2, and the $N = 8$ supersymmetry is broken down to $N = 1$ at the Planck scale. The implications for grand unified theories are briefly discussed.

PACS numbers: 04.60.+n, 11.30.Pb, 12.25.+e

Since its discovery in 1976, progress in supergravity[1] has evolved along rather diverse lines. There are those who, awed by the majesty of extended supersymmetry, have looked to the $N = 8$ theory as a possible unification of all interactions, while the more phenomenologically minded have concentrated on $N = 1$ supersymmetry as a means of solving the gauge hierarchy problem and accomodating chiral fermion representations in grand unified theories. Consequently, the hope has sometimes been expressed that these two approaches could be linked if $N = 8$ supersymmetry were to break down to $N = 1$ at the Planck scale. In this paper, we demonstrate that this hope can indeed be realized.

Our starting point is the observation[2,3] that the field equations of $N = 1$ supergravity in $d = 11$ dimensions[4] admit a candidate ground-state solution corresponding to the product of four-dimensional anti–de Sitter space (AdS) and the seven-sphere with its standard metric, i.e., the coset space SO(8)/SO(7). This seven-sphere admits 28 Killing vectors and eight Killing spinors (i.e., eight spinors which are covariantly constant with respect to the de Sitter covariant derivative appearing in the transformation law for the gravitino). It gives rise, à la Kaluza-Klein, to an effective four-dimensional theory with local SO(8) invariance and $N = 8$ supersymmetry. The SO(8) Yang-Mills coupling constant e is given by $e^2 \sim m^2 M_P^{-2}$, where m^{-1} is the radius of S^7 and M_P is the Planck mass. Thus m is of order M_P if e is of order 1. The massless sector describes one spin 2, eight spin $\frac{3}{2}$, 28 spin 1, 56 spin $\frac{1}{2}$, and 70 spin 0 (35 scalars plus 35 pseudoscalars) and may probably be identified with the gauged $N = 8$ theory of de Wit and Nicolai,[5] where, in addition to the obvious local SO(8) with 28 elementary gauge bosons, one finds a hidden SU(8) with 63 composite gauge bosons. This SU(8) has formed the basis for possible grand unification schemes.[6]

By extending some earlier work on "squashing" three-spheres and symmetry breaking,[7] it has recently been suggested[2,3] that other solutions of the $d = 11$ field equations which are topologically still S^7 but which deviate from the maximally symmetric S^7 geometry could give rise to a spontaneously broken version of this gauged $N = 8$ theory. The idea is that distortion of the seven-sphere corresponds to nonvanishing vacuum expectation values for the scalar fields and hence to a Higgs and super-Higgs effect in $d = 4$. In this paper we exhibit such a solution corresponding to the fact that S^7 admits not one but two Einstein metrics. In addition to the "round" S^7 discussed above there is a "squashed" S^7 with isometry group $Sp(2) \otimes Sp(1) \cong SO(5) \otimes SU(2)$ and for which the de Sitter connection has holonomy group G_2. Remarkably, we find that it admits but one Killing spinor and hence yields an effective four-dimensional theory with one unbroken supersymmetry. Thus $N = 8$ supersymmetry is broken down to $N=1$ at the Planck scale.

We adopt the following conventions. The $d=11$ space has signature $(-+++\ldots)$, and $d=11$ indices $M, N, \ldots$ will be decomposed into $d=4$ indices $\mu, \nu, \ldots$ and $d=7$ indices $m, n, \ldots$. The $d=11$ Dirac matrices satisfy

$$\{\Gamma_A, \Gamma_B\} = 2\eta_{AB} \tag{1}$$

and may be written

$$\Gamma_A = (\gamma_\alpha \otimes \underline{1},\ \gamma_5 \otimes \Gamma_a), \tag{2}$$

where γ_α and Γ_a are the Dirac matrices in $d=4$ and $d=7$, respectively. Spontaneous compactification works as follows: We set the fermion fields to zero and examine the boson field equation of the $d=11$ theory[4]:

$$R_{MN} - \tfrac{1}{2} g_{MN} R = \tfrac{1}{3}\left[F_{MPQR} F_N{}^{PQR} - \tfrac{1}{8} g_{MN} F^2 \right], \tag{3}$$

$$\nabla_M F^{MNPQ} = -\tfrac{1}{576}\, \epsilon^{M_1 \ldots M_8 NPQ} F_{M_1 \ldots M_4} F_{M_5 \ldots M_8}. \tag{4}$$

Equation (4) is solved by the Freund-Rubin[8] choice for which all components F_{MNPQ} vanish except

$$F_{\mu\nu\rho\sigma} = 3m\epsilon_{\mu\nu\rho\sigma}, \tag{5}$$

where m is a constant. Substituting into Eq. (3) gives

$$R_{\mu\nu} = -12m^2 g_{\mu\nu} \tag{6}$$

and

$$R_{mn} = 6m^2 g_{mn}. \tag{7}$$

Thus the eleven-dimensional space becomes a product of a four-dimensional Einstein space with negative cosmological constant and a seven-dimensional Einstein space with positive cosmological constant.

There are still infinitely many solutions of (6) and (7) but the ground state should presumably be distinguished by its symmetries. With this in mind, we proceed as in Refs. 2 and 3 to restrict the solution further by requiring that the vacuum be supersymmetric, i.e., by requiring the existence of covariantly constant spinors ϵ for which

$$\delta\psi_M = \bar{D}_M \epsilon = 0, \tag{8}$$

where

$$\bar{D}_M = D_M - \frac{i}{144}(\Gamma^{NPQR}{}_M + 8\Gamma^{PQR}\delta_M{}^N) F_{NPQR}\epsilon. \tag{9}$$

Substituting Eq. (5) into Eq. (9) yields

$$\bar{D}_\mu = D_\mu + m\gamma_\mu\gamma_5, \tag{10}$$

$$\bar{D}_m = D_m - \tfrac{1}{2} m\Gamma_m. \tag{11}$$

When $m \neq 0$, the requirement of $N=8$ supersymmetry singles out the unique choice of $AdS \otimes S^7$ with the standard SO(8)-invariant metric on S^7. This is because there will be one unbroken supersymmetry (i.e., one massless gravitino) for each Killing spinor in $d=7$, i.e., each spinor satisfying

$$\bar{D}_m \eta = 0. \tag{12}$$

The integrability condition is

$$[\bar{D}_m, \bar{D}_n]\eta = -\tfrac{1}{4} R_{mn}{}^{ab}\Gamma_{ab}\eta + \tfrac{1}{2} m^2 \Gamma_{mn}\eta. \tag{13}$$

Since spinors in $d=7$ have eight components, there will be eight covariantly constant spinors if and only if

$$R_{mnpq} = m^2(g_{mp}g_{nq} - g_{mq}g_{np}), \tag{14}$$

and we obtain the maximally symmetric special case of Eq. (7) corresponding to the seven-sphere with its standard metric of constant curvature. Similarly the vanishing of $[\bar{D}_\mu, \bar{D}_\nu]$ on spinors in $d=4$ implies that space-time is AdS = SO(3, 2)/SO(3, 1). As discussed in Refs. 2 and 3, this vacuum solution then yields the effective $d=4$ theory with local SO(8) invariance and $N=8$ supersymmetry.

One can now contemplate other solutions of Eq. (7), i.e., other seven-dimensional Einstein metrics which might admit fewer than eight Killing spinors. Of particular interest would be one with the same S^7 topology since this would correspond to a spontaneously broken version of the previous theory. Remarkably S^7 does indeed admit another Einstein metric for which the sphere is squashed in a special homogeneous manner. It may be described as the distance sphere in $P_2(H)$ the quaternionic projective plane, i.e., as the level surface formed by all geodesics of length r emanating from a point in $P_2(H)$.[9] When r is very small, the distance sphere approaches the standard "round" S^7 [which is an S^3 bundle over S^4; in fact this corresponds to the $K=1$ SU(2) Yang-Mills instanton in four-dimensional Euclidean space] but as r increases the length of the S^3 fibers shrinks relative to the size of the S^4 base, and the eigenvalues α_n $(0 \leq n \leq 6)$ of the Ricci tensor split into two sets $\alpha_0 = \alpha_1 = \alpha_2 = \alpha_3 = \alpha$, $\alpha_4 = \alpha_5 = \alpha_6 = \beta$, $\alpha \neq \beta$. However, for a certain value of r the two sets become equal again, and the metric becomes an Einstein metric. The symmetry of this distorted sphere is the group which leaves a point in $P_2(H)$ fixed, namely $Sp(2) \otimes Sp(1)$ [where $Sp(n)$ is the group of $n \times n$ quaternionic matrices], which is isomorphic to $SO(5) \otimes SU(2)$.

Explicitly, the Fubini-Study Einstein metric on $P_2(H)$, written in terms of quaternionic coordinates (q_1, q_2), is

$$ds^2 = (1 + \sum_k \bar{q}_k q_k)^{-1} \sum_i d\bar{q}_i dq_i - (1 + \sum_k \bar{q}_k q_k)^{-2} \sum_{i,j} \bar{q}_i dq_i d\bar{q}_j q_j . \tag{15}$$

Defining

$$q_1 = \tan\chi \cos(\tfrac{1}{2}\mu) U, \quad q_2 = \tan\chi \sin(\tfrac{1}{2}\mu) V, \tag{16}$$

where U and V are quaternions of unit modulus, it is straightforward to show that the 1-forms σ_i, Σ_i defined by

$$U^{-1} dU = i\sigma_1 + j\sigma_2 + k\sigma_3, \quad V^{-1} dV = i\Sigma_1 + j\Sigma_2 + k\Sigma_3, \tag{17}$$

satisfy the algebra $d\sigma_1 = -\sigma_2 \wedge \sigma_3$, etc., of SU(2). ($i, j, k$ are the imaginary quaternions.) Finally, defining

$$\nu_i = \sigma_i + \Sigma_i, \quad \omega_i = \sigma_i - \Sigma_i, \tag{18}$$

the metric on $P_2(H)$ becomes

$$ds^2 = d\chi^2 + \tfrac{1}{4}\sin^2\chi[d\mu^2 + \tfrac{1}{4}\sin^2\mu\,\omega_i{}^2 + \tfrac{1}{4}\cos^2\chi(\nu_i + \cos\mu\,\omega_i)^2]. \tag{19}$$

The distance sphere of radius $r(=\tan\chi)$, centered on $q_1 = q_2 = 0$, inherits the metric obtained from (19) by setting $\chi = \text{const}$. Thus up to an overall scaling constant, the squashed seven-sphere has metric

$$ds^2 = d\mu^2 + \tfrac{1}{4}\sin^2\mu\,\omega_i{}^2 + \tfrac{1}{4}\lambda^2(\nu_i + \cos\mu\,\omega_i)^2, \tag{20}$$

where λ is a constant parameter describing the degree of distortion. The round, SO(8)-invariant, sphere corresponds to $\lambda^2 = 1$.

Introducing the orthonormal basis $e^0 = d\mu$, $e^1 = \frac{1}{2}\sin\mu\,\omega_1, \ldots, e^4 = \frac{1}{2}\lambda(\nu_1 + \cos\mu\,\omega_1)$, etc., a straightforward calculation shows that the curvature form $\theta_{mn} = \frac{1}{2}R_{mnpq}e^p \wedge e^q$ is given by

$$\begin{aligned}
\theta_{01} &= (1 - \tfrac{3}{4}\lambda^2)e^0 \wedge e^1 + \tfrac{1}{2}(1 - \lambda^2)e^5 \wedge e^6, \\
\theta_{04} &= \tfrac{1}{4}\lambda^2 e^0 \wedge e^4 + \tfrac{1}{4}(1 - \lambda^2)[e^3 \wedge e^5 - e^2 \wedge e^6], \\
\theta_{12} &= (1 - \tfrac{3}{4}\lambda^2)e^1 \wedge e^2 + \tfrac{1}{2}(1 - \lambda^2)e^4 \wedge e^5, \\
\theta_{14} &= \tfrac{1}{4}\lambda^2 e^1 \wedge e^4 + \tfrac{1}{4}(1 - \lambda^2)[e^3 \wedge e^6 + e^2 \wedge e^5], \\
\theta_{15} &= \tfrac{1}{4}\lambda^2 e^1 \wedge e^5 - \tfrac{1}{4}(1 - \lambda^2)[e^0 \wedge e^6 + e^2 \wedge e^4], \\
\theta_{16} &= \tfrac{1}{4}\lambda^2 e^1 \wedge e^6 + \tfrac{1}{4}(1 - \lambda^2)[e^0 \wedge e^5 - e^3 \wedge e^4], \\
\theta_{45} &= \tfrac{1}{4}\lambda^{-2} e^4 \wedge e^5 + \tfrac{1}{2}(1 - \lambda^2)[e^0 \wedge e^3 + e^1 \wedge e^2],
\end{aligned} \tag{21}$$

where the remaining 14 components are obtained by performing simultaneous cyclic permutations of the triplets (1, 2, 3) and (4, 5, 6). The Ricci tensor $R_{mn} = \text{diag}(\alpha, \alpha, \alpha, \alpha, \beta, \beta, \beta)$ with

$$\alpha = 3 - \tfrac{3}{2}\lambda^2, \quad \beta = \lambda^2 + 1/2\lambda^2. \tag{22}$$

The Einstein condition, $\alpha = \beta$, gives two solutions for λ^2. $\lambda^2 = 1$ gives the round sphere discussed in Refs. 2 and 3, while $\lambda^2 = \frac{1}{5}$ gives the squashed sphere of this paper.

Substituting Eq. (21) with $\lambda^2 = \frac{1}{5}$ into Eq. (13) we find

$$[\bar{D}_m, \bar{D}_n]\eta = -\tfrac{1}{4}C_{mn}\eta, \tag{23}$$

where

$$C_{mn} = C_{mn}{}^{ab}\Gamma_{ab} \tag{24}$$

and $C_{mn}{}^{ab}$ is the Weyl tensor. Letting m run over $(0, i, \hat{i})$ where $i = 1, 2, 3$, and $\hat{i} = 4, 5, 6, = \hat{1}, \hat{2}, \hat{3}$, we find the 14 linearly independent components

$$C_{0i} = \tfrac{4}{5}[\Gamma_{0i} + \tfrac{1}{2}\epsilon_{ijk}\Gamma_{\hat{j}\hat{k}}], \tag{25}$$

$$C_{ij} = \tfrac{4}{5}[\Gamma_{ij} + \Gamma_{\hat{i}\hat{j}}], \tag{26}$$

$$C_{i\hat{j}} = \tfrac{4}{5}[-\Gamma_{i\hat{j}} - \tfrac{1}{2}\Gamma_{j\hat{i}} + \tfrac{1}{2}\delta_{ij}\Gamma_{k\hat{k}} - \tfrac{1}{2}\epsilon_{ijk}\Gamma_{0\hat{k}}], \tag{27}$$

where $C_{i\hat{j}}$ is trace free. The subgroup of SO(7) generated by these 14 linear combinations of Γ_{ab} corresponds to the holonomy[10] group of the connection of Eq. (11). Using the standard classification of Lie groups,[11] we find the exceptional group G_2.

To solve $C_{0i}\eta = 0 = C_{ij}\eta$ we resorted to an explicit representation,

$$\Gamma_0 = \gamma_0 \otimes \underline{1}, \quad \Gamma_i = \gamma_i \otimes \underline{1}, \quad \Gamma_{\hat{i}} = i\gamma_5 \otimes \tau_i, \tag{28}$$

and found that there is one solution. It automatically satisfies $C_{i\hat{j}}\eta = 0$. The existence of just one Killing spinor is related to the fact that G_2 is the stability subgroup of a seven-dimensional spinor. Thus spontaneous compactification on the squashed S^7 gives an effective $d = 4$ theory with one unbroken supersymmetry. This means in particular that seven of the eight gravitinos will acquire masses of order M_P.

Carrying out the complete calculation of the four-dimensional Lagrangian will not be an easy task, however, and it remains to be seen what influence the isometry group of $SO(5) \otimes SU(2)$ and the holonomy group of G_2 will have on the classification of particle states. We simply note that both contain $SU(2) \otimes SU(2)$ as a subgroup. There is a subtle interplay between the obvious SO(8) symmetry and hidden SU(8) symmetry in the unbroken phase and we expect an analogous interplay in the spontaneously broken phase. It will be interesting to see what effect this might have on the SU(8) unification schemes. As discussed by Witten,[12] in addition to the gauge hierarchy problem and chiral fermion problem in grand unified theories, there is another reason why one unbroken supersymmetry at the energy of compactification is an attractive feature of Kaluza-Klein theories: It can provide a mechanism for stabilizing the Kaluza-Klein vacuum.

Of course the surviving $N = 1$ supersymmetry must also be eventually spontaneously broken. In this connection, we note that a third solution of $d = 11$ field equations with S^7 topology has recently been found.[13] The seven-sphere is not squashed but the F_{mnpq} components of F_{MNPQ} are now nonvanishing and provide a parallelizing torsion. This solution also admits a Higgs interpretation corresponding to nonzero vacuum expectation values for the pseudoscalars.[2,3] It also involves the group G_2 (Ref. 13) but according to D'Auria, Fre, and van Nieuwenhuizen[14] all eight supersymmetries are broken. Since our one Killing spinor η may be used to build the required totally antisymmetric torsion, namely $\bar{\eta}\Gamma_{abc}\eta$, we expect more general solutions with both squashing and torsion and this suggests an $(N = 8) \to (N = 1) \to (N = 0)$ hierarchy.

We are grateful to C. J. Isham and K. S. Stelle for helpful discussions; one of us (C.N.P.) also thanks J. P. Bourguignon and N. J. Hitchin.

---

[a]Present address: CERN, CH-1211 Geneva 23, Switzerland.

[1]D. Z. Freedman, P. van Nieuwenhuizen, and S. Ferrara, Phys. Rev. D 13, 3214 (1976); S. Deser and B. Zumino, Phys. Lett. 62B, 335 (1976).

[2]M. J. Duff and C. N. Pope, in "Supergravity 82," edited by S. Ferrara, J. G. Taylor, and P. van Nieuwenhuizen (to be published). Reported by M. J. Duff, in *Supergravity 81*, edited by S. Ferrara and J. G. Taylor (Cambridge Univ. Press, London, 1982), and by M. J. Duff and D. J. Toms, in *Unification of the Fundamental Interactions II*, edited by S. Ferrara and J. Ellis (Plenum, New York, 1982), and CERN Report No. TH. 3259, 1982 (to be published).

[3]M. J. Duff, in Proceedings of the Marcel Grossman Meeting, Shanghai, August 1982, CERN Report No. TH. 3451, 1982 (to be published).

[4]E. Cremmer and B. Julia, Nucl. Phys. B159, 141 (1979).

[5]B. de Wit and H. Nicolai, Phys. Lett. 108B, 285 (1982).

[6]J. Ellis, M. K. Gaillard, L. Maiani, and B. Zumino, in *Unification of the Fundamental Interactions*, edited by S. Ferrara, J. Ellis, and P. van Nieuwenhuizen (Plenum, New York, 1980).

[7]R. Coquereaux and M. J. Duff, to be published.

[8]P. G. O. Freund and M. A. Rubin, Phys. Lett. 97B, 233 (1980).

[9]J. P. Bourguignon and A. Karcher, Ann. Sci. Normale Sup. 11, 71 (1978).

[10]S. Kobayashi and N. Nomizu, *Foundations of Differential Geometry* (Interscience, New York, 1969), Vol. 1.

[11]R. Gilmore, *Lie Group, Lie Algebras and Some of Their Applications* (Interscience, New York, 1974).

[12]E. Witten, Nucl. Phys. B186, 412 (1981).

[13]F. Englert, Phys. Lett. 119B, 339 (1982), and in Proceedings of the Supergravity Workshop, Trieste, September 1982 (to be published).

[14]R. D'Auria, P. Fre, and P. van Nieuwenhuizen CERN Report No. TH-3453, 1982 (to be published).

Reprinted from Phys. Lett. B **129** (1983) 39–42

COMPACTIFICATION OF d = 11 SUPERGRAVITY ON K3 × T³

M.J. DUFF [1], B.E.W. NILSSON and C.N. POPE [1]
Theory Group and Center for Theoretical Physics, The University of Texas at Austin, Austin, TX 78712, USA

Received 14 April 1983
Revised manuscript received 14 June 1983

When $d = 11$ supergravity spontaneously compactifies to $d = 4$, the number of unbroken sypersymmetries, $0 \leq N \leq 8$, is determined by the holonomy group, $\mathcal{H}$, of the $d = 7$ ground-state connection. Here we present a new solution: Minkowski spacetime × K3 × T^3, for which $\mathcal{H}$ = SU(2) and $N = 4$. The massless sector in $d = 4$ is given by $N = 4$ supergravity coupled to 22 $N = 4$ vector multiplets. Aside from its intrinsic interest, this example throws new light on Kaluza–Klein supergravity. In particular, we note that the 192 + 192 massless degrees of freedom obtained from K3 × T^3 exceed the 128 + 128 of the $N = 8$ theory obtained from T^7 or S^7.

The field equations of $N = 1$ supergravity in $d = 11$ dimensions admit of candidate ground-state solutions in which seven dimensions are compactified. Setting $\psi_M = 0$ ($M,N = 1,..., 11$), these equations are

$$R_{MN} - \tfrac{1}{2} g_{MN} R = \tfrac{1}{3} (F_{MPQR} F_N{}^{PQR} - \tfrac{1}{8} g_{MN} F_{PQRS} F^{PQRS}) \quad (1)$$

$$\nabla_M F^{MPQR} = -\tfrac{1}{576} \epsilon^{M_1 \ldots M_8 PQR} F_{M_1 \ldots M_4} F_{M_5 \ldots M_8} \, . \quad (2)$$

The Freund–Rubin [1] choice for which F_{MNPQ} vanished except for

$$F_{\mu\nu\rho\sigma} = 3m \, \epsilon_{\mu\nu\rho\sigma} \quad (3)$$

yields the product of a four-dimensional Einstein space time

$$R_{\mu\nu} = -12m^2 g_{\mu\nu} \quad (4)$$

with Minkowski signature and a seven-dimensional Einstein space with euclidean signature

$$R_{mn} = 6m^2 g_{mn} \, , \quad (5)$$

where $\mu, \nu = 1, ..., 4$ and $m, n = 5,..., 11$. Eq. (5) implies compactification when $m \neq 0$ and is consistent with, but does not imply, compactification when $m = 0$.

[1] On leave from The Blackett Laboratory, Imperial College, London, UK.

As discussed in refs. [2–4], the number of unbroken supersymmetries, N, in the resulting four-dimensional theory, is determined by the number of Killing spinors on the $d = 7$ manifold i.e. the number of spinors satisfying

$$\bar{D}_m \eta \equiv (\partial_m - \tfrac{1}{4} \omega_m{}^{ab} \Gamma_{ab} - \tfrac{1}{2} m e_m{}^a \Gamma_a) \eta = 0 \, , \quad (6)$$

where Γ_a are the $d = 7$ Dirac matrices,

$$\{\Gamma_a, \Gamma_b\} = -2\delta_{ab} \, . \quad (7)$$

$\Gamma_{ab} = \Gamma_{[a}\Gamma_{b]}$, and $\omega_m{}^{ab}$ and $e_m{}^a$ are the spin connection and siebenbein of the ground state solution to eq. (5). Such Killing spinors satisfy the integrability condition

$$[\bar{D}_m, \bar{D}_n] \, \eta = -\tfrac{1}{4} C_{mn}{}^{ab} \Gamma_{ab} \eta = 0 \, , \quad (8)$$

where $C_{mn}{}^{ab}$ is the Weyl tensor. The subgroup of Spin (7) generated by these linear combinations of the Spin (7) generators Γ_{ab} corresponds to the holonomy group $\mathcal{H}$ of the connection of eq. (6). Thus the maximum number of unbroken supersymmetries, N, is equal to the number of spinors left invariant by $\mathcal{H}$.

It is the exception, rather than the rule, that the ground state admits Killing spinors and, to date, only three examples have been discussed in the literature, two with $N = 8$ and $C_{mn}{}^{ab} = 0$ and one with $N = 1$ and $C_{mn}{}^{ab} \neq 0$. The case $m = 0$ and $N = 8$ singles out the seven torus of Cremmer and Julia [5] for which

$\mathcal{H} = \mathbf{1}$; the case $m \neq 0$ and $N = 8$ singles out the round seven-sphere of Duff and Pope [2,3] for which $\mathcal{H} = \mathbf{1}$ and then there is the squashed seven-sphere of Awada et al. [4] for which $m \neq 0, N = 1$ and $= G_2$. (In fact the full $d = 4$ theory obtained from the squashed S^7, including the massive states, corresponds to a spontaneously broken phase of the one obtained from the round S^7 [6].) Although T^7 and the round S^7 exhaust all possible $N = 8$ solutions (since they are the only Einstein spaces to be conformally flat[‡1] : $C_{mn}{}^{ab} = 0$) it is of interest to ask whether there are any others with $0 < N < 8$, and this leads naturally to the study of possible holonomy groups $\mathcal{H}$ for the connection of eq. (6). For $\mathcal{H} = \mathbf{1}$ we have $N = 8$ and for $\mathcal{H} = G_2$ we have $N = 1$ because these are the groups which, in seven dimensions, leave invariant 8 and 1 spinors respectively.

In this paper we examine another solution of eq. (5) with $m = 0$ for which $\mathcal{H} = SU(2)$. Since this SU(2) leaves invariant 4 spinors we find an effective $d = 4$ theory with $N = 4$ supersymmetry. The solution is given by the Ricci flat metric on $K3 \times T^3$ where K3 is Kummer's quartic surface in CP^3. A review of the K3 literature may be found in ref. [7]. The Ricci flat metric on K3 is not known explicitly but there is an existence proof. Moreover it is known to have 58 parameters, to have a self-dual Riemann tensor, and no symmetries. Topologically, K3 has Euler number $\chi = 24$ Hirzebruch signature $\tau = 16$ and Betty numbers

$$b_0 = 1, \quad b_1 = 0, \quad b_2 = 22, \quad b_3 = 0, \quad b_4 = 1\,. \tag{9}$$

This information will be sufficient for us to determine the Kaluza–Klein ansätze necessary to isolate the massless particle content of the resulting $N = 4$ supergravity theory. The number of massless particles of each spin is given by the number of zero-eigenvalue modes of the corresponding mass matrices. These are given by differential operators on the seven-dimensional ground state manifold (second order for bosons and first order for fermions) and are discussed in detail in ref. [3] for the case $m \neq 0$, where they were applied to S^7. To apply them to the $K3 \times T^3$ solution of this paper, we need only set $m = 0$. The results are given in table 1, where we compare with the reduction on T^7.

The single $g_{\mu\nu}$ comes from the single zero mode of the scalar laplacian, the three B_μ from the three Killing vectors on T^3 (K3 has no Killing vectors), and the 64 scalars S from the zero modes of the Lichnerowicz operator acting on symmetric rank-two tensors: 58 from K3 (the 58 parameters) and 6 from T^3 (the 6 parameters of the metric on $S^1 \times S^1 \times S^1$). We note that these 6 are Killing tensors but that the 58 are not.

As far as the fermions are concerned, we first note that since K3 is half-flat the holonomy group is SU(2) rather than the $SU(2) \times SU(2)$ of a generic four manifold and hence it admits two covariantly constant spinors (i.e. Killing spinors) which are left or right handed according as K3 is self-dual or anti-self-dual [8]. The four ψ_μ come from the four Killing spinors on $K3 \times T^3$ (2 on $K3 \times$ 2 on T^3). To obtain the 92 spin-$\frac{1}{2}$ fields χ we note that there are 40 zero-modes of the Rarita–Schwinger operator on K3: 38 of which are Γ-trace-free and 2 of which are not but are covariantly constant, while on T^3 there are 6 such zero-modes which are covariantly constant but not Γ-trace-free. We note that these 6 are Killing vector-spinors but that the 40 are not. With these conditions the

Table 1

$d = 11$	$d = 4$	spin	T^7	$K3 \times T^3$
g_{MN}	$g_{\mu\nu}$	2	1	1
	B_μ	1	7	3
	S	0	28	64
ψ_M	ψ_μ	3/4	8	4
	χ	1/2	56	92
A_{MNP}	$A_{\mu\nu\rho}$	–	1	1
	$A_{\mu\nu}$	0	7	3
	A_μ	1	21	25
	A	0	35	67

‡1 Strictly speaking, the Weyl tensor characterizes the *restricted* holonomy group of $\bar{D}_m$; i.e., it describes the rotation of a spinor parallel transported around a closed loop which is homotopic to zero. If the space is not simply connected there may be global obstructions to the existence of covariantly constant spinors, in addition to any local obstruction implied by the Weyl tensor. In addition to the ground state solutions T^7 and S^7, there will also exist solutions of the form T^7/Γ (generalizations of Klein bottles) and S^7/Γ (generalizations of lens spaces), where Γ is a discrete group. These spaces, like their T^7 or S^7 covering spaces, have $C_{abcd} = 0$, but these global considerations imply that they admit fewer than 8 covariantly constant spinors, and hence provide another means of obtaining $0 < N < 8$ supersymmetry.

We thank Don N. Page for discussions on these points.

92 modes, given by 40 on K3 × 2 Killing spinors on T^3 plus 6 on T^3 × 2 Killing spinors on K3, will be zero-modes of the spin-$\frac{1}{2}$ mass matrix [3].

The numbers of $A_{\mu\nu\rho}$, $A_{\mu\nu}$, A_μ and A fields are given by the zero-modes of the Hodge–de Rham operator acting on 0, 1, 2 and 3 forms respectively; i.e. by the Betti numbers b_0, b_1, b_2 and b_3 of K3 × T^3. But for a product manifold M = M′ × M″,

$$b_p = \sum_{r=0}^{p} b'_r b''_{p-r}, \tag{10}$$

where b_p, b'_p, b''_p are the p'th Betti numbers of M, M′ and M″ respectively. Hence from eq. (9), and the Betti numbers $b_p = \binom{3}{p}$ for T^3, we obtain the numbers given in table 1. A detailed discussion of boson and fermion zero-modes on K3 (and their relation to axial and conformal anomalies) may be found in refs. [8,9].

To summarize, the spin content is given by 1 spin 2, 4 spin $\frac{3}{2}$, 28 spin 1, 92 spin $\frac{1}{2}$, 67 scalars and 67 pseudoscalars. This corresponds to an $N = 4$ supergravity multiplet (1, 4, 6, 4, 1 + 1) coupled to 22 $N = 4$ spin-one matter multiplets (1, 4, 3 + 3).

Several comments are now in order especially since K3 × T^3 provides a counterexample to many claims to be found in the Kaluza–Klein literature.

(1) The number of massless degrees of freedom (per $d = 4$ spacetime point) of this $N = 4$ theory obtained from K3 × T^3, namely 192 + 192, exceeds the 128 + 128 of the $N = 8$ theory obtained from T^7 (or S^7). Thus one's naive expectation that the $N = 8$ theory maximizes the number of zero-modes is seen not to be fulfilled. Note that per $d = 4$ spacetime point, the $d = 11$ theory has (128 + 128) × ∞^7 degrees of freedom and so there is no contradiction in obtaining more than (128 + 128) when one isolates the massless states from the infinite tower of massive states. We do not know whether 192 + 192 is the maximum.

(2) Note that K3 × T^3 is neither a group manifold nor a coset space. Indeed K3 has no symmetries at all, yet this does not prevent a sensible Kaluza–Klein theory with a large number of massless particles. Of course, the 28 massless spin 1 are only abelian gauge fields, the gauge group being $[U(1)]^3 \times [GL(1, \mathbf{R})]^{25}$. Note also that K3 × T^3 provides the first example of a supersymmetric Kaluza–Klein theory for which the extra dimensional ground-state manifold is not parallelizable.

(3) The ansatz for the massless scalars coming from g_{MN} is not in general given by products of Killing vectors. When $m = 0$, the criterion for masslessness corresponds to zero-modes of the Lichnerowicz operator Δ_L. These are in one-to-one correspondence with the *number of parameters* of the ground state metric g_{mn} because Δ_L describes the first variation of the Einstein tensor and so its zero-modes preserve eq. (5) when $m = 0$. Thus T^7 yields 28 and K3 × T^3 yields 64. This ceases to be true when $m \neq 0$, however, because the ground state solution of eq. (4) is now anti de Sitter space. Massless scalars must now obey the conformal wave equation and hence, on the round S^7 for example, the mass matrix of ref. [3] is $(\Delta_L - 16\,m^2)$ which has 35 zero-modes rather than $(\Delta_L - 12\,m^2)$ which follows from the first variation of eq. (5) and which has no zero-modes. Hence it was found that 35 massless scalars come from g_{MN} even though the S^7 solution of eq. (5) has no parameters.

(4) How do the many parameters of K3 × T^3 show up in the effective four-dimensional theory? The answer is in the expectation values of the scalar fields. Compactification on Ricci flat manifolds yields no effective potential for the scalars and their expectation values are arbitrary. This contrasts with compactification on Einstein manifolds with $m \neq 0$. {Note incidentally that in this respect ungauged $N = 8$ supergravity [5] obtained from T^7 has many *more* parameters than gauged $N = 8$ supergravity [10] obtained from S^7, contrary to the claim that gauging *increases* the parameters from one (Newton's constant) to two (Newton's constant plus gauge coupling constant).}

Although we have focussed our attention on $N = 1$ supergravity in $d = 11$, solutions of the kind discussed here also exist for $N = 1$ in $d = 10$, 9 and 8 for which spacetime is Minkowski space and for which the extra dimensions are K3 × T^2, K3 × S^1 and K3 respectively. Owing to the Weyl condition in $d = 10$, we have $N = 2$ rather than $N = 4$. Omitting the details, we quote the results. Starting from the 64 + 64 components of $N = 1$ in $d = 10$, we obtain the 96 + 96 components in $d = 4$ of $N = 2$ supergravity coupled to 3 $N = 2$ vector multiplets and 20 $N = 2$ scalar multiplets. Starting from the 56 + 56 components of $N = 1$ in $d = 9$, we obtain the 92 + 92 components in $d = 4$ of $N = 2$ supergravity coupled to 2 $N = 2$ vector multiplets and 20 $N = 2$ scalar multiplets. Starting from the 48 + 48 components of $N = 1$ in $d = 8$, obtain the 88 + 88 components of $N = 2$ supergravity coupled to 1 $N = 2$ vector multiplet and 20 $N = 2$ scalar multiplets. Note that each drop in dimension from 10 to 8 corresponds to one less vector multiplet in $d = 4$.

Returning to the case of $d = 11$, we recall that solutions of eqs. (1) and (2) may be found for which the F_{mnpq} components of F_{MNPQ} are also non-zero. However, the solution of ref. [11] is known to break all 8 supersymmetries [12,13] and this is in fact an inevitable feature [14] of $F_{mnpq} \neq 0$ solutions. In order to find out whether other supersymmetry-preserving solutions exist, one can look to the holonomy group. For example the SU(3) subgroup of G_2 leaves invariant 2 spinors but we do not know of any solutions of eq. (5) with $\mathcal{H} = SU(3)$. Thus the outstanding problem is to classify all $d = 7$ Einstein metrics of non-negative curvature and their holonomy groups.

We are grateful to S. Weinberg for stimulating discussions on Kaluza–Klein theories and to J.A. Wheeler and S. Weinberg for their hospitality at the Center for Theoretical Physics and at the Theory Group. This publication was assisted by NSF Grant PHY 8205717 and by organized research funds of The University of Texas at Austin. Supported in part by the Robert A. Welch Foundation.

References

[1] P.G.O. Freund and M.A. Rubin, Phys. Lett. 97B (1980) 233.

[2] M.J. Duff, CERN preprint TH 3451 (1982), to be published in: Third Marcel Grossman Meeting (Shanghai, August 1982).

[3] M.J. Duff and C.N. Pope, Imperial College preprint ICTP/82-83/7, to be published in: Supergravity 82, eds. S. Ferrera, J.G. Taylor and P. van Niewenhuizen.

[4] M.A. Awada, M.J. Duff and C.N. Pope, Phys. Rev. Lett. 50 (1983) 294.

[5] E. Cremmer and B. Julia, Nucl. Phys. B159 (1979) 141.

[6] M.J. Duff, B.E.W. Nilsson and C.N. Pope, University of Texas preprint UTTG-5-83; and Phys. Rev. Lett. to be published.

[7] D.N. Page, Phys. Lett. 80B (1978) 55.

[8] S.W. Hawking and C.N. Pope, Nucl. Phys. B146 (1978) 381.

[9] S.M. Christensen and M.J. Duff, Nucl. Phys. B154 (1979) 301.

[10] B. de Wit and H. Nicolai, Phys. Lett. 108B (1982) 285; Nucl. Phys. B 208 (1982) 323.

[11] F. Englert, Phys. Lett. 119B (1982) 339.

[12] R. D'Auria, P. Fré and P. van Nieuwenhuizen, Phys. Lett. 122B (1983) 225.

[13] B. Biran, F. Englert, B. de Wit and H. Nicolai, Phys. Lett. 124B (1983) 45.

[14] F. Englert, M. Rooman and P. Spindel, Phys. Lett. 127B (1983) 47.

Reprinted from Phys. Lett. B **139** (1984) 39–41

EXACT PP-WAVE SOLUTIONS OF 11-DIMENSIONAL SUPERGRAVITY

C.M. HULL [1]

Department of Mathematics, Massachusetts Institute of Technology, Cambridge, MA 02139, USA

Received 11 January 1984

Exact solutions of the 11-dimensional supergravity field equations with non-trivial antisymmetric tensor and Rarita–Schwinger fields are given. Some of these are found to admit Killing spinors. The dimensional reduction to pp-wave or anti-gravitating multi-centre solutions of the four-dimensional theory is discussed.

The lagrangian of 11-dimensional supergravity [1] with the notation and conventions of ref. [2] and $\kappa = 1$, is

$$\begin{aligned}\mathcal{L} = & -\tfrac{1}{4} VR(\omega) - \tfrac{1}{48} V F_{MNPQ} F^{MNPQ} \\ & - \tfrac{1}{2}\mathrm{i} V \bar{\psi}_M \Gamma^{MNP} D_N(\tfrac{1}{2}(\omega + \hat{\omega}))\psi_P \\ & + \tfrac{1}{192} V \bar{\psi}_M (\Gamma^{MNWXYZ} + 12 g^{M[W}\Gamma^{XY}g^{Z]N}) \\ & \times \psi_N (F_{WXYZ} + \hat{F}_{WXYZ}) \\ & + (2/12^4)\epsilon^{M_1M_2\ldots M_{11}} F_{M_1M_2M_3M_4} F_{M_5M_6M_7M_8} \\ & \times A_{M_9M_{10}M_{11}} , \end{aligned} \tag{1}$$

where ω_{MAB} is the spin-connection including contorsion and $\hat{F}$ and $\hat{\omega}$ are supercovariantizations of F and ω. The lagrangian varies into a total divergence under the local supersymmetry transformations [1]

$$\delta e_M{}^A = -\mathrm{i}\bar{\epsilon}\Gamma^A \psi_M , \tag{2}$$

$$\delta A_{MNP} = \tfrac{3}{2}\bar{\epsilon}\Gamma_{[MN}\psi_{P]} , \tag{3}$$

$$\begin{aligned}\delta\psi_M = & D_M(\hat{\omega})\epsilon \\ & + \tfrac{1}{144}\mathrm{i}(\Gamma^{NPQR}{}_M - 8\Gamma^{PQR}\delta^N{}_M)\hat{F}_{NPQR}\epsilon \equiv \hat{D}_M\epsilon .\end{aligned} \tag{4}$$

Consider the 11-dimensional generalization of the pp-wave metric [3]

$$\mathrm{d}s^2 = 2\mathrm{d}u\,\mathrm{d}v - 2H(u,x^i)\mathrm{d}u^2 - \sum_{i=1}^{9}(\mathrm{d}x^i)^2 , \tag{5}$$

where the "light cone" co-ordinates are (v, u, x^i) with $i, j = 1, \ldots, 9$. Such metrics have been studied as solutions to pure 11-dimensional Einstein theory by J. Richer (unpublished). They admit a covariantly constant null vector k^M

$$k^M\partial/\partial x^M = \partial/\partial v, \quad k_{M;N} = 0, \quad g_{MN}k^Mk^N = 0 . \tag{6}$$

The space also admits covariantly constant spinors. It is convenient to introduce another null vector l^M

$$l^M\partial/\partial x^M = \partial/\partial u + H\partial/\partial v , \tag{7}$$

so that $k^M g_{MN} l^N = 1$.

A simple ansatz for a solution to the field equations obtained by varying (1), analogous to those discussed in refs. [4–6] is given by the metric (5) together with

$$\psi_M = k_M \chi , \tag{8}$$

$$F_{MNPQ} = k_{[M}\xi_{NPQ]} , \tag{9}$$

for some $\chi(u, x^i)$, $\xi_{MNP}(u, x^i)$ which are taken to be independent of the retarded time, v. Ansatz (8) is due to Urrutia [4] and leads to the vanishing of the contorsion and the Rarita–Schwinger stress-energy tensor and to trivial supercovariantizations, $\hat{F}_{MNPQ} = F_{MNPQ}$ and $\hat{\omega}_{MAB} = \omega_{MAB}$. The non-vanishing components of the curvature tensor for the space–time are

$$R_{MiNj} = k_M k_N H_{,ij} , \tag{10}$$

and so the scalar curvature R vanishes, which implies, through the gravitational field equation, that the spin-one field strength must be null ($F^2 = 0$) so that one

[1] Supported by a SERC/NATO Fellowship.

0.370-2693/84/$ 03.00 © Elsevier Science Publishers B.V.
(North-Holland Physics Publishing Division)

must have

$$\xi_{MNP}k^M = 0 \,. \tag{11}$$

The spin-one field equation and Bianchi identity are satisfied if

$$\xi^{MNP}{}_{;P} = 0, \quad \xi_{[MNP;Q]} = 0 \,, \tag{12}$$

so that ξ_{MNP} is harmonic, while the Einstein equation gives

$$\sum_{i=1}^{9} H_{,ii} = \tfrac{1}{12}\xi_{MNP}\xi^{MNP} \,. \tag{13}$$

The Rarita–Schwinger equation then reduces to

$$\Gamma^{MNP}D_N\psi_P = \Gamma^{MNP}k_P D_N\chi = 0 \,, \tag{14}$$

where the covariant derivative is given by

$$D_N\chi = \partial_N\chi - \tfrac{1}{2}\sum_{i=1}^{9}\Gamma^i H_{,i}(k\cdot\Gamma)k_N\chi \,. \tag{15}$$

Following ref. [6], it is convenient to introduce the projection operators

$$P_1 = \tfrac{1}{2}(k\cdot\Gamma)(l\cdot\Gamma), \quad P_2 = \tfrac{1}{2}(l\cdot\Gamma)(k\cdot\Gamma) \,, \tag{16}$$

and define

$$\zeta = P_1\chi \,, \quad \eta = P_2\chi \,. \tag{17}$$

Then (14) has solutions similar to those found in refs. [4,6],

$$\zeta = \zeta(x^i, u) \,, \quad \eta = \eta(u) \,, \tag{18}$$

with

$$\sum_{i=1}^{9}\Gamma^i\partial_i\zeta = 0 \,. \tag{19}$$

Then (5), (8) and (9) provide a solution to the full field equations, provided that $H(u, x^i)$, $\chi = \zeta(u, x^i) + \eta(u)$, $\xi_{MNP}(u, x^i)$ satisfy (11), (12), (13), (18) and (19). The solution can be interpreted as a supergravity wave advancing in the null direction given by k^M. If H, ξ_{MNP} and χ are independent of u, it is a plane wave, while for u-dependent functions one has a beam of radiation whose amplitude varies over the wave front. Just as a soliton is an extended field configuration sharing many of the properties of a massive particle, these wave solutions are in some ways analogous to massless particles.

A Killing spinor ϵ is one that parameterizes a supersymmetry transformation leaving the fields invariant, up to gauge transformations [6]. For (4) to vanish the spinor must satisfy

$$\hat{D}_M\epsilon = 0 \,. \tag{20}$$

As in ref. [6], it can be shown that the integrability conditions for (20) imply that

$$\xi_{MNP,i} = 0 \rightarrow \xi_{MNP} = \xi_{MNP}(u) \,. \tag{21}$$

If (21) holds, there are solutions $\epsilon = P_1\epsilon(u)$, $\epsilon(u)$ being given by the solutions of

$$\partial\epsilon/\partial u = (\mathrm{i}/72)\xi_{MNP}\Gamma^{MNP}\epsilon \,. \tag{22}$$

For the corresponding supersymmetry transformations of the Bose fields (2) and (3) to vanish, one must further impose

$$\chi_{,i} = 0 \,, \quad \psi_M = \psi_M(u) \,. \tag{23}$$

Thus, if (21) and (23) hold, the space is supersymmetric. It is puzzling that in such spaces, the gravitational wave amplitude given by H can have arbitrary dependence on the transverse directions x^i, while the spin-1 and spin-3/2 amplitudes cannot.

Using the methods of ref. [6], necessary and sufficient conditions for the Rarita–Schwinger field to be non-trivial (i.e., not pure gauge) can be given in terms of higher order integrability conditions. As in ref. [6], it is found that, if (21) holds, nearly all possible solutions are non-trivial, whereas if (21) does not hold only a restricted class of solutions is non-trivial.

Consider now the dimensional reduction to four dimensions, leading to a solution of the ungauged $N = 8$ supergravity theory [2]. If one takes the space–time co-ordinates as (u, v, x^1, x^2) with all the fields independent of the seven internal co-ordinates (x^3, $x^4 \dots x^9$), one obtains a pp-wave solution of the $N = 8$ supergravity with all the fields non-trivial. If the 11-dimensional space was supersymmetric, the corresponding 4-space will admit a real 16-dimensional space of Killing spinors.

Alternatively, introducing a time co-ordinate t and a space co-ordinate x^{10} by

$$\sqrt{2}\,\mathrm{d}t = \mathrm{d}v + (1 - H)\,\mathrm{d}u \,, \tag{24}$$

$$\sqrt{2}\,\mathrm{d}x^{10} = \mathrm{d}v - (1 + H)\,\mathrm{d}u \,, \tag{25}$$

one can regard (t, x^1, x^2, x^3) as the space–time coordinates, provided that all the fields are independent of the seven internal co-ordinates $(x^4, x^5, ..., x^{10})$. Choosing H to be of the form

$$H(x^1, x^2, x^3) = 1 + \sum_{s=1}^{s=n} M_s/|\boldsymbol{r} - \boldsymbol{r}_s| \, , \tag{26}$$

gives some solutions of the $N = 8$ supergravity field equations generalizing those found in the $N = 4$ theory by Gibbons [7]. This gives a static "multi-centre" space with $\boldsymbol{n}$ localized field configurations at the points $\boldsymbol{r} = \boldsymbol{r}_s$ in equilibrium with the gravitational and scalar attraction between them being exactly balanced by electric repulsion, a phenomenon called antigravity [8]. However, although these solutions are derived from regular 11-dimensional spaces, the 4-dimensional metrics are singular, with naked singularities at the points $\boldsymbol{r} = \boldsymbol{r}_s$ [7]. They can also be obtained by boosting a black hole solution in the x^{10} direction in the limit in which the velocity tends to that of light [7]. These could only be supersymmetric if ξ_{MNP} and χ were constant fields. Regular, antigravitating soliton solutions of $N = 8$ supergravity can, however, be obtained from certain static solutions of the 11-dimensional theory, as will be discussed elsewhere.

I would like to thank G.W. Gibbons for many useful discussions.

References

[1] E. Cremmer, B. Julia and J. Scherk, Phys. Lett. 76B (1978) 409.
[2] E. Cremmer and B. Julia, Nucl. Phys. B159 (1979) 141.
[3] M.W. Brinkman, Proc. Natl. Acad. Sci. USA 9 (1923) 1; J. Ehlers and W. Kundt, in: Gravitation. An introduction to current research, ed. L. Witten (Wiley, New York, 1962).
[4] L.F. Urrutia, Phys. Lett. 102B (1981) 393.
[5] P.C. Aichelburg and T. Dereli, Phys. Rev. D18 (1978) 1754; Phys. Lett. 80B (1979) 357.
[6] C.M. Hull, Killing spinors and exact plane wave solutions of extended supergravity, DAMTP preprint (1983), to be published in Phys. Rev. D.
[7] G.W. Gibbons, Nucl. Phys. B207 (1982) 337.
[8] J. Scherk, Phys. Lett. 88B (1979) 265.

Reprinted from Phys. Rev. D **31** (1985) 2503–6

$N=4$ remaining supersymmetry in a Kaluza-Klein monopole background in $D=11$ supergravity theory

Seung Kee Han*
International Centre for Theoretical Physics, Trieste, Italy

I. G. Koh
International Centre for Theoretical Physics, Trieste, Italy and Physics Department, Sogang University, Seoul, Korea
(Received 18 September 1984)

Upon seven-torus compactification of eleven-dimensional supergravity, a Kaluza-Klein monopole is embedded into one U(1) group of the isometry group $U(1)^7$. Four independent Killing spinors remain unbroken in this background.

Higher-dimensional general relativity (Kaluza-Klein theories)[1] is a promising candidate to unify gauge theories and gravity. However, dimensions of Kaluza-Klein theories are completely arbitrary unless constrained by supersymmetry. Requirement of the absence of states with spin higher than 2 puts an upper limit to the dimension of $D \leq 11$ for Riemmanian space and $D \leq 24$ for quasi-Riemmanian space.[2] Although the quantum behavior of Kaluza-Klein theories is not well studied yet, the supersymmetry will soften ultraviolet divergences compared to nonsupersymmetric cases. Furthermore, the topologically nontrivial solutions will play an important role in nonperturbative effects.[3]

The finite-energy Kaluza-Klein monopole solutions have been studied recently.[4–7] They are spherically symmetric and static, and are the usual magnetic monopoles in the asymptotic region of four space-time dimensions. This monopole solution is regular at the origin and the space-time geometry is intrinsically interwoven with the internal space in which direction the monopole is embedded.

It is interesting to examine the remaining supersymmetry in the background of Kaluza-Klein monopoles. In eleven-dimensional supergravity which is maximal in the pseudo-Riemmanian Kaluza-Klein theory, the fate of supersymmetry upon compactifications has been rather extensively studied. Some of them are the seven-torus[8] ($N=8$), round seven-sphere[9] ($N=8$), left squashed ($N=1$) and right squashed ($N=0$) seven-sphere,[10] M^{ppr} ($N=2$) and M^{pqr} ($N=0, p\neq q$) manifold,[11] where the number of remaining supersymmetries is given inside the parentheses. A table of known compactifications and surviving supersymmetry can be found in Ref. 12. Most interestingly, the $K_3 \times T_3$ solution[13] is an example with four surviving supersymmetries, but without isometry group corresponding to K^3. It turns out that there is an $N=4$ supersymmetric solution in the seven-torus compactification with a Kaluza-Klein monopole embedded into one U(1) group of the isometry $U(1)^7$ group. It is noteworthy that the Kaluza-Klein monopole solution is a unique example of $N=4$ supersymmetry with an isometry group.

The bosonic parts in $D=11$, $N=1$ supergravity relevant for the background are

$$R_{MN} - \tfrac{1}{2} g_{MN} R = \tfrac{1}{3}(F_{MPQR} F_N{}^{PQR} - \tfrac{1}{8} g_{MN} F_{PQRS} F^{PQRS}) ,$$
$$\nabla_M F^{MPQR} = -\tfrac{1}{576} \epsilon^{M_1 \cdots M_8 PQR} F_{M_1 \cdots M_4} F_{M_5 \cdots M_8} . \tag{1}$$

The background solutions with Kaluza-Klein monopole are obtained by the following vacuum expectation values (VEV's):

$$\langle F_{\mu\nu\rho\sigma} \rangle = \langle F_{mnpq} \rangle = 0 ,$$
$$\langle \Psi_\mu \rangle = \langle \Psi_m \rangle = 0 . \tag{2}$$

The VEV's for the elfbein are[4]

$$\begin{aligned} e^{(t)} &= dt , \\ e^{(r)} &= e^{b/2} dr , \\ e^{(\theta)} &= e^{b/2} r \, d\theta , \\ e^{(\phi)} &= e^{b/2} r \sin\theta \, d\phi , \\ e^{(5)} &= e^{-b/2}[dx^5 + n\kappa g(\cos\theta - 1) d\phi] , \\ e^{(6)} &= dx^6, \ldots, \quad e^{(11)} = dx^{11} . \end{aligned} \tag{3}$$

Here $e^b = 1 + |n| R/2r$, $R = 2\kappa g$, and n is the monopole charge. The indices inside the parentheses are for the frame indices. The spherically symmetric and static monopole solutions carrying the magnetic charge in more than one U(1) direction of the isometry group $U(1)^7$ simultaneously are shown to be absent.[6]

The number of independent supersymmetries will be determined by the Killing spinor equation[8–12]

$$D_M \eta = (\partial_M - \tfrac{1}{4} \omega_M{}^{AB} \Gamma_{AB}) \eta = 0 , \tag{4}$$

where

$$\Gamma_{AB} = \tfrac{1}{2}(\Gamma_A \Gamma_B - \Gamma_B \Gamma_A) .$$

Here early alphabet letters are used as frame labels, while

mid-alphabet letters are employed for world indices.

The relevant spin connections are

$$\begin{aligned}
\omega_\theta{}^{(r)}{}_{(\theta)} &= \frac{-(4r+|n|R)}{2(2r+|n|R)}\,,\\
\omega_\phi{}^{(r)}{}_{(\phi)} &= \frac{-(4r+|n|R)}{2(2r+|n|R)}\sin\theta\,,\\
\omega_\phi{}^{(r)}{}_{(5)} &= \frac{-n|n|R^2}{2(2r+|n|R)^2}(\cos\theta-1)\,,\\
\omega_5{}^{(r)}{}_{(5)} &= -\frac{|n|R}{(2r+|n|R)^2}\,,\\
\omega_\phi{}^{(\theta)}{}_{(\phi)} &= -\cos\theta+\frac{n^2R^2(\cos\theta-1)}{2(2r+|n|R)^2}\,,\\
\omega_5{}^{(\theta)}{}_{(\phi)} &= \frac{nR}{(2r+|n|R)^2}\,,\\
\omega_\phi{}^{(\theta)}{}_{(5)} &= \frac{nR}{2(2r+|n|R)}\sin\theta\,,\\
\omega_\theta{}^{(\phi)}{}_{(5)} &= -\frac{nR}{2(2r+|n|R)}\,.
\end{aligned} \tag{5}$$

The flat indices are raised and lowered with the metric $\eta_{AB}=\text{diag}(-1,+1,\ldots,+1)$, and $\omega_M{}^{AB}=-\omega_M{}^{BA}$. Of course, the flat-space limit without monopoles is obtained by putting $n=0$. Also since the Kaluza-Klein monopole solutions [equivalently Taub-Nut (Newman-Unti-Tamburino) solutions[14]] are regular at the origin, all spin connections are also regular.

The convenient choices for the Γ matrix with the Clifford algebra $\{\Gamma_A,\Gamma_B\}=-2\eta_{AB}$ are

$$\begin{aligned}
&\Gamma_0=\gamma_0\times\mathbf{1}_8,\quad \Gamma_i=\gamma_i\times\mathbf{1}_8,\quad i=1,2,3,\\
&\Gamma_5=\gamma_5\times\begin{bmatrix}0 & -\mathbf{1}_4\\ \mathbf{1}_4 & 0\end{bmatrix},\\
&\Gamma_{5+j}=\gamma_5\times\begin{bmatrix}\alpha_j & 0\\ 0 & -\alpha_j\end{bmatrix},\quad j=1,2,3\\
&\Gamma_{8+j}=\gamma_5\times\begin{bmatrix}0 & \beta_j\\ \beta_j & 0\end{bmatrix},\quad j=1,2,3\,.
\end{aligned} \tag{6}$$

Here

$$\gamma_0=\begin{bmatrix}-1 & 0\\ 0 & 1\end{bmatrix},\quad \gamma_i=\begin{bmatrix}0 & \sigma_i\\ -\sigma_i & 0\end{bmatrix},\quad \gamma_5=-i\gamma_0\gamma_1\gamma_2\gamma_3=\begin{bmatrix}0 & 1\\ 1 & 0\end{bmatrix},$$

and σ_i is the Pauli spin matrix. $\mathbf{1}_8$ and $\mathbf{1}_4$ are 8×8 and 4×4 identity matrices. The α_j and β_j are defined as

$$\begin{aligned}
&\alpha_1=\begin{bmatrix}0 & \sigma_1\\ -\sigma_1 & 0\end{bmatrix},\quad \alpha_2=\begin{bmatrix}0 & -\sigma_3\\ \sigma_3 & 0\end{bmatrix},\\
&\alpha_3=\begin{bmatrix}i\sigma_2 & 0\\ 0 & i\sigma_2\end{bmatrix},\quad \beta_1=\begin{bmatrix}0 & i\sigma_2\\ i\sigma_2 & 0\end{bmatrix},\\
&\beta_2=\begin{bmatrix}0 & 1\\ -1 & 0\end{bmatrix},\quad \beta_3=\begin{bmatrix}-i\sigma_2 & 0\\ 0 & i\sigma_2\end{bmatrix}.
\end{aligned} \tag{7}$$

Observe that Γ_5, Γ_{5+j}, and Γ_{8+j} are σ_2-, σ_3-, and σ_1-like in 4×4 block matrices. In these representations, Γ_0 and Γ_2 are symmetric, while all other Γ matrices are antisymmetric. Thus the charge conjugation matrix C is

$$C=\gamma_0\gamma_2\times\mathbf{1}_8\,. \tag{8}$$

The spinor representation of the tangent space group SO(1,10) is of 32 components, which are represented by

$$\eta_T=(\Psi_1^T,\ldots,\Psi_8^T)\,, \tag{9}$$

where $\Psi_j^T=(\lambda_j^T,\chi_j^T)$ with two-component spinor λ_j and χ_j. The Majorana condition $\eta=-C\bar{\eta}^T$ gives the relationship between λ_j and χ_j as

$$\chi_j=-\sigma_2\lambda_j^*\,. \tag{10}$$

The χ_j is dependent, and there are only 16 independent components corresponding to eight λ_j.

Since spin connections $\omega_M{}^{AB}$ with $M=t,r,6,\ldots,11$ vanish, their corresponding Killing spinor equations are trivial as

$$\partial_t\eta=\partial_r\eta=\partial_6\eta=\cdots=\partial_{11}\eta\,. \tag{11}$$

Thus the Killing spinor η is independent of coordinates $t,r,x^6,\ldots,x^{11}$. The Killing spinor equation in the θ direction reads

$$\begin{aligned}
&\partial_\theta\Psi_j+\frac{i}{2}\omega_\theta{}^{(r)(\theta)}\begin{bmatrix}\sigma_3 & 0\\ 0 & \sigma_3\end{bmatrix}\Psi_j+\frac{1}{2}\omega_\theta{}^{(\phi)(5)}\begin{bmatrix}\sigma_3 & 0\\ 0 & -\sigma_3\end{bmatrix}\Psi_{j+4}=0\,,\\
&\partial_\theta\Psi_{j+4}+\frac{i}{2}\omega_\theta{}^{(r)(\theta)}\begin{bmatrix}\sigma_3 & 0\\ 0 & \sigma_3\end{bmatrix}\Psi_{j+4}-\frac{1}{2}\omega_\theta{}^{(\phi)(5)}\begin{bmatrix}\sigma_3 & 0\\ 0 & -\sigma_3\end{bmatrix}\Psi_j=0,\quad j=1,\ldots,4.
\end{aligned} \tag{12}$$

Notice that all eight Ψ_j are not independent, but Ψ_j $(j=1,\ldots,4)$ are related to Ψ_{j+4} $(j=1,\ldots,4)$.

The Killing spinor equation in the ϕ direction reads

$$\partial_\phi \Psi_j - \frac{i}{2}\omega_\phi{}^{(r)(\phi)} \begin{bmatrix} \sigma_2 & 0 \\ 0 & \sigma_2 \end{bmatrix} \Psi_j + \frac{i}{2}\omega_\phi{}^{(\theta)(\phi)} \begin{bmatrix} \sigma_1 & 0 \\ 0 & \sigma_1 \end{bmatrix} \Psi_j + \frac{1}{2}\omega_\phi{}^{(r)(5)} \begin{bmatrix} \sigma_1 & 0 \\ 0 & -\sigma_1 \end{bmatrix} \Psi_{j+4} + \frac{1}{2}\omega_\phi{}^{(\theta)(5)} \begin{bmatrix} \sigma_2 & 0 \\ 0 & -\sigma_2 \end{bmatrix} \Psi_{j+4} = 0 ,$$

$$\partial_\phi \Psi_{j+4} - \frac{i}{2}\omega_\phi{}^{(r)(\phi)} \begin{bmatrix} \sigma_2 & 0 \\ 0 & \sigma_2 \end{bmatrix} \Psi_{j+4} + \frac{i}{2}\omega_\phi{}^{(\theta)(\phi)} \begin{bmatrix} \sigma_1 & 0 \\ 0 & \sigma_1 \end{bmatrix} \Psi_{j+4} - \frac{1}{2}\omega_\phi{}^{(r)(5)} \begin{bmatrix} \sigma_1 & 0 \\ 0 & -\sigma_1 \end{bmatrix} \Psi_j - \frac{1}{2}\omega_\phi{}^{(\theta)(5)} \begin{bmatrix} \sigma_2 & 0 \\ 0 & -\sigma_2 \end{bmatrix} \Psi_j = 0, \quad j = 1, \dots, 4. \tag{13}$$

Finally the Killing spinor equation in the fifth direction is

$$\partial_5 \Psi_j + \tfrac{1}{2}\omega_5{}^{(r)(5)} \begin{bmatrix} \sigma_1 & 0 \\ 0 & -\sigma_1 \end{bmatrix} \Psi_{j+4} + \frac{i}{2}\omega_5{}^{(\theta)(\phi)} \begin{bmatrix} \sigma_1 & 0 \\ 0 & \sigma_1 \end{bmatrix} \Psi_j = 0,$$

$$\partial_5 \Psi_{j+4} - \tfrac{1}{2}\omega_5{}^{(r)(5)} \begin{bmatrix} \sigma_1 & 0 \\ 0 & -\sigma_1 \end{bmatrix} \Psi_j + \frac{i}{2}\omega_5{}^{(\theta)(\phi)} \begin{bmatrix} \sigma_1 & 0 \\ 0 & \sigma_1 \end{bmatrix} \Psi_{j+4} = 0, \tag{14}$$

Solutions for these equations are obtained in a straightforward way. The Ψ_j $(j=1,\dots,4)$ are given by

$$\lambda_j = \begin{bmatrix} (A_j e^{i/2\phi} + B_j e^{-i/2\phi}) e^{i/2\theta} \\ (A_j e^{i/2\phi} - B_j e^{-i/2\phi}) e^{-i/2\theta} \end{bmatrix} ,$$

$$\chi_j = -\sigma_2 \lambda_j^* . \tag{15}$$

The Ψ_{j+4} $(j=1,\dots,4)$ are related to Ψ_j by

$$\lambda_{j+4} = \pm i \lambda_j ,$$

$$\chi_{j+4} = \mp i \chi_j , \tag{16}$$

where the upper (lower) sign corresponds to the positive (negative) monopole charge. Thus there are four independent Killing spinors Ψ_j $(j=1,\dots,4)$ for Eq. (4), and we obtain $N=4$ remaining supersymmetries in the Kaluza-Klein monopole backgrounds (2) and (3). In the limit of vanishing monopole charge ($n=0$), the mixing terms between Ψ_j and Ψ_{j+4} $(j=1,\dots,4)$ due to $\omega_\theta{}^{(\phi)(5)}$ in Eq. (12), $\omega_\phi{}^{(r)(5)}$ and $\omega_\phi{}^{(\theta)(5)}$ in Eq. (13), and $\omega_5^{(r)(5)}$ in Eq. (14) disappear. Then Ψ_{j+4} $(j=1,\dots,4)$ become independent of Ψ_j, and we recover the well-known $N=8$ solutions of seven-torus compactifications.[8]

It may be worthwhile to comment upon the consistency condition of Eq. (4), which is

$$[D_M, D_N]\eta = -\tfrac{1}{4} R_{MN}{}^{AB} \Gamma_{AB} \eta \propto C_{MN} \eta = 0 . \tag{17}$$

Here Γ_{AB} are the 55 SO(1,10) generators. The nonvanishing C_{MN} for the positive monopole charge are

$$C_{12} = -C_{35} = -\Gamma_{12} + \Gamma_{35} = 4T_1 ,$$
$$C_{13} = C_{25} = -\Gamma_{13} - \Gamma_{25} = 4T_2 , \tag{18}$$
$$C_{23} = -C_{15} = \Gamma_{23} - \Gamma_{15} = 4T_3 ,$$

and form the SO(3) subalgebra $[T_i, T_j] = \epsilon_{ijk} T_k$. Similar results hold also for the negatively charged monopole.

Now let us briefly consider the problem of remaining supersymmetries in D-dimensional supergravity theories other than the 11-dimensional one. The solutions for equations of motion in the torus compactifications are still given by Eq. (3) with the monopole embedded into one U(1) group of the isometry group $U(1)^{D-4}$. Vacuum expectation values of all the bosonic fields vanish, except vielbeins which have the identical vacuum expectation value as in Eq. (3). But if theories contain several fermionic fields other than the gravitino field, the remaining supersymmetry can be drastically altered in this Kaluza-Klein monopole background. For example in $D=6$, $N=2$ supergravity,[15] the supersymmetry transformation law for the gaugino λ_j of Yang-Mills supermultiplets in the Sp(1) direction is [Eq. (22) of Ref. 15 or Eq. (21) of Ref. 16]

$$\delta\lambda^j = -\frac{1}{2\sqrt{2}} e^{\phi/\sqrt{2}} F_{MN}{}^j \Gamma^{MN} \epsilon - \tfrac{1}{2} e^{-\phi/\sqrt{2}} C^{ij} T^i \epsilon , \tag{19}$$

where $C^{ij} = g'[A^i_\alpha (T^j \Phi)^\alpha - \delta^{ij}]$. Due to δ^{ij} in the C^{ij} term, all the supersymmetries are completely broken,

$$\langle \delta\lambda^j \rangle \neq 0 , \tag{20}$$

in the Kaluza-Klein monopole backgrounds. Thus one cannot expect some remaining supersymmetry generally in the Kaluza-Klein monopole background.

It is remarkable that the Kaluza-Klein monopole solution in $D=11$ supergravity is unique in giving $N=4$

remaining supersymmetries with nonvanishing isometry group.

We wish to thank M. A. Awada, M. J. Duff, O. Foda, J. Kowalski-Glikman, B. E. W. Nilsson, H. Nishino, and C. N. Pope for useful conversations. Special thanks are due to Z. F. Ezawa for many illuminating and helpful discussions. We are grateful to Professor Abdus Salam, the International Atomic Energy Agency, and UNESCO for kind hospitality at the International Centre for Theoretical Physics (ICTP), Trieste, Italy. This research is supported in part by the Korea Science and Engineering Foundation.

---

*On leave of absence from Physics Department, Chungbuk National University Cheongju, Chungbuk, Korea.

[1]T. Kaluza, Sitzungsber. Preuss. Akad. Wiss. Phys. Math. K1, 966 (1921); O. Klein, Z. Phys. **37**, 895 (1926).

[2]S. Weinberg, in *Fifth Workshop on Grand Unification*, Brown University, 1984, edited by K. Kang, H. Fried, and P. Frampton (World Scientific, Singapore, 1984).

[3]D. J. Gross, Princeton report, 1983 (unpublished).

[4]R. Sorkin, Phys. Rev. Lett. **51**, 87 (1983); D. J. Gross and M. J. Perry, Nucl. Phys. **B226**, 29 (1983); D. Pollard, J. Phys. A **16**, 565 (1983).

[5]M. J. Perry, Phys. Lett. **137B**, 171 (1984).

[6]Z. F. Ezawa and I. G. Koh, Phys. Lett. **140B**, 205 (1984).

[7]P. Nelson, Nucl. Phys. **B238**, 638 (1984); Q. Shafi and C. Wetterich, Bern Report No. BA-83-45, 1983 (unpublished); O. Foda, ICTP Report No. IC/84/66, 1984 (unpublished); S.-C. Lee, Phys. Lett. B (to be published).

[8]E. Cremmer and B. Julia, Nucl. Phys. **B159**, 141 (1979).

[9]M. J. Duff and D. J. Toms, in *Unifications of the Fundamental Interactions*, edited by J. Ellis, S. Ferrara, and P. van Nieuwenhuizen (Plenum, New York, 1982); M. J. Duff and C. N. Pope, in *Supersymmetry and Supergravity 82*, edited by S. Ferrara, J. G. Taylor, and P. van Nieuwenhuizen (World Scientific, Princeton, N.J., 1983).

[10]M. A. Awada, M. J. Duff, and C. N. Pope, Phys. Rev. Lett. **50**, 294 (1983); M. J. Duff, B. E. W. Nilsson, and C. N. Pope, *ibid.* **50**, 2043 (1983); F. A. Bais, H. Nicolai, and P. van Nieuwenhuizen, Nucl. Phys. **B228**, 333 (1983).

[11]E. Witten, Nucl Phys. **B186**, 412 (1981); L. Castellani, R. D'Auria, and P. Fré, *ibid.* **B239**, 610 (1984).

[12]L. Castellani, L. J. Romans, and N. P. Warner, Nucl. Phys. **B241**, 429 (1984).

[13]M. J. Duff, B. E. W. Nilsson, and C. N. Pope, Nucl. Phys. **B233**, 433 (1984).

[14]E. T. Newman, L. Tamburino, and T. Unti, J. Math. Phys. **4**, 915 (1963); C. W. Misner, *ibid.* **4**, 924 (1963).

[15]H. Nishino and E. Sezgin, Phys. Lett. **144B**, 187 (1984).

[16]I. G. Koh and H. Nishino, Phys. Lett. B (to be published).

Chapter 2

The eleven-dimensional supermembrane

Membrane theory has a strange history which goes back even further than strings. The idea that the elementary particles might correspond to modes of a vibrating membrane was put forward originally in 1962 by Dirac [1]. When string theory came along in the 1970's, there were some attempts to revive the membrane idea but without much success. Things did not change much until 1986 when Hughes, Liu and Polchinski [2] showed that it was possible to combine membranes with supersymmetry: the *supermembrane* was born.

Consequently, while all the progress in string theory was going on, a small splinter group was posing the question: Once you have given up 0-dimensional particles in favour of 1-dimensional strings, why not 2-dimensional membranes or in general p-dimensional objects ? Just as a 0-dimensional particle sweeps out a 1-dimensional *worldline* as it evolves in time, so a 1-dimensional string sweeps out a 2-dimensional *worldsheet* and a p-brane sweeps out a d-dimensional *worldvolume*, where $d = p + 1$. Of course, there must be enough room for the p-brane to move about in spacetime, so d must be less than or equal to the number of spacetime dimensions D. In fact supersymmetry places further severe restrictions both on the dimension of the extended object and the dimension of spacetime in which it lives [3]. One can represent these as points on a graph where we plot spacetime dimension D vertically and the p-brane dimension $d = p+1$ horizontally. This is the *brane scan* of table 2.1. In the early eighties Green and Schwarz [4] had shown that spacetime supersymmetry allows classical superstrings moving in spacetime dimensions $3, 4, 6$ and 10. (Quantum considerations rule out all but the ten-dimensional case as being truly fundamental. Of course some of these ten dimensions could be curled up to a very tiny size in the way suggested by Kaluza and Klein [5, 6]. Ideally six would be compactified in this way so as to yield the four spacetime dimensions with which we are familiar.) It was now realized, however, that these 1-branes in $D = 3, 4, 6$ and 10 should be viewed as but special cases of this more general class of supersymmetric extended object.

A simple way to understand the allowed points on the brane-scan is to demand equal numbers of boson and fermion degrees of freedom on the worldvolume. This matching of worldvolume bosons and fermions may, at first sight, seem puzzling since the Green–Schwarz approach begins with only spacetime supersymmetry. The explanation is as follows. As the p-brane moves through spacetime, its trajectory is described by the functions $X^M(\xi)$ where X^M are the spacetime coordinates ($M = 0, 1, \ldots, D-1$) and ξ^i are the worldvolume coordinates ($i = 0, 1, \ldots, d-1$). It is often convenient to make the so-called *static gauge choice* by making the $D = d + (D-d)$ split

$$X^M(\xi) = (X^\mu(\xi), Y^m(\xi)) \tag{2.1}$$

where $\mu = 0, 1, \ldots, d-1$ and $m = d, \ldots, D-1$, and then setting

$$X^\mu(\xi) = \xi^\mu. \tag{2.2}$$

Thus the only physical worldvolume degrees of freedom are given by the $(D - d)$ $Y^m(\xi)$. So the number of on-shell bosonic degrees of freedom is

$$N_B = D - d. \tag{2.3}$$

To describe the super p-brane we augment the D bosonic coordinates $X^M(\xi)$ with anticommuting fermionic coordinates $\theta^\alpha(\xi)$. Depending on D, this spinor could be Dirac, Weyl, Majorana or Majorana–Weyl. However, there is a fermionic kappa symmetry which implies that half of the spinor degrees of freedom are redundant and may be eliminated by a physical gauge choice. The net result is that the theory exhibits a *d-dimensional worldvolume supersymmetry* [3] where the number of fermionic generators is exactly half of the generators in the original spacetime super-symmetry. This partial breaking of supersymmetry is a key idea. Let M be the number of real components of the minimal spinor and N the number of supersymmetries in D spacetime dimensions and let m and n be the corresponding quantities in d worldvolume dimensions. Let us first consider $d > 2$. Since kappa symmetry always halves the number of fermionic degrees of freedom and going on-shell halves it again, the number of on-shell fermionic degrees of freedom is

$$N_F = \frac{1}{2}mn = \frac{1}{4}MN. \tag{2.4}$$

Worldvolume supersymmetry demands $N_B = N_F$ and hence

$$D - d = \frac{1}{2}mn = \frac{1}{4}MN. \tag{2.5}$$

A list of dimensions, number of real dimensions of the minimal spinor and possible supersymmetries is given in table 1.1 of chapter 1, from which we see that there are only 8 solutions of (2.5) all with $N = 1$, as indicated by the $d \geq 3$ points labelled S in table 2.1. We note in particular that $D_{\max} = 11$ since $M \geq 64$ for $D \geq 12$ and hence (2.5) cannot be satisfied. Similarly $d_{\max} = 6$ since $m \geq 16$ for $d \geq 7$. The case $d = 2$ is special because of the ability to treat left and right moving modes independently. If we require the sum of both left and right moving bosons and fermions to be equal, then we again find the condition (2.5). This provides a further 4 solutions all with $N = 2$, corresponding to Type II superstrings in $D = 3, 4, 6$ and 10 (or 8 solutions in all if we treat Type IIA and Type IIB separately). Both the gauge-fixed Type IIA and Type IIB superstrings will display $(8, 8)$ supersymmetry on the worldsheet. If we require only left (or right) matching, then (2.5) is replaced by

$$D-2=n=\frac{1}{2}MN \tag{2.6}$$

which allows another 4 solutions in $D=3,4,6$ and 10, all with $N=1$. The gauge-fixed theory will display $(8,0)$ worldsheet supersymmetry. The heterotic string falls into this category. The results [3] are indicated by the $d=2$ points labelled S in table 2.1. Point particles are usually omitted from the brane scan [3, 7, 8], but we have included them in table 2.1 by the $d=1$ points labelled S.

An equivalent way to arrive at the above conclusions is to list all scalar supermultiplets and to interpret the dimension of the target space, D, by

$$D-d=\text{ number of scalars.} \tag{2.7}$$

A useful reference is [9] which provides an exhaustive classification of all unitary representations of supersymmetry with maximum spin 2. In particular, we can understand $d_{\max}=6$ from this point of view since this is the upper limit for scalar supermultiplets. In summary, according to the above classification, Type II p-branes described by scalar supermultiplets do not exist for $p>1$.

There are four types of solution with $8+8$, $4+4$, $2+2$ or $1+1$ degrees of freedom respectively. Since the numbers 1, 2, 4 and 8 are also the dimension of the four division algebras, these four types of solution are referred to as real, complex, quaternion and octonion respectively. The connection with the division algebras can in fact be made more precise [10, 11].

$D\uparrow$												
11	.			S			T					
10	.	V	S/V	V	V	V	S/V	V	V	V	V	
9	.	S				S						
8	.				S							
7	.			S			T					
6	.	V	S/V	V	S/V	V	V					
5	.	S		S								
4	.	V	S/V	S/V	V							
3	.	S/V	S/V	V								
2	.	S										
1	.											
0	.	.	.	.	.	.	.	.	.	.	.	.
	0	1	2	3	4	5	6	7	8	9	10	11
												$d\rightarrow$

Table 2.1. The brane scan, where S, V and T denote scalar, vector and antisymmetric tensor multiplets.

Curiously enough, the maximum spacetime dimension permitted is eleven, where Bergshoeff, Sezgin and Townsend found their supermembrane [12, 13] which couples to eleven-dimensional supergravity [14, 5]. (The 3-form gauge field of

$D = 11$ supergravity had long been suggestive of a membrane interpretation [15]). Moreover, it was then possible to show [16] by simultaneous dimensional reduction of the spacetime and worldvolume that the membrane looks like a string in ten dimensions. In fact, it yields precisely the Type *IIA* superstring. This suggested that the eleven-dimensional theory was perhaps the more fundamental after all. See chapter 4.

Notwithstanding these and subsequent results, the supermembrane enterprise was, until recently, largely ignored by the mainstream physics community. Those who had worked on eleven-dimensional supergravity and then on supermembranes spent the early eighties arguing for *spacetime* dimensions greater than four, and the late eighties and early nineties arguing for *worldvolume* dimensions greater than two. The latter struggle [17] was by far the more bitter!

As we shall see in this volume, supermembranes now play a vital part in string duality [18], D-branes [19, 20, 21] and M-theory. Reviews on supermembranes may be found in [22, 23, 25, 8, 24, 17].

Another curious twist in the history of supermembranes concerns their interpretation as *solitons*: non-singular solutions of the classical field equations corresponding to lumps of field energy that are prevented from dissipating by a topological conservation law. The classical example of such a soliton is provided by magnetic monopole solutions of four-dimensional grand unified theories. Although the original Hughes–Liu–Polchinski [2] supermembrane was found as a soliton of a six-dimensional gauge theory, the subsequent development went in the opposite direction with membranes being treated as fundamental objects in their own right. One of the problems that membrane theory then had to confront was the question of quantization [26]: no one knows how to quantize fundamental p-branes with $p > 2$. All the techniques that worked so well for fundamental strings simply do not go through. A useful framework for analyzing these problems is the lightcone gauge [27, 28, 29], sometimes called the infinite momentum frame. For $p > 1$, this gauge choice does not eliminate all the unphysical degrees of freedom, and one finds in the case of the $D = 11$ supermembrane a quantum mechanical matrix model corresponding to a dimensionally reduced $D = 10$ Yang–Mills theory with gauge group $SU(N)$ as $N \to \infty$. Moreover, as was shown in [29], this theory does not possess a discrete spectrum: hence the negative title: *Supermembranes: a fond farewell?* We have included the paper by de Wit, Nicolai and Hoppe in our collection, however, because this infinite momentum frame has recently been resurrected in the context of the Matrix Model approach to M-theory [36, 31] discussed in chapter 6. In this new interpretation, the continuous spectrum is not a drawback but a virtue!

The next development came when Townsend [30] pointed out that not merely the $D = 6$ supermembrane but all the points marked S on the $\mathcal{H}, \mathcal{C}, \mathcal{R}$ sequences of the branescan correspond to topological defects of some globally supersymmetric field theory which preserve half of the spacetime supersymmetries. This naturally raises the question of whether the supergravity theories, in particular the $D = 11$ supergravity, might also admit the $\mathcal{O}$ sequence supermembranes as classical solutions preserving half the spacetime supersymmetry. Whether in global or local supersymmetry, states preserving half the supersymmetries (and known as *BPS states*) occupy a special status because, in appropriate units, their mass is equal to their charge. It follows from the supersymmetry algebra that they therefore belong to short multiplets and are thus protected from quantum corrections. Studying BPS states can thus give us vital information about the exact theory even at strong coupling. In the $D = 11$ supersymmetry algebra, the anticommutator of two supersymmetry generators Q_α is given by [32]

$$\{Q_\alpha, Q_\beta\} = (C\Gamma_M)_{\alpha\beta} P^M + (C\Gamma_{MN})_{\alpha\beta} Z^{MN} + (C\Gamma_{MNPQR})_{\alpha\beta} Z^{MNPQR} \quad (2.8)$$

where C is the charge conjugation matrix and $\Gamma_{M_1 \ldots M_n}$ is the antisymmetric product of n Dirac matrices. We see that the right-hand side involves not only the momentum P^M but also the 2-form charge Z^{MN} and the 5-form charge Z^{MNPQR}.

Indeed, in 1991 the eleven dimensional supermembrane was recovered as a solution of the $D = 11$ supergravity theory preserving one half of the supersymmetry [33]. The zero modes of the membrane solution belong to a $(d = 3, n = 8)$ supermultiplet consisting of eight scalars and eight spinors which correspond to the eight Goldstone bosons and their superpartners associated with breaking of the eight translations transverse to the membrane worldvolume. However, this elementary solution is a singular solution of the supergravity equations coupled to a supermembrane source and carries a Noether electric charge. It should not therefore be called a *soliton* which would be a non-singular solution of the source-free equations carrying a topological charge. (In this respect it resembles the fundamental string solution of ten-dimensional supergravity coupled to a string source of Dabholkar *et al* [34] to which it in fact reduces under a double dimensional reduction of spacetime and worldvolume, followed by a truncation from $N = 2$ to $N = 1$.) The true soliton solution of $D = 11$ supergravity is the $D = 11$ superfivebrane, discussed in chapter 3. In hindsight, we see that this is just what is expected from the supersymmetry algebra (2.8): the spatial components of the momentum P_M are carried by the plane wave of chapter 1, those of the 2-form charge Z_{MN} by the 2-brane and those of the 5-form charge Z_{MNPQR} by the 5-brane. The time components of the 2-form and 5-form are associated with the Type IIA sixbrane (which is equivalent [37] to the Kaluza–Klein monopole of chapter 1) and the Type IIA eightbrane of chapter 3 that arise on compactification to $D = 10$ [38, 39].

In all this earlier work, the $D = 11$ supermembrane was treated as *closed*, but following the arrival of $D = 10$ Dirichlet branes [19], surfaces of dimension p on which open strings can end, it was pointed out by Strominger [35] and Townsend [36] that the fivebrane can act as a surface on which open membranes can end.

References

[1] Dirac P A M 1962 An extensible model of the electron *Proc. R. Soc.* A **268** 57

[2] Hughes J, Liu J and Polchinski J 1986 Supermembranes *Phys. Lett.* B **180** 370

[3] Achucarro A, Evans J, Townsend P and Wiltshire D 1987 Super p-branes *Phys. Lett.* B **198** 441

[4] Green M B and Schwarz J 1984 Covariant description of superstrings *Phys. Lett.* B **136** 367

[5] Duff M J 1995 Kaluza–Klein theory in perspective *Proc. Nobel Symp. Oskar Klein Centenary, Stockholm, 1994* ed E Lindstrom (Singapore: World Scientific) NI-94-015, CTP-TAMU-22/94, hep-th/9410046

[6] Duff M J, Nilsson B E W and Pope C N 1986 Kaluza–Klein supergravity *Phys. Rep.* **130** 1

[7] Duff M J and Lu J X 1993 Type *II* p-branes: the brane scan revisited *Nucl. Phys.* B **390** 276

[8] Duff M J, Khuri R R and Lu J X 1995 String solitons *Phys. Rep.* **259** 213

[9] Strathdee J 1987 Extended Poincaré supersymmetry *Int. J. Mod. Phys.* A **2** 273

[10] Sierra G 1987 An application to the theories of Jordan algebras and Freundanthal triple systems to particles and strings *Class. Quantum Grav.* **4** 227

[11] Evans J M 1988 Supersymmetric Yang–Mills theories and division algebras *Nucl. Phys.* B **298** 92

[12] Bergshoeff E, Sezgin E and Townsend P 1987 Supermembranes and eleven-dimensional supergravity *Phys. Lett.* B **189** 75

[13] Bergshoeff E, Sezgin E and Townsend P 1988 Properties of the eleven-dimensional super membrane theory *Ann. Phys., NY* **185** 330

[14] Cremmer E, Julia B and Scherk J 1978 Supergravity theory in 11 dimensions *Phys. Lett.* B **76** 409

[15] Julia B 1982 Extra-dimensions: Recent progress using old ideas *Proc. Second M Grossmann Meeting, Trieste 1979* ed Ruffini (Amsterdam: North-Holland) p 79

[16] Duff M J, Howe P S, Inami T and Stelle K S 1987 Superstrings in $d = 10$ from supermembranes in d = 11 *Phys. Lett.* B **191** 70

[17] Duff M J 1996 *Supermembranes* Lectures given at the Theoretical Advanced Study Institute in Elementary Particle Physics (TASI 96), Boulder, 1996, hep-th/9611203

[18] Duff M J 1996 Electric/magnetic duality and its stringy origins *Int. J. Mod. Phys.* A **11** 4031

[19] Polchinski J 1995 Dirichlet-branes and Ramond–Ramond charges *Phys. Rev. Lett.* **75** 4724

[20] Polchinski J 1996 TASI lectures on D-branes, hep-th/9611050

[21] Polchinski J 1998 *String Theory* vol I and II (Cambridge: Cambridge University Press)

[22] Duff M J 1988 Supermembranes: The first fifteen weeks *Class. Quantum Grav.* **5** 189

[23] Duff M J 1989 Classical and quantum supermembranes *Class. Quantum Grav.* **6** 1577

[24] Townsend P K 1993 Three lectures on supersymmetry and extended subjects in *Proc. 13th GIFT Seminar on Theoretical Physics: Recent Problems in Mathematical Physics, Salamanca, Spain, 1992* ed L Ibort and M Rodriguez (Dordrecht: Kluwer)

[25] Duff M J, Pope C N and Sezgin E 1990 Supermembranes and physics in 2 + 1 dimensions *Proc. Supermembrane Workshop, ICTP, Trieste, 1989* (Singapore: World Scientific)

[26] Duff M J, Inami T, Pope C N, Sezgin E and Stelle K W 1988 Semiclassical quantization of the supermembrane *Nucl. Phys.* B **297** 515

[27] De Wit B, Hoppe J and Nicolai H 1988 On the quantum mechanics of supermembranes *Nucl. Phys.* B **305** 545

[28] De Wit B, Luscher M and Nicolai H 1989 The supermembrane is unstable *Nucl. Phys.* B **320** 135

[29] De Wit B and Nicolai H 1990 Supermembranes: a fond farewell? *Supermembranes and Physics in* $2+1$ *Dimensions* ed M J Duff, C N Pope and E Sezgin (Singapore: World Scientific) pp196–208

[30] Townsend P 1988 Supersymmetric extended solitons *Phys. Lett.* B **202** 53

[31] Banks T, Fischler W, Shenker S H and Susskind L 1997 M-theory as a matrix model: a conjecture *Phys. Rev.* D **55** 512–28

[32] Townsend P K 1996 P-brane democracy *Particles, Strings and Cosmology* ed J Bagger, G Domokos, A Falk and S Kovesi-Domokos (Singapore: World Scientific) p 271

[33] Duff M J and Stelle K 1991 Multimembrane solutions of $d = 11$ supergravity *Phys. Lett.* B **253** 113

[34] Dabholkar A, Gibbons G W, Harvey J A and Ruiz Ruiz F 1990 Superstrings and solitons *Nucl. Phys.* B **340** 33

[35] Strominger A 1996 Open p-branes *Phys. Lett.* B **383** 44

[36] Townsend P K 1996 D-branes from M-branes *Phys. Lett.* B **373** 68–75

[37] Duff M J and Lu J X 1994 Black and super p-branes in diverse dimensions *Nucl. Phys.* B **416** 301

[38] Hull C M 1998 Gravitational duality, branes and charges *Nucl. Phys.* B **509** 216

[39] Townsend P K 1998 M(embrane) theory on T^9 *Nucl. Phys. Proc. Suppl.* **68** 11

Reprinted from Phys. Lett. B **189** (1987) 75–8

SUPERMEMBRANES AND ELEVEN-DIMENSIONAL SUPERGRAVITY

E. BERGSHOEFF[1], E. SEZGIN
International Centre for Theoretical Physics, I-34100 Trieste, Italy

and

P.K. TOWNSEND
DAMPT, University of Cambridge, Silver Street, Cambridge CB3 9EW, UK

Received 6 February 1987

We construct an action for a supermembrane propagating in $d=11$ supergravity background. Using the constraints of $d=11$ curved superspace, we show that the action is invariant under Siegel-type transformations recently generalized by Hughes, Li and Polchinski. The transformation parameter is a world-volume scalar and $d=11$ spacetime spinor. We also discuss the general problem of the coupling of n-dimensional extended objects to d-dimensional supergravity.

1. Now that we have become accustomed to the notion that strings should replace particles, it is natural to investigate the properties of higher-dimensional extended objects, in particular of membranes since they are the simplest extended objects, and they might describe strings in an appropriate limit.

In 1962 Dirac [1] put forward a theory of an extended electron based on the idea of a relativistic membrane. In 1976, Collins and Tucker [2] studied the classical and quantum mechanics of free relativistic membranes. A year later a locally supersymmetric and reparametrization-invariant action for a spinning membrane was constructed by Howe and Tucker [3]. The action describes anti de Sitter supergravity coupled to a number of scalar multiplets in three dimensions. It is the membrane analog of the Neveu–Schwarz–Ramond formulation of the spinning string theory.

More progress towards the construction of a membrane theory was made by Sugamoto [4] in 1983. More recently, Hughes, Li and Polchinski [5] have constructed a Green–Schwarz-type action for a three-extended object propagating in flat six-dimensional spacetime. The consistency of the action requires the existence of a closed superspace five-form, in analogy with the Henneaux and Mezincescu [6] construction for the Green–Schwarz superstring action [7], where a closed superspace three-form is required. The novel feature of the theory of Hughes et al. is that the parameter of the Siegel-type transformation [8] is a scalar rather than a vector on the world volume.

The generalization of the Hughes et al. model to n-extended objects propagating in flat d-dimensional superspace is evident. All that is required is the existence of a closer super $(n+2)$-form given by

$$H=E^{\beta}E^{\alpha}E^{a_n}\ldots E^{a_1}(\gamma_{a_1\ldots a_n})_{\alpha\beta}\,, \tag{1}$$

where (E^{α}, E^{a}) are the basis one-forms in superspace. This form is closed provided that the following Γ-matrix identity holds:

$$(\gamma^{a})_{(\alpha\beta}(\gamma^{aa_1\ldots a_{n-1}})_{\gamma\delta)}=0\,. \tag{2}$$

The purpose of this note is to construct Hughes et al. -type actions describing the propagation of an n-extended object in d-dimensional *curved* superspace. We give a general formula for the action and the transformation rules, whose consistency requires, among other things (see below), the existence of a closed $(n+2)$-form in curved superspace. Thus we

[1] Supported in part by INFN, Sezione di Trieste, Trieste, Italy.

0370-2693/87/$ 03.50 © Elsevier Science Publishers B.V.
(North-Holland Physics Publishing Division)

expect that the n-extended objects under consideration can consistently propagate only in $d \leqslant 11$ supergravities whose superspace formulation involves a closed $(n+2)$-form. We further expect that such forms exist in supergravity theories in which a closed *bosonic* $(n+2)$-form occurs. As far as we know, the following possibilites exist (we include the Yang–Mills couplings whenever possible):

The dual formulation of $d=10$, $N=1$ supergravity involves a closed seven-form. Its dimensional reduction on a $(10-d)$-dimensional torus leads to real closed $(d-3)$-forms in d-dimensional supergravities. (These are $N=1$ supergravities in $d=8$, 9, 10; $N=2$ in $d=7$ and $N=2$ or 4 in $d=6$) [9]. Apart from these, there is: (i) *A real closed four-form in $d=11$, $N=1$* supergravity, (ii) a real closed three-form in non-chiral $d=10$, $N=2$ supergravity, (iii) a complex closed three-form in chiral $d=10$, $N=2$ supergravity.

Excluding Yang–Mills coupling, as is well known, closed super three-forms exist in $d=3$, 4, 6, and 10.

Considering the case of the membranes, from the above list it follows that the candidate dimensions are 7 and 11. Since the superspace formulation of $d=7$, $N=2$ supergravity is not known at present, we are led to consider the supermembrane propagating in eleven-dimensional spacetime.

Our main result is the construction of an action which describes a consistent coupling of $d=11$ supergravity to a supermembrane. In particular the Kalb–Ramond-like third rank antisymmetic tensor field of $d=11$ supergravity couples to the supermembane via a Wess–Zumino term.

In the following we focus our attention on the description of the supermembane action in $d=11$. The extension to the case of n-extended objects is given in the appendix.

2. We propose the following action for a supermembrane coupled to $d=11$ supergravity:

$$S=\int \mathrm{d}^2\xi\,(\tfrac{1}{2}\sqrt{-g}\,g^{ij}E_i{}^aE_j{}^b\eta_{ab} + \epsilon^{ijk}E_i{}^AE_j{}^BE_k{}^CB_{CBA}-\tfrac{1}{2}\sqrt{-g})\ . \tag{3}$$

Here $i=0, 1, 2$ labels the coordinates $\xi^i=(\tau,\sigma,\rho)$ of the world volume with metric g_{ij} and signature $(-, +, +)$. The super three-form B is needed for the superspace description of $d=11$ supergravity [10]. For the Levi-Civita symbol ϵ^{ijk} we use the same conventions as in ref. [11]. In (3) we have used the notation

$$E_i{}^A=(\partial_iZ^M)E_M{}^A\ , \tag{4}$$

where $Z^M(\xi)$ are the superspace coordinates, and $E_M{}^A(Z)$ is the supervielbein.

Note that the action has a cosmological constant with a fixed magnitude. This is so that the field equation of the metric g_{ij} gives the embedding equation

$$g_{ij}=E_i{}^aE_j{}^b\eta_{ab}\equiv T_{ij}\ . \tag{5}$$

We require that the action S is invariant under a fermionic gauge transformation of the form [5]

$$\delta E^a=0,\quad \delta E^\alpha=(1+\Gamma)^\alpha{}_\beta\kappa^\beta\ ,$$

$$\delta g_{ij}=2[X_{ij}-g_{ij}X^k{}_k/(n-1)]$$
$$(n=2 \text{ for membrane})\ , \tag{6}$$

where the transformation parameter $\kappa^\alpha(\xi)$ is a 32 component Majorana spinor, and a world-volume scalar, and

$$\delta E^A=\delta Z^ME_M{}^A, \tag{7}$$

$$\Gamma^\alpha{}_\beta=(1/6\sqrt{-g})\epsilon^{ijk}E_i{}^aE_j{}^bE_k{}^c(\gamma_{abc})^\alpha{}_\beta\ . \tag{8}$$

Here $\gamma^a(a=0, 1, ..., 10)$ are the Dirac matrices in eleven dimensions. X_{ij} is a function of $E_i{}^A$ which will be determined by the invariance of the action. The choice of δg_{ij} is due to the fact that, given a variation of the action of the form $\delta S=T_{ij}X^{ij}$, and writing this variation as

$$T_{ij}X^{ij}=g_{ij}X^{ij}+(T_{ij}-g_{ij})X^{ij}\ , \tag{9}$$

the second term on the right-hand side cancels $(\delta S/\delta g_{ij})\delta g_{ij}$. Thus we are left with the first term on the right-hand side, which equals the left-hand side upon the use of (5). Effectively, this means that whenever we encounter a variation of the form $T_{ij}X^{ij}$, we can use eq. (5), provided that we add X^{ij} to δg_{ij} as in (6).

The matrix $\Gamma^\alpha{}_\beta$ occurring in (8) satisfies the property

$$\Gamma^\alpha{}_\beta\Gamma^\beta{}_\delta=(T^i{}_{[i}T^j{}_jT^k{}_{k]})\delta^\alpha{}_\delta\equiv\Gamma^2\delta^\alpha{}_\delta\ . \tag{10}$$

The normalization in (8) is chosen such that upon the use of the equation $T_{ij}=g_{ij}$, the matrix $\Gamma^\alpha{}_\beta$ satisfies the relation $\Gamma^\alpha{}_\beta\Gamma^\beta{}_\gamma=\delta^\alpha{}_\gamma$.

Now using (6) the variation of the action (3) is (we consider a closed supermembrane and therefore discard the surface terms)

$$\delta S = \int d^2\xi\, [\sqrt{-g}\, g^{ij}(-\delta E^\beta E_i{}^\gamma T^a{}_{\gamma\beta}) E_{ja}$$

$$+ \sqrt{-g}\, g^{ij}(-\delta E^\beta E_i{}^c T^a{}_{c\beta}) E_{ja}$$

$$+ \epsilon^{ijk} E_i{}^A E_j{}^B E_k{}^C \delta E^\alpha H_{\alpha CBA}$$

$$- \tfrac{1}{2}\sqrt{-g}\, \delta g^{ij}(T^{ij} - \tfrac{1}{2} g^{ij} T - \tfrac{1}{2} g^{ij})] \,. \qquad (11)$$

The torsion two-form T^A and the four-form field strength H are defined by (our superspace conventions are those of Howe [12])

$$T^A = dE^A + E^B \Omega_B{}^A = \tfrac{1}{2} E^B E^C T_{CB}{}^A \,,$$

$$H = dB = \tfrac{1}{24} E^D E^C E^B E^A H_{ABCD} \,. \qquad (12)$$

We now organize the terms in (11) according to the number of one-forms E^α they contain. Those with three E^α and two E^α come only from the Wess–Zumino term. They must vanish seperately, and this requires the constaints

$$H_{\alpha\beta\gamma\delta} = H_{\alpha\beta\gamma d} = 0 \,. \qquad (13)$$

The cancellation of the terms linear in E^α lead to the constraints

$$T^a{}_{\alpha\beta} = (\gamma^a)_{\alpha\beta} \,, \qquad (14)$$

$$H_{\alpha\beta ab} = -\tfrac{1}{6} (\gamma_{ab})_{\alpha\beta} \,, \qquad (15)$$

while the cancellation of the terms not containing E^α require the constraint

$$\eta_{c(a} T^c{}_{b)\alpha} = \eta_{ab} \Lambda_\alpha \,, \qquad (16)$$

$$H_{\alpha abc} = -\tfrac{1}{2} \Lambda_\beta (\gamma_{abc})^\beta{}_\alpha \,. \qquad (17)$$

Here Λ_α is an arbitrary spinor superfield which is vanishing in $d=11$ [10].

It is important to realize that in obtaining (14)–(17) we have used the identity

$$\delta E^\alpha = \Gamma^\alpha{}_\beta \delta E^\beta + (1 - \Gamma^2) \kappa^\alpha \,. \qquad (18)$$

Using this identity in the variation of the kinetic term, the terms arising from $\Gamma^\alpha{}_\beta$ in (18) can be shown to cancel similar terms coming from the variation of the Wess–Zumino term, modulo terms which cancel by an appropriate variation of g_{ij}. [Using the argument below (8) once.] In the remaining terms coming from $(1-\Gamma^2)$, we use the argument given below (8) repeatedly to compute further contributions to δg_{ij}. Thus we find the result

$$X_{ij} = -\tfrac{1}{4} \epsilon_i{}^{kl} E_k{}^a E_l{}^b (\gamma_{ab})_{\alpha\beta} \delta E^\beta E_j{}^\alpha$$

$$+ \tfrac{1}{2} \kappa^\beta E_n{}^\alpha (\gamma^d)_{\alpha\beta} E^{nd} g_{i[j} (T^k{}_k T^l{}_{l]} + \delta^k{}_k T^l{}_{l]})$$

$$+ i \leftrightarrow j \,. \qquad (19)$$

In summary, the action (3) is invariant under (6) provided that (13)–(17) hold, and X^{ij} is given by (19). In addition, the following Bianchi identities must hold:

$$DT^A = -E^B \wedge R_B{}^A \,, \quad DH = 0 \,. \qquad (20)$$

The generalization of the results of this section to the general case of n-extended objects in d-dimensional supergravity is straightforward. The result is given in the appendix.

3. We observe that the superspace constraints of $d=11$ supergravity given by Cremmer and Ferrara [10] and Brink and Howe [10] do provide a solution to (13)–(17) *and* the Bianchi identities (20), with $\Lambda_\alpha = 0$.

In conclusion, we have shown that there exists a consistent coupling of a closed supermembrane to eleven-dimensional supergravity. (Note that it is natural to consider a *closed* supermembrane in eleven dimensions, since there are no matter multiplets in this dimension).

4. There are several directions in which the present work can be extended. We shall name a few.

Firstly, it is of interest to study the quantization of the supermembrane model in eleven dimensions. In particular, the question of whether massless gauge fields can possibly arise is a challenging one. Although usually one encounters difficulties in finding massless excitations of a membrane [13], it is encouraging that, here, we have a spacetime supersymmetric membrane action.

Secondly, it is natural to consider the dimensional reduction of our model from eleven- to ten-dimensional spacetime, *and* at the same time from three-dimensional world volume to a two-dimensional world sheet. It would be interesting to see what kind of $d=10$ string theories could possibly emerge in an infinitely thin membrane limit.

P.K.T. would like to thank Professor Abdus Salam for his kind hospitality and ICTP in Trieste where this work was carried out.

Appendix. In this appendix we construct the action for an n-extended object propagating in d-dimensional supergravity background. We also give the transformation rules, and the constraints on the background.

The action is

$$S=\int d^2\xi\ [\tfrac{1}{2}\sqrt{-g}\,g^{ij}E_i{}^aE_j{}^b\eta_{ab}$$

$$+\epsilon^{i_1\ldots i_n}E_{i_1}{}^{A_1}\ldots E_{i_{n+1}}{}^{A_{n+1}}B_{A_{n+1}\ldots A_1}$$

$$-\tfrac{1}{2}(n-1)\sqrt{-g}\,]\ . \tag{A1}$$

The transformation rules are those in (6), where the matrix $\Gamma^\alpha{}_\beta$ is now given by

$$\Gamma^\alpha{}_\beta=[\eta/(n+1)!\sqrt{-g}\,]$$

$$\times\epsilon^{i_1\ldots i_{n+1}}E_{i_1}{}^{a_1}\ldots E_{i_{n+1}}{}^{a_{n+1}}(\gamma_{a_1\ldots a_{n+1}})^\alpha{}_\beta\ , \tag{A2}$$

where η is given by

$$\eta=(-1)^{(n+1)(n-2)/4}\ . \tag{A3}$$

Invariance of the action (A1) is ensured by imposing the following set of constraints:

$$T^a{}_{\alpha\beta}=(\gamma^a)_{\alpha\beta}\ ,\quad \eta_{c(a}T^c{}_{b)\alpha}=\eta_{ab}\Lambda_\alpha\ , \tag{A4}$$

$$H_{\alpha a_{n+1}\ldots a_1}=(\eta/n!)\Lambda_\beta(\gamma_{a_1\ldots a_{n+1}})^\beta{}_\alpha\ ,$$

$$H_{\alpha\beta a_n\ldots a_1}=[\eta(-1)^n/(n+1)!](\gamma_{a_1\ldots a_n})_{\alpha\beta}\ , \tag{A5}$$

$$H_{\alpha\beta\gamma A_1\ldots A_{n-1}}=0\ , \tag{A5cont'd}$$

and by taking X_{ij} occurring in (6) to be

$$X_{ij}=(-\eta/2n!)$$

$$\times\epsilon_i{}^{k_1\ldots k_n}E_{k_1}{}^{a_1}\ldots E_{k_n}{}^{a_n}(\gamma_{a_1\ldots a_n})_{\alpha\beta}\delta E^\beta E_j{}^\alpha$$

$$+\tfrac{1}{2}\kappa^\beta[E_n{}^\alpha(\gamma^a)_{\alpha\beta}E^{na}+(n+1)\Lambda_\beta]$$

$$\times g_{i[j}(T^{k_1}{}_{k_1}\ldots T^{k_n}{}_{k_n]}+\delta^{k_1}{}_{k_1}T^{k_2}{}_{k_2}\ldots T^{k_n}{}_{k_n]}$$

$$+\delta^{k_1}{}_{k_1}\ldots\delta^{k_{n-1}}{}_{k_{n-1}}T^{k_n}{}_{k_n]})$$

$$-\tfrac{1}{2}\delta E^\beta\Lambda_\beta g_{ij}+(i\leftrightarrow j)\ . \tag{A6}$$

References

[1] P.A.M. Dirac, Proc. R. Soc. A 268 (1962) 57.
[2] P.A. Collins and R.W. Tucker, Nucl. Phys. B 112 (1976) 150.
[3] P.S. Howe and R.W. Tucker, J. Phys. A 10 (1977) L155.
[4] A. Sugamoto, Nucl. Phys. B 215 (1983) 381.
[5] J. Hughes, J. Liu and J. Polchinski, Supermembranes, University of Texas preprint (1986).
[6] M. Henneaux and L. Mezincescu, Phys. Lett. B 152 (1985) 340.
[7] M.B. Green and J.H. Schwarz, Nucl. Phys. B 243 (1984) 285.
[8] W. Siegel, Phys. Lett. B 128 (1983) 397.
[9] See e.g. A. Salam and E. Sezgin, eds., Supergravity in diverse dimensions, Vol. 1 (World Scientific, Singapore, 1987).
[10] E. Cremmer and S. Ferrara, Phys. Lett. B 91 (1980) 61; L. Brink and P. Howe, Phys. Lett. B 91 (1980) 384.
[11] E. Bergshoeff, E. Sezgin and P.K. Townsend, Phys. Lett. B 169 (1986) 191.
[12] P.S. Howe, Nucl. Phys. B 199 (1982) 309.
[13] K. Kikkawa and M. Yamasaki, Can the membrane be a unification model?, Osaka University preprint (August 1986).

Reprinted from Nucl. Phys. B 305 (1988) 545–81

ON THE QUANTUM MECHANICS OF SUPERMEMBRANES

B. de WIT

Institute for Theoretical Physics, University of Utrecht, POB 80.006, 3508 TA Utrecht, The Netherlands

J. HOPPE and H. NICOLAI*

Institut für Theoretische Physik, Universität Karlsruhe, POB 6380, D 7500 Karlsruhe, FRG

Received 11 April 1988

We study the quantum-mechanical properties of a supermembrane and examine the nature of its ground state. A supersymmetric gauge theory of area-preserving transformations provides a convenient framework for this study. The supermembrane can be viewed as a limiting case of a class of models in supersymmetric quantum mechanics. Its mass does not depend on the zero modes and vanishes only if the wave function is a singlet under supersymmetry transformations of the nonzero modes. We exhibit the complexity of the supermembrane ground state and examine various truncations of these models. None of these truncations has massless states.

1. Introduction

Some time ago an action for a membrane moving in a d-dimensional space-time was formulated, which is invariant under super-Poincaré transformations [1]. It is expressed in terms of the membrane coordinates $X^{\mu}(\zeta)$ and a set of anticommuting coordinates $\theta(\zeta)$, transforming as a d-dimensional vector and spinor, respectively; the parameters ζ^i $(i = 0, 1, 2)$ parametrize the world tube swept out by the membrane in space-time. As is well-known, similar actions exist for the superparticle, the superstring, as well as higher-extended objects ("p-branes") [1–4], and they are all characterized by the presence of a local (i.e., ζ-dependent) fermionic symmetry. This invariance requires the existence of a closed superspace form [5], appearing in the action in the form of a Wess-Zumino-Witten term, which is only possible for a specific number of space-time dimensions. Therefore, the supermembrane action can only be formulated in $d = 4$, 5, 7 and 11 dimensional space-times.

An intriguing result found in [1] is that a supermembrane can propagate in a curved superspace. In particular for $d = 11$, the membrane can couple consistently (i.e., without affecting the local fermionic symmetry) to a $d = 11$ supergravity background. Guided by the experience in string theory this result has been interpreted as an indication that the ground state of the supermembrane should be

* Address from May 1, 1988: II. Institut für Theoretische Physik, Universität Hamburg.

0169-6823/88/$03.50 © Elsevier Science Publishers B.V.
(North-Holland Physics Publishing Division)

degenerate and constitute the states of a massless $d = 11$ supergravity multiplet. In attempts to study this question the quantum fluctuations have been analyzed about solutions of the classical membrane equations [6, 7]. While the vacuum energy of these fluctuations vanishes for the solution considered in [6], it did not vanish for the solution described in [7], and neither did it constitute an integer as it does in the case of the string [8] (for the (open) bosonic membrane such a calculation was first undertaken in [9]). On the other hand, heuristic arguments were presented in [10], based on the vanishing of the vacuum energy for fluctuations about a solution with residual supersymmetry, which support the conjecture that the ground state has the structure of a massless $d = 11$ supermultiplet.

In this paper we will study the quantum mechanics of a supermembrane in more detail in the hope of constructing the ground-state wave function. We present an alternative formulation of the membrane as a gauge theory of the area-preserving transformations of the membrane surface. Here we are inspired by the fact that these transformations are the residual invariance of a relativistic membrane theory when quantized in the light-cone gauge [11]. It is possible to consider truncations of this gauge theory by truncating the infinite harmonic expansion of the membrane coordinates. At least for membranes with the topology of a sphere this can be done in such a way that the supersymmetry remains preserved. The group of area-preserving transformations is thereby reduced to SU(N).* These truncations lead to a class of matrix models in supersymmetric quantum mechanics [13, 14], which turn out to coincide with the models that have been presented in [15]. A priori, three different types of membrane ground states are possible. One possibility is that the ground state is a singlet under supersymmetry, which is thus annihilated by the supersymmetry charges. By virtue of the anticommutation relation which expresses the hamiltonian as the square of these charges, this ground state should be massless. However, this situation is not possible for the supermembrane: it follows from the explicit expression for the hamiltonian that all wave functions have an obvious degeneracy associated with the fermionic zero modes. Therefore the ground state must be degenerate and constitute a supermultiplet. There are then two possibilities. One is that the ground state is a massless supermultiplet, consisting of 2^7 bosonic and 2^7 fermionic states, in which case the supercharges associated with the nonzero modes must annihilate the ground-state wave functions. If this is not the case one has a massive supermultiplet. The ground-state degeneracy is then enormous, as a massive supermultiplet contains 2^{15} bosonic and 2^{15} fermionic states.

We restrict ourselves to supermembranes that move in a trivial space-time. Hence we consider no compactification as in [6] and neither do we study the possibility of membranes moving in nontrivial space-times such as in [16]. This means, in particular, that our considerations have no bearing on the results described in [16],

* This idea goes back to Goldstone [11]. The relation between SU(N) and the group of area-preserving transformations was exhibited in [12].

where the existence of infinitely many massless states of the supermembrane compactified to $AdS_4 \times S^7$ was demonstrated in a small-fluctuation analysis. Our work shows that the ground-state wave function of a supermembrane has a high degree of complexity. For instance, it is not possible for a massless ground state that the wave function factorizes into a bosonic and a fermionic part, if one of these factors is rotationally invariant. This is a distinct difference with the wave function for the superstring ground state. We then study the restrictions imposed by rotational invariance for the total ground-state wave function, but, unfortunately, this does not lead to useful simplifications. Although the condition that the wave function vanishes under the action of the supersymmetry charges has solutions, these solutions tend to be not square-integrable. This we demonstrate in a G_2-invariant truncation of the theory. We also consider a supermembrane propagating in a 4-dimensional space-time in a truncation where the group of area-preserving transformations is reduced to SU(2). Assuming that the wave function tends to zero at spatial infinity, we show that the energy of the supermembrane is lower than that of its bosonic version, but there is no solution with zero mass. However, the complexity of this problem makes it hard to reach a firm conclusion concerning the existence of massless solutions in the general case. We should also emphasize that, while the supersymmetric matrix models are well defined, it is not clear what will happen in the limit where the gauge group approaches the full infinite-dimensional group of area-preserving transformations. As is well-known, in quantum-mechanical systems based on an infinite number of degrees of freedom, degenerate ground states are not always contained in the same Hilbert space; this aspect is of immediate importance for possible applications of supermembrane theories. Also, while the models based on SU(N) yield, in the limit $N \to \infty$, the full group of area-preserving transformations corresponding to a membrane with the topology of a sphere, a corresponding result for other membrane topologies is not known.

In sect. 2 we start by formulating the membrane action in the light-cone gauge, emphasizing the role played by the area-preserving transformations. We introduce a gauge theory of these transformations, and verify the supersymmetry algebra. In sect. 3 we review the truncation to the finite-dimensional matrix models and discuss some properties of area-preserving transformations. Then, in sect. 4, we discuss attempts to solve the equations for the ground-state wave function of the supermembrane and demonstrate the absence of a massless ground state in two different truncations. In an appendix we analyze the implications of SO(9) invariance for a general wave function.

2. Lightcone formulation of the supermembrane

The starting point of this section is the lagrangian

$$\mathscr{L} = -\sqrt{-g(X,\theta)} - \varepsilon^{ijk}\left[\tfrac{1}{2}\partial_i X^\mu\left(\partial_j X^\nu + \bar\theta\Gamma^\nu\partial_j\theta\right) + \tfrac{1}{6}\bar\theta\Gamma^\mu\partial_i\theta\bar\theta\Gamma^\nu\partial_j\theta\right]\bar\theta\Gamma_{\mu\nu}\partial_k\theta, \quad (2.1)$$

where $X^\mu(\zeta)$ and $\theta(\zeta)$ denote the superspace coordinates of the membrane parametrized in terms of world-tube parameters ζ^i $(i = 0, 1, 2)$. The metric $g_{ij}(X, \theta)$ is the induced metric on the world tube,

$$g_{ij} = E_i^\mu E_j^\nu \eta_{\mu\nu}, \tag{2.2}$$

where E_i^μ are certain supervielbein components tangential to the world tube, defined by

$$E_i^\mu = \partial_i X^\mu + \bar{\theta}\Gamma^\mu \partial_i \theta, \tag{2.3}$$

and $\eta_{\mu\nu}$ is the flat $d = 11$ Minkowski metric. It is easy to see that E_i^μ is invariant under space-time supersymmetry transformations

$$\delta\theta = \varepsilon, \qquad \delta X^\mu = -\bar{\varepsilon}\Gamma^\mu\theta. \tag{2.4}$$

In fact this transformation also leaves the lagrangian (2.1) invariant (up to a total divergence) provided the following gamma matrix identity is satisfied

$$\bar{\psi}_{[1}\Gamma^\mu\psi_2\bar{\psi}_3\Gamma_{\mu\nu}\psi_{4]} = 0, \tag{2.5}$$

where we antisymmetrize over four arbitrary spinors ψ_1–ψ_4. This identity only holds in $d = 4$, 5, 7 and 11 space-time dimensions. In this paper we mainly restrict ourselves to $d = 11$, but this restriction is not important for the analysis to be presented below.

The field equations corresponding to the lagrangian (2.1) take the form

$$\partial_i\left(\sqrt{-g}\, g^{ij} E_j^\mu\right) - \varepsilon^{ijk} E_i^\nu\, \partial_j \bar{\theta}\Gamma^\mu{}_\nu \partial_k \theta = 0, \tag{2.6}$$

$$(\mathbf{1} + \Gamma) g^{ij} \not{E}_i \partial_j \theta = 0, \tag{2.7}$$

where Γ is defined by

$$\Gamma = \frac{\varepsilon^{ijk}}{6\sqrt{-g}} E_i^\mu E_j^\nu E_k^\rho \Gamma_{\mu\nu\rho}. \tag{2.8}$$

We note two important identities for Γ,

$$\Gamma^2 = \mathbf{1}, \qquad \Gamma \not{E}_i = \not{E}_i \Gamma = g_{ij} \frac{\varepsilon^{jkl}}{2\sqrt{-g}} E_k^\mu E_l^\nu \Gamma_{\mu\nu}. \tag{2.9}$$

The lagrangian (2.1) is manifestly invariant under reparametrizations of the world-tube coordinates ζ^i. It is also invariant under a *local* fermionic symmetry generated by

$$\theta = (\mathbf{1} - \Gamma)\kappa, \qquad \delta X^\mu = \bar{\kappa}(\mathbf{1} - \Gamma)\Gamma^\mu\theta, \tag{2.10}$$

where κ is an arbitrary ζ-dependent spinor. Observe that κ is always multiplied by the projection operator $(\mathbf{1} - \Gamma)$.

Of particular importance is the supersymmetry current associated with the transformations (2.4). It reads

$$J^i = -2\sqrt{-g}\, g^{ij} \not{E}_j \theta$$

$$- \varepsilon^{ijk}\left\{ E^\mu_j E^\nu_k \Gamma_{\mu\nu}\theta + \tfrac{4}{3}\left[\Gamma^\nu\theta\left(\bar{\theta}\Gamma_{\mu\nu}\partial_j\theta\right) + \Gamma_{\mu\nu}\theta\left(\bar{\theta}\Gamma^\nu\partial_j\theta\right)\right]\left[E^\mu_k - \tfrac{2}{5}\bar{\theta}\Gamma^\mu\partial_k\theta\right]\right\}. \tag{2.11}$$

As one can verify straightforwardly, this current is conserved by virtue of the field equations (2.6) and (2.7), provided the identity (2.5) holds.

In order to pass to the light-cone gauge we choose light-cone coordinates

$$X^\pm = \sqrt{\tfrac{1}{2}}\,(X^{10} \pm X^0). \tag{2.12}$$

Transverse coordinates will be denoted by $X^a(\zeta)$ $(a = 1, \ldots, 9)$. For the gamma matrices we make a similar decomposition,

$$\gamma_\pm = \sqrt{\tfrac{1}{2}}\,(\Gamma^{10} \pm \Gamma^0), \qquad \gamma_a = \Gamma_a, \qquad (a = 1, \ldots, 9), \tag{2.13}$$

so that $\{\gamma_+, \gamma_-\} = 2\mathbf{1}$, $\gamma_+^2 = \gamma_-^2 = 0$, $\{\gamma_\pm, \gamma_a\} = 0$. Furthermore we change notation and denote the parameters ζ^i by

$$(\zeta^0, \zeta^r) \to (\tau, \sigma^r), \qquad (r = 1, 2). \tag{2.14}$$

By a suitable reparametrization we now choose

$$X^+(\zeta) = X^+(0) + \tau, \tag{2.15}$$

so that $\partial_i X^+ = \delta_{i0}$. Furthermore we use the local fermionic symmetry to impose the gauge condition

$$\gamma_+\theta = 0. \tag{2.16}$$

With these gauge conditions we obtain the following result for the components of the induced metric

$$g_{rs} \equiv \bar{g}_{rs} = \partial_r \boldsymbol{X} \cdot \partial_s \boldsymbol{X},$$

$$g_{0r} \equiv u_r = \partial_r X^- + \partial_0 \boldsymbol{X} \cdot \partial_r \boldsymbol{X} + \bar{\theta}\gamma_- \partial_r \theta,$$

$$g_{00} = 2\partial_0 X^- + (\partial_0 \boldsymbol{X})^2 + 2\bar{\theta}\gamma_- \partial_0 \theta, \tag{2.17}$$

while the determinant of the metric can now be written as

$$g \equiv \det g = -\Delta \bar{g}, \tag{2.18}$$

where

$$\Delta = -g_{00} + u_r \bar{g}^{rs} u_s, \qquad \bar{g}^{rs} g_{st} = \delta^r_t, \qquad \bar{g} \equiv \det \bar{g}. \tag{2.19}$$

After imposing the gauge conditions (2.15) and (2.16) the lagrangian and supercharge densities take the form (using $\varepsilon^{0rs} = -\varepsilon^{rs}$, $g^{00} = -\Delta^{-1}$, $g^{0r} = \Delta^{-1}\bar{g}^{rs}u_s$)

$$\mathscr{L} = -\sqrt{\bar{g}\Delta} + \varepsilon^{rs}\,\partial_r X^a \bar{\theta}\gamma_- \gamma_a \partial_s \theta, \tag{2.20}$$

$$J^0 = 2\sqrt{\frac{\bar{g}}{\Delta}}\,[(\partial_0 X^a - u_r \bar{g}^{rs}\partial_s X^a)\gamma_a + \gamma_-]\theta + \varepsilon^{rs}\,\partial_r X^a \partial_s X^b \gamma_{ab}\theta. \tag{2.21}$$

In order to write down the corresponding hamiltonian density, we first determine the canonical momenta $\boldsymbol{P}$, P^+ and S conjugate to $\boldsymbol{X}$, X^- and θ, respectively. They are

$$\boldsymbol{P} \equiv \frac{\partial \mathscr{L}}{\partial(\partial_0 \boldsymbol{X})} = \sqrt{\frac{\bar{g}}{\Delta}}\,(\partial_0 \boldsymbol{X} - u_r \bar{g}^{rs}\partial_s \boldsymbol{X}),$$

$$P^+ \equiv \frac{\partial \mathscr{L}}{\partial(\partial_0 X^-)} = \sqrt{\frac{\bar{g}}{\Delta}},$$

$$S \equiv \frac{\partial \mathscr{L}}{\partial_L(\partial_0 \bar{\theta})} = -\sqrt{\frac{\bar{g}}{\Delta}}\,\gamma_- \theta, \tag{2.22}$$

where ∂_L denotes the *left* derivative. The hamiltonian density then takes the simple form

$$\begin{aligned} \mathscr{H} &\equiv \boldsymbol{P}\cdot\partial_0 \boldsymbol{X} + P^+ \partial_0 X^- + \bar{S}\partial_0 \theta - \mathscr{L} \\ &= \frac{\boldsymbol{P}^2 + \bar{g}}{2P^+} - \varepsilon^{rs}\,\partial_r X^a \bar{\theta}\gamma_- \gamma_a \partial_s \theta. \end{aligned} \tag{2.23}$$

The bosonic part of this expression was first found by Goldstone [11] (see also [12]), while its superextension was derived in [17].

One easily verifies that there are two primary constraints

$$\phi_r = \boldsymbol{P}\cdot\partial_r \boldsymbol{X} + P^+ \partial_r X^- + \bar{S}\partial_r \theta \approx 0, \tag{2.24}$$

$$\chi = S + P^+ \gamma_- \theta \approx 0, \tag{2.25}$$

where ≈ 0 indicates that the constraints are "weakly zero", so that they may have nonzero Poisson brackets with the phase-space variables. We recall that the time (i.e., τ) evolution in phase space is governed by the "total" hamiltonian [18])

$$H_T = \int d^2\sigma \left\{ \mathscr{H} + c^r\phi_r + \bar{d}\chi \right\}, \tag{2.26}$$

where c^r and d are Lagrange multipliers. One can easily verify that there are no secondary constraints at this point.

The gauge conditions (2.15) and (2.16) are still invariant under τ-dependent reparametrizations of σ^r

$$\sigma^r \to \sigma^r + \xi^r(\tau, \sigma). \tag{2.27}$$

Under such infinitesimal reparametrizations u^r changes into $u^r - \partial_0\xi^r + \partial_s\xi^r u^s - \xi^s\partial_s u^r$, which shows that one can impose yet another gauge condition, namely

$$u^r = 0. \tag{2.28}$$

In this gauge it follows that $c^r = 0$ according to the Hamilton equations corresponding to (2.26), so that $\partial_0 P^+ = 0$. Because P^+ transforms as a density under reparametrizations, it may be adjusted to a constant times some density $w(\sigma)$,

$$P^+ = P_0^+ w(\sigma), \tag{2.29}$$

where we will normalize $w(\sigma)$ according to

$$\int d^2\sigma\, w(\sigma) = 1. \tag{2.30}$$

Therefore the constant P_0^+ represents the membrane momentum in the direction associated with the coordinate X^-,

$$P_0^+ = \int d^2\sigma\, P^+. \tag{2.31}$$

The other momentum components are given by the integrals over $\boldsymbol{P}$ and $-\mathscr{H}$,

$$\boldsymbol{P}_0 = \int d^2\sigma\, \boldsymbol{P}, \qquad P_0^- = -\int d^2\sigma\, \mathscr{H}. \tag{2.32}$$

Hence the mass $\mathscr{M}$ of the membrane is given by

$$\mathscr{M}^2 = \int d^2\sigma \left\{ \frac{[\boldsymbol{P}^2]' + \bar{g}}{w} - 2P_0^+ \varepsilon^{rs}\, \partial_r X^a \bar{\theta}\gamma_-\gamma_a\partial_s\theta \right\}, \tag{2.33}$$

where the notation $[\boldsymbol{P}^2]'$ indicates that we are excluding the zero mode $\boldsymbol{P} = \boldsymbol{P}_0 w(\sigma)$ from the integrand. On the other hand, we observe that the zero modes $\boldsymbol{X}_0$ and θ_0, defined by

$$\boldsymbol{X}_0 = \int \mathrm{d}^2\sigma\, w(\sigma)\boldsymbol{X}, \qquad \theta_0 = \int \mathrm{d}^2\sigma\, w(\sigma)\theta, \tag{2.34}$$

do not appear in the equation for $\mathcal{M}^2$ either, at least if the membrane coordinates are single-valued functions of σ^r, which is the case if space-time is not compactified, or, for open membranes, if one assumes appropriate boundary conditions. The absence of $\boldsymbol{X}_0$, which is just the center-of-mass coordinate of the membrane, is rather obvious. The fact that $\mathcal{M}^2$ does not depend on θ_0 will play an extremely important role later on.

The coordinate X^- no longer appears explicitly in (2.33) and is determined by the gauge condition (2.28), or, equivalently, by the constraint (2.24) after imposing the gauge condition (2.29). The relevant formula is

$$\partial_r X^- = -\partial_0 \boldsymbol{X}\cdot\partial_r\boldsymbol{X} - \bar{\theta}\gamma_-\partial_r\theta\,. \tag{2.35}$$

Because X^- must be a globally defined function of σ^r this requires that

$$\oint(\partial_0\boldsymbol{X}\cdot\partial_r\boldsymbol{X} + \bar{\theta}\gamma_-\partial_r\theta)\,\mathrm{d}\sigma^r = 0 \tag{2.36}$$

for any closed curve on the membrane. Locally this condition implies

$$\varepsilon^{rs}(\partial_r\partial_0\boldsymbol{X}\cdot\partial_s\boldsymbol{X} + \partial_r\bar{\theta}\gamma_-\partial_s\theta) = 0\,. \tag{2.37}$$

Observe that, when space-time is not compactified so that $\boldsymbol{X}$ and θ are single-valued functions of σ, only the condition (2.37) is relevant.

The gauge conditions adopted above leave a residual reparametrization invariance consisting of time-independent area-preserving transformations. Infinitesimal transformations of this kind leave (2.29) invariant, and are thus defined by

$$\sigma^r \to \sigma^r + \xi^r(\sigma) \qquad \text{with} \qquad \partial_r(w(\sigma)\xi^r(\sigma)) = 0\,. \tag{2.38}$$

There exists an alternative formulation of the membrane theory, which emphasizes area-preserving reparametrizations from the start. Locally the area-preserving transformations can be written as

$$\xi^r(\sigma) = \frac{\varepsilon^{rs}}{w(\sigma)}\partial_s\xi(\sigma)\,. \tag{2.39}$$

If the membrane is topologically nontrivial, i.e. if the membrane surface has handles so that it contains uncontractible curves, then $\xi^r(\sigma)$ and $\xi(\sigma)$ will not necessarily be globally defined. However, we will restrict ourselves to the subgroup generated by functions $\xi(\sigma)$ that are globally defined. It is then convenient to introduce a Lie bracket of any two functions $A(\sigma)$ and $B(\sigma)$ by

$$\{A, B\} \equiv \frac{\varepsilon^{rs}}{w(\sigma)} \partial_r A(\sigma) \partial_s B(\sigma), \qquad (2.40)$$

which is antisymmetric in A and B and satisfies the Jacobi identity $\{A, \{B, C\}\} + \{B, \{C, A\}\} + \{C, \{A, B\}\} = 0$. Using this bracket, infinitesimal area-preserving reparametrizations act on X^a and θ according to

$$\delta X^a = \{\xi, X^a\}, \qquad \delta\theta = \{\xi, \theta\}. \qquad (2.41)$$

Now let us introduce a gauge field ω associated with time-dependent reparametrizations, transforming as

$$\delta\omega = \partial_0 \xi + \{\xi, \omega\}, \qquad (2.42)$$

and corresponding covariant derivatives

$$D_0 X^a = \partial_0 X^a - \{\omega, X^a\}, \qquad D_0\theta = \partial_0\theta - \{\omega, \theta\}. \qquad (2.43)$$

The following lagrangian density is then manifestly gauge invariant under the transformations (2.41) and (2.42),

$$w^{-1}\mathscr{L} = \tfrac{1}{2}(D_0 X)^2 + \bar{\theta}\gamma_- D_0\theta - \tfrac{1}{4}(\{X^a, X^b\})^2 + \bar{\theta}\gamma_-\gamma_a\{X^a, \theta\}, \qquad (2.44)$$

as well as under space-time supersymmetry transformations given by

$$\begin{aligned} \delta X^a &= -2\bar{\varepsilon}\gamma^a\theta, \\ \delta\theta &= \tfrac{1}{2}\gamma_+(D_0 X^a\gamma_a + \gamma_-)\varepsilon + \tfrac{1}{4}\{X^a, X^b\}\gamma_+\gamma_{ab}\varepsilon \\ \delta\omega &= -2\bar{\varepsilon}\theta. \end{aligned} \qquad (2.45)$$

The supercharge density associated with the transformations (2.45), equals

$$J^0 = w\left[2(D_0 X^a\gamma_a + \gamma_-) + \{X^a, X^b\}\gamma_{ab}\right]\theta. \qquad (2.46)$$

In the gauge $\omega = 0$ the latter result coincides with the charge density obtained previously (cf. (2.21) after imposing the gauge conditions (2.28) and (2.29)). To see that the supersymmetry transformations are associated with space-time, one may evaluate the supersymmetry commutator on X^a:

$$[\delta(\varepsilon_1), \delta(\varepsilon_2)]\, X^a = -2\bar{\varepsilon}_2\gamma_+\varepsilon_1 D_0 X^a - 2\bar{\varepsilon}_2\gamma^a\varepsilon_1 + \{\xi, X^a\}, \qquad (2.47)$$

where, on the right-hand side, we distinguish a τ-translation generated by D_0 (which, as we know, is related to a translation of the membrane coordinate X^+), a translation of X^a and an $\boldsymbol{X}$-dependent area-preserving gauge transformation with parameter $\xi = 2\bar{\varepsilon}_2\gamma^b\gamma_+\varepsilon_1 X^b$. In order to verify that the bosonic and fermionic degrees of freedom balance in the path integral associated with (2.44), one may impose a gauge condition $\omega = 0$, which leads to a (free) fermionic complex ghost field. Altogether one then counts 9 bosonic and $16 + 2 = 18$ (real) fermionic field components.

To establish full equivalence of (2.44) with the membrane lagrangian, we implement the gauge $\omega = 0$ and introduce canonical momenta $\boldsymbol{P}$ and S associated with $\boldsymbol{X}$ and θ,

$$\boldsymbol{P} = w\partial_0\boldsymbol{X}, \qquad S = -w\gamma_-\theta. \qquad (2.48)$$

The hamiltonian is then

$$H \equiv \int d^2\sigma \left\{ \boldsymbol{P}\cdot\partial_0\boldsymbol{X} + \bar{S}\partial_0\theta - \mathcal{L} \right\}$$

$$= \frac{1}{2}\int d^2\sigma \left\{ w^{-1}\boldsymbol{P}^2 + \tfrac{1}{2}w\left(\{X^a, X^b\}\right)^2 - 2w\bar{\theta}\gamma_-\gamma_a\{X^a, \theta\} \right\}, \qquad (2.49)$$

so that, after dropping the zero-mode $\boldsymbol{P}_0$, $2H$ coincides with eq. (2.33) for the membrane mass $\mathcal{M}$, provided θ is rescaled by a factor $\sqrt{P_0^+}$ (to make the comparison, use that $(\{X^a, X^b\})^2 = 2w^{-2}\bar{g}$).

Furthermore, the field equation for ω leads to the constraint

$$\varphi \equiv \{\partial_0\boldsymbol{X}, \cdot\boldsymbol{X}\} + \{\bar{\theta}\gamma_-, \theta\} \approx 0, \qquad (2.50)$$

or, in phase-space variables,

$$\varphi = \{w^{-1}\boldsymbol{P}, \cdot\boldsymbol{X}\} + \{w^{-1}\bar{S}, \theta\} \approx 0, \qquad (2.51)$$

This constraint is just (2.37), and we have thus established the equivalence of (2.44) with the initial lagrangian (2.1). The quantity φ is the "current" that couples to the gauge field ω, so it is obvious that $w\varphi$ represents the charge density associated with the area-preserving transformations. In addition there is the usual second-class constraint that expresses the fermionic momentum S into θ.

The Dirac brackets for the theory above are derived by standard methods and read

$$\big(X^a(\sigma), P^b(\sigma')\big)_{\rm DB} = \delta^{ab}\,\delta^2(\sigma-\sigma'),$$

$$\big(\theta_\alpha(\sigma), \bar\theta_\beta(\sigma')\big)_{\rm DB} = \frac{1}{4w}(\gamma_+)_{\alpha\beta}\,\delta^2(\sigma-\sigma'). \tag{2.52}$$

It is now possible to verify the full $d=11$ supersymmetry algebra. Decomposing the supersymmetry charges into two independent SO(9) spinors according to

$$Q = Q^+ + Q^- = \int d^2\sigma\, J^0, \tag{2.53}$$

where J^0 is given in (2.46) and $Q^\pm \equiv \frac{1}{2}\gamma_\pm\gamma_\mp Q$, we find the expressions

$$Q^+ = \int d^2\sigma\,\big(2P^a\gamma_a + w\{X^a, X^b\}\gamma_{ab}\big)\theta,$$

$$Q^- = -2\int d^2\sigma\, S = 2\gamma_-\theta_0. \tag{2.54}$$

Observe that Q^- acts only on the fermionic zero-modes θ_0, which, as we have pointed out before, do not appear in the expressions for the hamiltonian and the membrane mass.

It is now a straightforward exercise to determine the Dirac brackets for the supercharges. The result takes the following form

$$\big(Q^-_\alpha, \bar Q^-_\beta\big)_{\rm DB} = -2(\gamma_-)_{\alpha\beta},$$

$$\big(Q^+_\alpha, \bar Q^+_\beta\big)_{\rm DB} = 2(\gamma_+)_{\alpha\beta}H - 2(\gamma_a\gamma_+)_{\alpha\beta}\int d^2\sigma\, w\varphi X^a$$

$$+ (\gamma^a\gamma_+)_{\alpha\beta}\int d^2\sigma\,\partial_r S^r_a + (\gamma^{abcd}\gamma_+)_{\alpha\beta}\int d^2\sigma\,\partial_r S^r_{abcd},$$

$$\big(Q^+_\alpha, \bar Q^-_\beta\big)_{\rm DB} = -(\gamma_a\gamma_+\gamma_-)_{\alpha\beta}P^a_0 + (\gamma^{ab}\gamma_+\gamma_-)_{\alpha\beta}\int d^2\sigma\,\partial_r S^r_{ab}, \tag{2.55}$$

where the surface terms, given by

$$S^r_a = \varepsilon^{rs}\big\{2w^{-1}X_a P^b\partial_s X^b + 2X_a\bar\theta\gamma_-\partial_s\theta - \tfrac{3}{8}X^b\partial_s\big(\bar\theta\gamma_-\gamma_{ab}\theta\big)\big\},$$

$$S^r_{abcd} = \tfrac{1}{24}\varepsilon^{rs}X_{[a}\,\partial_s\big(\bar\theta\gamma_-\gamma_{bcd]}\theta\big),$$

$$S^r_{ab} = -\tfrac{1}{2}\varepsilon^{rs}X_{[a}\,\partial_s X_{b]}, \tag{2.56}$$

can contribute only if the membrane coordinates and momenta are not single-valued.

It is useful to separately consider the zero-mode contributions to Q^+, which define a conserved charge $Q^{+(0)}$. It reads

$$Q^{+(0)} = 2 P_0^a \gamma_a \theta_0 . \tag{2.57}$$

Together, $Q^{+(0)}$ and Q^- generate the algebra

$$\begin{aligned}
\left(Q_\alpha^-, \bar{Q}_\beta^-\right)_{\mathrm{DB}} &= -2(\gamma_-)_{\alpha\beta} , \\
\left(Q_\alpha^{+(0)}, \bar{Q}_\beta^{+(0)}\right)_{\mathrm{DB}} &= (\gamma_+)_{\alpha\beta} \boldsymbol{P}_0^2 , \\
\left(Q_\alpha^{+(0)}, \bar{Q}_\beta^-\right)_{\mathrm{DB}} &= -(\gamma_a \gamma_+ \gamma_-)_{\alpha\beta} P_0^a ,
\end{aligned} \tag{2.58}$$

where we have used that the hamiltonian for the zero modes is the center-of-mass hamiltonian $H^{(0)} = \frac{1}{2}\boldsymbol{P}_0^2$. For the remainder of the supercharge Q^+, which does not contain the zero modes anymore (provided that the membrane coordinates are single-valued), the Dirac bracket reads

$$\left(Q_\alpha^+, \bar{Q}_\beta^+\right)_{\mathrm{DB}} = (\gamma_+)_{\alpha\beta} \mathcal{M}^2 - 2(\gamma_a \gamma_+)_{\alpha\beta} \int \mathrm{d}^2\sigma \, w \varphi X^a + \cdots , \tag{2.59}$$

where the dots indicate the contribution from the surface terms. This relation plays a central role in the analysis of this paper.

So far we have been employing a $d = 11$ notation for the spinors θ. However, due to the gauge condition (2.16), the anticommuting coordinates are restricted to SO(9) spinors, satisfying

$$\gamma_1 \gamma_2 \cdots \gamma_9 \theta = \theta . \tag{2.60}$$

Furthermore we have

$$\bar{\theta}\gamma_- = i\sqrt{2}\,\theta^\dagger , \qquad \theta = \mathscr{C}^{-1}\theta^\dagger , \tag{2.61}$$

where $\mathscr{C}$ is the $d = 9$ charge conjugation matrix, which is symmetric and related to the $d = 11$ charge conjugation matrix by $\mathscr{C} = -C\gamma_{11}$; we also note that the SO(9) gamma matrices satisfy $\gamma_a^T = \mathscr{C}\gamma_a\mathscr{C}^{-1}$. Henceforth we will choose $\mathscr{C} = \mathbf{1}$, so that the SO(9) gamma matrices are symmetric.

In subsequent sections we shall study the ground-state wave function of the supermembrane. For that purpose it is convenient to have an explicit representation for the operators associated with the fermionic coordinates. As a first step towards constructing such a representation we decompose the real SO(9) spinor coordinates θ into a single complex 8-component spinor λ, which transforms linearly under the SO(7) $\times$ U(1) subgroup of SO(9). This decomposition is effected by expressing the two eigenspinors of γ_9, defined by $\gamma_9\,\theta^{(\pm)} = \pm\theta^{(\pm)}$, into a complex SO(7) spinor λ, according to

$$\theta^{(+)} = \frac{\lambda^\dagger + \lambda}{2^{5/4}}, \qquad \theta^{(-)} = i\frac{\lambda^\dagger - \lambda}{2^{5/4}}. \tag{2.62}$$

The bosonic coordinates X^a are then decomposed according to representations of this SO(7) $\times$ U(1) subgroup so we distinguish the components X^i of an SO(7) vector ($i = 1, 2, \ldots, 7$), while X^8 and X^9 are combined into a complex coordinate

$$Z = \sqrt{\tfrac{1}{2}}\,(X^8 + iX^9), \tag{2.63}$$

which transforms under U(1). Similarly, the momenta are decomposed in terms of an SO(7) vector P^i and a complex momentum $\mathscr{P}$ defined by

$$\mathscr{P} = \sqrt{\tfrac{1}{2}}\,(P^8 - iP^9). \tag{2.64}$$

The normalization factors in (2.62)–(2.64) are chosen such that the nonvanishing Dirac brackets are equal to

$$\begin{aligned}
\big(X^i(\sigma), P^j(\sigma')\big)_{\mathrm{DB}} &= \delta^{ij}\delta^2(\sigma - \sigma'),\\
\big(Z(\sigma), \mathscr{P}(\sigma')\big)_{\mathrm{DB}} &= \delta^2(\sigma - \sigma'),\\
\big(\lambda_\alpha(\sigma), \lambda^\dagger_\beta(\sigma')\big)_{\mathrm{DB}} &= -iw^{-1}\delta_{\alpha\beta}\,\delta^2(\sigma - \sigma').
\end{aligned} \tag{2.65}$$

The supercharges Q^+_α can also be written as a *complex* SO(7) spinor. When expressed in terms of the above coordinates these charges take the form

$$\begin{aligned}
Q &= \int d^2\sigma\Big[\big(P^i\Gamma_i + \tfrac{1}{2}w\{X^i, X^j\}\Gamma_{ij} - w\{Z, \bar{Z}\}\big)\lambda\\
&\qquad + \sqrt{2}\,\big(i\mathscr{P} + iw\{X^i, \bar{Z}\}\Gamma_i\big)\lambda^\dagger\Big],\\
Q^\dagger &= \int d^2\sigma\Big[\big(-P^i\Gamma_i + \tfrac{1}{2}w\{X^i, X^j\}\Gamma_{ij} + w\{Z, \bar{Z}\}\big)\lambda^\dagger\\
&\qquad + \sqrt{2}\,\big(-i\bar{\mathscr{P}} + iw\{X^i, Z\}\Gamma_i\big)\lambda\Big].
\end{aligned} \tag{2.66}$$

where Γ_i are the SO(7) gamma matrices*. In the same notation the hamiltonian reads

$$H = \int d^2\sigma \Big[\tfrac{1}{2} w^{-1} (P^i)^2 + w^{-1} |\mathscr{P}|^2 + \tfrac{1}{4} w (\{X^i, X^j\})^2 + w |\{Z, X^i\}|^2 + \tfrac{1}{2} w |\{Z, \bar{Z}\}|^2 + i w \lambda \Gamma_i \{X^i, \lambda^\dagger\} - \tfrac{1}{2}\sqrt{2}\, w \lambda \{Z, \lambda\} + \tfrac{1}{2}\sqrt{2}\, w \lambda^\dagger \{\bar{Z}, \lambda^\dagger\} \Big]. \quad (2.67)$$

The normalization of Q and $Q^\dagger$ is such that

$$(Q_\alpha, Q_\beta)_{\mathrm{DB}} = -\sqrt{2}\, \delta_{\alpha\beta} \int d^2\sigma\, w \bar{Z} \varphi,$$

$$(Q_\alpha, Q_\beta^\dagger)_{\mathrm{DB}} = -2i \delta_{\alpha\beta} H + 2i (\Gamma_i)_{\alpha\beta} \int d^2\sigma\, w X^i \varphi. \quad (2.68)$$

3. Area-preserving transformations and supersymmetric matrix models

The analysis presented in the foregoing section has led us to the constraint (2.24) (or, (2.35)–(2.37) and (2.50–51)), which generates the group G of area-preserving diffeomorphisms. All physical quantities, such as the expression (2.33) for the membrane mass, must be invariant under this group, and this statement applies equally to the classical theory (where (2.24) constrains the space of solutions) and to the quantum theory (where (2.24) must be imposed as a constraint on the physical states). The group G and its associated Lie algebra play an important role in the following and are also of interest in their own right [11, 12]. In this section, we summarize some properties of this group for spherical and toroidal membranes. Before going into the details we make some general remarks which also pertain to topologically more complicated membranes. We start by expanding the coordinates into a complete orthonormal basis of functions $Y^A(\sigma)$ on the membrane,

$$X(\sigma) = \sum_A X^A Y_A(\sigma), \qquad (A = 0, 1, 2, \ldots) \quad (3.1)$$

and likewise for the fermionic coordinates θ (or λ) and the momenta. The functions Y_A may be chosen real, in which case there are no restrictions on the modes, or complex, in which case there are further restrictions from the reality of $\boldsymbol{X}$. The following notation allows us to discuss both options in a uniform manner. First we

* Our conventions are as follows: $\{\Gamma_i, \Gamma_j\} = 2\delta_{ij}\mathbf{1}$, $\Gamma_{ij} \equiv \tfrac{1}{2}[\Gamma_i, \Gamma_j]$, $\Gamma_{ijk} \equiv \tfrac{1}{2}\{\Gamma_i, \Gamma_{jk}\}$, $\Gamma_1 \ldots \Gamma_7 = -i\mathbf{1}$. Also, $\Gamma_i = \Gamma_i^\dagger = -\Gamma_i^{\mathrm{T}} = -\Gamma_i^*$.

define

$$Y^A(\sigma) \equiv (Y_A(\sigma))^* \doteq \eta^{AB} Y_B(\sigma), \tag{3.2}$$

where the matrix η^{AB} satisfies $\eta^{AB}\eta_{BC} = \delta_C^A$ with $\eta_{AB} \equiv (\eta^{AB})^*$. The normalization of the functions Y_A is

$$\int d^2\sigma\, w(\sigma) Y^A(\sigma) Y_B(\sigma) = \delta_B^A, \tag{3.3a}$$

or, equivalently,

$$\int d^2\sigma\, w(\sigma) Y_A(\sigma) Y_B(\sigma) = \eta_{AB}, \tag{3.3b}$$

which shows that η_{AB} is symmetric. The reality condition on the expansion coefficients of $X(\sigma)$ then reads

$$X_A \equiv (X^A)^* = \eta_{AB} X^B. \tag{3.4}$$

Furthermore, completeness of the Y_A implies

$$\sum_A Y^A(\sigma) Y_A(\sigma') = \frac{1}{w(\sigma)} \delta^2(\sigma - \sigma'). \tag{3.5}$$

As explained in the previous section, area-preserving maps are expressed in terms of divergence-free vector fields, $\xi^r(\sigma)$; according to (2.39) these vector fields can be represented locally in terms of a scalar function $\xi(\sigma)$, which may or may not be globally defined.* We will concentrate on the subgroup of area-preserving maps generated by functions $\xi(\sigma)$ that are globally defined. As follows from (2.41), infinitesimal transformations can be expressed in terms of the Lie bracket defined in (2.40). Furthermore, the commutator of two infinitesimal transformations with parameters ξ_1 and ξ_2 yields an area-preserving transformation with parameter $\xi_3 = \{\xi_2, \xi_1\}$. Therefore the structure constants of the area-preserving maps that are globally defined, are given in terms of the Lie bracket (2.40). In order to make this more explicit, we decompose the Lie bracket of Y_A and Y_B according to

$$\{Y_A, Y_B\} = g_{AB}{}^C Y_C = g_{ABC} Y^C, \tag{3.6}$$

where indices of g_{ABC} are raised and lowered by means of η^{AB} and η_{AB}. Using the

* In the mathematics literature, the vector fields corresponding to functions $\xi(\sigma)$ that are globally defined, are called "hamiltonian vector fields"; if $\xi(\sigma)$ is not globally defined one speaks of "locally hamiltonian vector fields". See e.g. [19], p. 218. The latter contain harmonic vectors ξ^r and homotopically nontrivial reparametrizations.

normalization condition (3.3) it follows that g_{ABC} is defined by

$$g_{ABC} = \int d^2\sigma\, w(\sigma) Y_A(\sigma) \{ Y_B(\sigma), Y_C(\sigma) \}$$

$$= \int d^2\sigma\, \varepsilon^{rs} Y_A(\sigma) \partial_r Y_B(\sigma) \partial_s Y_C(\sigma). \tag{3.7}$$

Because the Lie bracket satisfies the Jacobi identity the structure constants will also satisfy this identity,

$$g_{[AB}{}^C g_{D]C}{}^E = 0. \tag{3.8}$$

In the space of functions that are globally defined, it follows directly from the definition (3.7) that the structure constants g_{ABC} are totally antisymmetric. As we will not consider compactified membranes, we will thus always be dealing with antisymmetric structure constants. Furthermore the zero-mode $Y_0(\sigma) = \text{constant}$ decouples from the other modes because

$$g_{0BC} = g_{A0C} = g_{AB0} = 0. \tag{3.9}$$

It is now straightforward to substitute the expansion (3.1) and similar ones for the fermionic coordinates into the expressions derived in sect. 2. The lagrangian corresponding to (2.44) thus reads

$$L = \tfrac{1}{2}(\partial_0 \boldsymbol{X}^0)^2 + \tfrac{1}{2}|D_0 \boldsymbol{X}^A|^2 + \bar{\theta}^0 \gamma_- \partial_0 \theta^0 + \bar{\theta}_A \gamma_- D_0 \theta^A$$

$$- \tfrac{1}{4} g_{AB}{}^E g_{CDE} X_a^A X_b^B X_a^C X_b^D - g_{ABC} X_a^A \bar{\theta}^B \gamma_- \gamma^a \theta^C, \tag{3.10}$$

where we have separately written the zero modes (corresponding to $A = 0$) and the nonzero modes with indices $A, B, \ldots$ ranging from 1 to ∞. The covariant derivatives in (3.10) are defined by

$$D_0 X_a^A = \partial_0 X_a^A - g_{BC}{}^A \omega^B X_a^C, \qquad D_0 \theta^A = \partial_0 \theta^A - g_{BC}{}^A \omega^B \theta^C, \tag{3.11}$$

where ω^A is the gauge field associated with time-dependent area-preserving transformations. The lagrangian (3.10) is invariant under time-dependent transformations, whose infinitesimal form is given by

$$\delta X_a^A = g_{BC}{}^A \xi^B X_a^C, \qquad \delta \theta^A = g_{BC}{}^A \xi^B \theta^C, \qquad \delta \omega^A = D_0 \xi^A. \tag{3.12}$$

so that the zero modes are invariant by virtue of (3.9) and the nonzero modes transform in the *adjoint* representation.

As in sect. 2, the hamiltonian associated with (3.10) in the gauge $\omega^A = 0$ leads to an equation for the membrane mass $\mathcal{M}$,

$$\mathcal{M}^2 = \boldsymbol{P}_A \cdot \boldsymbol{P}^A + \tfrac{1}{2} g_{AB}{}^E g_{CDE} X_a^A X_b^B X_a^C X_b^D + 2 g_{ABC} X_a^A \bar{\theta}^B \gamma_- \gamma^a \theta^C , \qquad (3.13)$$

which does not contain the zero modes. The relevant supercharge is the part of Q^+, defined in (2.54), that pertains to the nonzero modes,

$$Q = \left(2 P_a^A \gamma^a + g_{BC}{}^A X_b^B X_c^C \gamma^{bc} \right) \theta_A . \qquad (3.14)$$

As shown in (2.59) the Dirac bracket of Q with itself yields (3.13) and the constraint φ, whose components are

$$\varphi_A = g_{ABC} \left(\boldsymbol{P}^B \cdot \boldsymbol{X}^C + \bar{\theta}^B \gamma_- \theta^C \right) \approx 0 . \qquad (3.15)$$

The theory defined by (3.13)–(3.15) contains an infinite number of degrees of freedom. In order to make it well-defined, one would like to have some kind of regularization. This can be achieved by restricting the indices $A, B, C, \ldots$ to a finite range between 1 and some finite number Λ. The original theory would then be obtained in the limit $\Lambda \to \infty$. In general, this limiting procedure may destroy some of the symmetries of the theory, and it is not clear which of these will be restored in this limit. The most severe of these problems are cured if one can replace the full group G of area-preserving transformations by a finite-dimensional symmetry group G_Λ, which in the limit $\Lambda \to \infty$ coincides with G. The structure constant g^{ABC} can then be replaced by the structure constants f^{ABC} of the finite-dimensional group G_Λ, which satisfy

$$\lim_{\Lambda \to \infty} f^{ABC} = g^{ABC} . \qquad (3.16)$$

The existence of such a group G_Λ guarantees that supersymmetry is not affected, as this symmetry rests upon the existence of a Jacobi identity for the structure constants (it also depends on the space-time dimension through the condition (2.5)). The application of this regularization thus leads us to a class of $N = 16$ supersymmetric matrix models with hamiltonian

$$H = \mathrm{Tr}\left(\tfrac{1}{2} \boldsymbol{P}^2 + \tfrac{1}{4} [X_a, X_b]^2 + [X_a, \bar{\theta}] \gamma_- \gamma^a \theta \right) , \qquad (3.17)$$

where $\boldsymbol{P}$, $\boldsymbol{X}$ and $\boldsymbol{\theta}$ are matrices that take their values in the Lie algebra of G. Surprisingly enough, the quantum-mechanical version of these models coincides with the models proposed sometime ago in [15]*. However, it is not guaranteed that the group G_Λ will always exist. This has been demonstrated only for spherical membranes [11, 12]. In that case G_Λ is equal to the group SU(N), where N and Λ

* These models are reductions of supersymmetric Yang–Mills theories to 1 + 0 dimensions. The field ω introduced in sect. 2 corresponds to the timelike component of the gauge fields.

are related by $\Lambda = N^2 - 1$. Of course, subtle questions about the precise meaning of the limit $\Lambda \to \infty$ still remain and will require further study. However, we shall ignore such questions here and turn to a more detailed discussion of the area-preserving transformations for the sphere and the torus.

3.1. AREA-PRESERVING MAPS ON THE SPHERE

On the sphere one conventionally takes the spherical harmonics $Y_{lm}(\theta, \varphi)$ as basis functions, where we exclude the zero mode, so that the integers l and m satisfy $l \geq 1$, $|m| \leq l$. With this basis we have $w(\theta, \varphi) = (4\pi)^{-1}\sin\theta$. We choose the Condon-Shortley phase convention for the Y_{lm} (we follow the definitions of [20], except for the normalization of the Y_{lm} which differs by a factor $\sqrt{4\pi}$),

$$(Y_l^m)^* = (-)^m Y_l^{-m}, \tag{3.18}$$

so that

$$\eta_{(lm)(l'm')} = (-)^m \delta_{l-l'}\delta_{m+m'}, \tag{3.19}$$

where δ_l denotes the Kronecker symbol δ_{l0}. The Lie bracket of two spherical harmonics then reads

$$\begin{aligned}\{Y_{l_1m_1}, Y_{l_2m_2}\} &= \frac{4\pi}{\sin\theta}\left(\frac{\partial Y_{l_1m_1}}{\partial\theta}\frac{\partial Y_{l_2m_2}}{\partial\varphi} - \frac{\partial Y_{l_1m_1}}{\partial\varphi}\frac{\partial Y_{l_2m_2}}{\partial\theta}\right)\\ &= g_{l_1m_1,l_2m_2}{}^{l_3m_3}Y_{l_3m_2}.\end{aligned} \tag{3.20}$$

It should be obvious that $g_{l_1m_1,l_2m_2,l_3m_3} = 0$, unless $m_1 + m_2 + m_3 = 0$. Furthermore, one can verify that $l_1 + l_2 + l_3$ must be odd, for instance, by comparing the parity of both sides of (3.20), and that $l_3 \leq l_1 + l_2 - 1$. Using the antisymmetry of the structure constants it then follows that the structure constants only differ from zero if

$$|l_1 - l_2| + 1 \leq l_3 \leq l_1 + l_2 - 1, \qquad m_1 + m_2 + m_3 = 0. \tag{3.21}$$

Another way to see this is by writing the spherical harmonics as symmetric traceless homogeneous polynomials of three cartesian coordinates x_1, x_2, x_3:

$$Y_{lm}(\theta, \varphi) = r^{-l}a^{(lm)}_{i_1\ldots i_l}x_{i_1}\ldots x_{i_l}, \qquad (r^2 = x_1^2 + x_2^2 + x_3^2) \tag{3.22}$$

in which case the Lie bracket takes the form

$$\{A, B\} = 4\pi r\varepsilon_{ijk}x_i\partial_jA\partial_kB. \tag{3.23}$$

Substitution of (3.22) into (3.23) leads to the same restrictions on l_1, l_2 and l_3 as listed in (3.21). The representation (3.23) also shows that the structure constants for $l_1 = l_2 = l_3 = 1$ are proportional to those of SO(3).

In [12] it was shown that $g_{l_1m_1,l_2m_2,l_3m_3}$ are the $N \to \infty$ limit of SU(N) structure constants. Let us first indicate how SU(N) emerges in the truncation of the spherical harmonics to a finite set. This truncation is effected by restricting l to $l \leq N-1$, which leaves us with precisely N^2-1 functions Y_{lm}. To each Y_{lm}, which corresponds to the symmetric traceless homogeneous polynomials (3.22), we can generally assign an N-dimensional matrix by constructing the corresponding symmetric traceless product of SO(3) generators L_i in the N-dimensional representation (spin $s = \frac{1}{2}(N-1)$),

$$Y_{lm} \to T_{lm} = 4\pi\left(\frac{N^2-1}{4}\right)^{(1-l)/2} a^{(lm)}_{i_1\ldots i_l} L_{i_1}\ldots L_{i_l}. \tag{3.24}$$

As is well-known, the L_i satisfy the equations

$$\left[L_i, L_j\right] = i\varepsilon_{ijk}L_k, \qquad L_i^\dagger = L_i, \qquad L^2 = \frac{N^2-1}{4}\mathbf{1}, \tag{3.25}$$

as well as the pseudo-reality condition

$$L_i^* = L_i^{\mathrm{T}} = -\omega L_i \omega^{-1}. \tag{3.26}$$

The matrices (2.24) are traceless by virtue of the tracelessness of the tensors $a^{(lm)}$. The dimension of the representation is chosen such that the T_{lm} with $l \leq N-1$ form a *complete* set of traceless $N \times N$ matrices. This can easily be seen by writing them as the traceless part of $L_\pm^p L_3^q$, with $L_\pm$ the familiar raising and lowering operators, which are clearly independent, provided $p + q \leq N-1$. Using the symmetry property (3.26), it then follows that the T_{lm} with even (odd) $l \geq N$ can be expanded as a linear combination of the T_{lm} with even (odd) $l \leq N-1$. Note that the hermiticity of the T_{lm} follows from the phase convention adopted for the spherical harmonics, so that

$$(T_l^m)^\dagger = (-)^m T_l^{-m}. \tag{3.27}$$

From their completeness property it is obvious that the T_{lm} are the generators of SU(N) in the defining representation, and we obtain the structure constants from

$$\left[T_{l_1m_1}, T_{l_2m_2}\right] = if_{l_1m_1,l_2m_2}{}^{l_3m_3}T_{l_3m_3} \qquad (|m_i| \leq l_i \leq N-1). \tag{3.28}$$

Just as the structure constants of the area-preserving transformations, the SU(N) structure constants $f_{l_1m_1,l_2m_2,l_3m_3}$ are only different from zero if $l_1 + l_2 + l_3$ is odd (this follows from applying (3.26) to both sides of (3.28)), $l_3 \leq l_1 + l_2 - 1$ and $m_1 + m_2 + m_3 = 0$. Therefore we have the same restrictions on l_i and m_i as given in (3.21), except that one should keep in mind that, in the case of SU(N), there is the additional restriction that $l_i \leq N - 1$.

Due to (3.24) $f_{l_1 m_1, l_2 m_2, l_3 m_3}$ will converge to $g_{l_1 m_1, l_2 m_2, l_3 m_3}$, as $N \to \infty$ [12]. Eq. (3.24) also implies that the T_{lm} transform as tensor operators under rotations and once this is known the SU(N) structure constants defined by (3.28) are determined by group theory [21, 12] up to the calculation of the reduced matrix elements. One gets (without loss of generality, we have assumed that $l_1 \leq l_2 \leq l_3$ while $l_1 + l_2 - l_3$ is an odd positive integer)

$$f_{l_1 m_1, l_2 m_2, l_3 m_3} = -4\pi i \left(\prod_{i=1}^{3} \sqrt{2l_i + 1} \right) \begin{pmatrix} l_1 & l_2 & l_3 \\ m_1 & m_2 & m_3 \end{pmatrix} \times \begin{Bmatrix} l_1 & l_2 & l_3 \\ s & s & s \end{Bmatrix} (-)^N \frac{R_N(l_1) R_N(l_2)}{R_N(l_3)}, \tag{3.29}$$

where $\begin{pmatrix} l_1 & l_2 & l_3 \\ m_1 & m_2 & m_3 \end{pmatrix}$ and $\begin{Bmatrix} l_1 & l_2 & l_3 \\ s & s & s \end{Bmatrix}$ are the $3j$-symbol and the $6j$-symbol, respectively [20], with $s = \frac{1}{2}(N-1)$, while the function R_N is defined by

$$R_N(l) = \sqrt{\frac{(N+l)!\,(N^2-1)^{1-l}}{(N-l-1)!}}. \tag{3.30}$$

In the large-N limit, the expression to the right of the $3j$-symbol converges to

$$\begin{Bmatrix} l_1 & l_2 & l_3 \\ s & s & s \end{Bmatrix} (-)^N \frac{R_N(l_1) R_N(l_2)}{R_N(l_3)} \xrightarrow{N \to \infty} (1 + l_1 + l_2 + l_3) l_1! l_2! l_3! (-)^{l_3 - 1} \times \sqrt{\frac{(l_1 + l_2 - l_3)!\,(l_1 + l_3 - l_2)!\,(l_2 + l_3 - l_1)!}{(1 + l_1 + l_2 + l_3)!}} \sum_{n=0}^{l_1 + l_2 - l_3} \frac{n(-)^n}{F(n)} \tag{3.31}$$

where

$$F(n) = n!\,(l_1 + l_2 - l_3 - n)!\,(l_1 - n)!\,(l_2 - n)!\,(n + l_3 - l_1)!\,(n + l_3 - l_2)!. \tag{3.32}$$

The large-N limit of (3.29) coincides with the structure constants $g_{l_1 m_1, l_2 m_2, l_3 m_3}$ for the full group of area-preserving transformations. The mathematics underlying this result [12, 22] is quite intriguing, and could lead to the possibility of approximating other infinite-dimensional Lie algebras of symplectic diffeomorphisms on homogeneous manifolds by large-N matrix algèbras.

3.2. AREA-PRESERVING MAPS ON THE TORUS

Choosing torus coordinates $0 \leq \phi_1, \phi_2 < 2\pi$ the basis functions Y_A are labelled by two-dimensional vectors $\boldsymbol{m} = (m_1, m_2)$ with m_1, m_2 integer numbers. They are defined by

$$Y_{\boldsymbol{m}}(\boldsymbol{\phi}) = e^{i\boldsymbol{m}\cdot\boldsymbol{\phi}}, \tag{3.33}$$

where $\boldsymbol{m}\cdot\boldsymbol{\phi} = m_1\phi_1 + m_2\phi_2$. Again we will exclude the zero mode, so that $\boldsymbol{m} \neq 0$. Furthermore we have $w(\boldsymbol{\phi}) = (4\pi^2)^{-1}$, and $\eta_{\boldsymbol{mn}} = \delta_{\boldsymbol{m}+\boldsymbol{n}}$. For the Lie bracket of $Y_{\boldsymbol{m}}$ and $Y_{\boldsymbol{n}}$, one easily finds

$$\{Y_{\boldsymbol{m}}, Y_{\boldsymbol{n}}\} = -4\pi^2(\boldsymbol{m}\times\boldsymbol{n})\,Y_{\boldsymbol{m}+\boldsymbol{n}}, \tag{3.34}$$

where $\boldsymbol{m}\times\boldsymbol{n} \equiv m_1n_2 - m_2n_1$. The structure constants g_{ABC} follow directly from (3.7) and read

$$g_{\boldsymbol{mnk}} = -4\pi^2(\boldsymbol{m}\times\boldsymbol{n})\,\delta_{\boldsymbol{m}+\boldsymbol{n}+\boldsymbol{k}}. \tag{3.35}$$

The elements of the Lie algebra associated with G are thus labelled by the set of nonzero two-dimensional vectors $\boldsymbol{m}$ with integer coordinates. The commutator of two generators corresponding to two vectors of this lattice is then equal to the generator corresponding to the sum of the two vectors, multiplied by i times the oriented area of the parallelogram enclosed by the two vectors. Generators associated with parallel vectors thus commute. There exits an infinite variety of Cartan subalgebras, each infinite dimensional, consisting of the generators corresponding to the set of parallel vectors $\boldsymbol{m} = \lambda\boldsymbol{n}$, with $\boldsymbol{n}$ fixed and λ all nonzero integers.

The algebra corresponding to the structure constant (3.35) has been discussed in connection with the theory of incompressible fluids in [19]. Recently, it was emphasized that it contains subalgebras that are isomorphic to the Virasoro algebra [23]. One such subalgebra was explicitly given; its generators take the form

$$L_m = \frac{1}{4\pi^2}\sum_{k\neq 0}\frac{1}{k}\,Y_{(k,\,m+k)}. \tag{3.36}$$

More generally, solutions are obtained by taking a (logarithmically diverging) sum of the $Y_{\boldsymbol{m}}$ along a straight line in the 2-dimensional plane. For instance, one may take

$$L_m = \frac{1}{4\pi^2}\sum_{k\neq 0}\frac{1}{kp}\,Y_{(m,\,kp)}, \qquad \text{or} \qquad L_m = \frac{1}{4\pi^2}\sum_{k\neq 0}\frac{1}{kp}\,Y_{(kp,\,m+k)}, \tag{3.37}$$

where p is some nonzero integer. However, some caution is required with the infinite sums in (3.36)–(3.37), as the formal expressions for L_m do *not* correspond to differentiable functions of the torus coordinates ϕ_1 and ϕ_2. The Lie algebra based on (3.35) allows for a nontrivial central extension,

$$\{Y_m, Y_n\} = -4\pi^2(\boldsymbol{m}\times\boldsymbol{n})Y_{m+n} + \boldsymbol{c}\cdot\boldsymbol{m}\delta_{m+n}, \tag{3.38}$$

where $\boldsymbol{c}$ is a real two-dimensional vector. This result was also noted in [23]. Furthermore, one can enlarge the torus algebra to include fermionic generators X_r with (anti)commutation relations* (to avoid confusion with the usual symbol for the anticommutator, we replace $-(1/4\pi^2)\{\ ,\ \}$ by $[\ ,\ ]$)

$$\begin{aligned} [Y_m, Y_n] &= (\boldsymbol{m}\times\boldsymbol{n})Y_{m+n}, \\ \{X_r, X_s\} &= Y_{r+s}, \\ [Y_m, X_r] &= (\boldsymbol{m}\times\boldsymbol{r})X_{m+r}, \end{aligned} \tag{3.39}$$

where the fermionic generators X_r are labelled by the set of two-dimensional vectors $\boldsymbol{r} = (r_1, r_2)$, with r_1 and r_2 ranging either over the integers, or half integers (so that we get four different algebras, two of which are isomorphic to each other).

4. The supermembrane as a supersymmetric quantum-mechanical model

In this section we combine the previous results and study the properties of the supermembrane ground state. So far, we have not been able to prove or disprove the assertion that the supermembrane has massless states, although most of our results indicate that the ground state is massive. However, we stress that more work is needed before one can reach a definitive conclusion regarding this issue, and we hope that the results described here will pave the way for a more rigorous treatment of supermembranes which goes beyond perturbative (semi-classical) arguments.

The quantization of the supermembrane is straightforward in the SO(7) × U(1) formulation that we have presented in sect. 2. The coordinates are therefore $X^i(\sigma)$, $Z(\sigma)$, $\bar{Z}(\sigma)$ and $\lambda_\alpha(\sigma)$, with corresponding canonical momenta $P^i(\sigma)$, $\mathscr{P}(\sigma)$, $\bar{\mathscr{P}}(\sigma)$ and $\lambda^\dagger_\alpha(\sigma)$. The (anti)commutators of the operators associated with the coordinates and the momenta are given by the Dirac brackets (2.65) multiplied with an extra factor i. The operators P^i, $\mathscr{P}$, $\bar{\mathscr{P}}$ and $\lambda^\dagger$ can then be realized on wave functions (or rather functionals) $\Psi[X^i, Z, \bar{Z}, \lambda]$ by

$$\begin{aligned} P^i(\sigma) &= -i\frac{\partial}{\partial X^i(\sigma)}, & \mathscr{P}(\sigma) &= -i\frac{\partial}{\partial Z(\sigma)}, \\ \bar{\mathscr{P}}(\sigma) &= -i\frac{\partial}{\partial \bar{Z}(\sigma)}, & \lambda^\dagger(\sigma) &= \frac{1}{w}\frac{\partial}{\partial\lambda(\sigma)}. \end{aligned} \tag{4.1}$$

* This superextension of the algebra has been obtained in collaboration with Garreis (see [24]) and J. Wess.

It is now straightforward to write the relevant formulae from sect. 2 in this representation. Before doing so, we "regularize" the supermembrane theory by decomposing the coordinates and the momenta in terms of a finite set of function $Y^0(\sigma)$ and $Y^A(\sigma)$ with $A = 1, \ldots, \Lambda$. As explained in sect. 3, the structure constants g_{ABC} of the group of area-preserving transformations are then replaced by the structure constants f_{ABC} of a finite compact Lie group G, with dimension

$$\dim \mathrm{G} = \Lambda \,. \tag{4.2}$$

In the limit $\Lambda \to \infty$ the group G is assumed to coincide with the group of area-preserving transformations. This procedure turns the supermembrane into a model of supersymmetric quantum mechanics [13, 14] and leads precisely to the supersymmetric matrix models that have been constructed in [15]. An important consequence of this approach is that supersymmetry remains preserved, while the invariance under area-preserving maps is approximated by the invariance under G. For membranes topologically equivalent to S^2 the group G is equal to SU(N) and the limit $N \to \infty$ has been shown to yield the full group of area-preserving transformations [12]. However, in this section the precise nature of G does not play an important role.

The model that we will be considering in this section is thus based on a finite set of coordinates X_i^A, Z^A, $\bar{Z}^A$ and λ_α^A, together with their canonically conjugate momenta P_i^A, $\mathcal{P}^A$, $\bar{\mathcal{P}}^A$ and $\lambda_\alpha^{A\dagger}$. Here, the index A labels the adjoint representation of G. There are also the zero-mode (or center-of-mass) coordinates X_i^0, Z^0, $\bar{Z}^0$ and λ_α^0, but as we have already emphasized, these decouple entirely from the other coordinates, and do not contribute to the mass of the supermembrane states. The (anti)commutation relations corresponding to (2.65) are

$$\begin{aligned}
\left[X_i^A, P_{jB}\right] &= i\delta_{ij}\delta_B^A \,, \\
\left[Z^A, \mathcal{P}_B\right] &= \left[\bar{Z}^A, \bar{\mathcal{P}}_B\right] = i\delta_B^A \,, \\
\left\{\lambda_\alpha^A, \lambda_{\beta B}^\dagger\right\} &= \delta_{\alpha\beta}\delta_B^A \,,
\end{aligned} \tag{4.3}$$

while all other (anti)commutators vanish. The conjugate momenta can thus be represented by the operators

$$\begin{aligned}
P_{iA} &= -i\frac{\partial}{\partial X_i^A} \,, \qquad \mathcal{P}_A = -i\frac{\partial}{\partial Z^A} \,, \\
\bar{\mathcal{P}}_A &= -i\frac{\partial}{\partial \bar{Z}^A} \,, \qquad \lambda_{\alpha A}^\dagger = \frac{\partial}{\partial \lambda_\alpha^A} \,,
\end{aligned} \tag{4.4}$$

in agreement with (4.1), and the states of the theory correspond to the wave functions $\Psi(X_i^A, Z^A, \bar{Z}^A, \lambda_\alpha^A)$. The latter are elements of the Grassmann algebra generated by λ_α^A and may be expressed as*

$$\Psi = \sum_{k=0}^{8\Lambda} \Phi^{\alpha_1 \cdots \alpha_k}_{A_1 \cdots A_k}(X, Z, \bar{Z})\lambda^{A_1}_{\alpha_1}\lambda^{A_2}_{\alpha_2}\ldots\lambda^{A_k}_{\alpha_k}. \tag{4.5}$$

The norm of the state Ψ can then be defined through

$$\|\Psi\|^2 = \sum_{k=0}^{8\Lambda} \frac{1}{k!}\|\Phi^{\alpha_1 \cdots \alpha_k}_{A_1 \cdots A_k}\|^2, \tag{4.6}$$

with the usual L^2-norms for the coefficient functions $\Phi^{\alpha_1 \cdots \alpha_k}_{A_1 \cdots A_k}$. Of course, one also has the customary distinction between bosonic and fermionic states according to whether only even or odd powers of λ_α^A appear.

We next make the appropriate substitutions in the supercharge operators of sect. 2. The supercharges that pertain to the nonzero modes, follow directly from the $SO(7) \times U(1)$ covariant expressions in (2.66) and take the form

$$\begin{aligned} Q_\alpha &= \left\{ -i\Gamma^i_{\alpha\beta}\frac{\partial}{\partial X_i^A} + \tfrac{1}{2}f_{ABC}X_i^B X_j^C \Gamma^{ij}_{\alpha\beta} - f_{ABC}Z^B\bar{Z}^C\delta_{\alpha\beta}\right\}\lambda^A_\beta \\ &\quad + \sqrt{2}\left\{\delta_{\alpha\beta}\frac{\partial}{\partial Z^A} + if_{ABC}X_i^B\bar{Z}^C\Gamma^i_{\alpha\beta}\right\}\frac{\partial}{\partial\lambda_{\beta A}}, \\ Q^\dagger_\alpha &= \left\{ i\Gamma^i_{\alpha\beta}\frac{\partial}{\partial X_i^A} + \tfrac{1}{2}f_{ABC}X_i^B X_j^C \Gamma^{ij}_{\alpha\beta} + f_{ABC}Z^B\bar{Z}^C\delta_{\alpha\beta}\right\}\frac{\partial}{\partial\lambda_{\beta A}} \\ &\quad + \sqrt{2}\left\{-\delta_{\alpha\beta}\frac{\partial}{\partial \bar{Z}^A} + if_{ABC}X_i^B Z^C\Gamma^i_{\alpha\beta}\right\}\lambda^A_\beta. \end{aligned} \tag{4.7}$$

These charges define a supersymmetric quantum-mechanical model, whose hamiltonian follows from the $\{Q, Q^\dagger\}$ anticommutator. In order to exhibit this, let us evaluate the anticommutators of the supercharge operators Q and $Q^\dagger$. After a somewhat lengthy calculation, using the antisymmetry of f^{ABC} as well as the Jacobi

* Observe that we suppress the dependence on the zero-mode coordinates in (4.5). We will return to this shortly.

identities, one arrives at the following superalgebra

$$\{Q_\alpha, Q_\beta\} = 2\sqrt{2}\,\delta_{\alpha\beta}\bar{Z}^A\varphi_A\,,$$

$$\{Q^\dagger_\alpha, Q^\dagger_\beta\} = 2\sqrt{2}\,\delta_{\alpha\beta}Z^A\varphi_A\,,$$

$$\{Q_\alpha, Q^\dagger_\beta\} = 2\delta_{\alpha\beta}H - 2i\Gamma^i_{\alpha\beta}X^A_i\varphi_A\,. \tag{4.8}$$

This result is consistent with the Dirac brackets (2.65), with the operators H and φ_A corresponding to the contribution from the nonzero modes to the hamiltonian (2.67) and the constraint (2.51). The explicit expression for this hamiltonian, which is directly related to the membrane mass $\mathcal{M}$, reads

$$H = \tfrac{1}{2}\mathcal{M}^2 = H_\mathrm{b} + H_\mathrm{f}\,, \tag{4.9}$$

where

$$H_\mathrm{b} = -\frac{1}{2}\frac{\partial^2}{\partial X^i_A\,\partial X^A_i} - \frac{\partial^2}{\partial Z_A\,\partial\bar{Z}^A} + V(X, Z, \bar{Z})\,, \tag{4.10}$$

with positive potential V given by

$$V(X, Z, \bar{Z}) = \tfrac{1}{4}f^E_{AB}f_{CDE}\left\{X^A_iX^B_jX^C_iX^D_j + 4X^A_iZ^BX^C_i\bar{Z}^D + 2Z^A\bar{Z}^B\bar{Z}^CZ^D\right\}. \tag{4.11}$$

and

$$H_\mathrm{f} = if_{ABC}X^A_i\lambda^B_\alpha\Gamma^i_{\alpha\beta}\frac{\partial}{\partial\lambda_{\beta C}} + \tfrac{1}{2}\sqrt{2}\,f_{ABC}\left(Z^A\lambda^B_\alpha\lambda^C_\alpha - \bar{Z}^A\frac{\partial}{\partial\lambda_{\alpha B}}\frac{\partial}{\partial\lambda_{\alpha C}}\right). \tag{4.12}$$

The algebra (4.8) still contains the operators φ^A, which are the components of the constraint (2.51), and given by

$$\varphi^A = f^{ABC}\left(X_{iB}\frac{\partial}{\partial X^C_i} + Z_B\frac{\partial}{\partial Z^C} + \bar{Z}_B\frac{\partial}{\partial\bar{Z}^C} + \lambda_{\alpha B}\frac{\partial}{\partial\lambda^C_\alpha}\right). \tag{4.13}$$

Obviously, φ^A are just the generators of the group G, which must vanish on physical states, i.e.,

$$\varphi^A\Psi = 0\,. \tag{4.14}$$

Consequently the wave functions corresponding to physical states must be invariant under G (or the full group of area-preserving diffeomorphisms). On physical states one thus recovers the usual supersymmetry algebra. The expressions (4.9)–(4.14) precisely coincide with the results of [15], where quantum-mechanical models were discussed with up to 16 supercharges. Hence we have established that the supermembrane is a limiting case of this class of models.

The zero modes, which are not contained in the quantum-mechanical models of [15], lead also to corresponding supercharges, as we have already discussed in sect. 2. In the SO(7) × U(1) notation, there is one complex charge associated with Q^- and one with Q^+ (cf. (2.54), where we denote the latter by $Q^{+(0)}$ to indicate that it contains only contributions from the zero modes. In the representation (4.4) these charges read

$$Q^-_\alpha = \lambda^0_\alpha, \qquad Q^{-\dagger}_\alpha = \frac{\partial}{\partial \lambda^0_\alpha},$$

$$Q^{(0)}_\alpha = -i\Gamma^i_{\alpha\beta}\frac{\partial}{\partial X^0_i}\lambda^0_\beta + \sqrt{2}\,\frac{\partial}{\partial Z^0}\frac{\partial}{\partial \lambda^0_\alpha},$$

$$Q^{(0)\dagger}_\alpha = i\Gamma^i_{\alpha\beta}\frac{\partial}{\partial X^0_i}\frac{\partial}{\partial \lambda^0_\beta} + \sqrt{2}\,\frac{\partial}{\partial \bar{Z}^0}\lambda^0_\alpha. \tag{4.15}$$

It is easy to determine the supersymmetry algebra for the above charges, which is the quantum-mechanical analogue of (2.58) in SO(7) × U(1) notation. This algebra contains the hamiltonian

$$H^{(0)} = -\frac{1}{2}\frac{\partial^2}{\partial X^0_i\,\partial X^0_i} - \frac{\partial^2}{\partial Z^0\,\partial \bar{Z}^0}, \tag{4.16}$$

which is just the transverse kinetic energy of the membrane. The wave function associated with the zero modes is simply a plane-wave solution in terms of the transverse coordinates X^0_i, Z^0 and $\bar{Z}^0$ with a certain transverse momentum, multiplied by an arbitrary function of the fermionic zero modes λ^0. This wave function thus describes 128 bosonic states $1, \lambda^0_\alpha\lambda^0_\beta, \ldots$ and 128 fermionic states $\lambda^0_\alpha, \lambda^0_\alpha\lambda^0_\beta\lambda^0_\gamma, \ldots$. Under SO(9), these transform as the **44 ⊕ 84** and **128** representations. The 128 + 128 independent wave functions transform under the supercharge operators (4.15) as the states of a massless $d = 11$ supergravity multiplet. To see this, it is convenient to choose a Lorentz frame in which the transverse momentum vanishes, so that the charge $Q^{(0)}$ vanishes and one is only left with Q^-. Consequently, if the wave function (4.5) associated with the ground state of the nonzero-mode system is not degenerate, then the supermembrane ground state constitutes precisely a *massless* supermultiplet.

According to the above arguments, the zero modes are no longer relevant, and we have to determine the nature of the ground state corresponding to the hamiltonian H which governs the nonzero modes. According to (4.9), massless states Ψ must obey the Schrödinger equation

$$H\Psi = 0. \tag{4.17}$$

From the supersymmetry algebra, it follows that H can be written as

$$H = \tfrac{1}{16}\{Q_\alpha, Q_\alpha^\dagger\}. \tag{4.18}$$

The hamiltonian H is thus a positive operator, which vanishes if and only if the ground-state wave function Ψ is a singlet under supersymmetry, in which case

$$Q_\alpha\Psi = Q_\alpha^\dagger\Psi = 0. \tag{4.19}$$

Although this condition ensures that the ground state is massless, it does not immediately imply that the ground state constitutes the desired supermultiplet. In $d = 11$ dimensions one has to require separately that Ψ is also a singlet under SO(9).* For future purposes let us list the SO(9) generators in terms of the coordinates and momenta introduced above. It is convenient to decompose them into "orbital" and "spin" parts according to

$$J^{ab} = L^{ab} + S^{ab}, \tag{4.20}$$

where

$$\begin{aligned}
L_{ij} &= X_i^A \frac{\partial}{\partial X_j^A} - X_j^A \frac{\partial}{\partial X_i^A}, \\
L_{89} &= iZ^A \frac{\partial}{\partial Z^A} - i\bar{Z}^A \frac{\partial}{\partial \bar{Z}^A}, \\
L_{i+} &= X_i^A \frac{\partial}{\partial Z^A} - \bar{Z}^A \frac{\partial}{\partial X_i^A}, \\
L_{i-} &= X_i^A \frac{\partial}{\partial \bar{Z}^A} - Z^A \frac{\partial}{\partial X_i^A},
\end{aligned} \tag{4.21}$$

and

$$\begin{aligned}
S_{ij} &= \tfrac{1}{2}\lambda_\alpha^A \Gamma^{ij}_{\alpha\beta} \frac{\partial}{\partial \lambda_\beta^A}, \qquad S_{89} = -\tfrac{1}{2} i\lambda_\alpha^A \frac{\partial}{\partial \lambda_\alpha^A} + ic_0, \\
S_{i-} &= \frac{i}{2\sqrt{2}} \frac{\partial}{\partial \lambda_\alpha^A} \Gamma^i_{\alpha\beta} \frac{\partial}{\partial \lambda_{\beta A}}, \qquad S_{i+} = \frac{i}{2\sqrt{2}} \lambda_\alpha^A \Gamma^i_{\alpha\beta} \lambda_{\beta A}.
\end{aligned} \tag{4.22}$$

* In lower-dimensional space-times Ψ must transform nontrivially under the SO($d-2$) group of transverse rotations in order that the ground-state constitutes a supergravity multiplet.

Note the appearance of the "normal-ordering" constant $c_0 \equiv 2\Lambda$ in S_{89}. There is an associated hermitean U(1) charge operator J_{+-} which reads

$$J_{+-} \equiv iJ_{89} = \bar{Z}^A \frac{\partial}{\partial \bar{Z}^A} - Z^A \frac{\partial}{\partial Z^A} + \tfrac{1}{2}\lambda^A_\alpha \frac{\partial}{\partial \lambda^A_\alpha} - c_0 . \tag{4.23}$$

(with corresponding definitions for S_{+-} and L_{+-}). Defining the charge q of any operator $\mathcal{O}$ by means of $[J_{+-}, \mathcal{O}] = q\mathcal{O}$, we see the variables X_i^A, Z^A, $\bar{Z}^A$ and λ^A_α carry the U(1) charges 0, -1, $+1$ and $\frac{1}{2}$, respectively.

Our main task is now to solve (4.17), or equivalently (4.19), for some G-invariant wave function Ψ. We expect that the method of solving (4.17) for finite Λ cannot be used for purely bosonic membranes, because the ground-state energy of the bosonic membrane will diverge in the limit $\Lambda \to \infty$ and needs to be renormalized (see, e.g. [25]). Since this is a nonrenormalizable theory there is an inherent ambiguity in the calculation of the finite part of the infinite renormalization. On the other hand, if one succeeds in finding a state obeying (4.19) for the supermembrane, this state will remain a proper ground state in the limit $\Lambda \to \infty$. Nevertheless, we cannot a priori exclude the possibility that the lowest eigenvalue of H is strictly positive for finite Λ but only tends to zero as $\Lambda \to \infty$. At any rate, we expect that the Bose-Fermi symmetry leads to the usual softening of divergences associated with the large-Λ limit.

Up to this point, the analysis is completely analogous to the corresponding one for superstrings (a detailed discussion may be found in [26], sect. 11.7). The much more difficult part of the problem, however, resides in the nonzero mode part of Ψ. First of all, the hamiltonian (4.9) describes an interacting theory and not a free theory as in superstring theory. Secondly, the constraint (4.14) has no analog in string theory. There, one only demands invariance of the physical Hilbert space under rigid (i.e., length-preserving) translations that are generated by the operator $N_L - N_R$, which does not mix different oscillator modes. The group of area-preserving diffeomorphisms is much larger and, in particular, does not admit an invariant split into positively and negatively indexed modes.

In order to facilitate the calculations, one can make the additional assumption that Ψ is an SO(9) singlet. As alluded to above, this is in fact necessary if one wants to recover $d = 11$ supergravity as a "low-energy limit" from the supermembrane. For otherwise, the ground state would transform as $[(\mathbf{44} \oplus \mathbf{84})_b \oplus \mathbf{128}_f)]$ times a nonsinglet representation of SO(9) and would therefore describe states other than those of the $d = 11$ supergravity multiplet. Unfortunately, the requirement of SO(9) invariance does not lead to significant simplifications, so that this approach is not particularly useful. We refer the reader to the appendix for a more detailed analysis of the structure of SO(9)-invariant wave functions. However, one can show that the ground-state wave function cannot factorize into a bosonic and a fermionic function, i.e., it cannot be of the form $\Psi = \Psi_b \otimes \Psi_f$, with either Ψ_b or Ψ_f (or both) SO(9) or G invariant. The reason is that H_f, defined in (4.12), can be written as a product of two operators, a bosonic one equal to the bosonic coordinates, and a fermionic one, bilinear in the fermion operators, which both transform as a vector under SO(9) and in the adjoint representation of G. Sandwiching H_f between the ground-state wave functions, it follows from the SO(9) or G invariance of either Ψ_b or Ψ_f that $(\Psi, H_f\Psi)$ must vanish. Therefore, as a result of (4.17), $(\Psi, H_b\Psi) = (\Psi_b, H_b\Psi_b) = 0$. Because H_b is a positive operator, this implies that Ψ_b must vanish. This situation is in sharp contrast to superstrings where the (nonzero mode) ground-state factorizes into a bosonic and a fermionic SO(8) singlet, and where one has a mode-by-mode cancellation of the vacuum energies.

In general, the relevant equations $Q\Psi = Q^{\dagger}\Psi = 0$ are very difficult to solve. Therefore we will now consider two special cases to illustrate some of the difficulties. The first one is a truncation of the membrane theory, in which we discard the coordinates Z^A, $\bar{Z}^A$ and λ_8^A. We accordingly split the SO(7) spinor indices $\alpha, \beta, \ldots$ into $i, j, \ldots = 1, \ldots, 7$ and $\alpha, \beta, \ldots = 8$ and make use of the fact that (see, e.g. [27, 28])

$$(\Gamma^i)_{j8} = -i\delta^i_j, \qquad (\Gamma^i)_{jk} = ic_{ijk}, \tag{4.24}$$

where c_{ijk} are the octonionic structure constants obeying

$$c_{ijm}c^{klm} = 2\delta^{kl}_{ij} - \tfrac{1}{6}\varepsilon_{ijklmnp}c^{mnp}, \tag{4.25}$$

as well as a number of other relations which can be found in [28]. In this truncation the supercharges (4.7) take the form

$$Q_8 = \left\{ \frac{\partial}{\partial X_i^A} + \tfrac{1}{2}c^{ijk}f_{ABC}X_j^B X_k^C \right\} \lambda_i^A,$$

$$Q_8^{\dagger} = \left\{ -\frac{\partial}{\partial X_i^A} + \tfrac{1}{2}c^{ijk}f_{ABC}X_j^B X_k^C \right\} \frac{\partial}{\partial \lambda^i_A}. \tag{4.26}$$

The symmetry of this theory is now reduced to $N = 1$ supersymmetry, the G_2 subgroup of SO(9) and G. The equation $Q\Psi = Q^{\dagger}\Psi = 0$ can easily be solved and one finds two $G_2 \times G$ invariant solutions,

$$\Psi_1 = \left(\prod_{i,A} \lambda_i^A \right) \exp\left\{ \tfrac{1}{6}c^{ijk}f_{ABC}X_i^A X_j^B X_k^C \right\},$$

$$\Psi_2 = \exp\left\{ -\tfrac{1}{6}c^{ijk}f_{ABC}X_i^A X_j^B X_k^C \right\}. \tag{4.27}$$

It is amusing that in the membrane limit these two solutions become

$$\Psi_1[X(\sigma), \lambda(\sigma)] = \Big(\prod_{i,\sigma} \lambda_i(\sigma)\Big) \exp\Big\{\tfrac{1}{6}\int d^2\sigma\, \varepsilon^{rs} c_{ijk} X^i \partial_r X^j \partial_s X^k\Big\},$$

$$\Psi_2[X(\sigma), \lambda(\sigma)] = \exp\Big\{-\tfrac{1}{6}\int d^2\sigma\, \varepsilon^{rs} c_{ijk} X^i \partial_r X^j \partial_s X^k\Big\}, \tag{4.28}$$

so that the ground-state wave functionals are exponentials of a Wess-Zumino-Witten term, with corresponding torsion proportional to c_{ijk}. However, both solutions (4.27) fail to be square-integrable, and this problem persists for (4.28). Thus, there is no supersymmetric ground state, so that this truncation has no massless states. From the analogy with ordinary $N = 1$ supersymmetric quantum mechanics, this is what one would have intuitively expected for the full supermembrane, too, as the differential operator, which appears in (4.7), is $\pm\partial/\partial X + X^2$, rather than $\pm\partial/\partial X + X$ as in superstring theory [14]. However, the argument is vitiated by (amongst other things) the nonexistence of an SO(9)-invariant (or even SO(7)-invariant) three-index tensor analogous to c_{ijk}. Observe also that both solutions in (4.27) are singlets in their bosonic and fermionic factors. This does not contradict our findings above, because the wave functions do not tend to zero at spatial infinity, and for such functions the hamiltonian H_b is not a positive operator.

The second truncation which we will consider, consists in discarding the variables X_i^A and λ_i^A, thus retaining only Z^A, $\bar{Z}^A$ and $\lambda^A \equiv \lambda_8^A$. This corresponds to a membrane moving in a $d = 4$ dimensional space-time. The supercharges follow directly from (4.7) and read

$$Q = \sqrt{2}\,\frac{\partial}{\partial Z^A}\frac{\partial}{\partial \lambda_A} - f_{ABC} Z^A \bar{Z}^B \lambda^C,$$

$$Q^\dagger = -\sqrt{2}\,\frac{\partial}{\partial \bar{Z}^A}\lambda^A + f_{ABC} Z^A \bar{Z}^B \frac{\partial}{\partial \lambda_C}. \tag{4.29}$$

It is clear that the ground state cannot factorize into a bosonic and fermionic part and therefore we proceed from the ansatz*

$$\Psi = \Phi_0(Z, \bar{Z}) + \sum_{k\geq 1} \Phi_{A_1 \dots A_{2k}}(Z, \bar{Z})\lambda^{A_1} \dots \lambda^{A_{2k}}, \tag{4.30}$$

where the coefficient functions $\Phi^{A_1 \dots A_{2k}}$ are completely antisymmetric in the indices $A_1, \dots, A_{2k}$. To make life as simple as possible, we take G equal to SU(2), so that

* We could also choose Ψ such that only odd powers of λ appear.

$A, B, C, \ldots = 1, 2, 3$, and $f^{ABC} = \varepsilon^{ABC}$. The decomposition (4.30) then simplifies to

$$\Psi = \varphi_0(Z, \bar{Z}) + \varepsilon^{ABC}\varphi^A(Z, \bar{Z})\lambda^B\lambda^C. \tag{4.31}$$

(We choose a real basis for the adjoint representation of SU(2), so the position of indices is immaterial).

Requiring $Q\Psi = Q^\dagger\Psi = 0$, we get

$$\varepsilon^{ABC}Z^A\bar{Z}^B\varphi^C(Z, \bar{Z}) = 0, \tag{4.32}$$

which tells us that

$$\varphi^A = Z^A\varphi_1 + \bar{Z}^A\varphi_2, \tag{4.33}$$

and three more equations,

$$Z^A\frac{\partial\varphi_1}{\partial\bar{Z}^A} + 3\varphi_2 + \bar{Z}^A\frac{\partial\varphi_2}{\partial\bar{Z}^A} = 0, \tag{4.34}$$

$$2\sqrt{2}\,\varepsilon^{ABC}\left\{Z^B\frac{\partial\varphi_1}{\partial Z^C} + \bar{Z}^B\frac{\partial\varphi_2}{\partial Z^C}\right\} = \varepsilon^{ABC}Z^B\bar{Z}^C\varphi_0, \tag{4.35}$$

$$\sqrt{2}\,\frac{\partial\varphi_0}{\partial\bar{Z}^A} = 2[(Z\cdot\bar{Z})Z^A - Z^2\bar{Z}^A]\varphi_1 + 2[\bar{Z}^2Z^A - (Z\cdot\bar{Z})Z^A]\varphi_2. \tag{4.36}$$

Upon multiplication by Z^A and $\bar{Z}^A$, (4.36) leads to

$$\varphi_1 = \frac{1}{\sqrt{2}}\frac{1}{(Z\cdot\bar{Z})^2 - Z^2\bar{Z}^2}\bar{Z}^A\frac{\partial\varphi_0}{\partial\bar{Z}^A},$$

$$\varphi_2 = -\frac{1}{\sqrt{2}}\frac{1}{(Z\cdot\bar{Z})^2 - Z^2\bar{Z}^2}Z^A\frac{\partial\varphi_0}{\partial\bar{Z}^A}. \tag{4.37}$$

Substituting this result back into the previous equations, it turns out that (4.34) is identically satisfied, while (4.35) and (4.36) lead to

$$\varepsilon^{ABC}Z^A\bar{Z}^B\frac{\partial\varphi_0}{\partial Z^C} = \varepsilon^{ABC}Z^A\bar{Z}^B\frac{\partial\varphi_0}{\partial\bar{Z}^C} = 0, \tag{4.38}$$

$$\tilde{H}\varphi_0 \equiv \left\{H_b + \frac{\varepsilon^{ABC}Z^B\bar{Z}^C}{(Z\cdot\bar{Z})^2 - Z^2\bar{Z}^2}\varepsilon^{ADE}\left(\bar{Z}^D\frac{\partial}{\partial\bar{Z}^E} - Z^D\frac{\partial}{\partial Z^E}\right)\right\}\varphi_0 = 0. \tag{4.39}$$

Here H_b is the hamiltonian defined in (4.10), which in this case reads

$$H_b = -\frac{\partial}{\partial Z^A}\frac{\partial}{\partial \bar{Z}^A} + \tfrac{1}{2}\left[(Z\cdot\bar{Z})^2 - Z^2\bar{Z}^2\right]. \tag{4.40}$$

According to the constraint equations (4.14), the wave function must be SU(2) invariant, in which case eqs. (4.38) are obviously satisfied. Hence we are left with a Schrödinger equation for an SU(2)-invariant wave function φ_0, given by (4.39). The corresponding hamiltonian, $\tilde{H}$, consists of a linear combination of H_b, which is the hamiltonian for a bosonic membrane, and an extra term.

For the class of wave functions for which the hamiltonian is self-adjoint, we find that

$$(\varphi_0, H_b\varphi_0) = \int d^3Z\, d^3\bar{Z}\left\{\left|\frac{\partial\varphi_0}{\partial Z^A}\right|^2 + \tfrac{1}{2}\left[(Z\cdot\bar{Z})^2 - Z^2\bar{Z}^2\right]|\varphi_0|^2\right\}, \tag{4.41}$$

which is positive because

$$(Z\cdot\bar{Z})^2 - Z^2\bar{Z}^2 \geq 0. \tag{4.42}$$

Under the same conditions, we have

$$(\varphi_0, (\tilde{H} - H_b)\varphi_0) = \int d^3Z\, d^3\bar{Z}\,\frac{1}{2}\,\frac{\varepsilon^{ABC}Z^B\bar{Z}^C}{(Z\cdot\bar{Z})^2 - Z^2\bar{Z}^2}\,\varepsilon^{ADE}\left(\bar{Z}^D\frac{\partial}{\partial\bar{Z}^E} - Z^D\frac{\partial}{\partial Z^E}\right)|\varphi_0|^2. \tag{4.43}$$

Because

$$\varepsilon^{ADE}\left(\bar{Z}^D\frac{\partial}{\partial\bar{Z}^E} - Z^D\frac{\partial}{\partial Z^E}\right)\frac{\varepsilon^{ABC}Z^B\bar{Z}^C}{(Z\cdot\bar{Z})^2 - Z^2\bar{Z}^2} = 0, \tag{4.44}$$

the integrand in (4.43) can be written as a total divergence, which suggests that one can rewrite (4.43) as a surface integral. However, one has to take into account that the integrand has a singularity whenever $(Z\cdot\bar{Z})^2 = Z^2\bar{Z}^2$. This happens when Z^A becomes proportional to a real vector (or, in other words, whenever the two vectors $\mathrm{Re}\,Z^A$ and $\mathrm{Im}\,Z^A$ are aligned). Therefore, the integral (4.43) splits into two terms, one corresponding to the surface integral associated with large distances ($Z\cdot\bar{Z}\to\infty$), which yields a *positive* contribution, and another one corresponding to the contribution from the singularities, which turns out to be *negative*. To show this more explicitly, on may choose a parametrization in terms of the SU(2)-invariant variables

$$\zeta \equiv Z^A Z^A, \qquad \bar{\zeta} \equiv \bar{Z}^A \bar{Z}^A, \qquad \xi \equiv \sqrt{(Z \cdot \bar{Z})^2 - Z^2 \bar{Z}^2}. \tag{4.45}$$

It is not hard to see that $\tilde{H} - H_b$ is now equal to

$$\tilde{H} - H_b = \frac{2}{\xi} \sqrt{\xi^2 + |\zeta|^2} \frac{\partial}{\partial \xi}. \tag{4.46}$$

Furthermore, on SU(2)-invariant functions we have

$$d^3Z \, d^3\bar{Z} \propto \frac{\xi \, d\xi \, d\zeta \, d\bar{\zeta}}{\sqrt{\xi^2 + |\zeta|^2}}, \tag{4.47}$$

up to "angular" variables whose integral yields an irrelevant (positive) constant c. Substituting (4.46)–(4.47) into (4.43), and performing the integral over ξ, we then find

$$\left(\varphi_0, \left(\tilde{H} - H_b\right) \varphi_0\right) = -c \int d\zeta \, d\bar{\zeta} \, |\varphi_0(\xi = 0, \zeta, \bar{\zeta})|^2, \tag{4.48}$$

where we have dropped the contribution at $\xi = \infty$, which is proportional to $|\varphi_0|^2$ at spatial infinity. Therefore we have shown that for wave functions vanishing at infinity, the energy of a supermembrane will be *lower* than that of a corresponding bosonic membrane.

On the other hand, imposing the boundary condition that φ_0 vanishes when $Z \cdot \bar{Z} \to \infty$, one can see that no solution of (4.39) exists, as $\bar{H}$ is an elliptic differential operator (see e.g. [29], p. 320 ff.). Consequently, solutions that are subject to these boundary conditions do *not* have zero energy. We should emphasize, however, that the above boundary condition is not implied by square-integrability*, and we have not been able to establish the existence or nonexistence of a general square-integrable solution to (4.39).

It is now evident that the general case with arbitrary N is even harder to tackle because the number of coefficient functions in (4.30) as well as the number of SU(N) invariant variables analogous to (4.45) is further increased as N becomes larger. In particular, there seems no real advantage anymore to replacing the

* This is, for instance, demonstrated by the function $f(\xi, \zeta, \bar{\zeta}) = \zeta \exp[-\frac{1}{2}\xi^{1/4}|\zeta|^2 - \frac{1}{2}\xi^2]$, which does not satisfy the above boundary condition, as $\lim_{|\zeta| \to \infty} f(0, \zeta, \bar{\zeta}) = \infty$, but nevertheless $\int_0^\infty d\xi \int d^2\zeta \, |f(\xi, \zeta, \bar{\zeta})|^2 < \infty$!

second-order equation (4.17) by the first-order equation (4.19), since decoupling these equations will automatically lead to higher-order equations.

Note added

After this paper was completed we learnt that Claudson and Halpern (see [15]) consider wave functions similar to (4.27). Furthermore, we have meanwhile calculated the Witten index for the SU(2) model discussed at the end of sect. 4 along the lines of ref. [31] and found that it vanishes. This is consistent with the conclusion that there are no massless states.

Appendix

STRUCTURE OF SO(9)-INVARIANT WAVE FUNCTIONS

We here briefly describe how to construct SO(9)-invariant wave functions which do not factorize into bosonic and fermionic parts that are separately SO(9)-invariant. The basic idea is to first consider nontrivial SO(9) representations in either sector and then fold them together to form a singlet. This is completely obvious for the SO(7) subgroup of SO(9) and the nontrivial part of the analysis involves the generators $J_{i\pm}$ which are nonlinearly realized on the Grassmann algebra, cf. (4.22). As is well-known, any SO(9) representation can be characterized by its highest weight or, equivalently, by its Dynkin label (see e.g. [30]). In the present case this label consists of four positive integers $(a_1a_2a_3a_4)$, the first three of which indicate the SO(7) representation and the last of which is associated with the U(1) charge operators L_{+-} and S_{+-}. The highest-weight state $|(a_1a_2a_3a_4)\rangle$ must be annihilated by the raising operators L_{i+} and S_{i+}, i.e.

$$L_{i+}|(a_1a_2a_3a_4)\rangle_{\mathrm{b}} = 0, \qquad \text{or} \qquad S_{i+}|(a_1a_2a_3a_4)\rangle_{\mathrm{f}} = 0, \tag{A.1}$$

for a bosonic or fermionic representation, respectively. Of course, it must also be annihilated by the remaining raising operators of the SO(7) subgroup but this (and analogous statements) will be understood in the following. The representation is then generated by applying the lowering operators L_{i-} for the bosonic representations, or S_{i-} for the fermionic representations, until one reaches the lowest-weight state; in this procedure, the U(1) charge a_4 is changed by one unit at each step. From the discussion in sect. 4 we learn that the fermionic wave functions have a maximum U(1) charge which is equal to the normal-ordering constant $c_0 = 2\Lambda$, so we will restrict ourselves to representations with $|a_4| \leq c_0$.

We will now illustrate how this works by looking at various examples, first in the bosonic sector. So let us start with

$$|(000c_0)\rangle_{\mathrm{b}} = \bar{Z}^{A_1} \ldots \bar{Z}^{A_{c_0}}. \tag{A.2}$$

Obviously this state transforms under the symmetric tensor representation of the group G which is associated with the indices $A_1, \ldots, A_{c_0}$, but because G commutes with SO(9), this aspect is not very important. Clearly, the state (A.2) is an SO(7) singlet and annihilated by L_{i+} (use the explicit expressions in (4.21)). Acting on it with L_{i-}, we obtain

$$L_{i-}(\bar{Z}^{A_1}\dots\bar{Z}^{A_{c_0}}) = c_0 X_i^{(A_1}\bar{Z}^{A_2}\dots\bar{Z}^{A_{c_0})}. \tag{A.3}$$

The U(1) charge of (A.3) is $(c_0 - 1)$ while the G-representation content is evidently unaltered. Continuing in this fashion, we get

$$\begin{aligned} &L_{i-}L_{j-}(\bar{Z}^{A_1}\dots\bar{Z}^{A_{c_0}}) \\ &\quad = c_0(c_0-1)X_i^{(A_1}X_j^{A_2}\bar{Z}^{A_3}\dots\bar{Z}^{A_{c_0})} - c_0\delta_{ij}Z^{(A_1}\bar{Z}^{A_2}\bar{Z}^{A_3}\dots\bar{Z}^{A_{c_0})}, \end{aligned} \tag{A.4}$$

and so on. Hence, we just obtain a generalization of the usual SO(9) spherical harmonics. To also have an example with $a_4 = c_0 - 1$, one may start from any of the following states

$$\begin{aligned} |(* * * c_0 - 1)\rangle_{\mathrm{b}} &= X_i^{[B_1}\bar{Z}^{B_2]}\bar{Z}^{A_2}\dots\bar{Z}^{A_r}, \\ &X_i^{[B_1}X_j^{B_2}\bar{Z}^{B_3]}\bar{Z}^{A_2}\dots\bar{Z}^{A_r}, \qquad \text{or} \\ &X_i^{[B_1}X_j^{B_2}X_k^{B_3}\bar{Z}^{B_4]}\bar{Z}^{A_2}\dots\bar{Z}^{A_r}, \end{aligned} \tag{A.5}$$

where $(* * *)$ is the appropriate SO(7) label. Owing to the antisymmetry in the indices $B_1, B_2, \dots$ the states (A.5) are anihilated by L_{i+}.

The construction in the fermionic sector is similar. Since, by (4.23), the highest-weight state contains the maximal number of λ's, it is more convenient to start with the lowest-weight state. The analogue of (A.2) is then

$$|(000 - c_0)\rangle_{\mathrm{f}} = 1, \tag{A.6}$$

which is annihilated by S_{i-}. The action of S_{i+} now produces the state

$$S_{i+}|(000 - c_0)\rangle_{\mathrm{f}} = \frac{i}{2\sqrt{2}}\lambda^A\Gamma^i\lambda_A, \tag{A.7}$$

which has charge $-c_0 + 1$. The analogue of (A.5) is the set of states

$$\lambda^{B_1}\lambda^{B_2}, \qquad \lambda^{B_1}\Gamma^i\lambda^{B_2} - \frac{2}{c_0}\delta^{B_1B_2}\lambda^C\Gamma^i\lambda_C, \qquad \lambda^{B_1}\Gamma^{ij}\lambda^{B_2}, \qquad \lambda^{B_1}\Gamma^{ijk}\lambda^{B_2}. \tag{A.8}$$

An SO(9) singlet can now be formed by folding together the same bosonic and fermionic SO(9) representations. The resulting wavefunction can then be turned into a singlet with respect to G by contraction with an appropriate bosonic function of

SO(9) singlet variables such as $X_i^A X_i^B + Z^A \bar{Z}^B + Z^B \bar{Z}^A$, etc. For instance, from (A.2) and (A.6), we can construct the following SO(9)-singlet wavefunction

$$\begin{aligned}\Psi &= \left|(000c_0)\right\rangle_{\rm b} \otimes \left|(000-c_0)\right\rangle_{\rm f} \\ &+ \alpha L_{i-}\left|(000c_0)\right\rangle_{\rm b} \otimes S_{i+}\left|(000-c_0)\right\rangle_{\rm f} \\ &+ \beta L_{i-}L_{j-}\left|(000c_0)\right\rangle_{\rm b} \otimes S_{i+}S_{j+}\left|(000-c_0)\right\rangle_{\rm f} \\ &+ \gamma L_{i-}L_{i-}\left|(000c_0)\right\rangle_{\rm b} \otimes S_{j+}S_{j+}\left|(000-c_0)\right\rangle_{\rm f} \\ &+ \cdots . \end{aligned} \tag{A.9}$$

The coefficients $\alpha, \beta, \gamma, \ldots$ are determined from the requirement $J_{i\pm}\Psi = 0$. Using the SO(9) commutation relations and the known U(1) charges together with $L_{ij}|(000c_0)\rangle_{\rm b} = S_{ij}|(000-c_0)\rangle_{\rm f} = 0$ we find

$$\alpha = \frac{1}{c_0}, \qquad \beta = \frac{1}{2c_0(c_0-1)}, \qquad \gamma = \frac{\beta}{2c_0+5}. \tag{A.10}$$

After contraction with an appropriate bosonic wavefunction, (A.9) can also be expressed as

$$\begin{aligned}\Psi &= \Phi_{A_1\ldots A_{c_0}}(X, Z, \bar{Z}) \\ &\times \left\{ \bar{Z}^{A_1} \ldots \bar{Z}^{A_{c_0}} + \frac{i}{2\sqrt{2}} \lambda^B \Gamma^j \lambda_B X_j^{A_1} \bar{Z}^{A_2} \ldots \bar{Z}^{A_{c_0}} + \cdots \right\}. \end{aligned} \tag{A.11}$$

Another example is

$$\begin{aligned}\Psi' &= \Phi_{B_1 B_2 A_2 \ldots A_{c_0}}(X, Z, \bar{Z}) \\ &\times \left\{ \lambda^{B_1}\lambda^{B_2} \bar{Z}^{A_2} \ldots \bar{Z}^{A_{c_0}} + \frac{i}{2\sqrt{2}} \lambda^{B_1}\lambda^{B_2} \lambda^C \Gamma^j \lambda_C X_j^{A_2} \bar{Z}^{A_3} \ldots \bar{Z}^{A_{c_0}} + \cdots \right\}. \end{aligned} \tag{A.12}$$

It is not difficult to verify directly that indeed $J_{i\pm} = L_{i\pm} + S_{i\pm}$ vanish on Ψ and Ψ', at least to the order given. Obviously, there is a multitude of possibilities and very little hope of a complete classification. One can also prove that the supermembrane wave function for a massless ground-state cannot just be of the form (A.11). This follows directly from the observation that $H_{\rm f}\Psi$ contains no λ-independent term for Ψ given by (A.11), so that $\hat{H}_{\rm b}\Psi$ must vanish up to order λ^2 for a massless ground state. From the fact that H_b is positive, it then follows that Ψ must in fact vanish. This conclusion is already suggested by the fact that (A.11) is an eigenfunction of both $\boldsymbol{L}^2$ and $\boldsymbol{S}^2$, while the hamiltonian does not commute with these operators. A bothersome feature is that the degree of the SO(9) "spherical harmonic" is larger than or equal to $c_0 = 2\Lambda$ and therefore increases without bound as $\Lambda \to \infty$. It is hard to see what reasonably behaved wavefunction could ensure square-integrability of $\Psi, \Psi', \ldots$ or any linear combination thereof in this limit.

References

[1] E. Bergshoeff, E. Sezgin and P.K. Townsend, Phys. Lett. 189B (1987) 75; Ann. of Phys. 185 (1988) 330
[2] L. Brink and J.H. Schwarz, Phys. Lett. 100B (1981) 310;
W. Siegel, Phys. Lett. 128B (1983) 397; Class. Quantum Grav. 2 (1985) L95–L97
[3] M.B. Green and J.H. Schwarz, Phys. Lett. 136B (1984) 367; Nucl. Phys. B243 (1984) 285
[4] J. Hughes, J. Liu and J. Polchinsky, Phys. Lett. 180B (1986) 370
[5] M. Henneaux and L. Mezincescu, Phys. Lett. 152B (1985) 340
[6] M.J. Duff, T. Inami, C.N. Pope, E. Sezgin and K.S. Stelle, Nucl. Phys. B297 (1988) 515
[7] L. Mezincescu, R. Nepomechie and P. van Nieuwenhuizen, preprint ITP-SB-87-43
[8] L. Brink and H.B. Nielsen, Phys. Lett. 45B (1973) 332
[9] K. Kikkawa and M. Yamasaki, Progr. Theor. Phys. 76 (1986) 1379
[10] I. Bars, C.N. Pope and E. Sezgin, Phys. Lett. 198B (1987) 455;
I. Bars, preprint USC-87/HEP06 (1987)
[11] J. Goldstone, unpublished
[12] J. Hoppe, MIT Ph.D. Thesis, 1982, and in Proc. Int. Workshop on Constraint's theory and relativistic dynamics, ed. G. Longhi and L. Lusanna (World Scientific, 1987)
[13] H. Nicolai, J. Phys. A9 (1976) 1497, A10 (1977) 2143
[14] E. Witten, Nucl. Phys. B188 (1981) 513
P. Salomonson and J.W. van Holten, Nucl. Phys. B196 (1982) 509
M. De Crombrugghe and V. Rittenberg, Ann. of Phys. 151 (1983) 99
[15] M. Baake, P. Reinicke, and V. Rittenberg, J. Math. Phys. 26 (1985) 1070;
R. Flume, Ann. of Phys. 164 (1985) 189;
M. Claudson and M.B. Halpern, Nucl. Phys. B250 (1985) 689
[16] E. Bergshoeff, M.J. Duff, C.N. Pope and E. Sezgin, Phys. Lett. 199B (1987) 69;
E. Bergshoeff, E. Sezgin and Y. Tanii, ICTP preprint IC/88/5;
E. Bergshoeff, A. Salam, E. Sezgin and Y. Tanii, ICTP preprint IC/88/6
[17] E. Bergshoeff, E. Sezgin and Y. Tanii, Nucl. Phys. B298 (1988) 187
[18] P.A.M. Dirac, Lectures on quantum mechanics, Belfer Grad. School of Science Monograph Series 2, New York, 1964.
[19] V.I. Arnold, Mathematical methods of classical mechanics (Springer, 1978)
[20] A. Messiah, Quantum mechanics, vols. I and II (North-Holland, 1966)
[21] Judd, Operator techniques in atomic spectroscopy (McGraw-Hill, 1963)
[22] J. Hoppe, in preparation
[23] E. Floratos and J. Iliopoulos, Phys. Lett. 201B (1988) 237
[24] R. Garreis, Karlsruhe Ph.D. Thesis, 1988
[25] E. Brézin, C. Itzykson, G. Parisi and J.-B. Zuber, Commun. Math. Phys. 59 (1978) 35
[26] M.B. Green, J.H. Schwarz and E. Witten, Superstring theory, vol. 2 (Cambridge Univ. Press, 1987)
[27] M. Günaydin and F. Gürsey, J. Math. Phys. 14 (1973) 1651
[28] B. de Wit and H. Nicolai, Nucl. Phys. B231 (1984) 506
[29] R. Courant and D. Hilbert, Methods of mathematical physics, vol. 2 (Interscience Publ., 1962)
[30] R. Slansky, Phys. Reports 79 (1981) 1
[31] A.V. Smilga, Nucl. Phys. B266 (1986) 45

Reprinted from Phys. Lett. B **253** (1991) 113–18

Multi-membrane solutions of $D=11$ supergravity

M.J. Duff [1]
Center for Theoretical Physics, Physics Department, Texas A&M University, College Station, TX 77843, USA

and

K.S. Stelle
The Blackett Laboratory, Imperial College, London SW7 2BZ, UK

Received 29 September 1990

We find exact solutions to the field equations of eleven-dimensional supergravity corresponding to stable multi-membrane configurations. Their holonomy group is given by the SO(8) subgroup of an enlarged tangent space group SO(1, 2)×SO(16), and hence one half of the spacetime supersymmetries are broken. The solutions saturate a Bogomol'nyi bound between the mass per unit area and the Page charge, which also guarantees their stability.

Although the equations of motion of eleven-dimensional supergravity were written down by Cremmer, Julia and Scherk as long ago as 1978 [1], it was only recently that Bergshoeff, Sezgin and Townsend [2] constructed the eleven-dimensional supermembrane that couples to this background. In this paper we show that the supermembrane actually emerges as an exact solution of the supergravity field equations. Indeed, exact solutions for a superposition of arbitrarily many supermembranes can be obtained in this way.

It should be emphasized, however, that these membrane solutions are not "solitons" of the kind sought by Townsend [3], which would be non-singular configurations stabilized by an identically conserved topological charge. By constrast, our solutions have δ-function singularities on the worldvolume of the membrane and are stabilized by a charge conserved only by virtue of the field equations, which turns out to be the familiar Page charge [4,5] of eleven-dimensional supergravity. Nonetheless, in common with the soliton solutions, they break just one half of the spacetime supersymmetries and saturate a Bogomol'nyi bound between the mass per unit area and the conserved charge. Under a simultaneous dimensional reduction of the supermembrane in eleven dimensions to the superstring in ten dimensions [6], our solution goes over to the superstring solution of Dabholkar, Gibbons, Harvey and Ruiz-Ruiz [7].

We begin by making an ansatz for the $D=11$ gauge fields g_{MN} and A_{MNP} ($M=0, 1, ..., 10$) corresponding to the most general three–eight split invariant under $P_3\times SO(8)$, where P_3 is the $D=3$ Poincaré group. We split the $D=11$ coordinates

$$x^M=(x^\mu, y^m)\,, \tag{1}$$

where $\mu=0, 1, 2$ and $m=3, ..., 10$, and write the line-element as

$$ds^2=e^{2A}\,\eta_{\mu\nu}dx^\mu dx^\nu+e^{2B}\delta_{mn}dy^m dy^n\,, \tag{2}$$

and the three-form gauge field as

$$A_{\mu\nu\rho}=\pm\,\frac{1}{{}^3g}\,\varepsilon_{\mu\nu\rho}\,e^C\,, \tag{3}$$

where ${}^3g$ is the determinant of $g_{\mu\nu}$, $\varepsilon_{\mu\nu\rho}\equiv g_{\mu\alpha}g_{\nu\beta}g_{\rho\gamma}\varepsilon^{\alpha\beta\gamma}$ and $\varepsilon^{012}=+1$ i.e. $A_{012}=\pm e^C$. All other components of A_{MNP} and all components of the gravitino ψ_M are set to zero. P_3 invariance requires that the arbitrary functions A, B and C depend only on y^m; SO(8) in-

[1] Work supported in part by NSF grant PHY-90415132.

Elsevier Science Publishers B.V. (North-Holland)

variance then requires that this dependence be only through $r \equiv \sqrt{\delta_{mn} y^m y^n}$.

As we shall now show, the three arbitrary functions A, B and C are reduced to one by the requirement that the field configurations (2) and (3) preserve some unbroken supersymmetry. In other words, there must exist Killing spinors ε satisfying

$$\tilde{D}_M \varepsilon = 0 , \tag{4}$$

where $\tilde{D}_M$ is the supercovariant derivative appearing in the supersymmetry transformation rule of the gravitino

$$\delta\psi_M |_{\psi=0} = \tilde{D}_M \varepsilon , \tag{5}$$

$$\tilde{D}_M = \partial_M + \tfrac{1}{4}\omega_M{}^{AB}\Gamma_{AB} - \tfrac{1}{288}(\Gamma^{PQRS}{}_M + 8\Gamma^{PQR}\delta^S{}_M) F_{PQRS} , \tag{6}$$

where $F_{MNPQ} = 4\partial_{[M}A_{NPQ]}$. Here Γ_A are the $D=11$ Dirac matrices satisfying

$$\{\Gamma_A, \Gamma_B\} = 2\eta_{AB} , \tag{7}$$

A, B refer to the $D=11$ tangent space, $\eta_{AB} = \text{diag}\,(-, +, \ldots, +)$, and

$$\Gamma_{AB\ldots C} = \Gamma_{[A}\Gamma_B \ldots \Gamma_{C]} , \tag{8}$$

thus $\Gamma_{AB} = \frac{1}{2}(\Gamma_A\Gamma_B - \Gamma_B\Gamma_A)$, etc. The Γ's with world indices $P, Q, R\ldots$ in (6) have been converted using vielbeins $e_M{}^A$. We make a three–eight split

$$\Gamma_A = (\gamma_\alpha \otimes \Gamma_9, \mathbf{1} \otimes \Sigma_a) , \tag{9}$$

where γ_α and Σ_a are the $D=3$ and $D=8$ Dirac matrices respectively and where

$$\Gamma_9 \equiv \Sigma_3 \Sigma_4 \ldots \Sigma_{10} \tag{10}$$

so that $\Gamma_9^2 = 1$. The most general spinor field consistent with the $P_3 \times SO(8)$ takes the form

$$\varepsilon(x, y) = \epsilon \otimes \eta(r) , \tag{11}$$

where ϵ is a constant spinor of SO(1, 2) and η is an SO(8) spinor which may further be decomposed into chiral eigenstates via the projection operators $(1 \pm \Gamma_9)$.

In our background (2) and (3), the supercovariant derivative becomes

$$\tilde{D}_\mu = \partial_\mu - \tfrac{1}{2}\gamma_\mu e^{-A}\Sigma^m \partial_m e^A \Gamma_9 \mp \tfrac{1}{6}\gamma_\mu e^{-3A}\Sigma^m \partial_m e^C , \tag{12}$$

$$\tilde{D}_m = \partial_m + \tfrac{1}{4}e^{-B}(\Sigma_m\Sigma^n - \Sigma^n\Sigma_m)\partial_n e^B \mp \tfrac{1}{24}e^{-3A}(\Sigma_m\Sigma^n - \Sigma^n\Sigma_m)\partial_n e^C \Gamma_9 \mp \tfrac{1}{6}e^{-3A}\partial_m e^C \Gamma_9 . \tag{13}$$

Note that the γ_μ and Σ_m carry world indices. Hence we find that (4) admits two non-trivial solutions

$$(1 \pm \Gamma_9)\eta = 0 , \tag{14}$$

where the $\pm$ signs are correlated with the $\pm$ signs in our original ansatz (3),

$$\eta = e^{-C/6}\eta_0 , \tag{15}$$

where η_0 is a constant spinor, and

$$A = \tfrac{1}{3}C , \tag{16}$$

$$B = -\tfrac{1}{6}C + \text{constant} . \tag{17}$$

In each case, (14) means that one half of the maximal possible rigid supersymmetry survives.

To see the uniqueness of these solutions, we may appeal to holonomy arguments [5] i.e. the integrability conditions for (4) following from the commutators of the supercovariant derivatives (12) and (13). In this connection, it is important to realize that the holonomy of the supercovariant derivative $\tilde{D}_M$ is different from that of the ordinary Lorentz-covariant derivative D_M. After making a three–eight split of the kind we are considering, it is known that the SO(1, 2)×SO(8) subgroup of the $D=11$ tangent space group SO(1, 10) is enlarged to SO(1, 2)×SO(16) [8]. Essentially, this is because the SO(8) spin connections $\omega_M{}^{ab}\Sigma_{ab}$ are augmented by terms like $F_M{}^{\alpha\beta c}\gamma_{\alpha\beta}\Sigma_c$, $F_M{}^{\alpha bc}\gamma_\alpha\Gamma_9\Sigma_{bc}$ and $F_M{}^{abc}\Sigma_{abc}$ which conspire to produce the connection of an SO(16), under which η transforms as a 16-dimensional vector. The holonomy group $\mathscr{H}$ of a specific $\tilde{D}_M$ will be a subgroup of this enlarged tangent space group. In the trivial case where $\mathscr{H}$ is the identity, there are no restrictions on ϵ and the maximal number of 2×16 rigid supersymmetries are preserved, but spacetime is flat and $F_{MNPQ} = 0$. In the other trivial case where $\mathscr{H}$ coincides with the enlarged tangent space group, ε vanishes and no supersymmetry survives. Given our ansatz (2), (3), the only remaining possibilities are those given in (14)–(17), for which the holonomy group $\mathscr{H}$ is $\mathbf{1} \otimes (8)_\pm$ corresponding to two inequivalent embeddings of SO(8) in SO(16). Under SO(16) ⊃ SO(8), the 16 decomposes into an 8 plus 8 singlets. Since the number of unbroken supersymmetries is given by the number of singlets in this decomposition, we see that exactly half of the maximal rigid supersymmetries survive.

Thus, at this stage, the three unknown functions A, B and C have been reduced to one by choosing the case where half the supersymmetry survives. To determine this unknown function, we must substitute our ansatz into the field equations which follow from the action

$$S_G = \int d^{11}x\, \mathcal{L}_G \,, \tag{18}$$

where $\mathcal{L}_G$ is the supergravity lagrangian whose bosonic sector is given by

$$\kappa^2 \mathcal{L}_G = \tfrac{1}{2}\sqrt{-g}\, R - \tfrac{1}{96}\sqrt{-g}\, F_{MNPQ}F^{MNPQ} + \frac{1}{2(12)^4}\varepsilon^{MNOPQRSTUVW} F_{MNOP}F_{QRST}A_{UVW} \,. \tag{19}$$

Let us first consider the antisymmetric tensor field equation

$$\partial_M(\sqrt{-g}\, F^{MUVW}) + \tfrac{1}{1152}\varepsilon^{UVWMNOPQRST} F_{MNOP}F_{QRST} = 0 \,. \tag{20}$$

Substitution of (2), (3) and (16), (17) yields

$$\delta^{mn}\partial_m\partial_n e^{-C} = 0 \tag{21}$$

and hence, imposing the boundary condition that the metric be asymptotically minkowskian, we find

$$e^{-C} = 1 + \frac{K}{r^6}, \quad r > 0 \,, \tag{22}$$

where K is a constant, at this stage arbitrary. The same expression for C also solves the Einstein equations. Thus the two solutions are given by

$$ds^2 = \left(1 + \frac{K}{r^6}\right)^{-2/3} \eta_{\mu\nu}dx^\mu dx^\nu + \left(1 + \frac{K}{r^6}\right)^{1/3} \delta_{mn}dy^m dy^n \,,$$

$$A_{\mu\nu\rho} = \pm \frac{1}{3g}\varepsilon_{\mu\nu\rho}\left(1 + \frac{K}{r^6}\right)^{-1} \,. \tag{23}$$

In fact, these expressions do not solve the field equations everywhere because of the singularity at $r=0$. Instead of (21), for example, we have

$$\delta^{mn}\partial_m\partial_n e^{-C} = -6K\Omega_7\delta^8(y) \,, \tag{24}$$

where Ω_7 is the volume of the unit seven-sphere S^7. Similar remarks apply to the Einstein equations. In order that (23) be solutions everywhere, it is therefore necessary that the pure supergravity equations be augmented by source terms. This source is, of course, the supermembrane itself. To see this explicitly we consider the combined supergravity-supermembrane equations which follow from the action

$$S = S_G + S_M \,, \tag{25}$$

where S_M is the supermembrane action whose bosonic sector is given by

$$S_M = T\int d^3\xi \left(-\tfrac{1}{2}\sqrt{-\gamma}\,\gamma^{ij}\partial_i X^M \partial_j X^N g_{MN} + \tfrac{1}{2}\sqrt{-\gamma} \pm \frac{1}{3!}\varepsilon^{ijk}\partial_i X^M \partial_j X^N \partial_k X^P A_{MNP}\right) , \tag{26}$$

where T is the membrane tension. The Einstein equations are now

$$R_{MN} - \tfrac{1}{2}g_{MN}R = \kappa^2 T_{MN} \,, \tag{27}$$

where T_{MN} receives a contribution not only from the antisymmetric tensor kinetic term but also from the membrane itself,

$$\kappa^2 T^{MN} = \tfrac{1}{12}(F^M{}_{PQR}F^{NPQR} - \tfrac{1}{8}g^{MN}F_{PQRS}F^{PQRS}) - \kappa^2 T\int d^3\xi \sqrt{-\gamma}\,\gamma^{ij}\partial_i X^M \partial_j X^N \frac{\delta^{11}(x-X)}{\sqrt{-g}} \,, \tag{28}$$

while the antisymmetric tensor equation is now

$$\partial_M(\sqrt{-g}\, F^{MUVW}) + \tfrac{1}{1152}\varepsilon^{UVWMNOPQRST}F_{MNOP}F_{QRST} = \mp 2\kappa^2 T\int d^3\xi\, \varepsilon^{ijk}\partial_i X^U \partial_j X^V \partial_k X^W \delta^{11}(x-X) \,. \tag{29}$$

Furthermore, we have the membrane field equations

$$\partial_i(\sqrt{-\gamma}\,\gamma^{ij}\partial_j X^N g_{MN}) + \tfrac{1}{2}\sqrt{-\gamma}\,\gamma^{ij}\partial_i X^N \partial_j X^P \partial_M g_{NP} \pm \frac{1}{3!}\varepsilon^{ijk}\partial_i X^N \partial_j X^P \partial_k X^Q F_{MNPQ} = 0 \,, \tag{30}$$

$$\gamma_{ij} = \partial_i X^M \partial_j X^N g_{MN} \,. \tag{31}$$

It is not difficult to verify that the correct source term

in (24) and in the Einstein equations is obtained by the static gauge choice

$$X^{\mu}=\xi^{\mu}, \quad \mu=0,1,2, \tag{32}$$

and the solution

$$Y^{m}=\text{constant}, \tag{33}$$

provided

$$K=\frac{\kappa^2 T}{3\Omega_7}. \tag{34}$$

However, with the choice of $-$ sign in front of the Wess–Zumino term given in (26), we must choose the $-$ sign solution in (23). The $+$ sign solution also solves the combined supergravity-supermembrane equations but for the opposite choice of sign for the supermembrane Wess–Zumino term. One may also verify that (32), (33) satisfy the membrane field equations (30), (31).

Having established that the supergravity field configurations preserve half the supersymmetries, we must also verify that the membrane configurations (32), (33) preserve these supersymmetries. As discussed in ref. [9], the criterion is that in addition to the existence of Killing spinors ε satisfying (4), we must also have

$$(1\pm\Gamma)\varepsilon=0, \tag{35}$$

where the choice of sign is correlated with the sign of the Wess–Zumino term in (26), and where

$$\Gamma\equiv\frac{1}{3!\sqrt{-\gamma}}\varepsilon^{ijk}\partial_i X^M\partial_j X^N\partial_k X^P\Gamma_{MNP}. \tag{36}$$

Since $\Gamma^2=1$ and $\operatorname{tr}\Gamma=0$, $\frac{1}{2}(1\pm\Gamma)$ act as projection operators. From (32), (33), we see that for our solutions

$$\Gamma=\mathbf{1}\otimes\Gamma_9, \tag{37}$$

and hence (35) is indeed satisfied as a consequence of (14). Eq. (35) explains, from the membrane point of view, why the solutions we are seeking preserve just half the supersymmetries. It originates from the fermionic κ-symmetry of the supermembrane action. The fermionic zero-modes on the worldvolume are just the Goldstone fermions associated with the broken supersymmetry [10].

Under a simultaneous dimensional reduction of spacetime and worldvolume, the combined supergravity-supermembrane field equations (27)–(31) in $D=11$ reduce to the combined type IIA supergravity-superstring field equations in $D=10$ [6]. A further truncation yields the equations studied by Dabholkar et al. [7]. The tangent space group SO(1, 2)×SO(16) is thus reduced to SO(1, 1)×SO(8) ×SO(8). One might expect, therefore, that their solutions may be obtained from ours by simultaneous dimensional reduction, and this is indeed the case. Let us denote all $D=11$ variables by a carat, and then make the ten–one split

$$\hat{x}^{\hat{M}}=(x^M,x^2), \quad M=0,1,3,\ldots,9, \tag{38}$$

$$\hat{g}_{MN}=\mathrm{e}^{-\phi/6}g_{MN}, \quad \hat{g}_{22}=\mathrm{e}^{4\phi/3},$$

$$\hat{A}_{MN2}=B_{MN}, \tag{39}$$

and set to zero all other components of $\hat{g}_{\hat{M}\hat{N}}$ and $\hat{A}_{\hat{M}\hat{N}\hat{P}}$. Then we can read off the $D=10$ solutions from (23) and (32), (33):

$$\mathrm{d}s^2=g_{MN}\mathrm{d}x^M\mathrm{d}x^N$$

$$=\left(1+\frac{K}{r^6}\right)^{-3/4}\eta_{\mu\nu}\mathrm{d}x^\mu\mathrm{d}x^\nu$$

$$+\left(1+\frac{K}{r^6}\right)^{1/4}\delta_{mn}\mathrm{d}y^m\mathrm{d}y^n,$$

$$B_{01}=\pm\left(1+\frac{K}{r^6}\right)^{-1}, \quad \mathrm{e}^{\phi}=\left(1+\frac{K}{r^6}\right)^{-1/2},$$

$$X^\mu=\xi^\mu, \quad \mu=0,1, \quad Y^m=\text{constant}. \tag{40}$$

These agree (choosing the $-$ sign) with the supergravity-superstring solution of Dabholkar et al. [7], where the g_{MN}, B_{MN} and ϕ are the metric, antisymmetric tensor and dilation of $D=10$ supergravity and the X^M are the $D=10$ string variables. These authors showed that their string solution saturates a Bogomol'nyi bound for the mass per unit length. Using the same methods we may establish a similar bound for the mass per unit area of the membrane

$$\mathcal{M}=\int \mathrm{d}^8y\,\theta_{00}, \tag{41}$$

where θ_{MN} is the total energy–momentum pseudotensor of the combined gravity–matter system. One finds

$$\kappa^2\mathcal{M}\geqslant|P|, \tag{42}$$

where P is proportional to the central charge which appears in the $D=11-2=9$, $N=1$ supersymmetry algebra. The novel feature of this Bogomol'nyi bound from the point of view of $D=11$ supergravity is that P is nothing but the familiar Page charge [4,5] defined by

$$P=\tfrac{1}{2}\int_{S^7}({}^*F+\tfrac{1}{2}A\wedge F)\,. \tag{43}$$

Its conservation follows from (20). Under the simultaneous dimensional reduction (39), P reduces to

$$P=\tfrac{1}{2}\int_{S^7}{}^*\mathrm{e}^{-\phi}H\,, \tag{44}$$

where $H=\mathrm{d}B$, the quantity appearing in ref. [7]. Either way, one finds for our solutions that

$$P=\pm\kappa^2T\,. \tag{45}$$

Hence the bound is saturated and the mass per unit area is just the membrane tension. This provides another way, in addition to unbroken supersymmetry, to understand the stability of the solution. (Note that under simultaneous dimensional reduction the $D=10$ and $D=11$ Newton constants are related by $\kappa^2=(2\pi R)^{-1}\hat{\kappa}^2$ where R is the radius of the compactifying circle, but that the string and membrane tensions are also related by $T=2\pi R\hat{T}$. Hence $\kappa^2T=\hat{\kappa}^2\hat{T}$.)

So far we have concentrated on single membrane solutions of the supergravity field equations. However, there is a straightforward generalization to exact, stable multi-membrane configurations obtained by a linear superposition of solutions to eq. (21):

$$\mathrm{e}^{-C}=1+\sum_I\frac{K}{|\boldsymbol{r}-\boldsymbol{r}_I|^6}\,, \tag{46}$$

where $\boldsymbol{r}_I$ corresponds to the position of each membrane. The ability to superpose solutions of this kind is a well-known phenomenon in soliton and instanton physics and goes by the name of the "no-force condition". In the present context, it means the the mutual gravitational attraction of two widely separated membranes is exactly cancelled by an equal and opposite contribution from the antisymmetric tensor. This is closely related both to the saturation of the Bogomol'nyi bound and the existence of unbroken supersymmetry. In the supersymmetry context the no-force condition is sometimes called "antigravity". To see this explicitly, consider a stationary test membrane at some distance from a source membrane located at the origin. Let both satisfy $X^\mu=\xi^\mu$ so that, in particular, they have the same orientation. The lagrangian for this test membrane in the field of the source given by (2), (3) is, from (26)

$$\mathscr{L}_\mathrm{M}=-T[\sqrt{-\det(\mathrm{e}^{2A}\eta_{ij}+\mathrm{e}^{2B}\partial_iY^m\partial_jY_m)}-\mathrm{e}^C]\,, \tag{47}$$

corresponding to a potential V given by

$$V=T(\mathrm{e}^{3A}-\mathrm{e}^C)\,. \tag{48}$$

But this vanishes by the supersymmetry condition (16). On the other hand, if the test membrane had the opposite orientation, and hence the opposite Page charge P, then the sign change in the Wess–Zumino term in (45) would result in a net attractive force and the two membranes would annihilate one another. This is entirely analogous to the $D=10$ string solution [7].

Finally, we would like to emphasize that these solutions are not "solitons", because of the δ-function singularities on the membrane worldvolume. There has been a good deal of interest in interpreting supermembranes as solitons [10] or "cosmic p-branes" [3]. In particular, Strominger [11] has shown how the heterotic 5-brane [12] emerges as a soliton of the heterotic string. These solutions are all source-free and non-singular. By constrast, the singularity of our solution, like the string solution of Dabholkar et al. [7], really means that we are solving the coupled supergravity-supermembrane equations. They nevertheless share some of the same properties as the genuine solitons: the breaking of half the supersymmetries, the saturation of a Bogomol'nyi bound in which the mass per unit area is equal to the tension. Some interesting questions remain concerning the deeper physical significance of our solutions. One has grown used to the idea that superstrings and supermembranes are to be regarded as the fundamental objects, with the supergravity fields emerging as the massless states in the spectrum. The point of view most appropriate to the present work is opposite: the supermembrane or superstring emerges as a singular solution to the field theory. The preservation of just one half of the spacetime supersymmetry plays an important role in this relationship and it is known to be related to the κ-symmetry on the membrane worldvolume [10].

Eleven-dimensional supergravity is a theory for which no independent matter field theory exists, owing to its maximal supersymmetry. The $D = 11$ supermembrane is the only "matter" that is known to couple consistently to this maximal supergravity background. It may be that the coupled system owes its consistency to a so-far undiscovered off-shell symmetry which generalizes the κ-symmetry that is preserved [2,6] when the supergravity background is required to satisfy its field equations. To find such an off-shell symmetry remains an open problem for future work.

We have enjoyed numerous fruitful conversations with Jian Xin Lu.

References

[1] E. Cremmer, B. Julia and J. Scherk, Phys. Lett. B 76 (1978) 409.

[2] E. Bergshoeff, E. Sezgin and P.K. Townsend, Phys. Lett. B 189 (1987) 75.

[3] P.K. Towsend, Phys. Lett. B 202 (1988) 53.

[4] D.N. Page, Phys. Rev. D 28 (1983) 2976.

[5] M.J. Duff, B.E.W. Nilsson and C.N. Pope, Phys. Rep. 130 (1986) 1.

[6] M.J. Duff, P.S. Howe, T. Inami and K.S. Stelle, Phys. Lett. B 191 (1987) 70.

[7] A. Dabholkar, G. Gibbons, J. Harvey and F.R. Ruiz-Ruiz, preprint DAMTP/R 90-1, EFI-90-11, PUPT-1163 (1990).

[8] M.J. Duff, in: Quantum field theory and quantum statistics, Vol. 2, eds. I.A. Batalin, C.J. Isham and G. Vilkovisky (Adam Hilger, Bristol) p. 209;
H. Nicolai, Phys. Lett. B 187 (1987) 316.

[9] E. Bergshoeff, M.J. Duff, C.N. Pope and E. Sezgin, Phys. Lett. B 199 (1987) 69.

[10] J. Hughes, J. Liu and J. Polchinski, Phys. Lett. B 180 (1986) 370.

[11] A. Strominger, preprint UCSB-TH-90-16.

[12] M.J. Duff, Class. Quantum Grav. 5 (1988) 189.

Reprinted from Phys. Lett. B **383** (1996) 44–7

Open p-branes

Andrew Strominger
Department of Physics, University of California, Santa Barbara, CA 93106-9530, USA

Received 15 May 1996
Editor: M. Dine

Abstract

It is shown that many of the p-branes of type II string theory and $d = 11$ supergravity can have boundaries on other p-branes. The rules for when this can and cannot occur are derived from charge conservation. For example it is found that membranes in $d = 11$ supergravity and IIA string theory can have boundaries on fivebranes. The boundary dynamics are governed by the self-dual $d = 6$ string. A collection of N parallel fivebranes contains $\frac{1}{2}N(N-1)$ self-dual strings which become tensionless as the fivebranes approach one another.

Type II string theories contain a variety of BPS-saturated p-brane solitons carrying a variety of charges Q^i [1–3]. All of these are extended extremal black holes [2]. This means that they are extremal members of a one-parameter family of $M^i \geq Q^i$ solutions which, for $M^i > Q^i$, have regular event horizons and geodesically complete, nonsingular spacelike slices with a second asymptotic region. Furthermore the $M^i > Q^i$ solutions decay via Hawking emission to the BPS-saturated $M^i = Q^i$ states. Recently there has been spectacular progress, initiated by Polchinski, in describing the dynamics of those p-branes which carry RR charge by representing them as D-branes in a type I theory [3–15]. In this paper we will rederive some of these recent results from low-energy reasoning in a manner that will generalize to all p-branes and uncover new phenomena.

Viewing p-branes as extended holes in spacetime naturally leads one to consider configurations in which one p-brane threads through the hole at the core of the second p-brane [1]. For example consider a static configuration consisting of two like-charged, parallel NS-NS (i.e. symmetric) fivebranes in the IIB theory. The metric is given by

$$ds_{10}^2 = \eta_{\mu\nu}dy^\mu dy^\nu + (1 + \frac{\alpha'}{|x-x_1|^2} + \frac{\alpha'}{|x-x_2|^2})\delta_{jk}dx^j dx^k, \qquad (1)$$

where $\mu, \nu = 0, \ldots, 5$ and $j, k = 6, \ldots, 9$. This has two infinite throats located at $x = x_1$ and $x = x_2$. Next consider a RR closed string which comes out one throat and goes in the next:

$$X^0 = \tau, \quad X^1 = x_1 + (x_2 - x_1)\sigma. \qquad (2)$$

The existence of such a configuration may be obstructed by charge conservation. In particular an S^7 which surrounds a RR string has a non-zero integral $Q^{RR} = \int *H^{RR}$ where H^{RR} is the RR 3-form field

[1] At the extremal limit, many of the p-brane solutions are singular or strongly coupled at the core, so the spacetime metric is not a reliable guide to the geometry. One is still however led to consider the fate of a p-brane which threads a large, smooth non-extremal p-brane which subsequently evaporates down to extremality.

0370-2693/96/$12.00
PII S0370-2693(96)00712-5

strength. This would seem to prevent strings from ending, since in that case the S^7 may be contracted to a point by slipping it off the end. However in so doing one must first pass it through the fivebrane. Using the explicit construction of [1] [2] it may be seen that the low-energy effective field theory on the fivebrane worldvolume contains a coupling

$$\int d^6\sigma B^{\mathrm{RR}}_{\mu\nu}\mathcal{F}^{\mu\nu}, \tag{3}$$

where B^{RR} is the spacetime RR Kalb-Ramond field and $\mathcal{F}$ is the worldbrane $U(1)$ gauge field strength. This leads to the equation of motion

$$d * H^{\mathrm{RR}} = Q^{\mathrm{RR}}\delta^8 + *\mathcal{F} \wedge \delta^4, \tag{4}$$

where δ^4 (δ^8) is a transverse 4-form (8-form) delta function on the fivebrane (RR string) and $*\mathcal{F}$ denotes the Hodge dual within the worldvolume. The total integral of $d * H$ over any S^8 must vanish. Consider an S^8 which intersects the string at only one point. Such an S^8 point must intersect the fivebrane in an S^4. Integrating (4) over the S^8 we find

$$0 = Q^{\mathrm{RR}} + \int_{S^4} *\mathcal{F}. \tag{5}$$

We conclude that H^{RR} charge conservation can be maintained if an electric flux associated to the fivebrane $U(1)$ charge emanates from the point at which the string enters the fivebrane. In other words the end of the string looks like a charged particle on the worldbrane.

As most easily seen from the Green-Schwarz form of the string action, the stretched string preserves those supersymmetries generated by spinors ϵ obeying

$$\Gamma_{MN}\partial_+ X^M \partial_- X^N \epsilon = \epsilon. \tag{6}$$

(2) and (1) together preserves one quarter of the supersymmetries so this configuration is BPS-saturated to leading order. At next order one must include the back reaction of the string on the spacetime geometry and fields. Since there is no obstruction from charge conservation we presume that a fully supersymmetric configuration describing a RR string stretched between two NS-NS fivebranes exists and corresponds to a BPS state.

There is no coupling of the form (3) involving the NS-NS B field. Charge conservation therefore prohibits fundamental IIB strings from ending at NS-NS fivebranes. However $SL(2, Z)$ interchanges NS-NS and RR onebranes and fivebranes. Hence $SL(2, Z)$ invariance implies that a fundamental IIB string can end at a RR fivebrane. The latter (like all the RR solitons) can be realized as a D-brane. So this is not a surprise: we have reproduced results of [4,3].

Next let us consider what happens as the two fivebranes approach one another. The mass of the stretched string is given by a BPS bound and is a function on the two-fivebrane moduli space. It decreases with the string length. When the fivebranes become coincident, the stretched string has zero length and becomes a massless state carrying the $U(1)$ charges of both fivebranes. The result is therefore an $N = 4$ $U(2)$ gauge theory on the fivebrane [4,6]. Note that the dual relation to open string theory is not required for this conclusion. From this perspective the source of massless gauge bosons is similar to that in [17–19]: they arise from a degenerating one-cycle which threads two horizons.

A similar story applies to the RR threebrane. Reduction of the formulae in [2,20] leads to spacetime-worldbrane couplings of the form (3) for *both* the NS-NS and RR B fields. This is required by $SL(2, Z)$ invariance because the threebrane acts as a source for the self-dual 5-form and hence is itself self-dual. In [4] it was shown that fundamental strings can end on D-branes but here we see that D-strings may in some cases also end on D-branes. This dovetails nicely with S-duality of the $N = 4$, $d = 4$ gauge theory which lives on the threebrane: The ends of fundamental strings are electrically charged particles while the ends of D-strings are magnetically charged particles.

There may seem to be a puzzle for example for the RR 0-brane. Clearly charge conservation will prevent (except when there is a RR background [21]) a fundamental string from ending at a 0-brane. This may seem to conflict with the picture in [3] which involves an $SU(N)$ gauge theory for N 0-branes coming from strings ending at the 0-branes. However there is not really a conflict because our reasoning applies only to BPS states, and charge confinement in 0+1 $SU(N)$ gauge theories indeed eliminates the charged

[2] A correction to the zero mode wave function may be found in [16].

BPS states.

So far we have reproduced from a different perspective results previously obtained either directly in [4,3,6], as well as some $SL(2,Z)$ duals of those results. Our point of view gives the leading low-energy dynamics, but probably cannot easily reproduce the detailed prescription given in [4,3] for computing, e.g. finite momentum string-D-brane scattering as in [12]. However in considering higher p-branes this low-energy perspective will lead us to new phenomena.

As a further example we consider a membrane stretched between two fivebranes of eleven-dimensional supergravity [3]. (Of course reduction of this leads to examples in the IIA theory.) Unlike its IIB partner, the $d = 11$ (and IIA) fivebrane has chiral dynamics governed by the $d = 6$ tensor multiplet [23] containing 5 scalars and a self-dual antisymmetric tensor field strength $\mathcal{A}$ [1]. The membrane worldvolume condition for unbroken supersymmetry is [22,24]

$$\Gamma_{MNP}\epsilon^{\alpha\beta\gamma}\partial_\alpha X^M\partial_\beta X^N\partial_\gamma X^P\epsilon = \epsilon. \tag{7}$$

Again it is easily seen that a membrane stretched between two fivebranes preserves one quarter of the supersymmetries at leading order. For appropriate brane orientations the unbroken supersymmetries are generated by spinors obeying the two chirality conditions $\Gamma^{016}\epsilon = \Gamma^{012345}\epsilon = \epsilon$. The membrane can be surrounded by an S^7 for which there is a nonzero value of the charge $Q^M = \int_{S^7} *F$, where F here is the spacetime 4-form field strength. In the presence of a membrane and a fivebrane the equation of motion for F, as follows from formulae in [1,25,26], is

$$d*F = Q^M\delta^8 + \mathcal{A}\wedge\delta^5. \tag{8}$$

We see that the boundary of the membrane – which is a string lying in the fivebrane – must carry self-dual antisymmetric tensor charge $\int_{S^3}\mathcal{A} = -Q^M$. This string is the self-dual string of Duff and Lu [27].

Further insight into this construction can be gained by considering S- and T-duality. Polchinski [4,3] has shown that the worldbrane dynamics of the IIB RR fivebrane are described by open fundamental Dirichlet strings. $SL(2,Z)$ invariance then implies that worldbrane dynamics of the IIB NS-NS fivebrane are described by open RR strings (although this description is weakly coupled only at large g_s). Now periodically identify and T-dualize along one direction of the fivebrane. This gives a IIA theory [4]. The NS-NS (i.e. symmetric) fivebrane solution is represented by a conformal field theory involving only the transverse coordinates, and hence is unaffected by longitudinal T-duality (This is in contrast to RR p-branes, which lose (gain) a dimension under longitudinal (transverse) T-duality.). However the zero modes which propagate parallel to the fivebrane are affected, and the $N = 4$ $U(1)$ vector multiplet is transformed into an $N = 4$ antisymmetric tensor multiplet. At the same time the open strings which govern the IIB fivebrane dynamics are T-transformed into open membranes which govern the IIA fivebrane dynamics.

Next we consider the dynamics of N parallel $d = 11$ or IIA fivebranes. When the fivebranes are separated the low energy dynamics is governed by a globally supersymmetric $(0,2)$ $d = 6$ theory with N tensor multiplets. The moduli space of the $5N$ scalars is uniquely determined to be locally the symmetric space $T(5,N) \equiv SO(5,N)/(SO(5)\times SO(N))$. Since this is a chiral theory it is not possible for extra massless fields to appear when the fivebrane positions coincide. However tensionless strings can and do arise, because the tension of a BPS string which arises as the boundary of a membrane stretched between two fivebranes vanishes when the fivebrane coincides. These strings transform in the adjoint of the global $SO(N)$ which acts on the N tensor multiplets. Upon S^1 compactification along the fivebranes, winding states of the tensionless strings lead to the appearance of extra massless gauge bosons which – together with the reduced tensor multiplets which dualize to vector multiplets – fill out a $U(N)$ gauge theory [28], as predicted by T-duality.

Aspects of the preceding are quite similar to Witten's discussion [28] of K3 compactification of IIB string theory, whose moduli space is locally $T(5,21)$ and which contains (5) 5+16 (anti) self-dual antisymmetric tensor fields. In this case tensionless strings arise from threebranes wrapping degenerating 2-cycles. Indeed there is a dual IIA description of this IIB compactification, in the spirit of [29,8,9], as 16 toroidally compactified symmetric fivebranes and NS-NS orientifolds, where the extra 5+5 antisymmetric tensor fields arise from the supergravity multiplet [30]. In [29] it was shown that IIA on

[3] Preliminary observations on open membranes are made in [22].

K3 is equivalent to IIB on a D-manifold with 16 RR orientifolds and 16 RR fivebranes. IIB S-duality converts NS-NS to RR fields, so this is S-equivalent to a IIB configuration with 16 NS-NS orientifolds and 16 NS-NS fivebranes. Next T-dualize this last representation of IIA on K3 (yielding IIB on K3) along one of the noncompact directions. This will not affect the 4-geometry which involves only NS-NS fields. Hence IIB on K3 is equivalent to IIA on a "p-manifold" with 16 NS-NS orientifolds and 16 symmetric fivebranes. This provides the concrete connection to [28].

As pointed out in [28] the fact that self-dual strings (or open membranes [4]) become light as the fivebranes approach one another suggests that supergravity might be decoupled and the dynamics of self-dual strings and symmetric fivebranes consistently studied in isolation from the rest of string theory. This is also suggested by superconformal invariance of the tensor multiplet in $d = 6$ [23]. The relation by compactification to superconformal $d = 4$, $N = 4$ Yang-Mills makes this a particularly fascinating problem.

Further examples of p-branes with boundaries can be found. It may be directly checked in the IIB theory that charge conservation allows a threebrane to end on a membrane in the RR fivebrane. The membrane carries magnetic charge with respect to the fivebrane $U(1)$ gauge field. In general every RR p-brane has a $U(1)$ gauge field. Electric charges are always carried by zerobranes and arise from fundamental strings which terminate at the p-brane. Magnetic charges are carried by a $(p - 3)$-brane, which can arise as the boundary of a $(p - 2)$-brane. It is difficult to check charge conservation directly for $p > 5$ because the zero mode wave functions have not been worked out. However T-duality along a dimension transverse to an configuration of RR p-branes increases p, so we presume it is always possible (in IIA or IIB) for a RR $(p - 2)$-brane to end at a RR p-brane. All of these new multi-p-brane configurations can be used to construct p-manifold generalizations of the D-manifolds introduced in [8], and may arise in the process of dualization.

In conclusion string theory contains a rich variety of extended objects which interact in an intricate and beautiful fashion. Higher p-branes provide endpoints for branes of lower p, which latter in turn govern the dynamics of the former.

We thank J. Polchinski for useful discussions and for explaining the results of [29] prior to publication. This work was supported in part by DOE Grant No. DOE-91ER40618.

[4] The relation in the infrared between the self-dual open membranes and self-dual strings may involve Chern-Simons theory as in [31].

References

[1] C. Callan, J. Harvey and A. Strominger, Nucl. Phys. B 367 (1991) 60.
[2] G. Horowitz and A. Strominger, Nucl. Phys. B 360 (1991) 197.
[3] J. Polchinski, hep-th/9510017.
[4] J. Dai, R. Leigh and J. Polchinski, Mod. Phys. Lett. A 4 (1989) 2073.
[5] P. Horava, Phys. Lett. B 231 (1989) 251.
[6] E. Witten, hep-th/9510135.
[7] J. Polchinski and E. Witten, hep-th/9510169.
[8] M. Bershadsky, V. Sadov and C. Vafa, hep-th/9510225; hep-th/9511222.
[9] H. Ooguri and C. Vafa, hep-th/9501164.
[10] A. Sen, hep-th/9510229, 9511026.
[11] M. Li, hep-th/9510161.
[12] C. Callan and I. Klebanov, hep-th/9511173.
[13] C. Vafa, hep-th/9511088.
[14] E. Witten, hep-th/9511030.
[15] C. Bachas, hep-th/9511043.
[16] J. Harvey and A. Strominger, hep-th/9504047.
[17] C. Hull and P. Townsend, hep-th/9410167, Nucl. Phys. B 348 (1995) 109.
[18] E. Witten, Nucl. Phys. 443 (1995) 85.
[19] A. Strominger, hep-th/9504090, Nucl. Phys. B 451 (1995) 96.
[20] M. Duff and J. Lu, Phys. Lett. B 273 (1991) 409.
[21] J. Polchinski and A. Strominger, hep-th/9510227.
[22] E. Bergshoeff, E. Sezgin and P. Townsend, Phys. Lett. B 189 (1987) 75; Ann. Phys. B 185 (1988) 330.
[23] P. Howe, G. Sierra and P. Townsend, Nucl. Phys. B 221 (1983) 331.
[24] K. Becker, M. Becker and A. Strominger, hep-th/9507158.
[25] R. Guven, Phys. Lett. B 276 (1992) 49.
[26] D. Kaplan and J. Michelson, hep-th/9510053.
[27] M. Duff and J. Lu, hep-th/9306052, Nucl. Phys. B 416 (1994) 301.
[28] E. Witten, hep-th/9507121.
[29] T. Banks, M. Douglas, J. Polchinski and N. Seiberg, to appear.
[30] L. Romans, Nucl. Phys. B 276 (1986) 71.
[31] E. Witten, Comm. Math. Phys. 121 (1989) 351;
G. Moore and N. Seiberg, Phys. Lett. B 220 (1990) 422;
S. Carlip and I. Kogan, Phys. Rev. Lett. 64 (1990) 1487;
Mod. Phys. Lett. A 6 (1991) 171.

Reprinted from Phys. Lett. B **373** (1996) 68–75

D-branes from M-branes

P.K. Townsend
DAMTP, Silver St., Cambridge CB3 9EW, UK

Received 11 December 1995; revised manuscript received 12 January 1996
Editor: P.V. Landshoff

Abstract

The 2-brane and 4-brane solutions of ten dimensional IIA supergravity have a dual interpretation as Dirichlet-branes, or 'D-branes', of type IIA superstring theory and as 'M-branes' of an S^1-compactified eleven dimensional supermembrane theory, or M-theory. This eleven-dimensional connection is used to determine the ten-dimensional Lorentz covariant worldvolume action for the Dirichlet super 2-brane, and its coupling to background spacetime fields. It is further used to show that the 2-brane can carry the Ramond-Ramond charge of the Dirichlet 0-brane as a topological charge, and an interpretation of the 2-brane as a 0-brane condensate is suggested. Similar results are found for the Dirichlet 4-brane via its interpretation as a double-dimensional reduction of the eleven-dimensional fivebrane. It is suggested that the latter be interpreted as a D-brane of an open eleven-dimensional supermembrane.

1. Introduction

The importance of super p-branes for an understanding of the non-perturbative dynamics of type II superstring theories is no longer in doubt. For example, they are relevant to U-duality of toroidally-compactified type II superstrings [1,2], and symmetry enhancement at singular points in the moduli space of K_3 or Calabi-Yau compactified type II superstrings [3–6] as required by the type II/heterotic string-string duality [1,3,7]. Type II p-branes were first found as solutions of the effective $D = 10$ supergravity theory [8–11]. Because their worldvolume actions involve worldvolume gauge fields [12,13], in addition to the scalars and spinors expected on the basis of spontaneously broken translation invariance and supersymmetry, they were not anticipated in the original classification of super p-branes [14]. For the same reason, the fully $D = 10$ Lorentz covariant action for these type II super p-branes is not yet known. One purpose of this paper is to report progress on this front.

The type II p-branes are conveniently divided into those of Neveu/Schwarz-Neveu/Schwarz (NS-NS) type and those of Ramond-Ramond (RR) type according to the string theory origin of the $(p + 1)$-form gauge potential for which they are a source. The supergravity super p-branes found in the NS-NS sector comprise a string and a fivebrane. The string has a naked timelike singularity and can be identified as the effective field theory realization of the fundamental string [1]. The fivebrane solution is non-singular and has a 5-volume tension $\sim \lambda^{-2}$ expected of a soliton, where λ is the string coupling constant. Since a 5-brane is the magnetic dual of a string in $D = 10$

[1] Note that the existence of this solution is necessary for the consistency of any string theory with massless spin 2 excitations since a macroscopic string will then have long range fields which must solve the source free equations of the effective field theory.

0370-2693/96/$12.00
PII S0370-2693(96)00104-9

[15], this solution is an analogue of the BPS magnetic monopole of $D = 4$ super Yang-Mills (YM) theory.

In the RR sector the ten-dimensional ($D = 10$) IIA supergravity has p-brane solutions for $p = 0, 2, 4, 6$, while the IIB theory has RR p-branes solutions for $p = 1, 3, 5$ (see [2] for a recent review) [2]. With the exception of the 3-brane, which is self-dual, these p-branes come in $(p, \tilde{p})$ electric/magnetic pairs with $\tilde{p} = 6 - p$. The RR p-brane solutions all have a p-volume tension $\sim \lambda^{-1}$ [3] so, although non-perturbative, they are not typically solitonic. Moreover, they are all singular, with the exception of the 3-brane, and even this exceptional case is not typical of solitons because the solution has an event horizon [19]. Thus, the RR p-branes are intermediate between the fundamental string and the solitonic fivebrane. It now appears [20] that they have their place in string theory as Dirichlet-branes, or D-branes [21,22].

It was shown in [23] how all the p-brane solutions of $D = 10$ IIA supergravity (with $p \leq 6$) have an interpretation in $D = 11$, extending previous results along these lines for the string and fourbrane [24–26]. In particular, the 0-branes were identified with the Kaluza-Klein (KK) states of $D = 11$ supergravity and their 6-brane duals were shown to be $D = 11$ analogues of the KK monopoles. The remaining p-brane solutions have their $D = 11$ origin in either the membrane [25] or the fivebrane [27] solutions of $D = 11$ supergravity. It was subsequently shown that $D = 11$ supergravity is the effective field theory of the type IIA superstring at strong coupling [3] and then that various dualities in $D < 10$ can be understood in terms of the electric/magnetic duality in $D = 11$ of the membrane and fivebrane [28,29]. These results suggest the existence of a consistent quantum theory underlying $D = 11$ supergravity. This may be a supermembrane theory as originally suggested [30], or it may be some other theory that incorporates it in some way. Whatever it is, it now goes by the name 'M-theory' [31,32].

The point of the above summary is to show that the RR p-brane solutions of $D = 10$ IIA supergravity theory currently have two quite different interpretations. On the one hand they are interpretable as D-branes of type IIA string theory. On the other hand they are interpretable as solutions of S^1 compactified $D = 11$ supergravity. In the $p = 2$ and $p = 4$ cases these $D = 11$ solutions are also p-branes; since they are presumably also solutions of the underlying $D = 11$ M-theory we shall call them 'M-branes'. We shall first exploit the interpretation of the $p = 2$ super D-brane as a dimensionally reduced $D = 11$ supermembrane to deduce its $D = 10$ Lorentz covariant worldvolume action. The bosonic action has been found previously [22] by requiring one-loop conformal invariance of the open string with the string worldsheet boundary on the D-brane [3]. One feature of the derivation via $D = 11$ is that the coupling to background fields can also be found this way, and the resulting action has a straightforward generalization to general p. The coupling to the dilaton is such that the p-volume tension is $\sim \lambda^{-1}$, as expected for a D-brane [21]. The M-brane interpretation of the Dirichlet 4-brane is as the *double*-dimensional reduction of the eleven-dimensional fivebrane. We propose a bosonic action for the latter including a coupling to the bosonic fields of eleven-dimensional supergravity, and exploit it to deduce the coupling to background supergravity fields, including the dilaton, of the Dirichlet 4-brane. The result agrees with that deduced by generalization of the $p = 2$ case.

One intriguing feature of these results is that they suggest an interpretation of the eleven-dimensional fivebrane as a Dirichlet-brane of an open $D = 11$ supermembrane, and we further suggest that the string-boundary dynamics is controlled by the conjectured [34], and intrinsically non-perturbative, six-dimensional self-dual string theory (which is possibly related to the self-dual string soliton [35], although this solution involves six-dimensional gravitational fields which are not, according to current wisdom, among the fivebrane's worldvolume fields).

Finally, we show that a spherical $D = 10$ 2-brane can carry the same RR charge that is carried by the Dirichlet 0-branes; this charge is essentially the magnetic charge associated with the worldvolume vector potential. This suggests that the 0-branes can be

[2] There is also a IIB 7-brane [16] and a IIA 8-brane [17] (see also [18]), but these do not come in electric/magnetic pairs and have rather different physical implications; for example, they do not contribute to the spectrum of particles in any $D \geq 4$ compactification. Partly for this reason, only the $p \leq 6$ cases will be discussed here.

[3] The action of [22] is not obviously equivalent to the bosonic sector of the one found here and the omission of a discussion of this point was a defect of an earlier version of this paper; fortunately, the equivalence has since been established by Schmidhuber [33].

viewed as collapsed 2-branes. We point out that this is consistent with the $U(\infty)$ Supersymmetric Gauge Quantum Mechanics interpretation of the supermembrane worldvolume action [36,37], which further suggests an interpretation of the supermembrane as a condensate of 0-branes. Viewed from the $D = 11$ perspective these results can be taken as further evidence that $D = 11$ supergravity is the effective field theory of a supermembrane theory.

2. The $D = 10$ 2-brane as a $D = 11$ M-brane

Consider first the $D = 10$ 2-brane. From its D-brane description we know that the worldvolume action is based on the $D = 10$ Maxwell supermultiplet dimensionally reduced to three dimensions [38], i.e. the worldvolume field content is

$$\{X^a \ (a = 1, \ldots, 7)\ ,\ A_i \ (i = 0, 1, 2)\ ; \quad \chi^I \ (I = 1, \ldots, 8)\} \tag{2.1}$$

where the χ^I are eight $Sl(2;\mathbb{R})$ spinors and A_i is a worldvolume vector potential[4]. As for every other value of p, only the bosonic part of the 10-dimensional Lorentz covariant action constructed from these fields is currently known [22]. However, the alternative interpretation of the 2-brane as an M-brane allows us to find the complete action. In this interpretation, the IIA 2-brane is the direct (as against double) dimensional reduction of the $D = 11$ supermembrane. The worldvolume fields of the dimensionally reduced $D = 10$ supermembrane are, before gauge-fixing, $\{X^m \ (m = 0, 1, \ldots, 9);\ \varphi;\ \theta\}$, where θ is a 32-component Majorana spinor of the $D = 10$ Lorentz group and X^m is a 10-vector. After gauge fixing the physical fields are

$$\{X^a \ (a = 1, \ldots, 7)\ ,\ \varphi\,;\chi^I \ (I = 1, \ldots, 8)\}\,. \tag{2.2}$$

The difference between (2.1) and (2.2) is simply that the scalar φ of (2.2) is replaced in (2.1) by its 3-dimensional dual, the gauge vector A. By performing this duality transformation in the action *prior* to gauge fixing we can determine the fully $D = 10$ Lorentz covariant Dirichlet supermembrane action.

[4] Throughout this paper we shall use the letter A to denote worldvolume gauge fields, of whatever rank, and B to denote spacetime gauge fields, of whatever rank.

The first step of this procedure is to isolate the dependence of the $D = 11$ supermembrane action on X^{11}, which is here called φ. We shall first consider the case for which the $D = 11$ spacetime is the product of S^1 with $D = 10$ Minkowski spacetime, returning subsequently to consider the interaction with background fields. It is convenient to use the Howe-Tucker (HT) formulation of the action for which there is an auxiliary worldvolume metric γ_{ij}. It is also convenient to introduce the spacetime supersymmetric differentials

$$\Pi^m = dX^m - i\bar\theta\Gamma^m d\theta\,. \tag{2.3}$$

The action, given in [30], is

$$\begin{aligned} S = &-\tfrac{1}{2}\int d^3\xi\,\sqrt{-\gamma}\big[\gamma^{ij}\Pi_i^m\Pi_j^n\eta_{mn} \\ &+\gamma^{ij}(\partial_i\varphi - i\bar\theta\Gamma_{11}\partial_i\theta)(\partial_j\varphi - i\bar\theta\Gamma_{11}\partial_j\theta) - 1\big] \\ &-\tfrac{1}{6}\int d^3\xi\ \varepsilon^{ijk}\left[b_{ijk} + 3b_{ij}\partial_k\varphi\right]\,, \end{aligned} \tag{2.4}$$

where η is the $D = 10$ Minkowski metric, and

$$\begin{aligned} \varepsilon^{ijk}b_{ijk} = 3\varepsilon^{ijk}\Big\{&i\bar\theta\Gamma_{mn}\partial_i\theta\big[\Pi_i^m\Pi_j^n\eta_{mn} \\ &+ i\Pi_i^m(\bar\theta\Gamma^n\partial_j\theta) - \tfrac{1}{3}(\bar\theta\Gamma^m\partial_i\theta)(\bar\theta\Gamma^n\partial_j\theta)\big] \\ &+ (\bar\theta\Gamma_{11}\Gamma_m\partial_i\theta)(\bar\theta\Gamma_{11}\partial_j\theta)(\partial_k X^m - \tfrac{2}{3}i\bar\theta\Gamma^m\partial_k\theta)\Big\}\,, \end{aligned} \tag{2.5}$$

while

$$\varepsilon^{ijk}b_{ij} = -2\varepsilon^{ijk}\ i\bar\theta\Gamma_m\Gamma_{11}\partial_i\theta(\partial_j X^m - \tfrac{1}{2}i\bar\theta\Gamma^m\partial_j\theta)\,. \tag{2.6}$$

The second step, the replacement of the worldvolume scalar φ by its dual vector field, can be achieved by promoting $d\varphi$ to the status of an independent worldvolume one-form L while adding a Lagrange multiplier term AdL to impose the constraint $dL = 0$. Eliminating L by its algebraic equation of motion yields the dual action in terms of the fields X^m and the worldvolume field strength two-form $F = dA$. This action is

$$\begin{aligned} S = &-\tfrac{1}{2}\int d^3\xi\ \sqrt{-\gamma} \\ &\times\left[\gamma^{ij}\ \Pi_i^m\Pi_j^n\ \eta_{mn} + \tfrac{1}{2}\gamma^{ik}\gamma^{jl}\hat F_{ij}\hat F_{kl} - 1\right] \\ &-\tfrac{1}{6}\int d^3\xi\ \varepsilon^{ijk}\left[b_{ijk} - 3i(\bar\theta\Gamma_{11}\partial_i\theta)\hat F_{jk}\right]\,. \end{aligned} \tag{2.7}$$

where

$$\hat{F}_{ij} = F_{ij} - b_{ij} . \tag{2.8}$$

Thus (2.7) is the fully $D = 10$ Lorentz covariant worldvolume action for the $D = 10$ IIA Dirichlet supermembrane. The bosonic action, obtained by setting the fermions to zero in (2.7), is equivalent to the Born-Infeld-type action found by Leigh [22]. The equivalence follows from the recent observation of Schmidhuber [33] that dualizing the vector to a scalar in the action of Leigh yields the action of a $D = 11$ membrane, which was precisely the (bosonic) starting point of the construction presented here[5]. It is interesting to note that a sigma-model one-loop calculation in the string theory is reproduced by the classical supermembrane.

It can now be seen why it was advantageous to start from the HT form of the action; whereas the auxiliary metric is simply eliminated from (2.4), leading to the standard Dirac-Nambu-Goto (DNG) form of the action, its elimination from (2.7) is far from straightforward, although possible in principle. The point is that the γ_{ij} equation is now the very non-linear, although still algebraic, equation

$$\gamma_{ij} = \left(1 + \tfrac{1}{2}\gamma^{kp}\gamma^{lq}\hat{F}_{kl}\hat{F}_{pq}\right)^{-1}\left(g_{ij} + \gamma^{kl}\hat{F}_{ik}\hat{F}_{lj}\right) \tag{2.9}$$

where $g_{ij} = \Pi_i^m \Pi_j^n \eta_{mn}$. This equation can be solved as a series in $\hat{F}$ of the form

$$\gamma_{ij} = g_{ij}\left[1 - \tfrac{1}{2}g^{kp}g^{lq}\hat{F}_{kl}\hat{F}_{pq}\right] + g^{kl}\hat{F}_{ik}\hat{F}_{jl} + \mathcal{O}(\hat{F}^4) , \tag{2.10}$$

and the approximation $\gamma_{ij} = g_{ij}$ yields the quadratic part of the action in $\hat{F}$.

Invariance of the action (2.7) under supersymmetry requires $\hat{F}$ to be invariant. To see how this comes about, we observe that the two-form b in $\hat{F}$ is precisely the one that defines the WZ term in the Green-Schwarz superstring action; it has the property that the three-form $h = db$ is superinvariant, which implies that $\delta_\epsilon b = da$ for some one-form $a(\epsilon)$, where ϵ is the (constant) supersymmetry parameter. The modified two-form field strength $\hat{F}$ is therefore superinvariant if we choose $\delta_\epsilon A = a$. The κ-transformation of A is similarly determined by requiring κ-gauge invariance of the action, but it can also be deduced directly from those of the $D = 11$ supermembrane given in [30]. The result is most simply expressed in terms of the variations of the supersymmetric forms Π^m and $\hat{F}$, which are[6]

$$\begin{aligned}
\delta_\kappa \Pi^m &= -2i(\delta_\kappa\bar{\theta})\Gamma^m d\theta \\
\delta_\kappa \hat{F} &= i(\delta_\kappa\bar{\theta})\Gamma_m\Gamma_{11} d\theta \wedge \Pi^m \\
\delta_\kappa \theta &= (1 + \Gamma)\kappa ,
\end{aligned} \tag{2.11}$$

where

$$\begin{aligned}
\Gamma = &\frac{1}{6\sqrt{-\gamma}}\varepsilon^{ijk}\Pi_i^m\Pi_j^n\Pi_k^p\Gamma_{mnp} \\
&- \tfrac{1}{2}\gamma^{ik}\gamma^{jl}F_{kl}\Pi_i^m\Pi_j^n\Gamma_{mn}\Gamma_{11}
\end{aligned} \tag{2.12}$$

and $\kappa(\xi)$ is the $D = 10$ Majorana spinor parameter.

The coupling of the action (2.7) to background fields can also be deduced from its $D = 11$ origin. We shall consider here only the bosonic membrane coupled to bosonic background fields. Consider first the NS-NS fields. In the $D = 10$ membrane action obtained by direct dimensional reduction from $D = 11$, the NS-NS two-form potential B couples to the topological current $\varepsilon^{ijk}\partial_k\varphi$. In the dual action this coupling corresponds to the replacement of F by $F - B$. The coupling to the $D = 10$ spacetime metric is obvious so this leaves the dilaton; to determine its coupling we recall (see e.g. [26]) that the $D = 11$ metric is

$$ds_{11}^2 = e^{-\frac{2}{3}\phi}ds^2 + e^{\frac{4}{3}\phi}d\varphi^2 , \tag{2.13}$$

where ds^2 is the string-frame $D = 10$ metric and ϕ is the dilaton. A repetition of the steps described above, but now for the purely bosonic theory and carrying along the dependence on the NS-NS-spacetime fields, leads (after a redefinition of the auxiliary metric to the action

$$\begin{aligned}
S = -\tfrac{1}{2}\int d^3\xi\, e^{-\phi}\sqrt{-\gamma}\Big[&\gamma^{ij}g_{ij} \\
&+ \tfrac{1}{2}\gamma^{ik}\gamma^{jl}(F_{ij} - B_{ij})(F_{kl} - B_{kl}) - 1\Big] ,
\end{aligned} \tag{2.14}$$

where now $g_{ij} = \partial_i X^m \partial_j X^n g_{mn}$. The appearance of F through the modified field strength $F - B$ could also

[5] The equivalence with Born-Infeld for $p = 1$, i.e. the D-string, was shown in [39].

[6] As explained in detail in [40], it is not necessary to specify the transformation of the metric γ_{ij} if use is made of the '1.5 order' formalism.

have been deduced simply by the replacement of the flat superspace two-form potential b in $\hat{F}$ by its curved superspace counterpart, since setting the fermions to zero then yields precisely $F-B$. As for the RR fields, the coupling to the 3-form potential is of course the standard Lorentz coupling while the coupling to the 1-form potential has interesting implications which will be discussed at the conclusion of this article.

The above result, and the known form of the bosonic p-brane action in the absence of worldvolume gauge fields, suggests that the corresponding bosonic part of the worldvolume action of the Dirichlet super p-brane is

$$S = -\tfrac{1}{2}\int d^{(p+1)}\xi\, e^{-\phi}\sqrt{-\gamma}\Big[\gamma^{ij}g_{ij} + \tfrac{1}{2}\gamma^{ik}\gamma^{jl}(F_{ij}-B_{ij})(F_{kl}-B_{kl}) - (p-1)\Big]. \tag{2.15}$$

Since the vacuum expectation value of e^{ϕ} is the string coupling constant λ, it follows from this result that the p-brane tension is $\sim \lambda^{-1}$, as expected for D-branes. Of course, the steps leading to this result were particular to $p=2$ but we shall shortly arrive at the same result for $p=4$ via a different route. Although the action (2.15) is only guaranteed to be correct to quadratic order in F for $p \neq 2$, this will prove sufficient for present purposes.

3. The $D = 11$ 5-brane as a supermembrane D-brane

Consider now the Dirichlet 4-brane. In this case its M-brane interpretation is as a double-dimensional reduction of the $D=11$ 5-brane. The (partially) gauge-fixed field content of the latter consists [41,42] of the fields of the $N=4$ six-dimensional antisymmetric tensor multiplet, i.e.

$$\{X^a\ (a=1,\dots,5)\ ,\ A^+_{ij}\ (i,j=0,1,\dots,5)\ ;\ \chi^I (I=1,\dots,4)\} \tag{3.1}$$

where χ^I are chiral symplectic-Majorana spinors in the **4** of $USp(4)\cong \mathrm{Spin}(5)$, and A^+ is the two-form potential for a self-dual 3-form field strength $F=dA^+$. Because of the self-duality of F we cannot expect to find a worldvolume action (at least, not one quadratic in F). We might try to find an action that leads to all equations *except* the self-duality constraint which we can then just impose by hand, as advocated elsewhere in another context [43]. We shall adopt this strategy here, but it is important to appreciate an inherent difficulty in its present application. The problem is that the self-duality condition involves a metric and it is not clear which metric should be used, e.g. the induced metric or the auxiliary metric; the possibilities differ by higher order terms in F. Because of this ambiguity we should consider the action as determining only the lowest order, quadratic, terms in F. With this proviso, an obvious conjecture for the $D=11$ 5-brane action is

$$S = -\tfrac{1}{2}\int d^6\xi\,\sqrt{-\gamma}\Big[\gamma^{ij}\partial_i X^M\partial_j X^N\eta_{MN} + \tfrac{1}{2}\gamma^{il}\gamma^{jm}\gamma^{kn}F_{ijk}F_{lmn} - 4\Big], \tag{3.2}$$

where the fields X^M, $M=(0,1,\dots,10)$, are maps from the worldvolume to the $D=11$ Minkowski spacetime. This action has an obvious coupling to the bosonic fields (g_{MN}, B_{MNP}) of $D=11$ supergravity[7]. The coupled action is

$$S = -\tfrac{1}{2}\int d^6\xi\,\sqrt{-\gamma}\Big[\gamma^{ij}g^{(11)}_{ij} + \tfrac{1}{2}\gamma^{il}\gamma^{jm}\gamma^{kn}\big(F_{ijk}-B_{ijk}\big)\big(F_{lmn}-B_{lmn}\big) - 4\Big] \tag{3.3}$$

where $g^{(11)}_{ij}$ is the pullback of the 11-metric g_{MN} and B_{ijk} is the pullback of the 3-form potential B_{MNP}. Up to quartic terms in F_{ij}, and setting to zero the RR spacetime fields, the double dimensional reduction of (3.3) to $D=10$ reproduces the action (2.15) with $p=4$, as required for the M-brane interpretation of the Dirichlet 4-brane. In particular, the dilaton dependence is exactly as given in (2.15).

The worldvolume vector of the $D=10$ Dirichlet p-branes allows not only a coupling to the 2-form potential of string theory but also to the endpoints of an open string via a boundary action [21,22,44]. Let $X^m(\sigma,\tau)$ be the locus in spacetime of the string's worldsheet, with boundary at $\tau=0$. If this boundary lies in the worldvolume of a p-brane, then

$$X^m(\sigma,\tau)\Big|_{\tau=0} = X^m(\xi(\sigma)), \tag{3.4}$$

[7] Although consistency with the self-duality condition is now problematic.

where $X^m(\xi)$ is the locus in spacetime of the p-brane's worldvolume. It is also convenient to introduce the conjugate momenta to the worldsheet scalar fields at the worldsheet boundary, π_m, defined by

$$\pi_m(\sigma) = \sqrt{-g}\, g_{mn}\big(X(\sigma,\tau)\big)\, \frac{dX^n(\sigma,\tau)}{d\tau}\bigg|_{\tau=0}. \tag{3.5}$$

The $D = 10$ Lorentz covariant boundary action can then be written as

$$S_b(\text{string}) = \oint d\sigma \left[A_i(\xi(\sigma)) \frac{d\xi^i(\sigma)}{d\sigma} + X^m(\xi(\sigma))\pi_m(\sigma) \right]. \tag{3.6}$$

Similarly, the worldvolume antisymmetric tensor A^+ of the $D = 11$ 5-brane allows not only a coupling to the 3-form potential of $D = 11$ supergravity but also to the boundary of an open membrane. Let $X^M(\sigma, \rho, \tau)$ be the locus in the $D = 11$ spacetime of the membrane's worldvolume, with boundary at $\tau = 0$. If this boundary lies in the worldvolume of a fivebrane, with coordinates ξ^i, then

$$X^M(\sigma,\rho,\tau)\Big|_{\tau=0} = X^M\big(\xi(\sigma,\rho)\big), \tag{3.7}$$

where $X^M(\xi)$ is the locus in spacetime of the fivebrane's worldvolume. Defining, as before, the conjugate momenta π_M to the membrane scalar fields at the membrane's boundary, we can write down the following natural generalization of (3.6):

$$S_b(\text{membrane}) = \oint d\sigma d\rho \left[A^+_{ij}(\xi) \frac{d\xi^i}{d\sigma}\frac{d\xi^j}{d\rho} + X^m(\xi)\pi_M \right]. \tag{3.8}$$

Moreover, the double-dimensional reduction of this membrane boundary action reproduces the string boundary action (3.6). This suggests that we interpret the $D = 11$ 5-brane as a Dirichlet-brane of an underlying open supermembrane. It seems possible that the dynamics of the membrane boundary in the fivebrane's worldvolume might be describable by a six-dimensional superstring theory, which one would expect to have $N = 2$ (i.e. minimal) six-dimensional supersymmetry (e.g. on the grounds that it is a 'brane within a brane' [45]). However, since the 3-form field strength to which this boundary string couples is self-dual, this superstring theory would be, like the supermembrane itself, intrinsically non-perturbative. The existence of such a new superstring theory was conjectured previously [34] in a rather different context.

4. 0-branes from 2-branes and 2-branes from 0-branes.

One of the properties expected of the $D = 11$ supermembrane theory or M-theory is that it have $D = 11$ supergravity as its effective field theory. Various arguments for and against this have been given previously ([23] contains a recent brief review). A further argument in favour of this idea is suggested by the recent results of Witten concerning the effective action of n coincident Dirichlet p-branes [38]. He has shown that the (partially gauge-fixed) effective action in this case is the reduction from $D = 10$ to $(p+1)$ dimensions of the $U(n)$ $D = 10$ super Yang-Mills (YM) theory. Consider the 0-brane case for which the super YM theory is one-dimensional i.e. a model of supersymmetric gauge quantum mechanics (SGQM). If the 0-branes condense at some point then the effective action will be the $n \to \infty$ limit of a $U(n)$ SGQM. But this is just another description of the supermembrane! It is amusing to note that the continuity of the spectrum of the quantum supermembrane [46], in the zero-width approximation appropriate to its D-brane description, might now be understood as a consequence of the zero-force condition between an infinite number of constituent 0-branes. However, it is known that quantum string effects cause the D-brane to acquire a finite size core [47], consistent with its M-brane interpretation as a solution with an event horizon [26], and it was argued in [23] that this fact should cause the spectrum to be discrete.

Actually, the supermembrane was usually stated as being equivalent to an $SU(\infty)$ SGQM model [36,37], but the additional $U(1)$ is needed to describe the dynamics of the centre of mass motion. Note that a $U(1)$ SGQM is precisely the action for a Dirichlet 0-brane. This suggests that there might exist some classical closed membrane configuration for which the ground state, on quantization, could be identified with the 0-brane. For this to be possible it would be necessary for the closed membrane to carry the RR charge associated with the 0-branes. We now explain how this can occur.

From the $D = 11$ point of view the RR 0-brane charge is just the KK charge, i.e. the electric charge that couples to the KK vector field, which we shall here call B_m. The coupling of B_m to the $D = 10$ membrane can be found by dimensional reduction from $D = 11$. To leading order this coupling has the standard Noether form $B_m \mathcal{J}^m$, where

$$\mathcal{J}^m(x) = \int d^3\xi\, \sqrt{-\gamma}\gamma^{ij}\partial_i X^m \partial_j \varphi\, \delta^{10}\big(x - X(\xi)\big)\,. \tag{4.1}$$

is the KK current density. After dualization of the scalar field this becomes

$$\mathcal{J}^m(x) = \int d^3\xi\, \varepsilon^{ijk}\partial_i X^m F_{jk}\, \delta^{10}\big(x - X(\xi)\big)\,. \tag{4.2}$$

The total KK charge is $Q \equiv \int d^9x \mathcal{J}^0$. Choosing the $X^0 = \xi^0$ gauge one readily sees that

$$Q = \oint F\,, \tag{4.3}$$

i.e. the integral of the worldvolume 2-form field strength F over the closed membrane.

Thus, a closed membrane can carry the 0-brane RR charge as a type of magnetic charge associated with its worldvolume vector field, and its centre of mass motion is described by the 0-brane $U(1)$ SQGM. This can be interpreted as further evidence that the 0-brane is included in the (non-perturbative) supermembrane spectrum. However, from the $D = 11$ point of view the 0-brane is just a massless quantum of $D = 11$ supergravity and supersymmetry implies the existence of all massless quanta given any one of them. Thus, we have found a new argument that the spectrum of the $D = 11$ supermembrane (or, perhaps, M-theory) should include the massless states of $D = 11$ supergravity.

Acknowledgements

Helpful conversations with M.B. Green, J. Hoppe, R.G. Leigh, G. Papadopoulos and A. Tseytlin are gratefully acknowledged.

References

[1] C.M. Hull and P.K. Townsend, Nucl. Phys. B 438 (1995) 109.
[2] P.K. Townsend, *p*-Brane Democracy , hep-th/9507048, to appear in proceedings of the PASCOS/Hopkins workshop, March 1995.
[3] E. Witten, Nucl. Phys. B 443 (1995) 85.
[4] C.M. Hull and P.K. Townsend, Nucl. Phys. B 451 (1995) 525.
[5] A. Strominger, Nucl. Phys. B 451 (1995) 96.
[6] P. Aspinwall, Phys. Lett. B 357 (1995) 329.
[7] S. Kachru and C. Vafa, Nucl. Phys. B 450 (1995) 69.
[8] A. Dabholkar, G.W. Gibbons, J.A. Harvey and F. Ruiz-Ruiz, Nucl. Phys. B 340 (1990) 33.
[9] C.G. Callan, J.A. Harvey and A. Strominger, Nucl. Phys. B 359 (1991).
[10] M.J. Duff and J.X. Lu, Nucl. Phys. B 354 (1991) 141.
[11] G.T. Horowitz and A. Strominger, Nucl. Phys. B 360 (1991) 197.
[12] C.G. Callan, J.A. Harvey and A. Strominger, Nucl. Phys. B 367 (1991) 60.
[13] M.J. Duff and J.X. Lu, Phys. Lett. B 273 (1991) 409.
[14] A. Achúcarro, J.M. Evans, P.K. Townsend and D.L. Wiltshire, Phys. Lett. B 198 (1987) 441.
[15] R.I. Nepomechie, Phys. Rev. D 31 (1985) 1921.
[16] G.W. Gibbons, M.B. Green and M.J. Perry, Instantons and Seven-Branes in Type II Superstring Theory , hep-th/9511080.
[17] J. Polchinski and E. Witten, Evidence for heterotic-Type I string duality , hep-th/9510169.
[18] E. Bergshoeff, M. de Roo, M.B. Green, G. Papadopoulos and P.K. Townsend, Duality of Type II 7-branes and 8-branes , hep-th/9601150.
[19] G.W. Gibbons, G.T. Horowitz and P.K. Townsend, Class. Quantum Grav. 12 (1995) 297.
[20] J. Polchinski, Phys. Rev. Lett. 75 (1995) 4724.
[21] J. Dai, R.G. Leigh and J. Polchinski, Mod. Phys. Lett. A 4 (1989) 2073.
[22] R.G. Leigh, Mod. Phys. Lett. A 4 (1989) 2767.
[23] P.K. Townsend, Phys. Lett. B 350 (1995) 184.
[24] M.J. Duff, P.S. Howe, T. Inami and K.S. Stelle, Phys. Lett. B 191 (1987) 70.
[25] M.J. Duff and K.S. Stelle, Phys. Lett. B 253 (1991) 113.
[26] M.J. Duff, G.W. Gibbons and P.K. Townsend, Phys. Lett. B 332 (1994) 321.
[27] R. Güven, Phys. Lett. B 276 (1992) 49.
[28] P.K. Townsend, Phys. Lett. B 354 (1995) 247.
[29] J.M. Izquierdo, N.D. Lambert, G. Papadopoulos and P.K. Townsend, Nucl. Phys. 460 (1996) 560.
[30] E. Bergshoeff, E. Sezgin and P.K. Townsend, Phys. Lett. B 189 (1987) 75; Ann. Phys. (N.Y.) 185 (1988) 330.
[31] P. Hořava and E. Witten, Heterotic and Type I String Dynamics from Eleven Dimensions , hep-th/9510209.
[32] J.H. Schwarz, The Power of M-theory , hep-th/9510086.
[33] C. Schmidhuber, D-brane Actions , hep-th/9601003.
[34] E. Witten, Some Comments on String Dynamics , hep-th/9507012.

[35] M.J. Duff and J.X. Lu, Nucl. Phys. B 416 (1994) 301.
[36] J. Hoppe, Ph.D. Thesis, MIT (1982).
[37] B. de Wit, J. Hoppe and H. Nicolai, Nucl. Phys. B 305 [FS 23] (1988) 545.
[38] E. Witten, Bound states of D-strings and p-branes, hep-th/9510135.
[39] E. Bergshoeff, L.A.J. London and P.K. Townsend, Class. Quantum Grav. 9 (1992) 2545.
[40] P.K. Townsend, Three Lectures on Supermembranes, in Superstrings '88, eds. M. Green, M. Grisaru, R. Iengo, E. Sezgin and A. Strominger, (World Scientific 1989).
[41] G.W. Gibbons and P.K. Townsend, Phys. Rev. Lett. 71 (1983) 3754.
[42] D. Kaplan and J. Michelson, Zero modes for the $D = 11$ membrane and five-brane, hep-th/9510053.
[43] E. Bergshoeff, B. Janssen and T. Ortín, Class. Quantum Grav. 12 (1995) 1.
[44] C.G. Callan and I.R. Klebanov, D-brane Boundary State Dynamics, hep-th/9511173.
[45] M. Douglas, Branes within Branes, hep-th/9512077.
[46] B. de Wit, M. Lüscher and H. Nicolai, Nucl. Phys. B 320 (1989) 135.
[47] C. Bachas, D-brane Dynamics, hep-th/9511043.

Chapter 3

The eleven-dimensional superfivebrane

According to the classification of [1] described in chapter 2, no Type II p-branes with $p > 1$ could exist. Moreover, the only brane allowed in $D = 11$ was $p = 2$. These conclusions were based on the assumption that the only fields propagating on the worldvolume were scalars and spinors, so that, after gauge fixing, they fall only into *scalar* supermultiplets, denoted by S on the brane scan of table 2.1 of chapter 2. Indeed, these were the only kappa symmetric actions known at the time. Using soliton arguments, however, it was pointed out in [2, 3] that both Type IIA and Type IIB superfivebranes exist after all. Moreover, the Type IIB theory also admits a self-dual superthreebrane [4]. The no-go theorem is circumvented because in addition to the superspace coordinates X^M and θ^α there are also higher spin fields on the worldvolume: vectors or antisymmetric tensors. This raised the question: are there other super p-branes and if so, for what p and D? In [5] an attempt was made to answer this question by asking what new points on the brane scan are permitted by bose-fermi matching alone. Given that the gauge-fixed theories display worldvolume supersymmetry, and given that we now wish to include the possibility of vector and antisymmetric tensor fields, it is a relatively straightforward exercise to repeat the bose-fermi matching conditions of chapter 2 for vector and antisymmetric tensor supermultiplets.

Let us begin with vector supermultiplets. Once again, we may proceed in one of two ways. First, given that a worldvolume vector has $(d-2)$ degrees of freedom, the scalar multiplet condition (2.5) gets replaced by

$$D - 2 = \frac{1}{2}\, mn = \frac{1}{4}\, MN. \tag{3.1}$$

Alternatively, we may simply list all the vector supermultiplets in the classification of [6] and once again interpret D via (2.7). The results [5, 7] are shown by the points labelled V in table 2.1.

Next we turn to antisymmetric tensor multiplets. In $d = 6$ there is a supermultiplet with a second rank tensor whose field strength is self-dual: $(B^-_{\mu\nu}, \lambda^I, \phi^{[IJ]})$, $I = 1, \ldots, 4$. This has chiral $d = 6$ supersymmetry. Since there are five scalars, we have $D = 6 + 5 = 11$. There is thus a new point on the scan corresponding to the $D = 11$ superfivebrane. One may decompose this $(n_+, n_-) = (2, 0)$ supermultiplet under $(n_+, n_-) = (1, 0)$ into a tensor multiplet with one scalar and a hypermultiplet with four scalars. Truncating to just the tensor multiplet gives the zero modes of a fivebrane in $D = 6 + 1 = 7$. These two tensor multiplets are shown by the points labelled T in table 2.1.

Two comments are now in order:

1) The number of scalars in a vector supermultiplet is such that, from (2.7), $D = 3, 4, 6$ or 10 only, in accordance with [6].

2) Vector supermultiplets exist for all $d \leq 10$ [6], as may be seen by dimensionally reducing the $(n = 1, d = 10)$ Maxwell supermultiplet. However, in $d = 2$, vectors have no degrees of freedom and, in $d = 3$, vectors have only one degree of freedom and are dual to scalars. In this sense, therefore, these multiplets will already have been included as scalar multiplets in section chapter 2. There is consequently some arbitrariness in whether we count these as new points on the scan and in [5, 7] they were omitted. For example, it was recognized [5] that by dualizing a vector into a scalar on the gauge-fixed $d = 3$ worldvolume of the Type IIA supermembrane, one increases the number of worldvolume scalars, i.e. transverse dimensions, from 7 to 8 and hence obtains the corresponding worldvolume action of the $D = 11$ supermembrane. Thus the $D = 10$ Type IIA theory contains a hidden $D = 11$ Lorentz invariance [5, 8, 9]!

However, the whole subject of Type II supermembranes underwent a major sea change in 1995 when Polchinski [10] realized that Type II super p-branes carrying Ramond–Ramond charges admit the interpretation of *Dirichlet-branes* that had been proposed earlier in 1989 [11]. These D-branes are surfaces of dimension p on which open strings can end. The Dirichlet p-brane is defined by Neumann boundary conditions in $(p + 1)$ directions (the worldvolume itself) and Dirichlet boundary conditions in the remaining $(D - p - 1)$ transverse directions. In $D = 10$, they exist for even $p = 0, 2, 4, 6, 8$ in the Type IIA theory and odd $p = -1, 1, 3, 5, 7, 9$ in the Type IIB theory, in complete correspondence with the points marked V on the brane scan of table 3. The fact that these points preserve one half of the spacetime supersymmetry and are described by dimensionally reducing the $(n = 1, d = 10)$ Maxwell multiplet fits in perfectly with the D-brane picture.

As we have said, the existence of the eleven-dimensional superfivebrane was first established by Gueven [12] who found it as a soliton solution of $D = 11$ supergravity. In fact, he showed that it corresponds to the extreme limit of a *black* fivebrane, i.e. one exhibiting an event horizon. Black p-brane solutions of Type IIA and Type IIB supergravity had previously been found by Horowitz and Strominger [13], and it was subsequently shown that they preserve half the spacetime supersymmetry in the extreme mass=charge limit [5].

In chapter 2 we learned from [14] that the mass per unit area of the membrane $\mathcal{M}_3$ is equal to its tension:

$$\mathcal{M}_3 = T_3. \tag{3.2}$$

This *elementary* solution is a singular solution of the supergravity equations coupled to a supermembrane source and carries a Noether 'electric' charge

$$Q = \frac{1}{\sqrt{2}\kappa_{11}} \int_{S^7} (*K_4 + \frac{1}{2} C_3 \wedge K_4) = \sqrt{2}\kappa_{11} T_3 \tag{3.3}$$

where ${\kappa_{11}}^2$ is the $D = 11$ gravitational constant. Hence the solution saturates the Bogomol'nyi bound $\sqrt{2}\kappa_{11}\mathcal{M}_3 \geq Q$. This is a consequence of the preservation of half the supersymmetries which is also intimately linked with the worldvolume kappa symmetry. In this chapter, we learned from [12] that the mass per unit 5-volume of the fivebrane $\mathcal{M}_6$ is equal to its tension:

$$\mathcal{M}_6 = \tilde{T}_6. \tag{3.4}$$

This *solitonic* solution is a non-singular solution of the source-free equations and carries a topological 'magnetic' charge

$$P = \frac{1}{\sqrt{2}\kappa_{11}} \int_{S^4} K_4 = \sqrt{2}\kappa_{11}\tilde{T}_6. \tag{3.5}$$

Hence the solution saturates the Bogomol'nyi bound $\sqrt{2}\kappa_{11}\mathcal{M}_6 \geq P$. Once again, this is a consequence of the preservation of half the supersymmetries. These electric and magnetic charges obey a Dirac quantization rule [15, 16]

$$QP = 2\pi n \qquad n = \text{integer}. \tag{3.6}$$

Or, in terms of the tensions [17, 18],

$$2{\kappa_{11}}^2 T_3 \tilde{T}_6 = 2\pi n. \tag{3.7}$$

This naturally suggests a $D = 11$ membrane/fivebrane duality. Note that this reduces the three dimensionful parameters T_3, $\tilde{T}_6$ and κ_{11} to two. Moreover, it was then shown [19, 20, 21] that they are not independent: the tension of the singly charged fivebrane is given by

$$\tilde{T}_6 = \frac{1}{2\pi} {T_3}^2. \tag{3.8}$$

It was recognized in 1995 that membrane/fivebrane duality will in general require gravitational Chern–Simons corrections arising from a sigma-model anomaly on the fivebrane worldvolume [19]. This in turn predicts a spacetime correction to the $D = 11$ supergravity action

$$I_{11}(Lorentz) = T_3 \int C_3 \wedge \frac{1}{(2\pi)^4}[-\frac{1}{768}(trR^2) + \frac{1}{192}trR^4]. \tag{3.9}$$

Such a correction was also derived in a somewhat different way in [22, 23]. By the simultaneous dimensional reduction described in chapter 2, it translates into a corresponding prediction for the Type IIA string [24]. In both cases, supersymmetry also requires corrections to the Einstein action quartic in the curvature [25–28], as described in the paper by Green, Gutperle and Vanhove. These permit yet more consistency checks on the assumed dualities between M-theory in eleven dimensions and string theories in ten. Although in the $D = 10$ Type IIA theory they correspond to one string loop effects, it is important to emphasize that in $D = 11$ these corrections are intrinsically M-theoretic with no counterpart in ordinary $D = 11$ supergravity. It is perfectly true that the same invariants appear as one-loop supergravity counterterms [26, 28]. Their coefficient is either cubically divergent or zero according as one uses a regularization scheme with or without a dimensionful parameter. However, in M-theory their coefficient is finite and is proportional to the membrane tension, as befits intrinsically braney effects. Since this tension behaves as a fractional power of Newton's constant $T_3 \sim \kappa^{-2/3}$, these corrections could never be generated in perturbative $D = 11$ supergravity. Finding out what process in M-theory, which when doubly dimensionally reduced on S^1, yields a string one-loop amplitude, may well throw a good deal of light on what M-theory really is!

Having obtained the plane wave, supermembrane, Kaluza–Klein mono- pole and the superfivebrane in eleven dimensions, a bewildering array of other solitonic p-branes may be obtained by vertical and diagonal dimensional reduction [7, 29, 30]. (An important exception is the $D = 10$ Type IIA eightbrane [31] which corresponds to a solution of the *massive* Type IIA supergravity of Romans [32]. There is as yet no satisfactory $D = 11$ origin for either the eightbrane or the massive supergravity.) In particular, when wrapped around $K3$ which admits 19 self-dual and 3 anti-self-dual 2-forms, the $d = 6$ worldvolume fields of the $D = 11$ fivebrane reduce to the $d = 2$ worldsheet fields of the $D = 7$ heterotic string [33, 34]. As a consistency check, one reproduces both the Yang–Mills and Lorentz corrections to the Bianchi identity of the $D = 7$ heterotic string, starting from the Bianchi identity of the $D = 11$ fivebrane [19]. If we replace $K3$ by T^4, we obtain the worldsheet fields of the $D = 7$ Type IIA string. An important consequence of this will be the eleven-dimensional origin of string/string duality described in chapter 6.

A covariant kappa-symmetric action and/or field equations for the superfivebrane was not achieved until after similar actions for the Type II p-branes, spurred on by their interpretation as D-branes, were written down. Three groups were responsible for the covariant M-theory superfivebrane [35–37] while a non-covariant formulation was given by [39]; we have selected [38], which gives a more detailed account of the superembedding approach, as a representative. The superfivebrane was, in fact, already implicit in [40] where super p-branes were derived by embedding the worldvolume superspace into the spacetime superspace. Having obtained the covariant fivebrane action, it was then possible to verify explicitly that it yields the heterotic string action when wrapped around $K3$ [41].

References

[1] Achucarro A, Evans J, Townsend P and Wiltshire D 1987 Super p-branes *Phys. Lett.* B **198** 441

[2] Callan C G, Harvey J A and Strominger A 1991 World sheet approach to heterotic instantons and solitons *Nucl. Phys.* B **359** 611

[3] Callan C G, Harvey J A and Strominger A 1991 Worldbrane actions for string solitons *Nucl. Phys.* B **367** 60

[4] Duff M J and Lu J X 1991 The self-dual Type IIB superthreebrane *Phys. Lett.* B **273** 409

[5] Duff M J and Lu J X 1993 Type II p-branes: the brane scan revisited *Nucl. Phys.* B **390** 276

[6] Strathdee J 1987 Extended Poincaré supersymmetry *Int. J. Mod. Phys.* A **2** 273

[7] Duff M J, Khuri R R and Lu J X 1995 String solitons *Phys. Rep.* **259** 213

[8] Schmidhuber C 1996 D-brane actions *Nucl. Phys.* B **467** 146

[9] Townsend P K 1996 D-branes from M-branes *Phys. Lett.* B **373** 68

[10] Polchinski J 1995 Dirichlet-branes and Ramond–Ramond charges *Phys. Rev. Lett.* **75** 4724

[11] Dai J, Leigh R G and Polchinski J 1989 New connections between string theories *Mod. Phys. Lett.* A **4** 2073

[12] Gueven R 1992 Black p-brane solutions of $D = 11$ supergravity theory *Phys. Lett.* B **276** 49

[13] Horowitz G T and Strominger A 1991 Black strings and p-branes *Nucl. Phys.* B **360** 197

[14] Duff M J and Stelle K 1991 Multimembrane solutions of d = 11 supergravity *Phys. Lett.* B **253** 113

[15] Nepomechie R I 1985 Magnetic monopoles from antisymmetric tensor gauge fields *Phys. Rev.* D **31** 1921

[16] Teitelboim C 1986 Monopoles of higher rank *Phys. Lett.* B **167** 63

[17] Duff M J and Lu J X 1991 Elementary fivebrane solutions of $D = 10$ supergravity *Nucl. Phys.* B **354** 141

[18] Duff M J and Lu J X 1994 Black and super p-branes in diverse dimensions *Nucl. Phys.* B **416** 301

[19] Duff M J, Liu J T and Minasian R 1995 Eleven-dimensional origin of string/string duality: A one loop test *Nucl. Phys.* B **452** 261

[20] Schwarz J H 1995 The power of M-theory *Phys. Lett.* B **360** 13

[21] De Alwis S P 1996 A note on brane tension and M-theory, hep-th/9607011

[22] Witten E 1996 Five-branes and M-theory on an orbifold *Nucl. Phys.* B **463** 383–97

[23] Witten E 1996 Five-brane effective action in M-theory, hep-th/9610234

[24] Vafa C and Witten E 1995 A one-loop test of string duality *Nucl. Phys.* B **447** 261

[25] Green M B and Vanhove P 1997 D-instantons, strings and M-theory, hep-th/9704145

[26] Green M B, Gutperle M and Vanhove P 1997 One loop in eleven dimensions, hep-th/9706175

[27] Antoniadis I, Ferrara S, Minasian R and Narain K 1997 R^4 couplings in M and Type IIA theories on Calabi–Yau spaces, hep-th/9707013

[28] Russo J G and Tseytlin A A 1997 One loop four graviton amplitude in eleven dimensions, hep-th/9707134

[29] Lu H, Pope C N, Sezgin E and Stelle K S 1996 Stainless super p-branes *Phys. Lett.* B **371** 206

[30] Lu H and Pope C N 1996 p-brane solitons in maximal supergravities *Nucl. Phys.* B **465** 127

[31] Bergshoeff E, Green M B, Papadopoulos G and Townsend P K 1995 The IIA supereightbrane, hep-th/9511079

[32] Romans L 1986 Massive $N = 2a$ supergravity in ten dimensions *Phys. Lett.* B **169** 374

[33] Harvey J and Strominger A 1995 The heterotic string is a soliton *Nucl. Phys.* B **449** 535

[34] Townsend P K 1996 String/membrane duality in seven dimensions *Phys. Lett.* B **354** 206

[35] Howe P S and Sezgin E 1997 $D = 11$, $p = 5$ *Phys. Lett.* B **394** 62

[36] Pasti P, Sorokin D and Tonin M 1997 Covariant action for a $D = 11$ five-brane with the chiral field *Phys. Lett.* B **398** 41

[37] Bandos I, Lechner K, Nurmagambetov A, Pasti P, Sorokin D and Tonin M 1997 Covariant action for the super-five-brane of M-theory *Phys. Rev. Lett.* **78** 4332
[38] Howe P S, Sezgin E and West P C 1997 Covariant field equations of the M-theory fivebrane *Phys. Lett.* B **399** 49–59
[39] Agananic M, Park J, Popescu C and Schwarz J H 1997 World-volume action of the M theory five-brane *Nucl. Phys.* B **496** 191
[40] Howe P S and Sezgin E 1997 Superbranes *Phys. Lett.* B **390** 133
[41] Cherkis S and Schwarz J H 1997 Wrapping the M-theory fivebrane on $K3$ *Phys. Lett.* B **403** 225

Reprinted from Phys. Lett. B **276** (1992) 49–55

Black p-brane solutions of $D=11$ supergravity theory

R. Güven

Department of Mathematics, Boğaziçi University, Bebek, TR-80815 Istanbul, Turkey

Received 13 November 1991

Various classes of extended black hole solutions of $D=11$ supergravity theory are presented. It is shown that $D=11$ supergravity admits a class of "electric" black p-brane solutions for $p=2, 4, 6$ and a "magnetic" type of black five-brane solution. Each of these solutions is characterized by a mass and a charge parameter. The only supersymmetric members of these families are the extreme cases where Bogomol'nyi type of bounds are saturated by the parameters. The extreme cases also allow multi-source generalizations. Upon double dimensional reduction these families give rise to two new solutions of Type IIA string theory and one of these is a black five-brane.

The studies of the string theory solitons and black holes have recently revealed certain interesting features. One of these is the fact that ten-dimensional, effective string theories admit black p-brane solutions [1,2]. These solutions have the structure of a p-dimensional extended object surrounded by an event horizon and approach the Minkowski spacetime at spatial infinity. They are characterized by two parameters which may be interpreted as the mass per unit p-volume and the charge of the object. The extreme members of these extended black holes, which are obtained when the mass and the charge saturate a Bogomol'nyi type of bound, are particularly interesting as they contain the fundamental string solution [3,4] and the elementary five-brane [5,6] as special cases. At least one of these extreme cases is expected to have no higher-order corrections in string theory [7] and there are indications that the "branescan" of the known p-brane actions [8–10] can be generalized to include new, supersymmetric extended objects [2,11].

In this paper we wish to study the black p-branes in the framework of $D=11$ supergravity theory. Whereas strings are described by ten-dimensional actions, supermembranes couple to $D=11$ supergravity theory [9] and it is known that $D=11$ supergravity has fundamental multi-membrane solutions [12]. Upon double dimensional reduction [13] these membranes go over to the string solution of refs. [3,4]. Since the inverse route, the double dimensional oxidation [14], can be used to associate with each Type IIA string theory solution a solution of $D=11$ supergravity, one may expect that $D=11$ supergravity admits black membrane solutions whose extremal member is the fundamental membrane. Although double dimensional oxidation need not respect the singularity structure and the asymptotic behavior that is appropriate for a black hole, we shall see that this expectation is indeed fulfilled. Moreover, we shall see that $D=11$ supergravity possesses a class of "electric" black p-branes for $p=2, 4, 6$ and a "magnetic" type of black five-brane and these are all characterized by the charge and the mass parameters. The only supersymmetric members of these solutions turn out to be the extreme cases where Bogomol'nyi type of bounds are saturated by the parameters. The extreme cases also permit multi-source generalizations. Under double dimensional reduction, these families give rise to two new solutions of Type IIA string theory and one of these is a black five-brane.

After neglecting the fermionic degrees of freedom, the field variables of $D=11$ supergravity may be taken to be the metric $\hat{g}_{MN}$ and the four-form field $\hat{F}_{KLMN}$ and these are governed by the field equations [15]

$$\hat{R}_{MN}=\tfrac{1}{3}(\hat{F}^{KLP}{}_{M}\hat{F}_{KLPN}-\tfrac{1}{12}\hat{g}_{MN}\hat{F}_{KLPQ}\hat{F}^{KLPQ}), \tag{1}$$

$$\mathrm{d}\hat{F}=0, \quad \mathrm{d}\hat{*}\hat{F}+\tfrac{1}{6}\hat{F}\wedge\hat{F}=0, \tag{2}$$

0370-2693/92/$ 05.00

where $\hat{R}_{MN}$ is the Ricci tensor, $\hat{F}=\frac{1}{4}\hat{F}_{KLMN}\,dX^K\wedge dX^L\wedge dX^M\wedge dX^N$ and $\hat{*}$ denotes the $D=11$ Hodge dual. We use the conventions where the signature of $\hat{g}_{MN}$ is -9 and the Levi-Civita tensor $\hat{\epsilon}_{ABC...JK}$ has the component $\hat{\epsilon}_{012345678910}=1$ in the tangent space. The Ricci tensor is defined as $\hat{R}_{MN}=\hat{R}^K{}_{MNK}$, where $\hat{R}_{KLMN}$ is the Riemann tensor. The hats are used to distinguish the $D=11$ objects from the lower-dimensional objects that we shall encounter.

For black p-branes the $D=11$ metric is taken to be of the warped product form

$$d\hat{s}^2=ds^2-e^{2B}\delta_{ij}\,dx^i\,dx^j\,, \tag{3}$$

where $i, j=1, 2, ..., p$ and the $(11-p)$-dimensional metric

$$ds^2=e^{2\phi}\,dt^2-e^{2\psi-2\phi}\,dr^2-R^2\,d\Omega^2_{9-p} \tag{4}$$

is itself a warped product of a two-dimensional metric and the metric $d\Omega^2_{9-p}$ of a $(9-p)$-dimensional unit sphere S^{9-p}. Here the metric functions ϕ, ψ, R, B are assumed to depend only on the radial coordinate r. In order to obtain the $D=11$ Minkowski spacetime as a limit, one imposes the boundary conditions $\phi\to 0$, $\psi\to 0$, $B\to 0$ and $R\to r$ as $r\to\infty$. Even under more general conditions, the ADM mass of such solutions may be defined as [16]

$$M=3\oint_{\partial\Sigma}(\bar{D}_b g_{ac}-\bar{D}_a g_{bc})\,\bar{g}^{ac}\,d\sigma^b\,, \tag{5}$$

where g_{ab} is the $D=10$ spatial metric on the $t=\text{const.}$ hypersurfaces Σ; $\bar{g}_{ab}$ is the reference spatial metric obtained from the limiting spacetime and $\bar{D}_a$ denotes the covariant differentiation with respect to $\bar{g}_{ab}$. The integration in (5) is to be performed at the $r\to\infty$ boundary $\partial\Sigma$ of Σ with the surface element $d\sigma^b$.

In the solutions of interest a second conserved quantity is to be furnished by the four-form field $\hat{F}$. The standard conserved quantity associated with $\hat{F}$ is the Page charge [17]

$$Q_e=\int(\hat{*}\hat{F}+\tfrac{1}{6}\hat{A}\wedge\hat{F})\,, \tag{6}$$

where $\hat{A}$ is the potential three-form: $\hat{F}=d\hat{A}$. In analogy with the $D=4$ Einstein–Maxwell theory, (6) can be viewed as an "electric" type of charge. Non-zero "magnetic" charges

$$Q_m=\int\hat{F}\,, \tag{7}$$

on the other hand, can occur in the cases where $\hat{F}$ is closed but not globally exact. Clearly, there is an asymmetry in the definition of these two charges and this reflects the fact that a dual formulation of $D=11$ supergravity theory is not available [18]. Due to this property and in contrast to the string theory examples, magnetic solutions of the $D=11$ supergravity cannot be obtained from the electric ones by duality rotations.

$\hat{F}$ can easily anchor a non-zero Q_e or Q_m on the metrics of the form (3) when one notes that the last Betti numbers of S^{9-p} are non-zero for all $p<9$. Let ϵ_{9-p} be the volume $(9-p)$-form of S^{9-p} and let $V_{9-p}=\int\epsilon_{9-p}$. Since $\hat{F}$ is a four-form, $\hat{*}\hat{F}$ is a seven-form and it is natural to try either

$$\hat{*}\hat{F}=q_e\epsilon_7 \tag{8}$$

or

$$\hat{F}=q_m\epsilon_4\,, \tag{9}$$

where q_e, q_m are constants. In each of these two cases the four-form field equations (2) are satisfied without putting any restrictions on the metric functions. If the metric can be fixed by solving the Einstein equations (1), (8) and (9) will correspond to the solutions that have the respective charges $Q_e=q_eV_7$ and $Q_m=q_mV_4$. Clearly, (8) implies that $p=2$, which is the case of membranes, whereas (9) requires the extended object to be a five-brane.

In order to take into account other possible values of p, one should, of course, assign more general forms to $\hat{F}$. For the electric case this can be achieved by letting

$$\hat{*}\hat{F}=q_e\epsilon_{9-p}\wedge\bar{*}J\,, \tag{10}$$

where $J=\frac{1}{2}J_{jk}\,dx^j\wedge dx^k$ is a certain two-form on the p-dimensional euclidean space E^p that appears as one of the factors in the warped product (3). The symbol $\bar{*}$ denotes the Hodge dual on E^p. Taking the $D=11$ dual of (10) shows that $\hat{F}$ is always of the form $\hat{F}=(*F)\wedge J$ and satisfies $\hat{F}\wedge\hat{F}=0$. Here $F=q_e\epsilon_{9-p}$ and $*F$ is a closed two-form on the $(11-p)$-dimensional subspace which has the metric (4) and the dual $*$. Due to these properties, the field equations (2) reduce to

$$\mathrm{d}J=0\,,\quad \mathrm{d}{*}J=0\,, \tag{11}$$

according to which J can be any two-form that is both closed and co-closed on E^p. Formally, (10) applies for every p in the range $0\leqslant p\leqslant 9$ but $Q_e=0$ unless $p\geqslant 2$. Since the black hole metrics cannot have a non-zero spherical section for $p>7$, the really interesting interval is $2\leqslant p\leqslant 7$. At $p=2$, ${*}J=1$ and (10) reduces to (8). When $p>2$, it follows from

$$Q_e=q_e V_{9-p}\int {*}J \tag{12}$$

that $q_e V_{9-p}$ can be interpreted as the charge per unit $(p-2)$-volume of the extended object.

Considering next the Einstein equations (1), one finds that not every p in the range $2\leqslant p\leqslant 7$ is admissible; (3) and (10) are compatible only if p is even and J is the Kähler form on E^p:

$$J_{jl}J^{lk}=-\delta_j^k\,. \tag{13}$$

When these conditions are met, the Einstein equations become

$$(R^{9-p}\,e^{2\phi-\psi+pB}\phi')'$$
$$=p\cdot\tfrac{1}{54}q_e^2R^{p-9}\,e^{\psi+(4-p)B}\,, \tag{14}$$

$$(R^{9-p}\,e^{2\phi-\psi+pB}\,B')'$$
$$=(6-p)\cdot\tfrac{1}{108}q_e^2R^{p-9}\,e^{\psi+(4-p)B}\,, \tag{15}$$

$$(R^{8-p}\,e^{2\phi-\psi+pB}\,R')'+(p-8)R^{7-p}\,e^{\psi+pB}$$
$$=-p\cdot\tfrac{1}{108}q_e^2R^{p-9}\,e^{\psi+(4-p)B}\,, \tag{16}$$

$$p\,e^{-B}(e^{B-\psi}B')'+(9-p)R^{-1}(e^{-\psi}R')'$$
$$=0\,, \tag{17}$$

where prime denotes differentiation with respect to r and p takes on the values 2, 4 and 6.

Similarly, one may seek a generalization of the magnetic case by incorporating the lower-dimensional spheres into (9). Suppose W_{p-5} is a $(p-5)$-form which is closed and co-closed on E^p for $5\leqslant p\leqslant 7$. Then it can be checked that

$$\hat{F}=q_m\,\epsilon_{9-p}\wedge W_{p-5}\,, \tag{18}$$

satisfies the four-form field equations (2) for each p in this range. The Einstein equations, however, turn out to be much more restrictive and the only consistent case occurs at $p=5$. This brings us back to (9). For the five-branes the Einstein equations require that

$$(R^4\,e^{2\phi-\psi+5B}\,\phi')'=\tfrac{1}{54}q_m^2R^{-4}\,e^{\psi+5B}\,, \tag{19}$$

$$(R^4\,e^{2\phi-\psi+5B}\,B')'=\tfrac{1}{54}q_m^2R^{-4}\,e^{\psi+5B}\,, \tag{20}$$

$$(R^3\,e^{2\phi-\psi+5B}\,R')'-3R^2\,e^{\psi+5B}$$
$$=-\tfrac{1}{27}q_m^2R^{-4}\,e^{\psi+5B}\,, \tag{21}$$

$$5\,e^{-B}(e^{B-\psi}B')'+4R^{-1}(e^{-\psi}R')'=0\,. \tag{22}$$

These equations reduce to the $p=5$ case of (14)–(17) when $q_e=q_m=0$ but if the source terms are present, this correspondence clearly breaks down. The equations for $p=2$ and $p=5$ exhibit a duality in the sense that these are the only two cases which allow solutions obeying $\phi=B$ for non-zero q_e and q_m.

To display the black p-brane solutions of these equations in a convenient form let us define, for each p, the radial functions

$$\Delta_\pm=1-(r_\pm/r)^{8-p}\,, \tag{23}$$

where r_+, r_- are constants and $r_+>r_-$. Then the $D=11$ metric

$$\mathrm{d}\hat{s}^2=\Delta_+\Delta_-^{(p-3)/3}\,\mathrm{d}t^2$$
$$-\Delta_-^{(p-2)(p-6)/6(8-p)}[\Delta_+^{-1}\Delta_-^{-1}\,\mathrm{d}r^2+r^2\,\mathrm{d}\Omega_{9-p}^2]$$
$$-\Delta_-^{(6-p)/6}\,(\mathrm{d}x_1^2+\ldots+\mathrm{d}x_p^2)\,, \tag{24}$$

together with the dual of (10),

$$\hat{F}=\frac{q_e}{r^{9-p}}\,\mathrm{d}t\wedge\mathrm{d}r\wedge J\,, \tag{25}$$

constitute the solution for the electric p-branes if $p=2$, 4, 6 and the constants satisfy

$$r_+^{8-p}r_-^{8-p}=\left(\frac{q_e}{3(8-p)}\right)^2. \tag{26}$$

The magnetic five-brane solution, on the other hand, is obtained by choosing $p=5$ in (23) and letting

$$\mathrm{d}\hat{s}^2=\Delta_+\Delta_-^{-2/3}\,\mathrm{d}t^2-\Delta_+^{-1}\Delta_-^{-1}\,\mathrm{d}r^2-r^2\,\mathrm{d}\Omega_4^2$$
$$-\Delta_-^{1/3}\,(\mathrm{d}x_1^2+\ldots+\mathrm{d}x_5^2)\,. \tag{27}$$

In this case $\hat{F}$ is given by (9) and the constants obey

$$r_+^3r_-^3=(\tfrac{1}{9}q_m)^2\,. \tag{28}$$

When the ADM masses of these solutions are calculated according to (5), one finds that $r_+>r_-$ is sufficient to guarantee the positivity of the mass for all even p. When $p=5$ one must also require $r_+>0$ in order to get a positive mass in the $r_-=0$ limit. Let μV_{9-p} be the mass per unit p-volume of the object: $\mu V_{9-p}=M/v_p$ where v_p is the p-volume. Then for the electric p-branes

$$\mu=21r_+^6-3r_-^6\,, \quad p=2\,, \tag{29}$$

$$\mu=15r_+^4-11r_-^4\,, \quad p=4\,, \tag{30}$$

$$\mu=9(r_+^2+r_-^2)\,, \quad p=6\,, \tag{31}$$

and for the magnetic five-brane solution

$$\mu=12r_+^3-3r_-^3\,. \tag{32}$$

Since (26) and (28) also hold, it follows that the following Bogomol'nyi type of inequalities are satisfied by the mass and the charge parameters:

$$\mu^2\geqslant q_e^2\,, \quad p=2\,, \tag{33}$$

$$\mu^2\geqslant(\tfrac{1}{3}q_e)^2\,, \quad p=4\,, \tag{34}$$

$$\mu^2\geqslant(3q_e)^2\,, \quad p=6\,, \tag{35}$$

$$\mu^2\geqslant q_m^2\,, \quad p=5\,. \tag{36}$$

When $r_+=r_-$ these become strict equalities and one obtains the extreme solutions.

In the generic case, $r_+>r_-$, each of the solutions considered so far has the structure of an extended black hole whose event horizon is located at $r=r_+$. The singularity at $r=r_+$, which is manifest in (24) and (27), is just a coordinate singularity; each solution can easily be extended over this surface. Hence $r=r_+$ is always a regular event horizon. In the generic case another surface of interest occurs at $r=r_-$ and depending on p, this is either an inner horizon or a singular surface which is hidden behind the horizon. For the black six-branes, the spacetime is the global product of the $D=5$ Reissner–Nordström manifold with E^6 and $r=r_-$ is a regular inner horizon. For other black p-branes $r=r_-$ is a singular surface that lies behind the horizon. In the extreme cases, these two surfaces coalesce and become regular event horizons unless $p=4$. The extreme four-brane metric appears to be singular at $r=r_-$.

Under a simultaneous dimensional reduction of the spacetime and the world-volumes these solutions go over to the black $(p-1)$-brane solutions of the Type IIA superstring theory. After labelling the $D=11$ coordinates as $X^M=(X^{M'}, x^p)$, the reduction can be achieved through a ten–one split of the fields. The $D=11$ metric then gives rise to a $D=10$ metric $g_{M'N'}$ and a dilaton field φ according to the relations [13]

$$g_{M'N'}=e^{2\varphi/3}\,\hat{g}_{M'N'}\,, \quad e^{4\varphi/3}=-\hat{g}_{pp}\,. \tag{37}$$

The potential three-form $\hat{A}$, on the other hand, reduces to two distinct $D=10$ potentials

$$B_{L'M'}=\hat{A}_{L'M'p}\,, \quad A_{L'M'N'}=\hat{A}_{L'M'N'}\,. \tag{38}$$

By choosing an appropriate Kähler form on E^p one can check that the $p=2$ case of (24) reduces to a black string solution of ref. [2]. For the remaining two cases of (24) both $B_{L'M'}$ and $\hat{A}_{L'M'N'}$ turn out to be non-trivial and consequently, these give rise to new $D=10$ solutions. The reduction of (27) is a known black four-brane solution of the Type IIA string theory [2].

Let us next consider the supersymmetry of these solutions. In order to maintain supersymmetry in the bosonic sector of $D=11$ supergravity, the spacetime must admit Killing spinors which satisfy

$$\hat{D}_M\hat{\epsilon}+\tfrac{1}{144}\mathrm{i}(\hat{\Gamma}^{NOPQ}{}_M-8\hat{\Gamma}^{OPQ}\delta^N_M)\hat{F}_{NOPQ}\hat{\epsilon}=0\,, \tag{39}$$

where $\hat{\epsilon}$ is a Majorana spinor, $\hat{D}_M$ is the spinor covariant derivative and $\hat{\Gamma}_{OPQ}$, $\hat{\Gamma}_{NOPQ}$ denote the antisymmetrized products of the $D=11$ Dirac matrices. On the above black p-brane backgrounds we have verified that (39) is integrable only if $r_+=r_-$. Therefore, only the extreme black p-brane solutions are supersymmetric.

The extreme solutions not only admit Killing spinors but also allow generalizations to black multi-p-brane configurations. In constructing these generalizations it is convenient to introduce, in addition to E^p, a second euclidean space E^{10-p} of dimension $10-p$. Let $\boldsymbol{y}=(y^\alpha)$, $\alpha=1, 2, ..., 10-p$, be the cartesian coordinates on E^{10-p}. Similarly, let the coordinates x^i of E^p be represented as $\boldsymbol{x}$ and suppose dots denote the standard euclidean inner products. Consider, in this notation, the fields

$$\mathrm{d}\hat{s}^2=U^{-p/3}\,\mathrm{d}t^2 - U^{p/6}\,\mathrm{d}\boldsymbol{y}\cdot\mathrm{d}\boldsymbol{y}-U^{(p-6)/6}\,\mathrm{d}\boldsymbol{x}\cdot\mathrm{d}\boldsymbol{x}\,, \tag{40}$$

$$\hat{F}=3U^{-2}\,\mathrm{d}t\wedge\mathrm{d}U\wedge J\,, \tag{41}$$

where $U=U(y^\alpha)$ is a smooth, positive function on E^{10-p} and p takes again the values $p=2, 4, 6$. If (40) and (41) are substituted in (1) and (2), all field equations reduce to

$$\Delta U = 0 , \tag{42}$$

where Δ is the Laplace operator on E^{10-p}. Hence each solution of the Laplace equation (42) gives rise to a solution of the $D=11$ supergravity theory. For the present problem the relevant solution is

$$U = 1 + \sum_{l=1}^{K} a_l R_l^{p-8} , \tag{43}$$

where $R_l = |\mathbf{y} - \mathbf{y}_l|$ and a_l, $\mathbf{y}_l$ are constants. With this choice (40) and (41) describe the static equilibrium of K extreme black p-branes ($p=2, 6$) which are positioned at $\mathbf{y}=\mathbf{y}_l$ in the background space E^{10-p}. The mass and the electric charge of the lth hole of this configuration are both given in terms of a_l. Masses are positive if $a_l > 0$. In the particular case $p=2$, (40), (41) and (43) reduce to the multi-membrane solutions of ref. [12]. All the extreme cases of (24) are obtained by setting $K=1$ in (43) and changing the radial coordinate as $R_1^{8-p} = r^{8-p} - r_+^{8-p}$.

For these electric multi-p-branes let us choose the orthonormal basis one-forms as

$$\hat{w}^0 = U^{-p/6}\, dt , \tag{44}$$

$$\hat{w}^\alpha = U^{p/12}\, dy^\alpha , \tag{45}$$

$$\hat{w}^j = U^{(p-6)/12}\, dx^j . \tag{46}$$

In this frame the Killing spinor equation can be readily integrated. Suppose $\hat{\eta}$ is a $D=11$ Majorana spinor which has constant entries in the above frame. In the $U=1$, Minkowski spacetime limit $\hat{\eta}$ is clearly a Killing spinor. When $\partial U/\partial y^\alpha \neq 0$ one can check that

$$\hat{\epsilon} = U^{-p/12} \hat{\eta} , \tag{47}$$

where $\hat{\eta}$, subject to p algebraic conditions

$$\hat{\Gamma}_j \hat{\eta} = i \hat{\Gamma}_0 J_{jk} \hat{\Gamma}^k \hat{\eta} , \tag{48}$$

is the general form of the Killing spinor in the electric case. The presence of the algebraic conditions means that the supersymmetry of the Minkowski spacetime is partially broken in the above backgrounds. In a Majorana representation of the Dirac matrices (48) annihilates the 16, 24 and 28 of the 32 independent, real entries of $\hat{\eta}$ when p takes on the respective values $p=2$, 4 and 6. Hence, in this sense, only the $\frac{1}{2}$, $\frac{1}{4}$ and $\frac{1}{8}$ of the Minkowski supersymmetry survives for $p=2$, 4, 6.

The solution which describes the equilibrium configuration of K magnetic, black five-branes can be written as

$$d\hat{s}^2 = U^{-1/3}(dt^2 - d\mathbf{x}\cdot d\mathbf{x}) - U^{2/3}\, d\mathbf{y}\cdot d\mathbf{y} , \tag{49}$$

$$\hat{F} = \mp 3 (\tilde{*} dU) , \tag{50}$$

where U is the $p=5$ case of (43) and $\tilde{*}$ is the Hodge dual on $E^5(y^\alpha)$. In this solution a_l can be interpreted as the magnetic charge parameters of the black holes and the $K=1$ specialization gives the extreme member of (27). If one refers the spinors to the frame

$$\hat{w}^0 = U^{i m 1/6}\, dt , \tag{51}$$

$$\hat{w}^j = U^{-1/6}\, dx^j , \tag{52}$$

$$\hat{w}^\alpha = U^{1/3}\, dy^\alpha , \tag{53}$$

then the Killing spinors of the magnetic multi-holes have the form

$$\hat{\epsilon} = U^{-1/12} \hat{\eta} . \tag{54}$$

Once again, the constant spinor $\hat{\eta}$ turns out to be not completely arbitrary unless U is constant. Assuming that $\partial U/\partial y^\alpha \neq 0$, (39) implies

$$\hat{\eta} = \pm i \hat{\Gamma}_1 \hat{\Gamma}_2 \hat{\Gamma}_3 \hat{\Gamma}_4 \hat{\Gamma}_5 \hat{\eta} , \tag{55}$$

where $\hat{\Gamma}_1 ... \hat{\Gamma}_5$ refer to the $E^5(y^\alpha)$ part of the frame that is given by (53). As a consequence of (55) half of the entries of $\hat{\eta}$ vanish in the Majorana representation. The number of the surviving supersymmetry parameters is therefore the same for the extreme membranes and the five-branes. Another property shared by the $p=2$ and $p=5$ extreme solutions is the presence of an enlarged symmetry group which includes the boosts in the x^j directions. Notice that, unless $\phi = B$, the metric (3) gives rise to solutions that are invariant under $\mathbb{R} \times SO(10-p) \times E(p)$ where $E(p)$ is the p-dimensional euclidean group. When $\phi = B$ one gets precisely the $p=2$ and $p=5$ extreme solutions and the symmetry group extends to $SO(10-p) \times P(p)$ where $P(p)$ is the p-dimensional Poincaré group. The $p=2$ and $p=5$ multi-hole metrics are also invariant under $SO(10-p) \times P(p)$.

It is well known that (43) satisfies (42) everywhere except at $\mathbf{y}=\mathbf{y}_l$ where one encounters delta function singularities. The $\mathbf{y}=\mathbf{y}_l$ singularities are, however, of no consequence to the field equations and the spacetime geometry unless $p=4$. If one uses (44)–(46) as a basis, the orthonormal frame components of (1) and (2) all reduce to $U^{-(p+6)/6}\,\Delta U=0$. When one refers (1) and (2) to the basis (51)–(53), the field equations for the magnetic case become $U^{-5/3}\,\Delta U=0$. In each of these bases the field equations are satisfied at $y=y_l$. The chosen vielbeins are, of course, singular at $y=y_l$ but, provided $p\neq 4$, this can also be remedied. The important point is that the coordinates used in (40) and (49) break down when U encounters a zero or a singularity. The solutions, however, can be extended smoothly over $y=y_l$ by changing the coordinates. One can check in this way that $y=y_l$ are regular event horizons of the black holes if $p=2, 5, 6$. The exceptional case is $p=4$ where the singularities at $y=y_l$ deserve further study.

Let us return to the single black hole configurations and consider the uncharged solutions. It is obvious that (24) and (27) both reduce to two distinct families of vacuum solutions when $q_e=q_m=0$. The families obtained by setting $r_+=0$ have negative masses and do not describe black holes. The $r_-=0$ families are the global products of the $(11-p)$-dimensional Schwarzschild solutions with E^p. In the black p-branes with a Schwarzschild factor p need not be restricted to the above values; for each p in the range $0\leqslant p\leqslant 7$ there is a valid vacuum solution. There is in fact a third family of vacuum solutions where the metric is

$$\mathrm{d}\hat{s}^2=\Delta_-^{(1-p)/(1+p)}\,\mathrm{d}t^2-\Delta_-^{-1}\,\mathrm{d}r^2-r^2\,\mathrm{d}\Omega_{9-p}^2 -\Delta_-^{2/(p+1)}\,\mathrm{d}\boldsymbol{x}\cdot\mathrm{d}\boldsymbol{x}\,, \tag{56}$$

and $0\leqslant p\leqslant 9$. At $p=2$ the third family coalesces with the $r_+=0$ specialization of (24) and $p=0$ corresponds to the $D=11$ Schwarzschild solution. Assuming that $r_->0$, the metric (56) gives rise to a positive mass only if $p=0, 1, 7$. The $p=1$ member of this family has another interesting aspect: the spacetime is the product of the $D=10$ euclidean Schwarzschild manifold with the real line and the string extends over the euclidean Schwarzschild time. None of the vacuum solutions is supersymmetric.

So far we have examined the solutions which approach the Minkowski spacetime at spatial infinity. Let us finally note that there are charged black hole solutions of $D=11$ supergravity theory which display a different asymptotic behavior but share certain interesting features with the above families. The $D=4$ Einstein–Maxwell theory can be consistently embedded in $D=11$ supergravity [19] and remarkably, the $D=4$ Reissner–Nordström family gives rise to a class of $D=11$ black hole solutions through this embedding. The resulting spacetimes are of the form $M_5\times T^6$ where M_5 is a U(1) bundle over the $D=4$ spacetime and T^6 is the flat six-torus. The T^6 factors of these solutions can be replaced with E^6 without any penalty. It was described in ref. [20] how the bosonic sector of the $D=11$ supermembrane theory picks, among these black holes, only the supersymmetric, extreme Reissner–Nordström solution. Precisely the same situation is encountered also in the above black membrane family. Let us take the $p=2$ case of (24) and (25) as supermembrane background fields and label the membrane coordinates as (τ, ρ, σ). Then it can be verified that

$$t=\tau\,,\quad x_1=\rho\,,\quad x_2=\sigma\,,$$
$$X^M=X_0^M\quad\text{for}\quad X^M\neq t, x_1, x_2\,, \tag{57}$$

where X_0^M are constants, is a solution of the supermembrane field equations of ref. [9] only if $r_+=r_-$. The supermembrane theory therefore picks once again the extremal member. The $D=3$ metric induced on the membrane, in the extreme case, is

$$\mathrm{d}s_3^2=\Delta_+^{2/3}(\mathrm{d}\tau^2-\mathrm{d}\rho^2-\mathrm{d}\sigma^2) \tag{58}$$

and while the spacetime geometry is regular, the membrane metric is singular at the horizon. This supermembrane solution is known to have the fermionic κ-symmetry [12].

Another solution of the supermembrane field equations which exists only on the generic $p=2$ cases of (24), (25) and which resides only in the region $r_-<r<r_+$ can also be constructed easily but this time the κ-symmetry is not available.

I thank Professor Abdus Salam, The International Atomic Energy Agency and UNESCO for hospitality at the International Centre for Theoretical Physics, Trieste. I thank Metin Gürses and Ergin Sezgin for useful discussions. The research reported in this paper has been supported in part by the Turkish Scientific and Technical Research Council.

References

[1] G.W. Gibbons and K. Meada, Nucl. Phys. B 298 (1988) 741.

[2] G.T. Horowitz and A. Strominger, Nucl. Phys. B 360 (1991) 197.

[3] A. Dabholkar and J.A. Harvey, Phys. Rev. Lett. 63 (1989) 719.

[4] A. Dabholkar, G. Gibbons, J.A. Harvey and F. Ruiz Ruiz, Nucl. Phys. B 340 (1990) 33.

[5] C.G. Callan, J.A. Harvey and A. Strominger, Nucl. Phys. B 359 (1991) 611.

[6] M.J. Duff and J.X. Lu, Nucl. Phys. B 354 (1991) 141.

[7] C.G. Callan, J.A. Harvey and A. Strominger, Worldsheet approach to heterotic solitons and instantons, preprint EFI-91-03, PUPT-1233, UCSB-TH-91-74 (1991).

[8] J. Hughes, J. Liu and J. Polchinski, Phys. Lett. B 180 (1986) 370.

[9] E. Bergshoeff, E. Sezgin and P.K. Townsend, Phys. Lett. B 189 (1987) 75.

[10] A. Achucarro, J.M. Evans, P.K. Townsend and D.L. Wiltshire, Phys. Lett. B 198 (1987) 441.

[11] C.G. Callan, J.A. Harvey and A. Strominger, Worldbrane actions for string solitons, preprint EFI-91-12 (1991).

[12] M.J. Duff and K.S. Stelle, Phys. Lett. B 253 (1991) 113.

[13] M.J. Duff, P.S. Howe, T. Inami and K.S. Stelle, Phys. Lett. B 191 (1987) 70.

[14] P.K. Townsend, Phys. Lett. B 202 (1988) 53.

[15] E. Cremmer, B. Julia and J. Scherk, Phys. Lett. B 76 (1978) 409.

[16] See e.g. L. Bombelli, R.K. Koul, G. Kunstatter, J. Lee and R. Sorkin, Nucl. Phys. B 289 (1987) 735.

[17] D. Page, Phys. Rev. D 28 (1983) 2976.

[18] H. Nicolai, P.K. Townsend and P. van Nieuwenhuizen, Lett. Nuovo Cimento 30 (1981) 315.

[19] E. Cremmer and B. Julia, Nucl. Phys. B 159 (1979) 141; C.N. Pope, Class. Quantum Grav. 2 (1985) L77.

[20] R. Güven, Phys. Lett. B 212 (1988) 277; in: Supermembranes and physics in 2 + 1 dimensions, eds. M.J. Duff, C.N. Pope and E. Sezgin (World Scientific, Singapore, 1990).

Reprinted from Nucl. Phys. B **452** (1995) 261–82

Eleven-dimensional origin of string/string duality: a one-loop test ★

M.J. Duff, James T. Liu, R. Minasian

Center for Theoretical Physics, Department of Physics, Texas A&M University, College Station, TX 77843-4242, USA

Received 22 June 1995; accepted 11 July 1995

Abstract

Membrane/fivebrane duality in $D = 11$ implies Type IIA string/Type IIA fivebrane duality in $D = 10$, which in turn implies Type IIA string/heterotic string duality in $D = 6$. To test the conjecture, we reproduce the corrections to the 3-form field equations of the $D = 10$ Type IIA string (a mixture of tree-level and one-loop effects) starting from the Chern–Simons corrections to the 7-form Bianchi identities of the $D = 11$ fivebrane (a purely tree-level effect). K3 compactification of the latter then yields the familiar gauge and Lorentz Chern–Simons corrections to 3-form Bianchi identities of the heterotic string. We note that the absence of a dilaton in the $D = 11$ theory allows us to fix both the gravitational constant and the fivebrane tension in terms of the membrane tension. We also comment on an apparent conflict between fundamental and solitonic heterotic strings and on the puzzle of a fivebrane origin of S-duality.

1. Introduction

With the arrival of the 1984 superstring revolution [1], eleven-dimensional Kaluza–Klein supergravity [2] fell out of favor, where it more or less remained until the recent observation by Witten [3] that $D = 11$ supergravity corresponds to the strong coupling limit of the $D = 10$ Type IIA superstring, coupled with the realization that there is a web of interconnections between Type IIA and all the other known superstrings: Type IIB, heterotic $E_8 \times E_8$, heterotic $SO(32)$ and open $SO(32)$. In particular, string/string duality [4–10] implies that the $D = 10$ heterotic string compactified to $D = 6$ on T^4 is dual to the $D = 10$ Type IIA string compactified to $D = 6$ on K3 [11]. Moreover, this

★ Research supported in part by NSF Grant PHY-9411543.

SSDI 0550-3213(95)00368-1

automatically accounts for the conjectured strong/weak coupling S-duality in $D = 4$, $N = 4$ supersymmetric theories, since S-duality for one string is just target-space T-duality for the other [8]. In this paper we find further evidence for an eleven-dimensional origin of string/string duality and hence for S-duality.

$D = 10$ string/fivebrane duality and $D = 6$ string/string duality can interchange the roles of space-time and worldsheet loop expansions [4]. For example, tree-level Chern–Simons corrections to the Bianchi identities in one theory may become one-loop Green–Schwarz corrections to the field equations in the other. In a series of papers [4,7,12–17], it has been argued that this provides a useful way of putting various duality conjectures to the test. In particular, we can compare quantum space-time effects in string theory with the σ-model anomalies for the dual p-branes [18–22] even though we do not yet know how to quantize the p-branes! This is the method we shall employ in the present paper. We reproduce the corrections to the 3-form field equations of the $D = 10$ Type IIA string (a mixture of tree-level and one-loop effects) starting from the Chern–Simons corrections to the 7-form $\tilde{K}_7 = *K_4$ Bianchi identities of the $D = 11$ fivebrane (a purely tree-level effect):

$$d\tilde{K}_7 = -\tfrac{1}{2}{K_4}^2 + (2\pi)^4\tilde{\beta}'\tilde{X}_8 \,, \tag{1.1}$$

where the fivebrane tension is given by $\tilde{T}_6 = 1/(2\pi)^3\tilde{\beta}'$ and where the 8-form polynomial $\tilde{X}_8$ describes the $d = 6$ σ-model Lorentz anomaly of the $D = 11$ fivebrane:

$$\tilde{X}_8 = \frac{1}{(2\pi)^4}\left[-\frac{1}{768}(\mathrm{tr}R^2)^2 + \frac{1}{192}\mathrm{tr}R^4\right] . \tag{1.2}$$

K3 compactification of (1.1) then yields the familiar gauge and Lorentz Chern–Simons corrections to 3-form Bianchi identities of the heterotic string:

$$d\tilde{H}_3 = \tfrac{1}{4}\tilde{\alpha}'(\mathrm{tr}F^2 - \mathrm{tr}R^2) \,. \tag{1.3}$$

The present paper thus provides evidence not only for the importance of eleven dimensions in string theory but also (in contrast to Witten's paper) for the importance of supersymmetric extended objects with $d = p + 1 > 2$ worldvolume dimensions: the super p-branes [1].

2. Ten to eleven: it is not too late

In fact it should have come as no surprise that string theory makes use of eleven dimensions, as there were already tantalizing hints in this direction:

(i) In 1986, it was pointed out [25] that $D = 11$ supergravity compactified on $K3 \times T^{n-3}$ [26] and the $D = 10$ heterotic string compactified on T^n [27,28] have the same moduli spaces of vacua, namely

$$\mathcal{M} = \frac{SO(16+n, n)}{SO(16+n) \times SO(n)} \,. \tag{2.1}$$

[1] Super p-branes are reviewed in Refs. [23,24,9].

It was subsequently confirmed [29,30], in the context of the $D = 10$ Type IIA theory compactified on $K3 \times T^{n-4}$, that this equivalence holds globally as well as locally.

(ii) In 1987 the $D = 11$ supermembrane was discovered [31,32]. It was then pointed out [33] that the $(d = 2, D = 10)$ Green–Schwarz action of the Type IIA superstring follows by simultaneous worldvolume/space-time dimensional reduction of the $(d = 3, D = 11)$ Green–Schwarz action of the supermembrane.

(iii) In 1990, based on considerations of this $D = 11$ supermembrane which treats the dilaton and moduli fields on the same footing, it was conjectured [34,35] that discrete subgroups of all the old non-compact global symmetries of compactified supergravity [36–39] (e.g. $SL(2,\mathbb{R})$, $O(22,6)$, $O(24,8)$, E_7, E_8, E_9, E_{10}) should be promoted to duality symmetries of either heterotic or Type II superstrings. The case for a target space $O(22,6;\mathbb{Z})$ (*T-duality*) had already been made, of course [40]. Stronger evidence for a strong/weak coupling $SL(2,\mathbb{Z})$ (*S-duality*) in string theory was subsequently provided in [5,9,41–51]. Stronger evidence for their combination into an $O(24,8;Z)$ duality in heterotic strings was provided in [50,10,52,53] and stronger evidence for their combination into a discrete E_7 in Type II strings was provided in [11], where it was dubbed *U-duality*.

(iv) In 1991, the supermembrane was recovered as an elementary solution of $D = 11$ supergravity which preserves half of the space-time supersymmetry [54]. (*Elementary* solutions are singular and carry a Noether "electric" charge, in contrast to *solitons* which are non-singular solutions of the source-free equations and carry a topological "magnetic" charge.) The preservation of half the supersymmetries is intimately linked with the worldvolume kappa symmetry. It followed by the same simultaneous dimensional reduction in (ii) above that the elementary Type IIA string could be recovered as a solution of Type IIA supergravity. By truncation, one then obtains the $N = 1, D = 10$ elementary string [55].

(v) In 1991, the elementary superfivebrane was recovered as a solution of the dual formulation of $N = 1, D = 10$ supergravity which preserves half of the space-time supersymmetry [56]. It was then reinterpreted [57,58] as a non-singular soliton solution of the usual formulation. Moreover, it was pointed out that it also provides a solution of both the Type IIA and Type IIB field equations preserving half of the space-time supersymmetry and therefore that there exist both Type IIA and Type IIB superfivebranes. This naturally suggested a Type II string/fivebrane duality in analogy with the earlier heterotic string/fivebrane duality conjecture [23,59]. Although no Green–Schwarz action for the $d = 6$ worldvolumes is known, consideration of the soliton zero modes means that the gauged fixed actions must be described by a chiral antisymmetric tensor multiplet $(B^-_{\mu\nu}, \lambda^I, \phi^{[IJ]})$ in the case of IIA and a non-chiral vector multiplet $(B_\mu, \chi^I, A^I{}_J, \xi)$ in the case of IIB [57,58].

(vi) Also in 1991, black p-brane solutions of $D = 10$ superstrings were found [60] for $d = 1$ (IIA only), $d = 2$ (Heterotic, IIA and IIB), $d = 3$ (IIA only), $d = 4$ (IIB only) $d = 5$ (IIA only), $d = 6$ (Heterotic, IIA and IIB) and $d = 7$ (IIA only). Moreover, in the extreme mass-equals-charge limit, they each preserve half of the space-time supersymmetry [61]. Hence there exist all the corresponding super p-branes, giving rise to $D = 10$ particle/sixbrane, membrane/fourbrane and self-dual threebrane duality conjectures in addition to the existing string/fivebrane conjectures. The soliton zero modes are described by the supermultiplets listed in Table 1. Note that in contrast to the fivebranes, *both* Type IIA and Type IIB string worldsheet supermultiplets are non-chiral [2]. As such, they follow from T^4 compactification of the Type IIA fivebrane worldvolume supermultiplets.

(vii) In 1992, a fivebrane was discovered as a soliton of $D = 11$ supergravity preserving half the space-time supersymmetry [62]. Hence there exists a $D = 11$ superfivebrane and it forms the subject of the present paper. Once again, its covariant action is unknown but consideration of the soliton zero modes means that the gauged fixed action must be described by the same chiral antisymmetric tensor multiplet in (v) above [63,64,9]. This naturally suggests a $D = 11$ membrane/fivebrane duality.

(viii) In 1993, it was recognized [61] that by dualizing a vector into a scalar on the gauge-fixed $d = 3$ worldvolume of the Type IIA supermembrane, one increases the number of worldvolume scalars (i.e. transverse dimensions) from 7 to 8 and hence obtains the corresponding worldvolume action of the $D = 11$ supermembrane. Thus the $D = 10$ Type IIA theory contains a hidden $D = 11$ Lorentz invariance!

(ix) In 1994 [65] and 1995 [66], all the $D = 10$ Type IIA p-branes of (vi) above were related to either the $D = 11$ supermembrane or the $D = 11$ superfivebrane.

(x) Also in 1994, the (extreme electric and magnetic black hole [50,67]) Bogomol'nyi spectrum necessary for the E_7 U-duality of the $D = 10$ Type IIA string compactified to $D = 4$ on T^6 was given an explanation in terms of the wrapping of either the $D = 11$ membrane or $D = 11$ fivebrane around the extra dimensions [11].

(xi) In 1995, it was conjectured [64] that the $D = 10$ Type IIA superstring should be identified with the $D = 11$ supermembrane compactified on S^1, with the charged extreme black holes of the former interpreted as the Kaluza–Klein modes of the latter.

(xii) Also in 1995, the conjectured duality of the $D = 10$ heterotic string compactified on T^4 and the $D = 10$ Type IIA string compactified on K3 [11,3], combined with the above conjecture implies that the $d = 2$ worldsheet action of the $D = 6$ ($D = 7$) heterotic string may be obtained by K3 compactification [3] of the $d = 6$ worldvolume action of the $D = 10$ Type IIA fivebrane ($D = 11$ fivebrane) [68,69]. We shall shortly make use of this result.

Following Witten's paper [3] it was furthermore proposed [70] that the combination of perturbative and non-perturbative states of the $D = 10$ Type IIA string could be assembled into $D = 11$ supermultiplets. It has even been claimed [71] that both the $E_8 \times E_8$ and $SO(32)$ heterotic strings in $D = 10$ may be obtained by compactifying the $D = 11$ theory on Ξ_1 and Ξ_2 respectively, where Ξ_1 and Ξ_2 are one-dimensional structures obtained by squashing K3!

[2] This corrects an error in [61,9].

[3] The wrapping of the $D = 10$ *heterotic* fivebrane worldvolume around K3 to obtain a $D = 6$ *heterotic* string was considered in [7].

Table 1
Gauge-fixed $D = 10$ theories on the worldvolume, corresponding to the zero modes of the soliton, are described by the above supermultiplets and worldvolume supersymmetries. The $D = 11$ membrane and fivebrane supermultiplets are the same as Type IIA in $D = 10$

$d=7$	Type IIA	$(A_\mu, \lambda, 3\phi)$		$n=1$
$d=6$	Type IIA	$(B_{\mu\nu}^-, \lambda_R{}^I, \phi^{[IJ]})$	$I=1,\ldots,4$	$(n_+,n_-)=(2,0)$
	Type IIB	$(B_\mu, \chi^I, A^I{}_J, \xi)$	$I=1,2$	$(n_+,n_-)=(1,1)$
	Heterotic	(ψ^a, ϕ^α)	$a=1,\ldots,60$	
			$\alpha=1,\ldots,120$	$(n_+,n_-)=(1,0)$
$d=5$	Type IIA	$(A_\mu, \lambda^I, \phi^{[IJ]]})$	$I=1,\ldots,4$	$n=2$
$d=4$	Type IIB	$(B_\mu, \chi^I, \phi^{[IJ]})$	$I=1,\ldots,4$	$n=4$
$d=3$	Type IIA	(χ^I, ϕ^I)	$I=1,\ldots,8$	$n=8$
$d=2$	Type IIA	$(\chi_L{}^I, \phi_L{}^I), (\chi_R{}^I, \phi_R{}^I)$	$I=1,\ldots,8$	$(n_+,n_-)=(8,8)$
	Type IIB	$(\chi_L{}^I, \phi_L{}^I), (\chi_R{}^I, \phi_R{}^I)$	$I=1,\ldots,8$	$(n_+,n_-)=(8,8)$
	Heterotic	$(0, \phi_L{}^M), (\chi_R{}^I, \phi_R{}^I)$	$M=1,\ldots,24$	
			$I=1,\ldots,8$	$(n_+,n_-)=(8,0)$

3. $D = 11$ membrane/fivebrane duality

We begin with the bosonic sector of the $d = 3$ worldvolume of the $D = 11$ supermembrane:

$$S_3 = T_3 \int d^3\xi \left[-\frac{1}{2}\sqrt{-\gamma}\gamma^{ij}\partial_i X^M \partial_j X^N G_{MN}(X) + \frac{1}{2}\sqrt{-\gamma} \right.$$
$$\left. -\frac{1}{3!}\epsilon^{ijk}\partial_i X^M \partial_j X^N \partial_k X^P C_{MNP}(X) \right] , \tag{3.1}$$

where T_3 is the membrane tension, ξ^i $(i = 1, 2, 3)$ are the worldvolume coordinates, γ^{ij} is the worldvolume metric and $X^M(\xi)$ are the space-time coordinates $(M = 0, 1, \ldots, 10)$. Kappa symmetry [31,32] then demands that the background metric G_{MN} and background 3-form potential C_{MNP} obey the classical field equations of $D = 11$ supergravity, whose bosonic action is

$$I_{11} = \frac{1}{2\kappa_{11}{}^2} \int d^{11}x\sqrt{-G} \left[R_G - \frac{1}{2\cdot 4!}K^2_{MNPQ} \right] - \frac{1}{12\kappa_{11}{}^2} \int C_3 \wedge K_4 \wedge K_4 , \tag{3.2}$$

where $K_4 = dC_3$ is the 4-form field strength. In particular, K_4 obeys the field equation

$$d * K_4 = -\frac{1}{2}K_4{}^2 \tag{3.3}$$

and the Bianchi identity

$$dK_4 = 0 . \tag{3.4}$$

While there are two dimensionful parameters, the membrane tension T_3 and the eleven-dimensional gravitational constant κ_{11}, they are in fact not independent. To see this, we note from (3.1) that C_3 has period $2\pi/T_3$ so that K_4 is quantized according to

$$\int K_4 = \frac{2\pi n}{T_3} , \qquad n = \text{integer} . \tag{3.5}$$

Consistency of such C_3 periods with the space-time action, (3.2), gives the relation

$$\frac{(2\pi)^2}{\kappa_{11}{}^2 T_3{}^3} \in 4\mathbb{Z} \,. \tag{3.6}$$

The $D = 11$ classical field equations admit as a soliton a dual superfivebrane [62,6] whose worldvolume action is unknown, but which couples to the dual field strength $\tilde{K}_7 = *K_4$. The fivebrane tension $\tilde{T}_6$ is given by the Dirac quantization rule [6]

$$2\kappa_{11}{}^2 T_3 \tilde{T}_6 = 2\pi n \qquad n = \text{integer} \,. \tag{3.7}$$

Using (3.6), this may also be written as

$$\pi \frac{\tilde{T}_6}{T_3{}^2} \in \mathbb{Z} \,, \tag{3.8}$$

which we will find useful below. Although Dirac quantization rules of the type (3.7) appear for other p-branes and their duals in lower dimensions [6], it is the absence of a dilaton in the $D = 11$ theory that allows us to fix both the gravitational constant and the dual tension in terms of the fundamental tension.

From (3.3), the fivebrane Bianchi identity reads

$$d\tilde{K}_7 = -\frac{1}{2} K_4{}^2 \,. \tag{3.9}$$

However, such a Bianchi identity will in general require gravitational Chern–Simons corrections arising from a sigma-model anomaly on the fivebrane worldvolume [7,14,18–22]:

$$d\tilde{K}_7 = -\frac{1}{2} K_4{}^2 + (2\pi)^4 \tilde{\beta}' \tilde{X}_8 \,, \tag{3.10}$$

where $\tilde{\beta}'$ is related to the fivebrane tension by $T_6 = 1/(2\pi)^3 \tilde{\beta}'$ and where the 8-form polynomial $\tilde{X}_8$, quartic in the gravitational curvature R, describes the $d = 6$ σ-model Lorentz anomaly of the $D = 11$ fivebrane. Although the covariant fivebrane action is unknown, we know from Section 2 that the gauge fixed theory is described by the chiral antisymmetric tensor multiplet $(B^-_{\mu\nu}, \lambda^I, \phi^{[IJ]})$, and it is a straightforward matter to read off the anomaly polynomial from the literature. See, for example Refs. [72,73]. The contribution from the anti self-dual tensor is

$$\tilde{X}_B = \frac{1}{(2\pi)^4} \frac{1}{5760} \left[-10(\mathrm{tr}R^2)^2 + 28\,\mathrm{tr}R^4 \right] \tag{3.11}$$

and the contribution from the four left-handed (symplectic) Majorana–Weyl fermions is

$$\tilde{X}_\lambda = \frac{1}{(2\pi)^4} \frac{1}{5760} \left[\frac{10}{4} (\mathrm{tr}R^2)^2 + 2\,\mathrm{tr}R^4 \right] \,. \tag{3.12}$$

Hence $\tilde{X}_8$ takes the form quoted in the introduction:

$$\tilde{X}_8 = \frac{1}{(2\pi)^4} \left[-\frac{1}{768} (\mathrm{tr}R^2)^2 + \frac{1}{192} \mathrm{tr}R^4 \right] \,. \tag{3.13}$$

Thus membrane/fivebrane duality predicts a space-time correction to the $D = 11$ supermembrane action

$$I_{11}(\text{Lorentz}) = T_3 \int C_3 \wedge \frac{1}{(2\pi)^4}\left[-\frac{1}{768}(\mathrm{tr}R^2)^2 + \frac{1}{192}\mathrm{tr}R^4\right] . \tag{3.14}$$

Unfortunately, since the correct quantization of the supermembrane is unknown, this prediction is difficult to check. However, by simultaneous dimensional reduction [33] of $(d = 3, D = 11)$ to $(d = 2, D = 10)$ on S^1, this prediction translates into a corresponding prediction for the Type IIA string:

$$I_{10}(\text{Lorentz}) = T_2 \int B_2 \wedge \frac{1}{(2\pi)^4}\left[-\frac{1}{768}(\mathrm{tr}R^2)^2 + \frac{1}{192}\mathrm{tr}R^4\right] , \tag{3.15}$$

where B_2 is the string 2-form, T_2 is the string tension, $T_2 = 1/2\pi\alpha'$, related to the membrane tension by

$$T_2 = 2\pi R T_3 , \tag{3.16}$$

where R is the S^1 radius.

As a consistency check we can compare this prediction with previous results found by explicit string one-loop calculations. These have been done in two ways: either by computing directly in $D = 10$ the one-loop amplitude involving four gravitons and one B_2 [74–77], or by compactifying to $D = 2$ on an 8-manifold M and computing the B_2 one-point function [17]. We indeed find agreement. In particular, we note that

$$\tilde{X}_8 = \frac{1}{6}[2Y_8^{\mathrm{NS,R}} - Y_8^{\mathrm{R,R}}] , \tag{3.17}$$

where

$$\begin{aligned} Y_8^{\mathrm{NS,R}} &= \frac{1}{(2\pi)^4}\frac{1}{2880}\left[-\frac{25}{4}(\mathrm{tr}R^2)^2 + 31\,\mathrm{tr}R^4\right] , \\ Y_8^{\mathrm{R,R}} &= \frac{1}{(2\pi)^4}\frac{1}{2880}\left[10(\mathrm{tr}R^2)^2 - 28\,\mathrm{tr}R^4\right] . \end{aligned} \tag{3.18}$$

Upon compactification to $D = 2$, we arrive at

$$\begin{aligned} n_{\mathrm{NS,R}} &= \int_M Y_8^{\mathrm{NS,R}} , \\ n_{\mathrm{R,R}} &= \int_M Y_8^{\mathrm{R,R}} , \end{aligned} \tag{3.19}$$

where in the (NS,R) sector $n_{\mathrm{NS,R}}$ computes the index of the Dirac operator coupled to the tangent bundle on M and in the (R,R) sector $n_{\mathrm{R,R}}$ computes the index of the Dirac operator coupled to the spin bundle on M. We also find agreement with the well-known tree-level terms

$$\frac{1}{2\kappa_{10}{}^2}\int \frac{1}{2} B_2 \wedge K_4 \wedge K_4 , \tag{3.20}$$

where

$$\kappa_{11}{}^2 = 2\pi R \kappa_{10}{}^2 \ . \tag{3.21}$$

Thus using $D = 11$ membrane/fivebrane duality we have correctly reproduced the corrections to the B_2 field equations of the $D = 10$ Type IIA string (a mixture of tree-level and string one-loop effects) starting from the Chern–Simons corrections to the Bianchi identities of the $D = 11$ superfivebrane (a purely tree-level effect). It is now instructive to derive this same result from $D = 10$ string/fivebrane duality.

4. $D = 10$ Type IIA string/fivebrane duality

To see how a double worldvolume/space-time compactification of the $D = 11$ supermembrane theory on S^1 leads to the Type IIA string in $D = 10$ [33], let us denote all $(d = 3, D = 11)$ quantities by a hat and all $(d = 2, D = 10)$ quantities without. We then make a ten-one split of the space-time coordinates

$$\hat{X}^{\hat{M}} = (X^M, Y) \ , \quad M = 0, 1, \ldots, 9 \tag{4.1}$$

and a two-one split of the worldvolume coordinates

$$\hat{\xi}^{\hat{i}} = (\xi^i, \rho) \ , \quad i = 1, 2 \tag{4.2}$$

in order to make the partial gauge choice

$$\rho = Y \ , \tag{4.3}$$

which identifies the eleventh dimension of space-time with the third dimension of the worldvolume. The dimensional reduction is then affected by taking Y to be the coordinate on a circle of radius R and discarding all but the zero modes. In practice, this means taking the background fields $\hat{G}_{\hat{M}\hat{N}}$ and $\hat{C}_{\hat{M}\hat{N}\hat{P}}$ to be independent of Y. The string backgrounds of dilaton Φ, string σ-model metric G_{MN}, 1-form A_M, 2-form B_{MN} and 3-form C_{MNP} are given by [4]

$$\begin{aligned} \hat{G}_{MN} &= e^{-\Phi/3} \begin{pmatrix} G_{MN} + e^{\Phi} A_M A_N & e^{\Phi} A_M \\ e^{\Phi} A_N & e^{\Phi} \end{pmatrix} , \\ \hat{C}_{MNP} &= C_{MNP} \ , \\ \hat{C}_{MNY} &= B_{MN} \ . \end{aligned} \tag{4.4}$$

The actions (3.1) and (3.2) now reduce to

[4] The choice of dilaton prefactor, $e^{-\Phi/3}$, is dictated by the requirement that G_{MN} be the $D = 10$ string σ-model metric. To obtain the $D = 10$ fivebrane σ-model metric, the prefactor is unity because the reduction is then space-time only and not simultaneous worldvolume/space-time. This explains the remarkable "coincidence" [6] between $\hat{G}_{MN}$ and the fivebrane σ-model metric.

$$S_2 = T_2 \int d^2\xi \left[-\frac{1}{2}\sqrt{-\gamma}\gamma^{ij}\partial_i X^M \partial_j X^N G_{MN}(X) - \frac{1}{2!}\epsilon^{ij}\partial_i X^M \partial_j X^N B_{MN}(X) + \ldots \right] \tag{4.5}$$

and

$$I_{10} = \frac{1}{2\kappa_{10}{}^2} \int d^{10}x \sqrt{-G}\, e^{-\Phi} \left[R_G + (\partial_M \Phi)^2 - \frac{1}{2 \cdot 3!} H^2_{MNP} \right.$$
$$\left. - \frac{1}{2 \cdot 2!} e^{\Phi} F^2_{MN} - \frac{1}{2 \cdot 4!} e^{\Phi} J^2_{MNPQ} \right] - \frac{1}{2\kappa_{10}{}^2} \int \frac{1}{2} K_4 \wedge K_4 \wedge B_2 \,, \tag{4.6}$$

where the field strengths are given by $J_4 = K_4 + A_1 H_3$, $H_3 = dB_2$ and $F_2 = dA_1$. Let us now furthermore consider a simple space-time compactification of the fivebrane theory on the same S^1 to obtain the Type IIA fivebrane in $D = 10$. From (3.4) and (3.10), the field equations and Bianchi identities for the field strengths J_4, H_3, F_2 and their duals $\tilde{J}_6 = *J_4$, $\tilde{H}_7 = e^{-\Phi} * H_3$, $\tilde{F}_8 = *F_2$ now read

$$dJ_4 = F_2 H_3 \,, \quad d\tilde{J}_6 = H_3 J_4 \,, \tag{4.7}$$

$$dH_3 = 0 \,, \quad d\tilde{H}_7 = -\frac{1}{2} J_4{}^2 + F_2 \tilde{J}_6 + (2\pi)^4 \tilde{\beta}' \tilde{X}_8 \,, \tag{4.8}$$

$$dF_2 = 0 \,, \quad d\tilde{F}_8 = -H_3 \tilde{J}_6 \,. \tag{4.9}$$

Of course, the Lorentz corrections to the Bianchi identity for $\tilde{H}_7$ could have been derived directly from the Type IIA fivebrane in $D = 10$ since its worldvolume is described by the same antisymmetric tensor supermultiplet. Note that of all the Type IIA p-branes in Table 1, only the fivebrane supermultiplet is chiral, so only the $\tilde{H}_7$ Bianchi identity acquires corrections.

From (3.7), (3.16) and (3.21), or from first principles of string/fivebrane duality [78], the Dirac quantization rule for $n = 1$ is now

$$2\kappa_{10}{}^2 = (2\pi)^5 \alpha' \tilde{\beta}' \,. \tag{4.10}$$

So from either $D = 10$ string/fivebrane duality or from compactification of $D = 11$ membrane/fivebrane duality, the B_2 field equation with its string one-loop correction is

$$d(e^{-\Phi} * H_3) = -\frac{1}{2} J_4{}^2 + F_2 * J_4 + \frac{2\kappa_{10}{}^2}{2\pi\alpha'} \tilde{X}_8 \,, \tag{4.11}$$

which once again agrees with explicit string one-loop calculations [17,74].

5. $D = 7$ string/membrane duality

Simultaneous worldvolume/space-time compactification of the $D = 11$ fivebrane on K3 gives a heterotic string in $D = 7$ [68,69]. The five worldvolume scalars produce $(5_L, 5_R)$ worldsheet scalars, the four worldvolume fermions produce $(0_L, 8_R)$ worldsheet fermions and the worldvolume self-dual 3-form produces $(19_L, 3_R)$ worldsheet scalars, which together constitute the field content of the heterotic string. We may thus derive the Bianchi identity for this string starting from the fivebrane Bianchi identity, (1.1):

$$d\tilde{K}_7 = -\frac{1}{2}{K_4}^2 + (2\pi)^4\tilde{\beta}'\tilde{X}_8 \ . \tag{5.1}$$

We begin by performing a seven-four split of the eleven-dimensional coordinates

$$X^M = (x^\mu, y^i) \ , \quad \mu = 0, 1, \ldots, 6; \ i = 7, 8, 9, 10 \tag{5.2}$$

so that the original set of ten-dimensional fields $\{\mathcal{A}_n\}$ may be decomposed in a basis of harmonic p-forms on K3:

$$\mathcal{A}_n(X) = \sum \mathcal{A}_{n-p}(x)\omega_p(y) \ . \tag{5.3}$$

In particular, we expand C_3 as

$$C_3(X) = C_3(x) + \frac{1}{2T_3}\sum C_1^I(x)\omega_2^I(y) \ , \tag{5.4}$$

where ω_2^I, $I = 1, \ldots, 22$ are an integral basis of b_2 harmonic two-forms on K3. We have chosen a normalization where the seven-dimensional $U(1)$ field strengths $K_2^I = dC_1^I$ are coupled to even charges

$$\int K_2^I \in 4\pi\mathbb{Z} \ , \tag{5.5}$$

which follows from the eleven-dimensional quantization condition, (3.5).

Following Ref. [7], let us define the dual (heterotic) string tension $\tilde{T}_2 = 1/2\pi\tilde{\alpha}'$ by

$$\frac{1}{2\pi\tilde{\alpha}'} = \frac{1}{(2\pi)^3\tilde{\beta}'}V \ , \tag{5.6}$$

where V is the volume of K3, and the dual string 3-form $\tilde{H}_3$ by

$$\frac{1}{2\pi\tilde{\alpha}'}\tilde{H}_3 = \frac{1}{(2\pi)^3\tilde{\beta}'}\int_{K3}\tilde{K}_7 \ , \tag{5.7}$$

so that $\tilde{H}_3$ satisfies the conventional quantization condition

$$\int \tilde{H}_3 = 4\pi^2 n\tilde{\alpha}' \ , \tag{5.8}$$

which follows from the underlying $\tilde{K}_7$ quantization. The dual string Lorentz anomaly polynomial, $\tilde{X}_4$, is given by

$$\begin{aligned}\tilde{X}_4 = \int_{K3}\tilde{X}_8 &= \frac{1}{(2\pi)^4}\int_{K3}\left[-\frac{1}{768}(\mathrm{tr}R^2 + \mathrm{tr}R_0^2)^2 + \frac{1}{192}(\mathrm{tr}R^4 + \mathrm{tr}R_0^4)\right] \\ &= \frac{1}{(2\pi)^2}\frac{1}{192}\mathrm{tr}R^2 p_1(K3) \\ &= -\frac{1}{(2\pi)^2}\frac{1}{4}\mathrm{tr}R^2 \ ,\end{aligned} \tag{5.9}$$

where $p_1(K3)$ is the Pontryagin number of K3

$$p_1(K3) = -\frac{1}{8\pi^2}\int_{K3} \mathrm{tr} R_0^2 = -48 \ . \tag{5.10}$$

We may now integrate (5.1) over K3, using the Dirac quantization rule, (3.8), to find

$$d\tilde{H}_3 = -\frac{\tilde{\alpha}'}{4}\left[K_2^I K_2^J d_{IJ} + \mathrm{tr}R^2\right] \ , \tag{5.11}$$

where d_{IJ} is the intersection matrix on K3, given by

$$d_{IJ} = \int_{K3} \omega_2^I \wedge \omega_2^J \tag{5.12}$$

and has $b_2^+ = 3$ positive and $b_2^- = 19$ negative eigenvalues. Therefore we see that this form of the Bianchi identity corresponds to a $D = 7$ toroidal compactification of a heterotic string at a generic point on the Narain lattice [27,28]. Thus we have reproduced *exactly* the $D = 7$ Bianchi identity of the heterotic string, starting from a $D = 11$ fivebrane!

6. $D = 6$ string/string duality

Further compactification of (5.11) on S^1 clearly yields the six-dimensional Bianchi identity with two additional $U(1)$ fields coming from S^1, giving $\mathrm{tr}F^2$ with signature $(4, 20)$. Alternatively, this may be obtained from K3 compactification of the $D = 10$ fivebrane, with Bianchi identity

$$d\tilde{H}_7 = -\frac{1}{2}J_4{}^2 + F_2\tilde{J}_6 + (2\pi)^4\tilde{\beta}'\tilde{X}_8 \ . \tag{6.1}$$

Although in this section we focus just on this identity, we present the compactification of the complete bosonic $D = 10$ Type IIA action, (4.6), in Appendix A.

The reduction from ten dimensions is similar to that from eleven. There is one subtlety, however, which is that J_4 is the $D = 10$ gauge invariant combination, $J_4 = K_4 + A_1 H_3$. Compactifying (6.1) to six dimensions on K3, we may identify 22 $U(1)$ fields coming from the reduction of J_4 and one each coming from F_2 and $\tilde{J}_6$. Normalizing these 24 six-dimensional $U(1)$ fields according to (5.5), we obtain

$$d\tilde{H}_3 = -\frac{\tilde{\alpha}'}{4}\left[J_2^I J_2^J d_{IJ} - 2F_2\tilde{J}_2 - 16\pi^2\tilde{X}_4\right] \ , \tag{6.2}$$

where $J_2^I = dC_2^I + A_1 db^I$ and $J_4 = dC_3 + A_1 H_3$. The 22 scalars b^I are torsion moduli of K3. While we may be tempted to identify these two-forms with $U(1)$ field strengths, this would not be correct since $dJ_2^I = F_2 db^I \neq 0$ and $d\tilde{J}_2 = J_2^I db^J d_{IJ} \neq 0$. Thus the actual field strengths must be shifted according to

$$\hat{K}_2^I = J_2^I - F_2 b^I ,$$
$$\hat{J}_2 = \tilde{J}_2 - J_2^I b^J d_{IJ} + \tfrac{1}{2} F_2 b^I b^J d_{IJ} , \tag{6.3}$$

so that $d\hat{K}_2^I = d\hat{J}_2 = 0$. Inverting these definitions and inserting them into (6.2) gives finally

$$d\tilde{H}_3 = -\frac{\tilde{\alpha}'}{4}\left[\hat{K}_2^I \hat{K}_2^J d_{IJ} - 2F_2\hat{J}_2 + \mathrm{tr}R^2\right] . \tag{6.4}$$

In order to compare this result with the toroidally compactified heterotic string, it is useful to group the $U(1)$ field-strengths into a 24-dimensional vector

$$\mathcal{F}_2 = [F_2,\ \hat{J}_2,\ \hat{K}_2^I]^T , \tag{6.5}$$

in which case the $D = 6$ Bianchi identity now reads

$$d\tilde{H}_3 = -\frac{\tilde{\alpha}'}{4}\left[\mathcal{F}^T L \mathcal{F} + \mathrm{tr}R^2\right] , \tag{6.6}$$

where the matrix $L = [(-\sigma^1) \oplus d_{IJ}]$ has 4 positive and 20 negative eigenvalues. This is in perfect agreement with the reduction of the $D = 7$ result, (5.11), and corresponds to a Narain compactification on $\Gamma_{4,20}$.

Note that the heterotic string tension $1/2\pi\tilde{\alpha}'$ and the Type IIA string tension $1/2\pi\alpha'$ are related by the Dirac quantization rule [6,7]

$$2{\kappa_6}^2 = (2\pi)^3 n\alpha'\tilde{\alpha}' , \tag{6.7}$$

where ${\kappa_6}^2 = {\kappa_{10}}^2/V$ is the $D = 6$ gravitational constant. Some string theorists, while happy to endorse string/string duality, eschew the *soliton* interpretation. It is perhaps worth emphasizing, therefore, that without such an interpretation with its Dirac quantization rule, there is no way to relate the two string tensions.

7. Elementary versus solitonic heterotic strings

Our success in correctly reproducing the *fundamental* heterotic string σ-model anomaly polynomial

$$X_4 = \frac{1}{4}\frac{1}{(2\pi)^2}(\mathrm{tr}R^2 - \mathrm{tr}F^2) , \tag{7.1}$$

by treating the string as a (K3 compactified fivebrane) *soliton*, now permits a re-evaluation of a previous controversy concerning fundamental [79] versus solitonic [9,12,78] heterotic strings. In an earlier one-loop test of $D = 10$ heterotic string/heterotic fivebrane duality [14], X_4 was obtained by the following logic: the $d = 2$ gravitational anomaly for complex fermions in a representation $\mathcal{R}$ of the gauge group is [72,73]

$$I_4 = \frac{1}{2}\frac{1}{(2\pi)^2}\left(\frac{r}{24}\mathrm{tr}R^2 - \mathrm{tr}_{\mathcal{R}}F^2\right) , \tag{7.2}$$

where r is the dimensionality of the representation and R is the *two-dimensional* curvature. Since the $SO(32)$ heterotic string has 32 left-moving gauge Majorana fermions (or, if we bosonize, 16 chiral scalars) and 8 physical right-moving space-time Majorana fermions, Dixon, Duff and Plefka [14] set $\mathcal{R}$ to be the fundamental representation and put $r = 32 - 8 = 24$ to obtain $X_4 = I_4/2$, on the understanding that R is now to be interpreted as the pull-back of the *space-time* curvature. Exactly the same logic was used in [14] in obtaining the heterotic[5] fivebrane $\tilde{X}_8$

$$\tilde{X}_8 = \frac{1}{(2\pi)^4}\left[\frac{1}{24}\,\mathrm{tr}F^4 - \frac{1}{192}\,\mathrm{tr}F^2\,\mathrm{tr}R^2 + \frac{1}{768}\,(\mathrm{tr}R^2)^2 + \frac{1}{192}\,\mathrm{tr}R^4\right] \tag{7.3}$$

and in Sections 3 and 4 above in obtaining the Type IIA fivebrane $\tilde{X}_8$ of (3.13). This logic was however criticized by Izquierdo and Townsend [15] and also by Blum and Harvey [16]. They emphasize the difference between the *gravitational anomaly* (which vanishes for the *fundamental* heterotic string [79]) involving the *two-dimensional* curvature and the *σ-model* anomaly (which is given by X_4 [80]) involving the pull-back of the *space-time* curvature. Moreover, they go on to point out that the 32 left-moving gauge Majorana fermions (or 16 chiral scalars) of the fundamental heterotic string do not couple at all to the spin connections of this latter curvature. They conclude that the equivalence between X_4 and $I_4/2$ is a "curious fact" with no physical significance. They would thus be forced to conclude that the derivation of the Type IIA string field equations presented in the present paper is also a gigantic coincidence!

An attempt to make sense of all this was made by Blum and Harvey. They observed that the zero modes of *solitonic* strings (and fivebranes) necessarily couple to the space-time spin connections because they inherit this coupling from the space-time fields from which they are constructed. For these objects, therefore, they would agree that the logic of Dixon, Duff and Plefka (and, by inference, the logic of the present paper) is correct. But they went on to speculate that although *fundamental* and *solitonic* heterotic strings may both exist, they are *not* to be identified! Recent developments in string/string duality [3,8,11,69,68,81], however, have convinced many physicists that the fundamental heterotic string is a soliton after all and so it seems we must look for an alternative explanation.

The correct way to resolve the apparent conflict is, we believe, rather mundane. The solitonic string and p-brane solitons are invariably presented in a *physical gauge* where one identifies d of the D space-time dimensions with the $d = p + 1$ dimensions of the p-brane worldvolume. As discussed in [14], this is best seen in the Green–Schwarz formalism, which is in fact the only formalism available for $d > 2$. In such a physical gauge (which is only well-defined for vanishing *worldvolume* gravitational anomaly) the worldvolume curvatures and pulled-back space-time curvatures are mixed up. So, in this sense, the gauge fermions do couple to the space-time curvature after all.

[5] Note that the heterotic string X_4, the heterotic fivebrane $\tilde{X}_8$ and the Type IIA fivebrane $\tilde{X}_8$ are the only non-vanishing anomaly polynomials, since from Table 1, these are the only theories with chiral supermultiplets.

8. Fivebrane origin of S-duality? Discard worldvolume Kaluza–Klein modes?

In a recent paper [8], it was explained how S-duality in $D = 4$ follows as a consequence of $D = 6$ string/string duality: S-duality for one theory is just T-duality for the other. Since we have presented evidence in this paper that Type IIA string/heterotic string duality in $D = 6$ follows as a consequence of Type IIA string/Type IIA fivebrane duality in $D = 10$, which in turn follows from membrane/fivebrane duality in $D = 11$, it seems natural to expect a fivebrane origin of S-duality. (Indeed, a fivebrane explanation for S-duality has already been proposed by Schwarz and Sen [46] and by Binetruy [48], although they considered a T^6 compactification of the *heterotic* fivebrane rather than a $K3 \times T^2$ compactification of the *Type IIA fivebrane.*)

The explanation of [8] relied on the observation that the roles of the axion/dilaton fields S and the modulus fields T trade places in going from the fundamental string to the dual string. It was proved that, for a dual string compactified from $D = 6$ to $D = 4$ on T^2, $SL(2, \mathbb{Z})_S$ is a symmetry that interchanges the roles of the dual string worldsheet Bianchi identities and the field equations for the internal coordinates y^m $(m = 4, 5)$. However, in unpublished work along the lines of [34,35], Duff, Schwarz and Sen tried and failed to prove that, for a fivebrane compactified from $D = 10$ to $D = 6$, $SL(2, \mathbb{Z})_S$ is a symmetry that interchanges the roles of the fivebrane worldvolume Bianchi identities and the field equations for the internal coordinates y^m $(m = 4, 5, 6, 7, 8, 9)$. A similar negative result was reported by Percacci and Sezgin [82].

Another way to state the problem is in terms of massive worldvolume Kaluza–Klein modes. In the double dimensional reduction of the $D = 10$ fivebrane to $D = 6$ heterotic string considered in Section 6, we obtained the heterotic string worldsheet multiplet of 24 left-moving scalars, 8 right moving scalars and 8 chiral fermions as the *massless modes* of a Kaluza–Klein compactification on K3. Taken in isolation, these massless modes on the dual string worldsheet will display the usual T-duality when the string is compactified from $D = 6$ to $D = 4$ and hence the fundamental string will display the desired S-duality. However, no-one has yet succeeded in showing that this T-duality survives when the *massive* Kaluza–Klein modes on the fivebrane worldvolume are included. Since these modes are just what distinguishes a string $X^M(\tau, \sigma)$ from a fivebrane $X^M(\tau, \sigma, \rho^i)$ $(i = 1, 2, 3, 4)$, this was precisely the reason in [8] for preferring a $D = 6$ string/string duality explanation for $SL(2, \mathbb{Z})$ over a $D = 10$ string/fivebrane duality explanation. (Another reason, of course, is that the quantization of strings is understood, but that of fivebranes is not!) The same question about whether or not to discard massive worldvolume Kaluza–Klein modes also arises in going from the membrane in $D = 11$ to the Type IIA string in $D = 10$. For the moment therefore, this inability to provide a fivebrane origin for $SL(2, \mathbb{Z})$ remains the Achilles heel of the super p-brane programme [6].

---

[6] Another unexplained phenomenon, even in pure string theory, is the conjectured $SL(2, \mathbb{Z})$ duality of the $D = 10$ Type IIB string [11], which gives rise to U-duality in $D = 4$. In this connection, it is perhaps worth noting from Table 1 that the gauged-fixed worldvolume of the self-dual Type IIB superthreebrane is described by the $d = 4, n = 4$ Maxwell supermultiplet [83]. Now $d = 4, n = 4$ abelian gauge theories are

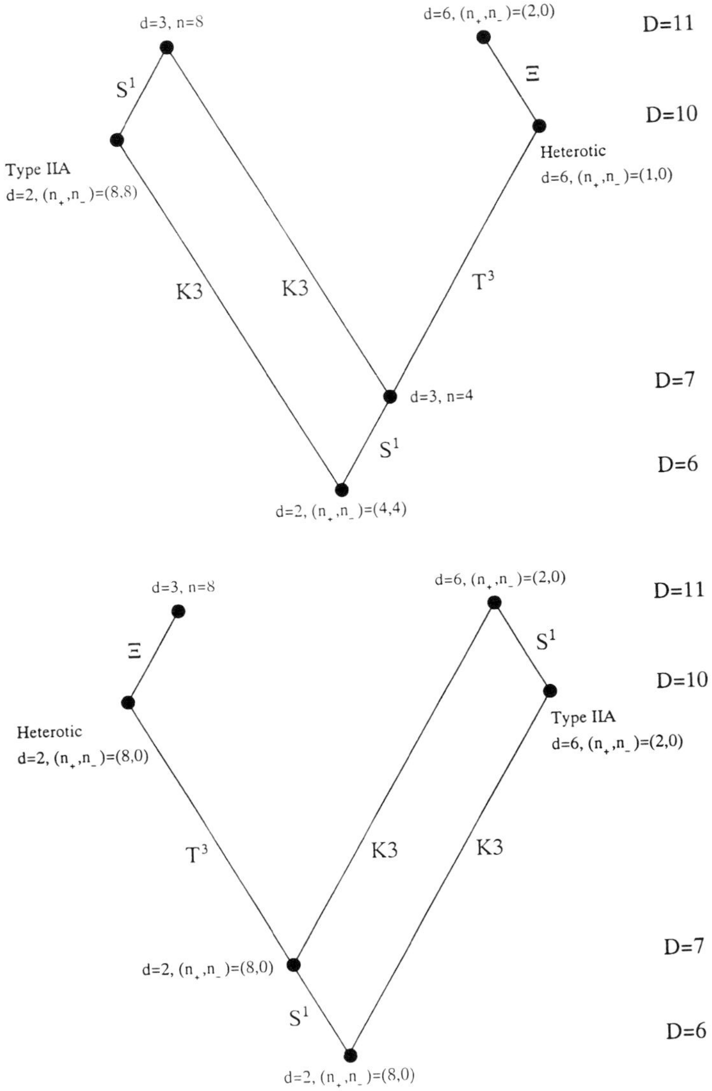

Fig. 1. Compactifications relating (a,top) the Type IIA fivebrane to the heterotic string and (b,bottom) the heterotic fivebrane to the Type IIA string. Worldvolume supersymmetries are indicated.

---

expected to display an $SL(2,\mathbb{Z})$ duality. See Refs. [84,85] for a recent discussion. Could this be the origin of the $SL(2,\mathbb{Z})$ of the Type IIB string which follows from a T^2 compactification of the threebrane? Note moreover, that the threebrane supermultiplet itself follows from T^2 compactification of either the Type IIA or Type IIB fivebrane supermultiplet. Compactifications of such $d = 6$ self-dual antisymmetric tensors have, in fact, recently been invoked precisely in the context of S-duality in abelian gauge theories [85]. Of course, the gauged-fixed action for the superthreebrane is presumably not simply the Maxwell action but some non-linear (possibly Born–Infeld [83]) version. Nevertheless, S-duality might still hold [86].

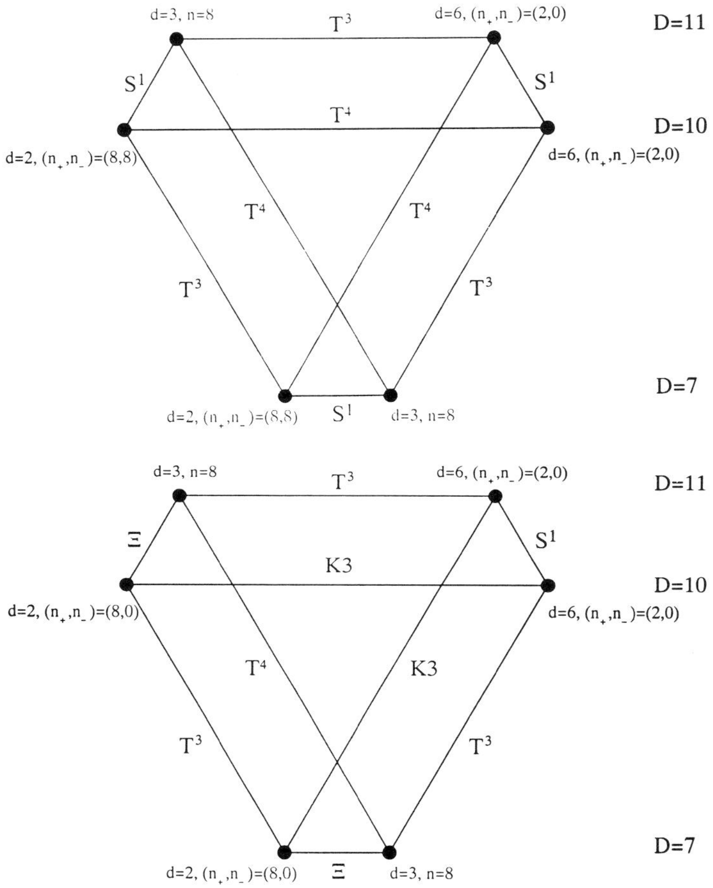

Fig. 2. Compactifications incorporating worldvolume reductions (a,top) and (b,bottom).

. Web of interconnections

We have discussed membrane and fivebranes in $D = 11$, heterotic strings and Type I fivebranes in $D = 10$, heterotic strings and membranes in $D = 7$, heterotic and ype II strings in $D = 6$ and how they are related by various compactifications. This omewhat bewildering mesh of interconnections is summarized in Fig. 1a. There are wo types of dimensional reduction to consider: lines sloping down left to right represent pace-time reduction $(d, D) \rightarrow (d, D-k)$ and lines sloping down right to left represent imultaneous space-time/worldsheet reduction $(d, D) \rightarrow (d-k, D-k)$. The worldsheet eductions may be checked against Table 1. Note that the simultaneous reduction on Ξ of he $D = 11$ membrane to yield the $D = 10$ heterotic string is still somewhat speculative 71], but we have included it since it nicely completes the diagram.

According to Townsend [68], a similar picture may be drawn relating the Type IIA tring and heterotic fivebrane, which we show in Fig. 1b, where we have once again

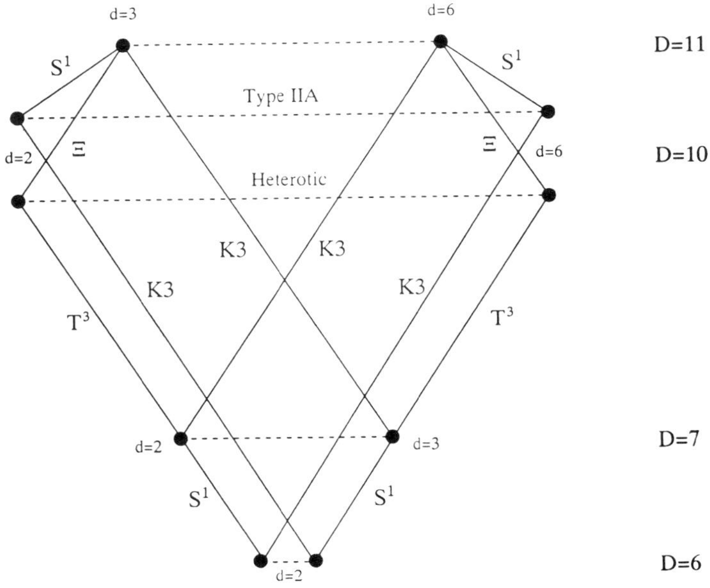

Fig. 3. A superposition of Figs. 1(a) and 1(b), illustrating strong/weak coupling dualities (denoted by the dotted lines).

speculated on a space-time reduction on Ξ of the $D = 11$ fivebrane to yield the $D = 10$ heterotic fivebrane. However, one must now explain how T^3 (or T^4) compactification of the $(120, 120)$ degrees of freedom of the gauge-fixed $D = 10$ heterotic fivebrane [59] can yield only the $(8, 8)$ of the $D = 7$ membrane (or the $(8_L, 8_L)$, $(8_R, 8_R)$ of the $D = 6$ Type IIA string). Townsend has given arguments to support this claim. There are more interrelationships one can illustrate by including horizontal lines representing worldsheet reduction only, $(d, D) \to (d - k, D)$, some of which are shown in Figs. 2a,b.

Note that these diagrams describe theories related by compactification and so relate weak coupling to weak coupling and strong to strong. In Fig. 3, we have superimposed Figs. 1a,b to indicate how the various theories are also related by *duality* (denoted by the dotted horizontal lines) which relates weak coupling to strong. We believe that these interrelationships, which have in particular enabled us to deduce supermembrane effects in agreement with explicit string one-loop calculations, strengthen the claim that eleven dimensions and supermembranes have a part to play in string theory: a triumph of *diversification* over *unification* [87].

Acknowledgements

We are grateful to Chris Hull, Joachim Rahmfeld, Paul Townsend and Edward Witten for conversations.

Appendix A. Reduction of the $D = 10$ Type IIA model on K3

In Section 6 we presented the reduction of the fivebrane Bianchi identity on K3. For completeness, we present the reduction of the bosonic part of the $D = 10$ Type IIA supergravity action, (4.6), which we write here in a form notation:

$$I_{10} = \frac{1}{2\kappa_{10}{}^2} \int d^{10}x\sqrt{-G}e^{-\Phi}\left[R_G + (\partial_M \Phi)^2\right] + \frac{1}{4\kappa_{10}{}^2} \int \left[F_2 \wedge *F_2 + e^{-\Phi} H_3 \wedge *H_3 + J_4 \wedge *J_4 - K_4 \wedge K_4 \wedge B_2\right] , \quad (A.1)$$

where the ten-dimensional bosonic fields are the metric G, dilaton Φ and the 1-, 2- and 3-form fields A_1, B_2 and C_3. Eleven-dimensional K_4 quantization, (3.5), as well as the usual Kaluza–Klein condition for F_2, give rise to the ten-dimensional conditions

$$\int K_4 = \frac{4\pi^2 nR}{T_2}, \quad \int H_3 = \frac{2\pi n}{T_2}, \quad \int F_2 = 2\pi nR . \quad (A.2)$$

Following the decomposition of the fields in Section 5, we write

$$A_1(X) = \frac{R}{2}A_1(x), \quad B_2(X) = B_2(x) + \frac{2\pi}{T_2}\sum b^I(x)\omega_2^I(y), \quad C_3(X) = \frac{R}{2}C_3(x) + \frac{\pi R}{T_2}\sum C_1^I(x)\omega_2^I(y), \quad (A.3)$$

in which case the four-form J_4 is given by

$$J_4(X) = \frac{R}{2}[K_4(x) + A_1(x)H_3(x)] + \frac{\pi R}{T_2}\sum[K_2^I(x) + A_1(x)db^I(x)]\omega_2^I(y) . \quad (A.4)$$

The constants are chosen so the six-dimensional $U(1)$ fields will be coupled to even charges

$$\int \mathcal{F}_2 \in 4\pi\mathbb{Z} . \quad (A.5)$$

For K3, with Betti numbers $b_0 = 1$, $b_1 = 0$, $b_2^+ = 3$ and $b_2^- = 19$, we may choose an integral basis of harmonic two-forms, ω_2^I with intersection matrix

$$d_{IJ} = \int_{K3} \omega_2^I \wedge \omega_2^J . \quad (A.6)$$

Since taking a Hodge dual of ω_2^I on K3 gives another harmonic two-form, we may expand the dual in terms of the original basis

$$\hat{*}\omega_2^I = \omega_2^J H^J{}_I \ , \tag{A.7}$$

where we use $\hat{*}$ to denote Hodge duals on K3. In this case, we find

$$\int_{K3} \omega_2^I \wedge \hat{*}\omega_2^J = d_{IK} H^K{}_J \ . \tag{A.8}$$

The matrix $H^I{}_J$ depends on the metric on K3, i.e. the $b_2^+ \cdot b_2^- = 57$ K3 moduli. Because of the fact that $\hat{*}\hat{*} = 1$, $H^I{}_J$ satisfies the properties [69]

$$\begin{aligned} H^I{}_J H^J{}_K &= \delta^I{}_K \ , \\ d_{IJ} H^J{}_K &= d_{KJ} H^J{}_I \ , \end{aligned} \tag{A.9}$$

so that

$$H^J{}_I d_{JK} H^K{}_L = d_{IL} \tag{A.10}$$

and hence is an element of $SO(3,19)/SO(3) \times SO(19)$.

Using these properties of K3, we may compactify the second line of (A.1) to obtain

$$\begin{aligned} I_6 = \frac{1}{2\kappa_6{}^2} \int \Big[& \tfrac{1}{2} e^{-\phi} H_3 \wedge *H_3 + \tfrac{1}{2} e^{-\phi} e^{\rho} db^I \wedge *db^J d_{IK} H^K{}_J \\ & + \frac{\tilde{\alpha}'}{4} \Big(e^{-\rho} F_2 \wedge *F_2 + e^{-\rho} J_4 \wedge *J_4 + J_2^I \wedge *J_2^J d_{IK} H^K{}_J \\ & - K_2^I \wedge K_2^J \wedge B_2 \, d_{IJ} - 2K_4 \wedge K_2^I b^J d_{IJ} \Big) \Big] \ . \end{aligned} \tag{A.11}$$

The six-dimensional dilaton is given by $\phi = \Phi + \rho$ where Φ is the ten-dimensional dilaton and ρ is the breathing mode of K3:

$$e^{-\rho} = \frac{1}{V} \int_{K3} \hat{*}1 \ . \tag{A.12}$$

In order to make contact with the compactified heterotic string, we wish to dualize the four-form J_4. Note, however, that since $d(e^{-\rho}\hat{*}J_4) = J_2^I db^J d_{IJ}$, the proper expression for dualizing J_4 is given by (6.3). Performing such a step and rewriting J_2^I as well, we finally arrive at

$$\begin{aligned} I_6 = \frac{1}{2\kappa_6{}^2} \int \Big[& \tfrac{1}{2} e^{-\phi} H_3 \wedge *H_3 + \tfrac{1}{2} e^{-\phi} e^{\rho} db^I \wedge *db^J d_{IK} H^K{}_J \\ & + \frac{\tilde{\alpha}'}{4} \Big(e^{-\rho} F_2 \wedge *F_2 + (\hat{K}_2^I + F_2 b^I) \wedge *(\hat{K}_2^J + F_2 b^J) d_{IK} H^K{}_J \\ & + e^{\rho} (\hat{J}_2 + \hat{K}_2^I b^J d_{IJ} + \tfrac{1}{2} F_2 b^I b^J d_{IJ}) \wedge *(\hat{J}_2 + \hat{K}_2^K b^L d_{KL} + \tfrac{1}{2} F_2 b^K b^L d_{KL}) \\ & - (\hat{K}_2^I \wedge \hat{K}_2^J d_{IJ} - 2F_2 \wedge \hat{J}_2) \wedge B_2 \Big) \Big] \ . \end{aligned} \tag{A.13}$$

This expression can be brought into a $SO(4,20)/SO(4) \times SO(20)$ invariant form. As in Section 6, we group the $U(1)$ field strengths into the 24-component vector

$$\mathcal{F}_2 = [\, F_2,\ \hat{J}_2,\ \hat{K}_2^I\,]^T \ , \tag{A.14}$$

which allows us to rewrite the bosonic lagrangian as

$$I_6 = \frac{1}{2\kappa_6{}^2}\int d^6x\sqrt{-G}e^{-\phi}\left(R + (\partial_\mu\phi)^2 - \frac{1}{2\cdot 3!}H^2_{\mu\nu\lambda} + \frac{1}{8}\mathrm{Tr}[\partial_\mu ML\partial_\mu ML]\right) \\ + \frac{1}{2\kappa_6{}^2}\int \frac{\tilde{\alpha}'}{4}\left(\mathcal{F}_2{}^T(LML)\wedge *\mathcal{F}_2 - \mathcal{F}_2{}^T\wedge L\mathcal{F}_2\wedge B_2\right) . \tag{A.15}$$

The matrix L is given by

$$L = \begin{bmatrix} -\sigma^1 & 0 \\ 0 & d_{IJ} \end{bmatrix} , \tag{A.16}$$

where $\sigma^1 = \begin{pmatrix} 0 & 1 \\ 1 & 0 \end{pmatrix}$. The matrix M contains the $1 + 57 + 22 = 80$ moduli of K3 with torsion, broken up in terms of e^ρ, $H^I{}_J$ and b^I respectively:

$$M = \begin{bmatrix} e^\rho & -\frac{1}{2}e^\rho(b^Ib^Jd_{IJ}) & e^\rho b^I \\ -\frac{1}{2}e^\rho(b^Ib^Jd_{IJ}) & e^{-\rho} + b^Ib^Jd_{IK}H^K{}_J + \frac{1}{4}e^\rho(b^Ib^Jd_{IJ})^2 & -b^KH^I{}_K - \frac{1}{2}e^\rho b^I(b^Kb^Ld_{KL}) \\ e^\rho b^J & -b^KH^J{}_K - \frac{1}{2}e^\rho b^J(b^Kb^Ld_{KL}) & H^I{}_Kd^{JK} + e^\rho b^Ib^J \end{bmatrix} . \tag{A.17}$$

In the last entry of M, d^{IJ} is the inverse of d_{IJ}. We verify that

$$M^T = M, \qquad MLM^T = L^{-1} \ . \tag{A.18}$$

This agrees with the bosonic action given in [81].

References

[1] M.B. Green, J.H. Schwarz and E. Witten, Superstring Theory (Cambridge Univ. Press, Cambridge, 1987).

[2] M.J. Duff, B.E.W. Nilsson and C.N. Pope, Kaluza-Klein supergravity, Phys. Rep. 130 (1986) 1.

[3] E. Witten, String theory dynamics in various dimensions, IASSNS-HEP-95-18, hep-th/9503124 (March 1995).

[4] M.J. Duff and J.X. Lu, Nucl. Phys. B 357 (1991) 534.

[5] M.J. Duff and R.R. Khuri, Nucl. Phys. B 411 (1994) 473.

[6] M.J. Duff and J.X. Lu, Nucl. Phys. B 416 (1994) 301.

[7] M.J. Duff and R. Minasian, Nucl. Phys. B 436 (1995) 507.

[8] M.J. Duff, Nucl. Phys. B 442 (1995) 47.

[9] M.J. Duff, R.R. Khuri and J.X. Lu, String solitons, NI-94-017, CTP-TAMU-67/92, McGill/94-53, CERN-TH.7542/94, hep-th/9412184, to be published in Phys. Rep. (December 1994).

[10] M.J. Duff, Classical/quantum duality, in Proceedings of the International High Energy Physics Conference, Glasgow (June 1994).

[11] C.M. Hull and P.K. Townsend, Nucl. Phys. B 438 (1995) 109.

[12] M.J. Duff and J.X. Lu, Phys. Rev. Lett. 66 (1991) 1402.

[13] M.J. Duff and J.X. Lu, Class. Quantum Grav. 9 (1992) 1.
[14] J.A. Dixon, M.J. Duff and J.C. Plefka, Phys. Rev. Lett. 69 (1992) 3009.
[15] J.M. Izquierdo and P.K. Townsend, Nucl. Phys. B 414 (1994) 93.
[16] J. Blum and J.A. Harvey, Nucl. Phys. B 416 (1994) 119.
[17] C. Vafa and E. Witten, A one-loop test of string duality, HUTP-95-A015, IASSNS-HEP-95-33, hep-th/9505053 (May 1995).
[18] J.A. Dixon, M.J. Duff and E. Sezgin, Phys. Lett. B 279 (1992) 265.
[19] J.A. Dixon and M.J. Duff, Phys. Lett. B 296 (1992) 28.
[20] R. Percacci and E. Sezgin, Int. J. Mod. Phys. A 8 (1993) 5367.
[21] E. Bergshoeff, R. Percacci, E. Sezgin, K.S. Stelle and P.K. Townsend, Nucl. Phys. B 398 (1993) 343.
[22] M. Cederwall, G. Ferretti, B.E.W. Nilsson and A. Westerberg, Higher dimensional loop algebras, non-abelian extensions and p-branes, Nucl. Phys. B 424 (1994) 97.
[23] M.J. Duff, Class. Quantum Grav. 5 (1988) 189.
[24] P.K. Townsend, Three lectures on supersymmetry and extended objects, in Proceedings of the 13th GIFT Seminar on Theoretical Physics: Recent problems in Mathematical Physics, Salamanca, Spain (15–27 June 1992).
[25] M.J. Duff and B.E.W. Nilsson, Phys. Lett. B 175 (1986) 417.
[26] M.J. Duff, B.E.W. Nilsson and C.N. Pope, Phys. Lett. B 129 (1983) 39.
[27] K.S. Narain, Phys. Lett. B 169 (1986) 41.
[28] K.S. Narain, M.H. Sarmadi and E. Witten, Nucl. Phys. B 279 (1987) 369.
[29] N. Seiberg, Nucl. Phys. B 303 (1988) 286.
[30] P.S. Aspinwall and D.R. Morrison, String theory on K3 surfaces, DUK-TH-94-68, IASSNS-HEP-94/23, hep-th/9404151 (April 1994).
[31] E. Bergshoeff, E. Sezgin and P.K. Townsend, Phys. Lett. B 189 (1987) 75.
[32] E. Bergshoeff, E. Sezgin and P.K. Townsend, Ann. Phys. 185 (1988) 330.
[33] M.J. Duff, P.S. Howe, T. Inami and K.S. Stelle, Phys. Lett. B 191 (1987) 70.
[34] M.J. Duff and J.X. Lu, Duality for strings and membranes, in Duff, Sezgin and Pope, eds., Supermembranes and Physics in $2+1$ Dimensions (World Scientific, Singapore, 1990); also published in Arnowitt, Bryan, Duff, Nanopoulos, Pope and Sezgin, eds., Strings 90 (World Scientific, Singapore, 1991).
[35] M.J. Duff and J.X. Lu, Nucl. Phys. B 347 (1990) 394.
[36] E. Cremmer, J. Scherk and S. Ferrara, Phys. Lett. B 74 (1978) 61.
[37] E. Cremmer and B. Julia, Nucl. Phys. B 159 (1979) 141.
[38] N. Marcus and J.H. Schwarz, Nucl. Phys. B 228 (1983) 145.
[39] M.J. Duff, $E_8 \times SO(16)$ symmetry of $D = 11$ supergravity?, in Batalin, Isham and Vilkovisky, eds., Quantum Field Theory and Quantum Statistics, Essays in Honour of E.S. Fradkin (Adam Hilger, 1986).
[40] A. Giveon, M. Porrati and E. Rabinovici, Phys. Rep. 244 (1994) 77.
[41] A. Font, L. Ibanez, D. Lust and F. Quevedo, Phys. Lett. B 249 (1990) 35.
[42] S.-J. Rey, Phys. Rev. D 43 (1991) 526.
[43] S. Kalara and D.V. Nanopoulos, Phys. Lett. B 267 (1991) 343.
[44] A. Sen, Nucl. Phys. B 404 (1993) 109.
[45] A. Sen, Phys. Lett. B 303 (1993) 22.
[46] J.H. Schwarz and A. Sen, Phys. Lett. B 312 (1993) 105.
[47] J.H. Schwarz and A. Sen, Nucl. Phys. B 411 (1994) 35.
[48] P. Binetruy, Phys. Lett. B 315 (1993) 80.
[49] A. Sen, Mod. Phys. Lett. A 8 (1993) 5079.
[50] M.J. Duff and J. Rahmfeld, Phys. Lett. B 345 (1995) 441.
[51] J.P. Gauntlett and J.A. Harvey, S duality and the spectrum of magnetic monopoles in heterotic string theory, EFI-94-11, hep-th/9407111 (July 1994).
[52] A. Sen, Nucl. Phys. B 434 (1995) 179.
[53] M.J. Duff, S. Ferrara, R.R. Khuri and J. Rahmfeld, Supersymmetry and dual string solitons, NI-94-034, CTP-TAMU-50/94, hep-th/9506057 (June 1995).
[54] M.J. Duff and K.S. Stelle, Phys. Lett. B 253 (1991) 113.
[55] A. Dabholkar, G. Gibbons, J.A. Harvey and F. Ruiz-Ruiz, Nucl. Phys. B 340 (1990) 33.
[56] M.J. Duff and J.X. Lu, Nucl. Phys. B 354 (1991) 141.
[57] C.G. Callan, J.A. Harvey and A. Strominger, Nucl. Phys. B 359 (1991) 611.

[58] C.G. Callan, J.A. Harvey and A. Strominger, Nucl. Phys. B 367 (1991) 60.
[59] A. Strominger, Nucl. Phys. B 343 (1990) 167; B 353 (1991) 565 (E).
[60] G.T. Horowitz and A. Strominger, Nucl. Phys. B 360 (1991) 197.
[61] M.J. Duff and J.X. Lu, Nucl. Phys. B 390 (1993) 276.
[62] R. Guven, Phys. Lett. B 276 (1992) 49.
[63] G.W. Gibbons and P.K. Townsend, Phys. Rev. Lett. 71 (1993) 3754.
[64] P.K. Townsend, The eleven-dimensional supermembrane revisited, hep-th/9501068 (1995).
[65] M.J. Duff, G.W. Gibbons and P.K. Townsend, Phys. Lett. B 332 (1994) 321.
[66] G.W. Gibbons, G.T. Horowitz and P.K. Townsend, Class. Quantum Grav. 12 (1995) 297.
[67] M.J. Duff, R.R. Khuri, R. Minasian and J. Rahmfeld, Nucl. Phys. B 418 (1994) 195.
[68] P.K. Townsend, String-membrane duality in seven dimensions, DAMTP-R/95/12, hep-th/9504095 (April 1995).
[69] J.A. Harvey and A. Strominger, The heterotic string is a soliton, EFI-95-16, hep-th/9504047 (April 1995).
[70] I. Bars, Stringy evidence for $D = 11$ structure in strongly coupled Type II-A superstring, USC-95/HEP-B2, hep-th/9503228 (March 1995).
[71] P.S. Aspinwall and D.R. Morrison, U-duality and integral structures, CLNS-95/1334, hep-th/9505025 (May 1995).
[72] L. Alvarez-Gaumé and E. Witten, Nucl. Phys. B 234 (1984) 269.
[73] L. Alvarez-Gaumé and P. Ginsparg, Ann. Phys. 161 (1985) 423.
[74] W. Lerche, B.E.W. Nilsson and A.N. Schellekens, Nucl. Phys. B 289 (1987) 609.
[75] W. Lerche, B.E.W. Nilsson, A.N. Schellekens and N.P. Warner, Nucl. Phys. B 299 (1988) 91.
[76] A.N. Schellekens and N.P. Warner, Phys. Lett. B 177 (1986) 317.
[77] A.N. Schellekens and N.P. Warner, Nucl. Phys. B 287 (1987) 317.
[78] M.J. Duff and J.X. Lu, Nucl. Phys. B 354 (1991) 129.
[79] D.J. Gross, J.A. Harvey, E. Martinec and R. Rohm, Nucl. Phys. B 256 (1985) 253.
[80] C.M. Hull and E. Witten, Phys. Lett. B 160 (1985) 398.
[81] A. Sen, String string duality conjecture in six dimensions and charged solitonic strings, TIFR-TH-95-16, hep-th/9504027 (April 1995).
[82] R. Percacci and E. Sezgin, Mod. Phys. Lett. A 10 (1995) 441.
[83] M.J. Duff and J.X. Lu, Phys. Lett. B 273 (1991) 409.
[84] E. Witten, On S-duality in abelian gauge theory, IASSNS-HEP-95-36, hep-th/9505186 (May 1995).
[85] E. Verlinde, Global aspects of electric-magnetic duality, CERN-TH/95-146, hep-th/9506011 (May 1995).
[86] G.W. Gibbons and D.A. Rasheed, Electric-magnetic duality rotations in non-linear electrodynamics, DAMTP preprint (June 1995).
[87] F.J. Dyson, Infinite in All Directions: Gifford lectures given at Aberdeen, Scotland (Harper & Row, New York, 1988).

Reprinted from Phys. Lett. B **409** (1997) 177–84

One loop in eleven dimensions

Michael B. Green [a,1], Michael Gutperle [a,2], Pierre Vanhove [b,3]

[a] *DAMTP, Silver Street, Cambridge CB3 9EW, UK*

[b] *Centre de Physique Théorique, Ecole Polytechnique, 91128 Palaiseau, France*

Received 3 July 1997
Editor: R. Gatto

Abstract

Four-graviton scattering in eleven-dimensional supergravity is considered at one loop compactified on one, two and three-dimensional tori. The dependence on the toroidal geometry determines the known perturbative and non-perturbative terms in the corresponding processes in type II superstring theories in ten, nine and eight dimensions. The ultra-violet divergence must be regularized so that it has a precisely determined finite value that is consistent both with T-duality in nine dimensions and with eleven-dimensional supersymmetry. © 1997 Elsevier Science B.V.

1. Introduction

The leading term in the M-theory effective action is the classical eleven-dimensional supergravity of [1]. Although terms of higher dimension must be strongly constrained by the large amount of supersymmetry they have not been systematically investigated. There is known to be an eleven-form, $\int C^{(3)} \wedge X_8$ (where X_8 is an eight-form made out of the curvature R and $C^{(3)}$ is the three-form potential), which is necessary for consistency with anomaly cancellation [2,3]. Eleven-dimensional supersymmetry relates this to a particular R^4 term [4] as well as a host of other terms and might well determine the complete effective action. Furthermore, the effective action of the compactified theory depends nontrivially on moduli fields associated with the geometry of the compact manifold. This dependence is very strongly restricted by consistency with the duality symmetries of string theory in ten and lower dimensions. For example, the R^4 term in M-theory compactified on a two-torus must be consistent with the structure of perturbative and non-perturbative terms in nine-dimensional IIA and IIB superstrings [5]. This provides strong evidence that it has the form [4]

$$S_{R^4} = \frac{1}{\kappa_{11}^{2/9}} \int d^9x \sqrt{G^{(9)}} h(\Omega, \bar{\Omega}; \mathcal{V}_2) t_8 t_8 R^4, \qquad (1)$$

where $G^{(9)}_{\mu\nu}$ is the M-theory metric in the space transverse to the torus, $\Omega = \Omega_1 + i\Omega_2$ is the complex structure of the torus and $\mathcal{V}_2$ is its volume in eleven-dimensional Planck units. The parameter κ_{11} has dimensions $[\text{length}]^{9/2}$ in arbitrary units (and the volume of the torus is given by $(\kappa_{11})^{4/9}\mathcal{V}_2$ in these units). The notation $t_8 t_8 R^4$ (to be defined below) indicates the particular contraction of four Riemann tensors that arises from integration over fermionic zero modes at one loop in superstring theory [6] as well as from integration over fermionic modes on a D-instanton [5].

[1] E-mail: M.B.Green@damtp.cam.ac.uk.
[2] M.Gutperle@damtp.cam.ac.uk.
[3] E-mail: vanhove@cpth.polytechnique.fr.

PII S0370-2693(97)00931-3

The function h has to be invariant under the action of the modular group, $Sl(2,\mathbb{Z})$, acting on Ω and in [5,4] various arguments were given for why it should have the form

$$h(\Omega,\bar{\Omega};\mathcal{V}_2) = \mathcal{V}_2^{-1/2} f(\Omega,\bar{\Omega}) + \frac{2\pi^2}{3}\mathcal{V}_2. \qquad (2)$$

The function f is a modular-invariant non-holomorphic Eisenstein series which is uniquely specified by the fact that it is an eigenfunction of the Laplace operator on the fundamental domain of $Sl(2,\mathbb{Z})$ with eigenvalue 3/4,

$$\Omega_2^2\, \partial_\Omega \partial_{\bar{\Omega}} f = \tfrac{3}{4} f \qquad (3)$$

(i.e. $f = \zeta(3)E_{\frac{3}{2}}$, where E_s is a Maass waveform of eigenvalue $s(s-1)$ [7]). Significantly, this is the kind of Ward identity that the threshold corrections in lower-dimensional $N = 2$ theories satisfy [8] and it suggests a very stringent set of geometrical constraints. The expansion of f for large Ω_2 has two power-behaved terms plus an infinite series of exponentially decreasing terms. These have exactly the correct coefficients to be identified with the tree-level and one-loop terms, together with an infinite series of D-instanton contributions in type IIB superstring theory. This identification makes use of T-duality to relate a multiply-charged D-instanton of type IIB with a multiply-wound world-line of a multiply-charged D-particle of type IIA. Indeed, semiclassical quantization around these D-instanton configurations may be carried out by functional integration around the D-particle background, as outlined in [4].

In this paper we will show how the sum of the perturbative and D-instanton contributions to the $t_8t_8R^4$ term are efficiently encoded in M-theory in the expression for the scattering of four gravitons at one loop in eleven-dimensional supergravity perturbation theory. The particles circulating around the loop are the 256 physical states that comprise the massless eleven-dimensional supergraviton. It may seem surprising that perturbation theory is of any significance since supergravity has terrible ultra-violet divergences in eleven dimensions. Furthermore, the absence of any scalar fields means that there is no small dimensionless coupling constant. However, there is strong reason to believe that the one-loop $t_8t_8R^4$ terms are protected from receiving higher-loop contributions by an eleven-dimensional nonrenormalization theorem since they are related by supersymmetry to the $C^{(3)} \wedge X_8$ term. Upon compactification to ten or fewer dimensions the Kaluza–Klein modes of the circulating fields are reinterpreted in terms of the windings of euclidean D-particle world-lines. The massive D-particles reproduce the D-instanton effects while the massless one (the massless ten-dimensional supergraviton) is equivalent to the perturbative one-loop string effects. The $t_8t_8R^4$ term obtained at tree level in string theory arises, somewhat miraculously, from windings of the D-particle world-lines in the eleventh dimension.

The one-loop diagram can, in principle, be obtained using covariant Feynman rules by summing over the contributions of the component fields circulating in the loop the graviton, gravitino and third-rank antisymmetric tensor fields. Alternatively, it can be expressed in terms of on-shell superfields. In that case the dynamics is defined by superspace quantum mechanics with the massless superparticle action which reduces in a fixed parameterization of the world-line to [9],

$$S_{\text{particle}} = \frac{1}{2}\int d\tau G_{\mu\nu}(\dot{X}^\mu - i\bar{\Theta}\Gamma^\mu\dot{\Theta})(\dot{X}^\nu - i\bar{\Theta}\Gamma^\nu\dot{\Theta}), \qquad (4)$$

where Θ is a 32 component $SO(10,1)$ spinor, $\mu = 1,\cdots,11$ and the reparameterization constraint requires the action density to vanish on physical states. For present purposes it will be sufficient to limit consideration to processes in which the external states do not carry momentum in the eleventh dimension and which are also not polarized in that direction although these are not essential conditions. This loop amplitude can be calculated by making use of the light-cone description of the super-particle in which the vertex operator for a graviton has the form

$$V^{(r)}(\zeta^{(r)},\tau^{(r)}) = \zeta^{(r)ik}(\dot{X}^i - \frac{1}{4p^+}\theta\gamma^{ij}\theta k_j^{(r)}) \times (\dot{X}^k - \frac{1}{4p^+}\theta\gamma^{kl}\theta k_l^{(r)})e^{ik^{(r)}\cdot X}, \qquad (5)$$

where $i,j,\cdots = 3,\cdots,11$, $\zeta^{(r)}_{\mu\nu}$ is the graviton wave function with momentum $k^{(r)}_\mu$ (where $(k^{(r)})^2 = 0 = k^{(r)\mu}\zeta^{(r)}_{\mu\nu}$), θ^a is a $SO(9)$ spinor in the light-cone gauge defined by $\Gamma^+\Theta = 0$ and $X^+ = p^+\tau$ (where $V^\pm \equiv V^1 \pm V^2$ with timelike V^1). This vertex operator is attached at a point $\tau^{(r)}$ on the world-line and is defined in a frame in which $k^{(r)+} = 0$, $\zeta^{(r)+\mu} = 0$. In a canonical treatment of this system the equations of motion determine that $X^i = p^i\tau + x^i$ and $\theta^a = S^a/\sqrt{p^+}$ and the (anti)commutation relations are $[p^i, x^j] = -i\delta^{ij}$, $\{S^a, S^b\} = \delta^{ab}$ (just as with the zero mode components of the corresponding relations in the ten-dimensional type IIA superstring theory).

The loop amplitude reduced to $(11-n)$ dimensions by compactification on an n-dimensional torus, T^n, has the form

$$\begin{aligned} A_4^{(n)} &= \frac{1}{\pi^{5/2}\mathcal{V}_n}\mathrm{Tr}\int d^{11-n}\mathbf{p} \\ &\times \int_0^\infty \frac{d\tau}{\tau}\left(\prod_{r=1}^4 \int_0^\tau d\tau^{(r)} V^{(r)}\right)\sum_{\{l_I\}} e^{-\tau(\mathbf{p}^2 + G^{(n)IJ} l_I l_J)} \\ &= \frac{1}{\pi^{\frac{n}{2}-3}\mathcal{V}_n}\tilde{K}\int_0^\infty d\tau\tau^{n/2-13/2}\sum_{\{l_I\}} e^{-\tau G^{(n)IJ} l_I l_J} \\ &\times \int_0^\tau \prod_{r=1}^4 d\tau^{(r)} F(\{k^{(r)}, \tau^{(r)}\}), \end{aligned} \tag{6}$$

where $\mathbf{p} = p^i$ is the $(11-n)$-dimensional loop momentum transverse to the compact directions and $G_{IJ}^{(n)}$ $(I, J = 1, \cdots n)$ is the metric on T^n which has volume $\mathcal{V}_n = \sqrt{\det G^{(n)}}$. The kinematic factor, $\tilde{K}$, in the second line involves eight powers of the external momenta and follows from the trace over the components of S^a. It may also be written as

$$\tilde{K} \sim \int d^{16}\eta \prod_{r=1}^4 \left(\zeta^{(r)}_{\mu_r\omega_r} k^{(r)}_{\nu_r} k^{(r)}_{\tau_r} \bar{\eta}\Gamma^{\mu_r\nu_r\rho_r}\eta\bar{\eta}\Gamma^{\omega_r\tau_r}{}_{\rho_r}\eta\right), \tag{7}$$

where η is a chiral $SO(9,1)$ Grassmann spinor. The overall normalization will be chosen so that $\tilde{K}$ is the linearized approximation to

$$t_8 t_8 R^4 \equiv t^{\mu_1\ldots\mu_8} t_{\nu_1\ldots\nu_8} R^{\nu_1\nu_2}_{\mu_1\mu_2}\cdots R^{\nu_7\nu_8}_{\mu_7\mu_8}, \tag{8}$$

where the tensor $t_8^{\mu_1\cdots\mu_8}$ was defined in [10]. The function F is a simple function of the external momenta. Since we are interested here in the leading term in the low-energy limit (the $t_8 t_8 R^4$ term) we can set the momenta $k^{(r)}$ to zero in the integrand so that $\int\prod d\tau^{(r)} F$ is replaced by τ^4 giving

$$A_4^{(n)} = \frac{\pi^{3/2}}{\mathcal{V}_n}\tilde{K}\int_0^\infty d\tau\tau^{n/2-5/2}\sum_{\{l_I\}} e^{-\pi\tau G^{(n)IJ} l_I l_J}. \tag{9}$$

Though this expression was obtained in a special frame we know that there is an $(11-n)$-dimensional covariant extension (including the case $n = 0$) that would follow directly from the covariant Feynman rules and should be easy to check by an explicitly calculation using the component form of the supergravity field theory action.

The expression for $A_4^{(n)}$ will contribute to the $t_8 t_8 R^4$ terms in the effective action for M-theory compactified on T^n. In order to determine the dependence of the amplitude on the geometry of the torus on which it is compactified it will be important to express $A_4^{(n)}$ in terms of the winding of the loop around T^n. This could be obtained directly from the definition of the loop amplitude as a functional integral or by performing a Poisson summation on the n integers, l_i, which amounts to inverting the metric in (9). The result is

$$A_4^{(n)} = \pi^{3/2}\tilde{K}\int_0^\infty d\hat{\tau}\hat{\tau}^{1/2}\sum_{\{\hat{l}_I\}} e^{-\pi\hat{\tau} G_{IJ}\hat{l}_I\hat{l}_J}, \tag{10}$$

where $\hat{\tau} = \tau^{-1}$.

The ultraviolet divergence of eleven-dimensional supergravity comes from the zero winding number term, $\{\hat{l}_I\} = 0$, in the limit that the loop shrinks to a point ($\hat{\tau} \to \infty$). We will formally write this divergent term as the ill-defined expression, $C \equiv \int d\hat{\tau}\hat{\tau}^{1/2}$.[4] The fact that the one-loop supergravity amplitude is infinite is a signal that point-particle dynamics alone cannot determine the short-distance physics of M-theory. A microscopic theory – such as Matrix theory [12] – should determine the correct finite value of C. This is somewhat analogous to the way in which divergent loop amplitudes in ten-dimensional super Yang–Mills are regularized by ten-dimensional string theory (for example, the F^4 terms in the effective action of the heterotic and open string theories [13,14]). Indeed, we will soon see that consistency with the duality symmetries of string theory together with the assumption that the eleven-dimensional theory can be obtained as a limit of the lower-dimensional theories

[4] The presence of a cubically divergent $t_8 t_8 R^4$ term in eleven-dimensional supergravity was first suggested in [11].

determines the precise finite renormalized value for the constant, C, that is also consistent with eleven-dimensional supersymmetry. It is a challenge to Matrix theory to reproduce this number.

In the following we will associate the integer $\hat{l}_r$ with the winding number of the loop around a compact dimension of circumference R_{12-r} (for $r \geq 1$). If a single direction is compactified on a circle of circumference $R_{11} \equiv \mathcal{V}_1$, the loop can be expressed as a sum over the winding number of the euclidean supergraviton world-line.

$$\frac{1}{\pi^{3/2}} A_4^{(1)} = C\tilde{K} + 2\tilde{K} \int_0^\infty d\hat{\tau}\hat{\tau}^{1/2} \sum_{\hat{l}_1 > 0} e^{-\pi\hat{\tau}\hat{l}_1^2 R_{11}^2}$$

$$= C\tilde{K} + \tilde{K}\zeta(3)\frac{1}{\pi R_{11}^3}. \qquad (11)$$

Rather strikingly, the finite R_{11}-dependent term gives a term in the effective ten-dimensional action that is precisely that obtained in [15,16] from the tree-level IIA string theory (here written in the M-theory frame). Although the regularized constant, C, is still undetermined, we will see later that it must be set equal to the coefficient of the one-loop $t_8 t_8 R^4$ term of the low energy effective action of string theory. The absence of any further perturbative or nonperturbative terms is in accord with the conjectures in [5,4].

Compactification on a torus ($n = 2$) gives a richer structure. In this case the one-loop amplitude has the form

$$\frac{1}{\pi^{3/2}} \mathcal{V}_2 A_4^{(2)} = \mathcal{V}_2 C\tilde{K}$$

$$+ \mathcal{V}_2^{-1/2}\tilde{K} \sum_{(\hat{l}_1, \hat{l}_2) \neq (0,0)} \int d\hat{\tau}\hat{\tau}^{1/2} e^{-\pi\hat{\tau}\frac{1}{\Omega_2}|\hat{l}_1 + \hat{l}_2\Omega|^2}$$

$$= \mathcal{V}_2 C\tilde{K} + \frac{1}{2\pi}\mathcal{V}_2^{-1/2}\tilde{K} \sum_{(\hat{l}_1, \hat{l}_2) \neq (0,0)} \frac{\Omega_2^{3/2}}{|\hat{l}_1 + \hat{l}_2\Omega|^3}$$

$$= \frac{1}{2\pi}\tilde{K}\left(2\pi C\mathcal{V}_2 + \mathcal{V}_2^{-1/2} f(\Omega, \bar{\Omega})\right), \qquad (12)$$

where the divergent zero winding term, $\hat{l}_1 = \hat{l}_2 = 0$ has again been separated from the terms with non-zero winding. The function f in this expression is precisely the (finite) Ω-dependent term in (2). In particular, in the limit $\mathcal{V}_2 \to 0$ M-theory should reduce to type IIB superstring theory in ten dimensions [17,18] with the complex scalar field, $\rho \equiv C^{(0)} + ie^{-\phi^B}$, identified with Ω (where $C^{(0)}$ is the $R \otimes R$ scalar and ϕ^B is the IIB dilaton). More precisely, the correspondence between the parameters in M-theory and in IIB is,

$$\mathcal{V}_2 \equiv R_{10}R_{11} = e^{\frac{1}{3}\phi^B} r_B^{-\frac{4}{3}},$$

$$\Omega_2 \equiv \frac{R_{10}}{R_{11}} = e^{-\phi^B} \qquad (13)$$

(where r_B is the radius of the tenth dimension expressed in the IIB sigma-model frame). Using the fact that $\sqrt{G^{(9)}}(\mathcal{V}_2)^{-\frac{1}{2}} t_8 t_8 R^4 = \sqrt{g^{B(9)}} r_B t_8 t_8 R^4$ (where $g^{B(d)}$ denotes the determinant of the IIB sigma-model metric in d dimensions) we see that (12) leads, in the ten-dimensional IIB limit ($r_B \to \infty$), to the expression suggested in [5]. This has the property that, when expanded in perturbation theory ($e^{-\phi^B} = \Omega_2 \to \infty$), it exactly reproduces both the tree-level and one-loop $t_8 t_8 R^4$ terms of the type IIB theory as well as an infinite series of D-instanton terms [5,4]. Importantly, the divergent term in (12)is proportional to $\mathcal{V}_2$ and does not contribute in the limit of relevance to ten-dimensional type IIB – thus the eleven-dimensional one-loop calculation reproduces the complete, finite, $t_8 t_8 R^4$ effective action in the type IIB theory.

As before, the coefficient of the tree-level term in the type IIB superstring perturbation theory is reproduced by the configurations with $\hat{l}_2 = 0$, in which the particle in the loop winds around the eleventh dimension but not the tenth (obviously there is a symmetry under the interchange of these directions so we could equally well consider the terms with $\hat{l}_1 = 0$). In order to expand (12) systematically for large Ω_2 it is necessary to undo the Poisson summation on $\hat{l}_1$ for the terms with $\hat{l}_2 \neq 0$. These terms are then expressed as a sum of multiply-wound D-particle world-lines where the winding number is $\hat{l}_2$ and the D-particle charge is the Kaluza–Klein charge, l_1. In the limit $\mathcal{V}_2 \to 0$ the terms with $l_1 = 0$ reproduce the one-loop $t_8 t_8 R^4$ term of ten-dimensional type IIB while the $l_1 \neq 0$ terms give the contribution of the sum of D-instantons. The precise contribution due to the world-line of a particular wrapped massive D-particle (of mass l_1 and winding $\hat{l}_2$) to this instanton sum is identical to that obtained by considering semiclassical quantization of four-graviton scattering in this background. Supersymmetry causes all quantum corrections to vanish. The additional fact that the one-loop string theory result is equivalent to the sum of windings of a massless D-particle (the supergraviton) is notable [4]. From the point of view of the string calculation this term arises from wrapping the string world-sheet in a degenerate manner around a circle.

We can now use the additional constraint of T-duality to pin down the precise value of C. This is determined by recalling that the one-loop terms in both the IIA and IIB theories are invariant under inversion of the circumference, $r_A \leftrightarrow r_B^{-1}$. This equates the coefficients of the $\mathcal{V}_2$ term and the $\Omega_2^{1/2}\mathcal{V}_2^{-1/2}$ terms in (12), and the result is that the coefficient C must be set equal to the particular value,

$$C = \frac{\pi}{3}. \tag{14}$$

The fact that the modular function in (2) is a Maass wave form satisfying (3) is easily deduced from the integral representation, (12). Developing a geometrical understanding of the origin of this equation would be of interest.

Upon compactifying on T^3 new issues arise. The full U-duality group is $Sl(3,\mathbb{Z}) \times Sl(2,\mathbb{Z})$. The seven moduli consist of the six moduli associated with the three-torus and $C_{123}^{(3)}$, the component of the antisymmetric three-form on the torus. The latter couples to the euclidean three-volume of the M-theory two-brane which can wrap around T^3. The perturbative eleven-dimensional one-loop expression can be expected to reproduce the effects of the Kaluza–Klein modes but not of the wrapped Membrane world-volume. However, these wrapped Membrane effects will be determined in the following by imposing U-duality and making use of the one-loop results for type II string theory compactified on T^2 [4]. We will write the complete four-graviton amplitude as

$$\mathcal{V}_3 A_4^{(3)} = \pi^{3/2}\tilde{K}H, \tag{15}$$

where the scalar function H depends on the seven moduli fields. There are several distinct classes of terms that will make separate contributions to the complete function $H = \sum_i H_i$.

The effects of the Kaluza–Klein modes are obtained from (10) with $n = 3$. In order to compare with string theory on T^2 we will choose R_{11} to be the special M-theory direction so that $R_{11} = e^{2\phi^A/3}$, where ϕ^A is the IIA dilaton (although the expression obviously has complete symmetry between all three compact directions). The sums over windings will be divided into various groups of terms. Firstly, there is the term with zero winding in all directions which is again divergent but will be set equal to the regularised value given by C in (14), which implies

$$H_1 = \frac{\pi}{3}\mathcal{V}_3 \equiv \frac{\pi}{3}T_2, \tag{16}$$

where T_2 is the imaginary part of the Kähler structure of T^2. The sum over $\hat{l}_1 \neq 0$ with $\hat{l}_2 = \hat{l}_3 = 0$ once again leads to the correct tree-level string contribution proportional to $\zeta(3)$,

$$H_2 = \zeta(3)\frac{1}{\pi R_{11}^3} = \zeta(3)\frac{1}{\pi}e^{-2\phi^A}. \tag{17}$$

The remaining sum is over all values of $\hat{l}_1$, $\hat{l}_2$ and $\hat{l}_3$ excluding the $\hat{l}_2 = \hat{l}_3 = 0$ terms. This is usefully reexpressed by converting the $\hat{l}_1$ sum to a sum over Kaluza–Klein modes by a Poisson resummation. The sum of these terms is

$$\begin{aligned} H_3 + H_4 = {} & \sqrt{\frac{\det G}{G_{11}}} \sum_{(\hat{l}_2,\hat{l}_3)\neq(0,0)} \sum_{l_1} \int_0^\infty d\hat{\tau} \\ & \times \exp\Bigg[2\pi i l_1 \hat{l}_2 \frac{G_{12}}{G_{11}} + 2\pi i l_1 \hat{l}_3 \frac{G_{13}}{G_{11}} - \pi l_1^2 \frac{1}{\hat{\tau} G_{11}} \\ & - \pi\hat{\tau}\Bigg(\hat{l}_2^2\Big(G_{22} - \frac{G_{12}^2}{G_{11}}\Big) + \hat{l}_3^2\Big(G_{33} - \frac{G_{13}^2}{G_{11}}\Big) \\ & + 2\Big(G_{23} - \frac{G_{12}G_{13}}{G_{11}}\Big)\hat{l}_2\hat{l}_3\Bigg)\Bigg] \\ = {} & T_2 \sum_{(\hat{l}_2,\hat{l}_3)\neq(0,0)} \sum_{l_1} \int_0^\infty d\hat{\tau} \\ & \times \exp\left[-\pi l_1^2 e^{-2\phi^A}\frac{1}{\hat{\tau}} + 2\pi i l_1 \hat{l}_i A^{(i)} - \pi\hat{\tau}\hat{l}_i\hat{l}_j g_{ij}^A\right]. \end{aligned} \tag{18}$$

In this expression G_{ij} is the metric on T^3 in M-theory coordinates with the convention that $i = 12 - \mu$ ($i = 1, 2, 3$) and the components of the IIA string sigma-model metric on the two-torus are given by

$$g_{ij}^A = R_{11}\left(G_{ij} - \frac{G_{1i}G_{1j}}{G_{11}}\right), \tag{19}$$

where $i, j = 2, 3$. The components of the $R \otimes R$ one-form potentials in the directions of the two-torus in (18) are defined by

$$A^{(i)} = \frac{G_{1i}}{G_{11}}. \tag{20}$$

The expression (18) depends on the $R \otimes R$ one-form, the complex structure of the two-torus,

$$U = \frac{1}{g^A_{22}}(g^A_{23} + i\sqrt{\det g^A}) \tag{21}$$

and the combination $T_2 e^{-2\phi^A}$. But it does *not* depend separately on the Kähler structure,

$$T = B_{12} + i\sqrt{\det g^A} = C^{(3)}_{123} + i\mathcal{V}_3, \tag{22}$$

where $\mathcal{V}_3 = R_9 R_{10} R_{11}$. In the last step we have used the usual identification of the $NS \otimes NS$ two-form with the M-theory three-form, $C^{(3)}$, and the fact that $r^A_2 r^A_3 = R_9 R_{10} R_{11}$, where r^A_i is the circumference of the dimension labelled i in the IIA sigma-model frame.

The expression (18) contains perturbative and non-perturbative contributions to the $t_8 t_8 R^4$ term in the IIA effective action. The perturbative term is obtained by setting $l_1 = 0$. The resulting double sum over $\hat{l}_2$ and $\hat{l}_3$ is logarithmically divergent, just as in the analogous problem considered in [8]. This is a reflection of the fact that the one-loop diagram in eight-dimensional supergravity is logarithmically divergent. As in [8,4], this divergence may be regularized in a unique manner that is consistent with modular invariance by adding a term, $\Sigma = \ln(T_2 U_2/\Lambda^2)$, giving

$$\begin{aligned} H_3 &= \sum_{(\hat{l}_2,\hat{l}_3)\neq(0,0)} \frac{U_2}{|\hat{l}_2 + \hat{l}_3 U|^2} - \ln(U_2 T_2/\Lambda^2) \\ &= -\left[\ln(U_2|\eta(U)|^4) + \ln(T_2)\right], \end{aligned} \tag{23}$$

where Λ^2 is adjusted to cancel the divergence coming from the sum. So we see that the piece of the perturbative string theory one-loop amplitude that depends on U is reproduced by configurations in which a massless particle propagating in the loop has a world-line that winds around the torus. This is the generalization of the way in which the IIB one-loop term was reproduced earlier by windings of a massless D-particle around a circle.

The terms with $l_1 \neq 0$ in (18) consist of a sum of non-perturbative D-instanton contributions,

$$\begin{aligned} H_4 = 2U_2\rho^A_2 &\sum_{\substack{(\hat{l}_2,\hat{l}_3)\neq(0,0)\\ l_1\neq 0}} \frac{|l_1|}{|\hat{l}_2 + \hat{l}_3 U|} \\ &\times K_1\left(2\pi\rho^A_2|\hat{l}_2 + \hat{l}_3 U||l_1|\right) e^{2i\pi l_1(\hat{l}_2 A^{(1)} + \hat{l}_3 A^{(2)})}, \end{aligned} \tag{24}$$

where $\rho^A_2 = r^A_2 e^{-\phi^A}$. Using the fact that $K_1(z) = \sqrt{\frac{\pi}{2z}}e^{-z}(1 + o(1/z))$ for large z we see that at weak IIA coupling, $e^{-\phi^A} \to \infty$, these terms are exponentially suppressed. The contribution of these instanton terms in the nine-dimensional case described earlier is obtained by letting $r^A_3 \to \infty$. In this case $U \to i\infty$ and only the $\hat{l}_3 = 0$ term in (24) survives. The double sum over l_1 and $\hat{l}_2$ becomes the nine-dimensional D-instanton sum contained in (12) which was explicitly given in [4].

So far we have ignored the contributions to the $t_8 t_8 R^4$ term arising from configurations in which the world-volume of the M-theory Membrane is wrapped around T^3. Such contributions are obviously not contained in the one-loop $D = 11$ supergravity amplitude. As with the contributions that came from circulating D-particles, the configurations that contribute to the $t_8 t_8 R^4$ term are described by the multiple windings of world-lines of nine-dimensional BPS states in ultra-short (256-dimensional) multiplets. Recall that these nine-dimensional states are winding states of fundamental strings with no momentum or oscillator excitations which are configurations of the wrapped M-theory Membrane with no Kaluza–Klein excitations. Such contributions are therefore labelled by two integers and depend only on the volume of the three-torus, $\det G$, and on $C^{(3)}$ but are independent of the other five components of the metric (i.e., they depend only on T and $\bar{T}$). These configurations of the IIA string world-sheet are just those that enter the functional integral for the $t_8 t_8 R^4$ term at one loop in string perturbation theory. Indeed, as explained in [4] (and in an analogous problem in [13,14]), the piece of the one-loop string amplitude that depends on T and $\bar{T}$ is given by a sum over non-degenerate wrapped world-sheets and contributes

$$H_5 = 2\sum_{m,n>0} \frac{1}{n}\left(e^{2\pi imnT} + e^{-2\pi imn\bar{T}}\right)$$
$$= -\left[\ln(|\eta(T)|^4) + \frac{\pi}{3}T_2 - \ln(T_2)\right], \qquad (25)$$

where m, n are the integers that label the windings of the world-sheet. The sum of H_1, H_3 and H_5 reproduces the full one-loop perturbative string theory result. Applying T-duality in one of the toroidal directions transforms this into the one-loop term of the IIB theory. The complete non-perturbative structure of the ten-dimensional $t_8t_8R^4$ terms of the IIB theory can then be recovered using the series of dualities described in [4].

The total contribution to the $t_8t_8R^4$ terms in the eight-dimensional effective action is given (in IIA string coordinates) by

$$S_{R^4} \sim \int d^8x\sqrt{g^{A(8)}}\, r_2^A r_3^A\, H\, t_8t_8R^4, \qquad (26)$$

where $H = \sum_{i=1}^{5} H_i$ and we have ignored an overall constant. This expression is invariant under the requisite $Sl(3,\mathbb{Z})\otimes Sl(2,\mathbb{Z})$ U- duality symmetry. The particle winding numbers $(\hat{l}_1, \hat{l}_2, \hat{l}_3)$ transform as a **3** of $Sl(3,\mathbb{Z})$ while the windings of the Membrane (m, n) transform as a **2** of $Sl(2,\mathbb{Z})$. The decoupling of the two factors in the U-duality group arises from the fact that the ultra-short BPS states in nine dimensions do not contain both a wrapped Membrane and Kaluza–Klein charges. Compactification on T^4 to seven dimensions is more complicated since the U-duality group is $Sl(5,\mathbb{Z})$, which is not a product of two factors. The $t_8t_8R^4$ terms in this case depends on the BPS spectrum in eight dimensions, which was discussed in [19].

In this paper we have studied properties of the one-loop amplitude in eleven-dimensional supergravity compactified on tori to lower dimensions. Upon compactifying to nine dimensions on T^2 this amplitude reproduces the complete perturbative and non-perturbative $t_8t_8R^4$ terms in the effective actions for the corresponding string theories if the ulta-violet divergence is chosen to have a particular finite regularized value (a value that can presumably be derived from Matrix Theory). This value is also in agreement with that obtained by supersymmetry which relates it to the $C^{(3)} \wedge X_8$ term [4]. In the limit in which the two-torus has zero volume, $\mathcal{V}_2 \to 0$, the regularized term does not contribute and the complete $t_8t_8R^4$ term of the ten-dimensional IIB theory is reproduced precisely by the one-loop supergravity calculation. It is noteworthy that the $t_8t_8R^4$ terms in the IIB theory only get string-theory perturbative contributions at tree-level and one loop, in addition to the non-perturbative D-instanton contributions. This is tantalizingly similar to the structure of the F^2 terms in $N = 2$ super Yang–Mills theory in $D = 4$ dimensions.

Upon compactification to eight dimensions on T^3 the one-loop eleven-dimensional supergravity amplitude reproduces the $Sl(3,\mathbb{Z})$-symmetric piece of the $t_8t_8R^4$ term that is associated with Kaluza–Klein instantons. The remaining piece that arises from the wrapped Membrane is uniquely determined by consistency with the T-duality that relates the IIA and IIB theories, together with one-loop string perturbation theory. We have not addressed the new issues that arise in compactification on manifolds of non-trivial holonomy or compactification to lower dimensions. For example, compactification on T^6 requires considerations of the wrapped world-volume of the M-theory five-brane.

In addition to the R^4 terms considered here there are many other terms of the same dimension involving the other fields of ten-dimensional string theory and M-theory. In the language of type IIB supergravity some of these terms conserve the R-symmetry charge (as with the R^4 term) while some of them violate it in a manner consistent with the instanton effects (such as the λ^{16} term described in [5]).

Since there is no scalar field there is no possibility of a well-defined perturbation expansion in powers of a small coupling constant in eleven-dimensional supergravity. Fortuitously, the relation of the $t_8t_8R^4$ term to the eleven-form, $C^{(3)} \wedge X_8$, via supersymmetry, implies that the one-loop expression is exact with no corrections from higher-loop diagrams (sin e normalization of the eleven-form is fixed by ano cancellation). This adds to the ever-increasing boo evidence that the constraints of maximal supergrav are profoundly restrictive.

We wish to acknowledge EC support under the Human Capital and Mobility programme. MBG is grateful to the University of Paris VI where this work was completed.

References

[1] E. Cremmer, B. Julia and J. Scherk, Phys. Lett. B 76 (1978) 409.

[2] C. Vafa and E. Witten, hep-th/9505053, Nucl. Phys. B 447 (1995) 261.

[3] M.J. Duff, J.T. Liu and R. Minasian, hep-th/9506126, Nucl. Phys. B 452 (1995) 261.

[4] M.B. Green and P. Vanhove, hep-th/9704145.

[5] M.B. Green and M. Gutperle, hep-th/9701093.

[6] M.B. Green and J.H. Schwarz, Phys. Lett. B 109 (1982) 444.

[7] A. Terras, Harmonic Analysis on Symmetric Spaces and Applications, Vol. I, Springer–Verlag (1985).

[8] L. Dixon, V. Kaplunovsky and J. Louis, Nucl. Phys. B 355 (1991) 649.

[9] R. Kallosh, hep-th/9705056.

[10] M.B. Green and J.H. Schwarz, Nucl. Phys. B 198 (1982) 441.

[11] E.S. Fradkin and A.A. Tseytlin, Nucl. Phys. B 227 (1983) 252.

[12] T. Banks, W. Fischler, S.H. Shenker and L. Susskind, hep-th/9610043, Phys. Rev. D 55 (1997) 5112.

[13] C. Bachas and E. Kiritsis, Proceedings of Trieste Spring School and Workshop, April 1996 hep-th/9611205.

[14] C. Bachas, C. Fabre, E. Kiritsis, N.A. Obers and P. Vanhove Heterotic/ type-I duality, D-brane instantons and generalized prepotentials (to appear).

[15] M.T. Grisaru, A.E.M Van de Ven and D. Zanon, Nucl. Phys. B 277 (1986) 388; Nucl. Phys. B 277 (1986) 409.

[16] D.J. Gross and E. Witten, Nucl. Phys. B 277 (1986) 1.

[17] P. Aspinwall, in: Proceedings of 'S-duality and mirror symmetry, Trieste 1995, hep-th/9508154; Nucl. Phys. Proc. 46 (1996) 30.

[18] J.H. Schwarz, hep-th/9607201.

[19] D. Berenstein, R. Corrado and J. Distler, hep-th/9704087.

Reprinted from Nucl. Phys. B **463** (1996) 383–97

Five-branes and M-theory on an orbifold

Edward Witten [1]

School of Natural Sciences, Institute for Advanced Study, Olden Lane, Princeton, NJ 08540, USA

Received 4 January 1996; accepted 19 January 1996

Abstract

We relate Type IIB superstrings compactified to six dimensions on K3 to an eleven-dimensional theory compactified on $(\mathbf{S}^1)^5/\mathbf{Z}_2$. Eleven-dimensional five-branes enter the story in an interesting way.

1. Introduction

By now, there is substantial evidence for the existence of an eleven-dimensional quantum theory with eleven-dimensional supergravity as its long wave-length limit. Moreover, the theory contains two-branes and five-branes at least macroscopically, and some of their properties are known; for instance, the κ-invariant Bergshoeff-Sezgin-Townsend action [1] describes the long wavelength excitations of a macroscopic two-brane.

The description by eleven-dimensional supergravity with two-branes and five-branes is expected to be valid when all characteristic length scales (of a space-time and the branes that it contains) are large compared to the Planck length. One also has some information about the behavior under certain conditions when some dimensions of space-time are *small* compared to the Planck scale. For instance, the eleven-dimensional "M-theory" (where M stands for magic, mystery, or membrane, according to taste) on $X \times \mathbf{S}^1$, with X any ten-manifold, is equivalent to Type IIA on X, with a Type IIA string coupling constant that becomes small when the radius of $\mathbf{S}^1$ goes to zero. Likewise, the M-theory on $Y \times$ K3, with Y a seven-manifold, is equivalent to the heterotic string on $Y \times \mathbf{T}^3$, and the M-theory on $X \times \mathbf{S}^1/Z_2$, with X a ten-manifold, is equivalent to the $E_8 \times E_8$

[1] Research supported in part by NSF Grant PHY92-45317.

PII S0550-3213(96)00032-6

heterotic string on X; in each case, the string coupling constant becomes small when the volume of the last factor goes to zero.

The evidence for the existence of the M-theory (beyond the consistency of the classical low energy theory) comes mainly from the success of statements deduced from the relations of the M-theory to strings. Even a few more similar examples might therefore significantly enrich the story. The purpose of the present paper is to add one more such example, by arguing that the M-theory on $Z \times (\mathbf{S}^1)^5/\mathbf{Z}_2$ is equivalent to the Type IIB superstring on $Z \times$ K3. Here Z is an arbitrary six-manifold, but as usual in such arguments, by scaling up the metric of Z, one can reduce to the case that $Z = \mathbf{R}^6$. In fact, once an equivalence is established between the M-theory on $Z \times (\mathbf{S}^1)^5/\mathbf{Z}_2$ and Type IIB on $Z \times$ K3 when Z is large, it can be followed into the region of small Z.

The equivalence of the M-theory on $\mathbf{R}^6 \times (\mathbf{S}^1)^5/\mathbf{Z}_2$ with Type IIB on $\mathbf{R}^6 \times$ K3 was also conjectured recently by Dasgupta and Mukhi [2] who independently pointed out a problem – involving anomaly cancellation and the distribution of the twisted sectors among fixed points – that will be addressed below. Some general comments about Type IIB on K3 as an M-theory orbifold were also made recently by Hull [3].

2. The low energy supergravity

Compactification of the Type IIB superstring on K3 gives a six-dimensional theory with a chiral supersymmetry which (upon toroidal compactification to four dimensions) is related to $N = 4$ supersymmetry in $D = 4$. We will call this six-dimensional chiral $N = 4$ supersymmetry (though the number of supercharges is only twice the minimum possible number in $D = 6$).

The supergravity multiplet of chiral $N = 4$ supergravity contains, in addition to the graviton, five self-dual tensors (that is two-forms with self-dual field strength) plus gravitinos. The graviton in six dimensions has nine helicity states, while the self-dual tensor has three, so the total number of bosonic helicity states is $9 + 5 \cdot 3 = 24$; the gravitinos likewise have 24 helicity states. The supergravity multiplet has gravitational anomalies (which cannot be canceled by the Green-Schwarz mechanism alone), so any consistent theory with chiral $N = 4$ supergravity in six dimensions must contain matter multiplets also.

There is actually only one possible matter multiplet in chiral $N = 4$ supersymmetry. It is the tensor multiplet, which contains five spin zero bosons, an anti-self-dual anti-symmetric tensor (that is a two-form field whose field strength is anti-self-dual) with three helicity states, and $5 + 3 = 8$ helicity states of chiral fermions. Cancellation of gravitational anomalies requires that the number of tensor multiplets be precisely 21.

Using only the low energy supergravity, one can deduce (for a survey of such matters see [4]) that the moduli space $\mathcal{M}$ of vacua is locally the homogeneous space $SO(21,5)/SO(21) \times SO(5)$. In the particular case of a chiral $N = 4$ theory obtained by compactification of Type IIB on K3, the global structure is actually (as asserted in Eq. (4.16) of [5]; see [6] for a more precise justification)

$\mathcal{M} = SO(21,5;\mathbf{Z})\backslash SO(21,5)/SO(21) \times SO(5)$. This depends on knowledge of conformal field theory T-duality on K3 [7] together with the $SL(2,\mathbf{Z})$ symmetry of ten-dimensional Type IIB superstring theory.

Note that since there is no scalar in the chiral $N = 4$ supergravity multiplet, the dilaton is one of the $5 \times 21 = 105$ scalars that come from the tensor multiplets. The $SO(21,5;\mathbf{Z})$ discrete symmetry mixes up the dilaton with the other 104 scalars, relating some but not all of the "strong coupling" regimes to regions of weak coupling or large volume.

2.1. *Five-branes and the tensor multiplet anomaly*

We will need some background about five-branes and gravitational anomalies.

We want to consider a certain model of *global* chiral $N = 4$ supergravity with the tensor multiplet. To do this, we begin in eleven-dimensional Minkowski space, with coordinates $x^1, \ldots, x^{11}$ (x^1 being the time), and gamma matrices $\Gamma^1, \ldots, \Gamma^{11}$ which obey

$$\Gamma^1 \Gamma^2 \ldots \Gamma^{11} = 1. \tag{2.1}$$

Now we introduce a five-brane with world-volume given by the equations

$$x^7 = \ldots = x^{11} = 0. \tag{2.2}$$

The presence of this five-brane breaks half of the 32 space-time supersymmetries. The 16 surviving supersymmetries are those that obey $\Gamma^7 \ldots \Gamma^{11} = 1$, or equivalently, in view of (2.1), $\Gamma^1 \ldots \Gamma^6 = 1$. Thus, the surviving supersymmetries are chiral in the six-dimensional sense; the world-volume theory of the five-brane has chiral $N = 4$ supersymmetry. This is *global* supersymmetry since – as the graviton propagates in bulk – there is no massless graviton on the five-brane world-volume.

Therefore, the massless world-volume fields must make up a certain number of tensor multiplets, this being the only matter multiplet allowed by chiral $N = 4$ supergravity. In fact, there is precisely one tensor multiplet. The five massless scalars are simply the fluctuations in $x^7, \ldots, x^{11}$; the massless world-volume fermions are the Goldstone fermions associated with the supersymmetries under which the five-brane is not invariant; and the anti-self-dual tensor has an origin that was described semiclassically in [8]. The assertion that the massless world-volume excitations of the five-brane consist of precisely one tensor multiplet can also be checked by compactifying the x^{11} direction on a circle, and comparing to the structure of a Dirichlet four-brane of Type IIA [9]. (In compactifying the M-theory to Type IIA, the five-brane wrapped around x^{11} turns into a four-brane; the tensor multiplet of $5 + 1$ dimensions reduces to a vector multiplet in $4 + 1$ dimensions, which is the massless world-volume structure of the Dirichlet four-brane.)

Now we want to allow fluctuations in the position of the five-brane and compute the quantum behavior at long wavelengths. At once we run into the fact that the tensor multiplet on the five-brane world-volume has a gravitational anomaly. Without picking a coordinate system on the five-brane world-volume, how can one cancel the anomaly in the one loop effective action of the massless world-volume fields (even at very long wavelengths where the one loop calculation is valid)?

This question was first discussed by Duff, Liu, and Minasian [10]; what follows is a sort of dual version of their resolution of the problem.[2] The tensor multiplet anomaly *cannot* be cancelled, as one might have hoped, by a world-volume Green-Schwarz mechanism. Instead one has to cancel a world-volume effect against a bulk contribution from the eleven-dimensional world, rather as in [11].

This theory has in the long-wavelength description a four-form F that is closed in the absence of five-branes, but which in the presence of five-branes obeys

$$dF = \delta_V \tag{2.3}$$

where δ_V is a delta-function supported on the five-form world-volume V. There is here a key point in the terminology: given a codimension n submanifold W of space-time, the symbol δ_W will denote not really a delta "function" but a closed n-form supported on W which integrates to one in the directions normal to W. For instance, in one dimension, if P is the origin on the x-axis, then $\delta_P = \delta(x)\,dx$ where $\delta(x)$ is the "Dirac delta function" and $\delta(x)\,dx$ is, therefore, a closed one-form that vanishes away from the origin and whose integral over the x-axis (that is, the directions normal to P) is 1. With δ_V thus understood as a closed five-form in the five-brane case, (2.3) is compatible with the Bianchi identity $d(dF) = 0$ and is, in fact, sometimes taken as a defining property of the five-brane as it asserts that the five-brane couples magnetically to F.

Now suppose that in the low energy expansion of the effective eleven-dimensional theory on a space-time M there is a term

$$\Delta L = \int_M F \wedge I_7 \tag{2.4}$$

where I_7 is a gravitational Chern-Simons seven-form. Exactly which Chern-Simons seven-form it should be will soon become clear. Under an infinitesimal diffeomorphism $x^I \to x^I + \epsilon v^I$ (ϵ being an infinitesimal parameter and v a vector field), I_7 does not transform as a tensor, but rather $I_7 \to I_7 + dJ_6$, where J_6 is a certain six-form (which depends upon v). The transformation of ΔL under a diffeomorphism is therefore

$$\Delta L \to \Delta L + \int_M F \wedge dJ_6 = \Delta L - \int_M dF \wedge J_6. \tag{2.5}$$

Thus, ΔL is generally covariant in the absence of five-branes, but in the presence of a five-brane, according to (2.3), one gets

[2] In the very similar case of ten-dimensional Type IIA five-branes, the dual version was worked out in unpublished work by J. Blum and J.A. Harvey.

$$\Delta L \to \Delta L - \int_V J_6. \tag{2.6}$$

But gravitational anomalies in n dimensions involve precisely expressions $\int J_n$ where J_n is as above (that is, J_n appears in the transformation law of a Chern-Simons $n+1$-form I_{n+1} by $I_{n+1} \to I_{n+1} + dJ_n$; see [12] for an introduction to such matters.) Thus with the correct choice of I_7, the anomaly of ΔL in the presence of a five-brane precisely cancels the world-volume anomaly of the tensor multiplet. This is thus a case in which an interaction in the bulk is needed to cancel on anomaly on the world-volume. Moreover (as explained in a dual language in [10]), the presence in eleven dimensions of the interaction ΔL can be checked by noting that upon compactification on a circle, this interaction reduces to the $H \wedge I_7$ term found in [13] for Type IIA superstrings; here H is the usual three-form field strength of the Type IIA theory.

What has been said to this point is sufficient for our purposes. However, I cannot resist a further comment that involves somewhat similar ideas. The seven-form F' dual to F does not obey $dF' = 0$ even in the absence of five-branes; from the eleven-dimensional supergravity one finds instead

$$dF' + \frac{1}{2} F \wedge F = 0. \tag{2.7}$$

One may ask how this is compatible with the Bianchi identity $d(dF') = 0$ once – in the presence of five-branes – one encounters a situation with $dF \neq 0$. The answer involves the anti-self-dual three-form field strength T on the five-brane world-volume. According to Eq. (3.3) of [14], this field obeys not – as one might expect – $dT = 0$, but rather $dT = F$. If then in the presence of a five-brane, (2.7) becomes

$$dF' + \frac{1}{2} F \wedge F - T \wedge \delta_V = 0, \tag{2.8}$$

then the Bianchi identity still works even in the presence of the five-brane. The $T \wedge \delta_V$ term in fact follows from the coupling in Eq. (3.3) of [14], which gives a five-brane contribution to the equation of motion of the three-form A. Thus, we get a new derivation of the relevant coupling and in particular of the fact that $dT = F$.

3. Type IIB on K3

We now come to the main focus of this paper. One would like to understand the "strong coupling behavior" of the Type IIB theory compactified on K3, or more precisely, the behavior as one goes to infinity in the moduli space $\mathcal{M}$ of vacua. As explained above, this theory has a $SO(21,5;\mathbf{Z})$ discrete symmetry, which gives many identifications of strong coupling or small volume with weak coupling or large volume, but there remain (as in, [5], Section 3, or [6]) inequivalent limits in which one can go to infinity in $\mathcal{M}$.

Any limit can be reached by starting at a given point $P \in \mathcal{M}$ and then considering the one-parameter family of vacua $P_t = e^{tx}P$ where x is a generator of $SO(21,5)$ and t is a positive real number. As $t \to \infty$, one approaches infinity in $\mathcal{M}$ in a direction that depends upon x. In any such limit, by looking at the lightest states, one aims to find a description by an effective ten-dimensional string theory or eleven-dimensional field theory. The duality group visible (though mostly spontaneously broken, depending on the precise choice of P) in this effective theory will include the subgroup Γ of $SO(21,5;\mathbf{Z})$ that commutes with x (and so preserves the particular direction in which one has gone to infinity).

As in [5,6], one really only needs to consider x's that lead to a maximal set of light states, and because of the discrete $SO(21,5;\mathbf{Z})$, there are only finitely many cases to consider. We will focus here on the one limit that seems to be related most directly to the M-theory.

Consider a subgroup $SO(16) \times SO(5,5)$ of $SO(21,5)$. Let x be a generator of $SO(5,5)$ that commutes with an $SL(5)$ subgroup. Then the subgroup of $SO(21,5;\mathbf{Z})$ that commutes with x – and so is visible if one goes to infinity in the direction determined by x – contains $SO(16) \times SL(5,\mathbf{Z})$.

Since it will play a role later, let us discuss just how $SL(5,\mathbf{Z})$ can be observed as a symmetry at infinity. Instead of making mathematical arguments, we will discuss another (not unrelated, as we will see) physical problem with $SO(21,5;\mathbf{Z})$ symmetry, namely the compactification on a five-torus of the $SO(32)$ heterotic string to five dimensions, with $SO(21,5;\mathbf{Z})$ as the T-duality group. The region at infinity in moduli space in which there is a visible $SL(5,\mathbf{Z})$ symmetry is simply the large volume limit, with the torus large in all directions. In what sense can $SL(5,\mathbf{Z})$ be "observed"? It is spontaneously broken (to a finite subgroup, generically trivial) by the choice of a metric on the five-torus, but, if one is free to move around in the moduli space of large volume metrics (remaining at infinity in $\mathcal{M}$) one can see that there is a spontaneously broken $SL(5,\mathbf{Z})$.

Now, actually, the relevant region at infinity in moduli space is parametrized by a large metric on the torus, a B-field, and a flat $SO(32)$ bundle described by five commuting Wilson lines W_j. (For the moment we take the flat bundle to be topologically trivial, a point we return to in Subsection 3.4.) If one is free to vary all of these, one can certainly observe the full $SL(5,\mathbf{Z})$. Suppose, though, that in some method of calculation, the Wilson lines are frozen at particular values, and one can only vary the metric and B-field. Then one will only observe the subgroup of $SL(5,\mathbf{Z})$ that leaves the Wilson lines invariant.

For instance, if the Wilson lines are trivial – a rather special situation with unbroken $SO(32)$ – one will see all of $SL(5,\mathbf{Z})$. Here is another case that will enter below though it will appear mysterious at the moment. As the W_j commute, they can be simultaneously diagonalized, with eigenvalues λ_j^a, $a = 1,\ldots,32$. Suppose that the λ_j^a are all ± 1, and have the property that for each fixed a, $\prod_j \lambda_j^a = -1$. There are 16 collections of five ± 1's whose product is -1 (namely $1,1,1,1,-1$ and four permutations of that sequence; $1,1,-1,-1,-1$ and nine permutations of that sequence; and $-1,-1,-1,-1,-1$). Let the λ^a_j be such that each such permutation appears exactly twice. This breaks $SL(5, \mathbf{Z})$ to the finite index subgroup Γ consisting of $SL(5, \mathbf{Z})$ matrices $M^j{}_k$, $j, k = 1, \ldots, 5$ such that $\sum_j M^j{}_k$ is odd for each fixed k. If the Wilson lines are frozen at the stated values, it is only Γ and not all of $SL(5, \mathbf{Z})$ that can be observed by varying the metric and B-field.

3.1. *Interpretation of the symmetry*

Let us go back to the Type IIB theory on K3 and the attempt to interpret the strong coupling limit that was described, the one with a visible $SL(5, \mathbf{Z})$. As in the example just discussed, the $SL(5, \mathbf{Z})$ symmetry is strongly suggestive of the mapping class group of a five-torus. Thus, one is inclined to relate this particular limit of Type IIB on K3 to the M-theory on $\mathbf{R}^6$ times a five-manifold built from $(\mathbf{S}^1)^5$. This cannot be $(\mathbf{S}^1)^5$ itself, because the M-theory on $\mathbf{R}^6 \times (\mathbf{S}^1)^5$ would have twice as much supersymmetry as we want. One is tempted instead to take an orbifold of $(\mathbf{S}^1)^5$ in such a way as to break half of the supersymmetry while preserving the $SL(5, \mathbf{Z})$.

A natural way to break half the supersymmetries by orbifolding is to divide by a $\mathbf{Z}_2$ that acts as -1 on all five circles. This is actually the only choice that breaks half the supersymmetry and gives a *chiral* $N = 4$ supersymmetry in six dimensions. In fact, dividing by this $\mathbf{Z}_2$ leaves precisely those supersymmetries whose generators obey $\Gamma^7 \Gamma^8 \ldots \Gamma^{11} \epsilon = \epsilon$. This condition was encountered in the discussion of the five-brane, and leaves the desired chiral supersymmetry. So M-theory compactified on $(\mathbf{S}^1)^5/\mathbf{Z}_2$ is our candidate for an eleven-dimensional interpretation of Type IIB superstrings on K3.

More precisely, the proposal is that M-theory on $(\mathbf{S}^1)^5/\mathbf{Z}_2$ has the property that when *any* $\mathbf{S}^1$ factor in $(\mathbf{S}^1)^5/\mathbf{Z}_2$ goes to zero radius, the M-theory on this manifold goes over to a weakly coupled Type IIB superstring. This assertion should hold not just for one of the five circles in the definition of $(\mathbf{S}^1)^5/\mathbf{Z}_2$, but for any of infinitely many circles obtained from these by a suitable symmetry transformation.

3.2. *Anomalies*

Let us work out the massless states of the theory, first (as in [15]) the "untwisted states," that is the states that come directly from massless eleven-dimensional fields, and then the "twisted states," that is, the states that in a macroscopic description appear to be supported at the classical singularities of $(\mathbf{S}^1)^5/\mathbf{Z}_2$.

The spectrum of untwisted states can be analyzed quickly by looking at antisymmetric tensors. The three-form A of the eleven-dimensional theory is odd under parity (because of the $A \wedge F \wedge F$ supergravity interaction). Since the $\mathbf{Z}_2$ by which we are dividing $(\mathbf{S}^1)^5$ reverses orientation, A is odd under this transformation. The zero modes of A on $(\mathbf{S}^1)^5$ therefore give, after the $\mathbf{Z}_2$ projection, five two-forms (and ten scalars, but no vectors or three-forms) on $\mathbf{R}^6$. The self-dual parts of these tensors are the expected five self-dual tensors of the supergravity multiplet, and the anti-self-dual parts are part of five tensor multiplets. The number of tensor multiplets from the untwisted sector is therefore five.

Just as in [15], the untwisted spectrum is anomalous; there are five tensor multiplets, while 21 would be needed to cancel the gravitational anomalies. 16 additional tensor multiplets are needed from twisted sectors.

The problem, as independently raised in [2], is that there appear to be 32 identical twisted sectors, coming from the 32 fixed points of the $\mathbf{Z}_2$ action on $(\mathbf{S}^1)^5$. How can one get 16 tensor multiplets from 32 fixed points? We will have to abandon the idea of finding a vacuum in which all fixed points enter symmetrically.

Even so, there seems to be a paradox. As explained in [15], since the eleven-dimensional theory has no gravitational anomaly on a smooth manifold, the gravitational anomaly of the eleven-dimensional massless fields on an orbifold is a sum of delta functions supported at the fixed points. In the case at hand, the anomaly can be canceled by 16 tensor multiplets (plus a Green-Schwarz mechanism), but there are 32 fixed points. Thus, each fixed point has an anomaly, coming from the massless eleven-dimensional fields, that could be canceled by $16/32 = 1/2$ tensor multiplets.[3] The paradox is that it is not enough to *globally* cancel the gravitational anomaly by adding sixteen tensor multiplets. One needs to cancel the anomaly *locally* in the eleven-dimensional world, somehow modifying the theory to add at each fixed point half the anomaly of a tensor multiplet. How can this be, given that the tensor multiplet is the only matter multiplet of chiral $N = 4$ supersymmetry, so that any matter system at a fixed point would be a (positive) integral number of tensor multiplets?

3.3. Resolution of the paradox

To resolve this paradox, the key point is that because the fixed points in $(\mathbf{S}^1)^5/\mathbf{Z}_2$ have codimension *five*, just like the codimension of a five-brane world-volume, there is another way to cancel anomalies apart from including massless fields on the world-volume. We can assume that the fixed points are magnetic sources of the four-form F. In other words, we suppose that (even in the absence of conventional five-branes) dF is a sum of delta functions supported at the orbifold fixed points. If so, then the bulk interaction $\Delta L = \int F \wedge I_7$ that was discussed earlier will give additional contributions to the anomalies supported on the fixed points.

Since a magnetic coupling of F to the five-brane cancels the anomaly of a tensor multiplet, if an orbifold fixed point has "magnetic charge" $-1/2$, this will cancel the anomaly from the eleven-dimensional massless fields (which otherwise could be canceled by $1/2$ a tensor multiplet). If an orbifold fixed point has magnetic charge $+1/2$, this doubles the anomaly, so that it can be canceled if there is in addition a "twisted sector" tensor multiplet supported at that fixed point. Note that it is natural that a $\mathbf{Z}_2$ orbifold point could have magnetic charge that is half-integral in units of the usual quantum of charge.

[3] The eleven-dimensional massless fields by obvious symmetries contribute the same anomaly at each fixed point.

A constraint comes from the fact that the sum of the magnetic charges must vanish on the compact space $(\mathbf{S}^1)^5/\mathbf{Z}_2$. Another constraint comes from the fact that if we want to maintain supersymmetry, the charge for any fixed point cannot be less than $-1/2$. Indeed, a fixed point of charge less than $-1/2$ would have an anomaly that could not be canceled by tensor multiplets; a negative number of tensor multiplets or a positive number of wrong chirality tensor multiplets (violating supersymmetry) would be required. An example of how to satisfy these constraints and ensure local cancellation of anomalies is to assign charge $-1/2$ to 16 of the fixed points, and charge $+1/2$ to the other 16. With one tensor multiplet supported at each of the last 16 fixed points, such a configuration has all anomalies locally cancelled in the eleven-dimensional sense.

Here is another way to cancel the anomalies locally. Assign magnetic charge $-1/2$ to each of the 32 fixed points, but include at each of 16 points on $(\mathbf{S}^1)^5/\mathbf{Z}_2$ a conventional five-brane, of charge 1. The total magnetic charge vanishes (as $32(-1/2)+16=0$) and since both a fixed point of charge $-1/2$ and a conventional five-brane are anomaly-free, all anomalies are cancelled locally. Each five-brane supports one tensor multiplet; the scalars in the tensor multiplets determine the positions of the five-branes on $(\mathbf{S}^1)^5/\mathbf{Z}_2$.

I would like to suggest that this last anomaly-canceling mechanism is the general one, and that the case that the magnetic charge is all supported on the fixed points is just a special case in which the five-branes and fixed points coincide. In fact, if a five-brane happens to move around and meet a fixed point, the charge of that fixed point increases by 1. This gives a very natural interpretation of the "twisted sector" modes of a fixed point of charge 1/2. Such a fixed point supports a tensor multiplet, which contains five scalars; we interpret the scalars as representing a possible perturbation in the five-brane position away from the fixed point.

If we accept this interpretation, there is no issue of what is the "right" configuration of charges for the fixed points; any configuration obeying the constraints (total charge 0 and charge at least $-1/2$ for each fixed point) appears somewhere on the moduli space. The only issue is what configuration of charges has the most transparent relation to string theory.

Let us parametrize the five circles in $(\mathbf{S}^1)^5$ by periodic variables x^j, $j=7,\ldots,11$, of period 1, with $\mathbf{Z}_2$ acting by $x^j \to -x^j$ so that the fixed points have all coordinates 0 or 1/2. We take $SL(5,\mathbf{Z})$ to act linearly, by $x^j \to M^j{}_k x^k$, with $M^j{}_k$ an $SL(5,\mathbf{Z})$ matrix. Thus $SL(5,\mathbf{Z})$ leaves invariant one fixed point P, the "origin" $x^j=0$, and acts transitively on the other 31. The only $SL(5,\mathbf{Z})$-invariant configuration of charges obeying the constraints is to assign magnetic charge $+31/2$ to P and $-1/2$ to each of the others. Then each of the 16 tensor multiplets would be supported at the origin. This configuration cancels the anomalies and is $SL(5,\mathbf{Z})$ invariant. However, it does not seem to be the configuration with the closest relation to string theory.

To see this, consider the limit in which one of the circles in $(\mathbf{S}^1)^5$ becomes small. To an observer who does not detect this circle, one is then left with $(\mathbf{S}^1)^4/\mathbf{Z}_2$, which is a K3 orbifold. Our hypothesis about M-theory on $(\mathbf{S}^1)^5/\mathbf{Z}_2$ says that this theory should go over to weakly coupled Type IIB on K3 when *any* circle shrinks. In $(\mathbf{S}^1)^4/\mathbf{Z}_2$, there are 16 fixed points; in quantization of Type IIB superstrings on this orbifold, one tensor multiplet comes from each of the 16 fixed points.

In M-theory on $(\mathbf{S}^1)^5/\mathbf{Z}_2$, there are 32 fixed points. When one of the circles is small, then – to an observer who does not resolve that circle – the 32 fixed points appear to coalesce pairwise to the 16 fixed points of the string theory on $(\mathbf{S}^1)^4/\mathbf{Z}_2$. To reproduce the string theory answer that one tensor multiplet comes from each singularity, we want to arrange the charges on $(\mathbf{S}^1)^5/\mathbf{Z}_2$ in such a way that each pair of fixed points differing only in the values of one of the coordinates contributes one tensor multiplet.

This can be done by arranging the charges in the following "checkerboard" configuration. If a fixed point has $\sum_j x^j$ integral, we give it charge $-1/2$. If $\sum_j x^j$ is a half-integer, we give it charge $+1/2$. Then any two fixed points differing only by the value of the x^j coordinate – for any given j – have equal and opposite charge, and contribute a total of one tensor multiplet.

Moreover, the four-form field strength F of the M-theory reduces in ten dimensions to a three-form field strength H. This vanishes for string theory on K3, so one can ask how the string theory can be a limit of an eleven-dimensional theory in a vacuum with non-zero F. If we arrange the charges in the checkerboard fashion, this puzzle has a natural answer. In the limit in which the jth circle shrinks to zero, equal and opposite charges are superposed and cancel, so the resulting ten-dimensional theory has zero H.

The checkerboard configuration is not invariant under all of $SL(5,\mathbf{Z})$, but only under the finite index subgroup Γ introduced just prior to Subsection 3.1 (the subgroup consisting of matrices $M^j{}_k$ such that $\sum_j M^j{}_k$ is odd for each k). Thus the reduction to ten-dimensional string theory can work not only if one shrinks one of the five circles in the definition of $(\mathbf{S}^1)^5/\mathbf{Z}_2$, but also if one shrinks any of the (infinitely many) circles obtainable from these by a Γ transformation.

Just as in the discussion in which Γ was introduced, in the checkerboard vacuum, one cannot see the full $SL(5,\mathbf{Z})$ if the only parameters one is free to vary are the metric and three-form A on $(\mathbf{S}^1)^5/\mathbf{Z}_2$. An $SL(5,\mathbf{Z})$ transformation w not in Γ is a symmetry only if combined with a motion of the other moduli – in fact a motion of some five-branes to compensate for the action of w on the charges of fixed points.

3.4. Check by comparison to other dualities

In the study of string theory dualities, once a conjecture is formulated that runs into no immediate contradiction, one of the main ways to test it is to try to see what implications it has when combined with other, better established dualities.

In the case at hand, we will (as was done independently by Dasgupta and Mukhi [2]) mainly compare our hypothesis about M-theory on $(\mathbf{S}^1)^5/\mathbf{Z}_2$ to the assertion that M-theory on $X \times \mathbf{S}^1$ is equivalent, for any five-manifold X, to Type IIA on X.

To combine the two assertions in an interesting way, we consider M-theory on $\mathbf{R}^5 \times \mathbf{S}^1 \times (\mathbf{S}^1)^5/\mathbf{Z}_2$. On the one hand, because of the $\mathbf{S}^1$ factor, this should be equivalent to Type IIA on $\mathbf{R}^5 \times (\mathbf{S}^1)^5/\mathbf{Z}_2$, and on the other hand, because of the $(\mathbf{S}^1)^5/\mathbf{Z}_2$ factor, it should be equivalent to Type IIB on $\mathbf{R}^5 \times \mathbf{S}^1 \times$ K3.

It is easy to see that, at least in general terms, we land on our feet. Type IIB on $\mathbf{R}^5 \times \mathbf{S}^1 \times \mathrm{K3}$ is equivalent by T-duality to Type IIA on $\mathbf{R}^5 \times \mathbf{S}^1 \times \mathrm{K3}$, and the latter is equivalent by Type IIA - heterotic duality to the heterotic string on $\mathbf{R}^5 \times \mathbf{S}^1 \times \mathbf{T}^4 = \mathbf{R}^5 \times \mathbf{T}^5$, and thence by heterotic - Type I duality to Type I on $\mathbf{R}^5 \times \mathbf{T}^5$.

On the other hand, Type IIA on $\mathbf{R}^5 \times (\mathbf{S}^1)^5/\mathbf{Z}_2$ is an orientifold which is equivalent by T-duality to Type I on $\mathbf{R}^5 \times \mathbf{T}^5$ [16,17].

So the prediction from our hypothesis about M-theory on $(\mathbf{S}^1)^5/\mathbf{Z}_2$ – that Type IIA on $(\mathbf{S}^1)^5/\mathbf{Z}_2$ should be equivalent to Type IIB on $\mathbf{S}^1 \times \mathrm{K3}$ – is correct. This is a powerful test.

Components of the moduli space

What remains to be said? The strangest part of our discussion about M-theory on $(\mathbf{S}^1)^5/\mathbf{Z}_2$ was the absence of a vacuum with symmetry among the fixed points. We would like to find a counterpart of this at the string theory level, for Type IIA on $(\mathbf{S}^1)^5/\mathbf{Z}_2$.

The Type IIA orientifold on $(\mathbf{S}^1)^5/\mathbf{Z}_2$ needs – to cancel anomalies – 32 D-branes located at 32 points in $(\mathbf{S}^1)^5$; moreover, this configuration of 32 points must be invariant under $\mathbf{Z}_2$. It is perfectly possible to place one D-brane at each of the 32 fixed points, maintaining the symmetry between them. Does this not contradict what we found in eleven dimensions?

The resolution of this puzzle starts by observing that the D-branes that are not at fixed points are paired by the $\mathbf{Z}_2$. So as the D-branes move around in a $\mathbf{Z}_2$-invariant fashion, the number of D-branes at each fixed point is conserved modulo two (if a D-brane approaches a fixed point, its mirror image does also). Thus, there is a $\mathbf{Z}_2$-valued invariant associated with each fixed point; allowing for the fact that the total number of D-branes is even, there are 31 independent $\mathbf{Z}_2$'s.

What does this correspond to on the Type I side? A configuration of 32 D-branes on $(\mathbf{S}^1)^5/\mathbf{Z}_2$ is T-dual to a Type I theory compactified on $(\mathbf{S}^1)^5$ with a flat $SO(32)$ bundle. However, the moduli space of flat $SO(32)$ connections on the five-torus is not connected – there are many components. One component contains the trivial connection and leads when one considers the deformations to the familiar Narain moduli space of the heterotic string on the five-torus. This actually corresponds to a D-brane configuration with an even number of D-branes at each fixed point. The Wilson lines W_j can be simultaneously block-diagonalized, with 16 two-by-two blocks. The ath block in W_j is

$$\begin{pmatrix} \cos\theta_{j,a} & \sin\theta_{j,a} \\ -\sin\theta_{j,a} & \cos\theta_{j,a} \end{pmatrix}, \tag{3.1}$$

with $\theta_{j,a}$, $j = 1, \ldots, 5$ being angular variables that determine the position on $(\mathbf{S}^1)^5$ of the ath D-brane (which also has an image whose coordinates are $-\theta_{j,a}$).

There are many other components of the moduli space of flat connections on the five-torus, corresponding to the 32 $\mathbf{Z}_2$'s noted above. Another component – in a sense at the opposite extreme from the component that contains the trivial connection – is the following. Consider a flat connection with the properties that the W_j can all be simultaneously diagonalized, with eigenvalues $\lambda_{j,a} = \pm 1$, $a = 1, \ldots, 32$. Since the positions of the D-branes are the phases of the eigenvalues of the W_j, this corresponds to a situation in which all D-branes are at fixed points. Pick the $\lambda_{j,a}$ such that each of the 32 possible sequences of five ± 1's arises as $\lambda_{j,a}$ for some value of a. Then there is precisely one D-brane at each of the 32 fixed points. This flat bundle - call it F - cannot be deformed *as a flat bundle* to the flat bundle with trivial connection; that is clear from the fact that the number of D-branes at fixed points is odd. Therefore, F does not appear on the usual Narain moduli space of toroidal compactification of the heterotic string to five dimensions. However, it can be shown that the bundle F is topologically trivial so that the flat connection on it can be deformed (but not via flat connections) to the trivial connection.[4] Thus, compactification using the bundle F is continuously connected to the usual toroidal compactification, but only by going through configurations that are not classical solutions.

The fact that the configuration with one D-brane at each fixed point is not on the usual component of the moduli space leads to a solution to our puzzle. In reconciling the two string theory descriptions of M-theory on $\mathbf{R}^5 \times \mathbf{S}^1 \times (\mathbf{S}^1)^5/\mathbf{Z}_2$, a key step was Type IIA - heterotic string duality relating Type IIA on $\mathbf{R}^5 \times \mathbf{S}^1 \times \mathrm{K3}$ to the heterotic string on $\mathbf{R}^6 \times \mathbf{S}^1 \times (\mathbf{S}^1)^4 = \mathbf{R}^5 \times \mathbf{T}^5$. This duality holds with the *standard* component of the moduli space on $\mathbf{T}^5$, so even though the symmetrical D-brane configuration exists, it is not relevant to our problem because it is related to a different component of the moduli space of flat $SO(32)$ bundles.

Working on the Type IIA orientifold on $(\mathbf{S}^1)^5/\mathbf{Z}_2$ which is T-dual to a flat $SO(32)$ bundle on the usual component of the moduli space means that the number of D-branes at each fixed point is even. With 32 D-branes and 32 fixed points, it is then impossible to treat symmetrically all fixed points. One can, however, pick any 16 fixed points, and place two D-branes at each of those, and none at the others. In the quantization, one then gets one five-dimensional vector multiplet from each fixed point that is endowed with a D-brane and none from the others.[5] Recalling that the vector multiplet is the dimensional reduction of the tensor multiplet from six to five dimensions, this result agrees with what we had in eleven dimensions: given any 16 of the 32 fixed points, there is a point in moduli space such that each of those 16 contributes precisely one matter multiplet, and the others contribute none.

It is possible that the absence of a vacuum with symmetrical treatment of all fixed points means that these theories cannot be strictly understood as orbifolds, but in any event, whatever the appropriate description is in eleven-dimensional M-theory, we have

---

[4] In a previous draft of this paper, it was erroneously claimed that the bundle F was topologically non-trivial, with non-vanishing Stieffel-Whitney classes. The error was pointed out by E. Sharpe and some topological details were clarified by D. Freed.

[5] This is most easily seen by perturbing to a situation in which the pair of D-branes is near but not at the fixed point. For orientifolds, there are no twisted sector states from a fixed point that does not have D-branes. After the $\mathbf{Z}_2$ projection, a pair of D-branes in the orientifold produces the same spectrum as a single D-brane in an unorientifolded Type IIA, and this is a single vector multiplet, as explained in detail in Section 2 of [18].

found a precisely analogous behavior in the ten-dimensional Type IIA orientifold.

Other similar checks

One might wonder about other similar checks of the claim about M-theory on $(\mathbf{S}^1)^5/\mathbf{Z}_2$. One idea is to look at M-theory on $\mathbf{R}^5 \times \mathbf{S}^1/\mathbf{Z}_2 \times (\mathbf{S}^1)^5/\mathbf{Z}_2$. The idea would be that this should turn into an $E_8 \times E_8$ heterotic string upon taking the $\mathbf{S}^1/\mathbf{Z}_2$ small, and into a Type IIB orientifold on $\mathbf{S}^1/\mathbf{Z}_2 \times K3$ if one shrinks the $(\mathbf{S}^1)^5/\mathbf{Z}_2$. However, because the two $\mathbf{Z}_2$'s do not commute in acting on spinors, it is hard to make any sense of this orbifold.

A similar idea is to look at M-theory on $\mathbf{R}^4 \times \mathrm{K3} \times (\mathbf{S}^1)^5/\mathbf{Z}_2$. When the last factor shrinks, this should become Type IIB on K3 × K3, while if the K3 factor shrinks then (allowing, as in a discussion that will appear elsewhere [19], for how the $\mathbf{Z}_2$ orbifolding acts on the homology of K3) one gets the heterotic string on $(\mathbf{S}^1)^8/\mathbf{Z}_2$. These should therefore be equivalent. But one does not immediately have tools to verify or disprove that equivalence.

Relation to extended gauge symmetry and non-critical strings

A rather different kind of check can be made by looking at the behavior when some D-branes – or eleven-dimensional five-branes – coincide.

Type IIA on K3 gets an extended $SU(2)$ gauge symmetry when the K3 develops an A_1 singularity.[6] This is not possible for Type IIB on K3, which has a chiral $N = 4$ supersymmetry that forbids vector multiplets. Rather, the weakly coupled Type IIB theory on a K3 that is developing an A_1 singularity develops [21] a non-critical string (that is, a string that propagates in flat Minkowski space and does not have the graviton as one of its modes) that couples to the anti-self-dual part of one of the antisymmetric tensor fields (the part that is in a tensor multiplet, not in the supergravity multiplet).

This six-dimensional non-critical string theory is a perhaps rather surprising example, apparently, of a non-trivial quantum theory in six-dimensional Minkowski space. Recently, it was argued by Strominger [22] that by considering almost coincident parallel five-branes in eleven dimensions, one gets on the world-volume an alternative realization of the same six-dimensional non-critical string theory.

We can now (as partly anticipated by Strominger's remarks) close the circle and deduce from the relation between M-theory on $\mathbf{T}^5/\mathbf{Z}_2$ and Type IIB on K3 *why* Type IIB on a K3 with an A_1 singularity gives the same unusual low energy dynamics as two nearby parallel five-branes in eleven dimensions. This follows from the fact that in the map from M-theory on $\mathbf{T}^5/\mathbf{Z}_2$ to Type IIB on K3, a configuration on $\mathbf{T}^5/\mathbf{Z}_2$ with two coincident five-branes is mapped to a K3 with an A_1 singularity. To see that these configurations are mapped to each other, it is enough to note that upon compactification on an extra circle of generic radius, they are precisely the configurations that give an enhanced $SU(2)$. This may be deduced as follows:

---

[6] We really mean a quantum A_1 singularity including a condition on a certain world-sheet theta angle [20].

(1) M-theory on $\mathbf{R}^5 \times \mathbf{S}^1 \times \mathbf{T}^5/\mathbf{Z}_2$ is equivalent to Type IIA on $\mathbf{R}^5 \times \mathbf{T}^5/\mathbf{Z}_2$, with the five-branes replaced by D-branes, and gets an enhanced $SU(2)$ gauge symmetry precisely when two five-branes, or D-branes, meet. Indeed, when two D-branes meet, their $U(1) \times U(1)$ gauge symmetry (a $U(1)$ for each D-brane) is enhanced to $U(2)$ (from the Chan-Paton factors of two coincident D-branes), or equivalently a $U(1)$ is enhanced to $SU(2)$.

(2) Type IIB on $\mathbf{R}^5 \times \mathbf{S}^1 \times$ K3 is equivalent to Type IIA on $\mathbf{R}^5 \times \mathbf{S}^1 \times$ K3 and therefore – because of the behavior of Type IIA on K3 – the condition on the K3 moduli that causes a $U(1)$ to be extended to $SU(2)$ is precisely that there should be an A_1 singularity.

Other orbifolds

Dasgupta and Mukhi also discussed M-theory orbifolds $\mathbf{R}^{11-n} \times (\mathbf{S}^1)^n/\mathbf{Z}_2$. The $\mathbf{Z}_2$ action on the fermions multiplies them by the matrix $\widetilde{\Gamma} = \Gamma^{11-n+1}\Gamma^{11-n+2}\ldots\Gamma^{11}$, and the orbifold can therefore only be defined if $\widetilde{\Gamma}^2 = 1$ (and not -1), which restricts us to n congruent to 0 or 1 modulo 4.

The case $n = 1$ was discussed in [15], $n = 4$ gives a K3 orbifold, and $n = 5$ has been the subject of the present paper. The next cases are $n = 8, 9$. For $n = 8$, as there are no anomalies, it would take a different approach to learn about the massless states from fixed points. For $n = 9$, Dasgupta and Mukhi pointed out the beautiful fact that the number of fixed points – $2^9 = 512$ – equals the number of left-moving massless fermions needed to cancel anomalies, and suggested that one such fermion comes from each fixed point. Since the left-moving fermions are singlets under the (chiral, right-moving) supersymmetry, this scenario is entirely compatible with the supersymmetry and is very plausible.

Reduced rank

Finally, let us note the following interesting application of part of the discussion above. Toroidal compactification of the heterotic (or Type I) string on a flat $SO(32)$ bundle that is not on the usual component of the moduli space (being T-dual to a configuration with an odd number of D-branes at fixed points) gives an interesting and simple way to reduce the rank of the gauge group while maintaining the full supersymmetry. Since $2n+1$ D-branes at a fixed point gives gauge group $SO(2n+1)$, one can in this way get gauge groups that are not simply laced. Models with these properties have been constructed via free fermions [23] and as asymmetric orbifolds [24].

Acknowledgements

I would like to thank M. Duff, J. Polchinski, and C. Vafa for helpful discussions.

References

[1] E. Bergshoeff, E. Sezgin, and P.K. Townsend, Supermembranes And Eleven-Dimensional Supergravity, Phys. Lett. B 189 (1987) 75.

[2] K. Dasgupta and S. Mukhi, Orbifolds Of *M*-Theory, hepth/9512196.

[3] C.M. Hull, String Dynamics at Strong Coupling, hepth/9512181.

[4] A. Salam and E. Sezgin, Supergravities In Diverse Dimensions (North-Holland/World-Scientific, 1989).

[5] E. Witten, String Theory Dynamics In Various Dimensions, Nucl. Phys. B 443 (1995) 85.

[6] P. Aspinwall and D. Morrison, *U*-Duality And Integral Structures, hepth/9505025.

[7] P. Aspinwall and D. Morrison, String Theory On K3 Surfaces, hepth/9404151.

[8] D. Kaplan and J. Michelson, Zero Modes For The $D = 11$ Membrane And Five-Brane, hepth/9510053.

[9] J. Polchinski, Dirichlet-Branes And Ramond-Ramond Charges, hepth/9510017.

[10] M.J. Duff, J.T. Liu and R. Minasian, Eleven Dimensional Origin Of String/String Duality: A One Loop Test, hepth/9506126.

[11] C.G. Callan Jr. and J.A. Harvey, Anomalies And Fermion Zero Modes On Strings And Domain Walls, Nucl. Phys. B 250 (1985) 427, for a treatment of anomaly cancellation for certain four-dimensional extended objects by current inflow from the bulk.

[12] L. Alvarez-Gaumé and P. Ginsparg, The Structure Of Gauge And Gravitational Anomalies, Ann. Phys. 161 (1985) 423.

[13] C. Vafa and E. Witten, A One Loop Test Of String Duality, hepth/9505053.

[14] P. Townsend, *D*-Branes From *M*-Branes, hepth/95012062.

[15] P. Horava and E. Witten, Heterotic And Type I String Dynamics From Eleven Dimensions, hepth/9510209.

[16] J. Dai, R. Leigh and J. Polchinski, New Connections Between String Theories, Mod. Phys. Lett. A 4 (1989) 2073.

[17] P. Horava, Strings On World Sheet Orbifolds, Nucl. Phys. B 327 (1989) 461; Background Duality Of Open String Models, Phys. Lett. B 321 (1989) 251.

[18] E. Witten, Bound States Of Strings And *p*-Branes, hepth/9510135.

[19] M. Duff, R. Minasian and E. Witten, Evidence For Heterotic/Heterotic Duality, to appear.

[20] P. Aspinwall, Extended Gauge Symmetries And K3 Surfaces, hep-th/9507012.

[21] E. Witten, Some Comments On String Dynamics, hep-th/9507121.

[22] A. Strominger, Open *P*-Branes, hep-th/9512059.

[23] S. Chaudhuri, G. Hockney and J.D. Lykken, Maximally Supersymmetric String Theories in $D < 10$, hepth/9505054.

[24] S. Chaudhuri and J. Polchinski, Moduli Space Of CHL Strings, hepth/9506048.

Reprinted from Phys. Lett. B **399** (1997) 49–59

Covariant field equations of the M-theory five-brane

P.S. Howe [a], E. Sezgin [b,1], P.C. West [c,2]
[a] *Department of Mathematics, King's College, London, UK*
[b] *Center for Theoretical Physics, Texas A&M University, College Station, TX 77843, USA*
[c] *Isaac Newton Institute for Mathematical Sciences, Cambridge, UK*

Received 10 February 1997
Editor: M. Dine

Abstract

The component form of the equations of motion for the 5-brane in eleven dimensions is derived from the superspace equations. These equations are fully covariant in six dimensions. It is shown that double-dimensional reduction of the bosonic equations gives the equations of motion for a 4-brane in ten dimensions governed by the Born-Infeld action. © 1997 Elsevier Science B.V.

1. Introduction

It is now widely believed that there is a single underlying theory which incorporates all superstring theories and which also has, as a component, a new theory in eleven dimensions which has been christened "M-theory". Opinion is divided as to whether M-theory is itself the fundamental theory or whether it is one corner of a large moduli space which has the five consistent ten-dimensional superstring theories as other corners. Whichever viewpoint turns out to be correct it seems certain that M-theory will play a crucial role in future developments. Not much is known about this theory at present, apart from the fact that it has eleven-dimensional supergravity as a low energy limit and that it has two basic BPS p-branes, the 2-brane and the 5-brane, which preserve half-supersymmetry. The former can be viewed as a fundamental (singular) solution to the supergravity equations whereas the latter is solitonic. It is therefore important to develop a better understanding of these branes and in particular the 5-brane, since the Green-Schwarz action for the 2-brane has been known for some time.

In a recent paper [1] it was shown that all branes preserving half-supersymmetry can be understood as embeddings of one superspace, the worldsurface, into another, the target superspace, which has spacetime as its body, and that the basic embedding condition which needs to be imposed is universal and geometrically natural. The results of [1] were given mainly at the linearized level; in a sequel [2] the eleven-dimensional 5-brane was studied in more detail and the full nonlinear equations of motion were derived. However, these were expressed in superspace notation. It is the purpose of this paper to interpret these equations in a more familiar form, in other words to derive their component equivalents. In the context of superembeddings the component formalism means the Green-Schwarz formalism since the leading term in the worldsurface θ-expansion of the embedding describes a map from a bosonic worldsurface to a target superspace.

[1] Research supported in part by NSF Grant PHY-9411543.
[2] Permanent address.

0370-2693/97/$17.00
PII S0370-2693(97)00257-8

Partial results for the bosonic sector of the eleven-dimensional fivebrane have been obtained in [3–6]. More recently, a non-covariant bosonic action has been proposed [7,8]. In this approach, only the five-dimensional covariance is manifest. In [9], a complete bosonic action has been constructed. The action contains an auxiliary scalar field, which can be eliminated at the expense of sacrificing the six-dimensional covariance, after which it reduces to the action of [7,8].

In this paper we will show that the covariant superfield equations of motion of the eleven-dimensional superfivebrane presented in [1,2] can be written in κ-invariant form, and that they do have the anticipated Born-Infeld structure. The κ-symmetry emerges from the worldsurface diffeomorphism invariance of the superspace equations, the parameter of this symmetry being essentially the leading component in the worldsurface θ-expansion of an odd diffeomorphism. We find neither the need to introduce a scalar auxiliary field, nor the necessity to have only five-dimensional covariance. As long as one does not insist on having an action, it is possible to write down six-dimensional covariant equations, as one normally expects in the case of chiral p-forms.

In order to show that our equations have the expected Born-Infeld form we perform a double-dimensional reduction and compare them with the equations of motion for a 4-brane in ten dimensions. In Section 4, we do this comparison in the bosonic sector, and flat target space, and show that the Born-Infeld form of the 4-brane equations of motion does indeed emerge. The work of Refs. [1,2] is briefly reviewed in the next section, and in Section 3 the equations of motion are described in Green-Schwarz language.

2. Equations of motion in superspace

The 5-brane is described by an embedding of the worldsurface M, which has (even|odd) dimension $(6|16)$ into the target space, $\underline{M}$, which has dimension $(11|32)$. In local coordinates $z^{\underline{M}} = (x^{\underline{m}}, \theta^{\underline{\mu}})$ for $\underline{M}$ and z^M for M the embedded submanifold is given as $z^{\underline{M}}(z)$ [3]. We define the embedding matrix $E_A{}^{\underline{A}}$ to be the derivative of the embedding referred to preferred bases on both manifolds:

$$E_A{}^{\underline{A}} = E_A{}^M \partial_M z^{\underline{M}} E_{\underline{M}}{}^{\underline{A}}, \tag{1}$$

where $E_M{}^A$ $(E_A{}^M)$ is the supervielbein (inverse supervielbein) which relates the preferred frame basis to the coordinate basis, and the target space supervielbein has underlined indices. The notation is as follows: indices from the beginning (middle) of the alphabet refer to frame (coordinate) indices, latin (greek) indices refer to even (odd) components and capital indices to both, non-underlined (underlined) indices refer to M $(\underline{M})$ and primed indices refer to normal directions. We shall also employ a two-step notation for spinor indices; that is, for general formulae a spinor index α (or α') will run from 1 to 16, but to interpret these formulae, we shall replace a subscript α by a subscript pair αi and a subscript α' by a pair ${}_i^\alpha$, where $\alpha = 1, \ldots, 4$ and $i = 1, \ldots, 4$ reflecting the $Spin(1,5) \times USp(4)$ group structure of the $N = 2, d = 6$ worldsurface superspace. (A lower (upper) α index denotes a left-handed (right-handed) $d = 6$ Weyl spinor and the $d = 6$ spinors that occur in the theory are all symplectic Majorana-Weyl.)

The torsion 2-form $T^{\underline{A}}$ on $\underline{M}$ is given as usual by

$$T^{\underline{A}} = dE^{\underline{A}} + E^{\underline{B}} \Omega_{\underline{B}}{}^{\underline{A}}, \tag{2}$$

where Ω is the connection 1-form. The pull-back of this equation onto the worldsurface reads, in index notation,

$$\begin{aligned} &\nabla_A E_B{}^{\underline{C}} - (-1)^{AB} \nabla_B E_A{}^{\underline{C}} + T_{AB}{}^C E_C{}^{\underline{C}} \\ &\quad = (-1)^{A(B+\underline{B})} E_B{}^{\underline{B}} E_A{}^{\underline{A}} T_{\underline{AB}}{}^{\underline{C}}, \end{aligned} \tag{3}$$

where the derivative ∇_A is covariant with respect to both spaces, i.e. with respect to both underlined and non-underlined indices, the connection on M being, at this stage at least, independent of the target space connection.

The basic embedding condition is

$$E_\alpha{}^{\underline{a}} = 0, \tag{4}$$

from which it follows that (using (3))

$$E_\alpha{}^{\underline{\alpha}} E_\beta{}^{\underline{\beta}} T_{\underline{\alpha\beta}}{}^{\underline{c}} = T_{\alpha\beta}{}^c E_c{}^{\underline{c}}. \tag{5}$$

If the target space geometry is assumed to be that of (on-shell) eleven-dimensional supergravity equation

[3] We shall also denote the coordinates of $M(\underline{M})$ by $z = (x, \theta)$ $(\underline{z} = (\underline{x}, \underline{\theta}))$ if it is not necessary to use indices

(4) actually determines completely the induced geometry of the worldsurface and the dynamics of the 5-brane. In fact, as will be discussed elsewhere, it is not necessary to be so specific about the target space geometry, but it will be convenient to adopt the on-shell geometry in the present paper. The structure group of the target superspace is $Spin(1,10)$ and the non-vanishing parts of the target space torsion are [10,11]

$$T_{\alpha\beta}{}^{\underline{c}} = -i(\Gamma^{\underline{c}})_{\underline{\alpha\beta}}, \tag{6}$$

$$T_{\underline{a}\beta}{}^{\underline{\gamma}} = -\tfrac{1}{36}(\Gamma^{\underline{bcd}})_{\underline{\beta}}{}^{\underline{\gamma}} H_{\underline{abcd}} - \tfrac{1}{288}(\Gamma_{\underline{abcde}})_{\underline{\beta}}{}^{\underline{\gamma}} H^{\underline{bcde}}, \tag{7}$$

where $H_{\underline{abcd}}$ is totally antisymmetric, and the dimension 3/2 component $T_{\underline{ab}}{}^{\underline{\gamma}}$. $H_{\underline{abcd}}$ is the dimension-one component of the closed superspace 4-form H_4 whose only other non-vanishing component is

$$H_{\underline{ab\gamma\delta}} = -i(\Gamma_{\underline{ab}})_{\underline{\gamma\delta}}. \tag{8}$$

With this target space geometry equation (5) becomes

$$E_\alpha{}^{\underline{\alpha}} E_\beta{}^{\underline{\beta}} (\Gamma^{\underline{c}})_{\underline{\alpha\beta}} = i T_{\alpha\beta}{}^{c} E_c{}^{\underline{c}}. \tag{9}$$

The solution to this equation is given by

$$E_\alpha{}^{\underline{\alpha}} = u_\alpha{}^{\underline{\alpha}} + h_\alpha{}^{\beta'} u_{\beta'}{}^{\underline{\alpha}}, \tag{10}$$

and

$$E_a{}^{\underline{a}} = m_a{}^b u_b{}^{\underline{a}}, \tag{11}$$

together with

$$T_{\alpha\beta}{}^{c} = -i(\Gamma^c)_{\alpha\beta} \rightarrow -i\eta_{ij}(\gamma^c)_{\alpha\beta}. \tag{12}$$

with $\eta_{ij} = -\eta_{ji}$ being the $USp(4)$ invariant tensor and the pair $(u_\alpha{}^{\underline{\alpha}}, u_{\alpha'}{}^{\underline{\alpha}})$ together making up an element of the group $Spin(1,10)$. Similarly, there is a $u_{a'}{}^{\underline{a}}$ such that the pair $(u_a{}^{\underline{a}}, u_{a'}{}^{\underline{a}})$ is the element of $SO(1,10)$ corresponding to this spin group element. (The inverses of these group elements will be denoted $(u_{\underline{\alpha}}{}^{\alpha}, u_{\underline{\alpha}}{}^{\alpha'})$ and $(u_{\underline{a}}{}^{a}, u_{\underline{a}}{}^{a'})$.) The tensor $h_\alpha{}^{\beta'}$ is given by [4]

$$h_\alpha{}^{\beta'} \rightarrow h_{\alpha i\beta}{}^{j} = \tfrac{1}{6}\delta_i{}^j(\gamma^{abc})_{\alpha\beta} h_{abc}, \tag{13}$$

where h_{abc} is self-dual, and

$$m_a{}^b = \delta_a{}^b - 2h_{acd}h^{bcd}. \tag{14}$$

This solution is determined up to local gauge transformations belonging to the group $Spin(1,5) \times USp(4)$, the structure group of the worldsurface. One also has the freedom to make worldsurface super-Weyl transformations but one can consistently set the conformal factor to be one and we shall do this throughout the paper.

It is useful to introduce a normal basis $E_{A'} = E_{A'}{}^{\underline{A}} E_{\underline{A}}$ of vectors at each point on the worldsurface. The inverse of the pair $(E_A{}^{\underline{A}}, E_{A'}{}^{\underline{A}})$ is denoted by $(E_{\underline{A}}{}^{A}, E_{\underline{A}}{}^{A'})$. The odd-odd and even-even components of the normal matrix $E_{A'}{}^{\underline{A}}$ can be chosen to be

$$E_{\alpha'}{}^{\underline{\alpha}} = u_{\alpha'}{}^{\underline{\alpha}}, \tag{15}$$

and

$$E_{a'}{}^{\underline{a}} = u_{a'}{}^{\underline{a}}. \tag{16}$$

Together with (10) and (11), it follows that the inverses in the odd-odd and even-even sectors are

$$E_{\underline{\alpha}}{}^{\alpha} = u_{\underline{\alpha}}{}^{\alpha}, \quad E_{\underline{\alpha}}{}^{\alpha'} = u_{\underline{\alpha}}{}^{\alpha'} - u_{\underline{\alpha}}{}^{\beta} h_\beta{}^{\alpha'}, \tag{17}$$

and

$$E_{\underline{a}}{}^{a} = u_{\underline{a}}{}^{b}(m^{-1})_b{}^a, \quad E_{\underline{a}}{}^{a'} = u_{\underline{a}}{}^{a'}. \tag{18}$$

Later, we will also need the relations [2]

$$u_\alpha{}^{\underline{\alpha}} u_\beta{}^{\underline{\beta}} (\Gamma^{\underline{a}})_{\underline{\alpha\beta}} = (\Gamma^a)_{\alpha\beta} u_a{}^{\underline{a}}, \tag{19}$$

$$u_{\alpha'}{}^{\underline{\alpha}} u_{\beta'}{}^{\underline{\beta}} (\Gamma^{\underline{a}})_{\underline{\alpha\beta}} = (\Gamma^a)_{\alpha'\beta'} u_a{}^{\underline{a}}, \tag{20}$$

$$u_\alpha{}^{\underline{\alpha}} u_{\beta'}{}^{\underline{\beta}} (\Gamma^{\underline{a}})_{\underline{\alpha\beta}} = (\Gamma^{a'})_{\alpha\beta'} u_{a'}{}^{\underline{a}}, \tag{21}$$

which follow from the fact that the u's form a 32×32 matrix that is an element of $Spin(1,10)$.

The field h_{abc} is a self-dual antisymmetric tensor, but it is not immediately obvious how it is related to a 2-form potential. In fact, it was shown in [2] that there is a superspace 3-form H_3 which satisfies

$$dH_3 = -\tfrac{1}{4}H_4, \tag{22}$$

where H_4 is the pull-back of the target space 4-form, and whose only non-vanishing component is H_{abc} where

$$H_{abc} = m_a{}^d m_b{}^e h_{cde}. \tag{23}$$

[4] We have rescaled the H_{abc} and h_{abc} of Refs. [1,2] by a factor of 6.

…uations of motion of the 5-brane can be obtained by systematic analysis of the torsion equation (3), subject to the condition (4) [2]. The bosonic equations are the scalar equation

$$\eta^{ab} K_{ab}{}^{c'} = \tfrac{1}{8} (\gamma^{c'})^{jk} (\gamma^a)^{\beta\gamma} Z_{a,\beta j,\gamma k}\,, \tag{24}$$

and the antisymmetric tensor equation

$$\hat{\nabla}^c h_{abc} = -\tfrac{1}{16} \eta^{jk} ((\gamma_{[a})^{\beta\gamma} Z_{b],\beta j,\gamma k} + \tfrac{1}{2} (\gamma_{ab}{}^c)^{\beta\gamma} Z_{c,\beta j,\gamma k})\,, \tag{25}$$

where

$$Z_{a\beta}{}^{\gamma'} = E_\beta{}^{\underline{\beta}} E_a{}^{\underline{a}} T_{\underline{a}\underline{\beta}}{}^{\underline{\gamma}} E_{\underline{\gamma}}{}^{\gamma'} - E_a{}^{\underline{\gamma}} \nabla_\beta E_{\underline{\gamma}}{}^{\gamma'}\,, \tag{26}$$

and

$$\hat{\nabla}_a h_{bcd} = \nabla_a h_{bcd} - 3 X_{a,[b}{}^e h_{cd]e}\,, \tag{27}$$

with

$$X_{a,b}{}^c = (\nabla_a u_b{}^{\underline{c}}) u_{\underline{c}}{}^c\,. \tag{28}$$

In the scalar equation we have introduced a part of the second fundamental form of the surface which is defined to be

$$K_{AB}{}^{C'} = (\nabla_A E_B{}^{\underline{C}}) E_{\underline{C}}{}^{C'}\,. \tag{29}$$

Finally, the spin one-half equation is simply

$$(\gamma^a)^{\alpha\beta} \chi_{a\beta}{}^{j} = 0\,, \tag{30}$$

where

$$\chi_a{}^{\alpha'} = E_a{}^{\underline{\alpha}} E_{\underline{\alpha}}{}^{\alpha'}\,. \tag{31}$$

We end this section by rewriting the equations of motion (24), (25) and (30) in an alternative form that will be useful for the purposes of the next section:

$$E_a{}^{\underline{\alpha}} E_{\underline{\alpha}}{}^{\beta'} (\Gamma^a)_{\beta'}{}^{\alpha} = 0\,, \tag{32}$$

$$\eta^{ab} \nabla_a E_b{}^{\underline{a}} E_{\underline{a}}{}^{b'} = -\tfrac{1}{8} (\Gamma^{b'a})_{\gamma'}{}^{\beta} Z_{a\beta}{}^{\gamma'}\,, \tag{33}$$

$$\hat{\nabla}^c h_{abc} = -\tfrac{1}{32} (\Gamma^c \Gamma_{ab})_{\gamma'}{}^{\beta} Z_{c\beta}{}^{\gamma'}\,. \tag{34}$$

It will also prove to be useful to rewrite (26) as

$$Z_{a\beta}{}^{\gamma'} = E_\beta{}^{\underline{\beta}} \left(T_{a\underline{\beta}}{}^{\underline{\gamma}} - K_{a\underline{\beta}}{}^{\underline{\gamma}} \right) E_{\underline{\gamma}}{}^{\gamma'}\,, \tag{35}$$

with the matrices T_a and K_a defined as

$$T_{a\underline{\beta}}{}^{\underline{\gamma}} = E_a{}^{\underline{a}} T_{\underline{a}\underline{\beta}}{}^{\underline{\gamma}}\,, \tag{36}$$

$$K_{a\underline{\beta}}{}^{\underline{\gamma}} = E_a{}^{\underline{\delta}} E_{\underline{\beta}}{}^{\gamma} (\nabla_\gamma E_{\underline{\delta}}{}^{\delta'}) E_{\delta'}{}^{\underline{\gamma}}\,. \tag{37}$$

3. Equations of motion in Green-Schwarz form

3.1. Preliminaries

In this section we derive the component equations of motion following from the superspace equations given in the last section. The idea is to expand the superspace equations as power series in θ^μ and to evaluate them at $\theta = 0$. We may choose a gauge in which the worldsurface supervielbein takes the form

$$\begin{aligned}
E_m{}^a(x,\theta) &= E_m{}^a(x) + O(\theta),\\
E_m{}^\alpha(x,\theta) &= E_m{}^\alpha(x) + O(\theta),\\
E_\mu{}^a(x,\theta) &= 0 + O(\theta),\\
E_\mu{}^\alpha(x,\theta) &= \delta_\mu{}^\alpha + O(\theta),
\end{aligned} \tag{38}$$

and the inverse takes the form

$$\begin{aligned}
E_a{}^m(x,\theta) &= E_a{}^m(x) + O(\theta),\\
E_a{}^\mu(x,\theta) &= E_a{}^\mu(x) + O(\theta),\\
E_\alpha{}^m(x,\theta) &= 0 + O(\theta),\\
E_\alpha{}^\mu(x,\theta) &= \delta_\alpha{}^\mu + O(\theta),
\end{aligned} \tag{39}$$

where $E_a{}^m(x)$ is the inverse of $E_m{}^a(x)$. The component field $E_m{}^\alpha(x)$ is the worldsurface gravitino, which is determined by the embedding, but which only contributes terms to the equations of motion which we shall not need for the purpose of this section. The field $E_a{}^\mu(x)$ is linearly related to the gravitino. From the embedding condition (4) we learn that

$$\partial_\mu z^{\underline{M}} E_{\underline{M}}{}^{\underline{a}} = 0 \quad \text{at } \theta = 0, \tag{40}$$

so that

$$E_a{}^{\underline{a}} = E_a{}^m \mathcal{E}_m{}^{\underline{a}} \quad \text{at } \theta = 0\,, \tag{41}$$

$$E_a{}^{\underline{\alpha}} = E_a{}^m \mathcal{E}_m{}^{\underline{\alpha}} \quad \text{at } \theta = 0\,, \tag{42}$$

where we have used the definitions

$$\mathcal{E}_m{}^{\underline{a}}(x) = \partial_m z^{\underline{M}} E_{\underline{M}}{}^{\underline{a}} \quad \text{at } \theta = 0\,, \tag{43}$$

$$\mathcal{E}_m{}^{\underline{\alpha}}(x) = \partial_m z^{\underline{M}} E_{\underline{M}}{}^{\underline{\alpha}} \quad \text{at } \theta = 0\,. \tag{44}$$

These are the embedding matrices in the Green-Schwarz formalism, often denoted by Π. From (11) we have

$$E_a{}^{\underline{a}} E_b{}^{\underline{b}} \eta_{\underline{a}\underline{b}} = m_a{}^c m_b{}^d \eta_{cd}\,, \tag{45}$$

this equation being true for all θ and in particular for $\theta = 0$. Therefore, if we put

$$e_a{}^m = ((m^{-1})_a{}^b E_b{}^m)(x), \tag{46}$$

we find that $e_m{}^a$ is the sechsbein associated with the standard GS induced metric

$$g_{mn}(x) = \mathcal{E}_m{}^{\underline{a}} \mathcal{E}_n{}^{\underline{b}} \eta_{\underline{ab}}. \tag{47}$$

There is another metric, which will make its appearance later, which we define as

$$G^{mn} = E_a{}^m(x) E_b{}^n(x)\, \eta^{ab} \tag{48}$$

$$= ((m^2)^{ab} e_a{}^m e_b{}^n)(x). \tag{49}$$

We also note the relation

$$u_a{}^{\underline{a}} = e_a{}^m \mathcal{E}_m{}^{\underline{a}}, \tag{50}$$

which follows from (11), (41) and (46).

For the worldsurface 3-form H_3 we have

$$H_{MNP} = E_P{}^C E_B{}^N E_A{}^M H_{ABC} \times (-1)^{((B+N)M+(P+C)(M+N))} \tag{51}$$

Evaluating this at $\theta = 0$ one finds

$$H_{mnp}(x) = (E_m{}^a E_n{}^b E_p{}^c H_{abc})(x) \tag{52}$$

so that, using (23) and (46), one finds

$$h_{abc}(x) = m_a{}^d e_d{}^m e_b{}^n e_c{}^p H_{mnp}(x). \tag{53}$$

We are now in a position to write down the equations of motion in terms of $\mathcal{E}_m{}^{\underline{\alpha}}$, $\mathcal{E}_m{}^{\underline{a}}$ and $H_{mnp}(x)$. The basic worldsurface fields are $x^{\underline{m}}$, $\theta^{\underline{\mu}}$ and $B_{mn}(x)$, where B_{mn} is the 2-form potential associated with H_{mnp} as $H_3 = dB_2 - \frac{1}{4}C_3$ and C_3 is the pull-back of the target space 3-form. We begin with the Dirac equation (30).

3.2. *The Dirac equation*

In order to extract the Dirac equation in κ-invariant component form, it is convenient to define the projection operators

$$E_{\underline{\alpha}}{}^{\alpha} E_{\alpha}{}^{\underline{\gamma}} = \tfrac{1}{2}(1+\Gamma)_{\underline{\alpha}}{}^{\underline{\gamma}}, \tag{54}$$

$$E_{\underline{\alpha}}{}^{\alpha'} E_{\alpha'}{}^{\underline{\gamma}} = \tfrac{1}{2}(1-\Gamma)_{\underline{\alpha}}{}^{\underline{\gamma}}. \tag{55}$$

The Γ-matrix, which clearly satisfies $\Gamma^2 = 1$, can be calculated from these definitions as follows. We expand

$$E_{\underline{\alpha}}{}^{\alpha} E_{\alpha}{}^{\underline{\gamma}} = \sum_{n=0}^{5} C^{\underline{a}_1\cdots\underline{a}_n} \left(\Gamma_{\underline{a}_1\cdots\underline{a}_n}\right)_{\underline{\alpha}}{}^{\underline{\gamma}}, \tag{56}$$

where C's are the expansion coefficients that are to be determined. Tracing this equation with suitable Γ-matrices, and using the relations (19)–(21), we find that the only non-vanishing coefficients are

$$C = \tfrac{1}{2}, \tag{57}$$

$$C^{\underline{abc}} = \tfrac{1}{6} h^{abc} u_a{}^{\underline{a}} u_b{}^{\underline{b}} u_c{}^{\underline{c}}, \tag{58}$$

$$C^{\underline{a}_1\cdots\underline{a}_6} = -\frac{1}{6!2} \epsilon^{a_1\cdots a_6} u_{a_1}{}^{\underline{a}_1} \cdots u_{a_6}{}^{\underline{a}_6}. \tag{59}$$

Substituting these back, and comparing with (54), we find

$$\Gamma = \frac{1}{6!\sqrt{-g}} \epsilon^{m_1\cdots m_6} \times (-\Gamma_{m_1\cdots m_6} + 40\Gamma_{m_1\cdots m_3} h_{m_4\cdots m_6}), \tag{60}$$

where we have used (50) and the definitions

$$\Gamma_m = \mathcal{E}_m{}^{\underline{a}} \Gamma_{\underline{a}}, \tag{61}$$

$$h_{mnp} = e_m{}^a e_n{}^b e_p{}^c h_{abc}. \tag{62}$$

The matrix Γ can also be written as

$$\Gamma = (-1 + \tfrac{1}{3}\Gamma^{mnp} h_{mnp})\Gamma_{(0)}, \tag{63}$$

where

$$\Gamma_{(0)} = \frac{1}{6!\sqrt{-g}} \epsilon^{m_1\cdots m_6} \Gamma_{m_1\cdots m_6}. \tag{64}$$

It is now a straightforward matter to derive the component for the Dirac equation (32). We use (19) to replace the worldsurface Γ-matrix by the target space Γ-matrix multiplied by factors of u, and recall (15), (17), (50) and (55) to find

$$\eta^{bc}(1-\Gamma)_{\underline{\gamma}}{}^{\underline{\beta}} (\Gamma_{\underline{a}})_{\underline{\alpha\beta}} E_c{}^{\underline{\gamma}} \mathcal{E}_b{}^{\underline{a}} = 0. \tag{65}$$

We recall that $\mathcal{E}_b{}^{\underline{a}} = e_b{}^m \mathcal{E}_m{}^{\underline{a}}$ and that $E_c{}^{\underline{\gamma}} = m_c{}^d e_d{}^n \mathcal{E}_n{}^{\underline{\gamma}}$. Using these relations, the Dirac equation can be written as

$$\mathcal{E}_a(1-\Gamma)\Gamma^b m_b{}^a = 0, \tag{66}$$

where $\Gamma^b = \Gamma^m e_m{}^a$ and the target space spinor indices are suppressed.

The Dirac equation obtained above has a very similar form to those of D-branes in ten dimensions [12–16], and indeed we expect that a double-dimensional reduction would yield the 4-brane Dirac equation.

The emergence of the projection operator $(1-\Gamma)$ in the Dirac equation in the case of D-branes, and the other known super p-branes is due to the contribution of Wess-Zumino terms in the action (see, for example, Ref. [17] for the eleven-dimensional supermembrane equations of motion). These terms are also needed for the κ-symmetry of the action. It is gratifying to see that the effect of Wess-Zumino terms is automatically included in our formalism through a geometrical route that is based on considerations of the embedding of a world superspace into target superspace.

3.3. The scalar equation

By scalar equation we mean the equation of motion for $x^{\underline{m}}(x)$, i.e. the coordinates of the target space, which are scalar fields from the worldsurface point of view. In a physical gauge, these describe the five scalar degrees of freedom that occur in the worldsurface tensor supermultiplet.

The scalar equation is the leading component of the superspace equation (33) which we repeat here for the convenience of the reader:

$$\eta^{ab}\nabla_a E_b{}^{\underline{a}} E_{\underline{a}}{}^{b'} = -\tfrac{1}{8}(\Gamma^{b'a})_{\gamma'}{}^{\beta} Z_{a\beta}{}^{\gamma'} . \tag{67}$$

The superspace equation for the covariant derivative

$$\nabla_a = E_a{}^m \nabla_m + E_a{}^{\mu}\nabla_{\mu}, \tag{68}$$

when evaluated at $\theta = 0$ involves the worldsurface gravitino $E_a{}^{\mu}(x)$ which is expressible in terms of the basic fields of the worldsurface tensor multiplet. Since it is fermionic it follows that the second term in the covariant derivative will be bilinear in fermions (at least), and we shall henceforth drop all such terms from the equations in order to simplify life a little. We shall temporarily make a further simplification by assuming that the target space is flat. The tensor Z, as we saw earlier, has two types of contribution, one (T_a) involving H_{abcd}, and the other (K_a) involving only terms which are bilinear or higher order in fermions. In accordance with our philosophy we shall henceforth ignore these terms.

To this order the right-hand side of the scalar equation vanishes as does the right-hand side of the tensor equation (25). Multiplying the scalar equation (67) with $E_{b'}{}^{\underline{c}}$, we see that it can be written in the form

$$\eta^{ab}(\nabla_a E_b{}^{\underline{c}} - K_{ab}{}^c E_c{}^{\underline{c}}) = 0 , \tag{69}$$

where $K_{ab}{}^c$ is defined below. Using the relation $E_b{}^{\underline{c}} = m_b{}^d u_d{}^{\underline{c}}$ and the definition of $X_{ab}{}^c$ in (28) we find that

$$\begin{aligned} K_{ab}{}^c &:= \nabla_a E_b{}^{\underline{d}} E_{\underline{d}}{}^c \\ &= (\hat{\nabla}_a m_b{}^d)(m^{-1})_d{}^c + X_{ab}{}^c . \end{aligned} \tag{70}$$

Using the relation

$$\eta^{ab}\hat{\nabla}_a m_b{}^c = 0 , \tag{71}$$

which we will prove later, we conclude that $\eta^{ab}K_{ab}{}^c = \eta^{ab}X_{ab}{}^c$. As a result, we can express the scalar equation of motion in the form

$$\eta^{ab}\hat{\nabla}_a E_b{}^{\underline{c}} = 0 . \tag{72}$$

where $\hat{\nabla}_a E_b{}^{\underline{c}} = \nabla_a E_b{}^{\underline{c}} - X_{ab}{}^d E_d{}^{\underline{c}}$. Relation (71) allows us to rewrite the scalar equation of motion in the form

$$m^{ab}\hat{\nabla}_a u_b{}^{\underline{c}} = 0 . \tag{73}$$

The next step is to find a explicit expression for the spin connection $\hat{\omega}_{a,b}{}^c$ associated with the hatted derivative. Using the definition of $X_{ab}{}^d$ given in (28), we find that this spin connection is given by

$$\hat{\omega}_{a,b}{}^c = \Omega_{a,b}{}^c + X_{a,b}{}^c = E_a{}^m(\partial_m u_b{}^{\underline{c}}) u_{\underline{c}}{}^c . \tag{74}$$

Recalling (50) and (46), we find that the hatted spin connection takes the form

$$\hat{\omega}_{a,b}{}^c = m_a{}^f e_f{}^n \left(\partial_n e_b{}^m g_{mp} e^{cp} + e_b{}^m \partial_n \mathcal{E}_m{}^{\underline{d}} \mathcal{E}_{p\underline{d}} e^{cp}\right) . \tag{75}$$

From this expression it is straightforward to derive the following result; given any vector V_m one has

$$\hat{\nabla}_a V_b = m_a{}^d e_d{}^n e_b{}^m \nabla_n V_m \tag{76}$$

where

$$\nabla_n V_m = \partial_n V_m - \Gamma_{nm}{}^p V_p \tag{77}$$

and

$$\Gamma_{nm}{}^{p} = \partial_n \mathcal{E}_m{}^{\underline{c}} \mathcal{E}_{s\underline{c}}\, g^{sp} \,. \qquad (78)$$

It is straightforward to verify that to the order to which we are working this connection is indeed the Levi-Civita connection for the induced metric g_{mn}.

We are now in a position to express the scalar equation in its simplest form which is in a coordinate basis using the hatted connection. Using the above result we find that (73) can be written as

$$G^{mn}\nabla_m \mathcal{E}_n{}^{\underline{a}} = 0. \qquad (79)$$

It remains to prove (71). Using the expression for $m_a{}^b$ given in (14) we find that

$$\eta^{ab}\hat{\nabla}_a m_b{}^c = -2\hat{\nabla}^b (h_{bde}h^{cde}) = -2h_{bde}\hat{\nabla}^b h^{cde}$$
$$= -\tfrac{2}{3}h_{bde}\hat{\nabla}^c h_{bde} = -\tfrac{1}{3}\hat{\nabla}^c (h_{bde}h^{bde}) = 0 \,. \qquad (80)$$

In carrying out the above steps we have used the h_{abc} equation of motion and the self-duality of this field.

In the case of a non-flat target space the derivation is quite a bit longer and the steps will be discussed elsewhere. One finds that the right hand side of the scalar equation in the form of (67) is given by

$$\eta^{ab}\nabla_a E_b{}^{\underline{a}} E_{\underline{a}}{}^{c'} = -\tfrac{1}{144}(1 - \tfrac{2}{3}\,\mathrm{tr}\, k^2)\epsilon^{c' e_1' e_2' e_3' e_4'} H_{e_1' e_2' e_3' e_4'}$$
$$+ \tfrac{2}{3} m_a{}^b H^{c'}{}_{bcd} h^{acd} \,, \qquad (81)$$

where

$$k_a{}^b := h_{acd}h^{bcd} \,. \qquad (82)$$

Using the steps given above this result can be expressed in the form

$$G^{mn}\nabla_m \mathcal{E}_n{}^{\underline{c}} = \frac{1}{\sqrt{-g}}(1 - \tfrac{2}{3}\,\mathrm{tr}\, k^2)\epsilon^{m_1\cdots m_6}$$
$$\times \Big(\frac{1}{6^2\cdot 4\cdot 5} H^{\underline{a}}{}_{m_1\cdots m_6} + \tfrac{2}{3} H^{\underline{a}}{}_{m_1 m_2 m_3} H_{m_4 m_5 m_6}\Big)$$
$$\times (\delta_{\underline{a}}{}^{\underline{c}} - \mathcal{E}_{\underline{a}}{}^{m}\mathcal{E}_m{}^{\underline{c}}) \,, \qquad (83)$$

where the target space indices on H_4 and H_7 have been converted to worldvolume indices with factors of $\mathcal{E}_m{}^{\underline{a}}$ and

$$H_{\underline{d}_1,\ldots,\underline{d}_4} = \frac{1}{7!}\epsilon_{\underline{d}_1,\ldots,\underline{d}_4 \underline{e}_1,\ldots,\underline{e}_7} H^{\underline{e}_1,\ldots,\underline{e}_7} \,. \qquad (84)$$

where H_7 is the seven-form that occurs in the dual formulation of eleven-dimensional supergravity. One can verify that the ratio between the two terms on the right hand side is precisely what one expects were this term to have been derived from the expected gauge invariant Wess-Zumino term of the form $C_6 + 4C_3 \wedge H_3$. We also note that the last factor in (83) implies that the RHS of the equation vanishes identically when multipled with $\mathcal{E}_{\underline{c}}{}^q$, as it should, indicating that only five of the eleven equations, which correspond to the Goldstone scalars, are independent.

3.4. *The tensor equation*

The tensor equation can be manipulated in a similar fashion. If we consider the simplest case of ignoring the fermion bilinears and assuming the target space to be flat we have, from (34)

$$\eta^{ab}\hat{\nabla}_a h_{bcd} = 0 \,. \qquad (85)$$

We can relate h to H using (53) and take the factor of m past the covariant derivative using (71) to get

$$m^{ab}\hat{\nabla}_a (e_b{}^m e_c{}^n e_c{}^p H_{mnp}) = 0 \,. \qquad (86)$$

Using similar steps to those given in the proved in the previous subsection and converting to a coordinate basis we find the desired form of the tensor equation in this approximation, namely

$$G^{mn}\nabla_m H_{npq} = 0 \,. \qquad (87)$$

In the case of a non-trivial target space a lengthy calculation is required to find the analogous result. One first finds that

$$\hat{\nabla}^c h_{abc} = \tfrac{1}{288} m_a^f m_b^g \epsilon_{fg e_1 e_2 e_3 e_4} H^{e_1 e_2 e_3 e_4}$$
$$- \tfrac{1}{72}\epsilon_{abde_1e_2e_3} m^d_f H^{fe_1e_2e_3}$$
$$+ 6h^{e_1e_2}{}_{[a} h_{bc]}{}^{e_3} m^{cc_1} H_{c_1e_1e_2e_3}$$
$$+ \tfrac{4}{3}h_{abc}h^{e_1e_2e_3} m^{cc_1} H_{c_1e_1e_2e_3} \equiv Y_{ab} \,. \qquad (88)$$

It is possible to rewrite Y_{ab} in the form

$$Y_{ab} = (\tilde{K} + m\tilde{K} + \tfrac{1}{4}mm\tilde{K})_{ab} \qquad (89)$$

where $\tilde{K}_{ab} = -\frac{1}{36.4!}\epsilon_{abcdef}H^{cdef}$, $(m\tilde{K})_{ab} = m^c_{|a}\tilde{K}_{b|c}$, $(mm\tilde{K})_{ab} = m^c_a m^d_b \tilde{K}_{cd}$. The scalar equation of motion can also be expressed in the form

$$G^{mn}\nabla_m H_{npq} = \frac{1}{1 - \frac{2}{3}\,\mathrm{tr}\, k^2} e_p^a e_q^b (4Y + 4mY + mmY)_{ab} \,, \qquad (90)$$

where mY and mmY are defined in a similar way to the $m\tilde{K}$ and $mm\tilde{K}$ terms above.

3.5. The κ-symmetry transformations

The κ-symmetry transformations are related to odd worldsurface diffeomorphisms. Under an infinitesimal worldsurface diffeomorphism $\delta z^M = -v^M$ the variation of the embedding expressed in a preferred frame basis is

$$\delta z^{\underline{A}} \equiv \delta z^{\underline{M}} E_{\underline{M}}{}^{\underline{A}} = v^A E_A{}^{\underline{A}} . \tag{91}$$

For an odd transformation ($v^a = 0$) one has

$$\delta z^{\underline{a}} = 0 , \quad \delta z^{\underline{\alpha}} = v^\alpha E_\alpha{}^{\underline{\alpha}} . \tag{92}$$

The vanishing of the even variation $\delta z^{\underline{a}}$ is typical of κ-symmetry and follows from the basic embedding condition (4).

The relation between the parameter v^α and the familiar κ transformation parameter $\kappa^{\underline{\alpha}}$ can be expressed as

$$v^\alpha = \kappa^{\underline{\gamma}} E_{\underline{\gamma}}{}^\alpha . \tag{93}$$

Therefore, recalling (54), the κ transformation rule (92) takes the form

$$\delta z^{\underline{\alpha}} = \kappa^{\underline{\gamma}} (1 + \Gamma)_{\underline{\gamma}}{}^{\underline{\alpha}} , \tag{94}$$

where we have absorbed a factor of two into the definition of κ. It is understood that these formulae are to be evaluated at $\theta = 0$, so that they are component results.

There remains the determination of the κ-symmetry transformation of the antisymmetric tensor field B_{mn}. It is more convenient to compute the κ transformations rule for the field $h_{abc}(x)$. (The relation between the two fields is described earlier.) Thus we need to consider

$$\delta h_{abc} = \kappa^{\underline{\gamma}} E_{\underline{\gamma}}{}^\alpha \nabla_\alpha h_{abc} \quad \text{at } \theta = 0 . \tag{95}$$

By including a Lorentz transformation we may write this transformation as

$$\delta h_{abc} = \kappa^{\underline{\gamma}} E_{\underline{\gamma}}{}^\alpha \hat{\nabla}_\alpha h_{abc} . \tag{96}$$

We have calculated $\hat{\nabla}_\alpha h_{abc}$, and the derivation of the result will be given elsewhere [18]. Using this result, we find

$$\delta h_{abc} = -\tfrac{i}{16} m_{|a|}{}^d \, \mathcal{E}_d (1 - \Gamma) \Gamma_{|bc|} \kappa , \tag{97}$$

where $\Gamma_a = \Gamma^m e_{ma}$ and the target space spinor indices are suppressed. One can check that the RHS is self-dual, modulo the Dirac equation (66).

4. Double-dimensional reduction

The procedure we shall adopt now is to use double-dimensional reduction [19] to obtain a set of equations for a 4-brane in ten dimensions and then to compare this set of equations with the equations that one derives by varying the Born-Infeld action. We shall take the target space to be flat and we shall ignore the terms bilinear in fermions on the right-hand-side of (24) and (25), that is, we drop the terms in these equations that involve the quantity Z defined by (26) and we also ignore terms involving the worldsurface gravitino. From the previous section, we read off the resulting equations of motion:

$$G^{mn} \nabla_m \mathcal{E}_n{}^{\underline{a}} = 0 , \tag{98}$$

$$G^{mn} \nabla_m H_{npq} = 0 . \tag{99}$$

We can further simplify matters by considering the corresponding bosonic problem, i.e. by neglecting $\underline{\theta}$ as well. In this limit, and recalling that we have assumed that the target space is flat, one has

$$\mathcal{E}_m{}^{\underline{a}} \longrightarrow \partial_m x^{\underline{a}} . \tag{100}$$

In order to carry out the dimensional reduction we shall, in this section, distinguish 6- and 11-dimensional indices from 5- and 10-dimensional indices by putting hats on the former. We have

$$x^{\hat{m}} = (x^m, y) \tag{101}$$

and

$$x^{\underline{\hat{m}}} = (x^{\underline{m}}, y) , \tag{102}$$

so that the sixth dimension of the worldsurface is identified with the eleventh dimension of the target space; moreover, this common dimension is taken to be a circle, and the reduction is effected by evaluating the equations of motion at $y = 0$. The metric is diagonal:

$$g_{\hat{m}\hat{n}} = (g_{mn}, 1) , \tag{103}$$

and the sechsbein can be chosen diagonal as well:

$$e_{\hat{m}}{}^{\hat{a}} = (e_m{}^a, 1) , \tag{104}$$

where both the five-dimensional metric and its associated fünfbein are independent of y. Since the fields do not depend on y, and since the connection has non-vanishing components only if all of its indices are five-dimensional, the equations of motion reduce to

$$G^{mn}\nabla_m\partial_n x^{\underline{a}} = 0, \tag{105}$$

$$G^{mn}\nabla_m F_{np} = 0, \tag{106}$$

where

$$F_{mn} = H_{mny}. \tag{107}$$

Since h in six dimensions is self-dual, and since H is related to h it follows that we only need to consider the py component of the tensor equation. It will be convenient to rewrite these equations in an orthonormal basis with respect to the five-dimensional metric; this basis is related to the coordinate basis by the fünfbein. Using a, b, etc., to denote orthonormal indices, the equations of motion become

$$G^{ab}\nabla_a\partial_b x^{\underline{a}} = 0, \tag{108}$$

$$G^{ab}\nabla_a F_{bc} = 0, \tag{109}$$

where

$$G^{ab} = (\hat{m}^2)^{ab}, \tag{110}$$

and where we have introduced a hat for the six-dimensional m-matrix for later convenience.

The claim is that these equations are equivalent to the equations of motion arising from the five-dimensional Born-Infeld Lagrangian,

$$\mathcal{L} = \sqrt{-\det K}, \tag{111}$$

where

$$K_{mn} = g_{mn} + F_{mn}, \tag{112}$$

g_{mn} being the induced metric. To prove this we first show that the Born-Infeld equations can be written in the form

$$L^{mn}\nabla_m\partial_n x^{\underline{a}} = 0, \tag{113}$$

$$L^{mn}\nabla_m F_{np} = 0, \tag{114}$$

where

$$L = (1 - F^2)^{-1}. \tag{115}$$

When matrix notation is used, as in the last equation, it is understood that the first index is down and the second up, and F^2 indicates that the indices are in the right order for matrix multiplication. L^{mn} is then obtained by raising the first index with the inverse metric as usual. To complete the proof we shall then show that G is proportional to L up to a scale factor.

The matrix K is $1 + F$ so that its inverse is

$$K^{-1} = (1 + F)^{-1} = (1 - F)L, \tag{116}$$

from which we find

$$(K^{-1})^{(mn)} = L^{mn},$$

$$(K^{-1})^{[mn]} = -(FL)^{mn}, \tag{117}$$

the right-hand side of the second equation being automatically antisymmetric. Varying the Born-Infeld Lagrangian with respect to the gauge field A_m ($F = dA$), gives

$$\partial_n(\sqrt{-\det K}(K^{-1})^{[mn]}) = 0. \tag{118}$$

Carrying out the differentiation of the determinant, switching to covariant derivatives, and using the Bianchi identity for F, one finds

$$\nabla_n(K^{-1})^{[mn]} + (K^{-1})^{[pq]}\nabla_p F_{qn}(K^{-1})^{[mn]} = 0. \tag{119}$$

Using the identity

$$(K^{-1})^{[mn]}F_{np} = \delta_p{}^m - L_p{}^m \tag{120}$$

and the expression for $(K^{-1})^{[mn]}$ in terms of L and F one derives from (119)

$$L_n{}^q\nabla_q(F^{np}L_p{}^m) + F^{pn}L_n{}^q\nabla_q L_p{}^m = 0. \tag{121}$$

On differentiating the product in this equation one finds that the two terms with derivatives of L vanish by symmetry. Multiplying the remaining term by $(L^{-1})_m{}^r$ then yields the claimed result, namely (114). A similar calculation is used to derive (113).

To complete the proof we need to show that G^{mn} is proportional to L^{mn}. We begin by setting

$$f_{ab} = h_{ab5}, \tag{122}$$

$$F_{ab} = e_a{}^m e_b{}^n F_{mn}. \tag{123}$$

We then find

$$h_{abc} = \tfrac{1}{2}\epsilon_{abcde}f^{de}\,, \tag{124}$$

$$F_{ab} = (m^{-1})_a{}^c f_{cb}\,, \tag{125}$$

where $m_a^b = \hat{m}_a{}^b$. The first equation follows from the self-duality of h_{abc}, while the second equation follows from (53), (104), (107) and (122).

We set

$$\hat{m}_{\hat{a}}{}^{\hat{b}} = (\hat{m}_a{}^b, \hat{m}_a{}^5, \hat{m}_5{}^b, \hat{m}_5{}^5) \tag{126}$$

$$= (m_a{}^b, M_a, M^b, N)\,. \tag{127}$$

Recalling that

$$\hat{m}_{\hat{a}}{}^{\hat{b}} = (1 - 2h^2)_{\hat{a}}{}^{\hat{b}}\,, \tag{128}$$

one finds

$$m_a{}^b = \delta_a{}^b(1 - 2t_1) + 8(f^2)_a{}^b\,, \tag{129}$$

$$M_a = -\epsilon_{abcde}f^{bc}f^{de}\,, \tag{130}$$

$$N = (1 + 2t_1)\,, \tag{131}$$

where $t_1 = \mathrm{tr}(f^2)$. Noting that $f_a{}^b M_b = 0$, as can be seen by symmetry arguments, it follows from (125) that

$$F_a{}^b M_b = 0\,. \tag{132}$$

Now, by a direct calculation, starting from (110) one finds that

$$G_{ab} = A\eta_{ab} + 16(f^2)_{ab}\,, \tag{133}$$

where

$$A = 1 - 4t_1 - 4(t_1)^2 + 16t_2\,, \tag{134}$$

and we have defined $t_2 = \mathrm{tr}(f^4)$. Now, multiplying $G_{ab} = (m^2)_{ab} + M_a M_b$ with $(F^2)^{bc}$, and recalling (125), one finds

$$GF^2 = f^2\,. \tag{135}$$

Using this relation in (133) we find

$$G = A(1 - 16F^2)^{-1}\,. \tag{136}$$

Therefore we have shown that (after a suitable rescaling of F), G is proportional to L and hence the equations of motion arising from the superspace formulation of the 5-brane, when reduced to a 4-brane in ten dimensions, coincide with those that one derives from the Born-Infeld Lagrangian.

5. Conclusions

The component form of the equations of motion for the 5-brane in eleven dimensions are derived from the superspace equations. They are formulated in terms of the worldsurface fields $x^{\underline{n}}, \theta^{\underline{\mu}}, B_{mn}$. These equations are fully covariant in six dimensions; they possess six-dimensional Lorentz invariance, reparametrization invariance, spacetime supersymmetry and κ symmetry. We have also derived the κ transformations of the component fields. The fivebrane equations are derived from the superspace embedding condition for p-branes which possess half the supersymmetry found previously [1] and used to find superspace equations for the 5-brane in eleven dimensions in [2]. In the superembedding approach advocated here, the κ-symmetry is nothing but the odd diffeomorphisms of the worldsurface and as such invariance of the equations of motion under κ-symmetry is guaranteed.

We have also carried out a double-dimensional reduction to obtain the 4-brane in ten dimensions. We find agreement with the known Born-Infeld formulation for this latter theory. The result in ten dimensions which emerges from eleven dimensions appears in an unexpected form and that generalises the Born-Infeld structure to incorporate the worldsurface chiral 2-from gauge field.

In a recent paper [7] it was suggested that it was impossible to find a covariant set of equations of motion for a self dual second rank tensor in six dimensions. However, in this paper we have presented just such a system whose internal consistency is ensured by the manner of its derivation. We would note that although the field h_{abc} which emerges form the superspace formalism obeys a simple duality condition, the field strength H_{mnp} of the gauge field inherits a version of this duality condition which is rather complicated. Using the solution of the chirality constraint on the 2-form, we expect that our bosonic equations of motion will reduce to those of [7,8].

In Ref. [9], an auxiliary field has been introduced to write down a 6D covariant action. It would be interesting to find if this field is contained in the formalism considered in this paper. We note, however, that in the approach of Ref. [9] one replaces the nonmanifest Lorentz symmetry with another bosonic symmetry that is equally nonmanifest, but necessary to eliminate the unwanted auxiliary field and that the proof the new symmetry involves steps similar to those needed to prove the nonmanifest Lorentz symmetry [9]. Further, it is not clear if a 6D covariant gauge fixing procedure is possible to gauge fix this extra symmetry.

In a forthcoming publication, we shall give in more detail the component field equation and the double-dimensional reduction [18]. We also hope to perform a generalized-dimensional reduction procedure to the worldsurface, but staying in eleven dimensions. In the approach of this paper, there is little conceptual difference in whether the worldsurface multiplet is a scalar multiplet (Type I branes), or vector multiplets (D branes), or indeed tensor multiplets (M branes) and we hope to report on the construction of all p-brane solutions from this view point.

We conclude by mentioning some open problems that are natural to consider, given the fact that we now know the 6D covariant field equations of the M-theory five-brane. It would be interesting to consider solitonic p-brane solutions of these equations, perform a semiclassical quantization, explore the spectral and duality properties of our system and study the anomalies of the chiral system. Finally, given the luxury of having manifest worldsurface and target space supersymmetries at the same time, it would be instructive to consider a variety of gauge choices, such as a static gauge, as was done recently for super D-branes [14], which would teach us novel and interesting ways to realize supersymmetry nonlinearly. This may provide useful tools in the search for the "different corners of M-theory".

6. Note added

While this paper was in the final stages of being written up, we saw two related papers appear on the net [20,21]. We hope to comment on the relationship between these papers and the work presented in a subsequent publication.

References

[1] P.S. Howe and E. Sezgin, Superbranes, hep-th/9607227.
[2] P.S. Howe and E. Sezgin, $D = 11, p = 5$, hep-th/9611008.
[3] P.K. Townsend, D-branes from M-branes, hep-th/9512062.
[4] O. Aharony, String theory dualities from M-theory, hep-th/9604103.
[5] E. Bergshoeff, M. de Roo and T. Ortin, The eleven-dimensional fivebrane, hep-th/9606118.
[6] E. Witten, Five-brane effective action in M-theory, hep-th/9610234.
[7] M. Perry and J.H. Schwarz, Interacting chiral gauge fields in six dimensions and Born-Infeld theory, hep-th/9611065.
[8] J.H. Schwarz, Coupling of self-dual tensor to gravity in six dimensions, hep-th/9701008.
[9] P. Pasti, D. Sorokin and M. Tonin, Covariant action for $D = 11$ five-brane with the chiral field, hep-th/9701037.
[10] E. Cremmer and S. Ferrara, Formulation of 11-dimensional supergravity in superspace, Phys. Lett. B 91 (1980) 61.
[11] L. Brink and P.S. Howe, Eleven-dimensional supergravity on the mass shell in superspace, Phys. Lett. B 91 (1980) 384.
[12] M. Cederwall, A. von Gussich, B.E.W. Nilsson and A. Westerberg, The Dirichlet super-three-brane in ten-dimensional Type IIB supergravity, hep-th/9610148.
[13] M. Aganagic, C. Popescu and J.H. Schwarz, D-brane actions with local kappa symmetry, hep-th/9610249.
[14] M. Aganagic, C. Popescu and J.H. Schwarz, Gauge-invariant and gauge-fixed D-brane actions, hep-th/9612080.
[15] M. Cederwall, A. von Gussich, B.E.W. Nilsson, P. Sundell and A. Westerberg, The Dirichlet super p-branes in ten-dimensional Type IIA and IIB supergravity, hep-th/9611159.
[16] E. Bergshoeff and P.K. Townsend, Super D-branes, hep-th/9611173.
[17] E. Bergshoeff, E. Sezgin and P.K. Townsend, Properties of eleven-dimensional supermembrane theory, Ann. Phys. 185 (1988) 330.
[18] P.S. Howe, E. Sezgin and P.C. West, in preparation.
[19] M.J. Duff, P.S. Howe, T. Inami and K.S. Stelle, Superstrings in $D = 10$ from supermembranes in $D = 11$, Phys. Lett. B 191 (1987) 70.
[20] I. Bandos, K. Lechner, A. Nurmagambetov, P. Pasti, D. Sorokin and M. Tonin, Covariant action for the super five-brane of M-theory, hep-th/9701149.
[21] M. Aganagic, J. Park, C. Popescu and J.H. Schwarz, Worldvolume action of the M-theory five-brane, hep-th/9701166.

Chapter 4

M-theory (before M-theory was cool)

This chapter addresses the question [1]: 'Should we have been surprised by the eleven-dimensional origin of string theory?' The importance of eleven dimensions is no doubt surprising from the point of view of perturbative string theory; from the point of view of membrane theory, however, there were already tantalizing hints in this direction:

(*i*) **K3 compactification**

In 1986, it was pointed out [2] that $D = 11$ supergravity on $R^{10-n} \times K3 \times T^{n-3}$ [3] and the $D = 10$ heterotic string on $R^{10-n} \times T^n$ [4] not only have the same supersymmetry but also the same moduli spaces of vacua, namely

$$\mathcal{M} = \frac{SO(16+n, n)}{SO(16+n) \times SO(n)}. \tag{4.1}$$

It took almost a decade for this 'coincidence' to be explained, but we now know that M-theory on $R^{10-n} \times K3 \times T^{n-3}$ is dual to the heterotic string on $R^{10-n} \times T^n$.

(*ii*) **Superstrings in D=10 from supermembranes in D=11**

As described in chapter 2, eleven dimensions received a big shot in the arm in 1987 when the $D = 11$ supermembrane was discovered [5]. It was then pointed out [6] that in an $R^{10} \times S^1$ topology the weakly coupled ($d = 2, D = 10$) Type IIA superstring follows by wrapping the ($d = 3, D = 11$) supermembrane around the circle in the limit that its radius R shrinks to zero. In particular, the Green–Schwarz action of the string follows in this way from the Green–Schwarz action of the membrane. It was necessary to take this $R \to 0$ limit in order to send to infinity the masses of the (at the time) unwanted Kaluza–Klein modes which had no place in weakly coupled Type IIA theory. The $D = 10$ dilaton, which governs the strength of the string coupling, is just a component of the $D = 11$ metric. A critique of superstring orthodoxy *circa* 1987, and its failure to accommodate the eleven-dimensional supermembrane, may be found in [7].

(*iii*) **Membrane at the end of the universe**

Being defined over the boundary of AdS_4, the $OSp(4|8)$ singleton[1] action [12]

[1] We recall that singletons are those strange representations of AdS first identified by Dirac [8] which admit no analogue in flat spacetime. They have been much studied by Fronsdal and collaborators [9, 10].

is a *three dimensional* superconformal theory with signature $(-, +, +)$ describing 8 scalars and 8 spinors. With the discovery of the eleven-dimensional supermembrane [5], it was noted that 8 scalars and 8 spinors on a three-dimensional worldvolume with signature $(-, +, +)$ is just what is obtained after gauge-fixing the supermembrane action! Moreover, kappa-symmetry of this supermembrane action forces the background fields to obey the field equations of $(N = 1, D = 11)$ supergravity. It was therefore suggested in 1987 [11] that on the $AdS_4 \times S^7$ supergravity background, the superconformal $OSp(4|8)$ singleton action describes a supermembrane whose worldvolume occupies the $S^1 \times S^2$ boundary of the AdS_4: *The membrane at the end of the universe* [13]. Noting that these singletons also appear in the Kaluza–Klein harmonic expansion of this supergravity background, this further suggested a form of bootstrap [11] in which the supergravity gives rise to the membrane on the boundary which in turn yields the supergravity in the bulk. This was thus a precursor of Maldacena's AdS/CFT correspondence [14], discussed in chapter 6, which conjectures a duality between physics in the bulk of AdS and a conformal field theory on the boundary. The other two supergroups in table 1.2 of chapter 1 also admit the so-called doubleton and tripleton supermultiplets [15] as shown in table 4.1.

Supergroup	**Supermultiplet**	**Field content**
$OSp(4\|8)$	$(n = 8, d = 3)$ singleton	8 scalars,8 spinors
$SU(2,2\|4)$	$(n = 4, d = 4)$ doubleton	1 vector,4 spinors,6 scalars
$OSp(6,2\|4)$	$((n_+, n_-) = (2,0), d = 6)$ tripleton	1 chiral 2-form,8 spinors, 5 scalars

Table 4.1. Superconformal groups and their singleton, doubleton and tripleton representations.

(*iv*) **Membranes and matrix models**

As mentioned in chapter 2, the $D = 11$ supermembrane in the lightcone gauge has a residual area preserving diffeomorphism symmetry. In 1988 it was shown to be described by a quantum mechanical matrix model [16] corresponding to a dimensionally reduced $D = 10$ Yang–Mills theory with gauge group $SU(k)$ as $k \to \infty$. This Hamiltonian has recently been resurrected in the context of the matrix model approach to M-theory [18, 17] discussed in chapter 6.

(*v*) **U-duality (when it was still non-U)**

Based on considerations of this $D = 11$ supermembrane, which on further compactification treats the dilaton and moduli fields on the same footing, it was conjectured [19] in 1990 that discrete subgroups of all the old non-compact global symmetries of compactified supergravity [20, 21] (e.g $SL(2, R)$, $O(6,6)$, E_7) should be promoted to duality symmetries of the supermembrane. Via the above wrapping around S^1, therefore, they should also be inherited by the Type IIA string [19].

(*vi*) **D=11 membrane/fivebrane duality**

In 1991, the supermembrane was recovered as an elementary solution of $D = 11$ supergravity which preserves half of the spacetime supersymmetry [22]. In 1992, the superfivebrane was discovered as a soliton solution of $D = 11$ supergravity also preserving half the spacetime supersymmetry [23]. This naturally suggests a $D = 11$ membrane/fivebrane duality.

(vii) **Hidden eleventh dimension**

We have seen how the $D = 10$ Type IIA string follows from $D = 11$. Is it possible to go the other way and discover an eleventh dimension hiding in $D = 10$? In 1993, it was recognized [24] that by dualizing a vector into a scalar on the gauge-fixed $d = 3$ worldvolume of the Type IIA supermembrane, one increases the number of worldvolume scalars (i.e. transverse dimensions) from 7 to 8 and hence obtains the corresponding worldvolume action of the $D = 11$ supermembrane. Thus the $D = 10$ Type IIA theory contains a hidden $D = 11$ Lorentz invariance! This device was subsequently used [25, 26] to demonstrate the equivalence of the actions of the $D = 10$ Type IIA membrane and the Dirichlet twobrane [27].

($viii$) **U-duality**

Of the conjectured Cremmer–Julia symmetries referred to in (v) above, the case for a target space $O(6,6;Z)$ (T-*duality*) in perturbative string theory had already been made, of course [28]. Stronger evidence for an $SL(2,Z)$ (S-*duality*) in string theory was subsequently provided in [29, 30] where it was pointed out that it corresponds to a *non-perturbative* electric/magnetic symmetry. In 1994, stronger evidence for the combination of S and T into a discrete duality of Type II strings, such as $E_7(Z)$ in $D = 4$, was provided in [31], where it was dubbed U-*duality*. Moreover, the BPS spectrum necessary for this U-duality was given an explanation in terms of the wrapping of either the $D = 11$ membrane or $D = 11$ fivebrane around the extra dimensions. This paper also conjectured a non-perturbative $SL(2,Z)$ of the Type IIB string in $D = 10$.

(ix) **Black holes**

In 1995, it was conjectured [32] that the $D = 10$ Type IIA superstring should be identified with the $D = 11$ supermembrane compactified on S^1, even for large R. The $D = 11$ Kaluza–Klein modes (which, as discussed in (ii) above, had no place in the *perturbative* Type IIA theory) were interpreted as charged extreme black holes of the Type IIA theory.

(x) **D=11 membrane/fivebrane duality and anomalies**

Membrane/fivebrane duality interchanges the roles of field equations and Bianchi identities and, as we saw in chapter 3, membrane/fivebrane duality thus predicts a spacetime correction to the $D = 11$ supergravity action [33, 34]. This prediction is intrinsically M-theoretic, with no counterpart in ordinary $D = 11$ supergravity. However, by simultaneous dimensional reduction [16] of $(d = 3, D = 11)$ to $(d = 2, D = 10)$ on S^1, it translates into a corresponding prediction for the Type IIA string. Thus using $D = 11$ membrane/fivebrane duality one can correctly reproduce the corrections to the 2-form field equations of the $D = 10$ Type IIA string (a mixture of tree-level and string one-loop effects) starting from the Chern–Simons corrections to the Bianchi identities of the $D = 11$ superfivebrane (a purely tree-level effect).

(xi) **Heterotic string from fivebrane wrapped around $K3$**

In 1995 it was shown that, when wrapped around $K3$ which admits 19 self-dual and 3 anti-self-dual 2-forms, the $d = 6$ worldvolume fields of the $D = 11$ fivebrane (or Type IIA fivebrane) $(B^-{}_{\mu\nu}, \lambda^I, \phi^{[IJ]})$ reduce to the $d = 2$ worldsheet fields of the heterotic string in $D = 7$ (or $D = 6$) [35, 36]. The 2-form yields $(19, 3)$ left and right moving bosons, the spinors yield $(0, 8)$ fermions and the scalars yield $(5, 5)$ which add up to the correct worldsheet degrees of freedom of the heterotic string [35, 36]. A consistency check is provided [33] by the derivation of the Yang–Mills and Lorentz Chern–Simons corrections to the Bianchi identity of the heterotic string starting from the fivebrane Bianchi identity. We also note that if we replace $K3$ by T^4 in the above derivation, the 2-form now yields $(3, 3)$ left and right moving bosons, the spinors now yield $(8, 8)$ fermions and the scalars again yield $(5, 5)$ which add up to the correct worldsheet degrees of freedom of the Type IIA string [1]. In this case, one recovers the trivial Bianchi identity of Type IIA.

(*xii*) **N=1 in D=4**

Also in 1995 it was noted [37–43] that $N = 1$ heterotic strings can be dual to $D = 11$ supergravity compactified on seven-dimensional spaces of G_2 holonomy which also yield $N = 1$ in $D = 4$ [44].

(*xiii*) **Non-perturbative effects**

Also in 1995 it was shown [45] that membranes and fivebranes of the Type IIA theory, obtained by compactification on S^1, yield e^{-1/g_s} effects, where g_s is the string coupling.

(*xiv*) **SL(2,Z)**

Also in 1995, strong evidence was provided for identifying the Type IIB string on $R^9 \times S^1$ with M-theory on $R^9 \times T^2$ [46, 43]. In particular, the conjectured $SL(2, Z)$ of the Type IIB theory discussed in (*viii*) above is just the modular group of the M-theory torus. Two alternative explanations of this $SL(2, Z)$ had previously been given: (a) identifying it with the S-duality [33] of the $d = 4$ Born–Infeld worldvolume theory of the self-dual Type IIB superthreebrane [47], and (b) using the four-dimensional heterotic/Type IIA/Type IIB triality [48] by noting that this $SL(2, Z)$, while non-perturbative for the Type IIB string, is perturbative for the heterotic string.

(*xv*) $\mathrm{E}_8 \times \mathrm{E}_8$ **heterotic string**

Also in 1995 (that *annus mirabilis*!), strong evidence was provided for identifying the $E_8 \times E_8$ heterotic string[2] on R^{10} with M-theory on $R^{10} \times S^1/Z_2$ [50].

This completes our summary of M-theory before M-theory was cool. The *phrase* M-theory (though, as we hope to have shown, not the *physics* of M-theory) first made its appearance in October 1995 [46, 50]. We shall return to M-theory in chapter 6.

References

[1] Duff M J 1996 *M*-theory: the theory formerly known as strings *Int. J. Mod. Phys.* A **11** 5623–42

[2] Duff M J and Nilsson B E W 1986 Four-dimensional string theory from the K3 lattice *Phys. Lett.* B **175** 417

[2] It is ironic that, having hammered the final nail in the coffin of $D = 11$ supergravity by telling us that it can never yield a *chiral* theory when compactified on a manifold [49], Witten pulls it out again by telling us that it does yield a chiral theory when compactified on something that is not a manifold!

[3] Duff M J, Nilsson B E W and Pope C N 1983 Compactification of $D = 11$ supergravity on $K3 \times T^3$ *Phys. Lett.* B **129** 39

[4] Narain K S 1986 New heterotic string theories in uncompactified dimensions < 10 *Phys. Lett.* B **169** 41

[5] Bergshoeff E, Sezgin E and Townsend P K 1987 Supermembranes and eleven-dimensional supergravity *Phys. Lett.* B **189** 75

[6] Duff M J, Howe P S, Inami T and Stelle K S 1987 Superstrings in $D = 10$ from supermembranes in $D = 11$ *Phys. Lett.* B **191** 70

[7] Duff M J 1990 Not the standard superstring review *Proc. International School of Subnuclear Physics, 'The Superworld II', Erice, 1987* ed A Zichichi (New York: Plenum) CERN-TH.4749/87

[8] Dirac P A M 1963 A remarkable representation of the 3+2 de Sitter group *J. Math. Phys.* **4** 901

[9] Fronsdal C 1982 Dirac supermultiplet *Phys. Rev.* D **26** 1988

[10] Flato M and Fronsdal C 1981 Quantum field theory of singletons. The rac *J. Math. Phys.* **22** 1100

[11] Duff M J 1988 Supermembranes: The first fifteen weeks *Class. Quantum Grav.* **5** 189

[12] Blencowe M P and Duff M J 1988 Supersingletons *Phys. Lett.* B **203** 229

[13] Bergshoeff E, Duff M J, Pope C N and Sezgin E 1988 Supersymmetric supermembrane vacua and singletons *Phys. Lett.* B **199** 69

[14] Maldacena J 1998 The large N limit of superconformal field theories and supergravity *Adv. Theor. Math. Phys.* **2** 231

[15] Gunaydin M 1990 Singleton and doubleton supermultiplets of space-time supergroups and infinite spin superalgebras *Proc. 1989 Trieste Conf. 'Supermembranes and Physics in* $2+1$ *Dimensions'* ed M Duff, C Pope and E Sezgin (Singapore: World Scientific)

[16] de Wit B, Hoppe J and Nicolai H 1988 On The Quantum Mechanics Of Supermembranes *Nucl. Phys.* B **305** 545

[17] Banks T, Fischler W, Shenker S H and Susskind L 1997 M-theory as a matrix model: a conjecture *Phys. Rev.* D **55** 5112–28

[18] Townsend P K 1996 D-branes from M-branes *Phys. Lett.* B **373** 68

[19] Duff M J and Lu J X 1990 Duality rotations in membrane theory *Nucl. Phys.* B **347** 394

[20] Cremmer E, Ferrara S and Scherk J 1978 $SU(4)$ invariant supergravity theory *Phys. Lett.* B **74** 61

[21] Cremmer E and Julia B 1979 The $SO(8)$ supergravity *Nucl. Phys.* B **159** 141

[22] Duff M J and Stelle K S 1991 Multimembrane solutions of $D = 11$ supergravity *Phys. Lett.* B **253** 113

[23] Gueven R 1992 Black p-brane solutions of $d = 11$ supergravity theory *Phys. Lett.* B **276** 49

[24] Duff M J and Lu J X 1993 Type II p-branes: the brane scan revisited *Nucl. Phys.* B **390** 276

[25] Townsend P 1996 D-Branes from M-Branes *Phys. Lett.* B **373** 68

[26] Schmidhuber C 1996 D-brane actions *Nucl. Phys.* B **467** 146

[27] Polchinski J 1995 Dirichlet branes and Ramond–Ramond charges *Phys. Rev. Lett.* **75** 4724

[28] Giveon A, Porrati M and Rabinovici E 1994 Target space duality in string theory *Phys. Rep.* **244** 77

[29] Font A, Ibanez L, Lust D and Quevedo F 1990 Strong-weak coupling duality and nonperturbative effects in string theory *Phys. Lett.* B **249** 35

[30] Rey S-J 1991 The confining phase of superstrings and axionic strings *Phys. Rev.* D **43** 526
[31] Hull C M and Townsend P K 1995 Unity of superstring dualities *Nucl. Phys.* B **438** 109
[32] Townsend P K 1995 The eleven-dimensional supermembrane revisited *Phys. Lett.* B **350** 184
[33] Duff M J, Liu J T and Minasian R 1995 Eleven dimensional origin of string-string duality: a one-loop test *Nucl. Phys.* B **452** 261
[34] Witten E 1996 Fivebranes and *M*-theory on an orbifold *Nucl. Phys.* B **463** 383
[35] Townsend P K 1995 String/membrane duality in seven dimensions *Phys. Lett.* B **354** 247
[36] Harvey J A and Strominger A 1995 The heterotic string is a soliton *Nucl. Phys.* B **449** 535
[37] Cadavid A C, Ceresole A, D'Auria R and Ferrara S 1995 11-dimensional supergravity compactified on Calabi–Yau threefolds *Phys. Lett.* B **357** 76
[38] Papadopoulos G and Townsend P K 1995 Compactification of D=11 supergravity on spaces with exceptional holonomy *Phys. Lett.* B **357** 300
[39] Schwarz J H and Sen A 1995 Type *IIA* dual of six-dimensional CHL compactifications *Nucl. Phys.* B **454** 427
[40] Harvey J A, Lowe D A and Strominger A 1995 $N = 1$ string duality *Phys. Lett.* B **362** 65
[41] Chaudhuri S and Lowe D A 1996 Type IIA-Heterotic duals with maximal supersymmetry *Nucl. Phys.* B **459** 113
[42] Acharya B S 1995 $N = 1$ Heterotic-supergravity duality and Joyce manifolds, hep-th/9508046
[43] Aspinwall P S 1996 Some relationships between dualities in string theory *Nucl. Phys. Proc. Suppl.* **46** 30
[44] Awada M A, Duff M J and Pope C N 1983 $N = 8$ supergravity breaks down to $N = 1$ *Phys. Rev. Lett.* **50** 294
[45] Becker K, Becker M and Strominger A 1995 Fivebranes, membranes and non-perturbative string theory *Nucl. Phys.* B **456** 130
[46] Schwarz J H 1995 The power of *M*-theory *Phys. Lett.* B **360** 13
[47] Duff M J and Lu J X 1991 The self-dual Type *IIB* superthreebrane *Phys. Lett.* B **273** 409
[48] Duff M J, Liu J T and Rahmfeld J 1996 Four-dimensional string/string/string triality *Nucl. Phys.* B **459** 125
[49] Witten E 1985 Fermion quantum numbers in Kaluza–Klein theory *Shelter Island II, Proc. 1983 Shelter Island Conf. on Quantum Field Theory and the Fundamental Problems of Physics* ed R Jackiw, N Khuri, S Weinberg and E Witten (Cambridge, MA: MIT Press) p 227
[50] Horava P and Witten E 1996 Heterotic and Type *I* string dynamics from eleven dimensions *Nucl. Phys.* B **460** 506

Reprinted from Phys. Lett. B **191** (1987) 70–4

SUPERSTRINGS IN $D=10$ FROM SUPERMEMBRANES IN $D=11$

M.J. DUFF [1], P.S. HOWE [2], T. INAMI [3] and K.S. STELLE [1]
CERN, CH-1211 Geneva 23, Switzerland

Received 3 March 1987

The type IIA superstring in ten dimensions is derived from the supermembrane in eleven dimensions by a simultaneous dimensional reduction of the world volume and the spacetime.

It is well known that $N=2a$ supergravity in ten dimensions (g_{mn}, A_m, Φ; ψ_m, χ; A_{mnp}, A_{mn}) may be obtained by dimensional reduction from $N=1$ supergravity in eleven dimensions ($\hat{g}_{\hat{m}\hat{n}}$; $\hat{\psi}_{\hat{m}}$; $\hat{A}_{\hat{m}\hat{n}\hat{p}}$). On the other hand, $n=2a$ supergravity is also the field theory limit of the type IIa superstring. Does this imply a connection between $D=11$ supergravity and strings? Bergshoeff, Sezgin and Townsend [1] have recently found a niche for $D=11$ supergravity within the framework of extended objects, but the extended object in question is a three-dimensional membrane rather than a two-dimensional string [‡1]. The purpose of this letter is to derive the type IIA superstring from this supermembrane by a dimensional reduction of the world volume from three to two dimensions and, simultaneously, a dimensional reduction of the spacetime from eleven to ten.

To describe the coupling of a closed three-membrane to a $d=11$ supergravity background, let us introduce world-volume coordinates $\hat{\xi}^{\hat{i}}$ ($\hat{i}=1, 2, 3$) and a world-volume metric $\hat{\gamma}_{\hat{i}\hat{j}}(\hat{\xi})$ with signature $(-, +, +)$. The target space is a supermanifold with superspace coordinates $\hat{z}^{\hat{M}}=(\hat{x}^{\hat{m}}, \hat{\theta}^{\hat{\mu}})$ where $\hat{m}=1, ..., 11$ and $\hat{\mu}=1, ..., 32$ with spacetime signature $(-, +, ..., +)$. We also define $\hat{E}_{\hat{i}}{}^{\hat{A}}=(\partial_{\hat{i}}\hat{z}^{\hat{M}})\hat{E}_{\hat{M}}{}^{\hat{A}}(\hat{z})$ where $\hat{E}_{\hat{M}}{}^{\hat{A}}$ is the supervielbein and $\hat{A}=(\hat{a}, \hat{\alpha})$ is the tangent space index ($\hat{a}=1, ..., 11$ and $\hat{\alpha}=1, ..., 32$). The action is then given by [1]

$$S=\int d^3\hat{\xi}\,\big(\tfrac{1}{2}\sqrt{-\hat{\gamma}}\,\hat{\gamma}^{\hat{i}\hat{j}}\hat{E}_{\hat{i}}{}^{\hat{a}}\hat{E}_{\hat{j}}{}^{\hat{b}}\eta_{\hat{a}\hat{b}} - \tfrac{1}{6}\epsilon^{\hat{i}\hat{j}\hat{k}}\hat{E}_{\hat{i}}{}^{\hat{A}}\hat{E}_{\hat{j}}{}^{\hat{B}}\hat{E}_{\hat{k}}{}^{\hat{C}}\hat{A}_{\hat{C}\hat{B}\hat{A}} - \tfrac{1}{2}\sqrt{-\hat{\gamma}}\big)\,. \qquad (1)$$

Note that there is a Wess–Zumino term involving the super three-form $\hat{A}_{\hat{A}\hat{B}\hat{C}}(\hat{z})$ and also a world-volume cosmological term. In addition to world-volume diffeomorphisms, target-space superdiffeomorphisms, Lorentz invariance and three-form gauge invariance, the action (1) is invariant under a fermionic gauge transformation [1] whose parameter $\hat{\kappa}^{\hat{\alpha}}(\hat{\xi})$ is a 32-component spacetime Majorana spinor and a world-volume scalar. This is a generalization to the case of membranes of the symmetry discovered by Siegel [4] for the superparticle and Green and Schwarz [5] for the superstring in the form given by Hughes, Liu and Polchinski [6]. We shall return to this later in eq. (23).

To see how the dimensional reduction works, let us first focus our attention on the purely bosonic sector for which the action (1) reduces to

$$S=\int d^3\hat{\xi}\,\big[\tfrac{1}{2}\sqrt{-\hat{\gamma}}\,\hat{\gamma}^{\hat{i}\hat{j}}\partial_{\hat{i}}\hat{x}^{\hat{m}}\partial_{\hat{j}}\hat{x}^{\hat{n}}\hat{g}_{\hat{m}\hat{n}}(\hat{x}) - \tfrac{1}{2}\sqrt{-\hat{\gamma}} + \tfrac{1}{6}\epsilon^{\hat{i}\hat{j}\hat{k}}\partial_{\hat{i}}\hat{x}^{\hat{m}}\partial_{\hat{j}}\hat{x}^{\hat{n}}\partial_{\hat{k}}\hat{x}^{\hat{p}}\hat{A}_{\hat{m}\hat{n}\hat{p}}(\hat{x})\big]\,. \qquad (2)$$

[1] On leave of absence from the Blackett Laboratory Imperial College, London SW7 2BZ, UK.

[2] On leave of absence from the Mathematics Department, King's College, The Strand, London WC2, UK.

[3] On leave of absence from RIFP, Kyoto University, Kyoto 606, Japan.

[‡1] It is interesting to note that the three–eight split SO(1, 10) ⊃ SO(1, 2) × SO(8) implied by the membrane had previously been invoked in refs. [2,3] to exhibit the hidden SO(16) of $D=11$ supergravity.

0370-2693/87/$ 03.50 © Elsevier Science Publishers B.V.
(North-Holland Physics Publishing Division)

Varying with the respect to the metric $\hat{\gamma}_{\tilde{i}\tilde{j}}$ yields the embedding equation

$$\hat{\gamma}_{\tilde{i}\tilde{j}}=\hat{g}_{\tilde{i}\tilde{j}}\equiv\partial_{\tilde{i}}\hat{x}^{\hat{m}}\partial_{\tilde{j}}\hat{x}^{\hat{n}}\hat{g}_{\hat{m}\hat{n}}(\hat{x}), \tag{3}$$

while varying with respect to $\hat{x}^{\hat{m}}$ yields the equation of motion

$$(1/\sqrt{-\hat{g}})\partial_{\tilde{i}}(\sqrt{-\hat{g}}\,\hat{g}^{\tilde{i}\tilde{j}}\partial_{\tilde{j}}\hat{x}^{\hat{m}})+\hat{\Gamma}_{\hat{n}\hat{p}}{}^{\hat{m}}\partial_{\tilde{i}}\hat{x}^{\hat{n}}\partial_{\tilde{j}}\hat{x}^{\hat{p}}\hat{g}^{\tilde{i}\tilde{j}}$$

$$=\tfrac{1}{6}\hat{F}^{\hat{m}}{}_{\hat{n}\hat{p}\hat{q}}\partial_{\tilde{i}}\hat{x}^{\hat{n}}\partial_{\tilde{j}}\hat{x}^{\hat{p}}\partial_{\tilde{k}}\hat{x}^{\hat{q}}\epsilon^{\tilde{i}\tilde{j}\tilde{k}}/\sqrt{-\hat{g}}\,. \tag{4}$$

where $F_{\hat{m}\hat{n}\hat{p}\hat{q}}$ is the field strength of $\hat{A}_{\hat{m}\hat{n}\hat{p}}$,

$$\hat{F}_{\hat{m}\hat{n}\hat{p}\hat{q}}\equiv4\partial_{[\hat{m}}\hat{A}_{\hat{n}\hat{p}\hat{q}]}\,. \tag{5}$$

We now make a two–one split of the world-volume coordinates

$$\xi^{\tilde{i}}=(\xi^{i},\rho),\quad i=1,2\,, \tag{6}$$

and a ten–one split of the spacetime coordinates

$$\hat{x}^{\hat{m}}=(x^{m},y),\quad m=1,\ldots,10\,, \tag{7}$$

in order to make the partial gauge choice

$$\rho=y\,, \tag{8}$$

which identifies the eleventh spacetime dimension with the third dimension of the world volume. The dimensional reduction is then affected by demanding that

$$\partial_{\rho}x^{m}=0\,, \tag{9}$$

and

$$\partial_{y}\hat{g}_{\hat{m}\hat{n}}=0=\partial_{y}\hat{A}_{\hat{m}\hat{n}\hat{p}}\,. \tag{10}$$

A suitable choice of ten-dimensional variables is now given by

$$\hat{g}_{\hat{m}\hat{n}}=\Phi^{-2/3}\begin{pmatrix}g_{mn}+\Phi^{2}A_{m}A_{n} & \Phi^{2}A_{m}\\ \Phi^{2}A_{n} & \Phi^{2}\end{pmatrix},$$

$$\hat{A}_{\hat{m}\hat{n}\hat{p}}=(\hat{A}_{mnp},\hat{A}_{mny})\;=(A_{mnp},A_{mn})\,. \tag{11}$$

From (3), the induced metic on the world sheet is now given by

$$\hat{g}_{\tilde{i}\tilde{j}}=\Phi^{-2/3}\begin{pmatrix}g_{ij}+\Phi^{2}A_{i}A_{j} & \Phi^{2}A_{i}\\ \Phi^{2}A_{j} & \Phi^{2}\end{pmatrix}, \tag{12}$$

where

$$g_{ij}\equiv\partial_{i}x^{m}\partial_{j}x^{n}g_{mn},\quad A_{i}\equiv\partial_{i}x^{m}A_{m}\,. \tag{13}$$

Note that

$$\sqrt{-\hat{g}}=\sqrt{-g}\,. \tag{14}$$

Substituting these expressions into the field equations (4) yields in the case $\hat{x}^{\hat{m}}=x^{m}$

$$(1/\sqrt{-g})\partial_{i}(\sqrt{-g}\,g^{ij}\partial_{j}x^{m})+\Gamma_{np}{}^{m}\partial_{i}x^{n}\partial_{j}x^{p}g^{ij}$$

$$=\tfrac{1}{2}F^{m}{}_{np}\partial_{i}x^{n}\partial_{j}x^{p}\epsilon^{ij}/\sqrt{-\hat{g}}\,, \tag{15}$$

where F_{mnp} is the field strength of A_{mn},

$$F_{mnp}\equiv3\partial_{[m}A_{np]}=\hat{F}_{mnpy}\,. \tag{16}$$

In the case $\hat{x}^{\hat{m}}=y$, (4) is an identity, as it must be for consistency. But (15) is just the ten-dimensional string equation of motion derivable from the action

$$S=\int d^{2}\xi\,(\tfrac{1}{2}\sqrt{-\gamma}\,\gamma^{ij}\partial_{i}x^{m}\partial_{j}x^{n}g_{mn}$$

$$+\tfrac{1}{2}\epsilon^{ij}\partial_{i}x^{m}\partial_{j}x^{n}A_{mn})\,. \tag{17}$$

Comparing with (2), we see that the overall effect is to reduce the eleven-dimensional membrane to a ten-dimensional string, to replace the three-form by a two-form in the Wess–Zumino term and to eliminate the world-volume cosmological constant. Note that the other ten-dimensional bosonic fields A_{mnp}, A_m and Φ have all decoupled. They have not disappeared from the theory, however, since their coupling still survives in the fermionic θ sector, to which we shall turn shortly. First, we make some remarks.

As is well known, the dimensional reduction (10) corresponds to a Kaluza–Klein compactification of spacetime on a circle in which one discards all the massive modes. The difference from conventional Kaluza–Klein is that by identifying the eleventh spacetime dimension with the third dimension on the world volume as in (8), the world volume is also compactified on the same circle. The condition (9) means that we are discarding the massive world-sheet modes at the same time. By retaining all the U(1) singlets but only the U(1) singlets, these truncations are guaranteed to be consistent [7] with the membrane equations of motion and, as we shall see, with the equations of motion of the background fields. As an extra check on consistency, we have been careful to substitute the Kaluza–Klein ansatz into the equations of motion rather than directly into the action. The signal for consistency is that the $\hat{x}^{\hat{m}}=y$ component of the field equations (4) is an identity. Having established consistency one may then, if so desired, substitute directly into the action (2) and integrate over ρ. The result is not quite the action (17) but an equivalent one which yields the same equations of motion. To see this, let us recall that since we are now treating $\hat{\gamma}_{\hat{i}\hat{j}}$ and $\hat{x}^{\hat{m}}$ as independent variables in (2), we should make independent Ansätze for both. Thus we write

$$\hat{\gamma}_{\hat{i}\hat{j}}=\phi^{-2/3}\begin{pmatrix}\gamma_{ij}+\phi^2 V_i V_j & \phi^2 V_i \\ \phi^2 V_j & \phi^2\end{pmatrix}, \tag{18}$$

where γ_{ij}, V_i and ϕ are, a priori, unrelated to g_{ij}, A_i and Φ of (12). Substituting into the action (2) and integrating over ρ yields

$$S=\int d^2\xi \,\{\tfrac{1}{2}\sqrt{-\gamma}\,\phi^{2/3}\Phi^{-2/3}$$
$$\times[\gamma^{ij}g_{ij}+\gamma^{ij}(A_i-V_i)(A_j-V_j)+\phi^{-2}\Phi^{-2}]$$
$$-\tfrac{1}{2}\sqrt{-\gamma}+\tfrac{1}{2}\epsilon^{ij}\partial_i x^m\partial_j x^n A_{mn}\} . \tag{19}$$

Since the equations of motion for γ_{ij}, V_i and ϕ are algebraic, we may eliminate all of them to yield the action

$$S=\int d^2\xi \,(\sqrt{-g}+\tfrac{1}{2}\epsilon^{ij}\partial_i x^m\partial_j x^n A_{mn}) , \tag{20}$$

which is the action we would have obtained by writing (2) in Nambu–Goto form

$$S=\int d^3\hat{\xi}\,(\sqrt{-\hat{g}}+\tfrac{1}{6}\epsilon^{\hat{i}\hat{j}\hat{k}}\partial_{\hat{i}}\hat{x}^{\hat{m}}\partial_{\hat{k}}\hat{x}^{\hat{p}}A_{\hat{m}\hat{n}\hat{p}}) . \tag{21}$$

Alternatively, we may eliminate just V_i and ϕ to obtain (17). It is interesting to note that the string action (17) we obtain by dimensional reduction is conformally invariant even though the membane theory we started from was not.

The foregoing discussion is readily generalized to a superspace setting. To facilitate a discussion of the fermionic symmetry, it is convenient to eliminate the world-volume metric as an independent variable. In this way we avoid having to discuss the rather complicated transformation rule for the metric. The action (1) then takes on its Nambu–Goto form

$$S=\int d^3\xi \,(\sqrt{-\det \hat{E}_{\hat{i}}{}^{\hat{a}}\hat{E}_{\hat{j}}{}^{\hat{b}}\eta_{\hat{a}\hat{b}}}$$
$$-\tfrac{1}{6}\epsilon^{\hat{i}\hat{j}\hat{k}}\hat{E}_{\hat{i}}{}^{\hat{A}}E_{\hat{j}}{}^{\hat{B}}E_{\hat{k}}{}^{\hat{C}}\hat{A}_{\hat{C}\hat{B}\hat{A}}) . \tag{22}$$

It is invariant under the transformation [1]

$$\delta\hat{z}^{\hat{a}}\equiv\delta\hat{z}^{\hat{M}}\hat{E}_{\hat{M}}{}^{\hat{a}}=0 ,$$

$$\delta\hat{z}^{\hat{\alpha}}\equiv\delta\hat{z}^{\hat{M}}\hat{E}_{\hat{M}}{}^{\hat{\alpha}}=\hat{\kappa}^{\beta}(1+\hat{\Gamma})_{\beta}{}^{\hat{\alpha}} , \tag{23}$$

where

$$\hat{\Gamma}_{\hat{\beta}}{}^{\hat{\alpha}}=(1/6\sqrt{-\hat{g}})\epsilon^{\hat{i}\hat{j}\hat{k}}\hat{E}_{\hat{i}}{}^{\hat{a}}\hat{E}_{\hat{j}}{}^{\hat{b}}\hat{E}_{\hat{k}}{}^{\hat{c}}(\hat{\Gamma}_{\hat{a}\hat{b}\hat{c}})_{\hat{\beta}}{}^{\hat{\alpha}} . \tag{24}$$

In (24) $\hat{g}_{\hat{i}\hat{j}}$ is the metric on the world volume induced from the bosonic metric on superspace,

$$\hat{g}_{\hat{i}\hat{j}}=\hat{E}_{\hat{i}}{}^{\hat{a}}\hat{E}_{\hat{j}}{}^{\hat{b}}\eta_{\hat{a}\hat{b}} . \tag{25}$$

In order for (22) to be invariant under this transformation, it is necessary that the background supergeometry be constrained. The constraints found in ref. [1] are

$$\hat{T}_{\hat{\alpha}\hat{\beta}}{}^{\hat{c}}=-\mathrm{i}(\hat{\Gamma}^{\hat{c}})_{\hat{\alpha}\hat{\beta}}, \quad \hat{F}_{\hat{\alpha}\hat{\beta}\hat{c}\hat{d}}=\mathrm{i}(\hat{\Gamma}_{\hat{c}\hat{d}})_{\hat{\alpha}\hat{\beta}} ,$$

$$\hat{F}_{\hat{\alpha}\hat{\beta}\hat{\gamma}\hat{\delta}}=\hat{F}_{\hat{\alpha}\hat{\beta}\hat{\gamma}\hat{d}}=0 , \tag{26}$$

$$\hat{T}_{\hat{\alpha}\hat{b}\hat{c}}=\eta_{\hat{b}\hat{c}}\hat{\Lambda}_{\hat{\alpha}}, \quad \hat{F}_{\hat{\alpha}\hat{b}\hat{c}\hat{d}}=(\Gamma_{\hat{b}\hat{c}\hat{d}})_{\hat{\alpha}}{}^{\hat{\beta}}\hat{\Lambda}_{\hat{\beta}} . \tag{27}$$

Although these equations are not the standard equations of on-shell $D=11$ supergravity in superspace [8], they are equivalent to them. That is to say, by suitable redefinitions of the superconnections and parts of the supervielbein, we may set $\hat{\Lambda}_{\hat{\alpha}}$, $\hat{T}_{\hat{\alpha}\hat{\beta}}{}^{\hat{\gamma}}$ and $\hat{T}_{\hat{a}\hat{b}}{}^{\hat{c}}$ to zero,

$$\hat{T}_{\hat{a}\hat{b}}{}^{\hat{c}}=\hat{T}_{\hat{\alpha}\hat{\beta}}{}^{\hat{\gamma}}=\hat{F}_{\hat{\alpha}\hat{b}\hat{c}\hat{d}}=0 . \tag{28}$$

Eqs. (26) and (28) are the on-shell supergravity equations, as may be checked using the Bianchi identities. Since we are always free to make such redefinitions, we may take the superspace constraints to be (26) and (28). This is therefore a stronger result than that given in ref. [1]; the fact that conventional constraints can be imposed was noted in the context of $N=1$ $D=10$ supersymmetric particles and strings in ref. [9].

The Kaluza–Klein Ansatz for the $N=1$ $D=11$ supervielbein is

$$\hat{E}_{\hat{M}}{}^{\hat{A}}=\begin{pmatrix}\hat{E}_M{}^a & \hat{E}_M{}^{\alpha} & \hat{E}_M{}^{11} \\ \hat{E}_y{}^a & \hat{E}_y{}^{\alpha} & \hat{E}_y{}^{11}\end{pmatrix}, \tag{29}$$

$$=\begin{pmatrix}E_M{}^a & E_M{}^{\alpha}+A_M\chi^{\alpha} & \Phi A_M \\ 0 & \chi^{\alpha} & \Phi\end{pmatrix}, \tag{30}$$

where $E_M{}^A=(E_M{}^a, E_M{}^\alpha)$ is the $N=2a$ $D=10$ supervielbein, A_M the superspace U(1) gauge field, and Φ and χ^a are superfields whose leading components are the dilaton and the dilatino respectively. In writing (23), we have made a partial $D=11$ local Lorentz gauge choice to set $\hat{E}_y{}^a=0$. For the superspace three-index potential $\hat{A}_{\hat{M}\hat{N}\hat{P}}$ we have

$$\hat{A}_{MNP}=A_{MNP}\ , \tag{31}$$

$$\hat{A}_{MNy}=A_{MN}\ . \tag{32}$$

All of the $D=10$ superfields $E_m{}^A$, χ^α, A_m, Φ, A_{MN}, A_{MNP} are taken to be independent of y. Note also that ten-dimensional spinor indices run from 1 to 32 so that α and $\hat{\alpha}$ can be identified. With $\hat{z}^{\hat{M}}=(z^M, y)$ we also impose

$$\partial_\rho z^M=0\ , \tag{33}$$

and fix

$$y=\rho\ . \tag{34}$$

Substituting the Ansätze (30), (31) and (32) into (22) and using (33) and (34) yields the action for a type IIA superstring coupled to a supergravity background

$$S=\int d^2\xi\ (\Phi\sqrt{-\det E_i{}^a E_j{}^b\eta_{ab}} - \tfrac{1}{2}\epsilon^{ij}\partial_i z^m\partial_j z^N A_{NM})\ . \tag{35}$$

Purely for convenience in superspace calculations, we have omitted an overall factor of $\Phi^{-2/3}$ in the Ansatz (29); the factor of Φ in (35) can be removed by a suitable rescaling of the supervielbein. To find the fermionic symmetry of the dimensionally reduced action (35), one substitutes the Kaluza–Klein Ansätze into (23). It is straightforward to show that

$$\hat{\Gamma}_\beta{}^\alpha=\Gamma_\beta{}^\alpha = (1/2\sqrt{-g})\epsilon^{ij}E_i{}^aE_j{}^b(\Gamma_{ab}\Gamma_{11})_\beta{}^\alpha \tag{36}$$

and that

$$\delta\hat{z}^\alpha=\delta z^\alpha\equiv\delta z^M E_M{}^\alpha=\kappa^\beta(1+\Gamma)_\beta{}^\alpha\ . \tag{37}$$

However, y also transforms under (23):

$$\delta_y=-\kappa^\beta(1+\Gamma)_\beta{}^\alpha A_\alpha \tag{38}$$

and a compensating infinitesimal world-volume diffeomorphism with parameter

$$(0, 0, \kappa^\beta(1+\Gamma)_\beta{}^\alpha A_\alpha) \tag{39}$$

must be made in order to maintain the gauge $y=\rho$.

Since (22) is invariant under (23) when the $D=11$ field equations are satisfied, it follows that the reduced action (35) will be invariant under (37) if the $N=2$a $D=10$ supergravity field equations are satisfied. This is because the compactification of the $N=1$ $D=11$ field theory on a circle is known to yield the $N=2$a $D=10$ field theory, though to the best of our knowledge this is the first time it has been done in superspace. Note that all of the $N=2$a supergravity fields are now coupled, including A_{mnp}, A_m and Φ which decoupled from the purely bosonic sector.

The transformation (37) can be recast into the Green–Schwarz [5,6] form by introducing

$$\lambda^{i\alpha}=\tfrac{1}{2}(\epsilon^{ij}/\sqrt{-g})E_j{}^a\kappa^\beta(\Gamma_a\Gamma_{11})_\beta{}^\alpha\ , \tag{40}$$

so that (37) becomes

$$\delta z^\alpha=E_i{}^a\{[\lambda_+^{i\beta}+(\epsilon^{ij}/\sqrt{-g})\lambda^\beta_{+j}] + [\lambda_-^{i\beta}-(\epsilon^{ij}/\sqrt{-g})\lambda^\beta_{-j}]\}(\Gamma_a)_\beta{}^\alpha\ , \tag{41}$$

where

$$\lambda_\pm^{i\alpha}=\tfrac{1}{2}\lambda^{i\beta}(1\pm\Gamma_{11})_\beta{}^\alpha\ . \tag{42}$$

In conclusion, we have succeeded in deriving (for the first time [‡2]) the action of the type IIA superstring coupled to an $N=2$a $D=10$ supergravity background starting from the action of the supermembrane coupled to the background of $N=1$ supergravity in $D=11$. The dimensional reduction corresponds to a compactification of both the spacetime and the world volume on the same circle and then discarding the massive modes. Classically, this is equivalent to letting the membrane tension α_3' tend to infinity and the radius of the circle R tend to zero in such a way that the string tension

$$\alpha_2'=2\pi R\alpha_3'$$

remains finite. The type IIA superstring is known to be a consistent quantum theory; the most urgent question for the supermembrane is whether it too is a consistent quantum theory in its own right.

We are grateful for discussions with Chris Pope and Ergin Sezgin.

[‡2] The type IIB action is given in ref. [10].

References

[1] E. Bergshoeff, E. Sezgin and P.K. Townsend, Phys. Lett. B 189 (1987) 75.
[2] M.J. Duff, in: Quantum field theory and quantum statistics, eds. I.H. Batalin, C. Isham and G. Vilkovisky (Adam Hilger, London, 1986).
[3] H. Nicolai, Phys. Lett. B 187 (1987) 316.
[4] W. Siegel, Phys. Lett. B 128 (1983) 397.
[5] M.B. Green and J.H. Schwarz, Phys. Lett. B 136 (1984) 367.
[6] J. Hughes, J. Liu and J. Polchinski, Phys. lett. B 180 (1986) 370.
[7] M.J. Duff, B.E.W. Nilsson and C.N. Pope, Phys. Rep. 130 (1986) 1.
[8] E. Cremmer and S. Ferrara, Phys. Lett. B 91 (1980) 61; L. Brink and P. Howe, Phys. Lett. B 91 (1980) 384.
[9] J. Shapiro and C. Taylor, Rutgers University preprint (1986).
[10] M.T. Grisaru, P. Howe, L. Mezincescu, B.E.W. Nilsson and P.K. Townsend, Phys. Lett. B 162 (1985) 116.

Reprinted from Nucl. Phys. B **347** (1990) 394–419

DUALITY ROTATIONS IN MEMBRANE THEORY

M.J. DUFF * and J.X. LU

Center for Theoretical Physics, Physics Department, Texas A&M University, College Station, TX 77843, USA

Received 14 May 1990

In analogy with a previous treatment of strings, it is shown that membrane theories exhibit global noncompact symmetries which have their origin in duality transformations on the three-dimensional worldvolume which rotate field equations into Bianchi identities. However, in contrast to the string, the worldvolume metric also transforms under duality by a conformal factor. In this way the Cremmer–Julia hidden symmetries of supergravity are seen to be a consequence of supermembrane duality. Moreover, the string duality follows from that of the membrane by simultaneous dimensional reduction. Generalization to higher-dimensional objects is straightforward.

1. Introduction

The purpose of this paper is to explore the phenomenon of "duality" in membrane theories. The word duality has come to acquire many meanings but here a duality transformation will mean a symmetry that rotates field equations into Bianchi identities on the worldvolume of the extended object. In a previous paper [1] devoted to duality rotations in string theory, we saw that in the case of a bosonic string compactified on an n-torus, these continuous transformations were described by an $SO(n, n)$ symmetry of the equations of motion in the presence of the massless background fields. The discrete subgroup, $SO(n, n; \mathbb{Z})$ which survives as a symmetry of the spectrum, contains the typical $R \leftrightarrow \alpha'/R$ transformations whose fixed points correspond to points of enhanced gauge symmetries and which have led to speculations about a minimum length in string theory [2]. As a preliminary to investigating questions of enhanced symmetry or minimum length in membrane theory, therefore, we first wish to discuss the continuous duality transformations.

In fact, our original motivation for studying duality on the worldvolume of extended objects sprang from the old observation that four-dimensional supergravity theories exhibit global noncompact continuous symmetries corresponding to duality symmetries that rotate space-time field equations into Bianchi identities

* Work supported in part by NSF grant PHY-9045132.

[3,4]. These were frequently referred to as "hidden" symmetries since their presence is far from obvious by inspection of the four-dimensional lagrangian. Indeed, they were originally best understood as a consequence of compactifying a higher-dimensional theory [4]. Nowadays, of course, supergravity theories are regarded as being merely the field theory limit of a superstring or supermembrane, and so it is natural to conjecture that these symmetries have their origin on the worldvolume of the appropriate extended object. Some evidence in favor of these ideas may be found in refs. [5,6]. For example, considerations of bosonic strings on group manifolds led us to expect a global $SO(n,n)$ in the lower dimension, where $n = \dim G$. In the case of the four-dimensional heterotic string this is replaced by global $SO(6,n)$. This latter symmetry is known to be present in four-dimensional $N = 4$ supergravity coupled to a Yang–Mills supermultiplet. The $6n$ scalars in the Yang–Mills sector are described by a nonlinear σ-model given by the coset $SO(6,n)/SO(6) \times SO(n)$. (There are also two scalars in the supergravity sector described by $SU(1,1)/U(1)$. The stringy origin of this coset is described in sect. 6.) Cosets of this kind were also encountered by Narain [7] in his torus compactification of the heterotic string but where $n = \text{rank}\ G$. By group manifold considerations, we were led to the larger symmetry with $n = \dim G$. It should be emphasized that these larger symmetries with $n = \dim G$ are broken by gauge interactions but they nevertheless exactly describe the nonlinear σ-model of the scalars. Similarly, one might expect that the global $E_{7(+7)}$ Cremmer–Julia [4] symmetry of $N = 8$ supergravity in four-dimensions has its origin on the worldvolume of the eleven-dimensional supermembrane [8] compactified on a seven-torus [4] or a seven-sphere [9]. Once again, in the case of the seven-sphere the $E_{7(+7)}$ will be broken by $SO(8)$ gauge interactions but the $E_{7(+7)}/SU(8)$ coset still describes the nonlinear σ-model of the scalars [10].

The proof of these conjectures (in the case of the string) was supplied by Cecotti et al. [11], who first pointed out that the two-dimensional worldsheet origin of these symmetries is quite similar to the way they appear in four-dimensional space-time i.e. through generalized duality transformations of the kind discussed in detail by Gaillard and Zumino [12]. Thus our task in this paper is to generalize the arguments of Cecotti et al. to the three-dimensional worldvolume of the membrane. We shall confine our attention in this paper to torus compactification and focus mainly on the bosonic sector. Moreover, we shall follow the approach described in ref. [1] for treating string duality which lends itself to generalization to membranes and other higher-dimensional objects. Let us therefore first recall string duality.

2. Review of string duality

The n-dimensional string is described by a two-dimensional σ-model with worldsheet coordinates $\xi^i = (\tau, \sigma)$, worldvolume metric $\gamma_{ij}(\xi)$ and target-space coordinates $x^{\mu}(\xi)$, with

$$\mathscr{L} = \tfrac{1}{2}\sqrt{-\gamma}\,\gamma^{ij}\,\partial_i x^{\mu}\,\partial_j x^{\nu}\,g_{\mu\nu} + \tfrac{1}{2}\varepsilon^{ij}\,\partial_i x^{\mu}\,\partial_j x^{\nu}\,b_{\mu\nu}\,. \tag{2.1}$$

We shall consider the case where x^{μ} correspond to the compactified coordinates with the background fields $g_{\mu\nu}$ and $b_{\mu\nu}$ ($\mu = 1, \ldots, n$) being x-independent. These backgrounds will admit the interpretation of scalar fields in space-time and hence will still depend on the space-time coordinates X^M. Let us define

$$\mathscr{F}^{i\mu} \equiv \sqrt{-\gamma}\,\gamma^{ij}\,\partial_j x^{\mu}\,, \qquad \tilde{\mathscr{F}}^{i\mu} \equiv \varepsilon^{ij}\,\partial_j x^{\mu}\,, \qquad (2.2), (2.3)$$

$$\tilde{\mathscr{G}}^{i}{}_{\mu} \equiv g_{\mu\nu}\mathscr{F}^{i\nu} + b_{\mu\nu}\tilde{\mathscr{F}}^{i\nu} = \partial\mathscr{L}/\partial\,\partial_i x^{\mu}\,. \tag{2.4}$$

Then there is a symmetry between the equations of motion,

$$\partial_i \tilde{\mathscr{G}}^{i}{}_{\mu} = 0\,, \tag{2.5}$$

and the Bianchi identities,

$$\partial_i \tilde{\mathscr{F}}^{i\mu} \equiv 0\,. \tag{2.6}$$

The invariance is summarized by the equations

$$\delta\tilde{\mathscr{F}}^{i\mu} = A^{\mu}{}_{\rho}\tilde{\mathscr{F}}^{i\rho} + B^{\mu\beta}\tilde{\mathscr{G}}^{i}{}_{\beta}\,, \qquad \delta\tilde{\mathscr{G}}^{i}{}_{\alpha} = C_{\alpha\rho}\tilde{\mathscr{F}}^{i\rho} + D_{\alpha}{}^{\beta}\tilde{\mathscr{G}}^{i}{}_{\beta}\,, \tag{2.7}$$

where A, B, C and D are constant parameters. We must also establish invariance of the equation of motion obtained by varying with respect to γ_{ij}, namely the vanishing of the energy–momentum tensor

$$\sqrt{-\gamma}\,\theta_{ij} = \frac{2\delta S}{\partial\gamma^{ij}} = \sqrt{-\gamma}\left[\partial_i x^{\mu}\,\partial_j x^{\nu}\,g_{\mu\nu} - \tfrac{1}{2}\gamma_{ij}\gamma^{kl}\,\partial_k x^{\mu}\,\partial_l x^{\nu}\,g_{\mu\nu}\right] = 0\,. \tag{2.8}$$

To fix the group structure more precisely, we introduced in ref. [1] a "dual" σ-model for which the roles of field equations and Bianchi identities are interchanged. One begins by noting that the equations of motion (2.5) may equivalently be derived from a first-order lagrangian with independent variables x^{μ} and $F_i{}^{\mu}$,

$$\begin{aligned}\mathscr{L}_x = {} & -\tfrac{1}{2}\sqrt{-\gamma}\,\gamma^{ij}F_i{}^{\mu}F_j{}^{\nu}g_{\mu\nu} - \tfrac{1}{2}\varepsilon^{ij}F_i{}^{\mu}F_j{}^{\nu}b_{\mu\nu} \\ & + \partial_i x^{\mu}\left(\sqrt{-\gamma}\,\gamma^{ij}F_j{}^{\nu}g_{\mu\nu} + \varepsilon^{ij}F_j{}^{\nu}b_{\mu\nu}\right).\end{aligned} \tag{2.9}$$

Varying with respect to $\partial_i x^\mu$, we have

$$\partial \mathscr{L}_x / \partial \partial_i x^\mu = \sqrt{-\gamma}\, \gamma^{ij} F_j{}^\nu g_{\mu\nu} + \varepsilon^{ij} F_j{}^\nu b_{\mu\nu}\,, \tag{2.10}$$

and with respect to $F_i{}^\mu$, we have

$$\begin{aligned} \partial \mathscr{L}_x / \partial F_i{}^\mu = & -\sqrt{-\gamma}\, \gamma^{ij} F_j{}^\nu g_{\mu\nu} - \varepsilon^{ij} F_j{}^\nu b_{\mu\nu} \\ & + \sqrt{-\gamma}\, \gamma^{ij} \partial_j x^\nu\, g_{\mu\nu} + \varepsilon^{ij} \partial_j x^\nu b_{\mu\nu} = 0\,. \end{aligned} \tag{2.11}$$

This latter equation says

$$F_i{}^\mu = \partial_i x^\mu\,, \tag{2.12}$$

and then eq. (2.10) yields the same equation of motion as in the second-order formalism.

Next consider a different first-order lagrangian with independent variables y_μ and $F_i{}^\mu$,

$$\mathscr{L}_y = \tfrac{1}{2}\sqrt{-\gamma}\, \gamma^{ij} F_i{}^\mu F_j{}^\nu g_{\mu\nu} + \tfrac{1}{2}\varepsilon^{ij} F_i{}^\mu F_j{}^\nu b_{\mu\nu} + \partial_i y_\mu \varepsilon^{ij} F_j{}^\mu\,. \tag{2.13}$$

We have

$$\partial \mathscr{L}_y / \partial \partial_i y_\mu = \varepsilon^{ij} F_j{}^\mu\,, \tag{2.14}$$

$$\partial \mathscr{L}_y / \partial F_i{}^\mu = \sqrt{-\gamma}\, \gamma^{ij} F_j{}^\nu g_{\mu\nu} + \varepsilon^{ij} F_j{}^\nu b_{\mu\nu} - \varepsilon^{ij} \partial_j y_\mu = 0\,. \tag{2.15}$$

This is an algebraic equation for $F_i{}^\mu$ with the solution

$$F_i{}^\mu = p^{\mu\nu} \left(1/\sqrt{-\gamma}\,\right) \gamma_{ij} \varepsilon^{jk}\, \partial_k y_\nu + q^{\mu\nu} \partial_i y_\nu\,, \tag{2.16}$$

where $p^{\mu\nu} = p^{\nu\mu}$ and $q^{\mu\nu} = -q^{\nu\mu}$ are related to $g_{\mu\nu} = g_{\nu\mu}$ and $b_{\mu\nu} = -b_{\nu\mu}$ by

$$p_{\mu\nu} = g_{\mu\nu} + b_{\mu\alpha} g^{\alpha\beta} b_{\nu\beta}\,, \qquad p_{\mu\nu} q^{\alpha\nu} = b_{\mu\beta} g^{\alpha\beta}\,, \tag{2.17}$$

where $p_{\mu\nu}$ is the inverse of $p^{\mu\nu}$. From eq. (2.14) the equation for y_μ is

$$\partial_i \left(\varepsilon^{ij} F_j{}^\mu \right) = 0\,, \tag{2.18}$$

which implies (2.12), at least locally.

Putting these results together, we find

$$\varepsilon^{ij}\,\partial_j y_\mu = \frac{\partial \mathcal{L}_x}{\partial\,\partial_i x^\mu} = g_{\mu\nu}\sqrt{-\gamma}\,\gamma^{ij}\,\partial_j x^\nu + b_{\mu\nu}\varepsilon^{ij}\,\partial_j x^\nu\,,$$

$$\varepsilon^{ij}\,\partial_j x^\mu = \frac{\partial \mathcal{L}_y}{\partial\,\partial_i y_\mu} = p^{\mu\nu}\sqrt{-\gamma}\,\gamma^{ij}\,\partial_j y_\nu + q^{\mu\nu}\varepsilon^{ij}\,\partial_j y_\nu\,. \tag{2.19}$$

Thus the field equations of the original lagrangian $\mathcal{L}_x$ are the Bianchi identities for the "dual" lagrangian $\mathcal{L}_y$, and vice versa.

To see the SO(n,n) symmetry explicitly, define

$$\mathcal{G}^i{}_\mu \equiv \sqrt{-\gamma}\,\gamma^{ij}\,\partial_j y_\mu\,. \tag{2.20}$$

Then eq. (2.19) may be written

$$\tilde{\mathcal{G}}^i{}_\mu = g_{\mu\nu}\mathcal{F}^{i\nu} + b_{\mu\nu}\tilde{\mathcal{F}}^{i\nu}\,, \qquad \tilde{\mathcal{F}}^{i\mu} = p^{\mu\nu}\mathcal{G}^i{}_\nu + q^{\mu\nu}\tilde{\mathcal{G}}^i{}_\nu\,, \tag{2.21}$$

or, in compact notation,

$$\Omega_{MN}\tilde{\Phi}^{iN} = G_{MN}\Phi^{iN}\,, \tag{2.22}$$

where $M, N = 1, \ldots, 2n$ and where

$$\tilde{\Phi}^{iN} = \begin{pmatrix} \tilde{\mathcal{F}}^{i\nu} \\ \tilde{\mathcal{G}}^i{}_\beta \end{pmatrix}, \qquad \Phi^{iN} = \begin{pmatrix} \mathcal{F}^{i\nu} \\ \mathcal{G}^i{}_\beta \end{pmatrix}. \tag{2.23}$$

The $2n \times 2n$ matrices Ω and G are given by

$$\Omega_{MN} = \begin{pmatrix} 0 & \delta_\mu{}^\beta \\ \delta^\alpha{}_\nu & 0 \end{pmatrix}, \tag{2.24}$$

$$G_{MN} = \begin{pmatrix} g_{\mu\nu} + b_{\mu\alpha}g^{\alpha\beta}b_{\nu\beta} & b_{\mu\alpha}g^{\alpha\beta} \\ g^{\alpha\beta}b_{\nu\beta} & g^{\alpha\beta} \end{pmatrix}. \tag{2.25}$$

The desired SO(n,n) symmetry is now manifest since the group SO(n,n) may be defined by parameters $A^M{}_N$ for which

$$\delta\Omega_{MN} = -A^P{}_M\Omega_{PN} - A^P{}_N\Omega_{MP} = 0\,. \tag{2.26}$$

Moreover,

$$G_{MP}\Omega^{PQ}G_{QN}=\Omega_{MN}, \tag{2.27}$$

and hence G is an element of SO(n,n). Thus eq. (2.22) is manifestly SO(n,n) invariant with $\tilde{\Phi}^{iN}$ and Φ^{iN} both transforming as the $2n$-dimensional vector representation

$$\delta\tilde{\Phi}^{iM}=A^{M}{}_{P}\tilde{\Phi}^{iP}, \qquad \delta\Phi^{iM}=A^{M}{}_{P}\Phi^{iP}. \tag{2.28}$$

The explicit transformation rules are those of eq. (2.7) with the restrictions

$$\begin{aligned} B^{\mu\beta}&=B^{[\mu\beta]}=A^{\mu\beta},\\ C_{\alpha\nu}&=C_{[\alpha\nu]}=A_{\alpha\nu},\\ D_{\alpha}{}^{\beta}&=A_{\alpha}{}^{\beta}=-A^{\beta}{}_{\alpha}, \end{aligned} \tag{2.29}$$

corresponding to the $n(2n-1)$ parameters of SO(n,n). Thus

$$\delta\tilde{\mathscr{F}}^{i\mu}=A^{\mu}{}_{\nu}\tilde{\mathscr{F}}^{i\nu}+B^{\mu\beta}\tilde{\mathscr{G}}^{i}{}_{\beta}, \qquad \delta\tilde{\mathscr{G}}^{i}{}_{\alpha}=C_{\alpha\nu}\tilde{\mathscr{F}}^{i\nu}-A^{\beta}{}_{\alpha}\tilde{\mathscr{G}}^{i}{}_{\beta}, \tag{2.30}$$

and similarly for $\mathscr{F}^{i\mu}$ and $\mathscr{G}^{i}{}_{\alpha}$. The transformation rule for G_{MN} is

$$\delta G_{MN}=-A^{P}{}_{M}G_{PN}-A^{P}{}_{N}G_{MP}, \tag{2.31}$$

and hence

$$\begin{aligned} \delta g_{\mu\nu}&=-A^{\rho}{}_{\mu}g_{\rho\nu}-A^{\rho}{}_{\nu}g_{\mu\rho}-b_{\mu\alpha}B^{\alpha\beta}g_{\beta\nu}-g_{\mu\alpha}B^{\alpha\beta}b_{\beta\nu},\\ \delta b_{\mu\nu}&=-A^{\rho}{}_{\mu}b_{\rho\nu}-A^{\rho}{}_{\nu}b_{\mu\rho}-b_{\mu\alpha}B^{\alpha\beta}b_{\beta\nu}-g_{\mu\alpha}B^{\alpha\beta}g_{\beta\nu}+C_{\mu\nu}. \end{aligned} \tag{2.32}$$

The action of SO(n,n) on the background fields $g_{\mu\nu}$ and $b_{\mu\nu}$ is nonlinear. A linear realization may be obtained by rewriting eq. (2.22) as

$$\tilde{\Phi}^{i}{}_{A}=\Phi^{i}{}_{A}, \tag{2.33}$$

where

$$\tilde{\Phi}^{i}{}_{A}=\Omega_{AB}E_{N}{}^{B}\tilde{\Phi}^{iN}, \qquad \Phi^{i}{}_{A}=E_{NA}\Phi^{iN} \tag{2.34}$$

and where $E_{M}{}^{A}$ is the "vielbein" for which

$$G_{MN}=E_{M}{}^{A}E_{NA}. \tag{2.35}$$

Multiplying (2.33) by $E_M{}^A$ and using the fact that $E_M{}^A$ is also an element of SO(n,n),

$$E_M{}^A \Omega_{AB} E_N{}^B = \Omega_{MN}\,, \tag{2.36}$$

we recover (2.22). As usual, the price to pay for a linear realization of G is a local symmetry H where H $\subset$ G. In this case

$$E_M{}^A \to \Lambda^A{}_B E_M{}^B \tag{2.37}$$

where Λ is an element of SO$(n) \times$ SO(n), the maximal compact subgroup of SO(n,n), whose elements commute with Ω. The n^2 physical scalar fields described by G_{MN} parametrize the coset SO$(n,n)/$SO$(n) \times$ SO(n) and their self-interaction is described by the corresponding nonlinear σ-model.

3. Membrane duality

The bosonic sector of the n-dimensional supermembrane is described by a three-dimensional σ-model with worldvolume coordinates $\xi^i = (\tau, \sigma, \rho)$, worldvolume metric $\gamma_{ij}(\xi)$ and target-space coordinates $x^\mu(\xi)$, with

$$\mathscr{L} = \tfrac{1}{2}\sqrt{-\gamma}\,\gamma^{ij}\,\partial_i x^\mu\,\partial_j x^\nu\, g_{\mu\nu} + \frac{1}{3!}\varepsilon^{ijk}\,\partial_i x^\mu\,\partial_j x^\nu\,\partial_k x^\rho\, b_{\mu\nu\rho} - \tfrac{1}{2}\sqrt{-\gamma} \tag{3.1}$$

with background fields $g_{\mu\nu}$ and $b_{\mu\nu\rho}(\mu = 1, \ldots, n)$. Define

$$\mathscr{F}^{i\mu} \equiv \sqrt{-\gamma}\,\gamma^{ij}\,\partial_j x^\mu\,, \qquad \tilde{\mathscr{F}}^{i\mu\nu} \equiv \varepsilon^{ijk}\,\partial_j x^\mu\,\partial_k x^\nu \qquad (3.2), (3.3)$$

$$\tilde{\mathscr{G}}^i{}_\mu = g_{\mu\nu}\mathscr{F}^{i\nu} + \tfrac{1}{2}b_{\mu\nu\rho}\tilde{\mathscr{F}}^{i\nu\rho} = \partial\mathscr{L}/\partial\,\partial_i x^\mu\,. \tag{3.4}$$

In the case when the background fields $g_{\mu\nu}$ and $b_{\mu\nu\rho}$ are independent of the coordinate x^μ, its equation of motion is just

$$\partial_i \tilde{\mathscr{G}}^i{}_\mu = 0\,, \tag{3.5}$$

whereas the Bianchi identity is

$$\partial_i \tilde{\mathscr{F}}^{i\mu\nu} \equiv 0\,. \tag{3.6}$$

Thus there is a duality symmetry that rotates field equations into Bianchi identities. The invariance is summarized by the equations

$$\delta\tilde{\mathscr{F}}^{i\mu\nu} = \tfrac{1}{2}A^{\mu\nu}{}_{\rho\sigma}\tilde{\mathscr{F}}^{i\rho\sigma} + B^{\mu\nu\rho}\tilde{\mathscr{G}}^{i}{}_{\rho},$$

$$\delta\tilde{\mathscr{G}}^{i}{}_{\alpha} = \tfrac{1}{2}C_{\alpha\rho\sigma}\tilde{\mathscr{F}}^{i\rho\sigma} + D_{\alpha}{}^{\rho}\tilde{\mathscr{G}}^{i}{}_{\rho}, \tag{3.7}$$

where A, B, C and D are constant parameters. Thus the first major difference from the string is that the duality transformations are nonlinear in the "field strength" $\partial_i x^\mu$ by virtue of $\tilde{\mathscr{F}}^{i\mu\nu}$, given in (3.3). We must also establish invariance of the equation of motion obtained by varying with respect to γ_{ij}, namely the vanishing of the energy–momentum tensor

$$\sqrt{-\gamma}\,\theta_{ij} = \frac{2\delta S}{\delta\gamma^{ij}} = \sqrt{-\gamma}\left[\partial_i x^\mu\,\partial_j x^\nu\,g_{\mu\nu} - \tfrac{1}{2}\gamma_{ij}\gamma^{kl}\,\partial_k x^\mu\,\partial_l x^\nu\,g_{\mu\nu} + \tfrac{1}{2}\gamma_{ij}\right] = 0\,. \tag{3.8}$$

This equation is just the statement that γ_{ij} is the induced metric on the world-volume

$$\gamma_{ij} = \partial_i x^\mu\,\partial_j x^\nu\,g_{\mu\nu}\,. \tag{3.9}$$

Here we encounter the second major difference from the string case, where the worldsheet metric is invariant under duality transformations. For the membrane, we must allow the possibility that γ_{ij} also transforms under duality.

The equations of motion (3.4), (3.5) and (3.8) may equivalently be derived from a first-order lagrangian with independent variables x^μ and $F_i{}^\mu$

$$\mathscr{L}_x = -\tfrac{1}{2}\sqrt{-\gamma}\,\gamma^{ij}F_i{}^\mu F_j{}^\nu g_{\mu\nu} - \tfrac{1}{3}\varepsilon^{ijk}F_i{}^\mu F_j{}^\nu F_k{}^\rho b_{\mu\nu\rho}$$

$$+\,\partial_i x^\mu\left(\sqrt{-\gamma}\,\gamma^{ij}F_j{}^\nu g_{\mu\nu} + \tfrac{1}{2}\varepsilon^{ijk}F_i{}^\nu F_k{}^\rho b_{\mu\nu\rho}\right) - \tfrac{1}{2}\sqrt{-\gamma}\,. \tag{3.10}$$

Varying with respect to $\partial_i x^\mu$, we have

$$\frac{\partial\mathscr{L}_x}{\partial\,\partial_i x^\mu} = \sqrt{-\gamma}\,\gamma^{ij}F_j{}^\nu g_{\mu\nu} + \tfrac{1}{2}\varepsilon^{ijk}F_j{}^\nu F_k{}^\rho b_{\mu\nu\rho}\,, \tag{3.11}$$

and with respect to $F_i{}^\mu$, we have

$$\partial\mathscr{L}_x/\partial F_i{}^\mu = -\sqrt{-\gamma}\,\gamma^{ij}F_j{}^\nu g_{\mu\nu} - \varepsilon^{ijk}F_j{}^\nu F_k{}^\rho b_{\mu\nu\rho}$$

$$+\sqrt{-\gamma}\,\dot{\gamma}^{ij}\,\partial_j x^\nu\,g_{\mu\nu} + \varepsilon^{ijk}\,\partial_j x^\nu F_k{}^\rho b_{\mu\nu\rho} = 0\,. \tag{3.12}$$

This latter equation says

$$F_i{}^\mu = \partial_i x^\mu \tag{3.13}$$

and then eq. (3.11) yields the same equations of motion as in the second-order formulation.

Now consider a different first-order lagrangian with independent variables $y_{\mu\nu}$ and $F_i{}^\mu$

$$\mathscr{L}_y = \tfrac{1}{2}\sqrt{-\gamma}\,\gamma^{ij}F_i{}^\mu F_j{}^\nu g_{\mu\nu} + \frac{1}{3!}\varepsilon^{ijk}F_i{}^\mu F_j{}^\nu F_k{}^\rho b_{\mu\nu\rho}$$
$$+ \varepsilon^{ijk}\,\partial_i y_{\mu\nu}F_j{}^\mu F_k{}^\nu - \tfrac{1}{2}\sqrt{-\gamma}\,. \tag{3.14}$$

We have

$$\partial\mathscr{L}_y/\partial\,\partial_i y_{\mu\nu} = \varepsilon^{ijk}F_j{}^\mu F_k{}^\nu \tag{3.15}$$

$$\partial\mathscr{L}_y/\partial F_i{}^\mu = \sqrt{-\gamma}\,\gamma^{ij}F_j{}^\nu g_{\mu\nu} + \tfrac{1}{2}\varepsilon^{ijk}F_j{}^\nu F_k{}^\rho b_{\mu\nu\rho} - 2\varepsilon^{ijk}\,\partial_j y_{\mu\nu}F_k{}^\nu = 0\,. \tag{3.16}$$

This is an algebraic equation for $F_i{}^\mu$ with the (implicit) solution

$$F_i{}^\mu = 2p^{\mu\nu}\left(1/\sqrt{-\gamma}\right)\gamma_{ij}\varepsilon^{jkl}\,\partial_k y_{\nu\sigma}F_l{}^\sigma - 2q^{\rho\sigma\mu}\,\partial_i y_{\rho\sigma} \tag{3.17}$$

where $p^{\mu\nu} = p^{\nu\mu}$ and $q^{\rho\sigma\mu} = q^{[\rho\sigma\mu]}$ are related to $g_{\mu\nu} = g_{\nu\mu}$ and $b_{\mu\nu\rho} = b_{[\mu\nu\rho]}$ by

$$p_{\mu\nu} = g_{\mu\nu} + \tfrac{1}{4}b_{\mu\alpha\beta}\,g^{\alpha\beta\gamma\delta}b_{\nu\gamma\delta}\,,$$
$$p_{\mu\nu}q^{\alpha\beta\nu} = \tfrac{1}{2}b_{\mu\gamma\delta}\,g^{\alpha\beta\gamma\delta}\,, \tag{3.18}$$

where $p_{\mu\nu}$ is the inverse of $p^{\mu\nu}$ and where

$$g^{\alpha\beta\gamma\delta} = \left(g^{\alpha\gamma}g^{\beta\delta} - g^{\alpha\delta}g^{\beta\gamma}\right). \tag{3.19}$$

In proving eq. (3.17), we have made use of the identities

$$\varepsilon^{ijk}\varepsilon^{lmn} = \gamma\big(\gamma^{il}\gamma^{jm}\gamma^{kn} + \gamma^{in}\gamma^{jl}\gamma^{km} + \gamma^{im}\gamma^{jn}\gamma^{kl}$$
$$-\gamma^{il}\gamma^{jn}\gamma^{km} - \gamma^{im}\gamma^{jl}\gamma^{kn} - \gamma^{in}\gamma^{jm}\gamma^{kl}\big),$$
$$\varepsilon^{ijk}\varepsilon^{lm}{}_k = \gamma\left(\gamma^{il}\gamma^{jm} - \gamma^{im}\gamma^{jl}\right),$$
$$\varepsilon^{ijk}\varepsilon^{l}{}_{jk} = 2\gamma\gamma^{il}\,, \qquad \varepsilon^{ijk}\varepsilon_{ijk} = 6\gamma \tag{3.20}$$

and have also used the field equation (3.9). From eq. (3.15) the equation for $y_{\mu\nu}$ is

$$\partial_i\left(\varepsilon^{ijk}F_j{}^{\mu}F_k{}^{\nu}\right)=0\,, \tag{3.21}$$

which implies (3.13), at least locally. Putting these results together, we find

$$2\varepsilon^{ijk}\,\partial_j y_{\mu\nu}\,\partial_k x^{\nu}=\frac{\partial\mathscr{L}_x}{\partial\,\partial_i x^{\mu}}=g_{\mu\nu}\sqrt{-\gamma}\,\gamma^{ij}\,\partial_j x^{\nu}+\tfrac{1}{2}b_{\mu\nu\rho}\varepsilon^{ijk}\,\partial_j x^{\nu}\,\partial_k x^{\rho}\,,$$

$$\varepsilon^{ijk}\,\partial_j x^{\mu}\,\partial_k x^{\nu}=\frac{\partial\mathscr{L}_y}{\partial\,\partial_i y_{\mu\nu}}=2p^{\mu\nu\rho\sigma}\sqrt{-\gamma}\,\gamma^{ij}\,\partial_j y_{\rho\sigma}+2q^{\mu\nu\rho}\varepsilon^{ijk}\,\partial_j y_{\rho\sigma}\,\partial_k x^{\sigma} \tag{3.22}$$

where $p^{\alpha\beta\gamma\delta}$ is defined by

$$g^{\alpha\beta\gamma\delta}=p^{\alpha\beta\gamma\delta}+q^{\alpha\beta\mu}p_{\mu\nu}q^{\gamma\delta\nu}\,. \tag{3.23}$$

Thus the field equations of the original lagrangian $\mathscr{L}_x$ are the Bianchi identities for the "dual" lagrangian $\mathscr{L}_y$, and vice versa.

4. A specific case: $n = 4$

In the case of the string, the n $\tilde{\mathscr{G}}^i{}_{\mu}$ and the n $\tilde{\mathscr{F}}^{i\mu}$ transform as the same $2n$-dimensional vector representation of the same orthogonal group, SO(n,n), for all n. In the case of the membrane, however, we shall see that each n tells a different story. The n $\tilde{\mathscr{G}}^i{}_{\mu}$ and the $n(n-1)/2$ $\tilde{\mathscr{F}}^{i\mu\nu}$ will transform as an $(n(n+1)/2)$-dimensional representation of some noncompact group. Then the $n(n+1)/2$ $g_{\mu\nu}$ and the $n(n-1)(n-2)/3!$ $b_{\mu\nu\rho}$ will parametrize some $(n(n^2+5)/3!)$-dimensional coset. (In fact, we shall see in sect. 5 that this is valid only for $n\leqslant 4$ and that the cases $n\geqslant 5$ require a separate treatment). The group and the coset will be different for each n.

In this section we shall focus explicitly on $n=4$. Then the 4 $\tilde{\mathscr{G}}^i{}_{\mu}$ and 6 $\tilde{\mathscr{F}}^{i\mu\nu}$ will transform as a 10-dimensional representation of the duality group which will turn out to be SL$(5,\mathbb{R})$, and the 10 $g_{\mu\nu}$ and 4 $b_{\mu\nu\rho}$ will parametrize the 14-dimensional coset SL$(5,\mathbb{R})/$SO(5). To see this explicitly, define

$$\mathscr{G}^i{}_{\mu\nu}\equiv 4\sqrt{-\gamma}\,\gamma^{ij}\,\partial_j y_{\mu\nu}\,. \tag{4.1}$$

Then eq. (3.22) may be written

$$\tilde{\mathscr{G}}^i{}_{\mu}=g_{\mu\nu}\mathscr{F}^{i\nu}+\tfrac{1}{2}b_{\mu\nu\rho}\tilde{\mathscr{F}}^{i\nu\rho}$$

$$\tilde{\mathscr{F}}^{i\mu\nu}=\tfrac{1}{2}p^{\mu\nu\rho\sigma}\mathscr{G}^i{}_{\rho\sigma}+q^{\mu\nu\rho}\tilde{\mathscr{G}}^i{}_{\rho}\,. \tag{4.2}$$

Now define $\tilde{\Phi}^i{}_{MN} = -\tilde{\Phi}^i{}_{NM}(M, N = 1, \ldots, 5)$ via

$$\tilde{\Phi}^i{}_{\mu\nu} = \frac{1}{2g}\varepsilon_{\mu\nu\alpha\beta}\mathscr{F}^{i\alpha\beta}\,, \qquad \tilde{\Phi}^i{}_{\mu 5} = \tilde{\mathscr{G}}^i{}_{\mu} \tag{4.3}$$

and $\Phi^{iMN} = -\Phi^{iNM}$ via

$$\Phi^{i\mu\nu} = \tfrac{1}{2}g^{-1/5}\varepsilon^{\mu\nu\rho\sigma}\mathscr{G}^i{}_{\rho\sigma}\,, \qquad \Phi^{i\mu 5} = g^{-1/5}\mathscr{F}^{i\mu}\,, \tag{4.4}$$

where g is the determinant of $g_{\mu\nu}$. Then eq. (4.2) may be written

$$\tilde{\Phi}^i{}_{MN} = (G_{MP}G_{NQ} - G_{MQ}G_{NP})\Phi^{iPQ} \tag{4.5}$$

where G_{MN} is the 5×5 symmetric matrix for which

$$G_{\mu\nu} = g^{-2/5}g_{\mu\nu}\,,$$

$$G_{\mu 5} = -\frac{1}{3!}g^{-2/5}g_{\mu\alpha}\varepsilon^{\alpha\beta\gamma\delta}b_{\beta\gamma\delta} = G_{5\mu}\,,$$

$$G_{55} = g^{3/5}\left(1 + \frac{1}{3!}b_{\alpha\beta\gamma}g^{\alpha\mu}g^{\beta\nu}g^{\gamma\rho}b_{\mu\nu\rho}\right). \tag{4.6}$$

We note that

$$\det G_{MN} = 1\,, \tag{4.7}$$

and hence that G_{MN} is an element of SL(5, $\mathbb{R}$). Eq. (4.5) is thus manifestly SL(5, $\mathbb{R}$) invariant with $\tilde{\Phi}^{iMN}$ transforming as the 10-dimensional representation

$$\delta\tilde{\Phi}^i{}_{MN} = -A^P{}_M\tilde{\Phi}^i{}_{PN} - A^P{}_N\tilde{\Phi}^i{}_{MP} \tag{4.8}$$

where $A^P{}_M$ is a traceless 5×5 constant matrix. Similarly, we have

$$\delta\Phi^{iMN} = A^M{}_P\Phi^{iPN} + A^N{}_P\Phi^{iMP}. \tag{4.9}$$

The explicit transformation rules are those of (3.7) with the restrictions

$$A^{\mu\nu}{}_{\rho\sigma} = \delta^\mu{}_\rho A^\nu{}_\sigma - \delta^\nu{}_\rho A^\mu{}_\sigma + \delta^\nu{}_\sigma A^\mu{}_\rho - \delta^\mu{}_\sigma A^\nu{}_\rho - \delta^\mu{}_\rho\delta^\nu{}_\sigma A^\alpha{}_\alpha + \delta^\mu{}_\sigma\delta^\nu{}_\rho A^\alpha{}_\alpha$$

$$B^{\mu\nu\rho} = B^{[\mu\nu\rho]} = -\varepsilon^{\mu\nu\rho\sigma}A^5{}_\sigma\,, \qquad C_{\mu\nu\rho} = C_{[\mu\nu\rho]} = -(1/g)\varepsilon_{\mu\nu\rho\sigma}A^\sigma{}_5$$

$$D_\mu{}^\rho = -A^\rho{}_\mu + \delta^\rho{}_\mu A^\alpha{}_\alpha\,, \tag{4.10}$$

corresponding to the 24 parameters of SL(5, ℝ). Thus

$$\delta\tilde{\mathscr{F}}^{i\mu\nu} = -A^{\alpha}{}_{\alpha}\tilde{\mathscr{F}}^{i\mu\nu} + A^{\mu}{}_{\sigma}\tilde{\mathscr{F}}^{i\sigma\nu} + A^{\nu}{}_{\sigma}\tilde{\mathscr{F}}^{i\mu\sigma} + B^{\mu\nu\lambda}\tilde{\mathscr{G}}^{i}{}_{\lambda}$$

$$\delta\tilde{\mathscr{G}}^{i}{}_{\mu} = A^{\alpha}{}_{\alpha}\tilde{\mathscr{G}}^{i}{}_{\mu} - A^{\sigma}{}_{\mu}\tilde{\mathscr{G}}^{i}{}_{\sigma} + \tfrac{1}{2}C_{\mu\rho\lambda}\tilde{\mathscr{F}}^{i\rho\lambda}\,. \tag{4.11}$$

Whereas,

$$\delta\mathscr{F}^{i\mu} = \left(-\tfrac{1}{3}A^{\alpha}{}_{\alpha} - \frac{2}{3}\frac{1}{3!}B^{\alpha\beta\gamma}b_{\alpha\beta\gamma}\right)\mathscr{F}^{i\mu} + A^{\mu}{}_{\sigma}\mathscr{F}^{i\sigma} - \tfrac{1}{2}B^{\mu\alpha\beta}\mathscr{G}^{i}{}_{\alpha\beta}$$

$$\delta\mathscr{G}^{i}{}_{\mu\nu} = \left(\tfrac{5}{3}A^{\alpha}{}_{\alpha} - \frac{2}{3}\frac{1}{3!}B^{\alpha\beta\gamma}b_{\alpha\beta\gamma}\right)\mathscr{G}^{i}{}_{\mu\nu} - A^{\alpha}{}_{\mu}\mathscr{G}^{i}{}_{\alpha\nu} - A^{\alpha}{}_{\nu}\mathscr{G}^{i}{}_{\mu\alpha} - C_{\rho\mu\nu}\mathscr{F}^{i\rho}\,. \tag{4.12}$$

The transformation rule for G_{MN} is given by

$$\delta G_{MN} = -A^{P}{}_{N}G_{MP} - A^{P}{}_{M}G_{PN}\,, \tag{4.13}$$

and hence

$$\begin{aligned}\delta g_{\mu\nu} &= \tfrac{4}{3}A^{\alpha}{}_{\alpha}g_{\mu\nu} - A^{\sigma}{}_{\mu}g_{\sigma\nu} - A^{\sigma}{}_{\nu}g_{\mu\sigma}\\ &\quad + \frac{2}{3}\frac{1}{3!}B^{\alpha\beta\gamma}b_{\alpha\beta\gamma}g_{\mu\nu} - \tfrac{1}{2}g_{\mu\rho}B^{\rho\alpha\beta}b_{\alpha\beta\nu} - \tfrac{1}{2}b_{\mu\alpha\beta}B^{\alpha\beta\rho}g_{\rho\nu}\,,\\ \delta b_{\mu\nu\rho} &= 2A^{\alpha}{}_{\alpha}b_{\mu\nu\rho} - A^{\sigma}{}_{\mu}b_{\sigma\nu\rho} - A^{\sigma}{}_{\nu}b_{\mu\sigma\rho} - A^{\sigma}{}_{\rho}b_{\mu\nu\sigma}\\ &\quad - \frac{1}{3!}B^{\alpha\beta\gamma}b_{\alpha\beta\gamma}b_{\mu\nu\rho} + B^{\alpha\beta\gamma}g_{\mu\alpha}g_{\nu\beta}g_{\rho\gamma} + C_{\mu\nu\rho}\,.\end{aligned} \tag{4.14}$$

This action of SL(5, ℝ) on the background fields is nonlinear. A linear realization of SL(5, ℝ) may be obtained by rewriting eq. (4.5) as

$$\tilde{\Phi}^{i}{}_{AB} = 2\Phi^{i}{}_{AB}\,, \tag{4.15}$$

where

$$\tilde{\Phi}^{i}{}_{AB} = E^{M}{}_{A}E^{N}{}_{B}\tilde{\Phi}^{i}{}_{MN}\,, \qquad \Phi^{i}{}_{AB} = E_{MA}E_{NB}\Phi^{iMN}\,, \tag{4.16}$$

with $E_{M}{}^{A}$ the "fünfbein" for which

$$G_{MN} = E_{M}{}^{A}E_{NA}\,, \tag{4.17}$$

where $E^{M}{}_{A}$ is its inverse. Multiplying eq. (4.15) by $E_{M}{}^{A}E_{N}{}^{B}$ we recover eq. (4.5).

As usual the price to pay for a linear realization of G is a local symmetry H where H ⊂ G. In this case

$$E_M{}^A \to \Lambda^A{}_B E_M{}^B , \tag{4.18}$$

where Λ is an element of SO(5), the maximal compact subgroup of SL(5, ℝ). The 14 physical scalar fields described by G_{MN} parametrize the coset SL(5, ℝ)/SO(5) and their self-interaction is described by the corresponding nonlinear σ-model.

Finally, we should also discuss the transformation properties of the membrane metric. From eqs. (3.2) and (3.9) we have

$$-\gamma\gamma^{ij} = g_{\mu\nu}\mathscr{F}^{i\mu}\mathscr{F}^{j\nu} . \tag{4.19}$$

Thus from the transformation rules for $g_{\mu\nu}$ of (4.14) and $\mathscr{F}^{i\mu}$ of (4.12) we may deduce

$$\delta\gamma_{ij} = \tfrac{1}{3}\left(A^\alpha{}_\alpha - \frac{1}{3!}B^{\alpha\beta\gamma}b_{\alpha\beta\gamma} - \tfrac{1}{2}\varepsilon \right)\gamma_{ij} , \tag{4.20}$$

where

$$\varepsilon \equiv \frac{1}{\sqrt{-\gamma}}\varepsilon^{lmn}\,\partial_l x^\mu\,\partial_m x^\nu\,\partial_n x^\rho\, g_{\mu\alpha}g_{\nu\beta}g_{\rho\gamma}B^{\alpha\beta\gamma} . \tag{4.21}$$

Thus γ_{ij} transforms conformally with a worldvolume coordinate-dependent conformal factor. The invariant

$$\tilde{\Phi}^i{}_{IJ}\Phi^{jIJ} = 0 \tag{4.22}$$

as a consequence of eq. (3.9) which just restates, in a manifestly SL(5, ℝ) invariant way, the vanishing of the energy–momentum tensor (3.8).

5. Comparison with $d = 11$ supergravity

By compactifying the membrane on T^4, we discovered in the last section that the duality symmetry is SL(5, ℝ) and that the 14 background fields $g_{\mu\nu}$ and $b_{\mu\nu\rho}$ parametrize the 14-dimensional coset SL(5, ℝ)/SO(5). But this duality symmetry and this coset are precisely those obtained by Cremmer and Julia [4] from compactification of $d = 11$ supergravity on T^4. Once we have accepted that duality symmetries of supergravity can in principle have their origin on the worldvolume of the supermembrane, this correspondence should not be surprising since we know that the κ-symmetry of the $d = 11$ supermembrane [8] forces the background fields of the three-dimensional σ-model to be solutions of the $d = 11$ supergravity equations [13]. With the exception of the graviton, which is a singlet under duality,

TABLE 1
The hidden global symmetries G and local symmetries H that result from compactifying $D = 11$ supergravity on T^n. For $n \leqslant 5$, these are compatible with the coset parametrized by the membrane background fields $g_{\mu\nu}$ and $b_{\mu\nu\rho}$ only. Extra space-time scalars must be included for $n \geqslant 6$

n	G	H	dim G/H	$n(n^2+5)/3!$	
1	$\mathbb{R}$	1	1	1	√
2	$GL(2, \mathbb{R})$	$SO(2)$	3	3	√
3	$SL(3, \mathbb{R}) \times SL(2, \mathbb{R})$	$SO(3) \times SO(2)$	7	7	√
4	$SL(5, \mathbb{R})$	$SO(5)$	14	14	√
5	$SO(5, 5)$	$SO(5) \times SO(5)$	25	25	√
6	$E_{6(+6)}$	$USp(8)$	42	41	×
7	$E_{7(+7)}$	$SU(8)$	70	63	×
8	$E_{8(+8)}$	$SO(16)$	128	92	×

all the space-time fields in the supergravity multiplet, gravitinos, vectors, spinors, scalars and antisymmetric tensors, will transform under duality. By retaining only the backgrounds $g_{\mu\nu}$ and $b_{\mu\nu\rho}$, we have so far been treating only the space-time scalars. The scalars alone are nevertheless sufficient to determine the duality symmetries. If we repeat the $n = 4$ analysis of sect. 4 for $n < 4$, one finds the duality symmetries listed in table 1 with the $n(n^2+5)/3!$ $g_{\mu\nu}$ and $b_{\mu\nu\rho}$ parametrizing the corresponding coset. These once again agree with those of Cremmer and Julia. Moreover the $n(n+1)/2$ "field strengths" $\tilde{\mathscr{G}}^i{}_\mu$ and $\tilde{\mathscr{F}}^{i\mu\nu}$ transform as the same representation of G as do the space-time vector fields*.

For $n \geqslant 6$, however, there is a mismatch with the number of space-time scalars as shown in table 1. The mismatch with the number of space-time vectors occurs already for $n \geqslant 5$ as shown in table 2. The reason for this is easy to explain and does not present a serious problem. By focussing only on those space-time scalars arising from $g_{\mu\nu}$ and $b_{\mu\nu\rho}$, we have ignored those arising from other sources [4]. For $n = 6$, we get 1 extra by dualizing the space-time three-form $b_{MNP}(M, N, P = 1, \ldots, 5)$; for $n = 7$ we get 7 extra by dualizing the space-time two-form $b_{MN\rho}(M, N = 1, \ldots, 4)$; for $n = 8$ we get 28 extra by dualizing the one-form $b_{M\nu\rho}$ and 8 extra by dualizing $g_{M\nu}(M = 1, 2, 3)$. This correctly accounts for the mismatch in table 1. Similarly, by focussing on the fields strengths $\tilde{\mathscr{G}}^i{}_\mu$ and $\tilde{\mathscr{F}}^{i\mu\nu}$, we have been ignoring others that couple to space-time vectors arising from dualization. For $n = 5$, we get 1 extra by dualizing the three-form b_{MNP}, for $n = 6$ we get 6 extra by dualizing the two-form $b_{MN\rho}$. This correctly accounts for the mismatch in table 2. The cases $n = 7, 8$ are special. The 28 space-time vectors field strengths combine with 28 dual field strengths to form a 56 of $E_{7(+7)}$ whereas for $n = 8$ all vectors are

* We are grateful to Ergin Sezgin for pointing out the importance of this.

TABLE 2
Representations of the duality symmetry under which the spin-1 space-time fields of compactified $D = 11$ supergravity transform. For $n \leqslant 4$, these are compatible with the representations of the membrane "field strengths" $\mathscr{G}^i{}_\mu$ and $\mathscr{F}^{i\mu\nu}$ only. Extra field strengths must be included for $n \geqslant 5$

n	G	spin-1 reps.	$n(n+1)/2$	
1	$\mathbb{R}$	1	1	√
2	GL(2, $\mathbb{R}$)	3	3	√
3	SL(3, $\mathbb{R}$) × SL(2, $\mathbb{R}$)	(3, 2)	6	√
4	SL(5, $\mathbb{R}$)	10	10	√
5	SO(5, 5)	16	15	×
6	$E_{6(+6)}$	27	21	×
7	$E_{7(+7)}$	56	28	×
8	$E_{8(+8)}$	—	36	×

dual to scalars. Thus we expect that all the hidden symmetries of Cremmer and Julia, including those for $n \geqslant 5$, will follow from membrane duality provided we start with an enlarged σ-model that includes the couplings to those background fields we could safely ignore for $n \leqslant 4$. We intend to return to this point elsewhere.

6. String duality from membrane duality

The dimension of the extended object ($p+1$ for a "p-brane") and the dimension of space-time (D) in which it moves, are severely limited by supersymmetry. Classically, one requires that in a physical gauge there be equal numbers of bosons and fermions on the worldvolume [14]. There are 12 possibilities displayed on the "brane-scan" of fig. 1. They fall into 4 sequences and the equations of motion for a lower member of the sequence may be obtained from those of a higher member in the same sequence by the process of "simultaneous dimensional reduction" [13]. This is illustrated by the diagonal lines in fig. 1 which terminate on the strings ($p = 1$) in $D = 3$, 4, 6 and 10. In particular, the Type IIA superstring in $D = 10$ follows from the supermembrane in $D = 11$. One suspects, therefore, that the string duality of sect. 2 should follow from the membrane duality of sect. 3 by the same simultaneous dimensional reduction. We shall now show that this is indeed the case, using the explicit $n = 4$ example of sect. 4.

Let us denote all membrane variables by a hat. Thus the equations of motion (3.5) and Bianchi identities (3.6) now read

$$\partial_{\hat{i}} \hat{\mathscr{G}}^{\hat{i}}{}_{\hat{\mu}} = 0, \qquad \partial_{\hat{i}} \hat{\mathscr{F}}^{\hat{i}\hat{\mu}\hat{\nu}} = 0. \tag{6.1), (6.2}$$

Similarly, we denote the membrane background fields by $\hat{g}_{\hat{\mu}\hat{\nu}}$ and $\hat{b}_{\hat{\mu}\hat{\nu}\hat{\rho}}$ and the SL(5, $\mathbb{R}$) parameters by $\hat{A}^{\hat{M}}{}_{\hat{N}}$. Here $\hat{i}$ runs over 1 to 3, $\hat{\mu}$ over 1 to 4 and $\hat{M}$ over 1

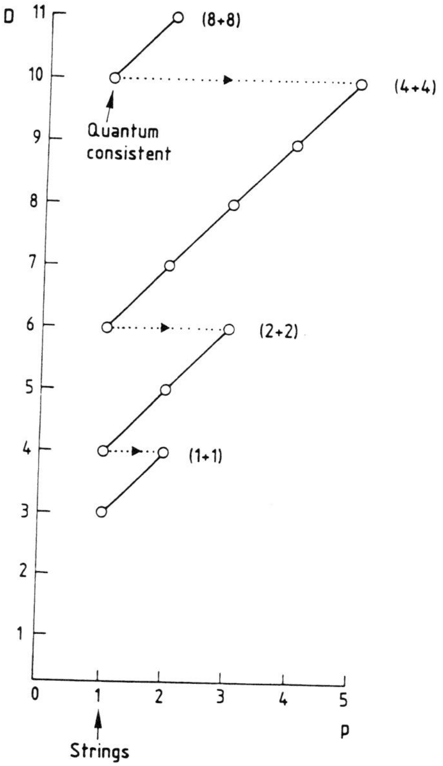

Fig. 1. The "brane-scan".

to 5. In this notation, eq. (3.9) becomes

$$\hat{\gamma}_{\hat{i}\hat{j}} = \partial_{\hat{i}}\hat{x}^{\hat{\mu}}\,\partial_{\hat{j}}\hat{x}^{\hat{\nu}}\,\hat{g}_{\hat{\mu}\hat{\nu}}\,. \tag{6.3}$$

Following ref. [13], we now make a two-one split of the worldvolume coordinates

$$\xi^{\hat{i}} = (\xi^{i}, \xi^{3})\,, \qquad i = 1,2\,, \tag{6.4}$$

and a three-one split of the target-space coordinates

$$\hat{x}^{\hat{\mu}} = (x^{\mu}, x^{4})\,, \qquad \mu = 1,2,3\,, \tag{6.5}$$

in order to make the partial gauge choice

$$\xi^{3} = x^{4}\,, \tag{6.6}$$

which identifies the fourth target-space dimension with the third worldvolume dimension. The dimensional reduction is then effected by demanding that

$$\partial x^{\mu}/\partial \xi^{3} = 0 . \tag{6.7}$$

(The other requirement of ref. [13], namely

$$\partial \hat{g}_{\hat{\mu}\hat{\nu}}/\partial x^{4} = 0 = \partial \hat{b}_{\hat{\mu}\hat{\nu}\hat{\rho}}/\partial x^{4} , \tag{6.8}$$

is here superfluous since we are already assuming that the background fields are independent of all $\hat{x}^{\hat{\mu}}$.) A suitable choice of four-dimensional variables is now

$$\hat{g}_{\hat{\mu}\hat{\nu}} = \phi^{-2/3}\begin{pmatrix} g_{\mu\nu} + \phi^{2}A_{\mu}A_{\nu} & \phi^{2}A_{\mu} \\ \phi^{2}A_{\nu} & \phi^{2} \end{pmatrix} , \tag{6.9}$$

$$\hat{b}_{\hat{\mu}\hat{\nu}\hat{\rho}} = \left(\hat{b}_{\mu\nu\rho}, \hat{b}_{\mu\nu 4}\right) = \left(b_{\mu\nu\rho}, b_{\mu\nu}\right) . \tag{6.10}$$

The background fields $g_{\mu\nu}$ and $b_{\mu\nu}$ will be identified with the string backgrounds of sect. 2 with $n = 3$, and ϕ will be the string dilaton. The fields A_{μ} and $b_{\mu\nu\rho}$ are the extra backgrounds that appear in the Type IIA but not the bosonic or heterotic strings. By working in a Green–Schwarz formalism and focussing only on the bosonic sector, we shall see that ϕ, A_{μ} and $b_{\mu\nu\rho}$ in fact all decouple from the equations of motion. Note, for example, that

$$\hat{g} = g . \tag{6.11}$$

If we now make the following identifications

$$\tilde{\mathscr{F}}^{i\mu} = \hat{\mathscr{F}}^{i\mu 4} , \qquad \tilde{\mathscr{G}}^{i}{}_{\mu} = \hat{\mathscr{G}}^{i}{}_{\mu} \tag{6.12), (6.13}$$

and substitute (6.9) and (6.10) into the membrane equations (6.1) and (6.2), we correctly recover the string equations (2.5) and (2.6). The $\hat{\mathscr{F}}^{\hat{\imath}\mu\nu}$ and $\hat{\mathscr{G}}^{\hat{\imath}}{}_{4}$ equations are identities, as they must be for consistency. Similarly, if we now make the identifications

$$A^{\mu}{}_{\nu} = \hat{A}^{\mu}{}_{\nu} - \delta^{\mu}{}_{\nu}\hat{A}^{\rho}{}_{\rho} , \qquad B^{\mu\nu} = \hat{A}^{5}{}_{\rho}\varepsilon^{\rho\mu\nu} , \qquad C_{\mu\nu} = (1/g)\hat{A}^{\rho}{}_{5}\varepsilon_{\rho\mu\nu} , \tag{6.14}$$

set to zero $\hat{A}^{5}{}_{4}$, $\hat{A}^{4}{}_{5}$, $\hat{A}^{4}{}_{4}$, $\hat{A}^{\mu}{}_{4}$ and $\hat{A}^{4}{}_{\mu}$, and substitute into the membrane transformation rules (4.11), (4.12) and (4.14), we correctly recover the string transformation rules (2.30) and (2.32). Once again, the extra equations not corresponding to string variables are just identities.

Thus in this explicit example, we have seen how the duality symmetry of the string equations of motion for $n = 3$, namely SO(3, 3), follows as a consequence of

TABLE 3
String duality from membrane duality via simultaneous dimensional reduction ($\rightarrow$). The string symmetries are actually larger than SO(n, n), where n is the number of compactified string dimensions

n	membrane duality	$\rightarrow$	string duality	SO(n, n)
1	GL(2, $\mathbb{R}$)	$\rightarrow$	SO(1, 1) $\times$ SO(1, 1)	SO(1, 1)
2	SL(3, $\mathbb{R}$) $\times$ SL(2, $\mathbb{R}$)	$\rightarrow$	SO(2, 2) $\times$ SO(1, 1)	SO(2, 2)
3	SL(5, $\mathbb{R}$)	$\rightarrow$	SO(3, 3) $\times$ SO(1, 1)	SO(3, 3)
4	SO(5, 5)	$\rightarrow$	SO(4, 4) $\times$ SO(1, 1)	SO(4, 4)
5	$E_{6(+6)}$	$\rightarrow$	SO(5, 5) $\times$ SO(1, 1)	SO(5, 5)
6	$E_{7(+7)}$	$\rightarrow$	SO(6, 6) $\times$ SO(2, 1)	SO(6, 6)
7	$E_{8(+8)}$	$\rightarrow$	SO(8, 8)	SO(7, 7)

the SL(5, $\mathbb{R}$) duality symmetry of the membrane for $n = 4$. It is not difficult to see, from a group theoretical point of view, how this would work for other values of n. The results are shown in table 3.

In fact, the string duality symmetries listed in the second column are larger than the SO(n, n) appearing in the third column and discussed in sect. 2. The reason is that the SO(n, n) refers to the coset parametrized by $g_{\mu\nu}$ and $b_{\mu\nu}$ only. However, even for the string, space-time scalars may arise from other sources. First there is the dilaton ϕ which, although decoupling from the bosonic sector, still survives in the fermi–fermi couplings. If we retain the $\hat{A}^4{}_4$ component of $\hat{A}^{\hat{\mu}}{}_{\hat{\nu}}$, there is an extra SO(1, 1) under which $g_{\mu\nu}$, $b_{\mu\nu}$ and ϕ transform by conformal factors. In four space-time dimensions we also have the axion b_{MN} (coming from $\hat{b}_{MN11}$ of the $D = 11$ supermembrane) which is dual to a scalar and which, together with ϕ parametrizes the coset SO(2, 1)/U(1). In three space-time dimensions we have 14 more scalars coming from $\hat{g}_{M\hat{\nu}}$ and $\hat{b}_{M\hat{\nu}11}$. These conspire with the dilaton and 49 $g_{\mu\nu}$ and $b_{\mu\nu}$ to parametrize SO(8, 8)/SO(8) $\times$ SO(8). For the heterotic string we promote each SO(n, n) to SO($16 + n, n$) corresponding to the extra 16 left-moving modes. Thus in $D = 3$, we would have SO(24, 8).

Of course, we could retain all the space-time background fields in the dimensional reduction including those that appear only in fermi–fermi couplings and thereby obtain the duality symmetries of the Type IIA superstring. Its duality symmetries would then be given by the first column in table 3 i.e. the same as those of the $D = 11$ supermembrane.

7. Higher extended objects

So far we have considered strings ($p = 1$) and membranes ($p = 2$), but similar duality symmetries will be present for other "p-branes" with $p \geqslant 3$. The lagrangian takes the form with $i = 1, \ldots, p+1$,

$$\mathcal{L} = \tfrac{1}{2}\sqrt{-\gamma}\,\gamma^{ij}\,\partial_i x^\mu\,\partial_j x^\nu\,g_{\mu\nu} - \tfrac{1}{2}(p-1)\sqrt{-\gamma} + \frac{1}{(p+1)!}\varepsilon^{ii_1\ldots i_p}\,\partial_i x^\mu\,\partial_{i_1} x^{\mu_1}\ldots\partial_{i_p} x^{\mu_p} b_{\mu\mu_1\ldots\mu_p} \tag{7.1}$$

with background fields $g_{\mu\nu}$ and $b_{\mu\mu_1,\ldots,\mu_p}$ ($\mu = 1,\ldots,n$). Define

$$\mathscr{F}^{i\mu} = \sqrt{-\gamma}\,\gamma^{ij}\,\partial_j x^\mu\,, \tag{7.2}$$

$$\tilde{\mathscr{F}}^{i\mu_1\ldots\mu_p} = \varepsilon^{ii_1\ldots i_p}\,\partial_{i_1} x^{\mu_1}\ldots\partial_{i_p} x^{\mu_p}\,, \tag{7.3}$$

$$\tilde{\mathscr{G}}^i{}_\mu = g_{\mu\nu}\mathscr{F}^{i\nu} + \frac{1}{p!} b_{\mu\mu_1\ldots\mu_p}\tilde{\mathscr{F}}^{i\mu_1\ldots\mu_p}\,. \tag{7.4}$$

Then once again there is a symmetry that rotates field equations

$$\partial_i \tilde{\mathscr{G}}^i{}_\mu = 0 \tag{7.5}$$

into Bianchi identities

$$\partial_i \tilde{\mathscr{F}}^{i\mu_1\ldots\mu_p} = 0\,. \tag{7.6}$$

The dual coordinate now has p indices: $y_{\mu_1\ldots\mu_p}$, and the analogue of eqs. (2.19) and (3.22) becomes

$$\begin{aligned}
&p!\varepsilon^{ii_1\ldots i_p}\,\partial_{i_1} y_{\mu\mu_2\ldots\mu_p}\,\partial_{i_2} x^{\mu_2}\ldots\partial_{i_p} x^{\mu_p}\\
&\quad = \partial\mathcal{L}_x/\partial\,\partial_i x^\mu\\
&\quad = g_{\mu\nu}\sqrt{-\gamma}\,\gamma^{ij}\,\partial_j x^\nu + \frac{1}{p!} b_{\mu\mu_1\ldots\mu_p}\varepsilon^{ii_1\ldots i_p}\,\partial_{i_1} x^{\mu_1}\ldots\partial_{i_p} x^{\mu_p}\,,\\
&\varepsilon^{ii_1\ldots i_p}\,\partial_{i_1} x^{\mu_1}\ldots\partial_{i_p} x^{\mu_p} = \partial\mathcal{L}_y/\partial\,\partial_i y_{\mu_1\ldots\mu_p}\\
&\quad = p!\,p^{\mu_1\ldots\mu_p\,\nu_1\ldots\nu_p}\sqrt{-\gamma}\,\gamma^{ij}\,\partial_j y_{\nu_1\ldots\nu_p}\\
&\quad\quad + p!\,q^{\mu_1\ldots\mu_p\nu}\varepsilon^{ii_1\ldots i_p}\,\partial_{i_1} y_{\nu\nu_2\ldots\nu_p}\,\partial_{i_2} x^{\nu_2}\ldots\partial_{i_p} x^{\nu_p}\,,
\end{aligned} \tag{7.7}$$

where $p^{\mu\nu} = p^{\nu\mu}$ and $q^{\mu_1 \ldots \mu_p} = q^{[\mu_1 \ldots \mu_p]}$ are related to $g_{\mu\nu} = g_{\nu\mu}$ and $b_{\mu_1 \ldots \mu_p} = b_{[\mu_1 \ldots \mu_p]}$ by

$$p_{\mu\nu} = g_{\mu\nu} + b_{\mu m} g^{mn} b_{\nu n} , \qquad p_{\mu\nu} q^{m\nu} = b_{\mu n} g^{mn} , \tag{7.8}$$

where $p_{\mu\nu}$ is the inverse of $p^{\mu\nu}$. Here we have adopted a condensed notation where the index m means

$$m \equiv [\mu_1 \ldots \mu_p] ,$$

and where a repeated m index means

$$a_m b^m \equiv (1/p!) a_{\mu_1 \ldots \mu_p} b^{\mu_1 \ldots \mu_p} . \tag{7.9}$$

The quantity g^{mn} is given by

$$g^{mn} \equiv g^{\mu_1 \ldots \mu_p \nu_1 \ldots \nu_p} = \sum_P (-1)^P g^{\mu_1 \nu_1} \ldots g^{\mu_p \nu_p} . \tag{7.10}$$

In this notation, the equations analogous to eqs. (2.21) and (4.2) may be written

$$\tilde{\mathscr{G}}^i{}_\mu = g_{\mu\nu} \mathscr{F}^{i\nu} + b_{\mu n} \tilde{\mathscr{F}}^{in} , \qquad \tilde{\mathscr{F}}^{im} = p^{mn} \mathscr{G}^i{}_n + q^{m\nu} \tilde{\mathscr{G}}^i{}_\nu \tag{7.11}$$

where

$$\tilde{\mathscr{G}}^i{}_\mu = p! \varepsilon^{i i_1 \ldots i_p} \partial_{i_1} y_{\mu \nu_2 \ldots \nu_p} \partial_{i_2} x^{\nu_2} \ldots \partial_{i_p} x^{\nu_p} ,$$

$$\mathscr{G}^i{}_n \equiv \mathscr{G}^i{}_{\nu_1 \nu_2 \ldots \nu_p} = (p!)^2 \sqrt{-\gamma}\, \gamma^{ij} \partial_j y_{\nu_1 \nu_2 \ldots \nu_p} , \tag{7.12}$$

and where p^{mn} is defined by

$$g^{mn} = p^{mn} + q^{m\mu} p_{\mu\nu} q^{n\nu} . \tag{7.13}$$

This may be rewritten as

$$\begin{pmatrix} \tilde{\mathscr{G}}^i{}_\mu \\ \tilde{\mathscr{F}}^{im} \end{pmatrix} = \begin{pmatrix} g_{\mu\nu} + b_{\mu m} g^{mn} b_{\nu n} & b_{\mu m} g^{mn} \\ g^{mn} b_{\nu n} & g^{mn} \end{pmatrix} \begin{pmatrix} \mathscr{F}^{i\nu} \\ \mathscr{G}^i{}_n \end{pmatrix} . \tag{7.14}$$

As remarked in ref. [1] in the context of strings, increasing the dimension of the target space, with coordinates x^μ, to include the extra y_m coordinates is strongly reminiscent of a Kaluza–Klein procedure. This analogy is seen to be even closer when we compare the matrix in eq. (7.14) with the typical Kaluza–Klein decomposition of the metric

$$G_{MN} = \begin{pmatrix} g_{\mu\nu} + A_\mu{}^m g_{mn} A_\nu{}^n & A_\mu{}^n g_{mn} \\ g_{mn} A_\nu{}^n & g_{mn} \end{pmatrix}. \tag{7.15}$$

Thus the role of the gauge field $A_\mu{}^m$ is played by the antisymmetric tensor $b_{\mu m}$. Untypical is the fact that the number of "internal" dimensions $\binom{n}{p}$ is determined by the number of "space-time" dimensions, n, with the curious twist that the role of the "internal" metric g_{mn} is played by g^{mn} which is built out of the inverse of the "space-time" metric $g^{\mu\nu}$ as in eq. (7.10). Indeed, if we introduce as in ref. [1] a target space with $n + \binom{n}{p}$ dimensions and coordinates

$$Z^M = (x^\mu, y_m) \tag{7.16}$$

the equations of motion and Bianchi identities may be united into a single equation, since after some rearrangement eq. (7.7) may be written

$$\Omega_{MM_1\ldots M_p}\varepsilon^{ii_1\ldots i_p}\partial_{i_1}Z^{M_1}\ldots\partial_{i_p}Z^{M_p} = G_{MN}\sqrt{-\gamma}\,\gamma^{ij}\partial_j Z^N, \tag{7.17}$$

where G_{MN} is the matrix appearing in (7.14) and $\Omega_{MM_1\ldots M_P} = \Omega_{M[M_1\ldots M_P]}$ is a numerical tensor whose nonvanishing components are given by

$$\Omega_\mu{}^{\alpha\beta\ldots\gamma}{}_{\nu\ldots\sigma} = \frac{1}{p!p}\delta^{\alpha\beta\ldots\gamma}_{\mu\nu\ldots\sigma}, \qquad \Omega^{\alpha\beta\ldots\gamma}{}_{\mu\nu\ldots\sigma} = \frac{1}{p!}\delta^{\alpha\beta\ldots\gamma}_{\mu\nu\ldots\sigma}, \tag{7.18}$$

where

$$\delta^{\alpha\beta\ldots\gamma}_{\mu\nu\ldots\sigma} = \sum_P (-1)^P \delta^\alpha{}_\mu \delta^\beta{}_\nu \ldots \delta^\gamma{}_\sigma. \tag{7.19}$$

Note that

$$\Omega_{[MM_1\ldots M_p]} = 0, \tag{7.20}$$

and so, multiplying both sides of eq. (7.17) by $\partial_i Z^M$ we learn that

$$\sqrt{-\gamma}\,\gamma^{ij}\partial_i Z^M G_{MN}\partial_j Z^N = 0 \tag{7.21}$$

as may also be verified explicitly.

Although eq. (7.17) is an elegant way of summarizing the combined field equations and Bianchi identities of an arbitrary p-brane in a target space of arbitrary D, one must not be lulled into thinking that all the hidden symmetries are thus rendered manifest. This is because, with the exception of the string, Ω is not an invariant tensor under the full duality transformations and $\partial_j Z^M$ does not transform as a vector. Rather, the manifest symmetry is only a subgroup of the full duality group, and turns out to be $SL(n, \mathbb{R}) \times \mathbb{R}^{\binom{n}{p}}$, under which γ_{ij} does not transform. To obtain the complete duality symmetry we must adopt a different route along the lines described in sect. 3. What will these p-brane symmetries be?

The first observation to make is that a p-brane couples to a $(p+1)$-form background $b_{MM_1\ldots M_P}$ $(M = 1, \ldots, D)$ whose field strength $F = \mathrm{d}b$ is a $(p+2)$-form. But in D space-time dimensions a $(p+2)$-form F is dual (in the sense of Poincaré duality) to a $(D-p-2)$-form $\tilde{F} = \mathrm{d}\tilde{b}$ where $\tilde{b}$ is $(D-p-3)$ form that couples to $(D-p-4)$-brane. Hence we expect the duality symmetry of a p-brane in D dimensions compactified on T^n to be the same as that of $(D-p-4)$-brane in D dimensions compactified on T^n. A good example is provided by the string in $D = 10$ and the 5-brane in $D = 10$. The former couples to the background fields of $D = 10$ supergravity with a 2-form b_{MN}, while the latter couples to the fields of the dual formulation of $D = 10$ supergravity in which the 2-form is replaced by a 6-form $\tilde{b}_{MNPQRS}$. Thus we anticipate that the 5-brane duality symmetries will be exactly the same as those for the string listed in table 3. Of course, to achieve this it will be necessary, as described in sect. 5, to augment the n-dimensional background scalar fields $g_{\mu\nu}$ and $\tilde{b}_{\mu\nu\rho\lambda\sigma\tau}$ with those arising from other sources and similarly for the field strengths $\tilde{\mathscr{G}}^i{}_\mu$ and $\tilde{\mathscr{F}}^{i\mu\nu\rho\lambda\sigma}$.

Note, incidentally that whereas in space-time one must replace the field-strengths $F_{MNN_1\ldots N_p}$ of the gauge $(p+1)$-forms by their duals in order to get equivalent degrees of freedom, for the compactified coordinates (where the $(p+1)$-form potentials are space-time scalars) one must replace the potentials $b_{\mu\mu_1\ldots\mu_p}$ themselves by their duals. Similar remarks apply when working with the space-time fields in the light-cone gauge rather than covariantly.

If we consider the full superspace (x^μ, θ^α) couplings and keep all the background fields, then the duality symmetry is preserved by the simultaneous dimensional reduction e.g. as discussed in sect. 6, the $D = 11$ membrane duality leads to the $D = 10$ Type IIA superstring duality. Thus there are really just four duality schemes corresponding to the four sequences on the "brane-scan" of fig. 1. Just as the largest duality group is $E_{8(+8)}$ for the octonionic sequence, a process of counting degrees of freedom and truncating leads to the three groups shown in

TABLE 4
Maximal (finite dimensional) duality symmetries for the 4 sequences of extended objects

Sequence	G	H	dim G/H
O	$E_{8(+8)}$	SO(16)	128
H	SO(8, 8)	SO(8) × SO(8)	64
C	SO(4, 4)	SO(4) × SO(4)	16
R	SO(2, 2)	SO(2) × SO(2)	4

table 4 for the quaternionic, complex and real sequences when compactified to $D=3$ space-time dimensions. Of course, one might conjecture that these may be enlarged even further to infinite-dimensional symmetries, for example E_9 [4] and E_{10} [5, 6, 19] in the octonionic case.

8. Conclusions

We have seen how the Cremmer–Julia hidden symmetries of supergravity have their origin in duality transformations on the three-dimensional worldvolume of the $D=11$ supermembrane, and how the string duality symmetries follow from those of the membrane by simultaneous dimensional reduction. The duality symmetries for a general p-brane in D space-time dimensions, will be the same as those of $(D-p-4)$-brane, to which it is related by Poincaré duality, an example being provided by the string and the 5-brane in $D=10$. Several questions now spring to mind.

First, we have succeeded in writing the equations in a manifestly duality invariant way. Since these symmetries are not symmetries of the p-brane action, however, their presence will never be obvious starting from the σ-model with physical background scalar fields $g_{\mu\nu}$ and $b_{\mu\mu_1\ldots\mu_p}$. However, it would be interesting to see whether we could write the p-brane action coupled to *all* the scalars of $E_M{}^A$ which describes both physical and unphysical modes. In this way, at least the local group H, which is the maximal compact subgroup of G, might then be manifest.

Secondly, all the continuous symmetries discussed here follow by demanding that the background fields are independent of the compactified coordinates. This corresponds to a naive dimensional reduction in which it is not even necessary to specify the topology of the extra dimensions. In reality, we must pick a specific topology and geometry e.g. the flat torus T^n, and keep all the Fourier modes. These symmetries will then be broken by the massive states. Moreover, there will be quantization conditions imposed by the torus topology. For the string, the τ components of both $\tilde{\mathscr{G}}^i{}_\mu$ and $\tilde{\mathscr{F}}^{i\mu}$ will be quantized, thus

$$P_\mu \equiv \tilde{\mathscr{G}}^\tau{}_\mu = y'_\mu = m_\mu\,, \qquad A^\mu \equiv \tilde{\mathscr{F}}^{\tau\mu} = x'^\mu = n^\mu\,, \tag{8.1}$$

where m_μ and n^μ are integers. From the point of view of the original σ-model $\mathscr{L}_x$, m_μ corresponds to the momentum modes and n^μ to the winding modes. Whereas from the point of view of the dual σ-model $\mathscr{L}_y$, the roles are reversed [1]. Thus there is a discrete subgroup of $SO(n,n)$ given by

$$\begin{pmatrix} n \\ m \end{pmatrix} \to S^{-1}\begin{pmatrix} n \\ m \end{pmatrix}$$

$$G \to S^{\mathrm{T}} G S \tag{8.2}$$

where G is given by eq. (2.25). This is the SO$(n, n; \mathbb{Z})$ referred to in sect. 1 and which leaves the string spectrum invariant. Thus the question arises whether there are also discrete subgroups of the membrane duality symmetries which leave the membrane spectrum invariant. The equations analogous to (8.1) are

$$P_\mu \equiv \tilde{\mathscr{G}}^{\tau}{}_{\mu} = 2\{x^\nu, y_{\mu\nu}\}, \qquad A^{\mu\nu} \equiv \tilde{\mathscr{F}}^{\tau\mu\nu} = \{x^\mu, x^\nu\}, \tag{8.3}$$

where the Lie bracket is defined by

$$\{X, Y\} = \varepsilon^{ab}\,\partial_a X\,\partial_b Y. \tag{8.4}$$

Unfortunately, this is a difficult question to answer since the membrane spectra are, as yet, unknown. (This is primarily because the usual light-cone gauge [15] action for a p-brane is highly nonlinear except for $p = 1$. An alternative gauge which linearizes the equations of motion has been proposed [16], at the expense of introducing a highly nonlinear constraint. Incidentally, on the subject of the light-cone gauge, we note that except for $p = 1$, it is inhomogeneous in the components of γ_{ij}, namely

$$\gamma_{ij} = \begin{pmatrix} -h & 0 \\ 0 & h_{ab} \end{pmatrix} \tag{8.5}$$

where $h \equiv \det h_{ab}$ and $a, b = 1, \ldots, p$. Consequently, the duality symmetry, under which γ_{ij} rescales by a conformal factor (4.20), would be obscured in this gauge. The conformal gauge of ref. [16], which for $p = 2$ looks like

$$\gamma_{ij} = \begin{pmatrix} A_1^2 & -A_1 A_2 & -A_1 A_3 \\ -A_1 A_2 & A_2^2 & -A_2 A_3 \\ -A_1 A_3 & -A_2 A_3 & A_3^2 \end{pmatrix} \tag{8.6}$$

would be much more suitable.) If such discrete subgroups do leave the membrane spectrum invariant, does the $R \to \alpha'/R$ idea continue to apply, and hence does the idea of a "minimum length" also hold for membranes? We would expect so, since as we saw in sect. 6, G (string) $\subset$ G (membrane). It would also be interesting to see whether the fixed points correspond to enhanced gauge symmetries.

Finally, is the similarity of the duality symmetry of the string in $D = 10$ and the 5-brane in $D = 10$ indicative of a deeper relationship? The two formulations of $D = 10$ supergravity, one with a 3-form field strength and one with a 7-form field strength, has long been something of an enigma from the point of view of superstrings. As field theories, each seems equally as good. In particular, provided we couple them to $E_8 \times E_8$ or SO(32) Yang–Mills, then both are anomaly-free [17, 18]. Since the 3-form version corresponds to the low-energy limit of the heterotic superstring (or Type-I superstring) it is naturally to conjecture, as was done some time ago [19], that there exists a "heterotic 5-brane" (or "Type-I

5-brane") whose low energy limit is the 7-form version. However, this would require a coupling of the super-5-brane to background Yang–Mills fields which, to date, has not been achieved. It was speculated that such a coupling might exist, making use of the property that a one-time and five-space dimensional worldvolume admits real self-dual three forms (and/or Weyl spinors).

This possibility has now become a virtual certainty thanks to the recent remarkable observation by Strominger [20] that the heterotic string admits the heterotic 5-brane as a soliton solution. The soliton interpretation of membranes was the motivation for the original supermembrane paper of Hughes et al. [21] and has also been pursued by Townsend [22]. Strominger went on to conjecture that the heterotic string and heterotic 5-brane might be "dual" in the sense of Olive and Montonen [23] and that the 5-brane describes the strong coupling limit of the string. Clearly, the many meanings of the word "duality" in theories of extended objects have still not been exhausted.

We are grateful to Paul Howe, Chris Pope and Ergin Sezgin for useful conversations. M.J.D. acknowledges the hospitality extended by members of the Theory Division at the Rutherford Laboratory, UK, where part of this work was carried out.

References

[1] M.J. Duff, Nucl. Phys. B335 (1990) 610

[2] K. Kikkawa and M. Yamasaki, Phys. Lett. B149 (1984) 357;
N. Sakai and I. Senda, Prog. Theor. Phys. 75 (1986) 692;
V.P. Nair, A. Shapere, A. Strominger and F. Wilczek, Nucl. Phys. B287 (1987) 414;
A. Shapere and F. Wilczek, Nucl. Phys. B320 (1989) 669;
A. Giveon, E. Rabinovici and G. Veneziano, Nucl. Phys. B322 (1989) 167;
M. Dine, P. Huet and N. Seiberg, Nucl. Phys. B322 (1989) 301;
P. Ginsparg and C. Vafa, Nucl. Phys. B289 (1987) 414;
B. Sathiapaian, Phys. Rev. Lett. 58 (1987) 1597;
R. Brandenberger and C. Vafa, Nucl. Phys. B316 (1989) 391;
C. Vafa, Mod. Phys. Lett. A4 (1989) 1615;
J.M. Molera and B. Ovrut, Phys. Rev. D40 (1989) 1146;
E. Alvarez and M.A.R. Osorio, Phys. Rev. D40 (1989) 1150;
J. Laver, J. Mas and H.P. Nilles, Phys. Rev. Lett. 226B (1989) 251

[3] S. Ferrara, J. Scherk and B. Zumino, Nucl. Phys. B121 (1977) 393;
E. Cremmer, J. Scherk and S. Ferrara, Phys. Lett. B68 (1977) 234;
E. Cremmer and J. Scherk, Nucl. Phys. B127 (1977) 259;
E. Cremmer, J. Scherk and S. Ferrara, Phys. Lett. B74 (1978) 61;
N. Marcus and J.H. Schwarz, Nucl. Phys. B228 (1983) 145;
J.P. Derendinger and S. Ferrara, *in* Supersymmetry and supergravity '84, ed. B. de Wit, P. Fayet and P. van Nieuwenhuizen (World Scientific, Singapore, 1985);
M. De Roo, Nucl. Phys. B255 (1985) 515; Phys. Lett. B156 (1985) 331;
E. Bergshoeff, I.G. Koh and E. Sezgin, Phys. Lett. B155 (1985) 646;
L. Castellani, A. Ceresole, S. Ferrara, R. D'Auria, P. Fre and E. Maina, Nucl. Phys. B268 (1986) 317; Phys. Lett. B161 (1985) 91

[4] E. Cremmer and B. Julia, Nucl. Phys. B139 (1979) 141;
E. Cremmer, *in* Supergravity '81 ed. S. Ferrara and J.G. Taylor (Cambridge Univ. Press, Cambridge, 1982);
B. Julia, *in* Superspace and supergravity, ed. S.W. Hawking and M. Rocek (Cambridge Univ. Press, Cambridge, 1980)
[5] M.J. Duff, Phys. Lett. B173 (1986) 289
[6] M.J. Duff, *in* Quantum field theory and quantum statistics, ed. I.A. Batalin, C.J. Isham and G. Vilkovisky (Hilger, London, 1986);
M.J. Duff, *in* Physics in higher dimensions, ed. T. Piran and S. Weinberg (World Scientific, Singapore, 1986);
M.J. Duff, *in* Architecture of the fundamental interactions at short distances, ed. P. Ramond and R. Stora (North-Holland, Amsterdam, 1987)
[7] K.S. Narain, Phys. Lett. B169 (1986) 61
[8] E. Bergshoeff, E. Sezgin and P.K. Townsend, Phys. Lett. B209 (1988) 451
[9] M.J. Duff, B.E.W. Nilsson and C.N. Pope, Phys. Rep. 130 (1986) 1
[10] B. de Wit and H. Nicolai, Phys. Lett. B108 (1982) 285
[11] S. Cecotti, S. Ferrara and L. Girardello, Nucl. Phys. B308 (1988) 436
[12] M.K. Gaillard and B. Zumino, Nucl. Phys. B193 (1981) 221
[13] M.J. Duff, P.S. Howe, T. Inami and K.S. Stelle, Phys. Lett. B191 (1987) 70
[14] A. Achucarro, J.M. Evans, P.K. Townsend and D.L. Wiltshire, Phys. Lett. B198 (1987) 441
[15] J. Hoppe, MIT PhD Thesis (1982)
[16] I. Bars, Membrane Symmetries and anomalies, Preprint USC-89/HEP 14, *in* Supermembranes and physics in $2+1$ dimensions, ed. M.J. Duff, C.N. Pope and E. Sezgin (World Scientific, Singapore) to be published
[17] M.B. Green and J.H. Schwarz, Phys. Lett. B149 (1984) 117
[18] S.J. Gates and H. Nishino, Phys. Lett. B173 (1986) 52
[19] M.J. Duff, Class. Quantum Gravity 5 (1988) 189
[20] A. Strominger, *in* Strings '90; Proc. of the Superstring Workshop, Texas A&M University, March 1990, ed. R. Arnowitt, R. Bryan, M.J. Duff, D.V. Nanopoulos, C.N. Pope and E. Sezgin (World Scientific, Singapore) to be published
[21] J. Hughes, J. Liu and J. Polchinski, Phys. Lett. B180 (1986) 370
[22] P.K. Townsend, Phys. Lett. B202 (1988) 53
[23] C. Montonen and D. Olive, Phys. Lett. B72 (1977) 117

Reprinted from Nucl. Phys. B **438** (1995) 109–37

Unity of superstring dualities

C.M. Hull [a], P.K. Townsend [b]

[a] *Physics Department, Queen Mary and Westfield College, Mile End Road, London E1 4NS, UK*
[b] *DAMTP, University of Cambridge, Silver Street, Cambridge CB3 9EW, UK*

Received 27 October 1994; accepted 2 December 1994

Abstract

The effective action for type II string theory compactified on a six-torus is $N = 8$ supergravity, which is known to have an E_7 duality symmetry. We show that this is broken by quantum effects to a discrete subgroup, $E_7(\mathbb{Z})$, which contains both the T-duality group $O(6, 6; \mathbb{Z})$ and the S-duality group $SL(2; \mathbb{Z})$. We present evidence for the conjecture that $E_7(\mathbb{Z})$ is an exact 'U-duality' symmetry of type II string theory. This conjecture requires certain extreme black hole states to be identified with massive modes of the fundamental string. The gauge bosons from the Ramond–Ramond sector couple not to string excitations but to solitons. We discuss similar issues in the context of toroidal string compactifications to other dimensions, compactifications of the type II string on $K_3 \times T^2$ and compactifications of 11-dimensional supermembrane theory.

1. Introduction

String theory in a given background can be formulated in terms of a sum over world-sheet fields, (super-) moduli and topologies of a world-sheet sigma-model with the background spacetime as its target space. Different backgrounds may define the same quantum string theory, however, in which case they must be identified. The transformations between equivalent backgrounds generally define a discrete group and such discrete gauge symmetries are referred to as duality symmetries of the string theory. An example is T-duality, which relates spacetime geometries possessing a compact abelian isometry group (see [1] and references therein). The simplest case arises from compactification of the string theory on a circle since a circle of radius R defines the same two-dimensional quantum field theory, and hence the same string theory, as that on a circle of radius α'/R. T-dualities are non-perturbative in the sigma-model coupling constant α' but valid order by order in the string coupling constant g. Some string theories may have

0550-3213/95/$09.50
SSDI 0550-3213(94)00559-1

additional discrete symmetries which are perturbative in α' but non-perturbative in g. An example is the conjectured S-duality of the heterotic string compactified on a six-torus [2–4]. In this paper we investigate duality symmetries of the type II string compactified to four dimensions and present evidence for a new 'U-duality' symmetry which unifies the S- and T-dualities and mixes sigma-model and string coupling constants.

Consider a compactified string for which the internal space is an n-torus with constant metric g_{ij} and antisymmetric tensor b_{ij}. The low-energy effective field theory includes a spacetime sigma-model whose target space is the moduli space $O(n, n)/[O(n) \times O(n)]$ of the torus, and the constants g_{ij} and b_{ij} are the expectation values of the n^2 scalar fields. There is a natural action of $O(n, n)$ on the moduli space. In general this takes one string theory into a different one, but a discrete $O(n, n; \mathbb{Z})$ subgroup takes a given string theory into an equivalent one. This is the T-duality group of the toroidally compactified string and the true moduli space of the string theory is the moduli space of the torus factored by the discrete T-duality group. There is a generalization to Narain compactifications on the '(p, q)-torus' $T(p, q)$ for which the left-moving modes of the string are compactified on a p-torus and the right-moving ones on a q-torus [5]. In this case the moduli space is $O(p, q)/[O(p) \times O(q)]$ factored by the T-duality group $O(p, q; \mathbb{Z})$. The $T(6, 22)$ case is relevant to the heterotic string compactified to four dimensions which has $O(6, 22; \mathbb{Z})$ as its T-duality group. At a generic point in the moduli space the effective field theory is $N = 4$ supergravity coupled to 22 abelian vector multiplets, giving a total of 28 abelian vector gauge fields [6] with gauge group $U(1)^{28}$. It follows from the compactness of the full gauge group for all 28 vector gauge fields that any electric or magnetic charges are quantized. The effective field theory has an $SL(2; \mathbb{R}) \times O(6, 22)$ invariance of the equations of motion which, due to the charge quantization and the fact that states carrying all types of charge can be found in the spectrum, is broken to the discrete subgroup $SL(2; \mathbb{Z}) \times O(6, 22; \mathbb{Z})$. The $O(6, 22; \mathbb{Z})$ factor extends to the T-duality group of the full string theory. It has been conjectured that the $SL(2; \mathbb{Z})$ factor also extends to a symmetry of the full string theory [3]. This is the S-duality group of the heterotic string. It acts on the dilaton field Φ and the axion field ψ (obtained by dualizing the four-dimensional two-form gauge field $b_{\mu\nu}$ that couples to the string) via fractional linear transformations of the complex scalar $\psi + ie^{-\Phi}$ and on the abelian field strengths by a generalized electric–magnetic duality. One of the $SL(2; \mathbb{Z})$ transformations interchanges the electric and magnetic fields and, when $\psi = 0$, takes Φ to $-\Phi$ which, since the expectation value of e^{Φ} can be identified with the string coupling constant g, takes g to $1/g$, and so interchanges strong and weak coupling.

Consider now the compactification of the type IIA or type IIB superstring to four dimensions on a six-torus. The low-energy effective field theory is $N = 8$ supergravity [7], which has 28 abelian vector gauge fields and 70 scalar fields taking values in $E_{7(7)}/[SU(8)/Z_2]$. The equations of motion are invariant under the action of $E_{7(7)}$ [7], which contains $SL(2; \mathbb{R}) \times O(6, 6)$ as a maximal subgroup. We shall show that certain quantum mechanical effects break $E_{7(7)}$ to a discrete subgroup which we shall call $E_7(\mathbb{Z})$, and this implies a breaking of the maximal $SL(2; \mathbb{R}) \times O(6, 6)$ subgroup to $SL(2; \mathbb{Z}) \times O(6, 6; \mathbb{Z})$. The $O(6, 6; \mathbb{Z})$ factor extends to the full string theory as the T-duality group, and it is natural to conjecture that the $SL(2; \mathbb{Z})$ factor also extends to the full string theory as an S-duality group. In fact, we shall present-evidence for the much stronger conjecture that the full $E_7(\mathbb{Z})$ group (to be defined below) extends to the full string theory as a new unified duality group, which we call U-duality. U-duality acts on the abelian gauge fields through a generalized electromagnetic duality and on the 70 scalar fields, the constant parts of which can each be thought of as a coupling constant of the theory. The zero-mode of the dilaton is related to the string coupling g, while 21 of the scalar zero-modes are the moduli of the metric on the 6-torus, and the others parameterise the space of constant antisymmetric tensor gauge fields on the six-torus. U-duality implies that all 70 coupling constants are on a similar footing despite the fact that the standard perturbative formulation of string theory assigns a special rôle to one of them.

Whereas T-duality is known to be an exact symmetry of string theory at each order in the string coupling constant g, the conjectured S-duality and U-duality are non-perturbative and so cannot be established within a perturbative formulation of string theory. However, it was pointed out in [4] in the context of the heterotic string that there are a number of quantities for which the tree level results are known to be, or believed to be, exact, allowing a check on S-duality by a perturbative, or semi-classical, calculation. We shall show that U-duality for the type II string passes the same tests.

First, for compactifications of the type II string that preserve at least $N = 4$ supersymmetry, the low-energy effective field theory for the massless modes is a supergravity theory whose form is determined uniquely by its local symmetries and is therefore not changed by quantum corrections. Duality of the string theory therefore implies the duality invariance of the equations of motion of the supergravity theory. This prediction is easily checked because the symmetries of the $N \geqslant 4$ supergravity/matter theories have been known for some time. In particular, the equations of motion of $N = 8$ supergravity are U-duality invariant. Second, another quantity that should be, and is, duality invariant is the set of values of electric and magnetic charges allowed by the Dirac–Schwinger–Zwanziger quantization condition. Third, the masses of states carrying electric or magnetic charges satisfy a Bogomolnyi bound which, for the compactifications considered here, is believed to be unrenormalized to arbitrary order in the string coupling constant. Duality invariance of the string theory requires this bound to be duality invariant. For soliton states the Bogomolnyi bound can be found from a classical bound on field configurations of the effective supergravity theory carrying electric or magnetic charges that generalizes the bound obtained in [8] for Maxwell–Einstein theory. We present this bound for $N = 8$ supergravity and show that it is U-duality invariant. Fourth, the spectrum of 'Bogomolnyi states' saturating the Bogomolnyi bound should also be duality invariant. These states include winding and momentum modes of the fundamental string and those found from quantization of solitons. We shall assume that soliton solutions of the type II string can be identified with those of its effective $N = 8$ supergravity theory, and these, as we shall see, include various types of extreme black hole [1].

One of the main concerns of this paper will be the Bogomolnyi states of the type II string theory that break half the supersymmetry. As we shall see, the soliton states in this category arise from quantization of a particular class of extreme black hole solution of $N = 8$ supergravity. It is essentially automatic that all soliton states of the type II string fall into representations of the U-duality group because this is a symmetry of the equations of motion of which the solitons are solutions [2]. A similar argument can be made for solitons of the heterotic string; for example, extreme black hole solutions of the low-energy field theory corresponding to the heterotic string fit into $SL(2;\mathbb{Z})$ representations [10]. There are two points to bear in mind, however. First, a duality transformation not only produces new soliton solutions from old but also changes the vacuum, as the vacuum is parameterised by the scalar expectation values and these change under duality. We shall assume, as in [4], that the new soliton state in the new vacuum can be continued back to give a new soliton state in the old vacuum with duality transformed electric and magnetic charges; this is certainly possible at the level of solutions of the low-energy effective action, since the extreme black hole solutions depend analytically on the scalar expectation values. Combining U-duality transformations with analytic continuations of the scalar field zero-modes in this way gives an $E_7(\mathbb{Z})$ invariance of the spectrum of soliton states in a given vacuum. (Note that whereas U-duality preserves masses, combining this with a scalar zero-mode continuation gives a transformation which changes masses and so is not an invariance of the hamiltonian.) Second, the four-dimensional metrics of many extreme black hole solitons are only defined up to a conformal rescaling by the exponential of a scalar field function that vanishes at spatial infinity. While the 'Einstein' metric is duality invariant, other metrics in the same conformal equivalence class will not be. In general one should therefore think of duality as acting on conformal equivalence classes of metrics, and the issue arises as to which metric within this class is the physically relevant one. As we shall see, for the solutions considered here each conformal class of metrics contains one that is (i) either completely regular or regular outside and on an event horizon and (ii) such that its spatial sections interpolate between topologically distinct vacua. The extreme black hole solutions corresponding to these metrics might reasonably be interpreted as solitons of the theory.

We now encounter an apparent contradiction with U-duality, and with S-duality, of the type II string theory because the fundamental string excitations include additional Bogomolnyi states which apparently cannot be assigned to duality multiplets containing solitons because the soliton multiplets are already complete. The only escape from this contradiction is to make the hypothesis that the

[1] See [9] for a discussion of the interpretation of extreme black holes as solitons.

[2] They also fall into supermultiplets because of the fermion zero-modes in the presence of an extreme black hole [11].

fundamental string states have already been counted among the soliton states. In order for this to be possible there must be soliton states carrying exactly the same quantum numbers as the fundamental Bogomolnyi states. This is indeed the case. The idea that particles with masses larger than the Planck mass, and hence a Compton wavelength less than their Schwarzschild radius, should be regarded as black holes is an old one [12,13], and it has recently been argued that Bogomolnyi states in the excitation spectrum of the heterotic string should be identified with extreme electrically charged dilaton black holes [14,15]. For the heterotic string, approximate solutions of the low-energy effective action include extreme black holes and self-gravitating BPS monopoles [16,17], and it is believed that these correspond to Bogomolnyi solitons of the heterotic string [4]. Any magnetically charged soliton will have an electrically charged soliton partner generated by the action of the $\mathbb{Z}_2$ electromagnetic duality subgroup of S-duality. Now, if the full string theory is S-duality invariant, and this $\mathbb{Z}_2$ subgroup acts on an electrically charged fundamental string state to give a magnetically charged soliton, as argued for the heterotic string in [4], then this fundamental string state must be identified with the corresponding electrically charged soliton. We shall return to these points later but it is worth noting here that solitons of the low-energy effective $N = 4$ or $N = 8$ supergravity theory fit into representations of the $S \times T$ or U-duality as these are symmetries of the supergravity equations of motion, and this is true *irrespective of whether the duality symmetry is actually a symmetry of the full heterotic or type II string theory*.

For compactifications of ten-dimensional string theories one expects solitons of the effective four-dimensional theory to have a ten-dimensional origin. For the type II string we are able to identify the four-dimensional solitons that break half the supersymmetry of $N = 8$ supergravity as six-torus 'compactifications' of the extreme black p-branes of either IIA or IIB ten-dimensional supergravity [18–22]. We note that, in this context, the Bogomolnyi bound satisfied by these states can be seen to arise from the algebra of Noether charges of the effective world-volume action [23]. Remarkably, the solitonic states that are required to be identified with fundamental string states are precisely those which have their ten-dimensional origin in the string soliton or extreme black 1-brane solution, which couples to the same two-form gauge field as the fundamental string. This suggests that we should identify the fundamental ten-dimensional string with the solitonic string. This is consistent with a suggestion made in [24], for other reasons, that the four-dimensional heterotic string be identified with an axion string.

A similar analysis can be carried out for non-toroidal compactification. A particularly interesting example is compactification of the type II superstring on $K_3 \times T^2$ [25] for which the effective four-dimensional field theory turns out to be identical to the effective field theory of the T^6-compactified heterotic string, and in particular has the same $SL(2; \mathbb{Z}) \times O(6, 22; \mathbb{Z})$ duality group. Furthermore, the spectrum of extreme black hole states is also the same. This raises the possibility that the two string theories might be *non-perturbatively* equivalent, even though they differ perturbatively. Such an equivalence would clearly have significant implications.

Finally we consider similar issues in the context of the 11-dimensional supermembrane [26]. This couples naturally to 11-dimensional supergravity [27] and hence to $N = 8$ supergravity after compactification on T^7 and to $N = 4$ supergravity coupled to 22 vector multiplets after compactification on $K_3 \times T^3$ [28]. At present it is not known how to make sense of a quantum supermembrane, so there is little understanding of what the massive excitations might be. However, some progress can be made using the methods sketched above for the string. We shall show that, if the elementary supermembrane is identified with the solitonic membrane solution [29] of 11-dimensional supergravity and account is taken of the solitonic five-brane solution [30], the results of this analysis for the four-dimensional theory are exactly the same as those of the type II string.

2. Charge quantization and the Bogomolnyi bound

Consider the four-dimensional lagrangian

$$L = \sqrt{-g}\left[\tfrac{1}{4}R - \tfrac{1}{2}g_{ij}(\phi)\partial_\mu\phi^i\,\partial^\mu\phi^j - \tfrac{1}{4}m_{IJ}(\phi)F^{\mu\nu I}F^J_{\mu\nu}\right.$$
$$\left. - \tfrac{1}{8}\epsilon^{\mu\nu\rho\sigma}a_{IJ}(\phi)F^I_{\mu\nu}F^J_{\rho\sigma}\right] \qquad (2.1)$$

for a spacetime 4-metric $g_{\mu\nu}$, scalars ϕ^i taking values in a sigma-model target space $\mathcal{M}$ with metric $g_{ij}(\phi)$, and k abelian vector fields A^I_μ with field strengths $F^I_{\mu\nu}$. The scalar functions $m_{IJ} + ia_{IJ}$ are entries of a positive definite $k \times k$ hermitian matrix. The bosonic sector of all supergravity theories without scalar potentials or non-abelian gauge fields can be put in this form. We shall be interested in those cases for which the equations of motion are invariant under some symmetry group G, which is necessarily a subgroup of $Sp(2k; \mathbb{R})$ [31] and an isometry group of $\mathcal{M}$. Of principal interest here are the special cases for which $\mathcal{M}$ is the homogeneous space G/H where H is the maximal compact subgroup of G. These cases include many supergravity theories, and all those with $N \geqslant 4$ supersymmetry. For $N = 4$ supergravity coupled to m vector multiplets $k = 6 + m$, $G = SL(2; \mathbb{R}) \times O(6, m)$ and $H = U(1) \times O(6) \times O(m)$. For $N = 8$ supergravity, $k = 28$, $G = E_{7(7)}$ and $H = SU(8)$. For the 'exceptional' $N = 2$ supergravity [32], $k = 28$, $G = E_{7(-25)}$ and $H = E_6 \times U(1)$.

Defining

$$G_{\mu\nu I} = m_{IJ} * F^J_{\mu\nu} + a_{IJ}F^J_{\mu\nu}, \qquad (2.2)$$

where $* F^I_{\mu\nu} = \tfrac{1}{2}\epsilon_{\mu\nu\rho\sigma}F^{\rho\sigma I}$, the A^I_μ field equations and Bianchi identities can be written in terms of the the $2k$-vector of two-forms

$$\mathcal{F} = \begin{pmatrix} F^I \\ G_I \end{pmatrix} \qquad (2.3)$$

as simply $d\mathcal{F}=0$. The group G acts on the scalars through isometries of $\mathcal{M}$ and on $\mathcal{F}$ as $\mathcal{F}\to\Lambda\mathcal{F}$ where $\Lambda\in G\subseteq Sp(2k;\mathbb{R})$ is a $2k\times 2k$ matrix preserving the $2k\times 2k$ matrix

$$\Omega=\begin{pmatrix}0 & \mathbb{1}\\ -\mathbb{1} & 0\end{pmatrix}. \tag{2.4}$$

An alternative way to represent the G/H sigma-model is in terms of a G-valued field $V(x)$ which transforms under rigid G-transformations by right multiplication and under local H-transformations by left multiplication:

$$V(x)\to h(x)V(x)\Lambda^{-1}, \qquad h\in H, \quad \Lambda\in G. \tag{2.5}$$

The local H-invariance can be used to set $V\in G/H$. Note that $\bar{\mathcal{F}}\equiv V\mathcal{F}$ is G-invariant. In most cases of interest, the scalar coset space can be parameterised by the complex scalars $z_{IJ}=a_{IJ}+im_{IJ}$, which take values in a generalized upper half-plane (m_{IJ} is positive definite), and the group G acts on z_{IJ} by fractional linear transformations. (This can be seen for $N=8$ supergravity as follows. In the symmetric gauge [7], the coset is parameterised by a scalar y_{IJ} which transforms under fractional linear transformations under G. However, z_{IJ} is related to y_{IJ} by a fractional linear transformation, $z_{IJ}=i(\mathbb{1}+\bar{y})/(\mathbb{1}-\bar{y})$, so that z in turn transforms under G by fractional linear transformations. Similar results follow for $N<8$ supergravities by truncation.)

We now define the charges

$$Q^I=\oint_\Sigma *F^I,\quad p_I=\frac{1}{2\pi}\oint_\Sigma F^I,\quad q_I=\oint_\Sigma G_I \tag{2.6}$$

as integrals of two-forms over a two-sphere Σ at spatial infinity. The charges p^I and g_I are the magnetic charges and the Noether electric charges, respectively. The charges Q^I are the electric charges describing the $1/r^2$ fall-off of the radial components of the electric fields, F^I_{0r}, and incorporate the shift in the electric charge of a dyon due to non-zero expectation values of axion fields [33]. Indeed, if the scalars ϕ^i tend to constant values $\bar{\phi}^i$ at spatial infinity, then

$$q_I=m_{IJ}(\bar{\phi})Q^J+a_{IJ}(\bar{\phi})p^J. \tag{2.7}$$

The charges (p^I,q_I) can be combined into a $2k$-vector

$$\mathcal{Z}=\oint_\Sigma\mathcal{F}=\begin{pmatrix}p^I\\ q_I\end{pmatrix}, \tag{2.8}$$

from which it is clear that $\mathcal{Z}\to\Lambda\mathcal{Z}$ under G.

The Dirac–Schwinger–Zwanziger (DSZ) quantization condition (with $\hbar=1$) for two dyons with charge vectors $\mathcal{Z}$ and $\mathcal{Z}'$ is

$$\mathcal{Z}^{\mathrm{T}}\Omega\mathcal{Z}'\equiv p^Iq'_I-p'^Iq_I=\nu, \tag{2.9}$$

for some integer ν. This quantization condition is manifestly G-invariant as $G\subseteq Sp(2k;\mathbb{R})$. However, it has implications for the quantum theory only if there exist both electric and magnetic charges. If, for example, there are no magnetic charges of one type then (2.9) places no constraint on the values of the corresponding electric charge. For the cases of interest to us here, we will show that there exist electric and magnetic charges of all types. We shall now proceed with our analysis of the general case assuming all types of charge exist and this, together with the quantization condition (2.9), implies that the Noether electric charges q_I lie in some lattice Γ and that the magnetic charges p^I lie in the dual lattice $\tilde{\Gamma}$. The group G is therefore broken to the discrete subgroup G($\mathbb{Z}$) which has the property that a vector $\mathscr{Z} \in \Gamma \oplus \tilde{\Gamma}$ is taken to another vector in the same self-dual lattice. The subgroup of Sp($2k$) preserving the lattice is Sp($2k$; $\mathbb{Z}$), so that the duality group is

$$\mathrm{G}(\mathbb{Z}) = \mathrm{G} \cap \mathrm{Sp}(2k; \mathbb{Z}). \tag{2.10}$$

For compact G, G($\mathbb{Z}$) is a finite group, while for non-compact G, it is an infinite discrete group. If we choose a basis for the fields A^I so that the electric charges, and hence the magnetic charges, are integers, then the lattice $\Gamma \oplus \tilde{\Gamma}$ is preserved by integer-valued matrices, so that Sp($2k$; $\mathbb{Z}$) consists of integer-valued $2k \times 2k$ matrices preserving Ω, and G($\mathbb{Z}$) is also represented by integer-valued $2k \times 2k$ matrices. Note that the group G($\mathbb{Z}$) is independent of the geometry of the lattice, as any two lattices Γ, Γ' are related by a GL(k, $\mathbb{R}$) transformation, so that the corresponding discrete groups G($\mathbb{Z}$), G′($\mathbb{Z}$) are related by GL(k, $\mathbb{R}$) conjugation and so are isomorphic. For $N = 4$ supergravity coupled to 22 vector multiplets, G($\mathbb{Z}$) is precisely the S × T duality group SL(2; $\mathbb{Z}$) × O(6, 22; $\mathbb{Z}$) of the toroidally compactified heterotic string, which was observed previously to be the quantum symmetry group of this effective field theory [4]. For $N = 8$ supergravity, G($\mathbb{Z}$) is a discrete subgroup of $\mathrm{E}_{7(7)}$ which we shall call $\mathrm{E}_{7(7)}(\mathbb{Z})$ and abbreviate to $\mathrm{E}_7(\mathbb{Z})$. It can be alternatively characterized as the subgroup of Sp(56; $\mathbb{Z}$) preserving the invariant quartic form of $\mathrm{E}_{7(7)}$. From the explicit form of this invariant given in [7], it is straightforward to see that $\mathrm{E}_7(\mathbb{Z})$ contains an SL(8, $\mathbb{Z}$) subgroup. We also have

$$\mathrm{E}_7(\mathbb{Z}) \supset \mathrm{SL}(2; \mathbb{Z}) \times \mathrm{O}(6, 6; \mathbb{Z}), \tag{2.11}$$

so that $\mathrm{E}_7(\mathbb{Z})$ contains the T-duality group of the toroidally compactified type II string. The minimal extension of the S-duality conjecture for the heterotic string would be to suppose that the SL(2; $\mathbb{Z}$) factor extends to an S-duality group of the type II string, but it is natural to conjecture that the full discrete symmetry group is the much larger U-duality group $\mathrm{E}_7(\mathbb{Z})$. $\mathrm{E}_7(\mathbb{Z})$ is strictly larger than SL(2; $\mathbb{Z}$) × O(6, 6; $\mathbb{Z}$), as it also contains an SL(8, $\mathbb{Z}$) subgroup. In the next section we shall verify some predictions of U-duality for the spectrum of states saturating a gravitational version of the Bogomolnyi bound, i.e. the 'Bogomolnyi states'. However, before turning to the spectrum we should verify that the Bogomolnyi bound is itself U-duality invariant, since otherwise a U-duality transformation could take a state in the Bogomolnyi spectrum to one that is not in this spectrum.

Consider first the cases of pure $N = 4$ supergravity (without matter coupling) and $N = 8$ supergravity, for which $\mathscr{M} = \mathrm{G/H}$ and $k = N(N-1)/2$. We define $Y_{mn} = t^I_{mn}(\bar{q}_I + i\bar{p}^{-I})$ where $(\bar{p}^I, \bar{q}_I)$ are the components of the $2k$-vector $\bar{\mathscr{Z}} = \bar{V}\mathscr{Z}$ and $\bar{V}$ is the constant asymptotic value of the G-valued field V at spatial infinity.

Here m, $n = 1, \ldots, N$ and $t^I_{mn} = -t^I_{nm}$ are the matrices generating the vector representation of SO(N). The Y_{mn} appear in a global supersymmetry algebra as central charges [34,8,9,35] and this allows a derivation of a Bogomolnyi bound. The antisymmetric complex $N \times N$ matrix Y_{mn} has $N/2$ complex skew eigenvalues λ_a, $a = 1, \ldots, N/2$, and the bound on the ADM mass of the Maxwell–Einstein theory [8] can be generalized to [35]

$$M_{\mathrm{ADM}} \geqslant \max |\lambda_a|. \tag{2.12}$$

Since $\mathcal{Z} \to \Lambda\mathcal{Z}$ and $\bar{V} \to \bar{V}\Lambda^{-1}$ under G, it follows that $\bar{\mathcal{Z}}$ and the λ_a are invariant under duality transformations, so the bound (2.12) is manifestly G-invariant. In the quantum theory this bound translates to a bound on the mass of the corresponding quantum state. Similar results apply to the case of $N = 4$ super-matter coupled to supergravity, with the difference that $t^I_{mn} = -t^I_{nm}$ are now certain scalar-field dependent matrices that 'convert' the SO(6, m) index I to the SO(6) composite index mn. Nevertheless, the charges λ_a remain duality invariant.

If the moduli of all the eigenvalues are equal, $|\lambda_{a_1}| = |\lambda_{a_2}| = \ldots = |\lambda_{gN/2}|$, then the bound (2.12) is equivalent to

$$M_{\mathrm{ADM}} \geqslant \sqrt{\frac{2}{N}}\sqrt{|\bar{\mathcal{Z}}|^2}, \tag{2.13}$$

where

$$|\bar{\mathcal{Z}}|^2 = \sum_a |\lambda_a|^2 = \frac{1}{2}\bar{Y}_{mn}\bar{Y}^{mn} = G_{IJ}\bar{p}^I\bar{p}^J + G_{IJ}\bar{q}_I\bar{q}_J \tag{2.14}$$

and $G_{IJ} = \frac{1}{2}t^I_{mn}t^J_{mn}$ is the identity matrix for pure supergravity, but is scalar dependent for the the matter-coupled $N = 4$ theory. However, in the general case of different eigenvalues, the bound (2.14) is strictly weaker than (2.12). If M_{ADM} is equal to the modulus of r of the eigenvalues λ_a, $M_{\mathrm{ADM}} = |\lambda_{a1}| = |\lambda_{a2}| = \ldots = |\lambda_{ar}|$, for some r with $0 \leqslant r \leqslant N/2$, then the soliton with these charges spontaneously breaks the N original supersymmetries down to r supersymmetries, so that for solitons for which $r = N/2$ precisely half of the N supersymmetries are preserved and the bound (2.12) is equivalent to (2.14). The duality invariance of the bound (2.14) for $N = 4$ was previously pointed out in [3,4].

3. Spectrum of Bogomolnyi states

There are many massive states in the spectrum of toroidally compactified string theories. The masses of those which do not couple to any of the U(1) gauge fields cannot be calculated exactly. This is also true in general of those that do couple to one of the U(1) gauge fields, but the masses of such particles are bounded by their charges, as just described. It is believed that the masses of string states that saturate the bound are not renormalized for theories with at least $N = 4$ supersymmetry. If this is so then these masses can be computed exactly. Such 'Bogomolnyi states' arise in the theory from winding and Kaluza–Klein modes of the fundamental string, and from quantization of non-perturbative soliton solutions of the string theory. The latter include extreme black holes and, for the heterotic string, self-gravitating BPS monopoles.

For generic compactifications of both heterotic and type II strings there are 28 abelian gauge fields and so a possible 56 types of electric or magnetic charge. We shall identify solitons of the effective supergravity theory carrying each type of charge, thereby justifying the quantization condition on these charges. These solitons are various types of extreme black holes. Initially, at least, we shall be interested in solitons carrying only one type of charge, in which case we should consistently truncate the supergravity theory to one with only one non-zero field strength, F. The coefficients of the F^2 terms can then be expressed in terms of a scalar field σ and a pseudoscalar field ρ (which are two functions of the ϕ^i) such that the truncated field theory has an action of the form

$$S = \int \mathrm{d}^4x\sqrt{-g}\left[\tfrac{1}{4}R + \tfrac{1}{4}\,\mathrm{e}^{-2a\sigma}F_{\mu\nu}F^{\mu\nu} + \tfrac{1}{4}\rho F_{\mu\nu} * F^{\mu\nu} + L(\sigma, \rho)\right], \tag{3.1}$$

where $L(\sigma, \rho)$ is the lagrangian for a scalar sigma-model and a is a constant. One can choose $a \geqslant 0$ without loss of generality since a is changed to $-a$ by the field redefinition $\sigma \to -\sigma$. For every value of a the equations of motion of (3.1) admit extreme multi-black hole solutions [36], parameterised by the asymptotic values of σ,ρ, which are arbitrary integration constants. There is an intrinsic ambiguity in the metric of the $a \neq 0$ extreme black hole solutions because a new metric can be constructed from the canonical metric (appearing in the action (3.1)) by a conformal rescaling by a power of e^{σ}. The general metric in this conformal equivalence class will not have an interpretation as a 'soliton' in the sense for which the $a = 0$ extreme Reissner–Nordstrom (RN) black hole is a soliton. One feature that is generally expected from a soliton is that it interpolates between different vacua: in the RN case these are the Minkowski spacetime near spatial infinity and the Robinson–Bertotti vacuum down an infinite Einstein–Rosen 'throat'. If we require of the $a \neq 0$ extreme black holes that they have a similar property then one must rescale the canonical metric $\mathrm{d}\hat{s}^2$ by $\mathrm{e}^{2a\sigma}$, after which one finds, for vanishing asymptotic values of σ and ρ, the solution

$$\begin{aligned}
\mathrm{d}s^2 &= \mathrm{e}^{2a\sigma}\,\mathrm{d}\hat{s}^2 \\
&= -\left(1 - \frac{(1+a^2)M}{r}\right)^{2(1-a^2)/(1+a^2)} \mathrm{d}t^2 \\
&\quad + \left(1 - \frac{(1+a^2)M}{r}\right)^{-2} \mathrm{d}r^2 + r^2\,\mathrm{d}\Omega_2^2, \\
\mathrm{e}^{-a\sigma} &= \left(1 - \frac{(1+a^2)M}{r}\right)^{a^2/(1+a^2)}, \quad \rho = 0,
\end{aligned} \tag{3.2}$$

where M is the ADM mass and $d\Omega_2^2$ is the metric on the unit 2-sphere. When $a = 1$ and σ is the dilaton field this rescaling of the canonical metric is exactly what is required to get the so-called 'string metric', so that the $a = 1$ black holes have a natural interpretation as string solitons. This might make it appear that the rescaling of the canonical metric by $e^{2a\sigma}$ is inappropriate to string theory when $a \neq 1$, but it must be remembered that the scalar field σ is not necessarily the dilaton but is, in general, a combination of the dilaton and modulus fields of the torus and gauge fields. Indeed, it was shown in [37] that for the $a = \sqrt{3}$ black holes this combination is such that the effective rescaling is just that of (3.2). For any value of a this metric has an internal infinity as $r \to (1 + a^2)M$ for constant t. For $a < 1$ the surface $r = (1 + a^2)M$ is an event horizon, but this event horizon is regular only if $2(1 - a^2)/(1 + a^2)$ is an integer, which restricts the values of a less than unity to $a = 0$ or $a = 1/\sqrt{3}$. The $a = 0$ case is the extreme RN black hole for which the soliton interpretation is widely accepted. The significance of the $a = 1/\sqrt{3}$ case has been explained in [38]. For $a \geqslant 1$ the surface $r = (1 + a^2)M$ is at infinite affine parameter along any geodesic, so one might admit all values of $a \geqslant 1$. On the other hand, the relevance of geodesic completeness is not clear in this context so one might still wish to insist that $2(1 - a^2)/(1 + a^2)$ be an integer so that the null surface $r = (1 + a^2)M$ is regular, in which case only the further values of $a = 1$ and $a = \sqrt{3}$ can be admitted. Curiously, the values

$$a = 0,\ 1/\sqrt{3},\ 1,\ \sqrt{3}, \tag{3.3}$$

which we find in this way by demanding that the solution (3.2) is a *bona fide* soliton, also arise from truncation of $N = 8$ supergravity. The possibility of the values $a = 0$ and $a = 1$ is guaranteed by the existence of consistent truncations of $N = 8$ supergravity to $N = 2$ and $N = 4$ supergravity, respectively. The possibility of the values $a = \sqrt{3}$ and $a = 1/\sqrt{3}$ is guaranteed by the existence of a consistent truncation of the maximal five-dimensional supergravity to simple five-dimensional supergravity since the subsequent reduction to four dimensions yields just these values.

Consider first the $a = 0$, electric and magnetic extreme RN black holes. Given any one such black hole with integral charge, an infinite number can be generated by acting with $G(\mathbb{Z})$, and these will include black holes carrying each of the 56 types of charge [35], and this is already sufficient to show that the continuous duality group $E_{7(7)}$ is broken to a discrete subgroup. These solutions break 3/4 of the supersymmetry in the $N = 4$ theories and 7/8 of the supersymmetry in the $N = 8$ case. For the remainder of the paper, we shall restrict ourselves to solitons which break half the supersymmetry, and the only extreme black hole solutions of this type are those with $a = \sqrt{3}$. This follows from consideration of the implications of supersymmetry for the moduli space of multi-black hole solutions. This multi-soliton moduli space is the target space for an effective sigma-model describing non-relativistic solitons [39]. This sigma-model must have 8 supersymmetries for solitons of a four-dimensional $N = 4$ supergravity theory that break half the supersymmetry, and this implies that the moduli space is hyper-Kähler. Similarly, the moduli space for multi-solitons of $N = 8$ supergravity that break half the supersymmetry is the target space for a sigma-model with 16 supersymmetries, and this implies that the moduli space is flat. However, the moduli space of multi-black hole solutions is flat if and only if $a = \sqrt{3}$ [40,41], so only these extreme black holes can be solutions of $N = 8$ supergravity that break half the supersymmetry. An alternative characterization of these extreme black holes is as 'compactifications' of the extreme black p-brane solitons of the ten-dimensional supergravity theory, which are known to break half the supersymmetry [42]. It follows that the moduli space of these solutions must be flat, and what evidence there is [43] confirms this prediction. This ten-dimensional interpretation of the solitons discussed here will be left to the following section where it will also become clear that they carry combinations of all 28 + 28 electric and magnetic charges associated with the 28 U(1) gauge fields.

This moduli space argument shows, incidentally, that whereas the flatness of the moduli space for solitons that break half the supersymmetry is protected by supersymmetry for $N = 8$ supergravity, this is not so for $N = 4$ theories. There is then no reason to expect the moduli space metric of extreme black hole solitons of the exact heterotic string theory (to all orders in α' and g) to be flat. Indeed, the $a = \sqrt{3}$ extreme black holes, which have a flat moduli space, are only approximate solutions of the heterotic string and are expected to receive higher order corrections. Furthermore, if BPS-type monopoles were to occur in the type II string theory, a possibility that is suggested by the occurrence of non-abelian gauge groups in some versions of the compactified type II string [44], they would have to break more than half the supersymmetry as their moduli space is not flat. This is in accord with the fact that the four-dimensional type II strings of [44] have at most $N = 4$ supersymmetry, so that solitons of these theories saturating a Bogomolnyi bound would have less than $N = 4$ supersymmetry. This provides further justification for our assumption that the solitons of the toroidally compactified type II string that break half the $N = 8$ supersymmetry are those of the effective $N = 8$ supergravity theory.

The complete set of soliton solutions of a supergravity theory fills out multiplets of the duality group G($\mathbb{Z}$), as mentioned in the introduction. We shall now explain this in more detail. Flat four-dimensional spacetime with the scalar fields ϕ^i taking constant values, ϕ_0^i, is a vacuum solution of the supergravity theory parameterised by these constants. The duality group acts non-trivially on such vacua as it changes the ϕ_0^i. The solitons for which the scalar fields tend asymptotically to the values ϕ_0^i provide the solitonic Bogomolnyi states about the vacuum state $|\phi_0\rangle$. A G($\mathbb{Z}$) transformation takes a Bogomolnyi state in this vacuum with charge vector $\mathcal{Z}$ to another Bogomolnyi state with charge vector $\mathcal{Z}'$ and equal mass but in a new vacuum $|\phi_0'\rangle$. As in [4], it will be assumed that one can smoothly continue from ϕ_0' to ϕ_0 without encountering a phase transition, to obtain a state with the charge vector $\mathcal{Z}'$, but a different mass in general, about the original vacuum $|\phi_0\rangle$. This assumption seems reasonable because the extreme black hole solutions depend analytically on the constants ϕ_0^i. We thus obtain a new Bogomolnyi soliton solution about the original vacuum but with a G($\mathbb{Z}$) transformed charge vector. The spectrum of all the Bogomolnyi states obtained in this way is G($\mathbb{Z}$) invariant by construction. In particular, the number of these Bogomolnyi states with charge vector $\mathscr{Z}$ will be the same as the number with charge vector $\mathscr{Z}'$ whenever $\mathscr{Z}$ is related to $\mathscr{Z}'$ by a $G(\mathbb{Z})$ transformation.

In addition to the Bogomolnyi states that arise from solitons, there are also the electrically charged Bogomolnyi states of the fundamental string. These states are purely perturbative and for the type II string they consist of the Kaluza–Klein (KK) and winding modes of the string. If they are also to fit into multiplets of the duality group they must have magnetically charged partners under duality, and these should be non-perturbative, i.e. solitonic. The soliton duality multiplets are, however, already complete for the reason just given. In order to have duality of the string theory we must therefore identify the fundamental states with electrically charged solitonic states. We shall see in the next section how this identification must be made.

It might be thought that all electrically charged soliton states should have an equivalent description in terms of fundamental states. This is presumably true of the heterotic string since there are fundamental string states carrying each of the 28 types of electric charge and these are related by the T-duality group $O(6, 22; \mathbb{Z})$. In contrast, the fundamental modes of the type II string carry only 12 of the possible 28 electric charges, because the 16 Ramond–Ramond (RR) U(1) gauge fields couple to the string through their field strengths only. The 12 string-mode electric charges are related by the T-duality group $O(6, 6; \mathbb{Z})$ of the type II string. It would be consistent with S- and T-duality to suppose that there are no charged states coupling to the 16 (RR) gauge fields, but this would not be consistent with U-duality, as we now show.

Recall that an n-dimensional representation of G gives an action of G on $\mathbb{R}^n$ which restricts to an action of $G(\mathbb{Z})$ on the lattice $\mathbb{Z}^n$. For both the heterotic and type II strings, the charge vector $\mathscr{Z}$ transforms under G as a 56-dimensional representation. For the heterotic string, $G = SL(2; \mathbb{R}) \times O(6, 22)$ and $\mathscr{Z}$ transforms according to its irreducible **(2, 28)** representation. This has the decomposition

$$(\mathbf{2}, \mathbf{28}) \rightarrow (\mathbf{2}, \mathbf{12}) + 16 \times (\mathbf{2}, \mathbf{1}) \tag{3.4}$$

in terms of representations of $SL(2; \mathbb{R}) \times O(6, 6)$. This is to be compared with the type II string for which $G = E_{7(7)}$ and $\mathscr{Z}$ transforms according to its irreducible **56** representation, which has the decomposition

$$\mathbf{56} \rightarrow (\mathbf{2}, \mathbf{12}) + (\mathbf{1}, \mathbf{32}) \tag{3.5}$$

under $SL(2; \mathbb{R}) \times O(6, 6)$. In both cases there is a common sector corresponding to the **(2, 12)** representation of $SL(2; \mathbb{R}) \times O(6, 6)$, plus an additional 32-dimensional representation corresponding, for the heterotic string, to the charges for the additional $U(1)^{16}$ gauge group and, for the type II strings, to the charges for the Ramond–Ramond (RR) sector gauge fields. It is remarkable that the latter fit into the irreducible spinor representation of $O(6, 6)$. These decompositions of the **56** representation of G induce corresponding decompositions of representations of $G(\mathbb{Z})$ into representations of $SL(2; \mathbb{Z}) \times O(6, 6; \mathbb{Z})$ on the charge lattice $\mathbb{Z}^{56}$. In particular, *U-duality requires the 16 + 16 electric and magnetic charges of the RR sector to exist and to transform irreducibly under the action of the T-duality group* $O(6, 6; \mathbb{Z})$, and we conclude that all charges in the RR sector must be carried by solitons. We shall later confirm this.

4. *p*-Brane interpretation of Bogomolnyi solitons

We have seen that the solitons of toroidally compactified superstrings fit into representations of the duality group $G(\mathbb{Z})$. Our concern here will be to identify states that break half the supersymmetry and carry just one of the 56 types of electric or magnetic charge. We shall call such states for which the charge takes the minimum value 'elementary'; acting on these with the duality group $G(\mathbb{Z})$ will generate a lattice of charged states. Here we wish to show how the elementary solitons arise from extreme black *p*-brane solitons of the ten-dimensional effective supergravity theory. These may be of electric or magnetic type. Electric *p*-brane solitons give electrically charged solitons of the four-dimensional dimensionally reduced field theory, while magnetic ones give magnetic monopoles, provided we use the form of the four-dimensional supergravity theory that comes directly from dimensional reduction without performing any duality transformations on the one-form gauge fields (although we convert two-form gauge fields to scalar fields in the usual way). If we had chosen a different dual form of action, the solutions would be the same, but some of the electric charges would be viewed as magnetic ones, and vice versa. This form of the action is manifestly invariant under T-duality: for the heterotic string, the action is the $O(6, 22)$ invariant one given in [45], which is related to the one of [46] by a duality transformation, and for the type II string, it is a new $O(6, 6)$ invariant form of the $N = 8$ supergravity action which is related to the $SL(8, \mathbb{R})$ invariant Cremmer–Julia action [7] by a duality transformation.

An extreme *p*-brane soliton of the ten-dimensional low-energy field theory has a metric of the form [21]

$$ds^2 = A(r)(-dt^2 + dx^i\, dx^i) + B(r)\, dr^2 + r^2\, d\Omega^2_{8-p}, \tag{4.1}$$

where x^i $(i = 1, \ldots, p)$ are p flat euclidean dimensions, $d\Omega^2_{8-p}$ is the metric on an $(8-p)$-sphere, r is a radial coordinate, t is a time coordinate and $A(r), B(r)$ are two radial functions that tend to unity as $r \to \infty$. These solitons couple either to an antisymmetric tensor gauge field A_r of rank $r = 7 - p$, in which case the *p*-brane is magnetically charged and $F = dA$ is proportional to the $(8-p)$-sphere volume form ϵ_{8-p}, or one of rank $r = p - 3$, in which case the brane is electric and $*F$ is proportional to ϵ_{8-p}. In some cases, the *p*-brane solutions will have corrections of higher order in α', but some of the solutions correspond to exact conformal field theories.

We shall be interested in four-dimensional solitons obtained by 'compactifying' *p*-brane solitons on the six-torus. Compactification on T^p is straightforward since one has only to 'wrap' the p-brane around the p-torus, which is achieved by making the appropriate identifications of the x^i coordinates. If $p < 6$, a soliton in four dimensions can then be found by taking periodic arrays on $T^{(6-p)}$ and making a periodic identification [3]. For example [47], a five-brane can be wrapped around a five-torus in six ways giving rise to six types of five-dimensional soliton, and these yield six types of black hole soliton in four dimensions on taking periodic arrays. Similarly, to 'compactify' a 0-brane (i.e. a 10-dimensional black hole) on a six-torus one first introduces a 6-dimensional periodic array of such black holes and periodically identifies. Instead of wrapping all p dimensions of a p-brane to obtain a point-like 0-brane in $3+1$ dimensions, one can wrap $p-q$ dimensions to obtain a q-brane soliton in $3+1$ dimensions; however, in what follows we shall restrict ourselves to 0-brane solitons in 4 dimensions.

The bosonic sectors of the ten-dimensional effective field theories of the heterotic and type IIA and type IIB superstrings each include a metric, g_{MN}, an antisymmetric tensor gauge field, b_{MN}, and a dilaton field, Φ. We shall first discuss this common sector of all three theories and then turn to the additional sectors characteristic of each theory. We expect the solutions we describe to be exact solutions of the classical type II theory, and their masses to be unrenormalized in the quantum theory, but for the heterotic string they are only approximate solutions (to lowest order in α') of the low-energy field theory.

Dimensional reduction of the common (g, b, Φ) sector on T^6 yields 6 Kaluza–Klein abelian gauge fields $(g_{\mu i} + \ldots)$ coming from g_{MN} and another 6 abelian gauge fields $(b_{\mu i} + \ldots)$ coming from b_{MN}. It is straightforward to identify the magnetically charged solitons associated with the KK gauge fields. These are the KK monopoles [48], consisting of the product of a self-dual Taub-NUT instanton, with topology $\mathbb{R}^4$, with a 5-torus and a time-like $\mathbb{R}$. As this is the product of a five-metric with a five-torus, this can also be viewed as a five-brane solution of the ten-dimensional theory wrapped around a five-torus [4]. There are six types of KK monopoles in four dimensions, one for each of the six KK gauge fields, because the five-brane can be wrapped around the six-torus in six different ways. As four-dimensional solutions the KK monopoles are extreme black holes with $a = \sqrt{3}$, as expected from the moduli space argument of the previous section. The elementary magnetically charged solitons associated with the $b_{\mu i}$ gauge fields can be identified with the six possible 'compactifications' of the extreme black five-brane [20,19] of the ten-dimensional (g, b, Φ) theory. We shall refer to these as abelian H-monopoles; they were first given in [49] and have been discussed further in [50,47,22]. It

[3] Alternatively, since the solution of extreme black p-branes always reduces to the solution of the Laplace equation in the transverse space, one has only to solve this equation on $\mathbb{R}^3 \times T^{(6-p)}$ instead of $\mathbb{R}^{(9-p)}$ to find solitons of the four-dimensional theory.

[4] For fixed r, t, the solution has topology $S^3 \times T^5$, and the S^3 can be regarded as a Hopf bundle of S^1 over S^2. Thus locally it is $S^2 \times T^6$, so that this solution might also be thought of as a twisted 6-brane.

is straightforward to check directly that the KK monopoles and the H-monopoles are indeed related by T-duality, as expected [51,52]. Note that we have not included KK modes of the 5-brane, i.e. configurations in which the 5-brane has momentum in some of the toroidal directions, as these either lead to extended objects in four dimensions or to localised solitons that carry more than one type of charge and so are not elementary.

The KK and abelian H-monopoles have electric duals. These electrically charged solitons have their ten-dimensional origin in the extreme black string [18] of the (g, b, Φ) theory, which is dual [53–55] to the extreme black five-brane. The 6 electric duals to the abelian H-monopoles are found by wrapping the solitonic string around the 6-torus, i.e. the 6 winding modes of the solitonic string. The electric duals of the KK monopoles come from Kaluza–Klein modes of the 1-brane, i.e. configurations in which the solitonic string has momentum in the toroidal directions. They can be thought of as pp-waves travelling in the compactified directions [11]. These 6 + 6 elementary electrically charged solitons couple to the 6 + 6 KK and $b_{\mu i}$ gauge fields. They are in one to one correspondence with the KK (i.e. torus momentum modes) and winding states of the fundamental string which couple to the same 12 gauge fields. This allows us, in principle, to identify the fundamental string states as soliton states and, as explained in earlier sections, U-duality of string theory forces us to do so.

Before turning to solitons of the additional sector of each string theory, we shall first explain here why these field theory solitons are exact solutions of type II string theory. Type II string theory in a (g, b, Φ) background is described by a non-linear sigma-model with (1, 1) world-sheet supersymmetry. The KK monopole background is described by a (4, 4) supersymmetric sigma-model plus a free (1, 1) supersymmetric field theory; this is conformally invariant [56] and so gives an exact classical solution of string theory. The pp-wave background is also an exact classical solution [57], so that the T-duals of these two solutions must be exact classical solutions too. In contrast, the heterotic string in a (g, b, Φ) background is described by a (1, 0) supersymmetric sigma-model, and at least some of the solutions described above only satisfy the field equations to lowest order in α'. In some cases, as we will describe later, these solutions can be modified to obtain exact classical heterotic string solutions. However, it is not known in general whether such backgrounds can be modified by higher order corrections to give an exact string solution.

We have now accounted for 12 + 12 of the required 28 + 28 types of charge of all three ten-dimensional superstring theories. We now consider how the additional 16 + 16 charges arise in each of these three theories, starting with the type II string. It is known that, after toroidal compactification, the type IIA and type IIB string theories are equivalent [58], but it is instructive to consider both of them. In either case, we showed in the last section that U-duality requires that the missing 16 + 16 types of charge transform as the irreducible spinor representation of the T-duality group. Since T-duality is a perturbative symmetry, if there were electrically charged states of this type in the fundamental string spectrum, there would also have to be magnetic ones. However, magnetic charges only occur in the soliton sector, so a prediction of U-duality is that the corresponding 16 electric charges are also to be found in the soliton sector and not, as one might have thought, in the elementary string spectrum. We shall confirm this.

First, we consider the type IIA theory. The ten-dimensional bosonic massless fields are the (g, b, Φ) fields of the common sector plus a one-form gauge potential, A_M, and a three-form gauge potential, A_{MNP}. These extra fields appear in the RR sector but couple to the string through their field strengths *only*. Upon compactification to four dimensions, A_M gives one abelian gauge field A_μ and A_{MNP} gives 15 abelian gauge fields A_μ^{ij}. These also couple to the string through their field strengths only and so there are no elementary string excitations that are electrically charged with respect to these 16 gauge fields, as expected. The solitonic p-brane solutions of the ten-dimensional field theory involving A_M or A_{MNP} and breaking only half the supersymmetry consists of a 0-brane, i.e. a (ten-dimensional) extreme black hole, a 2-brane (i.e. a membrane), a 4-brane and a 6-brane. The 0-brane and the 2-brane are of electric type. The 0-brane gives rise to an electrically charged four-dimensional black hole in the toroidally compactified theory by the procedure of taking periodic arrays of the ten-dimensional solution. The membrane gives a total of $6 \times 5/2 = 15$ electric black holes in four dimensions after 'wrapping' it around two directions of the six-torus and then taking periodic arrays to construct a four-dimensional solution. Similarly, the magnetic-type 4-brane and 6-brane can be wrapped around the six-torus (introducing periodic arrays where necessary) to give $15 + 1$ magnetically charged black holes in four dimensions. We have therefore found a total of 32 additional electric and magnetic charges. Combined with the previous 24 charges this gives a total of 56 elementary charged states carrying only one type of charge. From the low-energy field theory we know that these charges transform according to the **56** representation of E_7, and that acting on these elementary solitons with $E_7(\mathbb{Z})$ generates a 56-dimensional charge lattice. As anticipated, the extra $16 + 16$ electric and magnetic charges are inert under S-duality but are mixed by the T-duality group $O(6, 6; \mathbb{Z})$. In addition to the p-brane winding modes discussed above, there are also p-brane momentum modes; however, to give a 0-brane in 4 dimensions, the p-brane must wrap around the torus as well as having internal momentum, so that the resulting soliton would carry more than one type of charge and so would not be elementary; nevertheless, these solitons occur in the charge lattices generated by the elementary solitons.

A similar analysis can be made for the type IIB theory. In this case the extra massless bosonic fields in the ten-dimensional effective field theory are a scalar, a two-form potential, A_{MN}, and a four-form potential $A_{MNPQ}^{(+)}$ with self-dual five-form field strength. As for the type IIA string theory, these gauge fields couple through their field strengths only and so, again, there are no string excitations carrying the new electric charges. In the solitonic sector of the ten-dimensional field theory there is a neutral 5-brane, a self-dual 3-brane and a string, in addition to the string and neutral five-brane of the (g, b, Φ) sector. The new neutral five-brane gives 6 magnetic charges in four dimensions, the self-dual 3-brane gives 10 electric and 10 magnetic charges and the new string gives six electric charges. Note that the new solitonic string couples to the 16 U(1) gauge fields coming from A_{MN} and $A^{(+)}_{MNPQ}$. These 16 + 16 charges couple to A_{MN}, while the fundamental string and the solitonic string of the common sector both couple to b_{MN}; thus it may be consistent to identify the fundamental and common sector solitonic strings, but the new solitonic string cannot be identified with either. As for the type IIA string, all 56 charges generate the irreducible **56** dimensional representation of $E_7(\mathbb{Z})$.

Finally, we turn to the heterotic string. We have seen that the common sector solutions of the low-energy effective supergravity theory include 12 KK and abelian H-monopoles, and their 12 electric duals, and under T-duality these must have 16 + 16 electric and magnetic black hole partners coupling to the 16 remaining U(1) gauge fields. These have a ten-dimensional interpretation as the 0-branes and 8-branes of $N = 1$ ten-dimensional supergravity coupled to 16 abelian vector multiplets [21] (which can be taken to be those of the $U(1)^{16}$ subgroup of $E_8 \times E_8$ or $SO(32)/\mathbb{Z}_2$).

In addition to these black hole solutions, there are also BPS monopole solutions of the heterotic string arising from wrapping heterotic or gauge five-branes around the six-torus [16]. The BPS monopoles are not solutions of the effective supergravity theory with the abelian gauge group, but it has been argued (e.g. in [47]) that there should be modifications of these monopoles that are solutions of the abelian theory. The moduli spaces for multi-soliton solutions of BPS monopoles are hyper-Kähler [59] while those for extreme $a = \sqrt{3}$ black holes are flat [40,41], so that the black holes and BPS monopoles should not be related by duality. If the modified BPS monopoles also have a non-flat moduli space, then they too cannot be dual to black holes. However, it is also possible that they have a flat moduli space, and even that they are equivalent to black hole solutions. The modified BPS monopoles, if they exist, would have electric partners under S-duality which would be electric solitons. The magnetic partners under S-duality of electrically charged Bogomolnyi fundamental string states are expected to be magnetic monopole solitons, which might be either BPS-type solutions, or black holes (or both, if they are equivalent). In either case, the fundamental string states should be identified with the electrically charged solitons related to the magnetic monopoles by S-duality.

Whereas the solutions of the type II string we have discussed are exact conformal field theories, the solutions of the heterotic string are only approximate low-energy solutions. However, some of these heterotic solutions have a compact holonomy group, and for these one can set the Yang–Mills connection equal to the spin-connection so that the sigma-model becomes one with (1, 1) supersymmetry and the resulting background is an exact solution of string theory. Applying this to the five-brane gives the symmetric five-brane solution [50] and this can also be used to construct a 'symmetric KK monopole'. However, we do not know which of the other solutions of the low-energy effective theory can be corrected to give exact solutions, and the duality symmetry of non-abelian phases of the theory are not understood. Moreover the α' corrections to the four-dimensional supergravity action give a theory that is not S-duality invariant, but if string-loop corrections are also included, an S-duality invariant action should arise.

Table 1
Duality symmetries for type II string compactified to d dimensions

Space–time dimension d	Supergravity duality group G	String T-duality	Conjectured full string duality
10A	$SO(1, 1)/\mathbb{Z}_2$	$\mathbb{1}$	$\mathbb{1}$
10B	$SL(2; \mathbb{R})$	$\mathbb{1}$	$SL(2; \mathbb{Z})$
9	$SL(2; \mathbb{R})\times O(1, 1)$	$\mathbb{Z}_2$	$SL(2; \mathbb{Z})\times\mathbb{Z}_2$
8	$SL(3; \mathbb{R})\times SL(2; \mathbb{R})$	$O(2, 2; \mathbb{Z})$	$SL(3; \mathbb{Z})\times SL(2; \mathbb{Z})$
7	$O(5, 5)$	$O(3, 3; \mathbb{Z})$	$O(5, 5; \mathbb{Z})$
6	$SL(5; \mathbb{R})$	$O(4, 4; \mathbb{Z})$	$SL(5; \mathbb{Z})$
5	$E_{6(6)}$	$O(5, 5; \mathbb{Z})$	$E_{6(6)}(\mathbb{Z})$
4	$E_{7(7)}$	$O(6, 6; \mathbb{Z})$	$E_{7(7)}(\mathbb{Z})$
3	$E_{8(8)}$	$O(7, 7; \mathbb{Z})$	$E_{8(8)}(\mathbb{Z})$
2	$E_{9(9)}$	$O(8, 8; \mathbb{Z})$	$E_{9(9)}(\mathbb{Z})$
1	$E_{10(10)}$	$O(9, 9; \mathbb{Z})$	$E_{10(10)}(\mathbb{Z})$

5. Toroidal compactification to other dimensions

In this section, we extend the previous discussion to consider the duality symmetries of type II and heterotic strings toroidally compactified to d dimensions. The resulting low-energy field theory is a d-dimensional supergravity theory which has a rigid 'duality' group G, which is a symmetry of the equations of motion, and in odd dimensions is in fact a symmetry of the action. In each case the massless scalar fields of the theory take values in G/H, where H is the maximal compact subgroup of G. G has an $O(10-d, 10-d)$ subgroup for the type II string, and an $O(10-d, 26-d)$ subgroup for the heterotic string. In either string theory, it is known that this subgroup is broken down to the discrete T-duality group, $O(10-d, 10-d; \mathbb{Z})$ or $O(10-d, 26-d; \mathbb{Z})$. It is natural to conjecture that the whole supergravity duality group G is broken down to a discrete subgroup $G(\mathbb{Z})$ (defined below) in the d-dimensional string theory. We have already seen that this occurs for $d=4$ and will argue that for $d>4$ the symmetry G is broken to a discrete subgroup by a generalization of the Dirac quantization condition. In

Table 2
Duality symmetries for heterotic string compactified to d dimensions

Space–time dimension d	Supergravity duality group G	String T-duality	Conjectured full string duality
10	$O(16)\times SO(1, 1)$	$O(16; \mathbb{Z})$	$O(16; \mathbb{Z})\times\mathbb{Z}_2$
9	$O(1, 17)\times SO(1, 1)$	$O(1, 17; \mathbb{Z})$	$O(1, 17; \mathbb{Z})\times\mathbb{Z}_2$
8	$O(2, 18)\times SO(1, 1)$	$O(2, 18; \mathbb{Z})$	$O(2, 18; \mathbb{Z})\times\mathbb{Z}_2$
7	$O(3, 19)\times SO(1, 1)$	$O(3, 19; \mathbb{Z})$	$O(3, 19; \mathbb{Z})\times\mathbb{Z}_2$
6	$O(4, 20)\times SO(1, 1)$	$O(4, 20; \mathbb{Z})$	$O(4, 20; \mathbb{Z})\times\mathbb{Z}_2$
5	$O(5, 21)\times SO(1, 1)$	$O(5, 21; \mathbb{Z})$	$O(5, 21; \mathbb{Z})\times\mathbb{Z}_2$
4	$O(6, 22)\times SL(2, \mathbb{R})$	$O(6, 22; \mathbb{Z})$	$O(6, 22; \mathbb{Z})\times SL(2; \mathbb{Z})$
3	$O(8, 24)$	$O(7, 23; \mathbb{Z})$	$(8, 24; \mathbb{Z})$
2	$O(8, 24)^{(1)}$	$O(8, 24; \mathbb{Z})$	$O(8, 24)^{(1)}(\mathbb{Z})$

Tables 1 and 2, we list these groups for toroidally compactified superstring theories (at a generic point in the moduli space so that the gauge group is abelian).

For the type II string (Table 1), the supergravity duality groups G are given in [60,61] [5]. The Lie algebra of $E_{9(9)}$ is the $E_{8(8)}$ Kao–Moody algebra, while the algebra corresponding to the E_{10} Dynkin diagram has been discussed in [61,63]. The $d = 2$ duality symmetry contains the infinite-dimensional Geroch symmetry group of toroidally compactified general relativity. In $d = 9$, the conjectured duality group is a product of an SL(2; $\mathbb{Z}$) S-duality and a $\mathbb{Z}_2$ T-duality, while for $d < 8$ we conjecture a unified U-duality. For $d = 8$, the T-duality group O(2, 2; $\mathbb{Z}$) $\sim$ [SL(2; $\mathbb{Z}$) $\times$ SL(2; $\mathbb{Z}$)]/$\mathbb{Z}_2 \times \mathbb{Z}_2$ is a subgroup of the conjectured duality. In $d = 10$, the type IIA string has G = SO(1, 1)/$\mathbb{Z}_2$, while the type IIB has G = SL(2; $\mathbb{R}$), as indicated in the first two lines of Table 1. We shall abbreviate $E_{n(n)}(\mathbb{Z})$ to $E_n(\mathbb{Z})$ when no confusion can arise.

For the heterotic string (Table 2), the supergravity duality groups G for $d > 2$ can be found in articles collected in [64]. Pure $N = 4$ supergravity in $d = 4$ reduces to a theory with G = SO(8, 2) in $d = 3$ and to a theory with the supergravity duality group given by the affine group SO(8, 2)$^{(1)}$ in $d = 2$ [60]. Similar arguments suggest that the heterotic string should give a $d = 2$ supergravity theory with G given by the affine group O(8, 24)$^{(1)}$ symmetry. The heterotic string is conjectured to have an S $\times$ T duality symmetry in $d \geqslant 4$ and a unified U-duality in $d \leqslant 3$. Sen conjectured an O(8, 24; $\mathbb{Z}$) symmetry of $d = 3$ heterotic strings in [65]. The $d = 10$ supergravity theory has an O(16) symmetry acting on the 16 abelian gauge fields which is broken to the finite group O(16; $\mathbb{Z}$); we refer to this as the T-duality symmetry of the ten-dimensional theory.

The supergravity symmetry group G in d dimensions does not act on the d-dimensional spacetime and so survives dimensional reduction. Then G is necessarily a subgroup of the symmetry G′ in $d' < d$ dimensions, and dimensional reduction gives an embedding of G in G′, and G($\mathbb{Z}$) is a subgroup of G′($\mathbb{Z}$). We use this embedding of G into the duality group in $d' = 4$ dimensions to define the duality group G($\mathbb{Z}$) in $d > 4$ dimensions as G $\cap$ $E_7(\mathbb{Z})$ for the type II string and as G $\cap$ [O(6, 22; $\mathbb{Z}$) $\times$ SL(2, $\mathbb{Z}$)] for the heterotic string.

The symmetries in $d < 4$ dimensions can be understood using a type of argument first developed to describe the Geroch symmetry group of general relativity and used in [65] for $d = 3$ heterotic strings. The three-dimensional type II string can be regarded as a four-dimensional theory compactified on a circle and so is expected to have an $E_7(\mathbb{Z})$ symmetry. There would then be seven different $E_7(\mathbb{Z})$ symmetry groups of the three-dimensional theory corresponding to each of the seven different ways of first compactifying from ten to four dimensions, and then from four to three. The seven $E_7(\mathbb{Z})$ groups and the O(7, 7; $\mathbb{Z}$) T-duality group do not commute with each other and generate a discrete subgroup of E_8 which we define to be $E_8(\mathbb{Z})$. (Note that the corresponding Lie algebras, consisting of seven

[5] They have also been discussed in the context of world-volume actions of extended objects in supergravity backgrounds [62].

$E_{7(7)}$ algebras and an $O(7, 7)$, generate the whole of the $E_{8(8)}$ Lie algebra.) Similarly, in $d = 2$ dimensions, there are eight $E_8(\mathbb{Z})$ symmetry groups and an $O(8, 8; \mathbb{Z})$ T-duality group which generate $E_9(\mathbb{Z})$ as a discrete subgroup of $E_{9(9)}$, and in the heterotic string there are eight $O(8, 24; \mathbb{Z})$ symmetry groups from three dimensions and an $O(8, 24; \mathbb{Z})$ T-duality group which generate $O(8, 24; \mathbb{Z})^{(1)}$ as a discrete subgroup of $O(8, 24)^{(1)}$.

We now turn to the charge quantization condition and soliton spectrum in $d > 4$ dimensions. Consider first the example of type II string theory compactified to $d = 5$ dimensions. The low-energy theory is $d = 5$, $N = 8$ supergravity [66] which has 27 abelian vector gauge fields A^I_μ and an $E_{6(6)}$ rigid symmetry of the action. Recall that in five dimensions electric charge can be carried by particles or 0-brane solitons, while magnetic charge can be carried by strings or 1-brane solitons. The 27 types of electric charge q_I transform as a **27** of $E_{6(6)}$ while 27 types of magnetic charge p^I transform as a $\overline{\mathbf{27}}$. These charges satisfy the quantization condition $q_I p^I =$ integer [53,67] which is invariant under $E_{6(6)}$. As we shall see, all 54 types of charge occur and so the electric charges take values in a 27-dimensional lattice Λ and the magnetic ones take values in the dual lattice. This breaks the $E_{6(6)}$ symmetry down to the discrete subgroup which preserves the lattice. If the theory is now compactified to four dimensions, $E_{6(6)}$ survives as a subgroup of the $E_{7(7)}$ duality symmetry in $d = 4$ and the 27-dimensional lattice Λ survives as a sub-lattice of the 28-dimensional lattice of $d = 4$ electric charges (this will be checked for the elementary charged solitons below). Thus the subgroup of $E_{6(6)}$ preserving Λ is $E_{6(6)} \cap E_7(\mathbb{Z})$, which is precisely the discrete group $E_6(\mathbb{Z})$ defined above.

The five-dimensional theory has a Bogomolnyi bound involving the electric and magnetic charges [11,17] which is saturated by Bogomolnyi solutions that do not break all the supersymmetries, and the masses and charges of these states are expected to be unrenormalized in the quantum theory. This bound is invariant under $E_{6(6)}$ and $E_6(\mathbb{Z})$ and so Bogomolnyi solitons automatically fit into $E_6(\mathbb{Z})$ representations.

The $d = 5$ elementary solitons of the type IIA theory carrying precisely one type of electric or magnetic charge and breaking half the supersymmetry can be identified in a similar manner to that used in $d = 4$. The 27 elementary electrically charged solitons, which are all extreme black hole solutions in $d = 5$, and the 27 magnetic ones, which are all extreme black strings, arise from $d = 10$ solutions as follows. The 5-brane wrapped around the 5-torus gives one electrically charged 0-brane and 5 magentically charged strings. The $d = 10$ solitonic string gives 5 electric black holes and 1 black string. The 0-brane gives one black hole, the 4-brane gives 10 black holes and 5 black strings, the 4-brane gives 5 black holes and 10 black strings and the 6-brane gives 1 black string. In addition, there are 5 electric black holes, arising from pp-waves travelling in each of the 5 toroidal dimensions (these can also be viewed as momentum modes of the solitonic string), and their 5 magnetic duals, which are a magnetic string generalization of the KK magnetic monopole. These are the solutions consisting of the product of a 4-torus with self-dual Taub-NUT and two-dimensional Minkowski space. The non-compact six-dimensional subspace gives rise to a 5-dimensional magnetic string in the same way that a 4-dimensional KK magnetic monopole originates from a five-dimensional solution. Note that these solutions can be thought of as wrapped 4-brane solutions. This gives the 27 + 27 elementary charges, as required.

On further compactification to $d = 4$, the 27 electric charges give 27 electrically charged black holes in $d = 4$ and the magnetic strings give 27 magnetic black holes (together with 27 $d = 4$ black strings). There are two additional elementary charged states in $d = 4$, the pp-wave travelling in the fifth dimension and the KK monopole corresponding to the fifth dimension; these two solutions are uncharged from the five-dimensional point of view. This corresponds to the fact that the four-dimensional charges lie in a **56** of $E_{7(7)}$ and this decomposes into $E_{6(6)}$ representations as $\mathbf{56} \rightarrow \mathbf{27} + \overline{\mathbf{27}} + \mathbf{1} + \mathbf{1}$.

Similar arguments apply to other strings in $d \geqslant 4$ dimensions, where charge quantization effects break G to at most the string duality groups listed in the tables. In d dimensions there are electric point charges and magnetic $(d-4)$-brane solitons (which correspond to a subset of the four-dimensional black hole solitons on compactification), and the Dirac quantization of their charges [53,67] breaks the duality symmetry to the discrete subgroup $G(\mathbb{Z})$. There is a similar charge quantization condition on electric p-branes and magnetic $d - p - 4$ branes in d dimensions [53,67] which again break G to $G(\mathbb{Z})$. In each case, the Bogomolnyi solitons automatically fit into representations of the duality group (providing one can continue solutions from one vacuum to another as discussed in Section 3).

For $d < 4$, it is not clear how to understand the breaking of G to $G(\mathbb{Z})$ directly in terms of d-dimensional quantum effects. Thus, while we have shown that for $d \geqslant 4$ the group G is broken to at most $G(\mathbb{Z})$, there is less evidence for our conjectures for $d < 4$, although we do know that a subgroup of G is broken to the discrete T-duality group, and that the solitons will fit into representations of $G(\mathbb{Z})$.

6. Compactification of type II strings on $K_3 \times T^2$

The analysis of Bogomolnyi states can also be carried out for non-toroidal compactifications. An interesting example is compactification of the type II superstring on $K_3 \times T^2$ because, while K_3 has no non-trivial one-cycles and hence no string winding modes, it does have 22 non-trivial two-cycles around which a p-brane for $p > 1$ can wrap itself to produce a $(p-2)$-brane which will then produce monopole winding states on T^2 if $p < 5$ (taking periodic arrays where necessary). The effective four-dimensional supergravity can be found by the two-stage process of compactification to six dimensions on K_3 [25], followed by a straightforward reduction on T^2. It is an $N = 4$ supergravity with an $SL(2;\mathbb{R}) \times O(6, 22)$ symmetry and 28 U(1) gauge fields, exactly as for the compactification of the heterotic string on T^6 at a generic point in the moduli space. In fact, the four-dimensional supergravity theories are identical because the coupling of $N = 4$ supergravity to k abelian vector multiplets is uniqely determined by the choice of gauge group [46] [6]. The analysis of Section 2 again applies, with the result that the duality group is broken down to $SL(2; \mathbb{Z}) \times O(6, 22; \mathbb{Z})$ by the charge quantization condition, and the Bogomolnyi bound is again duality invariant. The soliton spectrum then automatically fits into representations of $SL(2; \mathbb{Z}) \times O(6, 22; \mathbb{Z})$, and the solitons correspond to precisely to the same extreme black holes as were discussed in Section 4 for the heterotic string. However, the ten-dimensional origin of the elementary charged solutions is now different and we now discuss these.

Consider first the common (g, b, Φ) sector. Because K_3 has no isometries and no non-trivial one-cycles all KK modes and string winding modes arise from the T^2-compactification. This yields modes carrying $2+2$ types of electric charge which couple to the $2+2$ gauge fields from the metric and antisymmetric tensor. The corresponding magnetic charges are the KK monopoles and the H-monopoles. The latter can be interpreted as the winding modes on T^2 of the six-dimensional solitonic string found from 'wrapping' the neutral ten-dimensional five-brane around the K_3 surface [68]. We have therefore identified the modes carrying just one type of the $4+4$ electric and magnetic charges in this sector. As before, we do not consider modes arising from the ten-dimensional solitonic string on the grounds that these are not independent of the fundamental string modes already considered.

Consider now the type IIA string. The additional 24 vector gauge fields in the four-dimensional effective field theory arise from the ten-dimensional RR gauge fields A_M and A_{MNP}. One of these vector gauge fields, A_μ, is the four-dimensional component of A_M. The remaining 23 come from expressing the ten-dimensional three-form A_{MNP} as the exterior product of a four-dimensional one-form gauge potential times each of the $22+1$ harmonic two-forms of $K_3 \times T^2$. We must now find the charged Bogomolnyi states to which these fields couple. Again we consider states carrying only one type of charge. The ten-dimensional 0-brane and six-brane solitons associated with A_M yield, respectively, one electric and one magnetic four-dimensional black hole coupling to A_μ. The electric 2-brane and the magnetic 4-brane solitons in ten dimensions produce the Bogomolnyi states carrying the other $23+23$ types of charge coupling to the other 23 gauge fields. Specifically, the 2-brane can be wrapped around the $22+1$ non-trivial two-cycles of $K_3 \times T^2$ to produce $22+1$ six-dimensional black holes of which one can then take periodic arrays to get $22+1$ four-dimensional electric black holes. The four-brane can be wrapped around the 22 homology two-cycles of K_3 to give 22 six-dimensional 2-branes, each of which can then be wrapped around T^2 to produce a four-dimensional magnetic black hole. Alternatively, the four-brane can be wrapped entirely around K_3 to give one six-dimensional black hole which then produces a further magnetic black hole in four dimensions on taking periodic arrays. We have now found a total of $24+24$ additional electric and magnetic

[6] Compactification of the heterotic string on $K_3 \times T^2$ leads to a four-dimensional effective field theory with only $N=2$ supersymmetry, for which the masses of the Bogomolnyi states might be expected to receive quantum corrections, so we shall not discuss this case here.

black holes. They each satisfy the Bogomolnyi bound because the ten-dimensional p-brane solitons do, and they each carry just one type charge. Combining these with the $4+4$ black holes from the (g, b, Φ) sector yields a total of $28+28$ elementary electric and magnetic extreme black holes which generate the $(\mathbf{2}, \mathbf{28})$ representation of $SL(2; \mathbb{Z}) \times O(6, 22; \mathbb{Z})$.

Consider instead the type IIB superstring. The RR gauge fields are A_{MN} and $A^{(+)}_{MNPQ}$ which produce $2+22$ four-dimensional gauge fields upon compactification on $K_3 \times T^2$ [25]. These fields couple to the soliton states in four dimensions obtained by wrapping the extra solitonic string and five-brane, and the self-dual three-brane, around the homology cycles of $K_3 \times T^2$, taking periodic arrays when necessary to get a four-dimensional soliton (alias extreme black hole). There are two homology one-cycles and two homology five-cycles so the extra solitonic string and five-brane produce $2+2$ four-dimensional electric and magnetic black holes. There are 44 three-cycles so the three-brane produces 44 four-dimensional solitons. Since the three-brane is self-dual 22 of these are electric and 22 magnetic. Again we have a total of $24+24$ additional charges. Combining these with the $4+4$ black holes from the (g, b, Φ) sector again yields a total of $28+28$ elementary electric and magnetic extreme black holes which generate the $(\mathbf{2}, \mathbf{28})$ representation of $SL(2; \mathbb{Z}) \times O(6, 22; \mathbb{Z})$.

Since the type II string compactified on $K_3 \times T^2$ and the generic toroidal compactification of the heterotic string have exactly the same four-dimensional low-energy field theory it is natural to conjecture that they might be equivalent string theories. If this is so then the Bogomolnyi states of the heterotic string discussed at the conclusion of the previous section would have a straightforward ten-dimensional interpretation after all. It would have some other remarkable consequences. For example, at special points of the heterotic string moduli space, there are extra massless fields and an enhanced (Yang–Mills) symmetry due to non-perturbative world-sheet effects (i.e. non-perturbative in α', but perturbative in g). If the compactified type II string is equivalent, it must have the same enhanced symmetry in vacua corresponding to the same points in the scalar-field coset space. This presumably does not arise from non-perturbative world-sheet effects, so would have to come from non-perturbative stringy effects, or from Wilson lines and their p-brane generalizations. This would mean that the sigma-model coupling constant α' of the heterotic string becomes one of the stringy coupling constants of the type II theory, as might have been expected from the fact that for the toridally compactified type II string, all coupling constants are on an equal footing and are mixed up under U-duality.

7. U-duality and the 11-dimensional supermembrane

As we have seen, the $E_7(\mathbb{Z})$ invariance of the spectrum of soliton states of $N=8$ supergravity is an automatic consequence of the $E_7(\mathbb{Z})$ invariance of the equations of motion. The non-trivial features are, firstly, that these states have an interpretation in terms of ten-dimensional KK solitons and solitonic p-branes and, secondly, that if the ten-dimensional field theory is considered to be the effective field theory of the type II string theory then $E_7(\mathbb{Z})$ invariance requires an identification of the solitonic string with the fundamental string. $N = 8$ supergravity can also be obtained by dimensional reduction of 11-dimensional supergravity on T^7. We shall now show that the elementary soliton states of $N = 8$ supergravity also have an interpretation in terms of KK solitons and solitonic p-branes of 11-dimensional supergravity. There are 7 KK magnetic monopoles and 7 electric duals, which are pp-waves of 11-dimensional supergravity [69] travelling in the internal dimensions. The 11-dimensional solitonic p-branes are the electric membrane and the magnetic five-brane. Each can be wrapped around the seven-torus to produce 21 four-dimensional solitons. Thus we have a total of 28 electric and 28 magnetic four-dimensional solitons each carrying one of the 56 types of charge which are a basis for the irreducible **56** representation of $E_7(\mathbb{Z})$. If we now wish to interpret 11-dimensional supergravity as an effective field theory of a fundamental $E_7(\mathbb{Z})$ supermembrane theory then we must identify the fundamental membrane with the solitonic one, just as we were forced to identify the fundamental string with the (appropriate) solitonic string.

Consider now the compactification to $\mathcal{M}_4$ on $K_3 \times T^3$ of 11-dimensional supergravity [28]. The effective field theory is the same as that of the type II superstring compactified on $K_3 \times T^2$, i.e. an $N = 4$ supergravity with 28 U(1) gauge fields and an $SL(2;\mathbb{Z}) \times O(6, 22; \mathbb{Z})$ symmetry. The soliton spectrum is also the same. In the monopole sector we have, firstly, 3 KK monopoles from the T^3 factor and, secondly, a further 25 monopoles from wrapping the solitonic five-brane around the $3 + 22$ homology five-cycles of $K_3 \times T^2$. This gives a total of 28 monopoles. The 28 electrically charged solitons are the electric duals of these monopoles which can be understood in terms of the KK and winding modes of either the solitonic membrane or a fundamental membrane. The entire set of 56 states can be assigned to the **(2, 28)** representation of $SL(2;\mathbb{R}) \times O(6, 22)$, inducing a corresponding representation of $SL(2;\mathbb{Z}) \times O(6, 22;\mathbb{Z})$.

These results are encouraging signs that it may be possible to define the quantum supermembrane theory entirely in terms of the solitonic membrane solution of 11-dimensional supergravity. Alternatively, one can envisage a dual formulation in terms of a fundamental 11-dimensional superfive-brane, in which case the solitonic five-brane might be identified with a fundamental five-brane. Of possible relevance in this connection is the fact that the membrane and five-brane solitons have a very different global structure. Both have a degenerate Killing horizon, but whereas the membrane horizon conceals a singularity in an interior region [70], much like the extreme Reissner–Nordstrom solution of four-dimensional Maxwell–Einstein theory, the five-brane is completely non-singular [38].

8. Comments

The equations of motion of four-dimensional effective supergravity theories of compactified superstring theories are invariant under a continuous duality group G that is broken by quantum effects to a discrete subgroup $G(\mathbb{Z})$. For the toroidally compactified heterotic string at a generic point in the moduli space, and for the $K_3 \times T^2$ compactified type II superstrings, this group is the $S \times T$ duality group $SL(2; \mathbb{Z}) \times O(6, 22; \mathbb{Z})$. For the toroidally compactified type II superstrings it is the U-duality group $E_7(\mathbb{Z})$ which contains the $S \times T$ duality group $SL(2; \mathbb{Z}) \times O(6, 22; \mathbb{Z})$. Whereas T-duality is known to be an exact symmetry of string theory at each order in the string coupling constant g, the conjectured S- and T-dualities are non-perturbative. We have provided evidence for U-duality of the type II string by considering those features of string theory that are expected to be given exactly by a semi-classical analysis, although it should be emphasized that this evidence depends only on the form of the effective supergravity theory and would apply equally to any consistent quantum theory of gravity for which this is the effective low-energy action. Nevertheless, by supposing this consistent quantum theory to be a string theory our arguments have led us to the remarkable conclusions that it is necessary to identify certain states of the string with extreme black holes, and the fundamental string with a solitonic string. We have also seen that the elementary Bogomolnyi states are extreme black hole solutions of the low-energy theory, and have shown how these arise from p-brane solitons of the ten-dimensional theory.

The zero-modes of the scalar fields of the low-energy field theory are all coupling constants of the string theory, so that $G(\mathbb{Z})$ symmetry relates different regimes in the perturbation theory in these coupling constants, interchanging strong and weak coupling and, in the case of U-duality, interchanging g with α', in the sense of mixing the quantum loop expansion in g with the sigma-model perturbation expansion in α' and the moduli of the compactification space. For the compactified type II superstring, any physical quantity (e.g. the S-matrix) can be expanded in terms of the 70 coupling constants associated with zero-modes of the 70 scalars. In the world-sheet approach to string theory, one first integrates over the sigma-model degrees of freedom on a Riemann surface of fixed genus, obtaining a result parameterised by the sigma-model coupling constants, and then sums over genus. As U-duality mixes up all 70 coupling constants, the final result may be expected to depend on all 70 scalars in a symmetric way, even though the calculation was very asymmetrical and in particular picked out the dilaton to play a special role. This would hugely constrain the theory, and the assumption of U-duality, together with $N = 8$ supersymmety, gives us a great deal of non-perturbative information, and might even enable us to solve the theory! This structure also suggests that there might be a new formulation of string theory which treats all the coupling constants on an equal footing.

We have seen that there are duality symmetries in all dimensions $d \leqslant 10$ and it is interesting to ask whether the $d < 10$ symmetries can correspond to symmetries of the 10-dimensional theory. If e.g., the four-dimensional dualities correspond to symmetries of the full ten-dimensional theory, these symmetries must interchange the various p-brane solitons of the string theory; such symmetries would probably have to be non-local. If the theories in $d < 4$ dimensions really do have the duality symmetries suggested in Section 5, and if these have analogues in higher dimensions, this would have remarkable consequences for string theory. For example, the

U-duality of the three-dimensional heterotic string includes transformations that would mix the string with 5-brane solitons in ten dimensions, and so would contain the transformations described as the 'duality of dualities' in [3]. For the type II string theory, this suggests that $E_{10}(\mathbb{Z})$ might be a discrete non-local symmetry of the ten-dimensional string theory!

One of the predictions of S-duality for the heterotic string is the presence of certain dyon bound states in the Bogomolnyi spectrum, which can be translated into a prediction concerning harmonic forms on the multi-monopole moduli space [71]. It would be interesting to consider the corresponding predictions for the type II string. Recently, some strong-coupling evidence for S-duality of the heterotic string has been found by studying partition functions of certain topological field theories [72], and again it would be interesting to seek similar strong-coupling tests for U-duality.

Finally, since $N = 8$ supergravity and its soliton spectrum are U-duality invariant, many of the properties previously thought to be unique to string theory are in fact already properties of the effective supergravity theory once account is taken of all soliton solutions. Since only stable states can appear in an exact S-matrix it is possible that the only states of the exact toroidally compactified type II string theory are the Bogomolnyi states and that these are in one to one correspondence with soliton states of the supergravity theory. This would support a previous suggestion [73] that a fundamental superstring theory might actually be equivalent to its effective field theory once solitons of the latter are taken into account. To pursue this further one would need to find higher-spin (> 2) soliton states corresponding to the higher-spin states of string theory. There seems to be no problem in principle with the existence of such higher-spin soliton states in $N = 8$ supergravity because the $a = 0$ and $a = 1$ extreme black holes must belong to massive supermultiplets with spins > 2 as they break more than half the supersymmetry. In this connection it is worth recalling the similarity of the mass/spin relation for Regge trajectories in four-dimensional string theory, $M^2 \propto J/\alpha'$, with that of the degenerate Kerr solutions of general relativity, $M^2 \propto J/G$.

Acknowledgements

We are grateful to Louise Dolan, Michael Duff, José Figueroa-O'Farrill, Jerome Gauntlett, Gary Gibbons, Michael Green, Bernard Julia, Wafic Sabra, Trevor Samols, Kellogg Stelle and Edward Witten for helpful comments.

References

[1] A. Giveon, M. Porrati and E. Rabinovici, Phys. Rep. 244 (1994) 77.
[2] A. Font, L. Ibanez, D. Lust and F. Quevedo, Phys. Lett. B 249 (1990) 35;
S.J. Rey, Phys. Rev. D 43 (1991) 526.
[3] J.H. Schwarz and A. Sen, Nucl. Phys. B 411 (1994) 35; Phys. Lett. B 312 (1993) 105.

[4] A. Sen, Nucl. Phys. B 404 (1993) 109; Phys. Lett. B 303 (1993); Int. J. Mod. Phys. A 8 (1993) 5079; Mod. Phys. Lett. A 8 (1993) 2023.
[5] K.S. Narain, Phys. Lett. B 169 (1986) 41.
[6] K.S. Narain, M.H. Sarmadi and E. Witten, Nucl. Phys. B 279 (1987) 369.
[7] E. Cremmer and B. Julia, Phys. Lett. B 80 (1978) 48; Nucl. Phys. B 159 (1979) 141.
[8] G.W. Gibbons and C.M. Hull, Phys. Lett. B 109 (1982) 190.
[9] G.W. Gibbons, *in* Supersymmetry, supergravity and related topics, eds. F. del Aguila, J.A. de Azcárraga and L.E. Ibañez (World Scientific, Singapore, 1985).
[10] R. Kallosh and T. Ortin, Phys. Rev. D 48 (1993) 742.
[11] G.W. Gibbons and M.J. Perry, Nucl. Phys. B 248 (1984) 629.
[12] S.W. Hawking, Mon. Not. R. Astron. Soc. 152 (1971) 75;
A. Salam, *in* Quantum gravity: an Oxford symposium, eds. C.J. Isham, R. Penrose and D.W. Sciama (Oxford U.P., Oxford, 1975);
G.'t Hooft, Nucl. Phys. B 335 (1990) 138.
[13] M.J. Duff, R.R. Khuri, R. Minasian and J. Rahmfeld, Nucl. Phys. B 418 (1994) 195.
[14] L. Susskind, preprint hep-th/9309145;
J.G. Russo and L. Susskind, preprint hep-th/9405117.
[15] M.J. Duff and J. Rahmfeld, preprint hep-th/9406105.
[16] J. Harvey, J. Liu, Phys. Lett. B 268 (1991) 40.
[17] G.W. Gibbons, D. Kastor, L. London, P.K. Townsend and J. Traschen, Nucl. Phys. B 416 (1994) 850.
[18] A. Dabholkar, G.W. Gibbons, J.A. Harvey and F. Ruiz-Ruiz, Nucl. Phys. B 340 (1990) 33;
M.J. Duff, G.W. Gibbons and P.K. Townsend, Phys. Lett. B 332 (1994) 321.
[19] C. Callan, J. Harvey and A. Strominger, Nucl. Phys. B 359 (1991) 611.
[20] M.J. Duff and J.X. Lu, Nucl. Phys. B 354 (1991) 141; Phys. Lett. B 273 (1991) 409.
[21] G.T. Horowitz and A. Strominger, Nucl. Phys. B 360 (1991) 197.
[22] M.J. Duff and J.X. Lu, Nucl. Phys. B 416 (1993) 301.
[23] J.A. de Azcárraga, J.P. Gauntlett, J.M. Izquierdo and P.K. Townsend, Phys. Rev. Lett. 63 (1989) 2443.
[24] E. Witten, Phys. Lett. B 153 (1985) 243.
[25] P.K. Townsend, Phys. Lett. B 139 (1984) 283;
M.B. Green, J.H. Schwarz and P.C. West, Nucl. Phys. B 254 (1985) 327.
[26] E. Bergshoeff, E. Sezgin and P.K. Townsend, Phys. Lett. B 189 (1987) 75; Ann. Phys. (NY) 185 (1988) 330.
[27] E. Cremmer, B. Julia and J. Scherk, Phys. Lett. B 76 (1978) 409.
[28] M.J. Duff, B.E.W. Nilsson and C.N. Pope, Phys. Lett. B 129 (1983) 39.
[29] M.J. Duff and K.S. Stelle, Phys. Lett. B 253 (1991) 113.
[30] R. Güven, Phys. Lett. B 276 (1992) 49.
[31] M.K. Gaillard and B. Zumino, Nucl. Phys. B 193 (1981) 221.
[32] M. Günaydin, G. Sierra and P.K. Townsend, Phys. Lett. B 133 (1983) 72; Nucl. Phys. B 242 (1984) 244.
[33] E. Witten, Phys. Lett. B 86 (1979) 283.
[34] E. Witten and D. Olive, Phys. Lett. B 78 (1978) 97.
[35] C.M. Hull, Ph.D. thesis, Cambridge 1983.
[36] G.W. Gibbons, Nucl. Phys. B 207 (1982) 337;
G.W. Gibbons and K. Maeda, Nucl. Phys. B 298 (1988) 741;
D. Garfinkle, G.T. Horowitz and A. Strominger, Phys. Rev. D 43 (1991) 3140;
C.F. Holzhey and F. Wilczek, Nucl. Phys. B 380 (1992) 447;
R. Kallosh, A. Linde, T. Ortin, A. Peet and A. Van Proeyen, Phys. Rev. D 46 (992) 5278.
[37] M.J. Duff, R.R. Khuri, R. Minasian and J. Rahmfeld, Nucl. Phys. B 418 (1994) 195.
[38] G.W. Gibbons, G.T. Horowitz and P.K. Townsend, Higher-dimensional resolution of dilatonic black hole singularities, Class. Quant. Grav., in press.
[39] J. Gauntlett, Nucl. Phys. B 400 (1993) 103.
[40] P. Ruback, Commun. Math. Phys. 107 (1986) 93.

[41] K. Shiraishi, J. Math. Phys. 34 (4) (1993) 1480.
[42] M.J. Duff and J.X. Lu, Nucl. Phys. B 390 (1993) 276.
[43] A.G. Felce and T.M. Samols, Phys. Lett. B 308 (1993) 30.
[44] Bluhm, P. Goddard and L. Dolan, Nucl. Phys. B 289 (1987) 364; B 309 (1988) 330.
[45] J. Maharana and J.H. Schwarz, Nucl. Phys. B 390 (1993) 3.
[46] M. de Roo, Nucl. Phys. B 255 (1985) 515.
[47] J. Gauntlett, J. Harvey and J. Liu, Nucl. Phys. B 409 (1993) 363;
J. Gauntlett and J. Harvey, S-duality and the spectrum of magnetic monopoles in heterotic string theory, preprint EFI-94-36.
[48] R. Sorkin, Phys. Rev. Lett. 51 (1983) 87;
D. Gross and M. Perry, Nucl. Phys. B 226 (1983) 29.
[49] R.R. Khuri, Phys. Lett. B 259 (1991) 261.
[50] R.R. Khuri, Nucl. Phys. B 387 (1992) 315.
[51] T. Banks, M. Dine, H. Dijkstra and W. Fischler, Phys. Lett. B 212 (1988) 45.
[52] C.M. Hull, in preparation.
[53] R. Nepomechie, Phys. Rev. D 31 (1985) 1921.
[54] A. Strominger, Nucl. Phys. B 343 (1990) 167.
[55] M.J. Duff and J.X. Lu, Phys. Rev. Lett. 66 (1991) 1402.
[56] C.M. Hull, Nucl. Phys. B 260 (1985) 182;
L. Alvarez-Gaumé and P. Ginsparg, Commun. Math. Phys. 102 (1985) 311.
[57] G.T. Horowitz and A.A. Tseytlin, IC preprints Imperial /TP/93-94/38, Imperial /TP/93-94/54.
[58] J. Dai, R.G. Leigh and J. Polchinski, Mod. Phys. Lett. A 4 (1989) 2073;
M. Dine, P. Huet and N. Seiberg, Nucl. Phys. B 322 (1989) 301.
[59] M. Atiyah and N. Hitchin, Phys. Lett. A 107 (1985) 21; Philos. Trans. R. Soc. London A 315 (1985) 459; The geometry and dynamics of magnetic monopoles (Princeton U.P., Princeton, NJ, 1988).
[60] B. Julia, *in* Supergravity and superspace, eds. S.W. Hawking and M. Rocek (Cambridge U.P., Cambridge, 1981).
[61] B. Julia, *in* Lectures in applied mathematics Vol. 21 (American Mathematical Society, Providence, R.I., 1985) p. 355.
[62] M.J. Duff and J.X. Lu, Nucl. Phys. B 347 (1990) 394.
[63] R.W. Gebert and H. Nicolai, preprint DESY-94-106 [hep-th 9406175].
[64] A. Salam and E. Sezgin, eds., Supergravity theories in diverse dimensions (World Scientific, Singapore, 1989)
[65] A. Sen, Tata Institute preprint TIFR-TH-94-19 [hep-th 9408083].
[66] E. Cremmer, *in* Supergravity and superspace, eds. S.W. Hawking and M. Rocek (Cambridge U.P., Cambridge, 1981).
[67] C. Teitelboim, Phys. Lett. B 67 (1986) 63, 69.
[68] A. Achúcarro, J. Evans, P.K. Townsend and D. Wiltshire, Phys. Lett. B 198 (1987) 441.
[69] C.M. Hull, Phys. Lett. B 139 (1984) 39.
[70] M.J. Duff, G.W. Gibbons and P.K. Townsend, Phys. Lett. B 332 (1994) 321.
[71] A. Sen, Phys. Lett. B 329 (1994) 217.
[72] C. Vafa and E. Witten, Harvard preprint HUTP-94-A017 [hep-th/9408074].
[73] P.K. Townsend, Phys. Lett. B 202 (1988) 53.

Reprinted from Phys. Lett. B **350** (1995) 184–8

The eleven-dimensional supermembrane revisited

P.K. Townsend
DAMTP, Univ. of Cambridge, Silver St., Cambridge, UK

Received 16 January 1995; revised manuscript received 28 March 1995
Editor: P.V. Landshoff

Abstract

It is argued that the type IIA 10-dimensional superstring theory is actually a compactified 11-dimensional supermembrane theory in which the fundamental supermembrane is identified with the solitonic membrane of 11-dimensional supergravity. The charged extreme black holes of the 10-dimensional type IIA string theory are interpreted as the Kaluza-Klein modes of 11-dimensional supergravity and the dual sixbranes as the analogue of Kaluza-Klein monopoles. All other p-brane solutions of the type IIA superstring theory are derived from the 11-dimensional membrane and its magnetic dual fivebrane soliton.

The effective field theory of the ten-dimensional type IIA superstring is $N = 2A$ supergravity. It has long been appreciated that this field theory is also the effective massless theory for eleven-dimensional supergravity compactified on S^1; the ten-dimensional dilaton thereby acquires a natural Kaluza-Klein (KK) interpretation. This leads one to wonder whether the type IIA string theory has an eleven-dimensional interpretation. An obvious candidate is the 11-dimensional supermembrane [1] since the double dimensional reduction of its worldvolume action yields the Green-Schwarz (GS) action of the type IIA superstring [2]. Despite this, the 11-dimensional interpretation of the *quantum* type IIA superstring is obscure because the dilaton vertex operator is radically different from the graviton vertex operator. In the GS action the dilaton comes from the R-R sector while the graviton comes from the NS-NS sector; there is therefore no obvious KK interpretation of the dilaton in string theory (in the bosonic string the dilaton is usually taken to couple to the worldsheet curvature but this makes the dilaton vertex operator even more dissimilar to the graviton vertex operator). It is possible, however, that this special status of the dilaton is an artefact of perturbation theory. It has recently been realized that some features of the effective field theories of compactified superstring theories, such as invariance under a generalized electromagnetic duality, may also be features of the full non-perturbative string theory even though this is not apparent in perturbation theory [3,4,5]. In this letter I similarly argue that the type IIA 10-dimensional superstring theory actually *is* a compactified 11-dimensional supermembrane theory.

Before further analysis of this conjecture, some discussion of the status of the 11-dimensional supermembrane is warranted. There is some reason to suppose that the supermembrane spectrum contains massless particles which can be identified as the graviton and other quanta of 11-dimensional supergravity [6]. The principal objection to this conclusion is that there are also reasons [7,8] to believe the spectrum to be continuous, which would preclude a particle interpretation. The physical reason for this is that there is no energy cost to a deformation of the membrane lead-

0370-2693/95/$9.50
SSDI 0370-2693(95)00397-5

ing to 'spikes' of arbitrary length but zero area, like a fakir's bed of nails (for the bosonic membrane there is an energy cost at the quantum level due to the Casimir effect, but this Casimir energy cancels for the supermembrane). The possibility of spikes of zero area is of course due to the supposition that the membrane has a core of zero width. A calculation [8] in the context of a first-quantized, regularized, zero-width supermembrane showed that the spectrum is indeed continuous, from zero, and this was widely interpreted as putting an end to the idea of a 'fundamental' supermembrane.

However, evidence was presented in [5] that the fundamental supermembrane should be identified with the solitonic membrane [9] of 11-dimensional supergravity. An additional reason for this identification is that κ-symmetry of the worldvolume action for a supermembrane requires the background fields to satisfy the *source-free* field equations of 11-dimensional supergravity [1]. This is paradoxical if the supermembrane is regarded as the source of the background fields, but the paradox would be resolved if the fundamental supermembrane were to be identified with a membrane solution of the source-free field equations, and the one of [9] is the only candidate. As originally presented this was seen as the exterior solution to a singular surface, which was interpreted as a membrane source, but the singularity can be interpreted equally well as a mere coordinate singularity at an event horizon, through which the source-free exterior solution can be analytically continued [10]. If one accepts the identification of the fundamental and solitonic supermembranes in the fully non-perturbative quantum theory, then it follows that the supermembrane acquires a core of finite size due to its gravitational field in the same way that a 'point' particle actually has a size of the order of its Schwarzschild radius once gravitational effects are included. In this case a 'spike' of a given length has a minimum area and therefore a minimum energy cost. Under these circumstances one would not expect a continuous spectrum.

A possible objection to this argument is that it could also be applied to string theory where, however, it is not needed because the spectrum is already discrete in perturbation theory. This may simply be a reflection of the fact that perturbation theory makes sense for strings because of the renormalizability of two-dimensional sigma-models whereas it does not make sense for membranes because of the non-renormalizability of three-dimensional sigma models. Also, there is no dilaton in 11-dimensions and so no obvious coupling constant with which to order a perturbation series. In any case, I shall assume in the following that the fully non-perturbative supermembrane spectrum is discrete for reasons along the above lines.

The determination of the spectrum of the 11-dimensional supermembrane, given that it is discrete, is impossible in practice, as it is for superstrings when account is taken of interactions and all non-perturbative effects. However, certain features of the spectrum can be reliably ascertained. Among these is the massless spectrum, for which the effective field theory is just 11-dimensional supergravity. This theory reduces to 10-dimensional $N = 2A$ supergravity upon compactification on S^1, but the spectrum in 10-dimensions will then also include the charged massive KK states. These states must also be present in the spectrum of the type IIA superstring if the latter is to be interpreted as a compactified supermembrane, as conjectured here. These states do not appear in perturbation theory but there are extreme black hole solutions of 10-dimensional $N = 2A$ supergravity that are charged with respect to the KK $U(1)$ gauge field [11]. Because these solutions preserve half of the supersymmetry there are good reasons (see e.g. [5] and references therein) to believe that their semi-classical quantization will be exact. I suggest that these states be identified as KK states. I shall now address possible objections to this identification.

First, the mass of a KK state is an integer multiple of a basic unit (determined by the S^1 radius) whereas the mass of an extreme black hole is apparently arbitrary. However, there are also 6-brane solutions of $N = 2A$ supergravity [11] that are the magnetic duals of the extreme black holes. It will be shown below that these 6-branes are completely non-singular when interpreted as solutions of the compactified 11-dimensional supergravity. It follows, if the 11-dimensional interpretation is taken seriously, that the 6-brane solitons must be included as solutions of the ten-dimensional theory and then, by the generalization of the Dirac quantization condition to p-branes and their duals [12], we conclude that in the quantum theory the electric charge of the extreme black holes is quantized. Since their mass is proportional to the modulus of their charge, with a universal constant of proportionality, their mass is also quantized. The *unit* of mass remains arbitrary, as was the S^1 radius.

Second, it may be objected that whereas the type IIA theory has only one set of charged states coupling to the $U(1)$ gauge field, the compactified supermembrane theory has *two*: the extreme black hole solutions of the effective 10-dimensional field theory after compactification on S^1 *and* the KK modes. The two sets of states have identical quantum numbers since the allowed charges must be the same in both cases. It has recently been argued in the context of compactifications of the heterotic [13] and the type II [5] superstrings that KK states should be *identified* with electrically charged extreme black holes (see also [14]). The reasons advanced for this identification do not obviously apply in the present context but once the principle is granted that this identification is possible it seems reasonable to invoke it more generally. Thus, I conjecture that the resolution of this second objection is that the KK and extreme black hole states of the S^1 compactified 11-dimensional supergravity are not independent in the context of the underlying supermembrane theory. This conjecture is similar to those made recently for the heterotic and type II superstrings but there is a crucial difference; in the string theory case the KK states also appear in the perturbative string spectrum since they result from compactification *from* the critical dimension, whereas the KK states discussed here do not appear in the perturbative string spectrum because they result from compactification *to* the critical dimension.

Little more can be said about the spectrum of *particle* states in ten dimensions since only those solutions of the effective field theory that do not break all supersymmetries can yield reliable information about the exact spectrum upon semi-classical quantization, and the only such particle-like solutions are the extreme electric black holes. However, there are also p-brane solitons of $N = 2A$ supergravity which preserve half the supersymmetry and are therefore expected to be exact solutions of type IIA string theory. These should also have an 11-dimensional interpretation. The extreme multi p-brane solutions associated with a R-R $(p+2)$-form field strength F_{p+2} of a 10-dimensional type II superstring have metric (in 'string sigma-model' form) and dilaton

$$ds_{10}^2 = V^{-\frac{1}{2}}(\boldsymbol{x})ds_{(p+1)}^2 + V^{\frac{1}{2}}\,d\boldsymbol{x}\cdot d\boldsymbol{x}\,,$$

$$e^{4\phi} = V^{(3-p)}(\boldsymbol{x})\,, \qquad (1)$$

where $ds_{(p+1)}^2$ is the Minkowski $(p+1)$-metric, $d\boldsymbol{x}\cdot d\boldsymbol{x}$ is the Euclidean metric on $\mathbb{R}^{(9-p)}$ and V is a harmonic function on $\mathbb{R}^{(9-p)}$ that approaches unity as $\rho^2 = \boldsymbol{x}\cdot\boldsymbol{x}$ tends to infinity; e.g. for the one R-R p-brane solution given in [11],

$$V = 1 + \frac{\mu}{\rho^{(7-p)}}\,, \qquad (2)$$

for some constant μ proportional to the mass per unit p-volume. The solutions (1) include the self-dual threebrane [11] of the type IIB superstring. They also include a R-R string and fivebrane of the type IIB superstring that appear not to have been considered previously in the string theory context, although they are special cases of the general p-brane solution of [11]. However, we are interested here in the type IIA p-branes. These comprise electric 0, 1 and 2 branes and magnetic 4, 5 and 6 branes, although the string and fivebrane are not of R-R type and so have a different form from (1).

The 6-brane soliton has already been mentioned; we now turn to its 11-dimensional interpretation. Consider the 11-metric

$$ds_{11}^2 = -dt^2 + dy\cdot dy + V(\boldsymbol{x})d\boldsymbol{x}\cdot d\boldsymbol{x} + V^{-1}(\boldsymbol{x})\left(dx^{11} - \boldsymbol{A}(\boldsymbol{x})\cdot d\boldsymbol{x}\right)^2, \qquad (3)$$

where $dy\cdot dy$ is the Euclidean metric on $\mathbb{R}^6$ (an infinite planar 6-brane) and $d\boldsymbol{x}\cdot d\boldsymbol{x}$ is the Euclidean metric on $\mathbb{R}^3$ (the uncompactified transverse space). This metric solves the 11-dimensional vacuum Einstein equations, and hence the field equations of 11-dimensional supergravity with zero four-form field strength, if $\nabla\times\boldsymbol{A} = \nabla V$, which implies that $\nabla^2 V = 0$. The one-soliton solution is

$$V = 1 + \frac{\mu}{\rho}\,, \qquad (4)$$

where $\rho = \sqrt{\boldsymbol{x}\cdot\boldsymbol{x}}$. The two-form $F = dA$ is then given by

$$F = \mu\varepsilon_2\,, \qquad (5)$$

where ε_2 is the volume form on the unit 2-sphere. The singularity of (3) at $\rho = 0$ is merely a coordinate singularity if x^{11} is identified modulo $4\pi\mu$ (to see this, set $\rho = \lambda^2$ and take the $\lambda \to 0$ limit). Thus

(3) is a *non-singular* solution of compactified 11-dimensional supergravity representing a magnetic KK 6-brane. It is an exact analogue in 11 dimensions of the KK monopole in 5 dimensions [16]. Considered as a solution of the effective field theory of ten-dimensional string theory, the 10-metric, in 'string sigma-model' form, is

$$ds_{10}^2 = e^{\frac{2\phi}{3}}\left[-dt^2 + dy \cdot dy + V\, d\boldsymbol{x} \cdot d\boldsymbol{x}\right], \tag{6}$$

where the 10-dimensional dilaton field ϕ is given by

$$e^{-2\phi} = V^{\frac{3}{2}}. \tag{7}$$

This is the sixbrane case of (1). In terms of the new radial coordinate $r = \rho + \mu$, the one-soliton solution is

$$\begin{aligned} ds_{10}^2 &= \left(1 - \frac{\mu}{r}\right)^{\frac{1}{2}}\left[-dt^2 + dy \cdot dy\right] \\ &\quad + \left(1 - \frac{\mu}{r}\right)^{-\frac{1}{2}} dr^2 + r^2\left(1 - \frac{\mu}{r}\right)^{\frac{3}{2}} d\Omega_2^2, \\ e^{-2\phi} &= \left(1 - \frac{\mu}{r}\right)^{-\frac{3}{2}}, \end{aligned} \tag{8}$$

where $d\Omega_2^2$ is the metric on the unit 2-sphere. This is just the 6-brane solution of 10-dimensional $N = 2A$ supergravity found by Horowitz and Strominger [5].

The remaining p-brane soliton solutions of $N = 2A$ supergravity are the string [17], membrane, fourbrane and fivebrane [11]. The string and fourbrane solitons have previously been shown [2,10] to be double-dimensional reductions of, respectively, the 11-dimensional membrane and the 11-dimensional fivebrane [18]. The 10-dimensional membrane and fivebrane differ from their 11-dimensional counterparts simply by the boundary conditions imposed on the harmonic function V that determines the solution, and the ten-dimensional soliton can be viewed as a periodic array of 11-dimensional solitons. Close to the horizon at the object's core the 10-dimensional solution approximates the 11-dimensional solution. A potential difficulty here is that the heterotic and type IIA superstring theories have the same fivebrane solution but we need the 11-dimensional interpretation only in the type II case. The resolution of this is that the fivebrane horizon is at infinite affine parameter in the ten-dimensional (string sigma-model) metric but at finite affine parameter (on timelike geodesics) in the 11-dimensional metric, so that both a ten and an eleven dimensional interpretation are possible. In contrast, the horizon of the 10-dimensional membrane is at *finite* affine parameter and one must pass to the 11-dimensional interpretation to avoid a singularity there. Moreover, as the horizon is approached the radius of the 11th dimension approaches infinity, so we have *dimensional decompactification at the horizon*. This behaviour may be contrasted with that of the sixbrane discussed above for which the (coordinate) singularity at the sixbrane core is due to the radius of the 11th dimension shrinking to zero.

Thus, all p-brane solitons of 10-dimensional $N = 2A$ supergravity have an 11-dimensional origin. Moreover, since the 11-dimensional fivebrane has a completely non-singular analytic extension through its horizon [15], the 10-dimensional magnetic 4, 5 and 6-brane solitons are all *completely non-singular* when interpreted as solutions of compactified 11-dimensional supergravity. The 11-dimensional membrane is singular, although the singularity is hidden behind an horizon. This is what one might expect in the context of a fundamental supermembrane theory. Together, these results for the type IIA electric and magnetic p-brane solitons can be taken as further evidence in favour of an 11-dimensional origin of the apparently 10-dimensional type IIA superstring theory. It is perhaps worth remarking that, not surprisingly, there is no similar interpretation of the p-brane solitons of type IIB superstring theory.

It may be objected here that while all of the p-brane *solitons* of the type IIA superstring may be solutions of an S^1-compactified supermembrane theory, the two theories differ in that one has an additional fundamental string while the other has an additional fundamental membrane. But this difference disappears once one identifies the fundamental string or membrane with the solitonic ones; both theories then have exactly the same spectrum of extended objects. In fact, it becomes a matter of convention whether one calls the theory a string theory, a membrane theory, or a p-brane theory for any of the other values of p for which there is a soliton solution; all are equal partners in a p-brane democracy. However, in the type IIA string these solitons must be interpreted as solutions of 11-dimensional supergravity and p-brane democracy has then to be interpreted as membrane fivebrane duality [5]. In contrast to the supermembrane, for which the worldvolume action is known [1], the six-dimensional worldvolume action for the 11-dimensional fivebrane has yet to be constructed, although it is known [19] that its six-dimensional physical field content is that of the self-dual antisymmetric tensor supermultiplet.

Discussions with M.J. Duff, C.M. Hull and K.S. Stelle are gratefully acknowledged

References

[1] E. Bergshoeff, E. Sezgin and P.K. Townsend, Phys. Lett. B 189 (1987) 75; Ann. Phys. (N.Y.) 185 (1988) 330.
[2] M.J. Duff, P.S. Howe, T. Inami and K.S. Stelle, Phys. Lett. B 191 (1987) 70.
[3] J.H. Schwarz and A. Sen, Nucl. Phys. B 411 (1994) 35; Phys. Lett. B 312 (1993) 105.
[4] A. Sen, Nucl. Phys. B 404 (1993) 109; Phys. Lett. B 303 (1993).
[5] C.M. Hull and P.K. Townsend, Unity of superstring dualities, Nucl. Phys. B, in press.
[6] I. Bars, C.N. Pope and E. Sezgin, Phys. Lett. B 198 (1987) 455.
[7] K.S. Stelle and P.K. Townsend, Are 2-branes better than 1? in Vol. 2 of Field Theory in Two Dimensions, and related topics, eds. G. Kunstatter, H.C. Lee, F.C. Khanna and I Umezawa (World Scientific 1988);
P.K. Townsend, Three lectures on supermembranes, i Superstrings '88, eds. M. Green, M. Grisaru, R. Iengo, I Sezgin and A. Strominger (World Scientific 1989).
[8] B. de Wit, M. Lüscher and H. Nicolai, Nucl. Phys. B 32 (1989) 135.
[9] M.J. Duff and K.S. Stelle, Phys. Lett. B 253 (1991) 113.
[10] M.J. Duff, G.W. Gibbons and P.K. Townsend, Phys. Lett. I 332 (1994) 321.
[11] G.T. Horowitz and A. Strominger, Nucl. Phys. B 360 (1991 197.
[12] R. Nepomechie, Phys. Rev. D 31 (1985) 1921;
C. Teitelboim, Phys. Lett. B 167 (1986) 69.
[13] M.J. Duff and J. Rahmfeld, Phys. Lett. B 345 (1995) 441
[14] L. Susskind, Some speculations about black hole entropy i string theory, preprint hep-th/9309145.
[15] G.W. Gibbons, G.T. Horowitz and P.K. Townsend, Class Quantum Grav. 12 (1995) 297.
[16] R. Sorkin, Phys. Rev. Lett. 51 (1983) 87;
D. Gross and M. Perry, Nucl. Phys. B 226 (1983) 29.
[17] A. Dabholkar, G.W. Gibbons, J.A. Harvey and F. Ruiz-Ruiz Nucl. Phys. B 340 (1990) 33.
[18] R. Güven, Phys. Lett. B 276 (1992) 49.
[19] G.W. Gibbons and P.K. Townsend, Phys. Rev. Lett. 71 (1993) 3754.

Chapter 5

Intersecting branes and black holes

The idea that elementary particles might behave like black holes is not a new one. Intuitively, one might expect that a pointlike object whose mass exceeds the Planck mass, and whose Compton wavelength is therefore less than its Schwarzschild radius, would exhibit an event horizon. In the absence of a consistent quantum theory of gravity, however, such notions would always remain rather vague. M-theory, on the other hand, not only predicts such massive states but may provide us with a consistent framework in which to discuss them. We might then be able to tackle some of the outstanding issues in black hole physics such as the information paradox and the microscopic origin of the Beckenstein–Hawking [1, 2] entropy formula. Moreover, in the M-theory framework, four-dimensional black holes may be regarded as originating from the elementary $D = 11$ building blocks of plane wave, membrane, Kaluza–Klein monopole or fivebrane by allowing these objects to wrap around some of the seven compactified directions. In particular, as we shall see in this chapter, black hole bound states have their M-theoretic origin in *intersecting* branes in eleven dimensions.

In [3] it was suggested that certain massive excitations of four-dimensional superstrings should indeed be identified with black holes. Of course, non-extreme black holes would be unstable due to the Hawking effect. To describe stable elementary particles, therefore, attention was focussed on extreme black holes whose masses saturate a Bogomol'nyi bound and these were identified with the BPS string states. There are also black holes which, though extreme, are not supersymmetric and which therefore do not obey any such bound. Nevertheless, evidence based on the mass and charge assignments was provided for also identifying these black holes with certain non-BPS string states. However, the paper remained agnostic concerning the relation between the other non-BPS string states and non-extreme black holes [4, 5] partly because superstring states always form supermultiplets whereas it has been argued that non-extreme black holes do not [6].

The motivation for identifying *elementary* electrically charged string states with black holes came from first noticing that the *solitonic* magnetically charged string states are extreme black holes [7] and then noting that they transform into one another under S duality. Of course, this involves extending the classical notion of a black hole from the weak coupling to the strong coupling regime. The words 'black hole' were therefore taken to describe a string state if there exists at least one string picture in which its mass exceeds the Planck mass for weak coupling. Further *dynamical* evidence for these identifications was supplied in [8, 9] where comparisons were made between the low energy scattering amplitudes.

By choosing appropriate combinations of dilaton and moduli fields to be the scalar field ϕ and appropriate combinations of the field strengths and their duals to be the Maxwell field F, the field equations of the four-dimensional low energy effective Lagrangians of M-theory can be consistently truncated to a form given by the Lagrangian

$$\mathcal{L} = \frac{1}{2\kappa^2}\sqrt{-g}\left[R - \frac{1}{2}(\partial\phi)^2 - \frac{1}{4}e^{-a\phi}F^2\right] + \dots. \tag{5.1}$$

A *consistent* truncation is defined to be one for which all solutions of the truncated theory are solutions of the original theory. The dots in (5.1) refer to terms involving a combination of pseudoscalar axion and dilaton fields which are in general required for consistency but which do not contribute to non-rotating black holes. Supersymmetric black hole solutions can be found for the four values of the dilaton–Maxwell coupling parameter $a = \sqrt{3}, 1, 1/\sqrt{3}, 0$. The cases $a = \sqrt{3}$, $a = 1$ and $a = 0$ correspond to the Kaluza–Klein black hole [10], the dilaton black hole [10] and the Reissner–Nordstrom black hole respectively. It was originally thought that only the $a = 1$ solution appeared in string theory. This is indeed true if the scalar ϕ refers purely to the dilaton [11, 12]. However, the case $a = \sqrt{3}$ was shown to be a solution of string theory in [7] by taking ϕ to be a linear combination of dilaton and moduli fields, and the case $a = 0$ was shown to be a dyonic solution of string theory in [3] by taking the field strength F to be a linear combination of a Maxwell field and its dual. The $a = 1/\sqrt{3}$ solution was discussed in [13] and its dyonic interpretation in [14].

For a truncation with $N = 2$ supersymmetries the fraction of supersymmetry preserved by these four values of a is $(1/2, 1/2, 1/2, 1/2)$; for $N = 4$ it is (1/2,1/2,1/4,1/4); for $N = 8$ it is $(1/2, 1/4, 1/8, 1/8)$ [14, 15, 3, 16, 17]. In each case, one may find BPS states in the superstring spectrum with the right masses and charges to be identified with these extreme black holes. On the basis of these mass and charge assignments, it was further suggested [3, 14] that we interpret these four values of a as $1-, 2-, 3-$ and 4-particle *bound states* with zero binding energy. For example, the Reissner–Nordstrom ($a = 0$) black hole combines four ($a = \sqrt{3}$) black holes: an electric Kaluza–Klein black hole, a magnetic Kaluza–Klein black hole, an electric winding black hole and a magnetic winding black hole. This zero-binding-energy bound-state conjecture can, in fact, be verified in the classical black hole picture by finding explicit 4-centred black hole solutions which coincide with the $a = \sqrt{3}, 1, 1/\sqrt{3}, 0$ solutions as we bring $1, 2, 3, 4$ centres together and take the remaining $3, 2, 1, 0$ centres out to infinity [18]. Such a construction is possible because of the appearance of four independent harmonic functions [19]. Moreover, this provides a novel realization of the *no-force* condition in that the charge carried by each black hole corresponds to a different $U(1)$. Thus the gravitational attraction cannot be cancelled by an electromagnetic repulsion but rather by a subtle repulsion due to scalar exchange. This phenomenon was also observed in [20]. In the above, for purposes of illustration, the special case has been chosen where all non-zero charges are equal to unity, but it is easily generalized to the case of different electric charges Q_1, P_2, Q_3, P_4 where the interpretation is that of a $(Q_1 + P_2 + Q_3 + P_4)$-particle bound state with zero binding energy [21].

A subsequent paper showed that this string-state/black-hole equivalence, and the corresponding bound state interpretation, are consistent not only with the mass and charge assignments, but also with the spin and supermultiplet structures [22]. String states are labelled by their superspin which tells us which supermultiplet we are talking about and by their ordinary spin which tells us which member of the supermultiplet we are talking about. In the string-state/black-hole dictionary, it is the bosonic Kerr angular momentum which yields the superspin whereas it is the fermionic angular momentum, provided by the Aichelburg–Embacher [23] fermionic hair, which yields the ordinary spin. As a further test, the gyromagnetic ratios of the black holes were calculated and found to be in agreement [24] with those of the string states [25]. Indeed, one of the motivations for believing the equivalence [3, 26] was based on the observation that both Kaluza–Klein string states [27] and extreme electrically charged Kaluza–Klein black holes [28] have the same (anomalous) gyromagnetic ratio $g = 1$.

In a similar fashion, it was then conjectured [29] that the Kaluza–Klein states arising from the compactification of $D = 11$ supergravity on a circle should be identified with the extreme electrically charged black hole solutions of $D = 10$ Type IIA theory [8].

These ideas extend quite naturally to the black p-branes of M-theory [8, 31, 33] which in the extreme limit may become super p-branes [32]. The same Lagrangian (5.1) appears in arbitrary dimensions $D \leq 11$ but where F is now a $(p+2)$-form. The parameter a can conveniently be expressed as [31, 35]

$$a^2 = \Delta - \frac{2(p+1)(D-p-3)}{D-2} \tag{5.2}$$

since Δ is a quantity that is preserved under dimensional reduction [35]. One may calculate the macroscopic entropy of the black p-branes and one finds that in the extreme limit it vanishes except when $a = 0$ and $p = 0$, i.e. for black holes in $D = 5$ with $\Delta = 4/3$ and in $D = 4$ with $\Delta = 1$ (the Reissner–Nordstrom solution). Moreover, branes with $\Delta = 4/n$ can also be regarded as bound states with zero binding energy of n fundamental $\Delta = 4$ branes [22, 34]. One again finds $1 \leq n \leq m$-centred p-brane solutions which reproduce the $\Delta = 4/n$ solutions of [35] as we allow n of the centres to coincide and take the remaining $(m-n)$ out to infinity.

In the case of $D = 4$ black holes, it remained a puzzle why these four values of a, namely $\sqrt{3}, 1, 1/\sqrt{3}, 0$ giving rise to $n = 1, 2, 3, 4$-particle bound states should be singled out. This puzzle was resolved by the realization [36] that the M-theoretic origin of the $a = \sqrt{3}, 1, 1/\sqrt{3}$ black holes preserving 2^{-n} of the supersymmetry was given by re-interpreting the $D = 11$ solutions of [37] as $n = 1, 2, 3$ *orthogonally intersecting* membranes or fivebranes in $D = 11$, which are then wrapped around the compact dimensions. Once one has introduced two or three intersecting membranes: $2 \perp 2$ or $2 \perp 2 \perp, 2$ and fivebranes: $5 \perp 5$ or $5 \perp 5 \perp, 5$, one can also envisage other supersymmetry-preserving intersections [38, 39]: $2 \perp 5$, $2 \perp 5 \perp 5$ and

$2 \perp 5 \perp 5$. Moreover, the missing $a = 0$ case preserving 1/8 supersymmetry admits the interpretation of *four* intersecting M-branes: $2 \perp 2 \perp 5 \perp 5$ [40] or alternatively as an intersecting membrane and fivebrane superposed by a Kaluza–Klein monopole [38, 39]. More complicated configurations, including branes intersecting at angles, are also possible [41–45].

What does all this have to do with the problem of finding a *microscopic* origin of the Beckenstein–Hawking black hole entropy S as one quarter of the area A of the event horizon? Macroscopically, the entropy of a black hole with scalar-Maxwell parameter a and inner and outer event horizons at $r_\pm$ is given by [12]

$$S = \frac{1}{4}A = \pi {r_+}^2 \left(\frac{r_+ - r_-}{r_+}\right)^{2a^2/(1+a^2)} \tag{5.3}$$

and so, according to this formula, the extreme ($r_+ = r_-$) black holes have zero entropy for $a \neq 0$ whereas the $a = 0$ case has $S = \pi {r_+}^2$. However, in the string-state/black-hole equivalence picture the entropy is supposed to be provided by the degeneracy of string states with the same mass and charges as the black hole. This degeneracy is certainly non-zero even for those states identified with the $a \neq 0$ black holes and so the test seems to fail. On the other hand, these solutions have non-trivial dilaton and hence involve strong coupling effects which render (5.3) untrustworthy. An attempt to remedy this was then made [46, 47] by showing that the entropy of the black holes evaluated at the *stretched horizon* [48] qualitatively matches the result expected from the degeneracy of string states. On the other hand, the $a = 0$ case in $D = 4$ and its $a = 0$ friend in $D = 5$ discussed above, can offer no such excuse. If this black hole picture is right, the entropy calculated from the logarithm of degeneracy of states has to yield one quarter the area of the event horizon!

The solution to this long-standing puzzle of explaining the microscopic origin of Beckenstein–Hawking entropy came from an unexpected quarter (no pun intended). We have already discussed the interpretation of black holes as bound states of wrapped p-branes. But if the charges in question are Ramond–Ramond charges, then these p-branes are just the Dirichlet branes [49]: surfaces of dimension p on which open strings can end. The problem of counting the number of string states is thus reduced to a solvable problem in conformal field theory! It was in this way that Strominger and Vafa [50] correctly provided the first *microscopic* derivation of the black hole entropy in the case $a = 0, D = 5$. The $a = 0, D = 4$ case followed soon after [51], and since then there has blossomed a whole industry involving generalizations to rotating black holes, non-extreme black holes, grey body factors and the like. All this is also suggestive of a resolution of the black hole information paradox, but this remains a controversial issue. The reader is referred to several reviews [52, 53, 54].

More recently, black hole solutions of *gauged* supergravity are attracting a good deal of attention due, in large part, to the correspondence between anti-de Sitter space and conformal field theories on its boundary as discussed in chapter 6. In [55], for example, new anti-de Sitter black hole solutions of gauged $N = 8$, $D = 4$, $SO(8)$ supergravity were presented. By focussing on the $U(1)^4$ Cartan subgroup, non-extremal 1, 2, 3 and 4 charge solutions were found. In the extremal limit, they may preserve up to 1/2, 1/4, 1/8 and 1/8 of the supersymmetry respectively. By contrast, the magnetic solutions preserve none. Since $N = 8$, $D = 4$ supergravity is a consistent truncation of $N = 1$, $D = 11$ supergravity, resulting from the S^7 compactification, it follows that these black holes will also be solutions of this theory. In [55], it was conjectured that a subset of the extreme electric black holes preserving 1/2 the supersymmetry may be identified with the S^7 Kaluza–Klein spectrum, with the non-abelian quantum numbers provided by the fermionic zero modes.

In [56] the non-linear S^7 Kaluza–Klein ansatz describing the embedding of the $U(1)^4$ truncation was presented. The charges for the black holes with toroidal horizons may be interpreted as the angular momenta of $D = 11$ membranes spinning in the transverse dimensions [57, 56]. The horizons of the black holes coincide with the worldvolume of the branes. It is curious that the same $U(1)^4$ black hole charges appear in the S^7 compactification of $D = 11$ supergravity as in the T^7 compactification, but for totally different reasons. Instead of arising from the intersection of different non-rotating branes, they arise from the different angular momenta of a single brane. This is indicative of deeper levels of duality yet to be uncovered.

References

[1] Beckenstein J 1973 Black holes and entropy *Phys. Rev.* D **7** 2333

[2] Hawking S W 1975 Particle creation by black holes *Commun. Math. Phys.* **43** 199

[3] Duff M J and Rahmfeld J 1995 Massive string states as extreme black holes *Phys. Lett.* B **345** 441

[4] Susskind L 1993 Some speculations about black hole entropy in string theory, hep-th/9309145

[5] Ellis J, Mavromatos N E and Nanopoulos D V 1992 Quantum mechanics and black holes in four-dimensional string theory *Phys. Lett.* B **278** 246

[6] Gibbons G W 1985 in *Supersymmetry, Supergravity and Related Topics* ed F del Aguila, J A Azcarraga and L E Ibanez (Singapore: World Scientific)

[7] Duff M J, Khuri R R, Minasian R and Rahmfeld J 1994 New black hole, string and membrane solutions of the four-dimensional heterotic string *Nucl. Phys.* B **418** 195

[8] Khuri R R and Myers R C 1995 Dynamics of extreme black holes and massive string states *Phys. Rev.* D **52** 6988

[9] Callan C G, Maldacena J M and A W Peet 1996 Extremal black holes as fundamental strings *Nucl. Phys.* B **475** 645

[10] Gibbons G W 1982 Antigravitating black hole solitons with scalar hair in $N = 4$ supergravity *Nucl. Phys.* B **207** 337

[11] Garfinkle D, Horowitz G T and Strominger A 1991 Charged black holes in string theory *Phys. Rev.* D **43** 3140

[12] Holzhey C F E and Wilczek F 1992 Black holes as elementary particles *Nucl. Phys.* B **380** 447

[13] Gibbons G W, Horowitz G T and Townsend P K 1995 Higher dimensional resolution of dilatonic black hole singularities *Class. Quantum Grav.* **12** 297

[14] Duff M J, Liu J T and Rahmfeld J 1996 Four-dimensional string/string/string triality *Nucl. Phys.* B **459** 125

[15] Kallosh R, Linde A, Ortin T, Peet A and Van Proeyen A 1992 Supersymmetry as a cosmic censor *Phys. Rev.* D **6** 5278

[16] Lu H and Pope C N 1996 *P*-brane solitons in maximal supergravities *Nucl. Phys.* B **465** 127
[17] Khuri R R and Ortin T 1996 Supersymmetric black holes in $N = 8$ supergravity *Nucl. Phys.* B **467** 355
[18] Rahmfeld J 1996 Extremal black holes as bound states *Phys. Lett.* B **372** 198
[19] Cvetic M and Tseytlin A 1996 Solitonic strings and BPS saturated dyonic black holes *Phys. Rev.* D **53** 589
[20] Kallosh R and Linde A 1996 Supersymmetric balance of forces and condensation of BPS states *Phys. Rev.* D **53** 5734
[21] Cvetič M and Youm D 1996 Dyonic BPS saturated black holes of heterotic string on a six-torus *Phys. Rev.* D **53** 584
[22] Duff M J and Rahmfeld J 1996 Bound states of black holes and other *P*-branes *Nucl. Phys.* B **481** 332
[23] Aichelburg P C and Embacher F 1986 Exact superpartners of $N = 2$ supergravity solitons *Phys. Rev.* D **34** 3006
[24] Duff M J, Liu J T and Rahmfeld J 1997 Dipole moments of black holes and string states *Nucl. Phys.* B **494** 161
[25] Russo J G and Susskind L 1995 Asymptotic level density in heterotic string theory and rotating black holes *Nucl. Phys.* B **437** 611
[26] Duff M J 1995 Kaluza–Klein theory in perspective *Proc. Nobel Symp. Oskar Klein Centenary, Stockholm, 1994* ed E Lindstrom (Singapore: World Scientific)
[27] Hosoya A, Ishikawa K, Ohkuwa Y and Yamagishi K 1984 Gyromagnetic ratio of heavy particles in the Kaluza–Klein theory *Phys. Lett.* B **134** 44
[28] Gibbons G W and Wiltshire D L 1986 Black holes in Kaluza–Klein theory *Ann. Phys., NY* **167** 201
[29] Townsend P K 1995 The eleven-dimensional supermembrane revisited *Phys. Lett.* B **350** 184–7
[30] Horowitz G T and Strominger A 1991 Black strings and *p*-branes *Nucl. Phys.* B **360** 197
[31] Duff M J and Lu J X 1994 Black and super p-branes in diverse dimensions *Nucl.Phys.* B **416** 301
[32] Duff M J and Lu J X 1993 Type *II* *p*-branes: the brane scan revisited *Nucl. Phys.* B **390** 276
[33] Duff M J, Lu H and Pope C N 1996 The black branes of *M*-theory *Phys. Lett.* B **382** 73–80
[34] Khviengia N, Khviengia Z, Lu H and Pope C N 1996 Intersecting *M*-branes and bound states *Phys. Lett.* B **388** 313
[35] Lu H, Pope C N, Sezgin E and Stelle K S 1995 Stainless super *p*-branes *Nucl. Phys.* B **456** 669
[36] Papadopoulos G and Townsend P K 1996 Intersecting *M*-branes *Phys. Lett.* B **380** 273–9
[37] Gueven R 1992 Black p-brane solutions of $d = 11$ supergravity theory *Phys.Lett.* B **276** 49–55
[38] Tseytlin A A 1997 Composite BPS configurations of *p*-branes in ten dimensions and eleven dimensions *Class. Quantum Grav.* **14** 2085
[39] Tseytlin A A 1996 Harmonic superpositions of *M*-branes *Nucl. Phys.* B **475** 149–63
[40] Klebanov I R and Tseytlin A A 1996 Intersecting *M*-branes as four-dimensional black holes *Nucl. Phys.* B **475** 179–92
[41] Gauntlett J, Kastor D and Traschen J 1996 Overlapping branes in *M*-theory *Nucl. Phys.* B **478** 544–60

[42] Berkooz M, Douglas M and Leigh R 1996 Branes intersecting at angles *Nucl. Phys.* B **480** 265

[43] Bergshoeff E, de Roo M, Eyras E, Janssen B and van der Schaar J P 1997 Intersections involving monopoles and waves in eleven dimensions, hep-th/9704120

[44] Bergshoeff E, de Roo M, Eyras E, Janssen B and van der Schaar J P 1997 Multiple intersections of D-branes and M-branes *Nucl. Phys.* B **494** 119

[45] Gauntlett J P, Gibbons G W, Papadopoulos G and Townsend P K 1997 Hyper-Kahler manifolds and multiply intersecting branes *Nucl. Phys.* B **500** 133

[46] Sen A 1995 Extremal black holes and elementary string states *Mod. Phys. Lett.* A **10** 2081

[47] Peet A 1995 Entropy and supersymmetry of D dimensional extremal electric black holes versus string states *Nucl. Phys.* B **456** 732

[48] Susskind L, Thorlacius L and Uglum J 1993 The stretched horizon and black hole complementarity *Phys. Rev.* D **48** 3743

[49] Polchinski J 1995 Dirichlet-branes and Ramond–Ramond charges *Phys. Rev. Lett.* **75** 4724

[50] Strominger A and Vafa C 1996 Microscopic origin of the Bekenstein–Hawking entropy *Phys. Lett.* B **379** 99

[51] Johnson C V, Khuri R R and Myers R C 1996 Entropy of 4D extremal black holes *Phys. Lett.* B **378** 78

[52] Maldacena J 1996 Black holes in string theory, hep-th/9607235

[53] Horowitz G T 1996 The origin of black hole entropy in string theory, gr-qc/9604051

[54] Polchinski J and Horowitz 1997 A correspondence principle for black holes and strings *Phys. Rev.* D **55** 6189

[55] Duff M J and Liu J T 1999 Anti-de Sitter black holes in gauged $N = 8$ supergravity, hep-th/9901149.

[56] Cvetic M, Duff M J, Hoxha P, Liu J T, Lu H, Lu J X, Martinez-Acosta R, Pope C N, Sati H and Tran T A 1999 Embedding AdS black holes in ten and eleven dimensions, hep-th/9903214

[57] Chamblin A, Emparan R, Johnson C V and Myers R C 1999 Charged AdS black holes in ten and eleven dimensions, hep-th/9902170

Reprinted from Phys. Lett. B **380** (1996) 273–9

Intersecting M-branes

G. Papadopoulos, P.K. Townsend
D.A.M.T.P., University of Cambridge, Silver Street, Cambridge CB3 9EW, UK

Received 30 March 1996
Editor: P.V. Landshoff

Abstract

We present the magnetic duals of Güven's electric-type solutions of $D = 11$ supergravity preserving $1/4$ or $1/8$ of the $D = 11$ supersymmetry. We interpret the electric solutions as n orthogonal intersecting membranes and the magnetic solutions as n orthogonal intersecting 5-branes, with $n = 2, 3$; these cases obey the general rule that p-branes can self-intersect on $(p-2)$-branes. On reduction to $D = 4$ these solutions become electric or magnetic dilaton black holes with dilaton coupling constant $a = 1$ (for $n = 2$) or $a = 1/\sqrt{3}$ (for $n = 3$). We also discuss the reduction to $D = 10$.

1. Introduction

There is now considerable evidence for the existence of a consistent supersymmetric quantum theory in 11 dimensions ($D = 11$) for which the effective field theory is $D = 11$ supergravity. This theory, which goes by the name of M-theory, is possibly a supermembrane theory [1]; in any case, the membrane solution of $D = 11$ supergravity [2], and its magnetic-dual 5-brane solution [3], (which we refer to jointly as 'M-branes') play a central role in what we currently understand about M-theory and its implications for non-perturbative superstring theory (see, for example, [4–10]). It is therefore clearly of importance to gain a fuller understanding of *all* the p-brane-like solutions of $D = 11$ supergravity.

For example, it was shown by Güven [3] that the membrane solution of [2] is actually just the first member of a set of three electric-type solutions parametrized, in the notation of this paper, by the integer $n = 1, 2, 3$. These solutions are

$$\begin{aligned} ds^2_{(11)} &= -H^{-2n/3}dt^2 + H^{(n-3)/3}ds^2(\mathbb{E}^{2n}) \\ &\quad + H^{n/3}ds^2(\mathbb{E}^{10-2n}) \\ F_{(11)} &= -3\, dt \wedge dH^{-1} \wedge J\,, \end{aligned} \tag{1}$$

where H is a harmonic function on $\mathbb{E}^{10-2n}$ with point singularities, J is a Kähler form on $\mathbb{E}^{2n}$ and $F_{(11)}$ is the 4-form field strength of $D = 11$ supergravity. The proportion of the $D = 11$ supersymmetry preserved by these solutions is 2^{-n}, i.e. $1/2, 1/4$ and $1/8$, respectively. The $n = 1$ case is the membrane solution of [2]. We shall refer to the $n = 2$ and $n = 3$ cases, which were interpreted in [3] as, respectively, a 4-brane and 6-brane, as the 'Güven solutions'. Their existence has always been something of a mystery since $D = 11$ supergravity does not have the five-form or seven-form potentials that one would expect to couple to a 4-brane or a 6-brane. Moreover, unlike the membrane which has a magnetic dual 5-brane, there are no known magnetic duals of the Güven solutions.

In our opinion, the p-brane interpretation given by Güven to his electric $n = 2, 3$ solutions is questionable because of the lack of $(p+1)$-dimensional Poincaré

PII S0370-2693(96)00506-0

invariance expected of such objects. This is to be contrasted with the $n = 1$ case, for which the solution (1) acquires a 3-dimensional Poincaré invariance appropriate to its membrane interpretation. In this paper we shall demystify the Güven solutions by re-interpreting them as *orthogonally intersecting membranes*. We also present their magnetic duals which can be interpreted as orthogonally intersecting 5-branes. The latter are new magnetic-type solutions of $D = 11$ supergravity preserving, respectively, $1/4$ and $1/8$ of the $D = 11$ supersymmetry. A novel feature of these solutions is that they involve the intersection of $D = 11$ fivebranes on 3-branes. We shall argue that this is an instance of a general rule: *p-branes can self-intersect on $(p-2)$-branes*.

Particle solutions in four dimensions ($D = 4$) can be obtained from M-brane solutions in $D = 11$ by wrapping them around 2-cycles or 5-cycles of the compactifying space. This is particularly simple in the case of toroidal reduction to $D = 4$. In this case, wrapped membranes and 5-branes can be interpreted [4] as, respectively, electric and magnetic $a = \sqrt{3}$ extreme black holes (in a now standard terminology which we elaborate below). Here we show that Güven's solutions, and their magnetic duals, have a $D = 4$ interpretation as either $a = 1$ (for $n = 2$) or $a = 1/\sqrt{3}$ (for $n = 3$) extreme electric or magnetic black holes. This $D = 11$ interpretation of the $a = 1, 1/\sqrt{3}$ extreme black holes in $D = 4$ is in striking accord with a recent interpretation [11] of them (following earlier suggestions [12], and using results of [13]) as bound states at threshold of two (for $a = 1$) or three (for $a = 1/\sqrt{3}$) $a = \sqrt{3}$ extreme black holes.

Rather than reduce to $D = 4$ one can instead reduce to $D = 10$ to find various solutions of IIA supergravity representing intersecting p-branes. We shall briefly mention these at the conclusion of this paper. There is presumably an overlap here with the discussion of intersections [14] and the 'branes within branes' [15–19] in the context of D-branes, but we have not made any direct comparison. The general problem of intersecting super p-branes was also discussed in [20] in the context of flat space extended solitons. We must also emphasize that the $D = 11$ supergravity solutions we discuss here have the interpretation we give them only after an integration over the position of the intersection in the 'relative transverse space'; we argue that this is appropriate for the interpretation as extreme black holes in $D = 4$.

2. Intersecting p-branes

We begin by motivating our re-interpretation of the $D = 11$ supergravity solutions (1). The first point to appreciate is that infinite planar p-branes, or their parallel multi p-brane generalizations, are not the only type of field configuration for which one can hope to find static solutions. Orthogonally intersecting p-branes could also be static. The simplest case is that of pairs of orthogonal p-branes intersecting in a q-brane, $q < p$. The next simplest case is three p-branes having a *common* q-brane intersection. Here, however, there is already a complication: one must consider whether the intersection of any two of the three p-branes is also a q-brane or whether it is an r-brane with $r > q$ (we shall encounter both cases below). There are clearly many other possibilities once one considers more than three intersecting p-branes, and even with only two or three there is the possibility of intersections of orthogonal p-branes for different values of p. A limiting case of orthogonal intersections of p-branes occurs when one p-brane lies entirely within the other. An example is the $D = 11$ solution of [21] which can be interpreted as a membrane lying within a 5-brane. For the purposes of this paper, orthogonal intersections of two or three p-branes for the same value of p will suffice.

Consider the case of n intersecting p-branes in D-dimensions for which the common intersection is a q-brane, with worldvolume coordinates ξ^μ, $\mu = 0, 1, \ldots, q$. The tangent vectors to the p-branes' worldvolumes that are *not* tangent to the q-brane's worldvolume span a space V, which we call the 'relative transverse space'; we denote its coordinates by x^a, $a = 1, \ldots, \ell$, where $\ell = \dim V$. Let y denote the coordinates of the remaining 'overall transverse space' of dimension $D - q - \ell$. The D-dimensional spacetime metric for a system of static and *orthogonal* p-branes intersecting in a q-brane should take the form

$$ds^2 = A(x,y)\,d\xi^\mu d\xi^\nu \eta_{\mu\nu} + B_{ab}(x,y)\,dx^a dx^b + C_{ij}(x,y)\,dy^i dy^j\,. \qquad (2)$$

Note the $(q+1)$-dimensional Poincaré invariance. We also require that $A \to 1$, and that B, C tend to the identity matrices, as $|y| \to \infty$, so the metric is asymptotic to the D-dimensional Minkowski metric in this limit.

A metric of the form (2) will have a standard interpretation as n intersecting p-branes only if the coefficients A, B, C functions are such that the metric approaches that of a single p-brane as one goes to infinity in V while remaining a finite distance from one of the n p-branes. The Güven solutions (1) do not have this property because they are translation invariant along directions in V. Specifically, they are special cases of (2) of the form

$$ds^2 = A(y)d\xi^\mu d\xi^\nu \eta_{\mu\nu} + B(y)dx^a dx^b \delta_{ab} + C(y)dy^i dy^j \delta_{ij} \,. \quad (3)$$

Because of the translational invariance in x directions, the energy density is the same at every point in V for fixed y. However, the translational invariance allows us to periodically identify the x coordinates, i.e. to take $V = \mathbb{T}^\ell$. In this case, the metric (3) could be viewed as that of a q-brane formed from the intersection of p-branes after averaging over the intersection points in V. If we insist that the p-branes have zero momentum in V-directions orthogonal to their q-brane, then this averaging is an immediate consequence of quantum mechanics. This delocalization effect should certainly be taken into account when the size of V is much smaller than the scale at which we view the dynamics in the y directions, i.e. for scales at which the effective field theory is $(D - \ell)$-dimensional. The q-brane solution of this effective field theory can then be lifted to a solution of the original D-dimensional theory; this solution will be of the form (3).

Thus, metrics of the form (3) can be interpreted as those of p-branes intersecting in a common q-brane. However, the solution does not determine, by itself, the combination of p-branes involved. That is, when interpreted as a q-brane intersection of n_α p_α-branes (for $\alpha = 1, 2, \ldots$) the numbers (n_α, p_α) are not uniquely determined by the numbers (D, q, ℓ). For example, the $n = 2$ Güven spacetime could be interpreted as intersections at a point of (i) 4 strings, or (ii) 2 strings and one membrane or (iii) a 0-brane and a 4-brane or (iv) 2 membranes. Additional information is needed to decide between these possibilities. In the context of M-theory, most of this additional information resides in the hypothesis that the 'basic' p-branes are the M-branes (i.e. the membrane and 5-brane), where 'basic' means that all other p-branes-like objects are to be constructed from them via orthogonal intersections, as described above. There is also additional information coming from the form of the 4-form field strength, which allows us to distinguish between electric, magnetic and dyonic solutions. With this additional information, the intersecting p-brane interpretation of the $n = 2, 3$ Güven solutions is uniquely that of 2 or 3 intersecting membranes.

It is convenient to consider the Güven solutions cases as special cases of n p-branes in D dimensions pairwise intersecting in a common q-brane, i.e. $\ell = n(p - q)$. To see what to expect of the magnetic duals of such solutions it is convenient to make a periodic identification of the x-coordinates in (3), leading to an interpretation of this configuration as a q-brane in $d \equiv D - n(p - q)$ dimensions. The magnetic dual of a q-brane in d dimensions is a $\tilde{q}$-brane, where $\tilde{q} = d - q - 4$. We must now find an interpretation of this $\tilde{q}$-brane as an intersection of n $\tilde{p}$-branes in D-dimensions, where $\tilde{p} = D - p - 4$. The consistency of this picture requires that the dimension of the space $\tilde{V}$ spanned by vectors tangent to the $\tilde{p}$-branes' worldvolumes that are *not* tangent to the $\tilde{q}$-brane's worldvolume be $D - d = D - n(p - q)$. This is automatic when $n = 2$ (but not when $n > 2$). As an example, consider the $n = 2$ Güven solution, interpreted as two orthogonal membranes with a 0-brane intersection. Periodic identification of the x-coordinates leads to a particle-like solution in an effective $D = 7$ supergravity theory. A particle in $D = 7$ is dual to a 3-brane. This 3-brane can now be interpreted as the intersection of two 5-branes. The vectors tangent to the 5-branes' worldvolumes that are *not* tangent to the 3-brane's worldvolume span a four-dimensional space, so the total dimension of the spacetime is $7 + 4 = 11$, as required.

Consider now the $n = 3$ Güven solution, interpreted as three orthogonal membranes intersecting at a common 0-brane. Periodic identification of the x-coordinates now leads to a particle-like solution in an effective $D = 5$ supergravity theory. A particle is dual to a string in $D = 5$, so we should look for a solution in $D = 11$ representing three orthogonal 5-branes whose common intersection is a string. The dimension of the space $\tilde{V}$ spanned by the vectors tangent to the 5-branes' worldvolumes that are *not* tangent to the string's worldsheet depends on whether the common intersection of all three 5-branes is also the intersection of any pair. If it were then $\tilde{V}$ would take its maximal dimension, $3(5-1) = 14$, leading to a total spacetime dimension of $5 + 14 = 19$. Since this is inconsistent with an interpretation in $D = 11$, we conclude that the pairwise intersection of the three 5-branes must be a q-brane with $q > 1$. In fact, the consistent choice is $q = 3$, i.e. each pair of 5-branes has a 3-brane intersection and the three 3-branes themselves intersect in a string [1]. In this case $\tilde{V}$ has dimension six, leading to a total spacetime dimension of eleven.

Note that all the cases of intersecting p-branes which we have argued should occur in M-theory have the property that p-brane pairs (for the same value of p) intersect on $(p-2)$-branes. Specifically, we have argued that 2-branes can intersect on 0-branes, that 5-branes can intersect on 3-branes and that these 3-brane intersections can themselves intersect on 1-branes. We shall conclude this section by explaining why we believe that this is a general rule, i.e. *p-branes can self-intersect on $(p-2)$-branes.*

Recall that the possibility of a membrane having a boundary on a 5-brane [16,17] arises from the fact that the 5-brane worldvolume contains a 2-form potential which can couple to the membrane's string boundary. The same argument does not obviously apply to *intersections* but it is plausible that it does, at least for those cases in which it is possible to view the q-brane intersection within a given p-brane as a dynamical object in its own right. Thus, it is reasonable to suppose that a condition for a p-brane to support a q-brane intersection is that the p-brane worldvolume field theory includes a $(q+1)$-form potential to which the q-brane can couple. We now observe that p-brane worldvolume actions always contain $(D-p-1)$ scalar fields. If one of these scalars is dualized then the worldvolume acquires a $(p-1)$-form potential, which can couple to a $(p-2)$-brane. Hence the rule stated above; the freedom of choice of which scalar to dualize corresponds to the possibility of an energy flow into the p-brane, at the intersection, in any of the directions orthogonal to its worldvolume.

[1] A useful analogy is that of three orthogonal planes in $\mathbb{E}^3$ which intersect pairwise on a line. The three lines intersect at a point.

3. Magnetic duals of Güven solutions

We now have sufficient information to find the magnetic duals of the series of electric solutions (1) of $D = 11$ supergravity. They should be of the form (3) with $q = 7 - 2n$ and they should preserve some fraction of the $D = 11$ supersymmetry. Solutions that preserve some supersymmetry can most easily be found by seeking bosonic backgrounds admitting Killing spinors. The Killing spinor equation can be found directly from the supersymmetry transformation law for the gravitino field ψ_M $(M = 0, 1, 2, \ldots, 10)$, and is

$$\left[D_M + \tfrac{1}{144}\left(\Gamma_M{}^{NPQR} - 8\delta_M^N\Gamma^{PQR}\right)F_{NPQR}\right]\zeta = 0\,, \qquad (4)$$

where D_M is the standard covariant derivative. Solutions ζ of this equation (if any) are the Killing spinors of the bosonic background, i.e. the $D = 11$ metric and 4-form field strength F_{MNPQ}. Backgrounds admitting Killing spinors for which the Bianchi identity for $F_{(11)}$ is also satisfied are automatically solutions of $D = 11$ supergravity. The proportion of the $D = 11$ supersymmetry preserved by such a solution equals the dimension of the space of Killing spinors divided by 32.

By substituting an appropriate ansatz for the metric and 4-form into (4) we have found a series of magnetic solutions parametrised by the integer $n = 1, 2, 3$. These are

$$\begin{aligned} ds^2_{(11)} &= H^{-n/3}(d\xi \cdot d\xi) + H^{-(n-3)/3} ds^2(\mathbb{E}^{2n}) \\ &\quad + H^{2n/3} ds^2(\mathbb{E}^3) \\ F_{(11)} &= \pm 3 \star dH \wedge J\,, \end{aligned} \qquad (5)$$

where $\star$ is the Hodge star of $\mathbb{E}^3$, J is the Kähler form on $\mathbb{E}^{2n}$ and $n = 1, 2, 3$. Our conventions for forms are such that

$$\begin{aligned} J &= \tfrac{1}{2} J_{ab} dx^a \wedge dx^b \\ F_{(11)} &= \tfrac{1}{4} F_{MNPR}\, dx^M \wedge dx^N \wedge dx^P \wedge dx^R\,. \end{aligned} \qquad (6)$$

The function H is harmonic on $\mathbb{E}^3$ with point singularities. Asymptotic flatness at 'overall transverse infinity' requires that $H \to 1$ there, so that

$$H = 1 + \sum_i \frac{\mu_i}{|x - x_i|}\,, \qquad (7)$$

for some constants μ_i. Note that these solutions have an $8-2n$ dimensional Poincaré invariance, as required.

In the $n = 1$ case the metric can be written as

$$ds^2_{(11)} = H^{-1/3} d\xi \cdot d\xi + H^{\frac{2}{3}} ds^2(\mathbb{E}^5) \qquad (8)$$

which is formally the same as the 5-brane solution of [3]. The difference is that the function H in our solution is harmonic on an $\mathbb{E}^3$ subspace of $\mathbb{E}^5$, i.e. our solution is a special case of the general 5-brane solution, for which H is harmonic on $\mathbb{E}^5$. The $n = 2, 3$ cases are *new* solutions of $D = 11$ supergravity with the properties expected from their interpretation as intersecting 5-branes. The solutions of the Killing spinor equation for the background given by (5) are

$$\zeta = H^{-n/12}\zeta_0, \quad \bar{\Gamma}_a\zeta_0 = \mp J_{ab}\gamma^*\bar{\Gamma}^b\zeta_0 , \qquad (9)$$

where $\{\bar{\Gamma}_a; a = 1, \ldots, 2n\}$ are the (frame) constant $D = 11$ gamma matrices along the $\mathbb{E}^{2n}$ directions, γ^* is the product of the three constant gamma matrices along the $\mathbb{E}^3$ directions and ζ_0 is a constant $D = 11$ spinor. It follows from (9) that the number of supersymmetries preserved by the magnetic intersecting 5-brane solutions is 2^{-n}, exactly as in the electric case.

4. $D = 4$ interpretation

We now discuss the interpretation of the solutions (1) and (5) in $D = 4$. The $D = 4$ field theory obtained by compactifying $D = 11$ supergravity on $\mathbb{T}^7$ can be consistently truncated to the massless fields of $N = 8$ supergravity. The latter can be truncated to

$$I = \int d^4x\sqrt{-g}\Big[R - 2(\partial\phi)^2 - \tfrac{1}{2}e^{-2a\phi}F^2\Big] , \qquad (10)$$

where F is an abelian 2-form field strength, provided that the scalar/vector coupling constant a takes one of the values[2] [4,22]

$$a = \sqrt{3}\,, \quad 1, \quad \frac{1}{\sqrt{3}}\,, \quad 0. \qquad (11)$$

The truncation of $N = 8$ supergravity to (10) is not actually a consistent one (in the standard Kaluza-Klein sense) since consistency requires that F satisfy $F \wedge F = 0$. However, this condition is satisfied for purely electric or purely magnetic field configurations, so purely electric or purely magnetic solutions of the field equations of (10) are automatically solutions of $N = 8$ supergravity, for the above values of a. In particular, the static extreme electric or magnetic black holes are solutions of $N = 8$ supergravity that preserve some proportion of the $N = 8$ supersymmetry. This proportion is $1/2,\ 1/4,\ 1/8,\ 1/8$ for $a = \sqrt{3},\ 1,\ 1/\sqrt{3},\ 0$, respectively.

It is known that the membrane and fivebrane solutions of $D = 11$ supergravity have a $D = 4$ interpretation as $a = \sqrt{3}$ extreme black holes. Here we shall extend this result to the $n = 2, 3$ cases by showing that the electric solutions (10) of $D = 11$ supergravity, and their magnetic duals (5) have a $D = 4$ interpretation as extreme black holes with scalar/vector coupling $a = \sqrt{(4/n) - 1}$. As we have seen, the $D = 11$ solutions for $n = 2, 3$, electric or magnetic, have a natural interpretation as particles in $D = 7$ and $D = 5$, respectively. It is therefore convenient to consider a two-step reduction to $D = 4$, passing by these intermediate dimensions. The $n = 3$ case is actually simpler, so we shall consider it first. We first note that for $a = 1/\sqrt{3}$ the action (10) can be obtained from that of simple supergravity in $D = 5$, for which the bosonic fields are the metric $ds^2_{(5)}$ and an abelian vector potential A with 2-form field strength $F_{(5)}$, by the ansatz

$$ds^2_{(5)} = e^{2\phi}ds^2 + e^{-4\phi}dx_5^2, \quad F_{(5)} = F , \qquad (12)$$

where ds^2, ϕ and F are the metric and fields appearing in the $D = 4$ action (10). Note that this ansatz involves the truncation of the $D = 4$ axion field A_5; it is the consistency of this truncation that requires $F \wedge F = 0$. As mentioned above, this does not present problems in the purely electric or purely magnetic cases, so these $D = 4$ extreme black hole solutions can be lifted, for $a = 1/\sqrt{3}$, to solutions of $D = 5$ supergravity. The magnetic black hole lifts to the $D = 5$ extreme black multi string solution [23]

$$ds^2_{(5)} = H^{-1}(-dt^2 + dx_5^2) + H^2 ds^2(\mathbb{E}^3)$$
$$F_{(5)} = {}^\star dH , \qquad (13)$$

where $\star$ is the Hodge star of $\mathbb{E}^3$ and H is a harmonic function on $\mathbb{E}^3$ with some number of point singularities, i.e. as in (7). We get a magnetic $a = 1/\sqrt{3}$ extreme black hole by wrapping this string around the x_5 direction.

---

[2] We may assume that $a \geq 0$ without loss of generality.

The electric $a = 1/\sqrt{3}$ extreme multi black hole lifts to the following solution of $D = 5$ supergravity:

$$ds^2_{(5)} = -H^{-2}dt^2 + Hds^2(\mathbb{E}^3 \times S^1)$$
$$F_{(5)} = dt \wedge dH^{-1} , \qquad (14)$$

where H is a harmonic function on $\mathbb{E}^3$. This solution is the 'direct' dimensional reduction of the extreme electrically-charged black hole solution of $D = 5$ supergravity [24]. The latter is formally the same as (14) but $\mathbb{E}^3 \times S^1$ is replaced by $\mathbb{E}^4$ and H becomes a harmonic function on $\mathbb{E}^4$.

To make the connection with $D = 11$ we note that the Kaluza-Klein (KK) ansatz

$$ds^2_{(11)} = ds^2_{(5)} + ds^2(\mathbb{E}^6), \quad F_{(11)} = F_{(5)} \wedge J , \qquad (15)$$

where J is a Kähler 2-form on $\mathbb{E}^6$, provides a consistent truncation of $D = 11$ supergravity to the fields of $D = 5$ simple supergravity. This allows us to lift solutions of $D = 5$ supergravity directly to $D = 11$. It is a simple matter to check that the $D = 5$ extreme black hole solution lifts to the $n = 3$ Güven solution and that the $D = 5$ extreme black string lifts to the magnetic $n = 3$ solution of (5).

The $a = 1$ case works similarly except that the intermediate dimension is $D = 7$. The KK/truncation ansatz taking us to $D = 7$ is

$$ds^2_{(11)} = e^{-(4/3)\phi} d\hat{s}^2_{(7)} + e^{(2/3)\phi} ds^2(\mathbb{T}^4)$$
$$F_{(11)} = F_{(7)} \wedge J . \qquad (16)$$

where $d\hat{s}^2_{(7)}$ is the string-frame $D = 7$ metric. Consistency of this truncation restricts $F_{(7)}$ to satisfy $F_{(7)} \wedge F_{(7)} = 0$, but this will be satisfied by our solutions. The ansatz then taking us to $D = 4$ is

$$d\hat{s}^2_{(7)} = d\hat{s}^2 + ds^2(\mathbb{T}^3), \quad F_{(7)} = F , \qquad (17)$$

where $d\hat{s}^2 = e^{2\phi} ds^2$ is the string-frame $D = 4$ metric. Combining the two KK ansätze, it is not difficult to check that the electric $a = 1$ extreme black hole lifts to the $n = 2$ Güven solution in $D = 11$ and that the magnetic $a = 1$ extreme black hole lifts to the new $n = 2$ magnetic $D = 11$ solution of this paper.

5. Comments

We have extended the $D = 11$ interpretation of $D = 4$ extreme black hole solutions of $N = 8$ supergravity with scalar/vector coupling $a = \sqrt{3}$ to two of the other three possible values, namely $a = 1$ and $a = 1/\sqrt{3}$. While the $a = \sqrt{3}$ black holes have a $D = 11$ interpretation as wrapped M-branes, the $a = 1$ and $a = 1/\sqrt{3}$ black holes have an interpretation as wrappings of, respectively, two or three *intersecting* M-branes. We have found no such interpretation for the $a = 0$ case, i.e. extreme Reissner-Nordström black holes; we suspect that their $D = 11$ interpretation must involve the gauge fields of KK origin (whereas this is optional for the other values of a).

The solution of $D = 11$ supergravity representing three intersecting 5-branes is essentially the same as the extreme black string solution of $D = 5$ supergravity. For both this solution and the $D = 11$ 5-brane itself the singularities of H are actually coordinate singularities at event horizons. Moreover, these solutions were shown in [23] to be geodesically complete, despite the existence of horizons, so it is of interest to consider the global structure of the solution representing two intersecting 5-branes. For this solution the asymptotic form of H near one of its singularities is $H \sim 1/r$, where r is the radial coordinate of $\mathbb{E}^3$. Defining a new radial coordinate ρ by $r = \rho^3$, we find that the asymptotic form of the metric near $\rho = 0$ is

$$ds^2_{(11)} \sim \rho^2 d\xi \cdot d\xi + \frac{1}{\rho} ds^2(\mathbb{E}^4) + 9\, d\rho^2 + \rho^2 d\Omega^2 \qquad (18)$$

where $d\Omega^2$ is the metric of the unit 2-sphere. 'Spatial' sections of this metric, i.e. those with $d\xi = 0$, are topologically $\mathbb{E}^4 \times S^2 \times \mathbb{R}^+$, where ρ is the coordinate of $\mathbb{R}^+$. Such sections are singular at $\rho = 0$ although it is notable that the volume element of $\mathbb{E}^4 \times S^2$ remains finite as $\rho \to 0$.

We have concentrated in this paper on solutions representing intersecting p-branes in $D = 11$, i.e. M-branes, but the main idea is of course applicable to supergravity theories in lower dimensions. In fact, the intersecting M-brane solutions in $D = 11$ can be used to deduce solutions of $D = 10$ IIA supergravity with a similar, or identical, interpretation by means of either direct or double dimensional reduction. Direct reduction yields solutions of $D = 10$ IIA supergravity with exactly the same interpretation as in $D = 11$, i.e. two (for $n = 2$) or three (for $n = 3$) membranes intersecting at a point, in the electric case, and, in the magnetic case, two 5-branes intersecting at a 3-brane (for $n = 2$) or three 5-branes intersecting at a string (for $n = 3$). On the other hand, double dimensional reduction of the electric $D = 11$ $n > 1$ solutions, i.e. wrapping one membrane around the S^1, gives solutions of $D = 10$ $N = 2A$ supergravity theory representing either a string and a membrane intersecting at a point (for $n = 2$) or a string and two membranes intersecting at a point (for $n = 3$). In the magnetic case, the wrapping can be done in two different ways. One way, which is equivalent to double-dimensional reduction, is to wrap along one of the relative transverse directions, in which case the $D = 10$ solutions represent either a 5-brane and a 4-brane intersecting at a 3-brane (for $n = 2$) or two 5-branes and a 4-brane intersecting at a string (for $n = 3$). The other way, which might reasonably be called 'triple dimensional' reduction, is to wrap along one of the directions in the *common* q-brane intersection, in which case one gets $D = 10$ solutions representing either two 4-branes intersecting at a membrane (for $n = 2$) or three 4-branes intersecting at a point (for $n = 3$). We expect that some of these IIA $D = 10$ solutions will have a superstring description via Dirichlet-branes.

Finally, we point out that the solutions (1) and (5) can both be generalized to the case in which $ds^2(\mathbb{E}^{2n})$ is replaced by any Ricci-flat Kähler manifold $\mathcal{M}^n$ of complex dimension n. Examples of compact manifolds $\mathcal{M}^n$ for $n = 1, 2, 3$ are $\mathcal{M}^1 = \mathbb{T}^2$, $\mathcal{M}^2 = K_3$, $\mathcal{M}^3$ a Calabi-Yau space. The new solutions of $D = 11$ supergravity obtained in this way generalize the corresponding KK vacuum solution of $D = 11$ supergravity to one representing an M-brane, or intersecting M-branes, wrapped around cycles in the the compactifying space. In any case, it is clear that the results of this paper are far from complete. It seems possible that a recent classification [25] of p-brane solutions of maximal supergravities in dimensions $D < 11$ might form a basis of a systematic M-theory interpretation, along the lines presented here, of all p-brane like solutions of $D = 11$ supergravity.

Acknowledgments

We would like to thank M.B. Green, J. Gauntlett and E. Witten for helpful discussions. G.P. is supported by a University Research Fellowship from the Royal Society.

References

[1] E. Bergshoeff, E. Sezgin and P.K. Townsend, Phys. Lett. B 189 (1987) 75; Ann. Phys. (N.Y.) 185 (1988) 330.
[2] M.J. Duff and K.S. Stelle, Phys. Lett. B 253 (1991) 113.
[3] R. Güven, Phys. Lett. B 276 (1992) 49.
[4] C.M. Hull and P.K. Townsend, Nucl. Phys. B 438 (1995) 109.
[5] P.K. Townsend, Phys. Lett. B 350 (1995) 184.
[6] J.H. Schwarz, The Power of M-theory, hep-th/9510086, M-theory extensions of T-duality hep-th/9601077.
[7] P. Hořava and E. Witten, Nucl. Phys. B 460 (1996) 506.
[8] K. Becker, M. Becker and A. Strominger, Nucl. Phys. B 456 (1995) 130.
[9] E. Witten, Fivebranes and M-theory on an Orbifold, hep-th/9512219.
[10] K. Becker and M. Becker, Boundaries in M-theory, hep-th/9602071.
[11] J. Rahmfeld, Phys. Lett. B 372 (1996) 198.
[12] M.J. Duff and J. Rahmfeld, Phys. Lett. 345 (1995) 441; M.J. Duff, J.T. Liu and J. Rahmfeld, Nucl. Phys. B 459 (1996) 125.
[13] M. Cvetič and A.A. Tseytlin, Phys. Lett. B 366 (1996) 95.
[14] A. Sen, Phys. Rev. D 53 (1996) 2874.
[15] M. Douglas, Branes within Branes, hep-th/9512077.
[16] A. Strominger, Open p-branes, hep-th/9512059.
[17] P.K. Townsend, D-branes from M-branes, hep-th/9512062.
[18] A.A. Tseytlin, Selfduality of Born-Infeld action and Dirichlet 3-brane of type IIB superstring theory, hep-th/9602064.
[19] M.B. Green and M. Gutperle, Comments on three-branes, hep-th/9602077.
[20] E.R.C. Abraham and P.K. Townsend, Nucl. Phys. B 351 (1991) 313.
[21] J.M. Izquierdo, N.D. Lambert, G. Papadopoulos and P.K. Townsend, Nucl. Phys. B 460 (1996) 560.
[22] R.R. Khuri and T. Ortín, Supersymmetric Black Holes in $N = 8$ Supergravity, hep-th/9512177.
[23] G.W. Gibbons, G.T. Horowitz and P.K. Townsend, Class. Quantum Grav. 12 (1995) 297.
[24] R.C. Myers and M.J. Perry, Ann. Phys. (N.Y.) 172 (1986) 304.
[25] H. Lü and C.N. Pope, p-brane solutions in maximal supergravities, hep-th/9512012.

Reprinted from Nucl. Phys. B **475** (1996) 149–63

Harmonic superpositions of M-branes

A.A. Tseytlin [1]

Theoretical Physics Group, Blackett Laboratory, Imperial College, London SW7 2BZ, UK

Received 22 April 1996; accepted 12 June 1996

Abstract

We present solutions describing supersymmetric configurations of 2 or 3 orthogonally intersecting 2-branes and 5-branes of $D = 11$ supergravity. The configurations which preserve 1/4 or 1/8 of maximal supersymmetry are $2\perp2$, $5\perp5$, $2\perp5$, $2\perp2\perp2$, $5\perp5\perp5$, $2\perp2\perp5$ and $2\perp5\perp5$ ($2\perp2$ stands for orthogonal intersection of two 2-branes over a point, etc.; p-branes of the same type intersect over (p-2)-branes). There exists a simple rule which governs the construction of composite supersymmetric p-brane solutions in $D = 10$ and 11 with a separate harmonic function assigned to each constituent 1/2-supersymmetric p-brane. The resulting picture of intersecting p-brane solutions complements their D-brane interpretation in $D = 10$ and seems to support possible existence of a $D = 11$ analogue of D-brane description. The $D = 11$ solution describing intersecting 2-brane and 5-brane reduces in $D = 10$ to a type II string solution corresponding to a fundamental string lying within a solitonic 5-brane (which further reduces to an extremal $D = 5$ black hole). We also discuss a particular $D = 11$ embedding of the extremal $D = 4$ dyonic black hole solution with finite area of horizon.

PACS: 04.50.+h; 04.20.Jb; 04.70.Bw; 11.25.Mj

1. Introduction

In view of recent suggestions that $D = 11$ supergravity may be a low-energy effective field theory of a fundamental 'M-theory' which generalises known string theories (see, e.g., [1]) it is important to gain better understanding of its classical p-brane solutions. It seems likely that supersymmetric BPS saturated p-brane solutions of low-dimensional theories can be understood as 'reductions' of basic $D = 11$ 'M-branes' – 2-brane

[1] E-mail: tseytlin@ic.ac.uk.
On leave from Lebedev Physics Institute, Moscow.

PII S0550-3213(96)00328-8

[2] and 5-brane [3] and their combinations [4]. The important questions are which combinations of M-branes do actually appear as stable supersymmetric solutions, how to construct them and how they are related to similar $D = 10$ p-brane configurations.

Here we shall follow and extend further the suggestion [4] that stable supersymmetric $D = 11$ p-brane configurations should have an interpretation in terms of orthogonal intersections of certain numbers of 2-branes and/or 5-branes. A possibility of existence of similar supersymmetric configurations was pointed out earlier (on the basis of charge conservation and supersymmetry considerations) in [5,6]. Discussions of related systems of D-branes in $D = 10$ string theories appeared in [7–9].

It should be noted that 'intersecting p-brane' solutions in [4] and below are isometric in all directions internal to all constituent p-branes (the background fields depend only on the remaining common transverse directions). They are different from possible virtual configurations where, e.g., a (p-2)-brane ends (in transverse space radial direction) on a p-brane [5] (such configurations may contribute to path integral but may not correspond to stable classical solutions). A configuration of, e.g., a p-brane and a p′-brane intersecting in (p+p′)-space may be also considered as a special anisotropic (cf. [10]) (p + p′)-brane. We expect (see also [4]) that there should exist more general solutions (with constituent p-branes effectively having different transverse spaces) which represent more complicated 'BPS bound states' of constituent p-branes and interpolate between such intersecting solutions and solutions with higher rotational symmetry for each p-brane.

The basic property of supersymmetric p-brane solutions of supergravity theories is that they are expressed in terms of harmonic functions of transverse spatial coordinates. This reflects the BPS saturated nature of these solutions and implies that there exist stable 'multicenter' configurations of multiple parallel p-branes of the same type. There may also exist stable supersymmetric solutions corresponding to combinations (intersections and bound states) of p-branes of the same or different types. While the rules of combining p-branes (in a way preserving supersymmetry and charge conservation) in $D = 10$ depend on a type (NS-NS or R-R) of the constituents [5], the following rules seem to be universal in $D = 11$ (these rules are consistent with $D = 10$ rules upon dimensional reduction):[2]

(i) p-branes of the same type can intersect only over a (p-2)-brane [4] (i.e. 2-branes can intersect over a 0-brane, 5-branes can intersect over a 3-brane, 3-branes can intersect over a string);

(ii) 2-brane can orthogonally intersect 5-brane over a string [5,6];

(iii) a configuration of n orthogonally intersecting M-branes preserves at least $1/2^n$ of maximal supersymmetry.[3]

Thus in addition to the basic (2- and 5-) M-branes preserving 1/2 of supersymmetry one should expect to find also the following composite configurations:

---

[2] Related conditions for supersymmetric combinations of D-branes in $D = 10$ are that the number of mixed Dirichlet-Neumann directions should be a multiple of 4 and that a (p-2)-brane can lie within a p-brane [8].

[3] In the case of general solutions involving parallel families of p-branes n stands for a number of intersecting families.

(i) $2\perp2$, $5\perp5$, $5\perp2$ preserving 1/4 of supersymmetry, and
(ii) $2\perp2\perp2$, $5\perp2\perp2$, $5\perp5\perp2$, $5\perp5\perp5$ preserving 1/8 of supersymmetry.

The allowed 1/16 supersymmetric configurations with *four* intersecting M-branes (i.e. $2\perp2\perp2\perp2$, $2\perp2\perp2\perp5$, $5\perp5\perp5\perp2$) have transverse space dimension $d < 3$ and thus (being described in terms of harmonic functions of transverse coordinates) are not asymptotically flat in transverse directions. The exception is $5\perp5\perp2\perp2$ for which the transverse dimension is 3 as in the $5\perp2\perp2$, $5\perp5\perp2$ and $5\perp5\perp5$ cases. Like the 'boosted' version of $5\perp5\perp5$ solution the $5\perp5\perp2\perp2$ background is 1/8-supersymmetric and upon compactification to $D = 4$ reduces to the dyonic $D = 4$ black hole [11,12] with four different charges and finite area of the horizon. This will be discussed in detail in [13]. Note also that the regular 3-charge dyonic $D = 5$ black hole [14] is described by $2\perp2\perp2$ or by 'boosted' $2\perp5$ solution.

In [4] the 'electric' $D = 11$ solutions of [3] with 1/4 and 1/8 of supersymmetry were interpreted as special $2\perp2$ and $2\perp2\perp2$ configurations and the corresponding 'magnetic' $5\perp5$ and $5\perp5\perp5$ solutions were found.

Below we shall generalise the solutions of [3,4] to the case when each intersecting p-brane is described by a separate harmonic function and will also present new solutions corresponding to case when intersecting M-branes are of different type, i.e. $5\perp2$, $5\perp2\perp2$ and $5\perp5\perp2$. The important $5\perp2$ solution reduces in $D = 10$ to a configuration which can be interpreted as a fundamental string lying within a solitonic (i.e. NS-NS) 5-brane (such $D = 10$ solution was given in [14]).[4]

The basic observation that clarifies the picture suggested in [4] and leads to various generalisations (both in $D = 11$ and $D = 10$) is that it is possible to assign an independent harmonic function to each intersecting p-brane (the solutions in [3,4] correspond to the 'degenerate' case when all harmonic functions are taken to be equal). For example, a generalisation of $2\perp2$ solution of [3,4] now parametrised by two independent harmonic functions describes, in particular, two orthogonally intersecting families of parallel 2-branes.

Combining the above $D = 11$ p-brane composition rules with the 'harmonic function rule' explained and illustrated on $D = 10$ examples in Section 2 below, it is easy to write down explicitly new solutions representing orthogonally intersecting (parallel families of) 2-branes and 5-branes mentioned above, i.e. $5\perp2$, $5\perp2\perp2$, $5\perp5\perp2$ (Section 3). A special version of $2\perp5$ solution superposed with a Kaluza-Klein monopole represents a particular $D = 11$ embedding of the extreme dyonic $D = 4$ black hole (Section 4).

[4] In addition to the intersecting $2\perp5$ configuration there should exist a supersymmetric $D = 11$ solution describing a 2-brane lying within a 5-brane (see [15] and Section 3.2). It should lead upon dimensional reduction (along 5-brane direction orthogonal to 2-brane) to a 2-brane within 4-brane configuration of type IIA theory (related by T-duality to a R-R string within 3-brane in type IIB theory) which is allowed from the point of view of D-brane description [8].

2. Harmonic function rule and $D = 10$ intersecting p-brane solutions

The metric and 4-form field strength of the basic extremal supersymmetric $p = 2$ [2] and $p = 5$ [3] p-brane solutions of $D = 11$ supergravity can be represented in the following form:

$$ds_{11}^2 = H_p^{(p+1)/9}(x)\left[H_p^{-1}(x)(-dt^2 + dy\,dy_p) + dx\,dx_{10-p}\right], \tag{2.1}$$

$$\mathcal{F}_{4(2)} = -3dt \wedge d(H_2^{-1}J), \qquad \mathcal{F}_{4(5)} = 3 * dH_5, \qquad \partial^2 H_p = 0, \tag{2.2}$$

where $dy\,dy_p \equiv dy_1^2 + ... + dy_p^2$, $dx\,dx_n \equiv dx_1^2 + ... + dx_n^2$ (y_a are internal coordinates of p-brane and x_i are transverse coordinates), $J = dy_1 \wedge dy_2$ is the volume form on R_y^2 and $*$ defines the dual form in R_x^5. H_p is a harmonic function on R_x^{10-p} which may depend only on part of x-coordinates (this may be viewed as a result of taking a periodic array of generic 1-center solutions; for simplicity, we shall still refer to such solution as a p-brane even though it will be 'delocalised' in some x-directions).

The structure of $\mathcal{F}_4$ in (2.2) is such that the contribution of the CS interaction term to the $\mathcal{F}_4$-equation of motion vanishes (i.e. $\mathcal{F}_4 \wedge \mathcal{F}_4 = 0$). This will also be the property of all intersecting solutions discussed below.

The structure of the metric (2.1) can be described as follows. If one separates the overall conformal factor which multiplies the transverse x-part then each of the squares of differentials of the coordinates belonging to a given p-brane is multiplied by the inverse power of the corresponding harmonic function. We suggest that this as a general rule ('*harmonic function rule*') which applies to any supersymmetric combination of orthogonally intersecting p-branes: if the coordinate y belongs to several constituent p-branes $(p_1, ..., p_n)$ then its contribution to the metric written in the conformal frame where the transverse part $dx\,dx$ is 'free' is multiplied by the product of the inverse powers of harmonic functions corresponding to each of the p-branes it belongs to, i.e. $H_{p_1}^{-1} ... H_{p_n}^{-1} dy^2$. The harmonic function factors thus play the role of 'labels' of constituent p-branes making the interpretation of the metric straightforward.

It can be checked explicitly that the specific backgrounds discussed below which can be constructed using this rule indeed solve the $D = 11$ supergravity equations of motion. While we did not attempt to give a general derivation of this rule directly from $D = 11$ field equations, it should be a consequence of the fact that intersecting configurations are required to be supersymmetric (i.e. it should follow from first-order equations implied by the existence of a Killing spinor). Since one should be able to superpose BPS states they must be parametrised (like their basic constituent p-branes) by harmonic functions. Taking the centers of each of the harmonic function at different points one can interpolate between the cases of far separated and coinciding p-branes, confirming the consistency of the 'harmonic function rule'.

This rule is also consistent (upon dimensional reduction) with analogous one which operates in $D = 10$ where it can be justified by conformal σ-model considerations (for specific NS-NS configurations) [12,14] or by T-duality [16] considerations (for R-R configurations).

2.1. $2\perp2$ $D=11$ *solution*

For example, the metric of two $D = 11$ 2-branes intersecting over a point constructed according to the above rule will be

$$ds^2_{11} = H^{1/3}_{2(1)}(x)H^{1/3}_{2(2)}(x)\big[-H^{-1}_{2(1)}(x)H^{-1}_{2(2)}(x)\,dt^2 \\ +H^{-1}_{2(1)}(x)dy\,dy^{(1)}_2 + H^{-1}_{2(2)}(x)dy\,dy^{(2)}_2 + dx\,dx_6\big]\,. \tag{2.3}$$

Here $y^{(1)}_1, y^{(1)}_2$ and $y^{(2)}_1, y^{(2)}_2$ are internal coordinates of the two 2-branes. The time part 'belongs' to both 2-branes and thus is multiplied by the product of the inverse powers of both harmonic functions. The corresponding field strength is

$$\mathcal{F}_{4(2\perp2)} = -3\,dt\wedge d(H^{-1}_{2(1)}J_1 + H^{-1}_{2(2)}J_2)\,. \tag{2.4}$$

The fact that the two harmonic functions can be centered at different (e.g. far separated) points together with supersymmetry and exchange symmetry with respect to the two 2-branes uniquely determines the form of the background, which indeed solves the $D = 11$ supergravity equations.

Setting $H_{2(2)} = 1$ one gets back to the special 2-brane solution (2.1), (2.2) where $H_2 = H_{2(1)}$ does not depend on two of the eight x-coordinates (called $y^{(2)}$ in (2.3)). Another special case $H_{2(1)} = H_{2(2)}$ corresponds to the '4-brane' solution of [3] interpreted in [4] as representing two intersecting 2-branes.

2.2. *Examples of intersecting p-brane solutions of $D = 10$ type II theories*

Before proceeding with the discussion of other composite $D = 11$ solutions let us demonstrate how the 'harmonic function rule' applies to various p-brane solutions of $D = 10$ type II superstring theories.

The basic $D = 10$ fundamental string solution [17] which has the following metric (we shall always use the string-frame form of the $D = 10$ metric):

$$ds^2_{10} = H^{-1}_1(x)(-dt^2 + dy^2) + dx\,dx_8\,. \tag{2.5}$$

The metric of the solution describing a fundamental string lying within the solitonic 5-brane [18,19] is given by [14]

$$ds^2_{10} = H^{-1}_1(x)(-dt^2 + dy^2_1) + dy^2_2 + \dots + dy^2_5 + H_5(x)dx\,dx_4 \\ = H_5(x)\big[H^{-1}_1(x)H^{-1}_5(x)(-dt^2 + dy^2_1) \\ +H^{-1}_5(x)(dy^2_2 + \dots + dy^2_5) + dx\,dx_4\big]\,. \tag{2.6}$$

Other NS-NS background fields have obvious 'direct sum' structure, i.e. the dilaton is given by $e^{2\phi} = H^{-1}_1 H_5$ and the antisymmetric 2-tensor has both 'electric' (fundamental string) and 'magnetic' (5-brane) components, $B_{ty_1} = H^{-1}_1$, $H_{mnk} = -\epsilon_{mnkl}\partial_l H_5$. The factorised harmonic function structure of this background has a natural explanation from the point of view of the associated conformal σ-model [14]. The solutions (2.5), (2.6) (as well as all solutions below which have a null hypersurface-orthogonal isometry) admit a straightforward 'momentum along string' generalisation $-dt^2 + dy_1^2 \to -dt^2 + dy_1^2 + K(x)(dt - dy_1)^2$ where K is an independent harmonic function (cf. [20]).

Applying $SL(2,\mathbb{Z})$ duality transformation of type IIB supergravity (which inverts the dilaton and does not change the Einstein-frame metric, i.e. modifies the string frame metric only by the conformal factor $e^{-\phi}$) one learns that the metric describing an R-R string lying within an R-R 5-brane has the same structure as (2.6), i.e. the structure consistent with the harmonic function rule (with the factor multiplying the square bracket now being $H_1^{1/2}H_5^{1/2}$). T-duality in the two 5-brane directions orthogonal to the string gives type IIB solution describing two 3-branes orthogonally intersecting over a string. Its metric has the form consistent with the 'harmonic function rule'

$$ds_{10}^2 = H_{3(1)}^{1/2}H_{3(2)}^{1/2}\Big[H_{3(1)}^{-1}H_{3(2)}^{-1}(-dt^2 + dy_1^2) + H_{3(1)}^{-1}dy\,dy_2^{(1)} + H_{3(2)}^{-1}dy\,dy_2^{(2)} + dx\,dx_4\Big], \tag{2.7}$$

where y_1 is the coordinate common to the two 3-branes.[5] The corresponding self-dual 5-tensor is

$$\mathcal{F}_{5(3\perp 3)} = dt \wedge (dH_{3(1)}^{-1} \wedge dy_1 \wedge dy_2^{(1)} \wedge dy_3^{(1)} + dH_{3(2)}^{-1} \wedge dy_1 \wedge dy_2^{(2)} \wedge dy_3^{(2)}) + *\,dH_{3(1)} \wedge dy_2^{(2)} \wedge dy_3^{(2)} + *dH_{3(2)} \wedge dy_2^{(1)} \wedge dy_3^{(1)}. \tag{2.8}$$

More general $1/8$ supersymmetric solutions describing the configurations $3\perp 3\perp 3$ and $3\perp 3\perp 3\perp 3$ will be discussed in [13].

While charge conservation prohibits the configuration with a fundamental string orthogonally intersecting solitonic 5-brane (and, by $SL(2,\mathbb{Z})$ duality, R-R string intersecting R-R 5-brane), the type IIB configuration of a fundamental string intersecting a R-R 5-brane (and its dual – R-R string intersecting a solitonic 5-brane) is allowed [5]. The corresponding solution is straightforward to write down. Its metric is given by (cf. (2.6); see also the discussion below)

$$ds_{10}^2 = H_5^{1/2}(x)\Big[-H_1^{-1}(x)H_5^{-1}(x)dt^2 + H_1^{-1}dy_1^2 + H_5^{-1}(x)(dy_2^2 + \dots + dy_6^2) + dx\,dx_3\Big]. \tag{2.9}$$

Here y_1 is the coordinate of the string intersecting 5-brane $(y_2, \dots, y_6)$ over a point.[6]

In general, metrics of $1/2$-supersymmetric p-branes of type II theories which carry R-R charges have the following form [21]:

$$ds_{10}^2 = H_p^{1/2}\,[H_p^{-1}(-dt^2 + dy\,dy_p) + dx\,dx_{9-p}], \tag{2.10}$$

[5] Adding a boost along the common string one finds upon reduction to $D = 5$ an extremal black hole with 3 charges and $x = 0$ as a regular horizon [14].

[6] For a multicenter choice of 5-brane harmonic function H_5 this metric describes a fundamental string intersecting several parallel 5-branes.

with the dilaton given by $e^{2\phi} = H_p^{(3-p)/2}$. It is straightforward to apply the 'harmonic function rule' and the supersymmetry and R-R charge conservation rules [5] to construct explicitly the solutions which describe multiple and intersecting R-R soliton configurations which are counterparts of the D-brane configurations discussed in [7,8]. The resulting procedure of constructing 'composite' supersymmetric backgrounds from 'basic' ones is in direct correspondence with a picture of 'free' parallel or intersecting D-brane hypersurfaces in flat space [22].

For example, the solution of type IIB theory representing a R-R string orthogonally intersecting 3-brane (T-dual to a 0-brane within a 4-brane in type IIA theory) is described by

$$ds_{10}^2 = H_1^{1/2} H_3^{1/2} [-H_1^{-1} H_3^{-1} dt^2 + H_1^{-1} dy_1^2 + H_3^{-1} (dy_2^2 + dy_3^2 + dy_4^2) + dx\, dx_5] . \tag{2.11}$$

By $SL(2,\mathbb{Z})$ duality the same (up to a conformal factor) metric represents a fundamental string intersecting a 3-brane.

An example of intersecting solution in type IIA theory is provided by a fundamental string orthogonally intersecting a 4-brane at a point (cf. (2.6), (2.11))

$$ds_{10}^2 = H_4^{1/2} \big[- H_1^{-1} H_4^{-1} dt^2 + H_1^{-1} dy_1^2 + H_4^{-1} (dy_2^2 + dy_3^2 + dy_4^2 + dy_5^2) + dx\, dx_4 \big] . \tag{2.12}$$

The required dilaton and antisymmetric tensors are given by direct sums of constituent fields. This background will be reproduced in Section 3.2 by dimensional reduction of orthogonally intersecting 2-brane and 5-brane solution of $D = 11$ supergravity.

Metrics describing configurations of different parallel type II p-branes lying within each other (with at least one of them being of R-R type) do not obey the 'harmonic function rule'. For example, the metric of the 'fundamental string – R-R string' bound state solution of type IIB theory (obtained by applying $SL(2,\mathbb{Z})$ transformation to the fundamental string background (2.10), see Schwarz in [1]) has the following structure:

$$ds_{10}^2 = \tilde{H}_1^{1/2} [H_1^{-1} (-dt^2 + dy_1^2) + dx\, dx_8] , \tag{2.13}$$

where H_1 and $\tilde{H}_1$ are 1-center harmonic functions with charges q and $\tilde{q} = qd^2/(c^2+d^2)$. The fundamental string limit corresponds to $\tilde{H}_1 = 1$ while the pure R-R string is recovered when $\tilde{H}_1 = H_1$. Other solutions related by T and $SL(2,\mathbb{Z})$ dualities (e.g. R-R string lying within 3-brane) have similar structure.

Let us note also that there exist a class of p-brane solutions [23–25] of the equations following from the action

$$S = \int d^D x \sqrt{g} \left[R - \tfrac{1}{2} (\partial\phi)^2 - \frac{1}{2(D-2-p)!} e^{-a\phi} F_{D-2-p}^2 \right] ,$$

with the metric being

$$ds_D^2 = H_p^{\alpha} \left[H_p^{-N} (-dt^2 + dy\, dy_p) + dx\, dx_{D-1-p} \right] ,$$

$$N = \frac{4}{\Delta}\,, \qquad \alpha = \frac{4(p+1)}{(D-2)\Delta}, \qquad \Delta \equiv a^2 + 2(p+1)\frac{D-3-p}{D-2}\,. \tag{2.14}$$

The power N is integer for supersymmetric p-branes with the amount of residual supersymmetry being at least $1/2^N$ of maximal (for $N = 4$ and $D = 4 + p$ the remaining fraction of supersymmetry is $1/8$). Lower dimensional ($D < 10$) solutions which have $N > 1$ can be re-interpreted as special limits of (reductions of) combinations of 1/2-supersymmetric 'basic' ($N = 1$) p-brane solutions in $D = 10, 11$. The higher than first power of the H_p^{-1} factor in the square bracket in (2.14) is a result of identifying the harmonic functions corresponding to basic constituent p-branes [26].

An example of a solution with $N = 2$ is the self-dual string in $D = 6$ [23]. It indeed can be reproduced as a special limit of the solitonic 5-brane plus fundamental string solution (2.6) with the four 'extra' 5-brane directions wrapped around a 4-torus (leading to the solution equivalent to the dyonic string of [27]) and the harmonic functions H_1 and H_5 set equal to each other.

3. Intersecting 2-branes and 5-branes in $D = 11$

3.1. $2\bot2\bot2$ and $5\bot5\bot5$ configurations

To write down the explicit form of intersecting 2- and 5-brane solutions in $D = 11$ it is useful first to simplify the notation: we shall use T (F) to denote the inverse power of harmonic function corresponding to a two-brane (five-brane), i.e. $T \equiv H_2^{-1}$, $F \equiv H_5^{-1}$. The lower index on T or F will indicate a number of a p-brane.

The solution which describes three 2-branes intersecting over a point is given by the straightforward generalisation of (2.3), (2.4):

$$\begin{aligned} ds_{11}^2 = (T_1T_2T_3)^{-1/3}\big[&- T_1T_2T_3\, dt^2 \\ &+T_1\, dy\, dy_2^{(1)} + T_2\, dy\, dy_2^{(2)} + T_3\, dy\, dy_2^{(3)} + dx\, dx_4\big]\,, \end{aligned} \tag{3.1}$$

$$\mathcal{F}_{4(2\bot2\bot2)} = -3\, dt \wedge d(T_1J_1 + T_2J_2 + T_3J_3)\,. \tag{3.2}$$

The three 2-branes are parametrised by 3 sets of coordinates $y_1^{(i)}, y_2^{(i)}$ and J_i are the volume forms on the corresponding 2-planes. Also, $\partial^2 T_i^{-1} = 0$, i.e. $T_i^{-1} = 1 + q_i/|x|^2$ in the simplest 1-center case. The special case of $T_1 = T_2 = T_3$ gives the '6-brane' solution of [3] correctly interpreted in [4] as representing three 2-branes orthogonally intersecting at one point. Other obvious special choices, e.g. $T_3 = 1$, lead to a particular case of $2\bot2$ solution (2.3), (2.4) with the harmonic functions not depending on two of the transverse coordinates.

This solution is regular at $x = 0$ and upon dimensional reduction to $D = 5$ along y_n-directions it becomes the 3-charge $D = 5$ Reissner-Nordström type black hole (discussed in the special case of equal charges in [28]) which is U-dual to NS-NS dyonic black hole constructed in [14].

Similar generalisation of the $5\bot5\bot5$ solution in [4] corresponding to the three 5-branes intersecting pairwise over 3-branes which in turn intersect over a string can be found by applying the 'harmonic function rule'

$$ds_{11}^2 = (F_1F_2F_3)^{-2/3}\left[F_1F_2F_3(-dt^2+dy_0^2)\right.$$
$$\left.+F_2F_3\,dy\,dy_2^{(1)} + F_1F_3\,dy\,dy_2^{(2)} + F_1F_2\,dy\,dy_2^{(3)} + dx\,dx_3\right], \quad (3.3)$$

$$\mathcal{F}_{4(5\perp5\perp5)} = 3\,(*dF_1^{-1}\wedge J_1 + *dF_2^{-1}\wedge J_2 + *dF_3^{-1}\wedge J_3)\,. \quad (3.4)$$

The coordinate y_0 is common to all three 5-branes, $y_1^{(1)}, y_2^{(1)}$ are common to the second and third 5-branes, etc. F_i depend on three x-coordinates. The duality $*$ is always defined with respect to the transverse x-subspace (R_x^3 in (3.3), (3.4)). The special case of $F_1 = F_2 = F_3$ gives the solution found in [4]. If $F_2 = F_3 = 1$ the above background reduces to the single 5-brane solution (2.1), (2.2) with the harmonic function $H_5 = F_1^{-1}$ being independent of the two of transverse coordinates (here denoted as $y_1^{(1)}, y_2^{(1)}$). The case of $F_3 = 1$ describes two 5-branes orthogonally intersecting over a 3-brane,

$$ds_{11}^2 = (F_1F_2)^{-2/3}\left[F_1F_2(-dt^2+dydy_3) + F_1dydy_2^{(1)} + F_2dydy_2^{(2)} + dxdx_3\right], \quad (3.5)$$

$$\mathcal{F}_{4(5\perp5)} = 3(*dF_1^{-1}\wedge J_1 + *dF_2^{-1}\wedge J_2)\,, \quad (3.6)$$

which again reduces to the corresponding solution of [4] when $F_1 = F_2$.

The $5\perp5\perp5$ configuration (3.3) has also the following generalisation obtained by adding a 'boost' along the common string:

$$ds_{11}^2 = (F_1F_2F_3)^{-2/3}\Big[F_1F_2F_3(du\,dv + K\,du^2)$$
$$+F_2F_3\,dy\,dy_2^{(1)} + F_1F_3\,dy\,dy_2^{(2)} + F_1F_2\,dy\,dy_2^{(3)} + dx\,dx_3\Big]\,. \quad (3.7)$$

Here $u, v = y_0 \mp t$ and K is a generic harmonic function of the three coordinates x_s. A non-trivial $K = Q/|x|$ describes a momentum flow along the string (y_0) direction. Upon compactification to $D = 4$ along isometric y_n-directions this background reduces [13] to extremal dyonic black hole with regular horizon which has the same metric as the solution of [11]. Thus the 'boosted' $5\perp5\perp5$ solution gives an embedding of the 1/8 supersymmetric dyonic black hole in $D = 11$ which is different from the one discussed in Section 4 below (see [13] for details).

3.2. *2-brane intersecting 5-brane*

Let us now consider other possible supersymmetric intersecting configurations not discussed in [4]. The most important one is a 2-brane orthogonally intersecting a 5-brane over a string (a possibility of such a configuration was pointed out in [5,6]). The corresponding background is easily constructed using the harmonic function rule

$$ds_{11}^2 = F^{-2/3}T^{-1/3}\left[FT(-dt^2+dy_1^2) + F(dy_2^2+\ldots+dy_5^2) + T\,dy_6^2 + dx\,dx_4\right], \quad (3.8)$$

$$\mathcal{F}_{4(5\perp2)} = -3\,dt\wedge dT\wedge dy_1\wedge dy_6 + 3*dF^{-1}\wedge dy_6\,, \quad (3.9)$$

where $y_1, \ldots, y_5$ belong to 5-brane and y_1, y_6 to 2-brane. This solution can be generalised further:

$$ds_{11}^2 = F^{-2/3}T^{-1/3}\big[FT(du\,dv + K\,du^2) + F(dy_2^2 + \ldots + dy_5^2) + T\,dy_6^2 + dx\,dx_4\big]\,, \tag{3.10}$$

where as in (3.7) $u, v = y_1 \mp t$ and K, like T^{-1} and F^{-1}, is a generic harmonic function of x_n. In the simplest 1-center case having a non-trivial K corresponds to adding a momentum flow along the string (y_1) direction.

Dimensional reduction of this solution to $D = 10$ along $x_{11} \equiv y_6$ (the direction of 2-brane orthogonal to 5-brane) leads to the NS-NS type II background corresponding to a fundamental string lying within a solitonic 5-brane. Using the relation between the $D = 11$ and (string frame) $D = 10$ metrics,

$$ds_{11}^2 = e^{4\phi/3}\left(dx_{11}^2 + e^{-2\phi}ds_{10}^2\right), \tag{3.11}$$

we indeed find the expected $D = 10$ background with the dilaton $e^{2\phi} = F^{-1}T$, the metric given by (2.6) (with $H_1 = T^{-1}$, $H_5 = F^{-1}$) and the antisymmetric 2-tensor field strength determined by the 3-tensor field strength (3.9).

Dimensional reduction along the string y_1 direction leads instead to the $D = 10$ solution corresponding to a fundamental string (along y_2) orthogonally intersecting a 4-brane (cf. (2.11)). Here the dilaton is $e^{2\phi} = H_1^{-1}H_4^{-1/2}$, $H_1 = T^{-1}$, $H_4 = F^{-1}$ and thus the resulting $D = 10$ metric has indeed the form (2.12) obtained by applying the harmonic function rule to combine the fundamental string (2.5) and R-R 4-brane (2.10) of type IIA theory. Another possibility is to compactify along one of the transverse directions, e.g., x_4 (assuming that harmonic functions are independent of it or forming a periodic array) in which case we find the type IIA solution describing a R-R 2-brane orthogonally intersecting solitonic 5-brane.

Compactification of all 6 isometric y-coordinates on a 6-torus leads to the extremal $D = 5$ black hole solution parametrised by 3 independent charges [14]. Thus the 'boosted' $2\perp5$ solution and $2\perp2\perp2$ solution discussed above represent two different $D = 11$ 'lifts' of the regular extremal 3-charge $D = 5$ black hole.

These black holes have a finite entropy[7] which is not surprising since (3.10) has a finite entropy directly as a $D = 11$ black brane background (assuming that internal directions of 2- and 5-branes are compactified). Setting $T^{-1} = 1 + \tilde{Q}/r^2$, $F^{-1} = 1 + P/r^2$, $K = Q/r^2$ $(r^2 = x_m x_m)$, one finds that $r = 0$ is a regular horizon (all radii are regular at $r \to 0$) with the area $A_9 = 2\pi^2 L^6\sqrt{Q\tilde{Q}P}$ (L is an equal period of y-coordinates). The corresponding thermodynamic entropy can then be understood as a statistical entropy (related to existence of degenerate $5\perp2$ BPS configurations with the same values of the charges) by counting relevant BPS states directly in $D = 11$ as suggested in [30].

[7] This makes possible to reproduce their entropy by counting the corresponding BPS states using D-brane description of the corresponding dual backgrounds with R-R charges [28,29] or using direct conformal field theory considerations [14].

The 2⊥5 metric (3.8) may be compared to the metric obtained by lifting to $D = 11$ the $D = 8$ dyonic membrane solutions [15],

$$ds_{11}^2 = T^{-1/3}\tilde{T}^{-1/3}\left[T(-dt^2 + dy_1^2 + dy_2^2) + \tilde{T}(dy_3^2 + dy_4^2 + dy_5^2) + dx\,dx_5\right], \tag{3.12}$$

where $T^{-1} = 1 + q/|x|^3$, and $\tilde{T}^{-1} = 1 + \tilde{q}/|x|^3$, $\tilde{q} = q\cos^2\xi$ (ξ is a free parameter). Since (3.12) reduces to the 2-brane metric if $\tilde{T} = 1$ and to the 5-brane metric if $\tilde{T} = T$ (and thus is similar to the metric (2.13) of a bound state of a NS-NS and R-R strings in type IIB theory) this background can presumably be interpreted as corresponding to a 2-brane lying within a 5-brane [15,4].

3.3. 2⊥2⊥5 and 5⊥5⊥2 configurations

Two other 1/8 supersymmetric configurations of three orthogonally intersecting M-branes are 2⊥2⊥5 and 5⊥5⊥2. The first one represents two 2-branes each intersecting 5-brane over a string with the two strings intersecting over a point (so that 2-branes intersect only over a point). The second one corresponds to a 2-brane intersecting each of the two 5-branes over a string with the 5-branes intersecting over a 3-brane (with the strings orthogonally intersecting 3-brane over a point).

In the first case we find

$$ds_{11}^2 = (T_1T_2)^{-1/3}F^{-2/3}\Big[-T_1T_2F\,dt^2 + T_1F\,dy_1^2 + T_1\,dy_2^2 + T_2F\,dy_3^2 + T_2\,dy_4^2 + F(dy_5^2 + dy_6^2 + dy_7^2) + dx\,dx_3\Big], \tag{3.13}$$

$$\mathcal{F}_{4(2\perp2\perp5)} = -3dt\wedge d(T_1\,dy_1\wedge dy_2 + T_2\,dy_3\wedge dy_4) + 3 * dF^{-1}\wedge dy_2\wedge dy_4, \tag{3.14}$$

where y_1, y_3, y_5, y_6, y_7 are 5-brane coordinates and y_1, y_2 and y_3, y_4 are coordinates of 2-branes.[8] In the second case

$$ds_{11}^2 = T^{-1/3}(F_1F_2)^{-2/3}\Big[-F_1F_2T\,dt^2 + F_1T\,dy_1^2 + F_2T\,dy_2^2 + F_1F_2(dy_3^2 + dy_4^2 + dy_5^2) + F_1\,dy_6^2 + F_2\,dy_7^2 + dx\,dx_3\Big], \tag{3.15}$$

$$\mathcal{F}_{4(5\perp5\perp2)} = -3dt\wedge d(T\,dy_1\wedge dy_2) + 3(*dF_1^{-1}\wedge dy_2\wedge dy_7 + *dF_2^{-1}\wedge dy_1\wedge dy_6). \tag{3.16}$$

[8] Note that this configuration is unique since (according to the rule that p-branes can intersect only over (p-2)-branes) the 2-branes cannot intersect over a string. For example, if one would try to modify (3.13) by combining dy_1^2 with dt^2 then y_1 would belong also to the second 2-brane.

Here y_1, y_2 belong to the 2-brane and y_1, y_3, y_4, y_5, y_6 and y_2, y_3, y_4, y_5, y_7 are coordinates of the two 5-branes intersecting over y_3, y_4, y_5.[9]

The backgrounds (3.13), (3.14) and (3.15), (3.16) have 'dual' structure. In the special case when $T_2 = 1$ in (3.13), (3.14) and $F_2 = 1$ in (3.15), (3.16) they become equivalent to the $2\perp5$ solution (3.8), (3.9) with the harmonic functions independent of one of the 4 transverse coordinates (y_4 in (3.13) and y_7 in (3.15)). Various possible dimensional reductions to $D = 10$ lead to expected p-brane intersection configurations of type IIA theory. For example, the reduction of $2\perp2\perp5$ (3.13) along the orthogonal direction y_2 of the first 2-brane leads to the configuration of a solitonic 5-brane with a fundamental string lying within it orthogonally intersected by 2-brane. Dimensional reduction along the direction y_1 common to the first 2-brane and 5-brane leads to the 4-brane orthogonally intersected by fundamental string and 2-brane, while the reduction along other 5-brane directions (y_5, y_6, y_7) gives $2\perp2\perp4$ type IIA configuration, etc.

4. $D = 11$ solution corresponding to $D = 4$ extremal dyonic black hole

The extreme dyonic $D = 4$ black hole string solutions with non-zero entropy [31,11] are described by the following NS-NS type II $D = 10$ background (compactified on 6-torus) [12]

$$ds^2_{10} = H_1^{-1}(x)\,[du\,dv + K(x)\,du^2] + dy\,dy_4 + H_5(x)V^{-1}(x)\,[dy_2 + a_s(x)\,dx^s]^2 + H_5(x)V(x)\,dx\,dx_3\,, \tag{4.1}$$

$$e^{2\phi} = H_1^{-1}H_5, \quad B = H_1^{-1}dt \wedge dy_1 - b_s dx^s \wedge dy_2, \quad db = - * dH_5, \quad da = - * dV\,, \tag{4.2}$$

where $u, v = y_1 \mp t$ and H_1, H_5, K, V are harmonic functions of x_s ($s = 1,2,3$). This background can be interpreted as representing a fundamental string (with an extra momentum along it, cf. (2.6)) lying within a solitonic 5-brane with all harmonic functions being independent of one of the four transverse directions (y_2) along which a Kaluza-Klein monopole [32] is introduced. Since the corresponding σ-model is invariant under T-duality, one cannot get rid of the off-diagonal KK monopole term in the metric by dualizing in y_2 direction. However, interpreting this background as a solution of type IIB theory one can apply the $SL(2,\mathbb{Z})$ duality to transform it first into a configuration of a R-R string lying on a R-R 5-brane 'distorted' by the Kaluza-Klein monopole. The metric one finds is then given by (4.1) rescaled by $e^{-\phi}$, i.e.

$$ds^2_{10IIB} = (H_1H_5)^{1/2}\Big[H_1^{-1}H_5^{-1}(du\,dv + K\,du^2) + H_5^{-1}\,dy\,dy_4 + V^{-1}(dy_2 + a_s\,dx^s)^2 + V\,dx\,dx_3\Big]\,. \tag{4.3}$$

---

[9] Another possibility could be to consider 2-brane intersecting each of the two 5-branes over the same string, i.e. $ds^2_{11} = T^{-1/3}(F_1F_2)^{-2/3}\big[F_1F_2T(-dt^2 + dy_1^2) + Tdy_2^2 + F_1F_2(dy_3^2 + dy_4^2) + F_1(dy_5^2 + dy_6^2) + F_2(dy_7^2 + dy_8^2) + dx\,dx_2\big]$. In this case, however, the transverse space is only 2-dimensional and thus the harmonic functions do not decay at infinity.

Since $B_{\mu\nu}$ in (4.2) is transformed into an R-R field one can now use T-duality along y_2 to exchange the off-diagonal term in the metric for an extra NS-NS $B_{\mu\nu}$ field. The resulting type IIA background has the metric

$$ds^2_{10IIA} = (H_1 H_5)^{1/2} V \Big[H_1^{-1} H_5^{-1} V^{-1} (du\, dv + K\, du^2) + H_5^{-1} V^{-1}\, dy\, dy_4 + H_5^{-1} H_1^{-1} d\tilde{y}_2^2 + dx\, dx_3 \Big] , \tag{4.4}$$

and the dilaton $e^{2\phi'} = H_1^{1/2} H_5^{-3/2} V$. As in other examples discussed above we can interpret this metric as describing a solitonic 5-brane (with the corresponding harmonic function now being V) which is lying within a R-R 6-brane (with the harmonic function H_5 and an extra dimension $\tilde{y}_2$), both being orthogonally intersected (over a string along y_1) by an R-R 2-brane (with coordinates $y_1, \tilde{y}_2$ and harmonic function H_1). Equivalent interpretation of this $D = 4$ dyonic black hole background was suggested in [33] where it was used to argue that statistical entropy found by D-brane counting of degenerate BPS states reproduces the finite thermodynamic entropy of the black hole.

Anticipating a possibility to compute the entropy by counting BPS states directly in $D = 11$ theory [30] it is of interest to lift the above type IIA $D = 10$ background to $D = 11$. Both forms of the $D = 10$ type IIA solution (4.1) and (4.4) lead to equivalent non-diagonal $D = 11$ metric.[10] From (4.1), (4.2) we find

$$ds^2_{11} = H_1^{1/3} H_5^{2/3} V \big[H_1^{-1} H_5^{-1} V^{-1} (du\, dv + K\, du^2) + H_5^{-1} V^{-1} dy\, dy_4 + H_1^{-1} V^{-1} dx_{11}^2 + V^{-2} (dy_2 + a_s\, dx^s)^2 + dx\, dx_3 \big] . \tag{4.5}$$

The corresponding 3-tensor field strength $\mathcal{F}_4$ is

$$\mathcal{F}_4 = 3\, dB \wedge dx_{11} = -3\, dt \wedge dH_1^{-1} \wedge dy_1 \wedge dx_{11} + 3 * dH_5 \wedge dy_2 \wedge dx_{11} . \tag{4.6}$$

Starting with (4.4) one obtains equivalent metric with $V \leftrightarrow H_5$, $x_{11} \to \tilde{y}_2$, $y_2 \to x_{11}$. The metric (4.5) can be interpreted as describing *intersecting 2-brane and 5-brane* (cf. (3.8) for $V = 1$, $F = H_5^{-1}$, $T = H_1^{-1}$, $x_{11} = y_6$, $y_2 = x_4$) *superposed with a KK monopole* along y_2 (for $H_1 = H_5 = 1$, $K = 1$ the metric becomes that of KK monopole times a 6-torus or type IIA 6-brane lifted to $D = 11$, see second reference in [1]).

The special cases of the background (4.1) when one or more harmonic functions are trivial are related to $a = 1/\sqrt{3}, \sqrt{3}, 1$ extremal $D = 4$ black holes. The 'irreducible' case when all 4 harmonic functions are non-trivial and equal ($H_1 = H_5 = K = V$) corresponds (for the 1-center choice of V) to the $a = 0$, $D = 4$ (Reissner-Nordström) black hole. The associated $D = 11$ metric (4.5) takes the form

$$ds^2_{11} = V^{-1}(x)\, du\, dv + du^2 + dy\, dy_4 + dx_{11}^2 + (dy_2 + a_s\, dx^s)^2 + V^2(x)\, dx\, dx_3 . \tag{4.7}$$

[10] Though the $D = 10$ metric (4.4) is diagonal, the R-R vector field supporting the 6-brane gives a non-vanishing $G_{11\mu}$ component of the $D = 11$ metric.

We conclude (confirming the expectation in [4]) that there exists an embedding of $a = 0$ RN black hole into $D = 11$ theory which has a non-trivial KK monopole type metric. This seems to represent an obstacle on the way of applying the $D = 11$ approach in order to give a statistical derivation of the $D = 4$ black hole entropy: one is to understand the effect of the presence of the KK monopole on counting of BPS states of systems of M-branes.[11]

The embedding of extreme 1/8 supersymmetric dyonic black holes into $D = 11$ theory discussed above is not, however, the only possible one. There exist two different 1/8 supersymmetric $D = 11$ solutions, namely, $5\perp5\perp5$ with a 'boost' along the common string (Section 3.1) and $2\perp2\perp5\perp5$, for which the $D = 11$ metric does not have KK monopole part but still reduces to an equivalent $D = 4$ dyonic black hole metric with regular horizon and finite entropy [13]. These M-brane configurations are likely to be a proper starting point for a statistical understanding of $D = 4$ black hole entropy directly from M-theory point of view.

5. Concluding remarks

As was discussed above, there are simple rules of constructing supersymmetric composite M-brane solutions from the basic building blocks – $D = 11$ 2-brane and 5-brane. This may be considered as an indication that there may exist a $D = 11$ analogue of D-brane description of R-R solitons in type II $D = 10$ string theories which applies directly to supersymmetric BPS configurations of $D = 11$ supergravity (in agreement with related suggestions in [5,6,35,36,30,37]).

We have also presented some explicit solutions corresponding to intersecting p-brane configurations of $D = 10$ type II theories. The resulting gravitational backgrounds complement the picture implied by D-brane approach. An advantage of viewing type IIA $D = 10$ configurations from $D = 11$ perspective is that this makes possible to treat various combinations of NS-NS and R-R p-branes on an equal footing, and in this sense goes beyond the D-brane description.

Acknowledgements

I am grateful to M. Cvetič, I. Klebanov and G. Papadopoulos for useful correspondence on related issues and acknowledge also the support of PPARC, ECC grant SC1*-CT92-0789 and NATO grant CRG 940870.

---

[11] For a discussion suggesting possible irrelevance of KK monopole for counting of BPS states in the D-brane approach see [34].

References

[1] C.M. Hull and P.K. Townsend, Nucl. Phys. B 438 (1995) 109;
P.K. Townsend, Phys. Lett. B 350 (1995) 184;
E. Witten, Nucl. Phys. B 443 (1995) 85;
J.H. Schwarz, Phys. Lett. B 367 (1996) 97, hep-th/9510086, hep-th/9601077;
P.K. Townsend, hep-th/9507048;
M.J. Duff, J.T. Liu and R. Minasian, Nucl. Phys. B 452 (1995) 261, hep-th/9506126;
K. Becker, M. Becker and A. Strominger, Nucl. Phys. B 456 (1995) 130;
I. Bars and S. Yankielowicz, hep-th/9511098;
P. Hořava and E. Witten, Nucl. Phys. B 460 (1996) 506;
E. Witten, hep-th/9512219.

[2] M.J. Duff and K.S. Stelle, Phys. Lett. B 253 (1991) 113.

[3] R. Güven, Phys. Lett. B 276 (1992) 49.

[4] G. Papadopoulos and P.K. Townsend, hep-th/9603087.

[5] A. Strominger, hep-th/9512059.

[6] P.K. Townsend, hep-th/9512062.

[7] E. Witten, hep-th/9510135;
M. Bershadsky, C. Vafa and V. Sadov, hep-th/9510225;
A. Sen, hep-th/9510229, hep-th/9511026;
C. Vafa, hep-th/9511088;
M. Douglas, hep-th/9512077.

[8] J. Polchinski, S. Chaudhuri and C.V. Johnson, hep-th/9602052.

[9] M.B. Green and M. Gutperle, hep-th/9604091.

[10] R.R. Khuri and R.C. Myers, hep-th/9512061.

[11] M. Cvetič and D. Youm, Phys. Rev. D 53 (1996) 584, hep-th/9507090.

[12] M. Cvetič and A.A. Tseytlin, Phys. Lett. B 366 (1996) 95, hep-th/9510097; hep-th/9512031.

[13] I.R. Klebanov and A.A. Tseytlin, Nucl. Phys. B 475 (1996) 179, hep-th/9604166.

[14] A.A. Tseytlin, Mod. Phys. Lett. A 11 (1996) 689, hep-th/9601177.

[15] J.M. Izquierdo, N.D. Lambert, G. Papadopoulos and P.K. Townsend, Nucl. Phys. B 460 (1996) 560, hep-th/9508177.

[16] E. Bergshoeff, C. Hull and T. Ortin, Nucl. Phys. B 451 (1995) 547, hep-th/9504081.

[17] A. Dabholkar and J.A. Harvey, Phys. Rev. Lett. 63 (1989) 478;
A. Dabholkar, G.W. Gibbons, J.A. Harvey and F. Ruiz-Ruiz, Nucl. Phys. B 340 (1990) 33.

[18] C.G. Callan, J.A. Harvey and A. Strominger, Nucl. Phys. B 359 (1991) 611.

[19] M.J. Duff and J.X. Lu, Nucl. Phys. B 354 (1991) 141.

[20] D. Garfinkle, Phys. Rev. D 46 (1992) 4286.

[21] G.T. Horowitz and A. Strominger, Nucl. Phys. B 360 (1991) 197.

[22] J. Polchinski, Phys. Rev. Lett. 75 (1995) 4724, hep-th/9510017.

[23] M.J. Duff and J.X. Lu, Nucl. Phys. B 416 (1994) 301, hep-th/9306052.

[24] G.W. Gibbons, G.T. Horowitz and P.K. Townsend, Class. Quant. Grav. 12 (1995) 297, hep-th/9410073.

[25] H. Lü, C.N. Pope, E. Sezgin and K.S. Stelle, Nucl. Phys. B 276 (1995) 669, hep-th/9508042.

[26] H. Lü and C.N. Pope, hep-th/9512012; hep-th/9512153.

[27] M.J. Duff, S. Ferrara, R.R. Khuri and J. Rahmfeld, Phys. Lett. B 356 (1995) 479, hep-th/9506057.

[28] A. Strominger and C. Vafa, HUTP-96-A002, hep-th/9601029.

[29] C.G. Callan and J.M. Maldacena, PUPT-1591, hep-th/9602043.

[30] R. Dijkgraaf, E. Verlinde and H. Verlinde, hep-th/9603126.

[31] R. Kallosh, A. Linde, T. Ortín, A. Peet and A. van Proeyen, Phys. Rev. D 46 (1992) 5278.

[32] R. Sorkin, Phys. Rev. Lett. 51 (1983) 87;
D. Gross and M. Perry, Nucl. Phys. B 226 (1983) 29.

[33] J. Maldacena and A. Strominger, hep-th/9603060.

[34] C. Johnson, R. Khuri and R. Myers, hep-th/9603061.

[35] K. Becker and M. Becker, hep-th/9602071.

[36] O. Aharony, J. Sonnenschein and S. Yankielowicz, hep-th/9603009.

[37] F. Aldabe, hep-th/9603183.

Reprinted from Phys. Lett. B **382** (1996) 73–80

The black branes of M-theory

M.J. Duff [1], H. Lü [2], C.N. Pope [2]
Center for Theoretical Physics, Texas A&M University, College Station, TX 77843, USA

Received 17 April 1996
Editor: L. Alvarez-Gaumé

Abstract

We present a class of black p-brane solutions of M-theory which were hitherto known only in the extremal supersymmetric limit, and calculate their macroscopic entropy and temperature.

1. Introduction

There is now a consensus that the best candidate for a unified theory underlying all physical phenomena is no longer ten-dimensional string theory but rather eleven-dimensional *M-theory*. The precise formulation of M-theory is unclear but membranes and fivebranes enter in a crucial way, owing to the presence of a 4-form field strength F_4 in the corresponding eleven-dimensional supergravity theory [1]. The membrane is characterized by a tension T_3 and an "electric" charge $Q_3 = \int_{S^7} *F_4$. For $T_3 > Q_3$, the membrane is "black" [2], exhibiting an outer event horizon at $r = r_+$ and an inner horizon at $r = r_-$, where $r = \sqrt{y^m y_m}$ and where y^m, $m = 1, 2, ..., 8$, are the coordinates transverse to the membrane. In the extremal tension=charge limit, the two horizons coincide, and one recovers the fundamental supermembrane solution which preserves half of the spacetime supersymmetries [3]. This supermembrane admits a covariant Green-Schwarz action [4]. Similar remarks apply to the fivebrane which is characterized by a tension T_6 and "magnetic charge" $P_6 = \int_{S^4} F_4$. It is also black when $T_6 > P_6$ and also preserves half the supersymmetries in the extremal limit [2]. There is, to date, no covariant fivebrane action, however. Upon compactification of M-theory to a lower spacetime dimension, a bewildering array of other black p-branes make their appearance in the theory, owing to the presence of a variety of $(p+2)$-form field strengths in the lower-dimensional supergravity theory [5,6]. Some of these p-branes may be interpreted as reductions of the eleven-dimensional ones or wrappings of the eleven-dimensional ones around cycles of the compactifying manifold [7–10]. In particular, one may obtain as special cases the four-dimensional black holes ($p = 1$). It has been suggested that, in the extremal limit, these black holes may be identified with BPS saturated string states [11–14]. Moreover, it is sometimes the case that multiply-charged black holes may be regarded as bound states at threshold of singly charged black holes [11,12,15,16]. Apart from their importance in the understanding of M-theory, therefore, these black p-branes have recently come to the fore as a way of providing a microscopic explanation of the Hawking entropy and temperature formulae [17–28] which have long been something of

[1] Research supported in part by NSF Grant PHY-9411543.
[2] Research supported in part by DOE Grant DE-FG05-91-ER40633.

PII S0370-2693(96)00521-7

an enigma. This latter progress has been made possible by the recognition that some p-branes carrying Ramond-Ramond charges also admit an interpretation as Dirichlet-branes, or D-branes, and are therefore amenable to the calculational power of conformal field theory [29].

The compactified eleven-dimensional supergravity theory admits a consistent truncation to the following set of fields: the metric tensor g_{MN}, a set of N scalar fields $\boldsymbol{\phi} = (\phi_1, \ldots, \phi_N)$, and N field strengths F_α of rank n. The Lagrangian for these fields takes the form [30,31]

$$e^{-1}\mathcal{L} = R - \tfrac{1}{2}(\partial\boldsymbol{\phi})^2 - \frac{1}{2n!}\sum_{\alpha=1}^{N} e^{\boldsymbol{a}_\alpha\cdot\boldsymbol{\phi}} F_\alpha^2 , \qquad (1.1)$$

where $\boldsymbol{a}_\alpha$ are constant vectors characteristic of the supergravity theory. The purpose of the present paper is to display a universal class of (non-rotating) black p-brane solutions to (1.1) and to calculate their classical entropy and temperature.

As discussed in Section 2, it is also possible to make a further consistent truncation to a single scalar ϕ and single field strength F:

$$e^{-1}\mathcal{L} = R - \tfrac{1}{2}(\partial\phi)^2 - \frac{1}{2n!}e^{a\phi}F^2 , \qquad (1.2)$$

where the parameter a can be conveniently re-expressed as

$$a^2 = \Delta - \frac{2d\tilde{d}}{D-2} , \qquad (1.3)$$

since Δ is a parameter that is preserved under dimensional reduction [32]. Special solutions of this theory have been considered before in the literature. Purely electric or purely magnetic black p-branes were considered in [5] for $D = 10$ dimensions and in [6] for general dimensions $D \leq 11$. All these had $\Delta = 4$. In the case of *extremal* black p-branes, these were generalized to other values of Δ in [32,30]. Certain non-extremal non-dilatonic ($a = 0$) black p-branes were also obtained in [33]

A particularly interesting class of solutions are the *dyonic* p-branes. Dyonic p-brane occur in dimensions $D = 2n$, where the n-index field strengths can carry both electric and magnetic charges. There are two types of dyonic solution. In the first type, each individual field strength in (1.1) carries either electric charge or magnetic charge, but not both. A particularly interesting example, owing to its non-vanishing entropy even in the extremal limit [34], is provided by the four-dimensional dyonic black hole. This is the $a = 0$ (Reissner-Nordstrom) solution, recently identified as a solution of heterotic string theory [11], but known for many years to be a solution of M-theory [35,36]. The construction of black dyonic p-branes of this type is identical to that for the solutions with purely electric or purely magnetic charges, discussed in Section 3.

In Section 4, we shall construct black dyonic p-branes of the second type, where there is one field strength, which carries both electric and magnetic charge. Special cases of these have also been considered before: the self-dual threebrane in $D = 10$ [5,37], the extremal self-dual string [6] and extremal dyonic string in $D = 6$ [41], a black self-dual string in $D = 6$ [33,19] and a different dyonic black hole in $D = 4$ [30]. See also [38] for the most general spherically symmetric extremal dyonic black hole solutions of the toroidally compactified heterotic string.

Black multi-scalar p-branes, the extremal limits of which may be found in [31], are discussed in Section 5.

The usual form of the metric for an isotropic p-brane in D dimensions is given by

$$ds^2 = e^{2A}(-dt^2 + dx^i dx^i) + e^{2B}(dr^2 + r^2 d\Omega^2) , \qquad (1.4)$$

where the coordinates (t, x^i) parameterise the d-dimensional world-volume of the p-brane. The remaining coordinates of the D dimensional spacetime are r and the coordinates on a $(D-d-1)$-dimensional unit sphere, whose metric is $d\Omega^2$. The functions A and B depend on the coordinate r only, as do the dilatonic scalar fields. The field strengths F_α can carry either electric or magnetic charge, and are given by

$$F_\alpha = \lambda_\alpha * \epsilon_{D-n} , \quad \text{or} \quad F_\alpha = \lambda_\alpha \epsilon_n , \qquad (1.5)$$

where ϵ_n is the volume form on the unit sphere $d\Omega^2$. The former case describes an elementary p-brane solution with $d = n - 1$ and electric charge $\lambda_\alpha = Q_\alpha$; the latter a solitonic p-brane solution with $d = D - n - 1$ and magnetic charge $\lambda_\alpha = P_\alpha$.

Solutions of supergravity theories with metrics of this form include extremal supersymmetric p-brane solitons, which saturate the Bogomol'nyi bound. The mass per unit p-volume of such a solution is equal to the sum of the electric and/or magnetic charges carried by participating field strengths. More general classes of "black" solutions exist in which the mass is an independent free parameter. In this paper, we shall show that there is a universal recipe for constructing such non-extremal generalisations of p-brane solutions, in which the metric (1.4) is replaced by

$$ds^2 = e^{2A}(-e^{2f}dt^2 + dx^i dx^i) + e^{2B}(e^{-2f}dr^2 + r^2 d\Omega^2)\,. \qquad (1.6)$$

Like A and B, f is a function of r. The ansätze for the field strengths (1.5) remain the same as in the extremal case. Remarkably, it turns out that the functions A, B and $\boldsymbol{\phi}$ take exactly the same form as they do in the extremal case, but for rescaled values of the electric and magnetic charges. The function f has a completely universal form:

$$e^{2f} = 1 - \frac{k}{r^{\tilde{d}}}\,, \qquad (1.7)$$

where $\tilde{d} = D - d - 2$. If k is positive, the metric has an outer event horizon at $r = r_+ = k^{1/\tilde{d}}$. When $k = 0$, the solution becomes extremal, and the horizon coincides with the location of the curvature singularity at $r = 0$.

The temperature of a black p-brane can be calculated by examining the behaviour of the metric (1.6) in the Euclidean regime in the vicinity of the outer horizon $r = r_+$. Setting $t = i\tau$ and $1 - kr^{-\tilde{d}} = \rho^2$, the metric (1.6) becomes

$$ds^2 = \frac{4r_+^2}{\tilde{d}^2}e^{2B(r_+)} \times \Big(d\rho^2 + \frac{\tilde{d}^2}{4r_+^2}e^{2A(r_+)-2B(r_+)}\rho^2 d\tau^2 + \cdots\Big)\,. \qquad (1.8)$$

We see that the conical singularity at the outer horizon ($\rho = 0$) is avoided if τ is assigned the period $(4\pi r_+/\tilde{d})e^{B(r_+)-A(r_+)}$. The inverse of this periodicity in imaginary time is the Hawking temperature,

$$T = \frac{\tilde{d}}{4\pi r_+}e^{A(r_+)-B(r_+)}\,. \qquad (1.9)$$

We may also calculate the entropy per unit p-volume of the black p-brane, which is given by one quarter of the area of the outer horizon. Thus we have

$$S = \tfrac{1}{4}r_+^{\tilde{d}+1}e^{(\tilde{d}+1)B(r_+)+(d-1)A(r_+)}\omega_{\tilde{d}+1}\,, \qquad (1.10)$$

where $\omega_{\tilde{d}+1} = 2\pi^{\tilde{d}/2+1}/(\frac{1}{2}\tilde{d})!$ is the volume of the unit $(\tilde{d}+1)$-sphere

In subsequent sections, we shall generalise various kinds of extremal p-brane solutions to obtain black single-scalar elementary and solitonic p-branes, black dyonic p-branes and black multi-scalar p-branes. The metric ansatz (1.6) gives rise to non-isotropic p-brane solutions for $d \geq 2$, in the sense that the Poincaré symmetry of the d-dimensional world volume is broken. When $d = 1$, however, the black hole solutions remain isotropic. In the extremal black hole solutions, the quantity $dA + \tilde{d}B$ vanishes, where A and B are defined in (1.4); whilst in the non-extremal cases, this quantity is non-vanishing. Isotropic p-brane solutions with $dA + \tilde{d}B \neq 0$ were discussed in [42].

2. Single-scalar black p-branes

The Lagrangian (1.1) can be consistently reduced to a Lagrangian for a single scalar and a single field strength

$$e^{-1}\mathcal{L} = R - \tfrac{1}{2}(\partial\phi)^2 - \frac{1}{2n!}e^{a\phi}F^2\,, \qquad (2.1)$$

where a, ϕ and F are given by [30]

$$a^2 = \Big(\sum_{\alpha,\beta}(M^{-1})_{\alpha\beta}\Big)^{-1}\,,$$

$$\phi = a\sum_{\alpha,\beta}(M^{-1})_{\alpha\beta}\,\boldsymbol{a}_\alpha\cdot\boldsymbol{\phi}\,,$$

$$(F_\alpha)^2 = a^2\sum_{\beta}(M^{-1})_{\alpha\beta}F^2\,, \qquad (2.2)$$

and $M_{\alpha\beta} = \boldsymbol{a}_\alpha\cdot\boldsymbol{a}_\beta$. The parameter a can conveniently be re-expressed as

$$a^2 = \Delta - \frac{2d\tilde{d}}{D-2}\,, \qquad (2.3)$$

where Δ is a parameter that is preserved under dimensional reduction [32]. Supersymmetric p-brane solutions can arise only when the value of Δ is given by $\Delta = 4/N$, with N field strengths participating in the solution. This occurs when the dot products of the dilaton vectors $\boldsymbol{a}_\alpha$ satisfy [31]

$$M_{\alpha\beta} = 4\delta_{\alpha\beta} - \frac{2d\tilde{d}}{D-2}. \qquad (2.4)$$

An interesting special case is provided by the four-dimensional black holes with $a^2 = 3, 1, 1/3, 0$, i.e. $N = 1, 2, 3, 4$ whose extremal limits admit the interpretation of 1, 2, 3, 4-particle bound states at threshold [11,15,16]. Their $D = 11$ interpretation has recently been discussed in [39,40].

To begin, let us consider the more general metric

$$ds^2 = -e^{2u}dt^2 + e^{2A}dx^i dx^i + e^{2v}dr^2 + e^{2B}r^2 d\Omega^2. \qquad (2.5)$$

It is straightforward to show that the Ricci tensor for this metric has the following non-vanishing components

$$\begin{aligned}
R_{00} &= e^{2(u-v)}\big(u'' - u'v' + v'^2 + (d-1)u'A' \\
&\quad + (\tilde{d}+1)u'(B' + \frac{1}{r})\big), \\
R_{ij} &= -e^{2(A-v)}\big(A'' - A'v' + A'u' + (d-1)A'^2 \\
&\quad + (\tilde{d}+1)A'(B' + \frac{1}{r})\big)\delta_{ij}, \\
R_{rr} &= -u'' + u'v' - u'^2 - (d-1)A'' + (d-1)A'v' \\
&\quad - (d-1)A'^2 - (\tilde{d}+1)B'' + \frac{\tilde{d}+1}{r}v' \\
&\quad - \frac{2(\tilde{d}+1)}{r}B' + (\tilde{d}+1)v'B' - (\tilde{d}+1)B'^2, \\
R_{ab} &= -e^{2(B-v)}\big(B'' + (B' + \frac{1}{r})[u' - v' + (d-1)A' \\
&\quad + (\tilde{d}+1)(B' + \frac{1}{r})] - \frac{1}{r^2}\big)g_{ab} + \tilde{d}g_{ab}, \qquad (2.6)
\end{aligned}$$

where a prime denotes a derivative with respect to r, and g_{ab} is the metric on the unit $(\tilde{d}+1)$-sphere. For future reference, we note that the ADM mass per unit p-volume for this metric is given by [43]

$$\begin{aligned}
m &= \big[(d-1)(e^{2A})'r^{\tilde{d}+1} + (\tilde{d}+1)(e^{2B})'r^{\tilde{d}+1} \\
&\quad - (\tilde{d}+1)(e^{2v} - e^{2B})r^{\tilde{d}}\big]\Big|_{r\to\infty}. \qquad (2.7)
\end{aligned}$$

The Ricci tensor for the metric (1.6) is given by (2.6) with $u = 2(A+f)$ and $v = 2(B-f)$. As in the case of isotropic p-brane solutions, the equations of motion simplify dramatically after imposing the ansatz

$$dA + \tilde{d}B = 0. \qquad (2.8)$$

Furthermore, the structure of the equations of motion implies that it is natural to take

$$f'' + \frac{\tilde{d}+1}{r}f' + 2f'^2 = 0, \qquad (2.9)$$

which has the solution given by (1.7). Note that we have chosen the asymptotic value of f to be zero at $r = \infty$. This is necessary in order that the metric (1.6) be Minkowskian at $r = \infty$. The equations of motion then reduce to the following three simple equations:

$$\begin{aligned}
&\phi'' + \frac{\tilde{d}+1}{r}\phi' + 2\phi'f' = -\frac{\epsilon a}{2}s^2 e^{-2f}, \\
&A'' + \frac{\tilde{d}+1}{r}A' + 2A'f' = \frac{\tilde{d}}{2(D-2)}s^2 e^{-2f}, \\
&d(D-2)A' + \tfrac{1}{2}\tilde{d}\phi'^2 + 2(D-2)A'f' \\
&\quad = \tfrac{1}{2}\tilde{d}s^2 e^{-2f}, \qquad (2.10)
\end{aligned}$$

where s is given by

$$s = \lambda e^{-\frac{1}{2}\epsilon a\phi + dA}\, r^{-(\tilde{d}+1)}, \qquad (2.11)$$

and $\epsilon = 1$ for elementary solutions and $\epsilon = -1$ for solitonic solutions. The last equation in (2.10) is a first integral of the first two equations, and hence determines an integration constant. The first two equations in (2.10) imply that we can naturally solve for the dilaton ϕ by taking $\phi = a(D-2)A/\tilde{d}$. The remaining equation can then be easily solved by making the ansatz that the function A takes the identical form as in the extremal case, but with a rescaled charge, i.e. it satisfies

$$\begin{aligned}
&A'' + \frac{\tilde{d}+1}{r}A' = \frac{\tilde{d}}{2(D-2)}\tilde{s}^2, \quad \text{with} \\
&\tilde{s} = \tilde{\lambda}e^{-\frac{1}{2}\epsilon a\phi + dA}\, r^{-(\tilde{d}+1)}. \qquad (2.12)
\end{aligned}$$

This has the solution $e^{-(D-2)\Delta A/(2\tilde{d})} = 1 + \tilde{\lambda}\sqrt{\Delta}/(2\tilde{d})\, r^{-\tilde{d}}$. Thus from (2.10) we have

$$2A'f' = (A'' + \frac{\tilde{d}+1}{r}A')(-1 + \frac{\lambda^2}{\tilde{\lambda}^2}e^{-2f}), \qquad (2.13)$$

implying

$$\frac{\lambda^2}{\tilde{\lambda}^2} - e^{2f} = c(1 + \frac{\tilde{\lambda}\sqrt{\Delta}}{2\tilde{d}}r^{-\tilde{d}}), \qquad (2.14)$$

where c is an integration constant. Substituting (1.7) into this, we deduce that

$$e^{-(D-2)\Delta A/(2\tilde{d})} = 1 + \frac{k}{r^{\tilde{d}}}\Big(\frac{\lambda^2}{\tilde{\lambda}^2} - 1\Big)^{-1}. \tag{2.15}$$

Thus it is natural to set $\tilde{\lambda} = \lambda \tanh\mu$, giving

$$e^{-(D-2)\Delta A/(2\tilde{d})} = 1 + \frac{k}{r^{\tilde{d}}}\sinh^2\mu. \tag{2.16}$$

The blackened single-scalar p-brane solution is therefore given by

$$\begin{aligned} ds^2 &= \Big(1 + \frac{k}{r^{\tilde{d}}}\sinh^2\mu\Big)^{-\frac{4\tilde{d}}{\Delta(D-2)}}(-e^{2f}dt^2 + dx^i dx^i) \\ &\quad + \Big(1 + \frac{k}{r^{\tilde{d}}}\sinh^2\mu\Big)^{\frac{4d}{\Delta(D-2)}}(e^{-2f}dr^2 + r^2 d\Omega^2), \\ e^{\frac{\epsilon\Delta}{2a}\phi} &= 1 + \frac{k}{r^{\tilde{d}}}\sinh^2\mu, \quad e^{2f} = 1 - \frac{k}{r^{\tilde{d}}}, \end{aligned} \tag{2.17}$$

with the two free parameters k and μ related to the charge λ and the mass per unit p-volume, m. Specifically, we find that

$$\begin{aligned} \lambda &= \frac{\tilde{d}k}{\sqrt{\Delta}}\sinh 2\mu, \\ m &= k\Big(\frac{4\tilde{d}}{\Delta}\sinh^2\mu + \tilde{d} + 1\Big). \end{aligned} \tag{2.18}$$

The extremal limit occurs when $k \longrightarrow 0$, $\mu \longrightarrow \infty$ while holding $ke^{2\mu} = \sqrt{\Delta}\lambda/\tilde{d} =$ constant. If k is non-negative, the mass and charge satisfy the bound

$$\begin{aligned} m - \frac{2\lambda}{\sqrt{\Delta}} &= \frac{k}{\Delta}\big[(\tilde{d}+1)\Delta - 2\tilde{d} + 2\tilde{d}e^{-2\mu}\big] \\ &\geq \frac{2k\tilde{d}^2(d-1)}{\Delta(D-2)} \geq 0, \end{aligned} \tag{2.19}$$

where the inequality is derived from $\Delta = a^2 + 2d\tilde{d}/(D-2) \geq 2d\tilde{d}/(D-2)$. The mass/charge bound (2.19) is saturated when k goes to zero, which is the extremal limit. In cases where $\Delta = 4/N$, the extremal solution becomes supersymmetric, and the bound (2.19) coincides with the Bogomol'nyi bound. Note however that in general there can exist extremal classical p-brane solutions for other values of Δ, which preserve no supersymmetry [30].

It follows from (1.9) and (1.10) that the Hawking temperature and entropy of the black p-brane (2.17) are given by

$$\begin{aligned} T &= \frac{\tilde{d}}{4\pi r_+}(\cosh\mu)^{-\frac{4}{\Delta}}, \\ S &= \tfrac{1}{4}r_+^{\tilde{d}+1}\,\omega_{\tilde{d}+1}\,(\cosh\mu)^{\frac{4}{\Delta}}. \end{aligned} \tag{2.20}$$

In the extremal limit, they take the form

$$\begin{aligned} T &\propto (e^{\mu})^{2(a^2 - \frac{2\tilde{d}^2}{D-2})/(\Delta\tilde{d})}, \\ S &\propto e^{\mu(4/\Delta - 2(\tilde{d}+1)/\tilde{d})}. \end{aligned} \tag{2.21}$$

Thus the entropy becomes zero in the extremal limit $\mu \to \infty$, unless the constant a is zero and $d = 1$, since the exponent can be rewritten as $\mu(4/\Delta - 2(\tilde{d}+1)/\tilde{d}) = -2\mu\big(2(d-1)\tilde{d}/(D-2) + (\tilde{d}+1)a^2/\tilde{d}\big)/\Delta$. In these special cases the dilaton ϕ vanishes and the entropy is finite and non-zero. The situation can arise for black holes with $\Delta = 4/3$ in $D = 5$, and $\Delta = 1$ in $D = 4$. The temperature of the extremal p-brane is zero, finite and non-zero, or infinite, according to whether $(a^2 - \frac{2\tilde{d}^2}{D-2})$ is negative, zero or positive.

3. Black dyonic *p*-branes

Dyonic p-brane occur in dimensions $D = 2n$, where the n-index field strengths can carry both electric and magnetic charges. There are two types of dyonic solution. In the first type, each individual field strength in (1.1) carries either electric charge or magnetic charge, but not both. The construction of black dyonic p-branes of this type is identical to that for the solutions with purely electric or purely magnetic charges, which we discussed in the previous section.

In this section, we shall construct black dyonic p-branes of the second type, where there is one field strength, which carries both electric and magnetic charge. The Lagrangian is again given by (2.1), with the field strength now taking the form

$$F = \lambda_1\,\epsilon_n + \lambda_2\,{*\epsilon_n}. \tag{3.1}$$

As in the case of purely elementary or purely solitonic p-brane solutions, we impose the conditions (2.8) and (2.9) on B and f respectively. The equations of motion then reduce to

$$\phi'' + \frac{n}{r}\phi' + 2\phi' f' = \tfrac{1}{2}a(s_1^2 - s_2^2)e^{-2f},$$

$$A'' + \frac{n}{r}A' + 2A'f' = \tfrac{1}{4}(s_1^2 + s_2^2)e^{-2f}\,,$$
$$d(D-2)A'^2 + \tfrac{1}{2}\tilde{d}\phi'^2 + 2(D-2)A'f'$$
$$= \tfrac{1}{2}\tilde{d}(s_1^2 + s_2^2)e^{-2f}\,, \qquad (3.2)$$

where

$$s_1 = \lambda_1 e^{\frac{1}{2}a\phi+(n-1)A}\, r^{-n}\,,$$
$$s_2 = \lambda_2 e^{-\frac{1}{2}a\phi+(n-1)A}\, r^{-n}\,. \qquad (3.3)$$

We can solve the Eqs. (3.2) for black dyonic p-branes by following analogous steps to those described in the previous section, relating the solutions to extremal dyonic solutions. In particular, we again find that the functions A, B and ϕ take precisely the same forms as in the extremal case, but with rescaled values of charges. Solutions for extremal dyonic p-branes are known for two values of a, namely $a^2 = n-1$ and $a = 0$ [30]. When $a^2 = n-1$, we find that the black dyonic p-brane solution is given by

$$e^{-\frac{1}{2}a\phi-(n-1)A} = 1 + \frac{k}{r^{n-1}}\sinh^2\mu_1\,,$$
$$e^{\frac{1}{2}a\phi-(n-1)A} = 1 + \frac{k}{r^{n-1}}\sinh^2\mu_2\,, \qquad (3.4)$$

with f given by (1.7). The mass per unit volume and the electric and magnetic charges are given by

$$m = k(2\sinh^2\mu_1 + 2\sinh^2\mu_2 + 1)\,,$$
$$\lambda_\alpha = (ak/\sqrt{2})\,\sinh(2\mu_\alpha)\,. \qquad (3.5)$$

For the non-negative values of k, the mass and the charges satisfy the bound

$$m - (\lambda_1 + \lambda_2) = k(n - 2 + e^{-2\mu_1} + e^{-2\mu_2}) \geq 0\,. \qquad (3.6)$$

The bound is saturated in the extremal limit $k \longrightarrow 0$. The solution (3.4) corresponds to the black dyonic string with $n = 3$ and $\Delta = 4$ in $D = 6$, and the dyonic black hole with $n = 2$ and $\Delta = 2$ in $D = 4$. In both cases, the extremal solution is supersymmetric and the bound (3.6) coincides with the Bogomol'nyi bound. Using (1.9) and (1.10), we find that the Hawking temperature and entropy of the non-extremal solutions are given by

$$T = \frac{\tilde{d}}{4\pi r_+}\big(\cosh\mu_1\cosh\mu_2\big)^{-\frac{2}{n-1}}\,,$$
$$S = \tfrac{1}{4}r_+^n\,\omega_n\,\big(\cosh\mu_1\cosh\mu_2\big)^{\frac{2}{n-1}}\,. \qquad (3.7)$$

When $a = 0$, the equations of motion degenerate and the dilaton ϕ decouples. We find the solution

$$\phi = 0\,, \quad e^{-(n-1)A} = 1 + \frac{k}{r^{n-1}}\sinh^2\mu\,, \qquad (3.8)$$

where again f is given by (1.7). The constant μ is related to the electric and magnetic charges by $\sqrt{\lambda_1^2 + \lambda_2^2} = k\sinh 2\mu$. In this case, unlike the $a^2 = n-1$ case, the solution is invariant under rotations of the electric and magnetic charges, and hence it is equivalent to the purely electric or purely magnetic solutions we discussed in the previous section. Note that in the dyonic solution (3.4), when the parameter $\mu_1 = \mu_2$, i.e. the electric and magnetic charges are equal, the dilaton field also decouples. For example, this can happen if one imposes a self-dual condition on the 3-form field strength in the dyonic string in $D = 6$. However, this is a different situation from the $a = 0$ dyonic solution, since in the latter case the electric and magnetic charges are independent free parameters. In fact the $a = 0$ dyonic solution with independent electric and magnetic charges occurs only in $D = 4$.

4. Black multi-scalar p-branes

To describe multi-scalar p-brane solutions, we return to the Lagrangian (1.1) involving N scalars and N field strengths. As we discussed previously, it can be consistently truncated to the single-scalar Lagrangian (2.1), in which case all the field strengths F_α are proportional to the canonically-normalised field strength F, and hence there is only one independent charge parameter. In a multi-scalar p-brane solution, the charges associated with each field strength become independent parameters. After imposing the conditions (2.8) and (2.9), the equations of motion reduce to

$$\varphi_\alpha'' + \frac{\tilde{d}+1}{r}\varphi_\alpha' + 2\varphi_\alpha' f' = -\tfrac{1}{2}\epsilon e^{-2f}\sum_{\beta=1}^{N} M_{\alpha\beta}\, S_\beta^2\,,$$

$$A'' + \frac{\tilde{d}+1}{r}A' + 2A'f' = \frac{\tilde{d}}{2(D-2)}e^{-2f}\sum_{\alpha=1}^{N} S_\alpha^2\,,$$

$$d(D-2)A'^2 + \tfrac{1}{2}\tilde{d}\sum_{\alpha,\beta=1}^{N}(M^{-1})_{\alpha\beta}\,\varphi'_\alpha\,\varphi'_\beta$$

$$+\,2(D-2)A'f' = \tfrac{1}{2}\tilde{d}e^{-2f}\sum_{\alpha=1}^{N} S_\alpha^2\,, \qquad (4.1)$$

where $\varphi_\alpha = \boldsymbol{a}_\alpha \cdot \boldsymbol{\phi}$ and $S_\alpha = \lambda_\alpha e^{-\frac{1}{2}\epsilon\varphi_\alpha + dA}\,r^{-(\tilde{d}+1)}$. We again find black solutions by taking A and φ_α to have the same forms as in the extremal case, with rescaled charges. Extremal solutions can be found in cases where the dot products of the dilaton vectors $\boldsymbol{a}_\alpha$ satisfy (2.4) [30]. Thus we find that the corresponding black solutions are given by

$$e^{\frac{1}{2}\epsilon\varphi_\alpha - dA} = 1 + \frac{k}{r^{\tilde{d}}}\sinh^2\mu_\alpha\,, \qquad e^{2f} = 1 - \frac{k}{r^{\tilde{d}}}\,,$$

$$ds^2 = \prod_{\alpha=1}^{N}\left(1 + \frac{k}{r^{\tilde{d}}}\sinh^2\mu_\alpha\right)^{-\frac{\tilde{d}}{D-2}}(-e^{2f}dt^2 + dx^i dx^i)$$

$$+\prod_{\alpha=1}^{N}\left(1 + \frac{k}{r^{\tilde{d}}}\sinh^2\mu_\alpha\right)^{\frac{d}{D-2}}(e^{-2f}dr^2 + r^2 d\Omega^2)\,. \qquad (4.2)$$

The mass per unit volume and the charges for this solution are given by

$$m = k(\tilde{d}\sum_{\alpha=1}^{N}\sinh^2\mu_\alpha + \tilde{d} + 1)\,,$$
$$\lambda_\alpha = \tfrac{1}{2}\tilde{d}k\sinh 2\mu_\alpha\,. \qquad (4.3)$$

For non-negative values of k, the mass and charges satisfy the bound

$$m - \sum_{\alpha=1}^{N}\lambda_\alpha = \tfrac{1}{2}k\tilde{d}\sum_{\alpha=1}^{N}(e^{-2\mu_\alpha} - 1) + k(\tilde{d}+1)$$

$$\geq \frac{k\tilde{d}(d-1)}{d} \geq 0\,. \qquad (4.4)$$

The bound coincides with the Bogomol'nyi bound. The Hawking temperature and entropy are given by

$$T = \frac{\tilde{d}}{4\pi r_+}\prod_{\alpha=1}^{N}(\cosh\mu_\alpha)^{-1}\,,$$

$$S = \tfrac{1}{4}r_+^{\tilde{d}+1}\,\omega_{\tilde{d}+1}\prod_{\alpha=1}^{N}\left(\cosh\mu_\alpha\right). \qquad (4.5)$$

In the extremal limit $k \longrightarrow 0$, the bound (4.4) is saturated, and the solutions become supersymmetric.

5. Conclusions

We have presented a class of black p-brane solutions of M-theory which were hitherto known only in the extremal supersymmetric limit and have calculated their macroscopic entropy and temperature. It would obviously be of interest to provide a microscopic derivation of the entropy and temperature using D-brane techniques and compare them with the macroscopic results found in this paper. Agreement would both boost the credibility of M-theory and further our understanding of black hole and black p-brane physics.

Acknowledgements

We have enjoyed useful conversations with Joachim Rahmfeld. We are very grateful to Arkady Tseytlin for drawing our attention to an error in the entropy formula (1.10) in the original version of this paper.

References

[1] E. Cremmer, B. Julia and J. Scherk, Supergravity theory in 11 dimensions, Phys. Lett. B 76 (1978) 409.
[2] R. Güven, Black p-brane solitons of $D = 11$ supergravity theory, Phys. Lett. B 276 (1992) 49.
[3] M.J. Duff and K.S. Stelle, Multi-membrane solutions of $D = 11$ supergravity, Phys. Lett. B 253 (1991) 113.
[4] E. Bergshoeff, E. Sezgin and P.K. Townsend, Supermembranes and eleven-dimensional supergravity, Phys. Lett. B 189 (1987) 75; Properties of the eleven-dimensional supermembrane theory, Ann. Phys. 185 (1988) 330.
[5] G.T. Horowitz and A. Strominger, Black strings and p-branes, Nucl. Phys. B 360 (1991) 197.
[6] M.J. Duff and J.X. Lu, Black and super p-branes in diverse dimensions, Nucl. Phys. B 416 (1994) 301.
[7] M.J. Duff, R.R. Khuri, R. Minasian and J. Rahmfeld, New black hole, string and membrane solutions of the four dimensional heterotic string, Nucl. Phys. B 418 (1994) 195.

[8] C.M. Hull and P.K. Townsend, Unity of superstring dualities, Nucl. Phys. B 294 (1995) 196.
[9] M.J. Duff, R.R Khuri and J.X. Lu, String solitons, Phys. Rep. 259 (1995) 213.
[10] A.A. Tseytlin, Extreme dyonic black holes in string theory, hep-th/9601177.
[11] M.J. Duff and J. Rahmfeld, Massive string states as extreme black holes, Phys. Lett. B 345 (1995) 441.
[12] A. Sen, Black hole solutions in heterotic string theory on a torus, Nucl. Phys. B 440 (1995) 421; Extremal black holes and elementary string states, Mod. Phys. Lett. A 10 (1995) 2081; A note on marginally stable bound states in Type II string theory, hep-th/9510229.
[13] R.R. Khuri and R.C. Myers, Dynamics of extreme black holes and massive string states, hep-th/9508045.
[14] A. Peet, Entropy and supersymmetry of D dimensional extremal electric black holes versus string states, hep-th/9506200.
[15] M.J. Duff, J.T. Liu and J. Rahmfeld, Four dimensional string/string/string triality, Nucl. Phys. B 459 (1996) 125.
[16] J. Rahmfeld, Extremal black holes as bound states, hep-th/9512089.
[17] A. Strominger and C. Vafa, Microscopic origin of Bekenstein-Hawking entropy, hep-th/9601029.
[18] C. Callan and J. Maldacena, D-brane approach to black hole quantum mechanics, hep-th/9602043.
[19] G.T. Horowitz and A. Strominger, Counting states of near extremal black holes, hep-th/9602051.
[20] A. Ghosh and P. Mitra, Entropy of extremal dyonic black holes, hep-th/9602057.
[21] J. Beckenridge, R. Myers, A. Peet and C. Vafa, D-branes and spinning black holes, hep-th/9602065.
[22] S.R. Das and S. Mathur, Excitations of D-strings, entropy and duality, hep-th/9611152.
[23] S.R. Das, Black hole entropy and string theory, hep-th/9602172.
[24] J. Beckenridge, D.A. Lowe, R. Myers, A. Peet, A. Strominger and C. Vafa, Macroscopic and microscopic entropy of near-extremal spinning black holes, hep-th/9603078.
[25] A. Strominger and J. Maldacena, Statistical entropy of four-dimensional extremal back holes, hep-th/9603060.
[26] C.V. Johnson, R.R. Khuri and R.C. Myers, Entropy of 4-D extremal black holes, hep-th/9603061.
[27] G.T. Horowitz, J. Maldacena and A. Strominger, Nonextremal black hole microstates and U-duality, hep-th/9603109.
[28] R. Dijkgraaf, E. Verlinde and H. Verlinde, BPS spectrum of the five-brane and black hole entropy, hep-th/96033126.
[29] J. Polchinski, Dirichlet-branes and Ramond-Ramond charges, hep-th/9511026.
[30] H. Lü and C.N. Pope, p-brane solitons in maximal supergravities, hep-th/9512012, to appear in Nucl. Phys. B.
[31] H. Lü and C.N. Pope, Multi-scalar p-brane solitons, hep-th/9512153, to appear in Int. J. Mod. Phys. A.
[32] H. Lü, C.N. Pope, E. Sezgin and K.S. Stelle, Stainless super p-branes, Nucl. Phys. B 456 (1995) 669.
[33] G.W. Gibbons, G.T. Horowitz and P.K. Townsend, Class. Quantum Grav. 12 (1995) 297.
[34] F. Larsen and F. Wilczek, Internal structure of black holes, hep-th/9511064.
[35] M.J. Duff and C.N. Pope, Consistent truncations in Kaluza-Klein theories, Nucl. Phys. B 255 (1985) 355.
[36] C.N. Pope, The embedding of the Einstein Yang-Mills equations in $D = 11$ supergravity, Class. Quant. Grav. 2 (1985) L77.
[37] M.J. Duff and J.X. Lu, The self-dual type IIB superthreebrane, Phys. Lett. B 273 (1991) 409.
[38] M. Cvetic and D. Youm, Dyonic BPS saturated black holes of heterotic string theory on a torus, hep-th/9507090.
[39] G. Papadopoulos and P.K. Townsend, Intersecting M-branes, hep-th/9603087.
[40] A.A. Tseytlin, Harmonic superpositions of M-branes, hep-th/9604035.
[41] M.J. Duff, S. Ferrara, R.R. Khuri and J. Rahmfeld, Supersymmetry and dual string solitons, Phys. Lett. B 356 (1995) 479.
[42] H. Lü, C.N. Pope and K.W. Xu, Liouville and Toda solitons in M-theory, hep-th/9604058.
[43] J.X. Lu, ADM masses for black strings and p-branes, Phys. Lett. B 313 (1993) 29.

Reprinted from Nucl. Phys. B **475** (1996) 179–92

Intersecting M-branes as four-dimensional black holes

I.R. Klebanov [a,1], A.A. Tseytlin [b,2]

[a] *Joseph Henry Laboratories, Princeton University, Princeton, NJ 08544, USA*
[b] *Theoretical Physics Group, Blackett Laboratory, Imperial College, London SW7 2BZ, UK*

Received 6 May 1996; accepted 18 June 1996

Abstract

We present two 1/8 supersymmetric intersecting p-brane solutions of 11-dimensional supergravity which upon compactification to four dimensions reduce to extremal dyonic black holes with finite area of horizon. The first solution is a configuration of three intersecting 5-branes with an extra momentum flow along the common string. The second describes a system of two 2-branes and two 5-branes. Related (by compactification and T-duality) solution of type IIB theory corresponds to a completely symmetric configuration of four intersecting 3-branes. We suggest methods for counting the BPS degeneracy of three intersecting 5-branes which, in the macroscopic limit, reproduce the Bekenstein-Hawking entropy.

PACS: 04.65.+e; 11.27.+d; 11.30.Pb

1. Introduction

The existence of supersymmetric extremal dyonic black holes with finite area of the horizon provides a possibility of a statistical understanding [1] of the Bekenstein-Hawking entropy from the point of view of string theory [2–4]. Such black hole solutions are found in four [5–7] and five [4,8] dimensions but not in $D > 5$ [9,10]. While the D-brane BPS state counting derivation of the entropy is relatively straightforward for the $D = 5$ black holes [4,11], it is less transparent in the $D = 4$ case, a complication being the presence of a solitonic 5-brane or Kaluza-Klein monopole in addition to a D-brane configuration in the descriptions used in [12,13].

[1] E-mail: klebanov@puhep1.princeton.edu.
[2] E-mail: tseytlin@ic.ac.uk.
On leave from Lebedev Physics Institute, Moscow.

PII S0550-3213(96)00338-0

One may hope to find a different lifting of the dyonic $D = 4$ black hole to $D = 10$ string theory that may correspond to a purely D-brane configuration. A related question is about the embedding of the $D = 4$ dyonic black holes into $D = 11$ supergravity (M-theory) which would allow to reproduce their entropy by counting the corresponding BPS states using the M-brane approach similar to the one applied in the $D = 5$ black hole case in [14].

As was found in [15], the (three-charge, finite area) $D = 5$ extremal black hole can be represented in M-theory by a configuration of orthogonally intersecting 2-brane and 5-brane (i.e. $2\perp5$) with a momentum flow along the common string, or by configuration of three 2-branes intersecting over a point ($2\perp2\perp2$). A particular embedding of (four-charge, finite area) $D = 4$ black hole into $D = 11$ theory given in [15] can be interpreted as a similar $2\perp5$ configuration 'superposed' with a Kaluza-Klein monopole.

Below we shall demonstrate that it is possible to get rid of the complication associated with having the Kaluza-Klein monopole. There exists a simple 1/8 supersymmetric configuration of *four* intersecting M-branes ($2\perp2\perp5\perp5$) with diagonal $D = 11$ metric. Upon compactification along six isometric directions it reduces to the dyonic $D = 4$ black hole with finite area and all scalars being regular at the horizon.

The corresponding $2\perp2\perp4\perp4$ solution of type IIA $D = 10$ superstring theory (obtained by dimensional reduction along a direction common to the two 5-branes) is T-dual to a $D = 10$ solution of type IIB theory which describes a remarkably symmetric configuration of *four* intersecting 3-branes.[3]

Our discussion will follow closely that of [15] where an approach to constructing intersecting supersymmetric p-brane solutions (generalising that of [19]) was presented.[4] The supersymmetric configurations of two or three intersecting 2- and 5-branes of $D = 11$ supergravity which preserve 1/4 or 1/8 of maximal supersymmetry are $2\perp2$, $5\perp5$, $2\perp5$, $2\perp2\perp2$, $5\perp5\perp5$, $2\perp2\perp5$ and $2\perp5\perp5$. Two 2-branes can intersect over a point, two 5-branes – over a 3-brane (which in turn can intersect over a string), 2-brane and 5-brane can intersect over a string [19]. There exists a simple 'harmonic function' rule [15] which governs the construction of composite supersymmetric p-brane solutions in both $D = 10$ and $D = 11$: a separate harmonic function is assigned to each constituent 1/2 supersymmetric p-brane.

[3] Similar D-brane configuration was independently discussed in [16,17]. Note that it is a combination of *four* and not *three* intersecting 3-branes that is related (for the special choice of equal charges) to the non-dilatonic ($a = 0$) RN $D = 4$ black hole. T-dual configuration of one 0-brane and three intersecting 4-branes of type IIA theory was considered in [18].

[4] Intersecting p-brane solutions in [19,15] and below are isometric in all directions internal to all constituent p-branes (the background fields depend only on the remaining common transverse directions). They are different from possible virtual configurations where, e.g., a (p-2)-brane ends (in transverse radial direction) on a p-brane [20]. A configuration of p-brane and p′-brane intersecting in p+p′-space may be also considered as a special anisotropic p+p′-brane. It seems unlikely that there exist more general static solutions (with constituent p-branes effectively having different transverse spaces [19,21]) which may 'interpolate' between intersecting p-brane solutions and solutions with one p-brane ending on another in the transverse direction of the latter.

Most of the configurations with *four* intersecting M-branes, namely, $2\perp2\perp2\perp2$, $2\perp2\perp2\perp5$ and $5\perp5\perp5\perp2$ are 1/16 supersymmetric and have transverse x-space dimension equal to two ($5\perp5\perp5\perp5$ configuration with 5-branes intersecting over 3-branes to preserve supersymmetry does not fit into 11-dimensional space-time). Being described in terms of harmonic functions of x they are thus not asymptotically flat in transverse directions. There exists, however, a remarkable exception – the configuration $2\perp2\perp5\perp5$ which (like $5\perp2\perp2$, $5\perp5\perp2$ and $5\perp5\perp5$) has transverse dimension equal to three and the fraction of unbroken supersymmetry equal to 1/8 (Section 3). Upon compactification to $D = 4$ it reduces to the extremal dyonic black hole with *four* different charges and finite area of the horizon.

Similar $D = 4$ black hole background can be obtained also from the 'boosted' version of the $D = 11$ $5\perp5\perp5$ solution [15] (Section 2) [5] as well from the $3\perp3\perp3\perp3$ solution of $D = 10$ type IIB theory (Section 4). The two $D = 11$ configurations $5\perp5\perp5$+'boost' and $2\perp2\perp5\perp5$ reduce in $D = 10$ to $0\perp4\perp4\perp4$ and $2\perp2\perp4\perp4$ solutions of $D = 10$ type IIA theory which are related by T-duality.

In Section 5 we shall suggest methods for counting the BPS entropy of three intersecting 5-branes which reproduce the Bekenstein-Hawking entropy of the $D = 4$ black hole. This seems to explain the microscopic origin of the entropy directly in 11-dimensional terms.

2. 'Boosted' $5\perp5\perp5$ solution of $D = 11$ theory

The $D = 11$ background corresponding to $5\perp5\perp5$ configuration [19] is [15]

$$ds_{11}^2 = (F_1F_2F_3)^{-2/3}\big[F_1F_2F_3(-dt^2 + dy_1^2) \tag{2.1}$$

$$+ F_2F_3(dy_2^2 + dy_3^2) + F_1F_3(dy_4^2 + dy_5^2) + F_1F_2(dy_6^2 + dy_7^2) + dx_s dx_s\big] ,$$

$$\mathcal{F}_4 = 3(*dF_1^{-1} \wedge dy_2 \wedge dy_3 + *dF_2^{-1} \wedge dy_4 \wedge dy_5 + *dF_3^{-1} \wedge dy_6 \wedge dy_7) . \tag{2.2}$$

Here $\mathcal{F}_4$ is the 4-form field strength and F_i are the inverse powers of harmonic functions of x_s ($s = 1, 2, 3$). In the simplest 1-center case discussed below $F_i^{-1} = 1 + P_i/r$ ($r^2 = x_s x_s$). The $*$-duality is defined with respect to the transverse 3-space. The coordinates y_n internal to the three 5-branes can be identified according to the F_i factors inside the square brackets in the metric: $(y_1, y_4, y_5, y_6, y_7)$ belong to the first 5-brane, $(y_1, y_2, y_3, y_6, y_7)$ to the second and $(y_1, y_2, y_3, y_4, y_5)$ to the third. 5-branes intersect over three 3-branes which in turn intersect over a common string along y_1. If $F_2 = F_3 = 1$ the above background reduces to the single 5-brane solution [22] with the harmonic function $H = F_1^{-1}$ independent of the two of transverse coordinates (here y_2, y_3). The

[5] The 'boost' along the common string corresponds to a Kaluza-Klein electric charge part in the $D = 11$ metric which is 'dual' to a Kaluza-Klein monopole part present in the $D = 11$ embedding of dyonic black hole in [15].

case of $F_3 = 1$ describes two 5-branes intersecting over a 3-brane.[6] The special case of $F_1 = F_2 = F_3$ is the solution found in [19].

Compactifying $y_1, .., y_7$ on circles we learn that the effective 'radii' (scalar moduli fields in $D = 4$) behave regularly both at $r = \infty$ and at $r = 0$ with the exception of the 'radius' of y_1. It is possible to stabilize the corresponding scalar by adding a 'boost' along the common string. The metric of the resulting more general solution [15] is (the expression for $\mathcal{F}_4$ remains the same)

$$ds_{11}^2 = (F_1F_2F_3)^{-2/3}\big[F_1F_2F_3(dudv + Kdu^2) \\ +F_2F_3(dy_2^2 + dy_3^2) + F_1F_3(dy_4^2 + dy_5^2) + F_1F_2(dy_6^2 + dy_7^2) + dx_s dx_s\big] . \tag{2.3}$$

Here $u = y_1 - t$, $v = 2t$ and K is a harmonic function of the three coordinates x_s. A non-trivial $K = 1 + Q/r$ describes a momentum flow along the string direction.[7] Q also has an interpretation of a 'boost' along y_1 direction.

The $D = 11$ metric (2.3) is regular at the $r = 0$ horizon and has a *non-zero* 9-area of the horizon (we assume that all y_n have period L)

$$A_9 = 4\pi L^7[r^2K^{1/2}(F_1F_2F_3)^{-1/2}]_{r\to 0} = 4\pi L^7\sqrt{QP_1P_2P_3} . \tag{2.4}$$

Compactification along $y_2, ...y_7$ leads to a solitonic $D = 5$ string. Remarkably, the corresponding 6-volume is constant so that one gets directly the Einstein-frame metric

$$ds_5^2 = H^{-1}(dudv + Kdu^2) + H^2 dx_s dx_s , \qquad H \equiv (F_1F_2F_3)^{-1/3} . \tag{2.5}$$

Further compactification along y_1 or u gives the $D = 4$ (Einstein-frame) metric which is isomorphic to the one of the dyonic black hole [6],

$$ds_4^2 = -\lambda(r)dt^2 + \lambda^{-1}(r)(dr^2 + r^2 d\Omega_2^2) , \tag{2.6}$$

$$\lambda(r) = \sqrt{K^{-1}F_1F_2F_3} = \frac{r^2}{\sqrt{(r+Q)(r+P_1)(r+P_2)(r+P_3)}} . \tag{2.7}$$

Note, however, that in contrast to the dyonic black hole background of [6] which has two electric and two magnetic charges here there is one electric (Kaluza-Klein) and 3 magnetic charges. From the $D = 4$ point of view the two backgrounds are related by U-duality. The corresponding 2-area of the $r = 0$ horizon is of course A_9/L^7.

In the special case when all 4 harmonic functions are equal ($K = F_i = H^{-1}$) the metric (2.3) becomes

---

[6] The corresponding 1/4 supersymmetric background also has 3-dimensional transverse space and reduces to a $D = 4$ black hole with two charges (it has $a = 1$ black hole metric when two charges are equal). The $5\perp 5$ configuration compactified to $D = 10$ gives $4\perp 4$ solution of type IIA theory which is T-dual to $3\perp 3$ solution of type IIB theory.

[7] The metric (2.3) with $F_i = 1$ (i.e. $ds^2 = -K^{-1}dt^2 + K|dy_1 + (K^{-1} - 1)dt|^2 + dy_n dy_n + dx_s dx_s$) reduces upon compactification along y_1 direction to the $D = 10$ type IIA R-R 0-brane background [23] with Q playing the role of the KK electric charge.

$$ds_{11}^2 = H^{-1}dudv + du^2 + dy_2^2 + ... + dy_6^2 + H^2 dx_s dx_s$$
$$= -H^{-2}dt^2 + H^2 dx_s dx_s + [dy_1 + (H^{-1} - 1)dt]^2 + dy_2^2 + ... + dy_6^2 , \quad (2.8)$$

and corresponds to a charged solitonic string in $D = 5$ or the Reissner-Nordström ($a = 0$) black hole in $D = 4$ ('unboosted' $5\perp5\perp5$ solution with $K = 1$ and equal F_i reduces to $a = 1/\sqrt{3}$ dilatonic $D = 4$ black hole [19]).

A compactification of this $5\perp5\perp5$+'boost' configuration to $D = 10$ along y_1 gives a type IIA solution corresponding to three 4-branes intersecting over 2-branes plus additional Kaluza-Klein (Ramond-Ramond vector) electric charge background, or, equivalently, to the $0\perp4\perp4\perp4$ configuration of three 4-branes intersecting over 2-branes which in turn intersect over a 0-brane. If instead we compactify along a direction common only to two of the three 5-branes we get $4\perp4\perp5$+'boost' type IIA solution.[8] Other related solutions of type IIA and IIB theories can be obtained by applying T-duality and $SL(2, \mathbb{Z})$ duality.

3. $2\perp2\perp5\perp5$ solution of $D = 11$ theory

Solutions with four intersecting M-branes are constructed according to the rules discussed in [15]. The $2\perp2\perp5\perp5$ configuration is described by the following background:

$$ds_{11}^2 = (T_1T_2)^{-1/3}(F_1F_2)^{-2/3}\big[- T_1T_2F_1F_2 \, dt^2$$
$$+ T_1F_1dy_1^2 + T_1F_2dy_2^2 + T_2F_1dy_3^2 + T_2F_2dy_4^2$$
$$+F_1F_2(dy_5^2 + dy_6^2 + dy_7^2) + dx_s dx_s\big] , \quad (3.1)$$

$$\mathcal{F}_4 = -3dt \wedge (dT_1 \wedge dy_1 \wedge dy_2 + dT_2 \wedge dy_3 \wedge dy_4)$$
$$+3(*dF_1^{-1} \wedge dy_2 \wedge dy_4 + *dF_2^{-1} \wedge dy_1 \wedge dy_3) . \quad (3.2)$$

Here T_i^{-1} are harmonic functions corresponding to the 2-branes and F_i^{-1} are harmonic functions corresponding to the 5-branes, i.e.

$$T_i^{-1} = 1 + \frac{Q_i}{r} , \qquad F_i^{-1} = 1 + \frac{P_i}{r} . \quad (3.3)$$

(y_1, y_2) belong to the first and (y_3, y_4) to the second 2-brane. $(y_1, y_3, y_5, y_6, y_7)$ and $(y_2, y_4, y_5, y_6, y_7)$ are the coordinates of the two 5-branes. Each 2-brane intersects each 5-brane over a string. 2-branes intersect over a 0-brane ($x = 0$) and 5-branes intersect over a 3-brane.

Various special cases include, in particular, the 2-brane solution [24] ($T_2 = F_1 = F_2 = 1$), as well as $5\perp5$ ($T_1 = T_2 = 1$) [19] and $2\perp5$ ($T_1 = F_2 = 1$), $2\perp2\perp5$ ($F_2 = 1$), $2\perp5\perp5$ ($T_2 = 1$) [15] configurations (more precisely, their limits when the harmonic functions do not depend on a number of transverse coordinates).

[8] This may be compared to another type IIA configuration (consisting of solitonic 5-brane lying within a R-R 6-brane, both being intersected over a 'boosted' string by a R-R 2-brane) which also reduces [12,15] to the dyonic $D = 4$ black hole.

As in the case of the $5\perp5\perp5+$'boost' solution (2.3), (2.2), the metric (3.1) is regular at the $r = 0$ horizon (in particular, all internal y_n-components smoothly interpolate between finite values at $r \to \infty$ and $r \to 0$) with the 9-area of the horizon being (cf. (2.4))

$$A_9 = 4\pi L^7 [r^2 (T_1 T_2 F_1 F_2)^{-1/2}]_{r\to 0} = 4\pi L^7 \sqrt{Q_1 Q_2 P_1 P_2}\,. \tag{3.4}$$

The compactification of y_n on 7-torus leads to a $D = 4$ background with the metric which is again the dyonic black hole one (2.6), now with

$$\lambda(r) = \sqrt{T_1 T_2 F_1 F_2} = \frac{r^2}{\sqrt{(r+Q_1)(r+Q_2)(r+P_1)(r+P_2)}}\,. \tag{3.5}$$

In addition, there are two electric and two magnetic vector fields (as in [6]) and also 7 scalar fields. The two electric and two magnetic charges are *directly* related to the 2-brane and 5-brane charges (cf. (3.2)).

When all 4 harmonic functions are equal ($T_i^{-1} = F_i^{-1} = H$) the metric (3.1) becomes (cf. (2.8))

$$ds^2_{11} = -H^{-2}dt^2 + H^2 dx_s dx_s + dy_1^2 + \ldots + dy_7^2\,, \tag{3.6}$$

i.e. describes a direct product of a $D = 4$ Reissner-Nordström black hole and a 7-torus.

Thus there exists an embedding of the dyonic $D = 4$ black holes into $D = 11$ theory which corresponds to a remarkably symmetric combination of M-branes only. In contrast to the embeddings with a Kaluza-Klein monopole [15] or electric charge ('boost') (2.3), (2.8) it has a diagonal $D = 11$ metric.

4. $3\perp3\perp3\perp3$ solution of type IIB theory

Dimensional reduction of the background (3.1), (3.2) to $D = 10$ along a direction common to the two 5-brane (e.g. y_7) gives a type IIA theory solution representing the R-R p-brane configuration $2\perp2\perp4\perp4$. This configuration is T-dual to $0\perp4\perp4\perp4$ one which is the dimensional reduction of the $5\perp5\perp5+$'boost' solution. This suggests also a relation between the two $D = 11$ configurations discussed in Sections 2 and 3.

T-duality along one of the two directions common to 4-branes transforms $2\perp2\perp4\perp4$ into the $3\perp3\perp3\perp3$ solution of type IIB theory. The explicit form of the latter can be found also directly in $D = 10$ type IIB theory (i.e. independently of the above $D = 11$ construction) using the method of [15], where the 1/4 supersymmetric solution corresponding to two intersecting 3-branes was given. One finds the following $D = 10$ metric and self-dual 5-form (other $D = 10$ fields are trivial):

$$\begin{aligned} ds^2_{10} = (T_1T_2T_3T_4)^{-1/2}\big[& -T_1T_2T_3T_4\, dt^2 \\ & + T_1T_2 dy_1^2 + T_1T_3 dy_2^2 + T_1T_4 dy_3^2 + T_2T_3 dy_4^2 + T_2T_4 dy_5^2 + T_3T_4 dy_6^2 + dx_s dx_s\big]\,, \end{aligned} \tag{4.1}$$

$$
\begin{aligned}
\mathcal{F}_5 = dt \wedge (& dT_1 \wedge dy_1 \wedge dy_2 \wedge dy_3 + dT_2 \wedge dy_1 \wedge dy_4 \wedge dy_5 \\
& + dT_3 \wedge dy_2 \wedge dy_4 \wedge dy_6 + dT_4 \wedge dy_3 \wedge dy_5 \wedge dy_6) \\
& + * dT_1^{-1} \wedge dy_4 \wedge dy_5 \wedge dy_6 + *dT_2^{-1} \wedge dy_2 \wedge dy_3 \wedge dy_6 \\
& + * dT_3^{-1} \wedge dy_1 \wedge dy_3 \wedge dy_5 + *dT_4^{-1} \wedge dy_1 \wedge dy_2 \wedge dy_4 \,. \qquad (4.2)
\end{aligned}
$$

The coordinates of the four 3-branes are (y_1, y_2, y_3), (y_1, y_4, y_5), (y_2, y_4, y_6) and (y_3, y_5, y_6), i.e. each pair of 3-branes intersect over a string and all 6 strings intersect at one point. T_i are the inverse harmonic functions corresponding to each 3-brane, $T_i^{-1} = 1 + Q_i/r$. Like the $2\perp2\perp5\perp5$ background of $D = 11$ theory this $D = 10$ solution is $1/8$ supersymmetric, has 3-dimensional transverse space and diagonal $D = 10$ metric.

Its special cases include the single 3-brane [23,25] with harmonic function independent of 3 of 6 transverse coordinates ($T_2 = T_3 = T_4 = 1$), $3\perp3$ solution found in [15] ($T_3 = T_4 = 1$) and also $3\perp3\perp3$ configuration ($T_4 = 1$). The $1/8$ supersymmetric $3\perp3\perp3$ configuration also has 3-dimensional transverse space [9] but the corresponding $D = 10$ metric

$$
\begin{aligned}
ds_{10}^2 = (T_1T_2T_3)^{-1/2}\big[& - T_1T_2T_3 \; dt^2 \\
& + T_1T_2dy_1^2 + T_1T_3dy_2^2 + T_1dy_3^2 + T_2T_3dy_4^2 + T_2dy_5^2 + T_3dy_6^2 + dx_sdx_s\big] \,, \qquad (4.3)
\end{aligned}
$$

is singular at $r = 0$ and has zero area of the $r = 0$ horizon. [10]

As in the two $D = 11$ cases discussed in the previous sections, the metric of the $3\perp3\perp3\perp3$ solution (4.1) has $r = 0$ as a regular horizon with finite 8-area, (cf. (2.4), (3.4))

$$
A_8 = 4\pi L^6 [r^2 (T_1T_2T_1T_2)^{-1/2}]_{r\to0} = 4\pi L^6 \sqrt{Q_1Q_2Q_3Q_4} \,. \qquad (4.4)
$$

A_8/L^6 is the area of the horizon of the corresponding dyonic $D = 4$ black hole with the metric (2.6) and

$$
\lambda(r) = \sqrt{T_1T_2T_3T_4} = \frac{r^2}{\sqrt{(r+Q_1)(r+Q_2)(r+Q_3)(r+Q_4)}} \,. \qquad (4.5)
$$

The gauge field configuration here involves 4 pairs of equal electric and magnetic charges. When all charges are equal, the $3\perp3\perp3\perp3$ metric (4.1) compactified to $D = 4$ reduces to the $a = 0$ black hole metric (while the $3\perp3\perp3$ metric (4.3) reduces to the $a = 1/\sqrt{3}$ black hole metric [26]).

---

[9] Similar configurations of three and four intersecting 3-branes, and, in particular, their invariance under the $1/8$ fraction of maximal supersymmetry were discussed in D-brane representation in [17,16].

[10] This is similar to what one finds for the 'unboosted' $5\perp5\perp5$ configuration (2.1), (2.2). As is well known from 4-dimensional point of view, one does need *four* charges to get a regular behaviour of scalars near the horizon and finite area.

5. Entropy of $D = 4$ Reissner-Nordström black hole

Above we have demonstrated the existence of supersymmetric extremal $D = 11$ and $D = 10$ configurations with finite entropy which are built solely out of the fundamental p-branes of the corresponding theories (the 2-branes and the 5-branes of the M-theory and the 3-branes of type IIB theory) and reduce upon compactification to $D = 4$ dyonic black hole backgrounds with regular horizon.

Namely, there exists an embedding of a four-dimensional dyonic black hole (in particular, of the non-dilatonic Reissner-Nordström black hole) into $D = 11$ theory which corresponds to a combination of M-branes only. This may allow an application of the approach similar to the one of [14] to the derivation of the entropy (3.4) by counting the number of different BPS excitations of the $2\perp 2\perp 5\perp 5$ M-brane configuration.

The $3\perp 3\perp 3\perp 3$ configuration represents an embedding of the 1/8 supersymmetric dyonic $D = 4$ black hole into type IIB superstring theory which is remarkable in that all four charges enter symmetrically. It is natural to expect that there should exist a microscopic counting of the BPS states which reproduces the Bekenstein-Hawking entropy in a (U-duality invariant) way that treats all four charges on an equal footing.

Although we hope to eventually attain a general understanding of this problem, in what follows we shall discuss the counting of BPS states for one specific example considered above: the M-theory configuration (2.3), (2.2) of the three intersecting 5-branes with a common line. Even though the counting rules of M-theory are not entirely clear, we see an advantage to doing this from M-theory point of view as compared to previous discussions in the context of string theory [12,13]: the 11-dimensional problem is more symmetric. Furthermore, apart from the entropy problem, we may learn something about the M-theory.

5.1. Charge quantization in M-theory and the Bekenstein-Hawking entropy

Upon dimensional reduction to four dimensions, the boosted $5\perp 5\perp 5$ solution (2.3), (2.2), reduces to the 4-dimensional black hole with three magnetic charges, P_1, P_2 and P_3, and an electric charge Q. The electric charge is proportional to the momentum along the intersection string of length L, $\mathcal{P} = 2\pi N/L$. The general relation between the coefficient Q in the harmonic function K appearing in (2.3) and the momentum along the $D = 5$ string (cf. (2.5)) wound around a compact dimension of length L is (see e.g. [27])

$$Q = \frac{2\kappa_{D-1}^2}{(D-4)\omega_{D-3}} \cdot \frac{2\pi N}{L} = \frac{\kappa_4^2 N}{L} = \frac{\kappa^2 N}{L^8}, \tag{5.1}$$

where $\kappa_4^2/8\pi$ and $\kappa^2/8\pi$ are Newton's constants in 4 and 11 dimensions. All toroidal directions are assumed to have length L.

The three magnetic charges are proportional to the numbers n_1, n_2, n_3 of 5-branes in the (14567), (12367), and the (12345) planes, respectively (see (2.1), (2.3)). The complete symmetry between n_1, n_2 and n_3 is thus automatic in the 11-dimensional approach. The precise relation between P_i and n_i is found as follows. The charge q_5 of a $D = 11$ 5-brane which is spherically symmetric in transverse $d + 2 \leq 5$ dimensions is proportional to the coefficient P in the corresponding harmonic function. For $d + 2 = 3$ appropriate to the present case (two of five transverse directions are isotropic, or, equivalently, there is a periodic array of 5-branes in these compact directions) we get

$$q_5 = \frac{\omega_{d+1} d}{\sqrt{2}\kappa} P \quad \longrightarrow \quad \frac{\omega_2 L^2}{\sqrt{2}\kappa} P = \frac{4\pi L^2}{\sqrt{2}\kappa} P \,. \tag{5.2}$$

At this point we need to know precisely how the 5-brane charge is quantized. This was discussed in [9], but we repeat the argument here for completeness. A different argument leading to equivalent results was presented earlier in [28]. Upon compactification on a circle of length L, the M-theory reduces to type IIA string theory where all charge quantization rules are known. We use the fact that double dimensional reduction turns a 2-brane into a fundamental string, and a 5-brane into a Dirichlet 4-brane. Hence, we have

$$T_2 \kappa^2 = T_1 \kappa_{10}^2 \,, \qquad T_5 \kappa^2 = T_4 \kappa_{10}^2 \,, \tag{5.3}$$

where the 10-dimensional gravitational constant is expressed in terms of the 11-dimensional one by $\kappa_{10}^2 = \kappa^2/L$. The charge densities are related to the tensions by

$$q_2 = \sqrt{2}\kappa T_2 \,, \qquad q_5 = \sqrt{2}\kappa T_5 \,, \tag{5.4}$$

and we assume that the minimal Dirac condition is satisfied, $q_2 q_5 = 2\pi$. These relations, together with the 10-dimensional expressions [29]

$$\kappa_{10} = g(\alpha')^2 \,, \qquad T_1 = \frac{1}{2\pi\alpha'} \,, \qquad \kappa_{10} T_4 = \frac{1}{2\sqrt{\pi\alpha'}} \,, \tag{5.5}$$

fix all the M-theory quantities in terms of α' and the string coupling constant, g. In particular, we find

$$\kappa^2 = \frac{g^3 (\alpha')^{9/2}}{4\pi^{5/2}} \,, \qquad L = \frac{g\sqrt{\alpha'}}{4\pi^{5/2}} \,. \tag{5.6}$$

The tensions turn out to be

$$T_2 = \frac{2\pi^{3/2}}{g(\alpha')^{3/2}} \,, \qquad T_5 = \frac{2\pi^2}{g^2 (\alpha')^3} \,. \tag{5.7}$$

Note that T_2 is identical to the tension of the Dirichlet 2-brane of type IIA theory, while T_5 – to the tension of the solitonic 5-brane. This provides a nice check on our results, since single dimensional reduction indeed turns the M-theory 2-brane into the Dirichlet 2-brane, and the M-theory 5-brane into the solitonic 5-brane. Note that the M-brane tensions satisfy the relation $2\pi T_5 = T_2^2$, which was first derived in [28] using toroidal compactification to type IIB theory in 9 dimensions. This serves as yet another consistency check.

It is convenient to express our results in pure M-theory terms. The charges are quantized according to [11]

$$q_2 = \sqrt{2}\,\kappa T_2 = n\sqrt{2}\,(2\kappa\pi^2)^{1/3}\,, \tag{5.8}$$

$$q_5 = \sqrt{2}\,\kappa T_5 = n\sqrt{2}\,(\frac{\pi}{2\kappa})^{1/3}\,, \tag{5.9}$$

i.e.

$$P_i = \frac{n_i}{2\pi L^2}(\frac{\pi\kappa^2}{2})^{1/3}\,. \tag{5.10}$$

The resulting expression for the Bekenstein-Hawking entropy of the extremal Reissner-Nordström type black hole, (2.6), (2.7), which is proportional to the area (2.4), is

$$S_{\rm BH} = \frac{2\pi A_9}{\kappa^2} = \frac{8\pi^2 L^7}{\kappa^2}\sqrt{P_1 P_2 P_3 Q} = 2\pi\sqrt{n_1 n_2 n_3 N}\,. \tag{5.11}$$

This agrees with the expression found directly in $D = 4$ [2,3,12,13].

In the case of the $2\bot2\bot5\bot5$ configuration we get (for each pair of 2-brane and 5-brane charges) $q_2 = \frac{4\pi L^5}{\sqrt{2}\,\kappa}Q$, $q_5 = \frac{4\pi L^2}{\sqrt{2}\,\kappa}P$. The Dirac condition on unit charges translates into $q_2 q_5 = 2\pi n_1 n_2$, where n_1 and n_2 are the numbers of 2- and 5-branes. We conclude that $Q_1 P_1 = \frac{\kappa^2}{4\pi L^7}n_1 n_2$. Then from (3.4) we learn that

$$S_{\rm BH} = \frac{2\pi A_9}{\kappa^2} = \frac{8\pi^2 L^7}{\kappa^2}\sqrt{Q_1 P_1 Q_2 P_2} = 2\pi\sqrt{n_1 n_2 n_3 n_4}\,. \tag{5.12}$$

Remarkably, this result does not depend on the particular choice of M-brane quantization condition (choice of $m_0 = \pi^2\kappa^{-2}T_2^{-3}$) or use of D-brane tension expression since the $2\bot2\bot5\bot5$ configuration contains equal number of 2-branes and 5-branes. This provides a consistency check. Note also that the $D = 4$ black holes obtained from the $2\bot2\bot5\bot5$ and from the $5\bot5\bot5$ M-theory configurations are not identical, but are related by U-duality. The equality of their entropies provides a check of the U-duality.

The same expression is obtained for the entropy of the $D = 10$ configuration $3\bot3\bot3\bot3$ (4.3) (or related $D = 4$ black hole). Each 3-brane charge q_3 is proportional to the corresponding coefficient Q in the harmonic function (cf. (5.2))

$$q_3 = \frac{1}{\sqrt{2}}(\frac{\omega_{d+1}d}{\sqrt{2}\,\kappa_{10}})Q \quad\rightarrow\quad \frac{\omega_2 L^3}{2\kappa_{10}}Q = \frac{2\pi L^3}{\kappa_{10}}Q\,, \tag{5.13}$$

where $\kappa_{10}^2/8\pi$ is the 10-dimensional Newton constant and the overall factor $\frac{1}{\sqrt{2}}$ is due to the dyonic nature of the 3-brane. The charge quantization in the self-dual case implies (see [9]) $q_3 = n\sqrt{\pi}$ (the absence of standard $\sqrt{2}$ factor here effectively compensates for the 'dyonic' $\frac{1}{\sqrt{2}}$ factor in the expression for the charge). [12] Thus, $Q_i = \frac{\kappa_{10}}{2\sqrt{\pi}L^3}n_i$, and the area (4.4) leads to the following entropy:

[11] In [30] it was argued that the 2-brane tension, T_2, satisfies $\kappa^2 T_2^3 = \pi^2/m_0$, where m_0 is a rational number that was left undetermined. The argument of [28], as well as our procedure [9], unambiguously fix $m_0 = 1/2$.

[12] This agrees with the D3-brane tension, $\kappa_{10}T_3 = \sqrt{\pi}$, since in the self-dual case $q_p = \kappa_{10}T_p$.

$$S_{\mathrm{BH}} = \frac{2\pi A_8}{\kappa_{10}^2} = \frac{8\pi^2 L^6}{\kappa^2}\sqrt{Q_1Q_2Q_3Q_4} = 2\pi\sqrt{n_1n_2n_3n_4}\,. \tag{5.14}$$

5.2. Counting of the microscopic states

The presence of the factor $\sqrt{N}$ in S_{BH} (5.11) immediately suggests an interpretation in terms of the massless states on the string common to all three 5-branes. Indeed, it is well known that, for a $(1+1)$-dimensional field theory with a central charge c, the entropy of left-moving states with momentum $2\pi N/L$ is, for sufficiently large N, given by [13]

$$S_{\mathrm{stat}} = 2\pi\sqrt{\tfrac{1}{6}cN}\,. \tag{5.15}$$

We should find, therefore, that the central charge on the intersection string is, in the limit of large charges, equal to

$$c = 6n_1n_2n_3\,. \tag{5.16}$$

The fact that the central charge grows as $n_1n_2n_3$ suggests the following picture. 2-branes can end on 5-branes, so that the boundary looks like a closed string [20,32,33]. It is tempting to associate the massless states with those of 2-branes attached to 5-branes near the intersection point. Geometrically, we may have a two-brane with three holes, each of the holes attached to different 5-dimensional hyperplanes in which the 5-branes lie. Thus, for any three 5-branes that intersect along a line, we have a collapsed 2-brane that gives massless states in the $(1+1)$-dimensional theory describing the intersection. What is the central charge of these massless states? From the point of view of one of the 5-branes, the intersection is a long string in $5+1$ dimensions. Such a string has 4 bosonic massless modes corresponding to the transverse oscillations, and 4 fermionic superpartners. Thus, we believe that the central charge arising from the collapsed 2-brane with three boundaries is $4(1+\frac{1}{2}) = 6$. [14]

The upshot of this argument is that each triple intersection contributes 6 to the central charge. Since there are $n_1n_2n_3$ triple intersections, we find the total central charge $6n_1n_2n_3$. One may ask why there are no terms of order n_1^3, etc. This can be explained by the fact that all parallel 5-branes are displaced relative to each other, so that the 2-branes produce massless states only near the intersection points.

One notable feature of our argument is that the central charge grows as a product of three charges, while in all D-brane examples one found only a product of two charges. We believe that this is related to the peculiar n^3 growth of the near-extremal entropy of n coincident 5-branes found in [9] (for coincident D-branes the near-extremal

[13] As pointed out in [31], this expression is reliable only if $N \gg c$. Requiring N to be much greater than $n_1n_2n_3$ is a highly asymmetric choice of charges. If, however, all charges are comparable and large, the entropy is dominated by the multiply wound 5-branes, which we discuss at the end of this section.

[14] Upon compactification on T^7, these massless modes are simply the small fluctuations of the long string in $4+1$ dimensions which is described by the classical solution (2.5). One should be able to confirm that the central charge on this string is equal to 6 by studying its low-energy modes.

entropy grows only as n^2). This is because the intersecting D-brane entropy comes from strings which can only connect objects pairwise. The 2-branes, however, can connect three different 5-branes. Based on our observations about entropy, we conjecture that the configurations where a 2-brane connects four or more 5-branes are forbidden (otherwise, for instance, the near-extremal entropy of n parallel 5-branes would grow faster than n^3). Perhaps such configurations do not give rise to massless states or give subleading contributions to the entropy.

The counting argument presented above applies to the configuration where there are n_1 parallel 5-branes in the (14567) hyperplane, n_2 parallel 5-branes in the (12367) hyperplane, and n_3 parallel 5-branes in the (12345) hyperplane. As explained in [31], if $n_1 \sim n_2 \sim n_3 \sim N$ we need to examine a different configuration where one replaces a number of disconnected branes by a single multiply wound brane. Let us consider, therefore, a single 5-brane in the (14567) hyperplane wound n_1 times around the y_1-circle, a single 5-brane in the (12367) hyperplane wound n_2 times around the y_1-circle, and a single 5-brane in the (12345) hyperplane wound n_3 times around the y_1-circle. Following the logic of [31], one can show that the intersection string effectively has winding number $n_1 n_2 n_3$: this is because the 2-brane which connects the three 5-branes needs to be transported $n_1 n_2 n_3$ times around the y_1-circle to come back to its original state.[15] Therefore, the massless fields produced by the 2-brane effectively live on a circle of length $n_1 n_2 n_3 L$. This implies [34] that the energy levels of the $(1+1)$-dimensional field theory are quantized in units of $2\pi/(n_1 n_2 n_3 L)$. In this theory there is only one species of the 2-brane connecting the three 5-branes; therefore, the central charge on the string is $c = 6$. The calculation of BPS entropy for a state with momentum $2\pi N/L$, as in [34,31], once again reproduces (5.11). While the end result has the form identical to that found for the disconnected 5-branes, the connected configuration is dominant when all four charges are of comparable magnitude [31]. Now the central charge is fixed, and the large entropy is due to the growing density of energy levels.

6. Black hole entropy in $D = 5$ and discussion

The counting arguments presented here are plausible, but clearly need to be put on a more solid footing. Indeed, it is not yet completely clear what rules apply to the 11-dimensional M-theory (although progress has been made in [14]). The rule associating massless states to collapsed 2-branes with three boundaries looks natural, and seems to reproduce the Bekenstein-Hawking entropy of extremal black holes in $D = 4$. Note also that a similar rule can be successfully applied to the case of the finite entropy $D = 5$ extremal dyonic black holes described in 11 dimensions by the 'boosted' $2 \perp 5$

[15] The role of $n_1 n_2 n_3$ as the effective winding number is suggested also by comparison of the $D = 5$ solitonic string metric, (2.5), with the fundamental string metric, $ds^2 = V^{-1}(dudv + Kdu^2) + dx_s dx_s$, where the coefficient in the harmonic function V is proportional to the tension times the winding number of the source string (see e.g. [27]). After a conformal rescaling, (2.5) takes the fundamental string form with $V = H^3 = (F_1 F_2 F_3)^{-1}$ so that near $r = 0$ the $dudv$ part of it is multiplied by $P_1 P_2 P_3 \sim n_1 n_2 n_3$. Thus, the source string may be thought of as wound $n_1 n_2 n_3$ times around the circle.

configuration [15]. Another possible $D = 11$ embedding of the $D = 5$ black hole is provided by $2\perp2\perp2$ configuration [15]. The relevant $D = 10$ type IIB configuration is $3\perp3$ (cf. (4.3)) with momentum flow along common string. In the case of $2 \perp 5$ configuration the massless degrees of freedom on the intersection string may be attributed to a collapsed 2-brane with a hole attached to the 5-brane and one point attached to the 2-brane. If the 5-brane is wound n_1 times and the 2-brane – n_2 times, the intersection is described by a $c = 6$ theory on a circle of length $n_1 n_2 L$. Following the arguments of [31], we find that the entropy of a state with momentum $2\pi N/L$ along the intersection string is

$$S_{\text{stat}} = 2\pi\sqrt{n_1 n_2 N}\,. \tag{6.1}$$

This seems to supply a microscopic M-theory basis, somewhat different from that in [14], for the Bekenstein-Hawking entropy of $D = 5$ extremal dyonic black holes.

We would now like to show that (6.1) is indeed equal to the expression for the Bekenstein-Hawking entropy for the 'boosted' $2 \perp 5$ configuration [15] (cf. (5.11)),

$$S_{\text{BH}} = \frac{2\pi A_9}{\kappa^2} = \frac{4\pi^3 L^6}{\kappa^2}\sqrt{QPQ'}\,. \tag{6.2}$$

Q and P are the parameters in the harmonic functions corresponding to the 2-brane and the 5-brane, and Q' is the parameter in the 'boost' function, i.e. $T^{-1} = 1 + Q/r^2$, $F^{-1} = 1 + P/r^2$, $K = 1 + Q'/r^2$. Note that here (cf. (5.1))

$$Q' = \frac{\kappa^2 N}{\pi L^7}\,, \qquad q_2 = \frac{4\pi^2 L^4}{\sqrt{2}\kappa} Q\,, \qquad q_5 = \frac{4\pi^2 L}{\sqrt{2}\kappa} P\,. \tag{6.3}$$

As in the case of the $2\perp2\perp5\perp5$ configuration, we can use the Dirac quantization condition, $q_2 q_5 = 2\pi n_1 n_2$, to conclude that $QP = \frac{\kappa^2}{4\pi^3 L^5} n_1 n_2$. This yields (6.1) when substituted into (6.2). A similar expression for the BPS entropy is found in the case of the completely symmetric $2\perp2\perp2$ configuration,

$$S_{\text{BH}} = \frac{4\pi^3 L^6}{\kappa^2}\sqrt{Q_1 Q_2 Q_3} = 2\pi\sqrt{n_1 n_2 n_3}\,, \tag{6.4}$$

where we have used the 2-brane charge quantization condition (5.8), which implies that $Q_i = n_i L^{-4}(\kappa/\sqrt{2\pi})^{4/3}$. Agreement of different expressions for the $D = 5$ black hole entropy provides another check on the consistency of (5.8), (5.9).

Our arguments for counting the microscopic states apply only to the configurations where M-branes intersect over a string. It would be very interesting to see how approach analogous to the above might work when this is not the case. Indeed, black holes with finite horizon area in $D = 4$ may also be obtained from the $2\perp2\perp5\perp5$ configuration in M-theory, and the $3\perp3\perp3\perp3$ one in type IIB, while in $D = 5$ – from the $2\perp2\perp2$ configuration. Although from the $D = 4, 5$ dimensional point of view these cases are related by U-duality to the ones we considered, the counting of their states seems to be harder at the present level of understanding. We hope that a more general approach to the entropy problem, which covers all the solutions we discussed, can be found.

Acknowledgements

We are grateful to V. Balasubramanian, C. Callan and M. Cvetič for useful discussions. I.R.K. was supported in part by DOE grant DE-FG02-91ER40671, the NSF Presidential Young Investigator Award PHY-9157482, and the James S. McDonnell Foundation grant No. 91-48. A.A.T. would like to acknowledge the support of PPARC, ECC grant SC1*-CT92-0789 and NATO grant CRG 940870.

References

[1] A. Sen, Mod. Phys. Lett. A 10 (1995) 2081, hep-th/9504147.
[2] F. Larsen and F. Wilczek, hep-th/9511064.
[3] M. Cvetič and A.A. Tseytlin, Phys. Rev. D 53 (1996) 5619, hep-th/9512031.
[4] A. Strominger and C. Vafa, hep-th/9601029.
[5] R. Kallosh, A. Linde, T. Ortín, A. Peet and A. van Proeyen, Phys. Rev. D 46 (1992) 5278.
[6] M. Cvetič and D. Youm, Phys. Rev. D 53 (1996) 584, hep-th/9507090.
[7] M. Cvetič and A.A. Tseytlin, Phys. Lett. B 366 (1996) 95, hep-th/9510097.
[8] A.A. Tseytlin, Mod. Phys. Lett. A 11 (1996) 689, hep-th/9601177.
[9] I.R. Klebanov and A.A. Tseytlin, Nucl. Phys. B 475 (1996) 164, hep-th/9604089 (revised).
[10] M. Cvetič and D. Youm, hep-th/9605051.
[11] C.G. Callan and J.M. Maldacena, hep-th/9602043.
[12] J.M. Maldacena and A. Strominger, hep-th/9603060.
[13] C. Johnson, R. Khuri and R. Myers, hep-th/9603061.
[14] R. Dijkgraaf, E. Verlinde and H. Verlinde, hep-th/9603126; hep-th/9604055.
[15] A.A. Tseytlin, Nucl. Phys. B 475 (1996) 149, hep-th/9604035.
[16] V. Balasubramanian and F. Larsen, hep-th/9604189.
[17] M.B. Green and M. Gutperle, hep-th/9604091.
[18] M. Cvetič and A. Sen, unpublished.
[19] G. Papadopoulos and P.K. Townsend, hep-th/9603087.
[20] A. Strominger, hep-th/9512059.
[21] G. Papadopoulos, hep-th/9604068.
[22] R. Güven, Phys. Lett. B 276 (1992) 49.
[23] G.T. Horowitz and A. Strominger, Nucl. Phys. B 360 (1991) 197.
[24] M.J. Duff and K.S. Stelle, Phys. Lett. B 253 (1991) 113.
[25] M.J. Duff and J.X. Lu, Phys. Lett. B 273 (1991) 409.
[26] G.W. Gibbons and K. Maeda, Nucl. Phys. B 298 (1988) 741.
[27] A. Dabholkar, J.P. Gauntlett, J.A. Harvey and D. Waldram, hep-th/9511053.
[28] J.H. Schwarz, Phys. Lett. B 367 (1996) 97, hep-th/9510086.
[29] J. Polchinski, S. Chaudhuri and C.V. Johnson, hep-th/9602052.
[30] M.J. Duff, J.T. Liu and R. Minasian, Nucl. Phys. B 452 (1995) 261, hep-th/9506126.
[31] J. Maldacena and L. Susskind, hep-th/9604042.
[32] P.K. Townsend, hep-th/9512062.
[33] K. Becker and M. Becker, hep-th/9602071.
[34] S. Das and S. Mathur, hep-th/9601152.

Chapter 6

M-theory and duality

In 1977, Montonen and Olive made a bold conjecture [1]. Might there exist a *dual* formulation of fundamental physics in which the roles of Noether charges and topological charges are reversed? In such a dual picture, the magnetic monopoles would be the fundamental objects and the quarks, W-bosons and Higgs particles would be the solitons! They were inspired by the observation that in certain *supersymmetric* grand unified theories, the masses M of all the particles whether elementary (carrying purely electric charge Q), solitonic (carrying purely magnetic charge P) or *dyonic* (carrying both) are described by a universal formula

$$M^2 = v^2(Q^2 + P^2) \tag{6.1}$$

where v is a constant. Note that the mass formula remains unchanged if we exchange the roles of P and Q. The Montonen–Olive conjecture was that this electric/magnetic symmetry is a symmetry not merely of the mass formula but is an exact symmetry of the entire quantum theory! The reason why this idea remained merely a conjecture rather than a proof has to do with the whole question of perturbative versus non-perturbative effects. According to Dirac, the electric charge Q is quantized in units of e, the charge on the electron, whereas the magnetic charge is quantized in units of $1/e$. In other words, $Q = me$ and $P = n/e$, where m and n are integers. The symmetry suggested by Montonen and Olive thus demanded that in the dual world, we not only exchange the integers m and n but we also replace e by $1/e$ (or $\hbar/e$ if we restore Planck's constant) and go from a regime of weak coupling to a regime of strong coupling. This was very exciting firstly because it promised a whole new window on non-perturbative effects and secondly because this would be an intrinsically quantum symmetry with no classical ($\hbar \to 0$) counterpart. On the other hand, it also made a proof very difficult and the idea was largely forgotten for the next few years.

Although the original supermembrane paper by Hughes, Liu and Polchinski [2] made use of the soliton idea, the subsequent impetus in supermembrane theory was to mimic superstrings and treat the p-branes as fundamental objects in their own right (analogous to particles carrying an electric Noether charge). Even within this framework, however, it was possible to postulate a Poincaré duality between one p-brane and another by relating them to the geometrical concept of *p-forms*.

(Indeed, this is how *p*-branes originally got their name.) Now the low energy limit of 10-dimensional string theory is a 10-dimensional supergravity theory with a 3-form field strength. However, 10-dimensional supergravity had one puzzling feature that had long been an enigma from the point of view of string theory. In addition to the above version there existed a *dual* version in which the field strength was a 7-form. This suggested [3], a dual version of string theory in which the fundamental objects are fivebranes! This became known as the *string/fivebrane duality conjecture*. The analogy was still a bit incomplete, however, because at that time no-one had thought of the fivebrane as a soliton.

Then in 1990, a major breakthrough for the string/fivebrane duality conjecture came along when Strominger [4] found that the equations of the 10-dimensional heterotic string admit a fivebrane as a soliton solution which also preserves half the spacetime supersymmetry and whose mass per unit 5-volume is given by the topological charge associated with the 3-form of the string. Moreover, this mass became larger, the smaller the strength of the string coupling, exactly as one would expect for a soliton. He went on to suggest a complete strong/weak coupling duality with the strongly coupled string corresponding to the weakly coupled fivebrane. By generalizing some earlier work of Nepomechie [5] and Teitelboim [6], moreover, it was possible to show that the electric charge of the fundamental string and the magnetic charge of the solitonic fivebrane obeyed a Dirac quantization rule. In this form, string/fivebrane duality was now much more closely mimicking the electric/magnetic duality of Montonen and Olive. However, since most physicists were already sceptical of electric/magnetic duality in four dimensions, they did not immediately embrace string/fivebrane duality in ten dimensions!

Furthermore, there was one major problem with treating the fivebrane as a fundamental object in its own right; a problem that has bedevilled supermembrane theory right from the beginning: no-one knows how to quantize fundamental *p*-branes with $p > 1$. All the techniques that worked so well for fundamental strings and which allow us, for example, to calculate how one string scatters off another, simply do not go through. Problems arise both at the level of the worldvolume equations where the old *bête noir* of non-renormalizability comes to haunt us and also at the level of the spacetime equations. Each term in string perturbation theory corresponds to a two-dimensional worldsheet with more and more holes: we must sum over all topologies of the worldsheet. But for surfaces with more than two dimensions we do not know how to do this. Indeed, there are powerful theorems in pure mathematics which tell you that it is not merely hard but impossible. Of course, one could always invoke the dictum that *God does not do perturbation theory*, but that does not cut much ice unless you can say what He does do! So there were two major impediments to string/fivebrane duality in 10 dimensions. First, the electric/magnetic duality analogy was ineffective so long as most physicists were sceptical of this duality. Secondly, treating fivebranes as fundamental raised all the unresolved issues of quantization.

The first of these impediments was removed, however, when Sen [7] revitalized the Montonen–Olive conjecture by establishing that certain dyonic states, which their conjecture demanded, were indeed present in the theory. Many duality sceptics were thus converted. Indeed this inspired Seiberg and Witten [8] to look for duality in more realistic (though still supersymmetric) approximations to the standard model. The subsequent industry, known as Seiberg–Witten theory, provided a wealth of new information on non-perturbative effects in four-dimensional quantum field theories, such as quark-confinement and symmetry-breaking, which would have been unthinkable just a few years ago.

The Montonen–Olive conjecture was originally intended to apply to four-dimensional grand unified field theories. In 1990, however, Font, Ibanez, Lust and Quevedo [9] and, independently, Rey [10] generalized the idea to four-dimensional superstrings, where in fact the idea becomes even more natural and goes by the name of S-duality.

In fact, superstring theorists had already become used to a totally different kind of duality called T-duality. Unlike S-duality which was a non-perturbative symmetry and hence still speculative, T-duality was a perturbative symmetry and rigorously established [11]. If we compactify a string theory on a circle then, in addition to the Kaluza–Klein particles we would expect in an ordinary field theory, there are also extra *winding* particles that arise because a string can wind around the circle. T-duality states that nothing changes if we exchange the roles of the Kaluza–Klein and winding particles provided we also exchange the radius of the circle R with its inverse $1/R$. In short, a string cannot tell the difference between a big circle and a small one!

Recall that, when wrapped around a circle, an 11-dimensional membrane behaves as if it were a 10-dimensional string. In a series of papers between 1991 and 1995, Duff, Khuri, Liu, Lu, Minasian and Rahmfeld [12–14, 16–18] argued that this may also be the way out of the problems of 10-dimensional string/fivebrane duality. If we allow four of the ten dimensions to be curled up and allow the solitonic fivebrane to wrap around them, it will behave as if it were a 6-dimensional solitonic string! The fundamental string will remain a fundamental string but now also in 6-dimensions. So the 10-dimensional string/fivebrane duality conjecture gets replaced by a 6-dimensional string/string duality conjecture. The obvious advantage is that, in contrast to the fivebrane, we do know how to quantize the string and hence we can put the predictions of string/string duality to the test. For example, one can show that the coupling constant of the solitonic string is indeed given by the inverse of the fundamental string's coupling constant, in complete agreement with the conjecture.

When we spoke of string/string duality, we originally had in mind a duality between one heterotic string and another, but the next major development in the subject came in 1994 when Hull and Townsend [19] suggested that, if the four-dimensional compact space is chosen suitably, a six-dimensional heterotic string can be dual to a six-dimensional Type IIA string! The barriers between the different string theories were beginning to crumble.

String/string duality has another unexpected pay-off [18]. If we compactify the six-dimensional spacetime on two circles down to four dimensions, the fundamental string and the solitonic string will each acquire a T-duality. But here is the miracle: the T-duality of the solitonic string is just the S-duality of the fundamental string, and vice-versa! This phenomenon, in which the non-perturbative replacement of e by $1/e$ in one picture is just the perturbative replacement of R

by $1/R$ in the dual picture, goes by the name of *Duality of Dualities*. Thus four-dimensional electric/magnetic duality, which was previously only a conjecture, now emerges automatically if we make the more primitive conjecture of six-dimensional string/string duality.

All this previous work on T-duality, S-duality, and string/string duality was suddenly pulled together by Witten [20] under the umbrella of eleven-dimensions. One of the biggest problems with $D = 10$ string theory is that there are *five* consistent string theories: Type I $SO(32)$, heterotic $SO(32)$, heterotic $E_8 \times E_8$, Type IIA and Type IIB. As a candidate for a unique *theory of everything*, this is clearly an embarrassment of riches. Witten put forward a convincing case that this distinction is just an artifact of perturbation theory and that non-perturbatively these five theories are, in fact, just different corners of a deeper theory. See table 6.1. Moreover, this deeper theory, subsequently dubbed *M-theory*, has $D = 11$ supergravity as its low energy limit! Thus the five string theories and $D = 11$ supergravity represent six different special points[1] in the moduli space of M-theory. The small parameters of perturbative string theory are provided by $< e^{\Phi} >$, where Φ is the dilaton field, and $< e^{\sigma_i} >$ where σ_i are the moduli fields which arise after compactification. What makes M-theory at once intriguing and yet difficult to analyze is that in $D = 11$ there is neither dilaton nor moduli and hence the theory is intrinsically non-perturbative. Consequently, the ultimate meaning of M-theory is still unclear, and Witten has suggested that in the meantime, M should stand for 'magic', 'mystery' or 'membrane', according to taste. Curiously enough, however, Witten still played down the importance of supermembranes. But it was only a matter of time before he too succumbed to the conclusion that we weren't doing just string theory any more! In the coming months, literally hundreds of papers appeared in the internet confirming that, whatever M-theory may be, it certainly involves supermembranes in an important way[64].

$E_8 \times E_8$ heterotic string	M theory
$SO(32)$ heterotic string	
$SO(32)$ Type I string	
Type IIA string	
Type IIB string	

Table 6.1. The five apparently different string theories are really just different corners of M-theory.

For example, the 6-dimensional string/string duality discussed above (and hence the 4-dimensional electric/magnetic duality) follows from 11-dimensional membrane/fivebrane duality [21, 22]. The fundamental string is obtained by wrapping the membrane around a one-dimensional space and then compactifying on a

[1] Some authors take the phrase *M-theory* to refer merely to this sixth corner of the moduli space. With this definition, of course, M-theory is no more fundamental than the other five corners. For us, *M-theory* means the whole kit and caboodle.

four-dimensional space; whereas the solitonic string is obtained by wrapping the fivebrane around the four-dimensional space and then compactifying on the one-dimensional space. Thus $S^1/Z_2 \times K3$, $S^1 \times K3$, $S^1/Z_2 \times T^4$ and $S^1 \times T^4$ yield heterotic/heterotic, Type IIA/heterotic, heterotic/Type IIA and Type IIA/Type IIA duality, respectively. Nor did it take long before the more realistic kinds of electric/magnetic duality envisioned by Seiberg and Witten [8] were also given an explanation in terms of string/string duality and hence M-theory [23, 24, 34, 35]. Even QCD now has a $D = 11$ interpretation [35].

It is interesting to ask whether we have exhausted all possible theories of extended objects with spacetime supersymmetry and fermionic gauge invariance on the worldvolume. This we claimed to have done in chapters 2 and 3 by demanding super-Poincaré invariance, but might there exist other Green–Schwarz type actions in which the supergroup is not necessarily super-Poincaré? Although we have not yet attempted to construct all such actions, one may nevertheless place constraints on the dimensions and signatures for which such theories are possible [25]. We simply impose bose-fermi matching but relax the requirement of a super-Poincaré algebra. Although the possibilities are richer, there are still several constraints. In particular, the maximum spacetime dimension is now $D = 12$ where we can have a worldvolume with $(2, 2)$ signature provided we have a $(10, 2)$ spacetime signature. This new case is particularly interesting since it belongs to the O sequence and furthermore admits Majorana–Weyl spinors. In fact, the idea of a twelfth timelike dimension in supergravity is an old one [26] and twelve-dimensional supersymmetry algebras have been discussed in the supergravity literature [27]. In particular, the chiral $(N_+, N_-) = (1, 0)$ supersymmetry algebra in $(S, T) = (10, 2)$ involves the anti-commutator

$$\{Q_\alpha, Q_\beta\} = (\Gamma^{MN})_{\alpha\beta} P_{MN} + (\Gamma^{MNPQRS})_{\alpha\beta} Z^+{}_{MNPQRS}. \tag{6.2}$$

The right-hand side yields a Lorentz generator and a six index object so it is certainly not super-Poincaré.

Despite all the objections one might raise to a world with two time dimensions, and despite the above problems of interpretation, the idea of a $(2, 2)$ object moving in a $(10, 2)$ spacetime has recently been revived [28] in the context of F-*theory* [29], which involves Type IIB compactifications where the axion and dilaton from the Ramond–Ramond sector are allowed to vary on the internal manifold. Given a manifold M that has the structure of a fiber bundle whose fiber is T^2 and whose base is some manifold B, then

$$F \text{ on } M \equiv \text{ Type } IIB \text{ on } B. \tag{6.3}$$

The utility of F-theory is beyond dispute and it has certainly enhanced our understanding of string dualities, Seiberg–Witten theory and much else, but should the twelve-dimensions of F-theory be taken seriously? And if so, should F-theory be regarded as more fundamental than M-theory? Given that there seems to be no supersymmetric field theory with $SO(10, 2)$ Lorentz invariance [30], and given that the on-shell states carry only ten-dimensional momenta [29], the more conservative interpretation is that the twelfth dimension is merely a mathematical artifact and that F-theory should simply be interpreted as a clever way of compactifying the IIB string [31]. Time (or should I say 'both times'?) will tell.

Even the chiral $E_8 \times E_8$ string, which according to Witten's earlier theorem could never come from eleven-dimensions, was given an eleven-dimensional explanation by Horava and Witten [32]. The no-go theorem is evaded by compactifying not on a circle (which has no ends), but on a line-segment (which has two ends). Witten went on to argue that if the size of this one-dimensional space is large compared to the six-dimensional Calabi–Yau manifold, then our world is approximately five-dimensional [33]. This may have important consequences for confronting M-theory with experiment. For example, it is known that the strengths of the four forces change with energy. In supersymmetric extensions of the standard model, one finds that the fine structure constants $\alpha_3, \alpha_2, \alpha_1$ associated with the $SU(3)\times SU(2)\times U(1)$ all meet at about 10^{16} GeV, entirely consistent with the idea of grand unification. The strength of the dimensionless number $\alpha_G = GE^2$, where G is Newton's constant and E is the energy, also almost meets the other three, but not quite. This near miss has been a source of great interest, but also frustration. However, in a universe of the kind envisioned by Witten [33], spacetime is approximately a narrow five dimensional layer bounded by four-dimensional walls. The particles of the standard model live on the walls but gravity lives in the five-dimensional bulk. As a result, it is possible to choose the size of this fifth dimension so that all four forces meet at this common scale. Note that this is much less than the Planck scale of 10^{19} GeV, so gravitational effects may be much closer in energy than we previously thought; a result that would have all kinds of cosmological consequences.

On the subject of cosmology, the S^7 compactification of M-theory and its massless sector of gauged $N = 8$ $D = 4$ supergravity have also featured in a recent cosmological context with attempts to reconcile an open universe with inflation [39–42].

Thus this eleven-dimensional framework now provides the starting point for understanding a wealth of new non-perturbative phenomena, including string/string duality, Seiberg–Witten theory, quark confinement, QCD, particle physics phenomenology and cosmology.

So what is M-theory?

There is still no definitive answer to this question, although several different proposals have been made. By far the most popular is M(atrix) theory [40]. The matrix models of M-theory are $SU(k)$ supersymmetric gauge quantum mechanical models with 16 supersymmetries. As we have seen in chapter 2, these models were first introduced in the context of M-theory in [42, 41, 43], where they appeared as regularizations of the $D = 11$ supermembrane in the lightcone gauge, sometimes called the *infinite momentum frame.* (The lightcone gauged-fixed supermembrane is a supersymmetric gauge quantum mechanical model with the group of area-preserving diffeomorphisms as its gauge group.) The matrix approach of [40] exploits the observation [44] that such models are also interpretable as the effective action of k coincident Dirichlet 0-branes, and that the continuous spectrum phenomenon is then just the no-force condition between them.

The theory begins by compactifying the eleventh dimension on a circle of radius R, so that the longitudinal momentum is quantized in units of $1/R$ with total $P_L =$

k/R with $k \to \infty$. The theory is *holographic* [40] in that it contains only degrees of freedom which carry the smallest unit of longitudinal momentum, other states being composites of these fundamental states. This is, of course, entirely consistent with their identification with the Kaluza–Klein modes. It is convenient to describe these k degrees of freedom as $k \times k$ matrices. When these matrices commute, their simultaneous eigenvalues are the positions of the 0-branes in the conventional sense. That they will in general be non-commuting, however, suggests that to properly understand M-theory, we must entertain the idea of a fuzzy spacetime in which spacetime coordinates are described by non-commuting matrices. In any event, this matrix approach has had success in reproducing many of the expected properties of M-theory such as $D = 11$ Lorentz covariance, $D = 11$ supergravity as the low-energy limit, and the existence of membranes and fivebranes. Other important contributions may be found in [53–57].

It was further proposed that when compactified on T^n, the quantum mechanical model should be replaced by an $(n+1)$-dimensional $SU(k)$ Yang–Mills field theory defined on the dual torus $\tilde{T}^n$. Another test of this M(atrix) approach, then, is that it should explain the U-dualities [45, 46, 19] of chapter 4. For $n = 3$, for example, this group is $SL(3, Z) \times SL(2, Z)$. The $SL(3, Z)$ just comes from the modular group of T^3 whereas the $SL(2, Z)$ is the electric/magnetic duality group of four-dimensional $N = 4$ Yang–Mills [47]. For $n > 3$, however, this picture looks suspicious because the corresponding gauge theory becomes non-renormalizable and the full U-duality group has still escaped explanation. There have been speculations [49, 56, 55] on what compactified M-theory might be, including a revival of the old proposal that it is really M(embrane)theory [56]. In other words, perhaps $D = 11$ supergravity together with its BPS configurations: plane wave, membrane, fivebrane, KK monopole and the $D = 11$ embedding of the Type IIA eightbrane, are all there is to M-theory and that we need look no further for new degrees of freedom. At the time of writing this is still being hotly debated.

The year 1998 marked a renaissance in anti de-Sitter space brought about by Maldacena's conjectured duality between physics in the bulk of AdS and a conformal field theory on the boundary [57]. In particular, M-theory on $AdS_4 \times S^7$ is dual to a non-abelian $(n = 8, d = 3)$ superconformal theory, Type IIB string theory on $AdS_5 \times S^5$ is dual to a $d = 4$ $SU(k)$ super Yang–Mills theory and M-theory on $AdS_7 \times S^4$ is dual to a non-abelian $((n_+, n_-) = (2, 0), d = 6)$ conformal theory. In particular, as has been spelled out most clearly in the $d = 4$ $SU(k)$ Yang–Mills case, there is seen to be a correspondence between the Kaluza–Klein mass spectrum in the bulk and the conformal dimension of operators on the boundary [58, 59, 60]. This duality thus holds promise not only of a deeper understanding of M-theory, but may also throw light on non-perturbative aspects of the theories that live on the boundary which can include four-dimensional gauge theories. Models of this kind, where a bulk theory with gravity is equivalent to a boundary theory without gravity, have also been advocated by 't Hooft [62] and by Susskind [63] who call them *holographic* theories. The reader may notice a striking similarity to the earlier idea of 'The membrane at the end of the universe' and interconnections between the two are currently being explored [61, 65]. For example, one immediately recognizes that the dimensions and supersymmetries of these three conformal theories are exactly the same as the singleton, doubleton and tripleton supermultiplets of chapter 4. Many theorists are understandably excited about the AdS/CFT correspondence because of what it can teach us about non-perturbative QCD. In the editor's opinion, however, this is, in a sense, a diversion from the really fundamental question: What is M-theory? So my hope is that this will be a two-way process and that superconformal field theories will also teach us more about M-theory.

Edward Witten is fond of imagining how physics might have developed on other planets in which major discoveries, such as general relativity, Yang–Mills and supersymmetry were made in a different order from that on Earth. In the same vein, I would like to suggest that on planets more logical than ours (Vulcan?), eleven dimensions would have been taken as the starting point from which ten-dimensional string theory was subsequently derived as a special case. Indeed, future (terrestrial) historians may judge the period 1984–95 as a time when theorists were like boys playing by the sea shore, and diverting themselves with the smoother pebbles or prettier shells of perturbative ten-dimensional superstrings while the great ocean of non-perturbative eleven-dimensional M-theory lay all undiscovered before them.

References

[1] Montonen C and Olive D 1977 Magnetic monopoles as gauge particles? *Phys. Lett.* B **72** 117

[2] Hughes J, Liu J and Polchinski J 1986 Supermembranes *Phys.Lett.* B **180** 370

[3] Duff M J 1988 Supermembranes: The first fifteen weeks *Class. Quantum Grav.* **5** 189

[4] Strominger A 1990 Heterotic solitons *Nucl. Phys.* B **343** 167

[5] Nepomechie R I 1985 Magnetic monopoles from antisymmetric tensor gauge fields *Phys. Rev.* D **31** 1921

[6] Teitelboim C 1986 Monopoles of higher rank *Phys. Lett.* B **167** 63

[7] Sen A 1994 Dyon-monopole bound states, self-dual harmonic forms on the multi-monopole moduli space and SL(2,Z) invariance in string theory *Phys. Lett.* B **329** 217

[8] Seiberg N and Witten E 1994 Electric/magnetic duality, monopole condensation and confinement in N=2 supersymmetric Yang–Mills theory *Nucl. Phys.* B **426** 19 (Erratum 1994 *Nucl. Phys.* B **430** 485)

[9] Font A, Ibanez L, Lust D and Quevedo F 1990 Strong-weak coupling duality and nonperturbative effects in string theory *Phys. Lett.* B **249** 35

[10] Rey S J 1991 The confining phase of superstrings and axionic strings *Phys. Rev.* D **43** 526

[11] Giveon A, Porrati M and Rabinovici E 1994 Target space duality in string theory *Phys. Rep.* **244** 77

[12] Duff M J and Lu J X 1991 Loop expansions and string/five-brane duality *Nucl. Phys.* B **357** 534

[13] Duff M J and Khuri R R 1994 Four-dimensional string/string duality *Nucl. Phys.* B **411** 473

[14] Duff M and Lu J 1994 Black and super p-branes in diverse dimensions *Nucl. Phys.* B **416** 301

[15] Duff M J, Khuri R R, Minasian R and Rahmfeld J 1994 New black hole, string and membrane solutions of the four-dimensional heterotic string *Nucl. Phys.* B **418** 195

[16] Duff M J and Minasian R 1995 Putting string/string duality to the test *Nucl. Phys.* B **436** 507
[17] Duff M J, Khuri R R and Lu J X 1995 String solitons *Phys. Rep.* **259** 213
[18] Duff M J 1995 Strong/weak coupling duality from the dual string *Nucl. Phys.* B **442** 47
[19] Hull C M and Townsend P K 1995 Unity of superstring dualities *Nucl.Phys.* B **438** 109–37
[20] Witten E 1995 String dynamics in various dimensions *Nucl. Phys.* B **443** 85–126
[21] Duff M J, Liu J T and Rahmfeld J 1996 Four-dimensional string/string/string triality *Nucl. Phys.* B **459** 125
[22] Duff M J, Minasian R and Witten E 1996 Evidence for heterotic/heterotic duality *Nucl. Phys.* B **465** 413
[23] Duff M J 1996 Electric/magnetic duality and its stringy origins *Int. J. Mod. Phys.* A **11** 4031
[24] Lerche W 1997 Introduction to Seiberg–Witten theory and its stringy origin *Nucl. Phys. Proc. Suppl.* B **55** 83
[25] Blencowe M and Duff M J 1988 Supermembranes and the signature of space-time *Nucl. Phys.* B **310** 387
[26] Julia B 1980 Symmetries of supergravity models *Proc. Conf. Mathematical Problems in Theoretical Physics, Lausanne 1979 (Lecture Notes in Physics 116)* ed K Osterwalder (Berlin: Springer) p 368
[27] van Holten J W and van Proyen A 1982 *J. Phys. A: Math. Gen.* **15** 3763
[28] Hewson S and Perry M 1997 The twelve-dimensional super $(2+2)$-brane *Nucl. Phys.* B **492** 249
[29] Vafa C 1996 Evidence for F-theory *Nucl. Phys.* B **469** 403
[30] Nishino H and Sezgin E 1996 Supersymmetric Yang–Mills equations in $10+2$ dimensions *Phys. Lett.* B **388** 569
[31] Sen A 1996 F-theory and orientifolds *Nucl. Phys.* B **475** 562
[32] Horava P and Witten E 1996 Heterotic and Type I string dynamics from eleven dimensions *Nucl. Phys.* B **460** 506
[33] Witten E 1996 Strong coupling expansion of Calabi–Yau compactification *Nucl.Phys.* B **471** 135
[34] Witten E 1997 Solutions of four-dimensional field theories via M- theory *Nucl. Phys.* B **500** 3
[35] Witten E 1997 Branes and the dynamics of QCD, hep-th/97006109
[36] Turok N and Hawking S W 1998 Open inflation, the four-form and the cosmological constant *Phys. Lett.* B **432** 271
[37] Bousso R and Chamblin A 1998 Open inflation from nonsingular instantons: wrapping the universe with a membrane, hep-th/9805167
[38] Bremer M S, Duff M J, Lu H, Pope C N and Stelle K S 1998 Instanton cosmology and domain walls from M-theory and string theory, hep-th/9807051
[39] Hawking S W and Reall H S 1998 Inflation, singular instantons and eleven-dimensional cosmology, hep-th/9807100
[40] Banks T, Fischler W, Shenker S H and Susskind L 1997 M-theory as a matrix model: a conjecture *Phys. Rev.* D **55** 5112–28
[41] De Wit B, Hoppe J and Nicolai H 1988 On the quantum mechanics of supermembranes *Nucl. Phys.* B **305** 545
[42] De Wit B, Luscher M and Nicolai H 1989 The supermembrane is unstable *Nucl. Phys.* B **320** 135

[43] De Wit B and Nicolai H 1990 Supermembranes: a fond farewell? *Supermembranes and Physics in* $2+1$ *Dimensions* ed M J Duff, C N Pope and E Sezgin (Singapore: World Scientific) p 196

[44] Townsend P K 1996 D-branes from M-branes *Phys. Lett.* B **373** 68

[45] Cremmer E and Julia B 1979 The $SO(8)$ supergravity *Nucl. Phys.* B **159** 141

[46] Duff M J and Lu J X 1990 Duality rotations in membrane theory *Nucl. Phys.* B **347** 394

[47] Susskind L 1996 T-duality in M(atrix) theory and S-duality in field theory, hep-th/9611164

[48] Taylor W 1997 D-brane field theory on compact spaces *Phys. Lett.* B **394** 283

[49] Seiberg N 1997 New theories in six dimensions and matrix description of M-theory on T^5 and T^5/Z_2 *Phys. Lett.* B **408** 98

[50] Becker K, Becker M, Polchinski J and Tseytlin A 1997 Higher order graviton scattering in m(atrix) theory *Phys. Rev.* D **56** 3174

[51] Seiberg N 1997 Why is the matrix model correct? *Phys. Rev. Lett.* **79** 3577

[52] Banks T, Seiberg N and Shenker S 1997 Branes from matrices *Nucl. Phys.* B **490**

[53] Sen A 1998 $D0$-branes on T^N and matrix theory *Adv. Theor. Math. Phys.* **2** 51

[54] Dijkgraaf R, Verlinde E and Verlinde H 1997 Matrix string theory *Nucl. Phys.* B **500** 43

[55] Martinec E 1997 Matrix theory and $N = (2,1)$ strings, hep-th/9706194

[56] Townsend P K 1998 M(embrane) theory on T^9 *Nucl. Phys. Proc. Suppl.* **68** 11

[57] Maldacena J 1998 The large N limit of superconformal field theories and supergravity *Adv. Theor. Math. Phys.* **2** 231, hep-th/9711200

[58] Gubser S S, Klebanov I R and Polyakov A M 1998 Gauge theory correlators from non-critical string theory *Phys. Lett.* B **428** 105, hep-th/9802109

[59] Witten E 1998 Anti-de Sitter space and holography *Adv. Theor. Math. Phys.* **2** 253, hep-th/9802150

[60] Witten E 1998 Anti-de Sitter space, thermal phase transition, and confinement in gauge theories *Adv. Theor. Math. Phys.* **2** 505, hep-th/9803131

[61] Duff M J 1999 Anti-de Sitter space, branes, singletons, superconformal field theories and all that *IJMP* A **14** 815

[62] t'Hooft G 1993 Dimensional reduction in quantum gravity, gr-qc/9310026

[63] Susskind L 1994 The world as a hologram, hep-th/9409089

[64] Keku M 1998 *Introduction to Superstrings and M-theory* 2nd edn (Berlin: Springer)

[65] Seiberg N and Witten E 1999 The $D1/D5$ system and singular CFT *J. High Energy Phys.* 9904

Reprinted from Nucl. Phys. B **443** (1995) 85–126

String theory dynamics in various dimensions

Edward Witten
School of Natural Sciences, Institute for Advanced Study, Olden Lane, Princeton, NJ 08540, USA

Received 23 March 1995; accepted 3 April 1995

Abstract

The strong coupling dynamics of string theories in dimension $d \geqslant 4$ are studied. It is argued, among other things, that eleven-dimensional supergravity arises as a low energy limit of the ten-dimensional Type IIA superstring, and that a recently conjectured duality between the heterotic string and Type IIA superstrings controls the strong coupling dynamics of the heterotic string in five, six, and seven dimensions and implies S-duality for both heterotic and Type II strings.

1. Introduction

Understanding in what terms string theories should really be formulated is one of the basic needs and goals in the subject. Knowing some of the phenomena that can occur for strong coupling – if one can know them without already knowing the good formulation! – may be a clue in this direction. Indeed, S-duality between weak and strong coupling for the heterotic string in four dimensions (for instance, see Refs. [1,2]) really ought to be a clue for a new formulation of string theory.

At present there is very strong evidence for S-duality in supersymmetric field theories, but the evidence for S-duality in string theory is much less extensive. One motivation for the present work was to improve this situation.

Another motivation was to try to relate four-dimensional S-duality to statements or phenomena in more than four dimensions. At first sight, this looks well-nigh implausible since S-duality between electric and magnetic charge seems to be very special to four dimensions. So we are bound to learn something if we succeed.

Whether or not a version of S-duality plays a role, one would like to determine the strong coupling behavior of string theories above four dimensions, just as S-duality – and its conjectured Type II analog, which has been called U-duality [3] – determines the

Elsevier Science B.V.
SSDI 0550-3213(95)00158-1

strong coupling limit after toroidal compactification to four dimensions.[1] One is curious about the phenomena that may arise, and in addition if there is any non-perturbative inconsistency in the higher-dimensional string theories (perhaps ultimately leading to an explanation of why we live in four dimensions) it might show up naturally in thinking about the strong coupling behavior.

In fact, in this paper, we will analyze the strong coupling limit of certain string theories in certain dimensions. Many of the phenomena are indeed novel, and many of them are indeed related to dualities. For instance, we will argue in Section 2 that the strong coupling limit of Type IIA supergravity in ten dimensions is eleven-dimensional supergravity! In a sense, this statement gives a rationale for "why" eleven-dimensional supergravity exists, much as the interpretation of supergravity theories in $d \leqslant 10$ as low energy limits of string theories explains "why" these remarkable theories exist. How eleven-dimensional supergravity fits into the scheme of things has been a puzzle since the theory was first predicted [5] and constructed [6].

Upon toroidal compactification, one can study the strong coupling behavior of the Type II theory in $d < 10$ using U-duality, as we will do in Section 3. One can obtain a fairly complete picture, with eleven-dimensional supergravity as the only "surprise."

Likewise, we will argue in Section 4 that the strong coupling limit of five-dimensional heterotic string theory is Type IIB in six dimensions, while the strong coupling limit of six-dimensional heterotic string theory is Type IIA in six dimensions (in each case with four dimensions as a K3), and the strong coupling limit in seven dimensions involves eleven-dimensional supergravity. These results are based on a relation between the heterotic string and the Type IIA superstring in six dimensions that has been proposed before [3,4]. The novelty in the present paper is to show, for instance, that vexing puzzles about the strong coupling behavior of the heterotic string in five dimensions disappear if one assumes the conjectured relation of the heterotic string to Type IIA in six dimensions. Also we will see – using a mechanism proposed previously in a more abstract setting [7] – that the "string–string duality" between heterotic and Type IIA strings in six dimensions implies S-duality in four dimensions, so the usual evidence for S-duality can be cited as evidence for string–string duality.

There remains the question of determining the strong coupling dynamics of the heterotic string above seven dimensions. In this context, there is a curious speculation[2] that the heterotic string in ten dimensions with SO(32) gauge group might have for its strong coupling limit the SO(32) Type I theory. In Section 5 we show that this relation, if valid, straightforwardly determines the strong coupling behavior of the heterotic string in nine and eight dimensions as well as ten, conjecturally completing the description of strong coupling dynamics except for $E_8 \times E_8$ in ten dimensions.

---

[1] By "strong coupling limit" I mean the limit as the string coupling constant goes to infinity keeping fixed (in the sigma model sense) the parameters of the compactification. Compactifications that are not explicitly described or clear from the context will be toroidal.

[2] This idea was considered many years ago by M. B. Green, the present author, and probably others, but not in print as far as I know.

The possible relations between different theories discussed in this paper should be taken together with other, better established relations between different string theories. It follows from T-duality that below ten dimensions the $E_8 \times E_8$ heterotic string is equivalent to the SO(32) heterotic string [8,9], and Type IIA is equivalent to Type IIB [10,11]. Combining these statements with the much shakier relations discussed in the present paper, one would have a web of connections between the five string theories and eleven-dimensional supergravity.

After this paper was written and circulated, I learned of a paper [12] that has some overlap with the contents of Section 2 of this paper.

2. Type II superstrings in ten dimensions

2.1. Type IIB in ten dimensions

In this section, we will study the strong coupling dynamics of Type II superstrings in ten dimensions. We start with the easy case, Type IIB. A natural conjecture has already been made by Hull and Townsend [3]. Type IIB supergravity in ten dimensions has an $SL(2,\mathbb{R})$ symmetry; the conjecture is that an $SL(2,\mathbb{Z})$ subgroup of this is an exact symmetry of the string theory.[3] This then would relate the strong and weak coupling limits just as S-duality relates the strong and weak coupling limits of the heterotic string in four dimensions.

This $SL(2,\mathbb{Z})$ symmetry in ten dimensions, if valid, has powerful implications below ten dimensions. The reason is that in $d < 10$ dimensions, the Type II theory (Type IIA and Type IIB are equivalent below ten dimensions) is known to have a T-duality symmetry $SO(10-d, 10-d; \mathbb{Z})$. This T-duality group does not commute with the $SL(2,\mathbb{Z})$ that is already present in ten dimensions, and together they generate the discrete subgroup of the supergravity symmetry group that has been called U-duality.[4] Thus, U-duality is true in every dimension below ten if the $SL(2,\mathbb{Z})$ of the Type IIB theory holds in ten dimensions.

In the next section we will see that U-duality controls Type II dynamics below ten dimensions. As $SL(2,\mathbb{Z})$ also controls Type IIB dynamics in ten dimensions, this fundamental duality between strong and weak coupling controls all Type II dynamics in all dimensions except for the odd case of Type IIA in ten dimensions. But that case will not prove to be a purely isolated exception: the basic phenomenon that we will find in Type IIA in ten dimensions is highly relevant to Type II dynamics below ten

[3] For earlier work on the possible role of the non-compact supergravity symmetries in string and membrane theory, see Ref. [13].

[4] For instance, in five dimensions, T-duality is $SO(5,5)$ and U-duality is E_6. A proper subgroup of E_6 that contains $SO(5,5)$ would have to be $SO(5,5)$ itself or $SO(5,5) \times \mathbb{R}^*$ ($\mathbb{R}^*$ is the non-compact form of $U(1)$), so when one tries to adjoin to $SO(5,5)$ the $SL(2)$ that was already present in ten dimensions (and contains two generators that map NS–NS states to RR states and so are not in $SO(5,5)$) one automatically generates all of E_6.

dimensions, as we will see in Section 3. In a way ten-dimensional Type IIA proves to exhibit the essential new phenomenon in the simplest context.

To compare to $N = 1$ supersymmetric dynamics in four dimensions [14], ten-dimensional Type IIA is somewhat analogous to supersymmetric QCD with $3N_c/2 > N_f > N_c + 1$, whose dynamics is controlled by an effective infrared theory that does not make sense at all length scales. The other cases are analogous to the same theory with $3N_c > N_f > 3N_c/2$, whose dynamics is controlled by an exact equivalence of theories - conformal fixed points - that make sense at all length scales.

2.2. Ramond–Ramond charges in ten-dimensional type IIA

It is a familiar story to string theorists that the string coupling constant is really the expectation of a field - the dilaton field ϕ. Thus, it can be scaled out of the low energy effective action by shifting the value of the dilaton.

After scaling other fields properly, this idea can be implemented in closed string theories by writing the effective action as $e^{-2\phi}$ times a function that is invariant under $\phi \to \phi + \text{constant}$. There is, however, an important subtlety here that affects the Type IIA and Type IIB (and Type I) theories. These theories have massless antisymmetric tensor fields that originate in the Ramond–Ramond (RR) sector. If A_p is such a p-form field, the natural gauge invariance is $\delta A_p = \mathrm{d}\lambda_{p-1}$, with λ_{p-1} a $(p-1)$-form - and no dilaton in the transformation laws. If one scales A_p by a power of e^{ϕ}, the gauge transformation law becomes more complicated and less natural.

Let us, then, consider the Type IIA theory with the fields normalized in a way that makes the gauge invariance natural. The massless bosonic fields from the (Neveu–Schwarz)^2 or NS–NS sector are the dilaton, the metric tensor g_{mn}, and the antisymmetric tensor B_{mn}. From the RR sector, one has a one-form A and a three-form A_3. We will write the field strengths as $H = \mathrm{d}B$, $F = \mathrm{d}A$, and $F_4 = \mathrm{d}A_3$; one also needs $F_4' = \mathrm{d}A_3 + A \wedge H$. The bosonic part of the low energy effective action can be written $I = I_{\mathrm{NS}} + I_{\mathrm{R}}$ where I_{NS} is the part containing NS–NS fields only and I_{R} is bilinear in RR fields. One has (in units with $\alpha' = 1$)

$$I_{\mathrm{NS}} = \frac{1}{2}\int \mathrm{d}^{10}x\,\sqrt{g}\,e^{-2\phi}\left(R + 4(\nabla\phi)^2 - \frac{1}{12}H^2\right) \tag{2.1}$$

and

$$I_{\mathrm{R}} = -\int \mathrm{d}^{10}x\,\sqrt{g}\left(\frac{1}{2\cdot 2!}F^2 + \frac{1}{2\cdot 4!}F_4'^2\right) - \frac{1}{4}\int F_4 \wedge F_4 \wedge B. \tag{2.2}$$

With this way of writing the Lagrangian, the gauge transformation laws of A, B, and A_3 all have the standard, dilaton-independent form $\delta X = \mathrm{d}\Lambda$, but it is not true that the classical Lagrangian scales with the dilaton like an overall factor of $e^{-2\phi}$.

Our interest will focus on the presence of the abelian gauge field A in the Type IIA theory. The charge W of this gauge field has the following significance. The Type IIA theory has two supersymmetries in ten dimensions, one of each chirality; call them

Q_α and $Q'_{\dot\alpha}$. The space-time momentum P appears in the anticommutators $\{Q, Q\} \sim \{Q', Q'\} \sim P$. In the anticommutator of Q with Q' it is possible to have a Lorentz-invariant central charge

$$\{Q_\alpha, Q'_{\dot\alpha}\} \sim \delta_{\alpha\dot\alpha} W. \tag{2.3}$$

To see that such a term does arise, it is enough to consider the interpretation of the Type IIA theory as the low energy limit of eleven-dimensional supergravity, compactified on $\mathbb{R}^{10} \times \mathbf{S}^1$. From that point of view, the gauge field A arises from the components $g_{m,11}$ of the eleven-dimensional metric tensor, W is simply the eleventh component of the momentum, and (2.3) is part of the eleven-dimensional supersymmetry algebra. [5]

In the usual fashion [17], the central charge (2.3) leads to an inequality between the mass M of a particle and the value of W:

$$M \geqslant c_0|W|, \tag{2.4}$$

with c_0 a "constant," that is a function only of the string coupling constant $\lambda = e^\phi$, and independent of which particle is considered. The precise constant with which W appears in (2.3) or (2.4) can be worked out using the low energy supergravity (there is no need to worry about stringy corrections as the discussion is controlled by the leading terms in the low energy effective action, and these are uniquely determined by supersymmetry). We will work this out at the end of this section by a simple scaling argument starting with eleven-dimensional supergravity. For now, suffice it to say that the λ dependence of the inequality is actually

$$M \geqslant \frac{c_1}{\lambda}|W| \tag{2.5}$$

with c_1 an absolute constant. States for which the inequality is saturated – we will call them BPS-saturated states by analogy with certain magnetic monopoles in four dimensions – are in "small" supermultiplets with 2^8 states, while generic supermultiplets have 2^{16} states.

In the elementary string spectrum, W is identically zero. Indeed, as A originates in the RR sector, W would have had to be a rather exotic charge mapping NS–NS to RR states. However, there is no problem in finding classical black hole solutions carrying the W charge (or any other gauge charge, in any dimension). It was proposed by Hull and Townsend [3] that quantum particles carrying RR charges arise by quantization of such black holes. Recall that, in any dimension, charged black holes obey an inequality $GM^2 \geqslant \text{const} \cdot W^2$ (G, M, and W are Newton's constant and the black hole mass and charge); with $G \sim \lambda^2$, this inequality has the same structure as (2.5). These two inequalities actually correspond in the sense that an extreme black hole, with the

[5] The relation of the supersymmetry algebra to eleven dimensions leads to the fact that both for the lowest level and even for the first excited level of the Type IIA theory, the states can be arranged in eleven-dimensional Lorentz multiplets [15]. If this would persist at higher levels, it might be related to the idea that will be developed below. It would also be interesting to look for possible eleven-dimensional traces in the superspace formulation [16].

minimum mass for given charge, is invariant under some supersymmetry [18] and so should correspond upon quantization to a "small" supermultiplet saturating the inequality (2.5).

To proceed, then, I will assume that there are in the theory BPS-saturated particles with $W \neq 0$. This assumption can be justified as follows. Hull and Townsend actually showed that upon toroidally compactifying to less than ten dimensions, the assumption follows from U-duality. In toroidal compactification, the radii of the circles upon which one compactifies can be arbitrarily big. That being so, it is implausible to have BPS-saturated states of $W \neq 0$ below ten dimensions unless they exist in ten dimensions; that is, if the smallest mass of a W-bearing state in ten dimensions were strictly bigger than $c|W|/\lambda$, then this would remain true after compactification on a sufficiently big torus.

If the ten-dimensional theory has BPS-saturated states of $W \neq 0$, then what values of W occur? A continuum of values of W would seem pathological. A discrete spectrum is more reasonable. If so, the quantum of W must be independent of the string coupling "constant" λ. The reason is that λ is not really a "constant" but the expectation value of the dilaton field ϕ. If the quantum of W were to depend on the value of ϕ, then the value of the electric charge W of a particle would change in a process in which ϕ changes (that is, a process in which ϕ changes in a large region of space containing the given particle); this would violate conservation of W.

The argument just stated involves a hidden assumption that will now be made explicit. The canonical action for a Maxwell field is

$$\frac{1}{4e^2}\int \mathrm{d}^n x\,\sqrt{g}\,F^2. \tag{2.6}$$

Comparing to (2.2), we see that in the case under discussion the effective value of e is independent of ϕ, and this is why the charge of a hypothetical charged particle is independent of ϕ. If the action were

$$\frac{1}{4}\int \mathrm{d}^n x\,\sqrt{g}\,\mathrm{e}^{\gamma\phi}F^2 \tag{2.7}$$

for some non-zero γ, then the current density would equal (from the equations of motion of A) $J_m = \partial^n(\mathrm{e}^{\gamma\phi}F_{mn})$. In a process in which ϕ changes in a large region of space containing a charge, there could be a current inflow proportional to $\nabla\phi\cdot F$, and the charge would in fact change. Thus, it is really the ϕ-independence of the kinetic energy of the RR fields that leads to the statement that the values of W must be independent of the string coupling constant and that the masses of charged fields scale as λ^{-1}.

Since the classical extreme black hole solution has arbitrary charge W (which can be scaled out of the solution in an elementary fashion), one would expect, if BPS-saturated charged particles do arise from quantization of extreme black holes, that they should possess every allowed charge. Thus, we expect BPS-saturated extreme black holes of mass

$$M = \frac{c|n|}{\lambda}, \tag{2.8}$$

where n is an arbitrary integer, and, because of the unknown value of the quantum of electric charge, c may differ from c_1 in (2.5).

Apart from anything else that follows, the existence of particles with masses of order $1/\lambda$, as opposed to the more usual $1/\lambda^2$ for solitons, is important in itself. It almost certainly means that the string perturbation expansion – which is an expansion in powers of λ^2 – will have non-perturbative corrections of order $\exp(-1/\lambda)$, in contrast to the more usual $\exp(-1/\lambda^2)$ [6]. The occurrence of such terms has been guessed by analogy with matrix models [20].

The fact that the masses of RR charges diverge as $\lambda \to 0$ – though only as $1/\lambda$ – is important for self-consistency. It means that these states disappear from the spectrum as $\lambda \to 0$, which is why one does not see them as elementary string states.

2.3. *Consequences for dynamics*

Now we will explore the consequences for dynamics of the existence of these charged particles.

The mass formula (2.8) shows that, when the string theory is weakly coupled, the RR charges are very heavy. But if we are bold enough to follow the formula into strong coupling, then for $\lambda \to \infty$, these particles go to zero mass. This may seem daring, but the familiar argument based on the "smallness" of the multiplets would appear to show that the formula (2.8) is exact and therefore can be used even for strong coupling. In four dimensions, extrapolation of analogous mass formulas to strong coupling has been extremely successful, starting with the original work of Montonen and Olive that led to the idea of S-duality. (In four-dimensional $N = 2$ theories, such mass formulas generally fail to be exact [21] because of quantum corrections to the low energy effective action. For $N = 4$ in four dimensions, or for Type IIA supergravity in ten dimensions, the relevant, leading terms in the low energy action are uniquely determined by supersymmetry.)

So for strong coupling, we imagine a world in which there are supermultiplets of mass $M = c|n|/\lambda$ for every λ. These multiplets necessarily contain particles of spin at least two, as every supermultiplet in Type IIA supergravity in ten dimensions has such states. (Multiplets that do not saturate the mass inequality contain states of spin $\geqslant 4$.) Rotation invariance of the classical extreme black hole solution suggests [7] (as does U-duality) that the BPS-saturated multiplets are indeed in this multiplet of minimum spin.

Thus, for $\lambda \to \infty$ we have light, charged fields of spin two. (That is, they are charged with respect to the ten-dimensional gauge field A.) Moreover, there are infinitely many of these. This certainly does not correspond to a local field theory in ten dimensions. What kind of theory will reproduce this spectrum of low-lying states? One is tempted to

---

[6] If there are particles of mass $1/\lambda$, then loops of those particles should give effects of order $e^{-1/\lambda}$, while loops of conventional solitons, with masses $1/\lambda^2$, would be of order $\exp(-1/\lambda^2)$.

[7] Were the classical solution not rotationally invariant, then upon quantizing it one would obtain a band of states of states of varying angular momentum. One would then not expect to saturate the mass inequality of an extreme black hole without taking into account the angular momentum.

think of a string theory or Kaluza–Klein theory that has an infinite tower of excitations. The only other option, really, is to assume that the strong coupling limit is a sort of theory that we do not know about at all at present.

One can contemplate the possibility that the strong coupling limit is some sort of a string theory with the dual string scale being of order $1/\lambda$, so that the charged multiplets under discussion are some of the elementary string states. There are two reasons that this approach does not seem promising: (i) there is no known string theory with the right properties (one needs Type IIA supersymmetry in ten dimensions, with charged string states coupling to the abelian gauge field in the gravitational multiplet); (ii) we do not have evidence for a stringy exponential proliferation of light states as $\lambda \to \infty$, but only for a single supermultiplet for each integer n, with mass $\sim |n|$.

Though meager compared to a string spectrum, the spectrum we want to reproduce is just about right for a Kaluza–Klein theory. Suppose that in the region of large λ, one should think of the theory not as a theory on $\mathbb{R}^{10}$ but as a theory on $\mathbb{R}^{10} \times \mathbf{S}^1$. Such a theory will have a "charge" coming from the rotations of $\mathbf{S}^1$. Suppose that the radius $r(\lambda)$ of the $\mathbf{S}^1$ scales as $1/\lambda$ (provided that distances are measured using the "string" metric that appears in (2.1) – one could always make a Weyl rescaling). Then for large λ, each massless field in the eleven-dimensional theory will give, in ten dimensions, for each integer n a single field of charge n and mass $\sim |n|\lambda$. This is precisely the sort of spectrum that we want.

So we need an eleven-dimensional field theory whose fields are in one-to-one correspondence with the fields of the Type IIA theory in ten dimensions. Happily, there is one: eleven-dimensional supergravity! So we are led to the strange idea that eleven-dimensional supergravity may govern the strong coupling behavior of the Type IIA superstring in ten dimensions.

Let us discuss a little more precisely how this would work. The dimensional reduction of eleven-dimensional supergravity to ten dimensions including the massive states has been discussed in some detail (for example, see Ref. [22]). Here we will be very schematic, just to touch on the points that are most essential. The bosonic fields in eleven-dimensional supergravity are the metric G_{MN} and a three-form A_3. The bosonic part of the action is

$$I = \frac{1}{2}\int \mathrm{d}^{11}x\,\sqrt{G}\left(R + |\mathrm{d}A_3|^2\right) + \int A_3 \wedge \mathrm{d}A_3 \wedge \mathrm{d}A_3. \tag{2.9}$$

Now we reduce to ten dimensions, taking the eleventh dimensions to be a circle of radius e^{γ}. That is, we take the eleven-dimensional metric to be $\mathrm{d}s^2 = G^{10}_{mn}\,\mathrm{d}x^m\,\mathrm{d}x^n + \mathrm{e}^{2\gamma}(\mathrm{d}x^{11} - A_m\,\mathrm{d}x^m)^2$ to describe a ten-dimensional metric G^{10} along with a vector A and scalar γ; meanwhile A_3 reduces to a three-form which we still call A_3, and a two-form B (the part of the original A_3 with one index equal to 11). Just for the massless fields, the bosonic part of the action becomes roughly

$$I = \frac{1}{2}\int \mathrm{d}^{10}x\,\sqrt{G^{10}}\left(\mathrm{e}^{\gamma}\left(R + |\nabla\gamma|^2 + |\mathrm{d}A_3|^2\right) + \mathrm{e}^{3\gamma}|\mathrm{d}A|^2 + \mathrm{e}^{-\gamma}|\mathrm{d}B|^2\right) + \ldots \tag{2.10}$$

This formula, like others below, is very rough and is only intended to exhibit the powers of e^γ. The point in its derivation is that, for example, the part of A_3 that does not have an index equal to "11" has a kinetic energy proportional to e^γ, while the part with such an index has a kinetic energy proportional to $e^{-\gamma}$.

The powers of e^γ in (2.10) do not, at first sight, appear to agree with those in (2.1). To bring them in agreement, we make a Weyl rescaling by writing $G^{10} = e^{-\gamma} g$. Then in terms of the new ten-dimensional metric g, we have

$$I = \frac{1}{2} \int d^{10}x \sqrt{g} \left(e^{-3\gamma} \left(R + |\nabla\gamma|^2 + |dB|^2 \right) + |dA|^2 + |dA_3|^2 + \ldots \right). \tag{2.11}$$

We see that (2.11) now does agree with (2.1) if

$$e^{-2\phi} = e^{-3\gamma}. \tag{2.12}$$

In the original eleven-dimensional metric, the radius of the circle is $r(\lambda) = e^\gamma$, but now, relating γ to the dilaton string coupling constant via (2.12), we can write

$$r(\lambda) = e^{2\phi/3} = \lambda^{2/3}. \tag{2.13}$$

The masses of Kaluza–Klein modes of the eleven-dimensional theory are of order $1/r(\lambda)$ when measured in the metric G^{10}, but in the metric g they are of order

$$\frac{e^{-\gamma/2}}{r(\lambda)} \sim \lambda^{-1}. \tag{2.14}$$

Manipulations similar to what we have just seen will be made many times in this paper.

Here are the salient points:

(1) The radius of the circle *grows* by the formula (2.13) as $\lambda \to \infty$. This is important for self-consistency; it means that when λ is large the eleven-dimensional theory is weakly coupled at its compactification scale. Otherwise the discussion in terms of eleven-dimensional field theory would not make sense, and we would not know how to improve on it. As it is, our proposal reduces the strongly coupled Type IIA superstring to a field theory that is weakly coupled at the scale of the low-lying excitations, so we get an effective determination of the strong coupling behavior.

(2) The mass of a particle of charge n, measured in the string metric g in the effective ten-dimensional world, is of order $|n|/\lambda$ from (2.14). This is the dependence on λ claimed in (2.5), which we have now in essence derived: the dependence of the central charge on ϕ is uniquely determined by the low energy supersymmetry, so by deriving this dependence in a Type IIA supergravity theory that comes by Kaluza–Klein reduction from eleven dimensions, we have derived it in general.

So far, the case for relating the strong coupling limit of Type IIA superstrings to eleven-dimensional supergravity consists of the fact that this enables us to make sense of the otherwise puzzling dynamics of the BPS-saturated states and that point (1) above worked out correctly, which was not obvious a priori. The case will hopefully get much stronger in the next section when we extend the analysis to work below ten dimensions and incorporate U-duality, and in Section 4 when we look at the heterotic string in seven dimensions. In fact, the most startling aspect of relating strong coupling string dynamics to eleven-dimensional supergravity is the Lorentz invariance that this implies between the eleventh dimension and the original ten. Both in Section 3 and in Section 4, we will see remnants of this underlying Lorentz invariance.

3. Type II dynamics below ten dimensions

3.1. U-duality and dynamics

In this section, we consider Type II superstrings toroidally compactified to $d < 10$ dimensions, with the aim of understanding the strong coupling dynamics, that is, the behavior when some parameters, possibly including the string coupling constant, are taken to extreme values.

The strong coupling behaviors of Type IIA and Type IIB seem to be completely different in ten dimensions, as we have seen. Upon toroidal compactification below ten dimensions, the two theories are equivalent under T-duality [10,11], and so can be considered together. We will call the low energy supergravity theory arising from this compactification Type II supergravity in d dimensions.

The basic tool in the analysis is U-duality. Type II supergravity in d dimensions has a moduli space of vacua of the form G/K, where G is a non-compact connected Lie group (which depends on d) and K is a compact subgroup, generally a maximal compact subgroup of G. G is an exact symmetry of the supergravity theory. There are also U(1) gauge bosons, whose charges transform as a representation of G. [8] The structure was originally found by dimensional reduction from eleven dimensions [23].

In the string theory realization, the moduli space of vacua remains G/K since this is forced by the low energy supergravity. Some of the Goldstone bosons parametrizing G/K come from the NS–NS sector and some from the RR sector. The same is true of the gauge bosons. In string theory, the gauge bosons that come from the NS–NS sector couple to charged states in the elementary string spectrum. It is therefore impossible for G to be an exact symmetry of the string theory – it would not preserve the lattice of charges. The U-duality conjecture says that an integral form of G, call it $G(\mathbb{Z})$, is a symmetry of the string theory. If so, then as the NS–NS gauge bosons couple to BPS-saturated charges, the same must be true of the RR gauge bosons – though the charges in question do not appear in the elementary string spectrum. The existence of such RR charges was our main assumption in the last section; we see that this assumption is essentially a consequence of U-duality.

[8] To make a G-invariant theory on G/K, the matter fields in general must be in representations of the unbroken symmetry group K. Matter fields that are in representations of K that do not extend to representations of G are sections of some homogeneous vector bundles over G/K with non-zero curvature. The potential existence of an integer lattice of charges forces the gauge bosons to be sections instead of a flat bundle, and that is why they are in a representation of G and not only of K.

The BPS-saturated states are governed by an exact mass formula – which will be described later in some detail – which shows how some of them become massless when one approaches various limits in the moduli space of vacua. Our main dynamical assumption is that the smallest mass scale appearing in the mass formula is always the smallest mass scale in the theory.

We assume that at a generic point in G/K, the only massless states are those in the supergravity multiplet. There is then nothing to say about the dynamics: the infrared behavior is that of d-dimensional Type II supergravity. There remains the question of what happens when one takes various limits in G/K – for instance, limits that correspond to weak coupling or large radius or (more mysteriously) strong coupling or very strong excitation of RR scalars. We will take the signal that something interesting is happening to be that the mass formula predicts that some states are going to zero mass. When this occurs, we will try to determine what is the dynamics of the light states, in whatever limit is under discussion.

We will get a complete answer, in the sense that for every degeneration of the Type II superstring in d dimensions, there is a natural candidate for the dynamics. In fact, there are basically only two kinds of degeneration; one involves weakly coupled string theory, and the other involves weakly coupled eleven-dimensional supergravity. In one kind of degeneration, one sees toroidal compactification of a Type II superstring from ten to d dimensions; the degeneration consists of the fact that the string coupling constant is going to zero. (The parameters of the torus are remaining fixed.) In the other degeneration one sees toroidal compactification of eleven-dimensional supergravity from eleven to d dimensions; the degeneration consists of the fact that the radius of the torus is going to infinity so that again the coupling constant at the compactification scale is going to zero.[9] (These are actually the degenerations that produce maximal sets of massless particles; others correspond to perturbations of these.)

Thus, with our hypotheses, one gets a complete control on the dynamics, including strong coupling. Every limit which one might have been tempted to describe as "strong coupling" actually has a weakly coupled description in the appropriate variables. The ability to get this consistent picture can be taken as evidence that the hypotheses are true, that U-duality is valid, and that eleven-dimensional supergravity plays the role in the dynamics that was claimed in Section 2.

It may seem unexpected that weakly coupled string theory appears in this analysis as a "degeneration," where some particles go to zero mass, so let me explain this situation. For $d < 9$, G is semi-simple, and the dilaton is unified with other scalars. The "string" version of the low energy effective action, in which the dilaton is singled out in the gravitational kinetic energy

$$\int \mathrm{d}^d x \sqrt{g}\, \mathrm{e}^{-2\phi} R \qquad (3.1)$$

[9] It is only in the eleven-dimensional description that the radius is going to infinity. In the ten-dimensional string theory description, the radius is fixed but the string coupling constant is going to infinity.

is unnatural for exhibiting such a symmetry. The G-invariant metric is the one obtained by a Weyl transformation that removes the $e^{-2\phi}$ from the gravitational kinetic energy. The transformation in question is of course the change of variables $g = e^{4\phi/(d-2)}g'$, with g' the new metric. This transformation multiplies masses by $e^{2\phi/(d-2)}$, that is, by

$$w_d = \lambda^{2/(d-2)} \tag{3.2}$$

(with λ the string coupling constant). Thus, while elementary string states have masses of order one with respect to the string metric, their masses are of order $\lambda^{2/(d-2)}$ in the natural units for discussions of U-duality. So, from this point of view, the region of weakly coupled string theory is a "degeneration" in which some masses go to zero.

It is amusing to consider that, in a world in which supergravity was known and string theory unknown, the following discussion might have been carried out, with a view to determining the strong coupling limit of a hypothetical consistent theory related to Type II supergravity. The string theory degeneration might then have been found, giving a clue to the existence of this theory. Similarly, the strong coupling analysis that we are about to perform might a priori have uncovered new theories beyond string theory and eleven-dimensional supergravity, but this will not be the case.

3.2. The Nature of Infinity

It is useful to first explain – without specific computations – why NS–NS (rather than RR) moduli play the primary role.

We are interested in understanding what particles become light – and how they interact – when one goes to infinity in the moduli space $G(\mathbb{Z})\backslash G/K$. The discussion is simplified by the fact that the groups G that arise in supergravity are the maximally split forms of the corresponding Lie groups. This simply means that they contain a maximal abelian subgroup A which is a product of copies of $\mathbb{R}^*$ (rather than U(1)). [10]

For instance, in six dimensions $G = SO(5,5)$, with rank 5. One can think of G as the orthogonal group acting on the sum of five copies of a two-dimensional real vector space H endowed with quadratic form

$$\begin{pmatrix} 0 & 1 \\ 1 & 0 \end{pmatrix}. \tag{3.3}$$

Then a maximal abelian subgroup of G is the space of matrices looking like a sum of five 2×2 blocks, of the form

$$\begin{pmatrix} e^{\lambda_i} & 0 \\ 0 & e^{-\lambda_i} \end{pmatrix} \tag{3.4}$$

for some λ_i. This group is of the form $(\mathbb{R}^*)^5$. Likewise, the integral forms arising in T- and U-duality are the maximally split forms over $\mathbb{Z}$; for instance the T-duality group

[10] Algebraists call A a "maximal torus," and T would be the standard name, but I will avoid this terminology because (i) calling $(\mathbb{R}^*)^n$ a "torus" might be confusing, especially in the present context in which there are so many other tori; (ii) in the present problem the letter T is reserved for the T-duality group.

upon compactification to $10 - d$ dimensions is the group of integral matrices preserving a quadratic form which is the sum of d copies of (3.3). This group is sometimes called $\mathrm{SO}(d, d; \mathbb{Z})$.

With the understanding that G and $G(\mathbb{Z})$ are the maximally split forms, the structure of infinity in $G(\mathbb{Z})\backslash G/K$ is particularly simple. A fundamental domain in $G(\mathbb{Z})\backslash G/K$ consists of group elements of the form $g = tu$, where the notation is as follows. u runs over a *compact* subset U of the space of generalized upper triangular matrices; compactness of U means that motion in U is irrelevant in classifying the possible ways to "go to infinity." t runs over A/W where A was described above, and W is the Weyl group.

Thus, one can really only go to infinity in the A direction, and moreover, because of dividing by W, one only has to consider going to infinity in a "positive" direction.

Actually, A has a very simple physical interpretation. Consider the special case of compactification from 10 to $10 - d$ dimensions on an orthogonal product of circles $\mathbf{S}^1_i$ of radius r_i. Then G has rank $d + 1$, so A is a product of $d + 1$ $\mathbb{R}^*$'s. d copies of $\mathbb{R}^*$ act by rescaling the r_i (making up a maximal abelian subgroup of the T-duality group $\mathrm{SO}(d, d)$), and the last one rescales the string coupling constant. So in particular, with this choice of A, if one starts at a point in moduli space at which the RR fields are all zero, they remain zero under the action of A.

Thus, one can probe all possible directions at infinity without exciting the RR fields; directions in which some RR fields go to infinity are equivalent to directions in which one only goes to infinity via NS–NS fields. Moreover, by the description of A just given, going to infinity in NS–NS directions can be understood to mean just taking the string coupling constant and the radial parameters of the compactification to zero or infinity.

3.3. The central charges and their role

Let us now review precisely why it is possible to predict particle masses from U-duality. The unbroken subgroup K of the supergravity symmetry group G is realized in Type II supergravity as an R-symmetry group; that is, it acts non-trivially on the supersymmetries. K therefore acts on the central charges in the supersymmetry algebra. The scalar fields parametrizing the coset space G/K enable one to write a G-invariant formula for the central charges (which are a representation of K) of the gauge bosons (which are a representation of G). For most values of d, the formula is uniquely determined, up to a multiplicative constant, by G-invariance, so the analysis does not require many details of supergravity. That is fortunate as not all the details we need have been worked out in the literature, though many can be found in Ref. [24].

For example, let us recall (following Ref. [3]) the situation in $d = 4$. The T-duality group is $\mathrm{SO}(6, 6)$, and S-duality would be $\mathrm{SL}(2)$ (acting on the axion-dilaton system and exchanging electric and magnetic charge). $\mathrm{SO}(6, 6) \times \mathrm{SL}(2)$ is a maximal subgroup of the U-duality group which is $G = E_7$ (in its non-compact, maximally split form) and has $K = \mathrm{SU}(8)$ as a maximal compact subgroup.

Toroidal compactification from ten to four dimensions produces in the NS–NS sector twelve gauge bosons coupling to string momentum and winding states, and transforming in the twelve-dimensional representation of SO(6, 6). The electric and magnetic charges coupling to any one of these gauge bosons transform as a doublet of SL(2), so altogether the NS–NS sector generates a total of 24 gauge charges, transforming as $(\mathbf{12}, \mathbf{2})$ of $\mathrm{SO}(6,6) \times \mathrm{SL}(2)$.

From the RR sector, meanwhile, one gets 16 vectors. (For instance, in Type IIA, the vector of the ten-dimensional RR sector gives 1 vector in four dimensions, and the three-form gives $6 \cdot 5/2 = 15$.) These 16 states give a total of $16 \cdot 2 = 32$ electric and magnetic charges, which can be argued to transform in an irreducible spinor representation of SO(6, 6) (of positive or negative chirality for Type IIA or Type IIB), while being SL(2) singlet. The fact that these states are SL(2) singlets means that there is no natural way to say which of the RR charges are electric and which are magnetic. Altogether, there are $24 + 32 = 56$ gauge charges, transforming as

$$(\mathbf{12}, \mathbf{2}) \oplus (\mathbf{32}, \mathbf{1}) \tag{3.5}$$

under $\mathrm{SO}(6,6) \times \mathrm{SL}(2)$; this is the decomposition of the irreducible **56** of E_7. Let us call the space of these charges V.

The four-dimensional theory has $N = 8$ supersymmetry; thus there are eight positive-chirality supercharges Q^i_α, $i = 1, \ldots, 8$, transforming in the **8** of $K = \mathrm{SU}(8)$. The central charges, arising in the formula

$$\{Q^i_\alpha, Q^j_\beta\} = \epsilon_{\alpha\beta} Z^{ij}, \tag{3.6}$$

therefore transform as the second rank antisymmetric tensor of SU(8), the **28**: this representation has complex dimension 28 or real dimension 56. Denote the space of Z^{ij}'s as W.

Indeed, the **56** of E_7, when restricted to SU(8), coincides with the **28**, regarded as a 56-dimensional real representation. (Equivalently, the **56** of E_7 when complexified decomposes as $\mathbf{28} \oplus \overline{\mathbf{28}}$ of SU(8).) There is of course a natural, SU(8)-invariant metric on W. As the **56** is a pseudoreal rather than real representation of E_7, there is no E_7-invariant metric on V. However, as V and W coincide when regarded as representations of SU(8), one can pick an embedding of SU(8) in E_7 and then define an SU(8)-covariant map $T : V \longrightarrow W$ which determines a metric on V.

There is no reason to pick one embedding rather than another, and indeed the space of vacua $E_7/\mathrm{SU}(8)$ of the low energy supergravity theory can be interpreted as the space of all SU(8) subgroups of E_7. Given $g \in E_7$, we can replace $T : V \longrightarrow W$ by

$$T_g = T g^{-1}. \tag{3.7}$$

This is not invariant under $g \longrightarrow gk$, with $k \in \mathrm{SU}(8)$, but it is so invariant up to an SU(8) transformation of W. So let $\psi \in V$ be a vector of gauge charges of some string state. Then

$$\psi \longrightarrow Z(\psi) = T_g \psi \tag{3.8}$$

gives a vector in W, representing the central charges of ψ. The map from "states" ψ to central charges $Z(\psi)$ is manifestly E_7-invariant, that is invariant under

$$\begin{aligned} \psi &\rightarrow g'\psi\,, \\ g &\rightarrow g'g\,. \end{aligned} \tag{3.9}$$

Also, under $g \rightarrow gk$, with $k \in \mathrm{SU}(8)$, Z transforms to $Tk^{-1}T^{-1}Z$, that is, it transforms by a "local SU(8) transformation" that does not affect the norm of the central charge. The formula (3.8) is, up to a constant multiple, the only formula with these properties, so it is the one that must come from the supergravity or superstring theory.

In supersymmetric theories with central charges, there is an inequality between the mass of a state and the central charge. For elementary string winding states and their partners under U-duality, the inequality is $M \geqslant |Z|$. (More generally, the inequality is roughly that M is equal to or greater than the largest eigenvalue of Z; for a description of stringy black holes with more than one eigenvalue, see Ref. [19]. Elementary string states have only one eigenvalue.)

So far, we have not mentioned the integrality of the gauge charges. Actually, states carrying the 56 gauge charges only populate a lattice $V_{\mathbb{Z}} \subset V$. If U-duality is true, then each lattice point related by U-duality to the gauge charges of an elementary string state represents the charges of a supermultiplet of mass $|Z(\psi)|$.

As an example of the use of this formalism, let us keep a promise made in Section 2 and give an alternative deduction, assuming U-duality, of the important statement that the masses of states carrying RR charges are (in string units) of order $1/\lambda$.[11] Starting from any given vacuum, consider the one-parameter family of vacua determined by the following one-parameter subgroup of $\mathrm{SO}(6,6) \times \mathrm{SL}(2)$: we take the identity in $\mathrm{SO}(6,6)$ (so that the parameters of the toroidal compactification are constant) times

$$g_t = \begin{pmatrix} \mathrm{e}^t & 0 \\ 0 & \mathrm{e}^{-t} \end{pmatrix} \tag{3.10}$$

in SL(2) (so as to vary the string coupling constant). We work here in a basis in which the "top" component is electric and the "bottom" component is magnetic.

Using the mass formula $M(\psi) = |Z(\psi)| = |Tg^{-1}\psi|$, the t dependence of the mass of a state comes entirely from the g action on the state. The NS–NS states, as they are in a doublet of SL(2), have "electric" components whose masses scale as e^{-t} and "magnetic" components with masses of e^t. On the other hand, as the RR states are SL(2) singlets, the mass formula immediately implies that their masses are independent of t.

These are really the masses in the U-dual "Einstein" metric. Making a Weyl transformation to the "string" basis in which the electric NS–NS states (which are elementary string states) have masses of order one, the masses are as follows: electric NS–NS, $M \sim 1$; magnetic NS–NS, $M \sim \mathrm{e}^{2t}$; RR, $M \sim \mathrm{e}^t$. But since we know that the magnetic

[11] The following argument was pointed out in parallel by C. Hull.

NS–NS states (being fairly conventional solitons) have masses of order $1/\lambda^2$, we identify $e^t = 1/\lambda$ (a formula one could also get from the low energy supergravity); hence the RR masses are of order $1/\lambda$ as claimed. [12]

The basic properties described above hold in any dimension above three. (In nine dimensions, some extra care is needed because the U-duality group is not semi-simple.) In three dimensions, new phenomena, which we will not try to unravel, appear because vectors are dual to scalars and charges are confined (for some of the relevant material, see Ref. [25]).

3.4. Analysis of dynamics

We now want to justify the claims made at the beginning of this section about the strong coupling dynamics.

To do this, we will analyze limits of the theory in which some of the BPS-saturated particles go to zero mass. Actually, for each way of going to infinity, we will look only at the particles whose masses goes to zero as fast as possible. We will loosely call these the particles that are massless at infinity.

Also, we really want to find the "maximal" degenerations, which produce maximal sets of such massless particles; a set of massless particles, produced by going to infinity in some direction, is maximal if there would be no way of going to infinity such that those particles would become massless together with others. A degeneration (i.e. a path to infinity) that produces a non-maximal set of massless particles should be understood as a perturbation of a maximal degeneration. (In field theory, such perturbations, which partly lift the degeneracy of the massless particles, are called perturbations by relevant operators.) We will actually also check a few non-maximal degenerations, just to make sure that we understand their physical interpretation.

To justify our claims, we should show that in any dimension d, there are only two maximal degenerations, which correspond to toroidal compactification of weakly coupled ten-dimensional string theory and to toroidal compactification of eleven-dimensional supergravity, respectively. The analysis is in fact very similar in spirit for any d, but the details of the group theory are easier for some values of d than others. I will first explain a very explicit analysis for $d = 7$, chosen as a relatively easy case, and then explain an efficient approach for arbitrary d.

In $d = 7$, the T-duality group is SO(3,3), which is the same as SL(4); U-duality extends this to $G = \mathrm{SL}(5)$. A maximal compact subgroup is $K = \mathrm{SO}(5)$.

In the NS–NS sector, there are six U(1) gauge fields that come from the compactification on a three-torus; they transform as a vector of SO(3,3) or second rank antisymmetric tensor of SL(4). In addition; four more U(1)'s, transforming as a spinor of SO(3,3) or a **4** of SL(4), come from the RR sector. These states combine with the

[12] We made this deduction here in four dimensions, but it could be made, using U-duality, in other dimensions as well. Outside of four dimensions, instead of using the known mass scale of magnetic monopoles to fix the relation between t and λ, one could use the known Weyl transformation (3.2) between the string and U-dual mass scales.

six from the NS–NS sector to make the second rank antisymmetric tensor, the **10** of SL(5).

In Type II supergravity in seven dimensions, the maximal possible R-symmetry is $K = \mathrm{SO}(5)$ or Sp(4). The supercharges make up in fact four pseudo-real spinors Q^i_α, $i = 1, \ldots, 4$, of the seven-dimensional Lorentz group SO(1,6), transforming as the **4** of Sp(4). The central charges transform in the symmetric part of $\mathbf{4} \times \mathbf{4}$, which is the **10** or antisymmetric tensor of SO(5). Thus, we are in a situation similar to what was described earlier in four dimensions: the gauge charges transform as the **10** of SL(5), the central charges transform in the **10** of SO(5), and a choice of vacuum in $G/K = \mathrm{SL}(5)/\mathrm{SO}(5)$ selects an SO(5) subgroup of SL(5), enabling one to identify these representations and map gauge charges to central charges.

A maximal abelian subgroup A of SL(5) is given by the diagonal matrices. A one-parameter subgroup of A consists of matrices of the form

$$g_t = \begin{pmatrix} e^{a_1 t} & 0 & 0 & 0 & 0 \\ 0 & e^{a_2 t} & 0 & 0 & 0 \\ 0 & 0 & e^{a_3 t} & 0 & 0 \\ 0 & 0 & 0 & e^{a_4 t} & 0 \\ 0 & 0 & 0 & 0 & e^{a_5 t} \end{pmatrix}, \tag{3.11}$$

where the a_i are constants, not all zero, with $\sum_i a_i = 0$. We want to consider the behavior of the spectrum as $t \to +\infty$. By a Weyl transformation, we can limit ourselves to the case that

$$a_1 \geqslant a_2 \geqslant \ldots \geqslant a_5. \tag{3.12}$$

Let ψ_{ij}, $i < j$ be a vector in the **10** of SL(5) whose components are zero except for the ij component, which is 1 (and the ji component, which is -1). We will also use the name ψ_{ij} for a particle with those gauge charges. The mass formula $M(\psi) = |Tg^{-1}\psi|$ says that the mass of ψ_{ij} scales with t as

$$M(\psi_{ij}) \sim e^{-t(a_i + a_j)}. \tag{3.13}$$

By virtue of (3.12), the lightest type of particle is ψ_{12}. For generic values of the a_i, this is the unique particle whose mass scales to zero fastest, but if $a_2 = a_3$ then ψ_{12} is degenerate with other particles. To get a maximal set of particles degenerate with ψ_{12}, we need a maximal set of a_i equal to a_2 and a_3. We cannot set all a_i equal (then they have to vanish, as $\sum_i a_i = 0$), so by virtue of (3.12), there are two maximal cases, with $a_1 = a_2 = a_3 = a_4$, or $a_2 = a_3 = a_4 = a_5$. So the maximal degenerations correspond to one-parameter subgroups

$$g_t = \begin{pmatrix} e^{t} & 0 & 0 & 0 & 0 \\ 0 & e^{t} & 0 & 0 & 0 \\ 0 & 0 & e^{t} & 0 & 0 \\ 0 & 0 & 0 & e^{t} & 0 \\ 0 & 0 & 0 & 0 & e^{-4t} \end{pmatrix} \tag{3.14}$$

or

$$g_t = \begin{pmatrix} e^{4t} & 0 & 0 & 0 & 0 \\ 0 & e^{-t} & 0 & 0 & 0 \\ 0 & 0 & e^{-t} & 0 & 0 \\ 0 & 0 & 0 & e^{-t} & 0 \\ 0 & 0 & 0 & 0 & e^{-t} \end{pmatrix} \tag{3.15}$$

with $t \to +\infty$. As we will see, the first corresponds to weakly coupled string theory, and the second to eleven-dimensional supergravity.

In (3.14), the particles whose masses vanish for $t \to +\infty$ are the ψ_{ij} with $1 \leqslant i < j \leqslant 4$. There are six of these, the correct number of light elementary string states of string theory compactified from ten to seven dimensions. Moreover, in (3.14), g_t commutes with a copy of SL(4) that acts on indices $1-2-3-4$. This part of the seven-dimensional symmetry group SL(5) is unbroken by going to infinity in the direction (3.14), and hence would be observed as a symmetry of the low energy physics at "infinity" (though most of the symmetry is spontaneously broken in any given vacuum near infinity). Indeed, SL(4) with six gauge charges in the antisymmetric tensor representation is the correct T-duality group of weakly coupled string theory in seven dimensions.

There is a point here that may be puzzling at first sight. The full subgroup of SL(5) that commutes with g_t is actually not SL(4) but SL(4) $\times\ \mathbb{R}^*$, where $\mathbb{R}^*$ is the one-parameter subgroup containing g_t. What happens to the $\mathbb{R}^*$? When one restricts to the *integral* points in SL(5), which are the true string symmetries, this $\mathbb{R}^*$ does not contribute, so the symmetry group at infinity is just the integral form of SL(4). A similar comment applies at several points below and will not be repeated.

Moving on now to the second case, in (3.15), the particles whose masses vanish for $t \to +\infty$ are the ψ_{1i}, $i > 1$. There are four of these, the correct number for compactification of eleven-dimensional supergravity on a four-torus $\mathbf{T}^4$ whose dimensions are growing with t. The gauge charges of light states are simply the components of the momentum along $\mathbf{T}^4$. The symmetry group at infinity is again SL(4). This SL(4) has a natural interpretation as a group of linear automorphisms of $\mathbf{T}^4$. [13] In fact, the gauge charges carried by the light states in (3.15) transform in the **4** of SL(4), which agrees with the supergravity description as that is how the momentum components along $\mathbf{T}^4$ transform under SL(4). As this SL(4) mixes three of the "original" ten dimensions with the eleventh dimension that is associated with strong coupling, we have our first evidence for the underlying eleven-dimensional Lorentz invariance.

Finally, let us consider a few non-maximal degenerations, to make sure we understand how to interpret them. [14] Degeneration in the direction

[13] That is, if $\mathbf{T}^4$ is understood as the space of real variables y^i, $i = 1, \ldots, 4$, modulo $y^i \to y^i + n^i$, with $n^i \in \mathbb{Z}$, then SL(4) acts by $y^i \to w^i{}_j y^j$. For this to be a diffeomorphism and preserve the orientation, the determinant of w must be one, so one is in SL(4). Given an n-torus $\mathbf{T}^n$, we will subsequently use the phrase "mapping class group" to refer to the SL(n) that acts linearly in this sense on $\mathbf{T}^n$.

[14] We will see in the next section that when the U-duality group has rank r, there are r naturally distinguished one-parameter subgroups. For SL(5), these are (3.14), (3.15), and the two introduced below.

$$g_t = \begin{pmatrix} e^{3t} & 0 & 0 & 0 & 0 \\ 0 & e^{3t} & 0 & 0 & 0 \\ 0 & 0 & e^{-2t} & 0 & 0 \\ 0 & 0 & 0 & e^{-2t} & 0 \\ 0 & 0 & 0 & 0 & e^{-2t} \end{pmatrix} \tag{3.16}$$

leaves as $t \to \infty$ the unique lightest state ψ_{12}. I interpret this as coming from partial decompactification to eight dimensions – taking one circle much larger than the others so that the elementary string states with momentum in that one direction are the lightest. This family has the symmetry group $SL(3) \times SL(2)$, which is indeed the U-duality group in eight dimensions, as it should be.

The family

$$g_t = \begin{pmatrix} e^{2t} & 0 & 0 & 0 & 0 \\ 0 & e^{2t} & 0 & 0 & 0 \\ 0 & 0 & e^{2t} & 0 & 0 \\ 0 & 0 & 0 & e^{-3t} & 0 \\ 0 & 0 & 0 & 0 & e^{-3t} \end{pmatrix} \tag{3.17}$$

gives three massless states ψ_{ij}, $1 \leqslant i < j \leqslant 3$, transforming as $(\mathbf{3}, \mathbf{1})$ of the symmetry group $SL(3) \times SL(2)$. I interpret this as decompactification to the Type IIB theory in ten dimensions – taking all three circles to be very large. The three light charges are the momenta around the three circles; $SL(3)$ is the mapping class group of the large three-torus, and $SL(2)$ is the U-duality group of the Type IIB theory in ten dimensions.

Partially saturated states

I will now justify an assumption made above and also make a further test of the interpretation that we have proposed.

First of all, we identified BPS-saturated elementary string states with charge tensors ψ_{ij} with (in the right basis) only one non-zero entry. Why was this valid?

We may as well consider NS–NS states; then we can restrict ourselves to the T-duality group $SO(3,3)$. The gauge charges transform in the vector representation of $SO(3,3)$. Given such a vector v_a, one can define the quadratic invariant $(v, v) = \sum_{a,b} \eta^{ab} v_a v_b$.

On the other hand, $SO(3,3)$ is the same as $SL(4)$, and v is equivalent to a second rank antisymmetric tensor ψ of $SL(4)$. In terms of ψ, the quadratic invariant is $(\psi, \psi) = \frac{1}{4}\epsilon^{ijkl}\psi_{ij}\psi_{kl}$. By an $SL(4)$ transformation, one can bring ψ to a normal form in which the independent non-zero entries are ψ_{12} and ψ_{34} only. Then

$$(\psi, \psi) = 2\psi_{12}\psi_{34}. \tag{3.18}$$

So the condition that the particle carries only one type of charge, that is, that only ψ_{12} or ψ_{34} is non-zero, is that $(\psi, \psi) = 0$.

Now let us consider the elementary string states. Such a state has in the toroidal directions left- and right-moving momenta p_L and p_R. p_L and p_R together form a vector of $SO(3,3)$, and the quadratic invariant is [8]

$$(p, p) = |p_L|^2 - |p_R|^2. \tag{3.19}$$

BPS-saturated states have no oscillator excitations for left- or right-movers, and the mass shell condition requires that they obey $|p_L|^2 - |p_R|^2 = 0$, that is, that the momentum or charge vector p is light-like. This implies, according to the discussion in the last paragraph, that in the right basis, the charge tensor ψ has only one entry. That is the assumption we made.

Now, however, we can do somewhat better and consider elementary string states of Type II that are BPS-saturated for left-movers only (or equivalently, for right-movers only). Such states are in "middle-sized" supermultiplets, of dimension 2^{12} (as opposed to generic supermultiplets of dimension 2^{16} and BPS-saturated multiplets of dimension 2^8). To achieve BPS saturation for the left-movers only, one puts the left-moving oscillators in their ground state, but one permits right-moving oscillator excitations; as those excitations are arbitrary, one gets an exponential spectrum of these half-saturated states (analogous to the exponential spectrum of BPS-saturated states in the heterotic string [26]). With oscillator excitations for right-movers only, the mass shell condition implies that $|p_L|^2 > |p_R|^2$, and hence the charge vector is not lightlike. The charge tensor ψ therefore in its normal form has both ψ_{12} and ψ_{34} non-zero. For such states, the mass inequality says that the mass is bounded below by the largest eigenvalue of $Tg^{-1}\psi$, with equality for the "middle-sized" multiplets.

With this in mind, let us consider the behavior of such half-saturated states in the various degenerations. In the "stringy" degeneration (3.14), a state with non-zero ψ_{12} and ψ_{34} has a mass of the same order of magnitude as a state with only ψ_{12} non-zero. This is as we would expect from weakly coupled string theory with toroidal radii of order one: the half-saturated states have masses of the same order of magnitude as the BPS-saturated massive modes. To this extent, string excitations show up in the strong coupling analysis.

What about the "eleven-dimensional" degeneration (3.15)? In this case, while the particles with only one type of charge have masses that vanish as e^{-3t} for $t \to \infty$, the particles with two kinds of charge have masses that *grow* as e^{+t}. The only light states that we can see with this formalism in this degeneration are the Kaluza–Klein modes of eleven-dimensional supergravity. There is, for instance, no evidence for membrane excitations; such evidence might well have appeared if a consistent membrane theory with eleven-dimensional supergravity as its low energy limit really does exist.

3.5. Framework for general analysis

It would be tiresome to repeat this analysis "by hand" in other values of the dimension. Instead, I will now [15] explain a bit of group theory that makes the analysis easy. One of the main points is to incorporate the action of the Weyl group. This was done above by choosing $a_1 \geqslant a_2 \geqslant \ldots \geqslant a_5$, but to exploit the analogous condition in E_7, for instance, a little machinery is useful.

---

[15] With some assistance from A. Borel.

In d dimensions, the U-duality group G has rank $r = 11 - d$. Given any one-parameter subgroup F of a maximal abelian subgroup A, one can pick a set of simple positive roots x_i such that the action of F on the x_i is

$$x_i \longrightarrow e^{c_i t} x_i \tag{3.20}$$

with c_i *non-negative*. In this restriction on the c_i, we have used the Weyl action. Conversely, for every set of non-negative c_i (not all zero), there is a one-parameter subgroup F that acts as (3.20).

The gauge charges are in some representation R of G; that is, for each weight in R there is a corresponding gauge charge. [16] Let $\rho = \sum_i e_i x_i$ be the highest weight in R. The e_i are positive integers. A particle whose only gauge charge is the one that corresponds to ρ has a mass that vanishes for $t \to +\infty$ as

$$M_\rho \sim \exp\left(-\sum_i c_i e_i t\right). \tag{3.21}$$

Any other weight in R is of the form $\rho' = \sum_i f_i x_i$, with $f_i \leqslant e_i$. A particle carrying the ρ' charge has mass of order

$$M_{\rho'} \sim \exp\left(-\sum_i c_i f_i t\right). \tag{3.22}$$

Thus $M_{\rho'} \geqslant M_\rho$ – the particle with only charge ρ always goes to zero mass at least as fast as any other – and $M_{\rho'} = M_\rho$ if and only if

$$c_i = 0 \text{ whenever } f_i < e_i. \tag{3.23}$$

Now, our problem is to pick the subgroup F, that is, the c_i, so that a maximal set of $M_{\rho'}$ are equal to M_ρ. If the c_i are all non-zero, then (as the highest weight state is unique) (3.23) implies that $\rho' = \rho$ and only one gauge charge is carried by the lightest particles. The condition in (3.23) becomes less restrictive only when one of the c_i becomes zero, and to get a maximal set of $M_{\rho'}$ degenerate with M_ρ, we must set as many of the c_i as possible to zero. As the c_i may not all vanish, the best we can do is to set $r - 1$ of them to zero. There are therefore precisely r one-parameter subgroups F_i to consider, labeled by which of the c_i is non-zero.

The x_i are labeled by the vertices in the Dynkin diagram of G, so each of the F_i is associated with a particular vertex P_i. Deleting P_i from the Dynkin diagram of G leaves the Dynkin diagram of a rank $r - 1$ subgroup H_i of G. It is the unbroken subgroup when one goes to infinity in the F_i direction.

[16] The particular representations R that actually arise in Type II string theory in $d \geqslant 4$ have the property (unusual among representations of Lie groups) that the non-zero weight spaces are all one-dimensional. It therefore makes sense to label the gauge charges by weights. (These representations are actually "minuscule" – the Weyl group acts transitively on the weights.) $d \leqslant 3$ would have some new features, as already mentioned above.

3.6. *Analysis in $d = 4$*

With this machinery, it is straightforward to analyze the dynamics in each dimension d. As the rank is $r = 11 - d$, there are $11 - d$ distinguished one-parameter subgroups to check. It turns out that one of them corresponds to weakly coupled string theory in d dimensions, one to toroidal compactification of eleven-dimensional supergravity to d dimensions, and the others to partial (or complete) decompactifications. In each case, the symmetry group when one goes to infinity is the expected one: the T-duality group $SO(10-d, 10-d)$ for the string degeneration; the mapping class group $SL(11-d)$ for supergravity; or for partial decompactification to d' dimensions, the product of the mapping class group $SL(d'-d)$ of a $(d'-d)$-torus and the U-duality group in d' dimensions.

I will illustrate all this in $d = 4$, where the U-duality group is E_7. Going to infinity in a direction F_i associated with one of the seven points in the Dynkin diagram leaves as unbroken subgroup H_i one of the following:

(1) $SO(6,6)$: this is the T-duality group for string theory toroidally compactified from ten to four dimensions. This is a maximal degeneration, with (as we will see) 12 massless states transforming in the **12** of $SO(6,6)$.

(2) $SL(7)$: this is associated with eleven-dimensional supergravity compactified to four dimensions on a seven-torus whose mapping class group is $SL(7)$. This is the other maximal degeneration; there are the expected seven massless states in the **7** of $SL(7)$.

(3) E_6: this and the other cases are non-maximal degenerations corresponding to partial decompactification. This case corresponds to partial decompactification to five dimensions by taking one circle to be much larger than the others; there is only one massless state, corresponding to a state with momentum around the large circle. E_6 arises as the U-duality group in five dimensions.

(4) $SL_2 \times SO(5,5)$: this is associated with partial decompactification to six dimensions. There are two light states, corresponding to momenta around the two large circles; they transform as $(\mathbf{2}, \mathbf{1})$ under $SL_2 \times SO(5,5)$. SL_2 acts on the two large circles and $SO(5,5)$ is the U-duality group in six dimensions.

(5) $SL_3 \times SL(5)$: this is associated with partial decompactification to seven dimensions. $SL(3)$ acts on the three large circles (and the three light charges), and $SL(5)$ is the U-duality in seven dimensions.

(6) $SL_4 \times SL(3) \times SL(2)$: this is associated with partial decompactification to eight dimensions. $SL(4)$ acts on the four large circles and light charges, and $SL(3) \times SL(2)$ is U-duality in eight dimensions.

(7) $SL_6 \times SL_2$: this is associated with decompactification to Type IIB in ten dimensions. SL_6 acts on the six large circles and light charges, and $SL(2)$ is the U-duality in ten dimensions.

In what follows, I will just check the assertions about the light spectrum for the first two cases, which are the important ones, and the third, which is representative of the others.

(1) F_1 can be described as follows. E_7 contains a maximal subgroup $SO(6,6) \times$

SL(2). F_1 can be taken as the subgroup of SL(2) consisting of matrices of the form

$$\begin{pmatrix} e^t & 0 \\ 0 & e^{-t} \end{pmatrix}. \tag{3.24}$$

The gauge charges are in the **56** of E_7, which decomposes under L_1 as $(\mathbf{12}, \mathbf{2}) \oplus (\mathbf{32}, \mathbf{1})$. The lightest states come from the part of the $(\mathbf{12}, \mathbf{2})$ that transforms as e^t under (3.24); these are the expected twelve states in the **12** of SO(6, 6).

(2) E_7 contains a maximal subgroup SL(8). F_2 can be taken as the subgroup of SL(8) consisting of group elements $g_t = \mathrm{diag}(e^t, e^t, \ldots, e^t, e^{-7t})$. The **56** of E_7 decomposes as $\mathbf{28} \oplus \mathbf{28}'$ – the antisymmetric tensor plus its dual. The states of highest eigenvalue (namely e^{8t}) are seven states in the **28** transforming in the expected **7** of the unbroken SL(7).

(3) E_7 has a maximal subgroup $E_6 \times \mathbb{R}^*$, and F_3 is just the $\mathbb{R}^*$. The **56** of E_7 decomposes as $\mathbf{27}^1 \oplus \mathbf{27}'^{-1} \oplus \mathbf{1}^3 \oplus \mathbf{1}^{-3}$, where the E_6 representation is shown in boldface and the $\mathbb{R}^*$ charge (with some normalization) by the exponent. Thus in the F_3 degeneration, there is a unique lightest state, the $\mathbf{1}^3$.

The reader can similarly analyze the light spectrum for the other F_i, or the analogous subgroups in $d \neq 4$.

4. Heterotic string dynamics above four dimensions

4.1. A puzzle in five dimensions

S-duality gives an attractive proposal for the strong coupling dynamics of the heterotic string after toroidal compactification to four dimensions: it is equivalent to the same theory at weak coupling. In the remainder of this paper, we will try to guess the behavior above four dimensions. This process will also yield some new insight about S-duality in four dimensions.

Toroidal compactification of the heterotic string from 10 to d dimensions gives $2(10-d)$ vectors that arise from dimensional reduction of the metric and antisymmetric tensor. Some of the elementary string states are electrically charged with respect to these vectors.

Precisely in five dimensions, one more vector arises. This is so because in five dimensions a two-form B_{mn} is dual to a vector A_m, roughly by $\mathrm{d}B = *\mathrm{d}A$. In the elementary string spectrum, there are no particles that are electrically charged with respect to A, roughly because A can be defined (as a vector) only in five dimensions. But it is easy to see where to find such electric charges. Letting H be the field strength of B (including the Chern–Simons terms) the anomaly equation

$$\mathrm{d}H = \mathrm{tr}\, F \wedge F - \mathrm{tr}\, R \wedge R \tag{4.1}$$

(F is the $E_8 \times E_8$ or SO(32) field strength and R the Riemann tensor) implies that the electric current of A is

$$J = *\mathrm{tr}\, F \wedge F - *\mathrm{tr}\, R \wedge R. \tag{4.2}$$

Thus, with $G = \mathrm{d}A$, (4.1) becomes

$$D^m G_{mn} = J_n, \tag{4.3}$$

showing that J_n is the electric current. So the charge density J_0 is the instanton density, and a Yang–Mills instanton, regarded as a soliton in $4+1$ dimensions, is electrically charged with respect to A.

Instantons (and their generalizations to include the supergravity multiplet [27,28]) are invariant under one half of the supersymmetries. One would therefore suspect that quantization of the instanton would give BPS-saturated multiplets, with masses given by the instanton action:

$$M = \frac{16\pi^2|n|}{\lambda^2}. \tag{4.4}$$

Here n is the instanton number or electric charge and λ is the string coupling.

To really prove existence of these multiplets, one would need to understand and quantize the collective coordinates of the stringy instanton. In doing this, one needs to pick a particular vacuum to work in. In the generic toroidal vacuum, the unbroken gauge group is just a product of U(1)'s. Then the instantons, which require a non-abelian structure, tend to shrink to zero size, where stringy effects are strong and the analysis is difficult. Alternatively, one can consider a special vacuum with an unbroken non-abelian group, but this merely adds infrared problems to the stringy problems. The situation is analogous to the study [29] of H-monopoles after toroidal compactification to four dimensions; indeed, the present paper originated with an effort to resolve the problems concerning H-monopoles. (The connection between instantons and H-monopoles is simply that upon compactification of one of the spatial directions on a circle, the instantons become what have been called H-monopoles.)

Despite the difficulty in the collective coordinate analysis, there are two good reasons to believe that BPS-saturated multiplets in this sector do exist. One, already mentioned, is the invariance of the classical solution under half the supersymmetries. The second reason is that if in five dimensions, the electrically charged states had masses bounded strictly above the BPS value in (4.4), the same would be true after compactification on a sufficiently big circle, and then the BPS-saturated H-monopoles required for S-duality could not exist.

Accepting this assumption, we are in a similar situation to that encountered earlier for the Type IIA string in ten dimensions: there is a massless vector, which couples to electric charges whose mass diverges for weak coupling. (The mass is here proportional to $1/\lambda^2$ in contrast to $1/\lambda$ in the other case.) Just as in the previous situation, we have a severe puzzle if we take the formula seriously for strong coupling, when these particles seem to go to zero mass.

If we are willing to take (4.4) seriously for strong coupling, then we have for each integer n a supermultiplet of states of charge n and mass proportional to $|n|$, going to zero mass as $\lambda \to \infty$. It is very hard to interpret such a spectrum in terms of local field theory in five dimensions. But from our previous experience, we know what to do: interpret these states as Kaluza–Klein states on $\mathbb{R}^5 \times \mathbf{S}^1$.

The $\mathbf{S}^1$ here will have to be a "new" circle, not to be confused with the five-torus $\mathbf{T}^5$ in the original toroidal compactification to five dimensions. (For instance, the *T*-duality group $SO(21,5)$ acts on $\mathbf{T}^5$ but not on the new circle.) So altogether, we seem to have eleven dimensions, $\mathbb{R}^5 \times \mathbf{S}^1 \times \mathbf{T}^5$, and hence we seem to be in need of an eleven-dimensional supersymmetric theory.

In Section 2, eleven-dimensional supergravity made a handy appearance at this stage, but here we seem to be in a quandary. There is no obvious way to introduce an eleventh dimension relevant to the heterotic string. Have we reached a dead end?

4.2. The heterotic string in six dimensions

Luckily, there is a conjectured relation between the heterotic string and Type II superstrings [3,4] which has just the right properties to solve our problem (though not by leading us immediately back to eleven dimensions). The conjecture is that the heterotic string toroidally compactified to *six* dimensions is equivalent to the Type IIA superstring compactified to six dimensions on a K3 surface.

The evidence for this conjecture has been that both models have the same supersymmetry and low energy spectrum in six dimensions and the same moduli space of vacua, namely $SO(20,4;\mathbb{Z})\backslash SO(20,4;\mathbb{R})/(SO(20)\times SO(4))$. For the toroidally compactified heterotic string, this structure for the moduli space of vacua is due to Narain [8]; for Type II, the structure was determined locally by Seiberg [30] and globally by Aspinwall and Morrison [31].

In what follows, I will give several new arguments for this "string–string duality" between the heterotic string and Type IIA superstrings:

(1) When one examines more precisely how the low energy effective actions match up, one finds that weak coupling of one theory corresponds to strong coupling of the other theory. This is a necessary condition for the duality to make sense, since we certainly know that the heterotic string for weak coupling is not equivalent to the Type IIA superstring for weak coupling.

(2) Assuming string–string duality in six dimensions, we will be able to resolve the puzzle about the strong coupling dynamics of the heterotic string in five dimensions. The strongly coupled heterotic string on $\mathbb{R}^5$ (times a five-torus whose parameters are kept fixed) is equivalent to a Type IIB superstring on $\mathbb{R}^5 \times \mathbf{S}^1$ (times a K3 whose parameters are kept fixed). The effective six-dimensional Type IIB theory is weakly coupled at its compactification scale, so this is an effective solution of the problem of strong coupling for the heterotic string in five dimensions.

(3) We will also see that – as anticipated by Duff in a more abstract discussion [7] – string–string duality in six dimensions implies *S*-duality of the heterotic string in four dimensions. Thus, all evidence for *S*-duality can be interpreted as evidence for string–string duality, and one gets at least a six-dimensional answer to the question "what higher-dimensional statement leads to *S*-duality in four dimensions?"

(4) The K3 becomes singular whenever the heterotic string gets an enhanced symmetry group; the singularities have an A–D–E classification, just like the enhanced symmetries.

(5) Finally, six-dimensional string–string duality also leads to an attractive picture for heterotic string dynamics in seven dimensions. (Above seven dimensions the analysis would be more complicated.)

I would like to stress that some of these arguments test more than a long distance relation between the heterotic string and strongly coupled Type IIA. For instance, in working out the five-dimensional dynamics via string–string duality, we will be led to a Type IIA theory with a *small* length scale, and to get a semi-classical description will require a T-duality transformation, leading to Type IIB. The validity of the discussion requires that six-dimensional string–string duality should be an exact equivalence, like the $SL(2,\mathbb{Z})$ symmetry for Type IIB in ten dimensions and unlike the relation of Type II to eleven-dimensional supergravity.

4.3. Low energy actions

Let us start by writing a few terms in the low energy effective action of the heterotic string, toroidally compactified to six dimensions. We consider the metric g, dilaton ϕ, and antisymmetric tensor field B, and we let C denote a generic abelian gauge field arising from the toroidal compactification. We are only interested in keeping track of how the various terms scale with ϕ. For the heterotic string, the whole classical action scales as $e^{-2\phi} \sim \lambda^{-2}$, so one has very roughly

$$I = \int d^6x \sqrt{g}\, e^{-2\phi} \left(R + |\nabla\phi|^2 + |dB|^2 + |dC|^2\right). \tag{4.5}$$

On the other hand, consider the Type IIA superstring in six dimensions. The low energy particle content is the same as for the toroidally compactified heterotic string, at least at a generic point in the moduli space of the latter where the unbroken gauge group is abelian. Everything is determined by $N = 4$ supersymmetry except the number of $U(1)$'s in the gauge group and the number of antisymmetric tensor fields; requiring that these match with the heterotic string leads one to use Type IIA rather than Type IIB. So in particular, the low energy theory derived from Type IIA has a dilaton ϕ', a metric g', an antisymmetric tensor field B', and gauge fields C'. [17] Here ϕ', g', and B' come from the NS–NS sector, but C' comes from the RR sector, so as we noted in Section 2, the kinetic energy of ϕ', g', and B' scales with the dilaton just like that in (4.5), but the kinetic energy of C' has no coupling to the dilaton. So we have schematically

$$I' = \int d^6x \sqrt{g'} \left(e^{-2\phi'} \left(R' + |\nabla\phi'|^2 + |dB'|^2\right) + |dC'|^2\right). \tag{4.6}$$

[17] We normalize B' and C' to have standard gauge transformation laws. Their gauge transformations would look different if one scaled the fields by powers of e^{ϕ}. This point was discussed in Section 2.

We need the change of variables that turns (4.5) into (4.6). In (4.5), the same power of e^{ϕ} multiplies R and $|dC'|^2$. We can achieve that result in (4.6) by the change of variables $g' = g''e^{2\phi'}$. Then (4.6) becomes

$$I' = \int d^6x \sqrt{g''} \left(e^{2\phi'} (R'' + |\nabla\phi'|^2) + e^{-2\phi'} |dB'|^2 + e^{2\phi'} |dC'|^2 \right). \tag{4.7}$$

Now the coefficient of the kinetic energy of B' is the opposite of what we want, but this can be reversed by a duality transformation. The field equations of B' say that $d * (e^{-2\phi'} dB') = 0$, so the duality transformation is

$$e^{-2\phi'} dB' = *dB''. \tag{4.8}$$

Then (4.7) becomes

$$I' = \int d^6x \sqrt{g''}\, e^{2\phi'} \left(R'' + |\nabla\phi'|^2 + |dB''|^2 + |dC'|^2 \right). \tag{4.9}$$

This agrees with (4.5) if we identify $\phi = -\phi'$. Putting everything together, the change of variables by which one can identify the low energy limits of the two theories is

$$\begin{aligned} \phi &= -\phi' , \\ g &= e^{2\phi} g' = e^{-2\phi'} g' , \\ dB &= e^{-2\phi'} * dB' , \\ C &= C' . \end{aligned} \tag{4.10}$$

Unprimed and primed variables are fields of the heterotic string and Type IIA, respectively.

In particular, the first equation implies that weak coupling of one theory is equivalent to strong coupling of the other. This makes it possible for the two theories to be equivalent without the equivalence being obvious in perturbation theory.

4.4. Dynamics in five dimensions

Having such a (conjectured) exact statement in six dimensions, one can try to deduce the dynamics below six dimensions. The ability to do this is not automatic because (just as in field theory) the dimensional reduction might lead to new dynamical problems at long distances. But we will see that in this particular case, the string–string duality in six dimensions does determine what happens in five and four dimensions.

We first compactify the heterotic string from ten to six dimensions on a torus (which will be kept fixed and not explicitly mentioned), and then take the six-dimensional world to be $\mathbb{R}^5 \times \mathbf{S}^1_r$, where $\mathbf{S}^1_r$ will denote a circle of radius r. We want to keep r fixed and take $\lambda = e^{\phi}$ to infinity. According to (4.10), the theory in this limit is equivalent to the Type IIA superstring on $\mathbb{R}^5 \times \mathbf{S}^1_{r'}$ times a K3 surface (of fixed moduli), with string coupling and radius λ' and r' given by

$$\lambda' = \lambda^{-1} ,$$

$$r' = \lambda^{-1} r. \tag{4.11}$$

In particular, the coupling λ' goes to zero in the limit for $\lambda \to \infty$. However, the radius r' in the dual theory is also going to zero. The physical interpretation is much clearer if one makes a T-duality transformation, replacing r' by

$$r'' = \frac{1}{r'} = \frac{\lambda}{r}. \tag{4.12}$$

The T-duality transformation also acts on the string coupling constant. This can be worked out most easily by noting that the effective five-dimensional gravitational constant, which is λ^2/r, must be invariant under the T-duality. So under $r' \to 1/r'$, the string coupling λ' is replaced by

$$\lambda'' = \frac{\lambda'}{r'} \tag{4.13}$$

so that

$$\frac{r'}{(\lambda')^2} = \frac{r''}{(\lambda'')^2}. \tag{4.14}$$

Combining this with (4.11), we learn that the heterotic string on $\mathbb{R}^5 \times \mathbf{S}^1_r$ and string coupling λ is equivalent to a Type II superstring with coupling and radius

$$\lambda'' = r^{-1}\,,$$
$$r'' = \frac{\lambda}{r}\,. \tag{4.15}$$

This is actually a Type IIB superstring, since the T-duality transformation turns the Type IIA model that appears in the string–string duality conjecture in six dimensions into a Type IIB superstring.

Eq. (4.15) shows that the string coupling constant of the effective Type IIB theory remains fixed as $\lambda \to \infty$ with fixed r, so the dual theory is not weakly coupled at all length scales. However, (4.15) also shows that $r'' \to \infty$ in this limit, and this means that at the length scale of the compactification, the effective coupling is weak. (The situation is similar to the discussion of the strongly coupled ten-dimensional Type IIA superstring in Section 2.) All we need to assume is that the six-dimensional Type II superstring theory, even with a coupling of order one, is equivalent at long distances to weakly coupled Type II supergravity. If that is so, then when compactified on a very large circle, it can be described at and above the compactification length by the weakly coupled supergravity, which describes the dynamics of the light degrees of freedom.

Moduli space of Type IIB vacua

The following remarks will aim to give a more fundamental explanation of (4.15) and a further check on the discussion.

Consider the compactification of Type IIB superstring theory on $\mathbb{R}^6 \times$ K3. This gives a chiral $N = 4$ supergravity theory in six dimensions, with five self-dual two-forms (that is, two-forms with self-dual field strength) and twenty-one anti-self-dual two-forms (that is, two-forms with anti-self-dual field strength). The moduli space of vacua of the low energy supergravity theory is therefore [32] G/K with $G = SO(21,5)$ and K the maximal subgroup $SO(21) \times SO(5)$.

The coset space G/K has dimension $21 \times 5 = 105$. The interpretation of this number is as follows. There are 80 NS–NS moduli in the conformal field theory on K3 (that, the moduli space of (4,4) conformal field theories on K3 is 80-dimensional). There are 24 zero modes of RR fields on K3. Finally, the expectation value of the dilaton – the string coupling constant – gives one more modulus. In all, one has $80 + 24 + 1 = 105$ states. In particular, the string coupling constant is unified with the others.

It would be in the spirit of U-duality to suppose that the Type IIB theory on $\mathbb{R}^6 \times K3$ has the discrete symmetry group $SO(21,5;\mathbb{Z})$. In fact, that follows from the assumption of $SL(2,\mathbb{Z})$ symmetry of Type IIB in ten dimensions [3] together with the demonstration in Ref. [31] of a discrete symmetry $SO(20,4;\mathbb{Z})$ for (4,4) conformal field theories on K3. For the $SO(20,4;\mathbb{Z})$ and $SL(2,\mathbb{Z})$ do not commute and together generate $SO(21,5;\mathbb{Z})$. The moduli space of Type IIB vacua on $\mathbb{R}^6 \times K3$ is hence

$$\mathcal{N} = SO(21,5;\mathbb{Z})\backslash SO(21,5;\mathbb{R})/(SO(21) \times SO(5)). \tag{4.16}$$

Now consider the Type IIB theory on $\mathbb{R}^5 \times \mathbf{S}^1 \times K3$. One gets one new modulus from the radius of the $\mathbf{S}^1$. No other new moduli appear (the Type IIB theory on $\mathbb{R}^6 \times K3$ has no gauge fields so one does not get additional moduli from Wilson lines). So the moduli space of Type IIB vacua on $\mathbb{R}^5 \times \mathbf{S}^1 \times K3$ is

$$\mathcal{M} = \mathcal{N} \times \mathbb{R}^+, \tag{4.17}$$

where $\mathbb{R}^+$ (the space of positive real numbers) parametrizes the radius of the circle.

What about the heterotic string on $\mathbb{R}^5 \times \mathbf{T}^5$? The T-duality moduli space of the toroidal vacua is precisely $\mathcal{N} = SO(21,5;\mathbb{Z})\backslash SO(21,5;\mathbb{R})/(SO(21) \times SO(5))$. There is one more modulus, the string coupling constant. So the moduli space of heterotic string vacua on $\mathbb{R}^5 \times \mathbf{T}^5$ is once again $\mathcal{M} = \mathcal{N} \times \mathbb{R}^+$. Now the $\mathbb{R}^+$ parametrizes the string coupling constant.

So the moduli space of toroidal heterotic string vacua on $\mathbb{R}^5 \times \mathbf{T}^5$ is the same as the moduli space of Type IIB vacua on $\mathbb{R}^5 \times \mathbf{S}^1 \times K3$, suggesting that these theories may be equivalent. The map between them turns the string coupling constant of the heterotic string into the radius of the circle in the Type IIB description. This is the relation that we have seen in (4.15) (so, in particular, strong coupling of the heterotic string goes to large radius in Type IIB).

To summarize the discussion, we have seen that an attractive conjecture – the equivalence of the heterotic string in six dimensions to a certain Type IIA theory – implies another attractive conjecture – the equivalence of the heterotic string in five dimensions to a certain Type IIB theory. The link from one conjecture to the other depended on a T-duality transformation, giving evidence that these phenomena must be understood in terms of string theory, not just in terms of relations among low energy field theories.

Detailed matching of states

Before leaving this subject, perhaps it would be helpful to be more explicit about how the heterotic and Type II spectra match up in five dimensions.

Compactification of the six-dimensional heterotic string theory on $\mathbb{R}^5 \times \mathbf{S}^1$ generates in the effective five-dimensional theory three U(1) gauge fields that were not present in six dimensions. There is the component g_{m6} of the metric, the component B_{m6} of the antisymmetric tensor field, and the vector A_m that is dual to the spatial components B_{mn} of the antisymmetric tensor field. Each of these couples to charged states: g_{m6} couples to elementary string states with momentum around the circle, B_{m6} to states that wind around the circle, and A_m to states that arise as instantons in four spatial dimensions, invariant under rotations about the compactified circle. The mass of these "instantons" is r/λ^2, with the factor of r coming from integrating over the circle and $1/\lambda^2$ the instanton action in four dimensions. The masses of these three classes of states are hence of order $1/r$, r, and r/λ^2, respectively, if measured with respect to the string metric. To compare to Type II, we should remember (4.10) that a Weyl transformation $g = \lambda^2 g'$ is made in going to the sigma model metric of the Type IIA description. This multiplies masses by a factor of λ, so the masses computed in the heterotic string theory but measured in the string units of Type IIA are

$$\begin{aligned} g_{m6} &: \ \frac{\lambda}{r} , \\ B_{m6} &: \ \lambda r , \\ A_m &: \ \frac{r}{\lambda} . \end{aligned} \tag{4.18}$$

Likewise, compactification of the six-dimensional Type IIA superstring on $\mathbb{R}^5 \times \mathbf{S}^1$ gives rise to three vectors g'_{m6}, B'_{m6}, and A'. The first two couple to elementary string states. The last presumably couples to some sort of soliton, perhaps the classical solution that has been called the symmetric five-brane [28]. Its mass would be of order $r'/(\lambda')^2$ in string units for the same reasons as before. The masses of particles coupling to the three vectors are thus in string units:

$$\begin{aligned} g'_{m6} &: \ \frac{1}{r'} , \\ B'_{m6} &: \ r' , \\ A'_m &: \ \frac{r'}{(\lambda')^2} . \end{aligned} \tag{4.19}$$

Now, (4.18) agrees with (4.19) under the expected transformation $\lambda = 1/\lambda'$, $r = \lambda r'$ provided that one identifies g_{m6} with g'_{m6}; B_{m6} with A'_m; and A_m with B'_{m6}. The interesting point is of course that B_{m6} and A_m switch places. But this was to be expected from the duality transformation $\mathrm{d}B \sim *\mathrm{d}B'$ that enters in comparing the two theories.

So under string–string duality the "instanton," which couples to A_m, is turned into the string winding state, which couples to B'_{m6}, and the string winding state that couples to B_{m6} is turned into a soliton that couples to A'_m.

4.5. Relation to S-duality

Now we would like to use six-dimensional string–string duality to determine the strong coupling dynamics of the heterotic string in four dimensions. Once again, we start with a preliminary toroidal compactification from ten to six dimensions on a fixed torus that will not be mentioned further. Then we take the six-dimensional space to be a product $\mathbb{R}^4 \times \mathbf{T}^2$, with $\mathbf{T}^2$ a two-torus. String–string duality says that this is equivalent to a six-dimensional Type IIA theory on $\mathbb{R}^4 \times \mathbf{T}^2$ (with four extra dimensions in the form of a fixed K3).

One might now hope, as in six and five dimensions, to take the strong coupling limit and get a useful description of strongly coupled four-dimensional heterotic string theory in terms of Type II. This fails for the following reason. In six dimensions, the duality related strong coupling of the heterotic string to weak coupling of Type IIA. In five dimensions, it related weak coupling of the heterotic string to coupling of order one of Type IIB (see (4.15)). Despite the coupling of order one, this was a useful description because the radius of the sixth dimension was large, so (very plausibly) the effective coupling at the compactification scale is small. A similar scaling in four dimensions, however, will show that the strong coupling limit of the heterotic string in four dimensions is related to a strongly coupled four-dimensional Type II superstring theory, and now one has no idea what to expect.

It is remarkable, then, that there is another method to use six-dimensional string–string duality to determine the strong coupling behavior of the heterotic string in four dimensions. This was forecast and explained by Duff [7] without reference to any particular example. The reasoning goes as follows.

Recall (such matters are reviewed in Ref. [33]) that the T-duality group of a two-torus is $SO(2,2)$ which is essentially the same as $SL(2) \times SL(2)$. Here the two $SL(2)$'s are as follows. One of them, sometimes called $SL(2)_U$, acts on the complex structure of the torus. The other, sometimes called $SL(2)_T$, acts on the combination of the area ρ of the torus and a scalar $b = B_{56}$ that arises in compactification of the antisymmetric tensor field B.

In addition to $SL(2)_U$ and $SL(2)_T$, the heterotic string in four dimensions is conjectured to have a symmetry $SL(2)_S$ that acts on the combination of the four-dimensional string coupling constant

$$\lambda_4 = \lambda \rho^{-1/2} \tag{4.20}$$

and a scalar a that is dual to the space-time components B_{mn} $(m, n = 1, \ldots, 4)$. We know that the heterotic string has $SL(2)_U$ and $SL(2)_T$ symmetry; we would like to know if it also has $SL(2)_S$ symmetry. If so, the strong coupling behavior in four dimensions is determined.

Likewise, the six-dimensional Type IIA theory, compactified on $\mathbb{R}^4 \times \mathbf{T}^2$, has $SL(2)_{U'} \times SL(2)_{T'}$ symmetry, and one would like to know if it also has $SL(2)_{S'}$ symmetry. Here $SL(2)_{U'}$ acts on the complex structure of the torus, $SL(2)_{T'}$ acts on the area ρ' and scalar b' derived from B'_{56}, and $\mathrm{SL}(2)_{S'}$ would conjecturally act on the string coupling constant λ'_4 and the scalar a' that is dual to the $\mathbb{R}^4$ components of B'.

If string–string duality is correct, then the metrics in the equivalent heterotic and Type IIA descriptions differ only by a Weyl transformation, which does not change the complex structure of the torus; hence $\mathrm{SL}(2)_U$ can be identified with $\mathrm{SL}(2)_{U'}$. More interesting is what happens to S and T. Because the duality between the heterotic string and Type IIA involves $\mathrm{d}B \sim *\mathrm{d}B'$, it turns a into b' and a' into b. Therefore, it must turn $\mathrm{SL}(2)_S$ into $\mathrm{SL}(2)_{T'}$ and $\mathrm{SL}(2)_{S'}$ into $\mathrm{SL}(2)_T$. Hence the known $\mathrm{SL}(2)_T$ invariance of the heterotic and Type IIA theories implies, if string–string duality is true, that these theories must also have $\mathrm{SL}(2)_S$ invariance!

It is amusing to check other manifestations of the fact that string–string duality exchanges $\mathrm{SL}(2)_S$ and $\mathrm{SL}(2)_T$. For example, the four-dimensional string coupling $\lambda_4 = \lambda\rho^{-1/2} = \lambda/r$ (r is a radius of the torus) turns under string–string duality into $1/r' = (\rho')^{-1/2}$. Likewise $\rho = r^2$ is transformed into $\lambda^2(r')^2 = (r'/\lambda')^2 = 1/(\lambda'_4)^2$. So string–string duality exchanges λ_4 with $\rho^{-1/2}$, as it must in order to exchange $\mathrm{SL}(2)_S$ and $\mathrm{SL}(2)_T$.

Some other models with S-duality

From string–string duality we can not only rederive the familiar S-duality, but attempt to deduce S-duality for new models. For instance, one could consider in the above a particular two-torus $\mathbf{T}^2$ that happens to be invariant under some $\mathrm{SL}(2)_U$ transformations, and take the orbifold with respect to that symmetry group of the six-dimensional heterotic string. This orbifold can be regarded as a different compactification of the six-dimensional model, so string–string duality – if true – can be applied to it, relating the six-dimensional heterotic string on this orbifold (and an additional four-torus) to a Type IIA string on the same orbifold (and an additional K3).

Orbifolding by a subgroup of $\mathrm{SL}(2)_U$ does not disturb $\mathrm{SL}(2)_T$, so the basic structure used above still holds; if six-dimensional string–string duality is valid, then $\mathrm{SL}(2)_S$ of the heterotic string on this particular orbifold follows from $\mathrm{SL}(2)_T$ of Type IIA on the same orbifold, and vice versa. This example is of some interest as – unlike previously known examples of S-duality – it involves vacua in which supersymmetry is completely broken. The S-duality of this and possible related examples might have implications in the low energy field theory limit, perhaps related to phenomena such as those recently uncovered by Seiberg [14].

4.6. Enhanced gauge groups

Perhaps the most striking phenomenon in toroidal compactification of the heterotic string is that at certain points in moduli space an enhanced non-abelian gauge symmetry appears. The enhanced symmetry group is always simply-laced and so a product of A, D, and E groups; in toroidal compactification to six dimensions, one can get any product of A, D, and E groups of total rank $\leqslant 20$.

How can one reproduce this with Type IIA on a K3 surface? [18] It is fairly obvious that one cannot get an enhanced gauge symmetry unless the K3 becomes singular; only then might the field theory analysis showing that the RR charges have mass of order $1/\lambda$ break down.

The only singularities a K3 surface gets are orbifold singularities. (It is possible for the distance scale of the K3 to go to infinity, isotropically or not, but that just makes field theory better.) The orbifold singularities of a K3 surface have an A–D–E classification. Any combination of singularities corresponding to a product of groups with total rank $\leqslant 20$ (actually at the classical level the bound is $\leqslant 19$) can arise.

Whenever the heterotic string on a four-torus gets an enhanced gauge group G, the corresponding K3 gets an orbifold singularity of type G. This assertion must be a key to the still rather surprising and mysterious occurrence of extended gauge groups for Type IIA on K3, so I will attempt to explain it.

The moduli space

$$\mathcal{M} = SO(20,4;\mathbb{Z})\backslash SO(20,4;\mathbb{R})/(SO(20)\times SO(4)) \tag{4.21}$$

of toroidal compactifications of the heterotic string to six dimensions – or K3 compactifications of Type II – can be thought of as follows. Begin with a 24 dimensional real vector space W with a metric of signature $(4,20)$, and containing a self-dual even integral lattice L (necessarily of the same signature). Let V be a four-dimensional subspace of W on which the metric of W is positive definite. Then $\mathcal{M}$ is the space of all such V's, up to automorphisms of L. Each V has a twenty-dimensional orthocomplement $V^{\perp}$ on which the metric is negative definite.

In the heterotic string description, V is the space of charges carried by right-moving string modes, and $V^{\perp}$ is the space of charges carried by left-moving string modes. Generically, neither V nor $V^{\perp}$ contains any non-zero points in L. When $V^{\perp}$ contains such a point P, we get a purely left-moving (antiholomorphic) vertex operator $\mathcal{O}_P$ of dimension $d_P = -(P,P)/2$. (Of course, $(P,P) < 0$ as the metric of W is negative definite on $V^{\perp}$.) d_P is always an integer as the lattice L is even. The gauge symmetry is extended precisely when $V^{\perp}$ contains some P of $d_P = 1$; the corresponding $\mathcal{O}_P$ generate the extended gauge symmetry.

In the K3 description, W is the real cohomology of K3 (including H^0, H^2, and H^4 together [31]). The lattice L is the lattice of integral points. V is the part of the cohomology generated by self-dual harmonic forms. The interpretation is clearest if we restrict to K3's of large volume, where we can use classical geometry. Then H^0 and H^4 split off, and we can take for W the 22-dimensional space H^2, and for V the three-dimensional space of self-dual harmonic two-forms.

Consider a K3 that is developing an orbifold singularity of type G, with r being the rank of G. In the process, a configuration of r two-spheres S_i (with an intersection matrix given by the Dynkin diagram of G) collapses to a point. These two-spheres

[18] This question was very briefly raised in Section 4.3 of Ref. [34] and has also been raised by other physicists.

are holomorphic (in one of the complex structures on the $K3$), and the corresponding cohomology classes $[S_i]$ have length squared -2. As they collapse, the $[S_i]$ become anti-self-dual and thus – in the limit in which the orbifold singularity develops – they lie in $V^\perp$. (In fact, as S_i is holomorphic, the condition for $[S_i]$ to be anti-self-dual is just that it is orthogonal to the Kähler class and so has zero area; thus the $[S_i]$ lie in $V^\perp$ when and only when the orbifold singularity appears and the S_i shrink to zero.) Conversely, the Riemann–Roch theorem can be used to prove that any point in $V^\perp$ of length squared -2 is associated with a collapsed holomorphic two-sphere.

In sum, precisely when an orbifold singularity of type G appears, there is in $V^\perp$ an integral lattice of rank r, generated by points of length squared -2, namely the S_i; the lattice is the weight lattice of G because of the structure of the intersection matrix of the S_i. This is the same condition on $V^\perp$ as the one that leads to extended symmetry group G for the heterotic string. In the K3 description, one U(1) factor in the gauge group is associated with each collapsed two-sphere. These U(1)'s should make up the maximal torus of the extended gauge group.

Despite the happy occurrence of a singularity – and so possible breakdown of field theory – precisely when an extended gauge group should appear, the occurrence of extended gauge symmetry in Type IIA is still rather surprising. It must apparently mean that taking the string coupling to zero (which eliminates the RR charges) does not commute with developing an orbifold singularity (which conjecturally brings them to zero mass), and that conventional orbifold computations in string theory correspond to taking the string coupling to zero first, the opposite of what one might have guessed.

4.7. Dynamics in seven dimensions

The reader might be struck by a lack of unity between the two parts of this paper. In Sections 2 and 3, we related Type II superstrings to eleven-dimensional supergravity. In the present section, we have presented evidence for the conjectured relation of Type II superstrings to heterotic superstrings. If both are valid, should not eleven-dimensional supergravity somehow enter in understanding heterotic string dynamics?

I will now propose a situation in which this seems to be true: the strong coupling limit of the heterotic string in seven dimensions. I will first propose an answer, and then try to deduce it from six-dimensional string–string duality.

The proposed answer is that the strong coupling limit of the heterotic string on $\mathbb{R}^7 \times \mathbf{T}^3$ gives a theory whose low energy behavior is governed by eleven-dimensional supergravity on $\mathbb{R}^7 \times$ K3! The first point in favor of this is that the moduli spaces coincide. The moduli space of vacua of the heterotic string on $\mathbb{R}^7 \times \mathbf{T}^3$ is

$$\mathcal{M} = \mathcal{M}_1 \times \mathbb{R}^+ \tag{4.22}$$

with

$$\mathcal{M}_1 = \mathrm{SO}(19,3;\mathbb{Z})\backslash\mathrm{SO}(19,3;\mathbb{R})/\mathrm{SO}(19) \times \mathrm{SO}(3). \tag{4.23}$$

Here $\mathcal{M}_1$ is the usual Narain moduli space, and $\mathbb{R}^+$ parametrizes the possible values of the string coupling constant. For eleven-dimensional supergravity compactified on $\mathbb{R}^7 \times \mathrm{K3}$, the moduli space of vacua is simply the moduli space of Einstein metrics on K3. This does *not* coincide with the moduli space of (4,4) conformal field theories on K3, because there is no second rank antisymmetric tensor field in eleven-dimensional supergravity. Rather the moduli space of Einstein metrics of volume 1 on K3 is isomorphic to $\mathcal{M}_1 = \mathrm{SO}(19,3;\mathbb{Z})\backslash \mathrm{SO}(19,3;\mathbb{R})/\mathrm{SO}(19) \times \mathrm{SO}(3)$. [19] Allowing the volume to vary gives an extra factor of $\mathbb{R}^+$, so that the moduli space of Einstein metrics on K3 coincides with the moduli space $\mathcal{M}$ of string vacua.

As usual, the next step is to see how the low energy effective theories match up. Relating these two theories only makes sense if large volume of eleven-dimensional supergravity (where perturbation theory is good) corresponds to strong coupling of the heterotic string. We recall that the bosonic fields of eleven-dimensional supergravity are a metric G and three-form A_3 with action

$$I = \frac{1}{2}\int \mathrm{d}^{11}x \sqrt{G}\left(R + |\mathrm{d}A_3|^2\right) + \int A_3 \wedge \mathrm{d}A_3 \wedge \mathrm{d}A_3. \tag{4.24}$$

To reduce on $\mathbb{R}^7 \times \mathrm{K3}$, we take the eleven-dimensional line-element to be $\mathrm{d}s^2 = \widetilde{g}_{mn}\mathrm{d}x^m\mathrm{d}x^n + \mathrm{e}^{2\gamma}h_{\alpha\beta}\mathrm{d}y^\alpha\mathrm{d}y^\beta$, with $m,n = 1,\ldots,7$, $\alpha,\beta = 1,\ldots,4$; here $\widetilde{g}$ is a metric on $\mathbb{R}^7$, h a fixed metric on K3 of volume 1, and e^γ the radius of the K3. The reduction of A_3 on $\mathbb{R}^7 \times \mathrm{K3}$ gives on $\mathbb{R}^7$ a three-form a_3, and 22 one-forms that we will generically call A. The eleven-dimensional Lagrangian becomes very schematically (only keeping track of the scaling with e^γ)

$$\int \mathrm{d}^7x \sqrt{\widetilde{g}}\left(\mathrm{e}^{4\gamma}(\widetilde{R} + |\mathrm{d}\gamma|^2 + |\mathrm{d}a_3|^2) + |\mathrm{d}A|^2\right). \tag{4.25}$$

To match this to the heterotic string in seven dimensions, we write $\widetilde{g} = \mathrm{e}^{-4\gamma}g$, with g the heterotic string metric in seven dimensions. We also make a duality transformation $\mathrm{e}^{6\gamma}\mathrm{d}a_3 = *\mathrm{d}B$, with B the two-form of the heterotic string. Then (4.25) turns into

$$\int \mathrm{d}^7x \sqrt{g}\,\mathrm{e}^{-6\gamma}\left(R + |\mathrm{d}\gamma|^2 + |\mathrm{d}B|^2 + |\mathrm{d}A|^2\right). \tag{4.26}$$

The important point is that the Lagrangian scales with an overall factor of $\mathrm{e}^{-6\gamma}$, similar to the overall factor of $\lambda^{-2} = \mathrm{e}^{-2\phi}$ in the low energy effective action of the heterotic string. Thus, to match eleven-dimensional supergravity on $\mathbb{R}^7 \times \mathrm{K3}$ with the heterotic string in seven dimensions, one takes the radius of the K3 to be

$$\mathrm{e}^\gamma = \mathrm{e}^{\phi/3} = \lambda^{1/3}. \tag{4.27}$$

In particular, as we hoped, for $\lambda \to \infty$, the radius of the K3 goes to infinity, and the eleven-dimensional supergravity theory becomes weakly coupled at the length scale of the light degrees of freedom.

---

[19] This space parametrizes three-dimensional subspaces of positive metric in $H^2(\mathrm{K3},\mathbb{R})$. The subspace corresponding to a given Einstein metric on K3 consists of the part of the cohomology that is self-dual in that metric.

Now, let us try to show that this picture is a consequence of string–string duality in six dimensions. We start with the heterotic string on $\mathbb{R}^6 \times \mathbf{S}^1 \times \mathbf{T}^3$, where $\mathbf{S}^1$ is a circle of radius r_1, and $\mathbf{T}^3$ is a three-torus that will be held fixed throughout the discussion.[20] If λ_7 and λ_6 denote the heterotic string coupling constant in seven and six dimensions, respectively, then

$$\frac{1}{\lambda_6^2} = \frac{r_1}{\lambda_7^2}. \tag{4.28}$$

We want to take r_1 to infinity, keeping λ_7 fixed. That will give a heterotic string in seven dimensions. Then, after taking r_1 to infinity, we consider the behavior for large λ_7, to get a strongly coupled heterotic string in seven dimensions.

The strategy of the analysis is of course to first dualize the theory, to a ten-dimensional Type II theory, and then see what happens to the dual theory when first r_1 and then λ are taken large. Six-dimensional string–string duality says that for fixed r_1 and λ, the heterotic string on $\mathbb{R}^6 \times \mathbf{S}^1 \times \mathbf{T}^3$ is equivalent to a Type IIA superstring on $\mathbb{R}^6 \times$ K3, with the following change of variables. The six-dimensional string coupling constant λ_6' of the Type IIA description is

$$\lambda_6' = \frac{1}{\lambda_6} = \frac{r_1^{1/2}}{\lambda_7}. \tag{4.29}$$

The metrics g and g' of the heterotic and Type IIA descriptions are related by

$$g = \mathrm{e}^{2\phi} g' = \lambda_6^2 g' = \frac{\lambda_7^2}{r_1} g'. \tag{4.30}$$

In addition, the parameters of the K3 depend on r_1 (and the parameters of the $\mathbf{T}^3$, which will be held fixed) in a way that we will now analyze.

There is no unique answer, since we could always apply an $SO(20, 4; \mathbb{Z})$ transformation to the K3. However, there is a particularly simple answer. The heterotic string compactified on $\mathbf{S}^1 \times \mathbf{T}^3$ has 24 abelian gauge fields. As the radius r_1 of the $\mathbf{S}^1$ goes to infinity, the elementary string states carrying the 24 charges behave as follows. There is one type of charge (the momentum around the $\mathbf{S}^1$) such that the lightest states carrying only that charge go to zero mass, with

$$M \sim \frac{1}{r_1}. \tag{4.31}$$

There is a second charge, the winding number around $\mathbf{S}^1$, such that particles carrying that charge have masses that blow up as r_1. Particles carrying only the other 22 charges have fixed masses in the limit.

Any two ways to reproduce this situation with a K3 will be equivalent up to a T-duality transformation. There is a particularly easy way to do this – take a fixed K3 and

[20] Generally, there are also Wilson lines on $\mathbf{T}^3$ breaking the gauge group to a product of U(1)'s; these will be included with the parameters of the $\mathbf{T}^3$ that are kept fixed in the discussion.

scale up the volume V, leaving fixed the "shape." This reproduces the above spectrum with a relation between V and r_1 that we will now determine.

We start with the Type IIA superstring theory in ten dimensions. The bosonic fields include the metric g'_{10}, dilaton ϕ'_{10}, gauge field A, and three-form A_3. The action is schematically

$$\int d^{10}x \sqrt{g'_{10}} \left(e^{-2\phi'_{10}} R'_{10} + |dA|^2 + |dA_3|^2 + \ldots \right). \tag{4.32}$$

Upon compactification on $\mathbb{R}^6 \times \mathrm{K3}$, massless modes coming from A and A_3 are as follows. A gives rise to a six-dimensional vector, which we will call a. A_3 gives rise to 22 vectors – we will call them C_I – and a six-dimensional three-form, which we will call a_3. If V is the volume of the $K3$, the effective action in six dimensions scales schematically as

$$\int d^6x \sqrt{g'} \left(\frac{1}{(\lambda'_6)^2} R' + V|da|^2 + V|da_3|^2 + |dC_I|^2 \right). \tag{4.33}$$

Visible in (4.33) are 23 vectors, namely a and the C_I. However, precisely in six dimensions a three-form is dual to a vector, by $V da_3 = *db$. So we can replace (4.33) with

$$\int d^6x \sqrt{g'} \left(\frac{1}{(\lambda'_6)^2} R' + V|da|^2 + \frac{1}{V}|db|^2 + |dC_I|^2 \right), \tag{4.34}$$

with 24 vectors. As the canonical kinetic energy of a vector is

$$\int d^6x \frac{1}{4e^2_{\mathrm{eff}}} |dA|^2, \tag{4.35}$$

with e_{eff} the effective charge, we see that we have one vector with effective charge of order $V^{-1/2}$, one with effective charge of order $V^{1/2}$, and 22 with effective charges of order one.

According to our discussion in Section 2, the mass of a particle carrying an RR charge is of order $e_{\mathrm{eff}}/\lambda'_6$. So for fixed λ'_6 and $V \to \infty$, one type of particle goes to zero mass, one to infinite mass, and 22 remain fixed – just like the behavior of the heterotic string as $r_1 \to \infty$. The lightest charge-bearing particle has a mass of order

$$M' = \frac{1}{V^{1/2}\lambda'_6}. \tag{4.36}$$

To compare this to the mass (4.31) of the lightest particle in the heterotic string description, we must remember the Weyl transformation (4.30) between the two descriptions. Because of this Weyl transformation, the relation between the two masses should be $M = \lambda_6^{-1} M' = \lambda'_6 M'$. So λ'_6 scales out, and the relation between the two descriptions involves the transformation

$$V = r_1^2. \tag{4.37}$$

The reason that the string coupling constant scales out is that it does not enter the map between the moduli space of heterotic string vacua on a four-torus and $(4,4)$ conformal field theories on K3; the relation (4.37) could have been deduced by studying the description of quantum K3 moduli space in Ref. [31] instead of using low energy supergravity as we have done.

Since we know from (4.37) and (4.29) how the parameters V and λ_6' of the Type IIA description are related to the heterotic string parameters, we can identify the ten-dimensional Type IIA string coupling constant λ_{10}', given by

$$\frac{V}{(\lambda_{10}')^2} = \frac{1}{(\lambda_6')^2}. \tag{4.38}$$

We get

$$\lambda_{10}' = \frac{r_1^{3/2}}{\lambda_7}. \tag{4.39}$$

Thus, for $r_1 \to \infty$, the Type IIA theory is becoming strongly coupled. At the same time, according to (4.37) one has $V \to \infty$, so the Type IIA theory is becoming decompactified.

In Section 2 we proposed a candidate for the strong coupling behavior of Type IIA on $\mathbb{R}^{10}$: it is given by eleven-dimensional supergravity on $\mathbb{R}^{10} \times \mathbf{S}^1$. To be more precise, the relation acted as follows on the massless modes. If the line element of the eleven-dimensional theory is $ds^2 = G_{ij}^{10} dx^i dx^j + r_{11}^2 (dx^{11})^2$, $i, j = 1, \ldots, 10$, with G^{10} a metric on $\mathbb{R}^{10}$ and r_{11} the radius of the circle, then r_{11} is related to the ten-dimensional Type IIA string coupling constant by

$$r_{11} = (\lambda_{10}')^{2/3} = \frac{r_1}{\lambda_7^{2/3}} \tag{4.40}$$

and the Type IIA metric g' is related to G^{10} by

$$g' = (\lambda_{10}')^{2/3} G^{10}. \tag{4.41}$$

As this result holds for any fixed metric g' on $\mathbb{R}^{10}$, it must, physically, hold on any ten-manifold M as long as the dimensions of M are scaled up fast enough compared to the growth of the ten-dimensional string coupling constant. I will assume that with λ_{10}' and V going to infinity as determined above, one is in the regime in which one can use the formulas (4.40), (4.41) that govern the strong coupling behavior on $\mathbb{R}^{10}$.

If this is so, then from (4.41) the volume V_{11} of the K3 using the metric of the eleven-dimensional supergravity is related to the volume V using the string metric of the Type IIA description by

$$V_{11} = (\lambda_{10}')^{-4/3} V = \lambda_7^{4/3} r_1^{-2} V = \lambda_7^{4/3}. \tag{4.42}$$

Now we have the information we need to solve our problem. The heterotic string on $\mathbb{R}^6 \times \mathbf{S}^1 \times \mathbf{T}^3$, with radius r_1 of the $\mathbf{S}^1$ and string coupling constant λ_7, is related to eleven-dimensional supergravity on $\mathbb{R}^6 \times \mathbf{S}^1 \times$ K3, where the radius of the $\mathbf{S}^1$ is given in

(4.40) and the volume of the K3 in (4.42). We are supposed to take the limit $r_1 \to \infty$ and then consider the behavior for large λ_7. The key point is that V_{11} is independent of r_1. This enables us to take the limit as $r_1 \to \infty$; all that happens is that $r_{11} \to \infty$, so the $\mathbb{R}^6 \times \mathbf{S}^1 \times \mathrm{K3}$ on which the supergravity theory is formulated becomes $\mathbb{R}^7 \times \mathrm{K3}$. (Thus we see Lorentz invariance between the "eleventh" dimension which came from strong coupling and six of the "original" dimensions.) The dependence on the heterotic string coupling λ_7 is now easy to understand: it is simply that the volume of the K3 is $V_{11} \sim \lambda_7^{4/3}$. That is of course the behavior of the volume expected from (4.27). So the relation that we have proposed between the heterotic string in seven dimensions and eleven-dimensional supergravity on $\mathbb{R}^7 \times \mathrm{K3}$ fits very nicely with the implications of string–string duality in six dimensions.

5. On heterotic string dynamics above seven dimensions

By now we have learned that the strong coupling dynamics of Type II superstrings is, apparently, tractable in any dimension and that the same appears to be true of the heterotic string in dimension $\leqslant 7$. Can we also understand the dynamics of the heterotic string above seven dimensions?

It might be possible to extend the use of six-dimensional string–string duality above seven dimensions (just as we extended it above six dimensions at the end of the last section). This will require more careful analysis of the K3's and probably more subtle degenerations than we have needed so far.

But is there some dual description of the heterotic string above seven dimensions that would give the dynamics more directly? For instance, can we find a dual of the heterotic string directly in ten dimensions?

Once this question is asked, an obvious speculation presents itself, at least in the case of SO(32). (For the $E_8 \times E_8$ theory in ten dimensions, I have no proposal to make.) There is another ten-dimensional string theory with SO(32) gauge group, namely the Type I superstring. Might they in fact be equivalent? [21]

The low energy effective theories certainly match up; this follows just from the low energy supersymmetry. Moreover, they match up in such a way that strong coupling of one theory would turn into weak coupling of the other. This is an essential point in any possible relation between them, since weak coupling of one is certainly not equivalent to weak coupling of the other. In terms of the metric g, dilaton ϕ, two-form B, and gauge field strength F, the heterotic string effective action in ten dimensions scales with the dilaton like

$$\int \mathrm{d}^{10}x \sqrt{g}\, \mathrm{e}^{-2\phi} \left(R + |\nabla\phi|^2 + F^2 + |\mathrm{d}B|^2\right). \tag{5.1}$$

[21] The SO(32) heterotic string has particles that transform as spinors of SO(32); these are absent in the elementary string spectrum of Type I and would have to arise as some sort of solitons if these two theories are equivalent.

If we transform $g = e^{\phi} g'$ and $\phi = -\phi'$, this scales like

$$\int d^{10}x \sqrt{g'} \left(e^{-2\phi'} \left(R' + |\nabla\phi'|^2 \right) + e^{-\phi'} F^2 + |dB|^2 \right). \tag{5.2}$$

This is the correct scaling behavior for the effective action of the Type I superstring. The gauge kinetic energy scales as $e^{-\phi'}$ instead of $e^{-2\phi'}$ because it comes from the disc instead of the sphere. The B kinetic energy scales trivially with ϕ' in Type I because B is an RR field. The fact that $\phi = -\phi'$ means that strong coupling of one theory is weak coupling of the other, as promised.

Though a necessary condition, this is scarcely strong evidence for a new string–string duality between the heterotic string and Type I. However, given that the heterotic and Type II superstrings and eleven-dimensional supergravity all apparently link up, one would be reluctant to overlook a possibility for Type I to also enter the story.

Let us try to use this hypothetical new duality to determine the dynamics of the heterotic string below ten dimensions. (Below ten dimensions, the SO(32) and $E_8 \times E_8$ heterotic strings are equivalent [9], so the following discussion applies to both.) We formulate the heterotic string, with ten-dimensional string coupling constant λ, on $\mathbb{R}^d \times \mathbf{T}^{10-d}$ with $\mathbf{T}^{10-d}$ a $(10-d)$-torus of radius r. This would be hypothetically equivalent to a toroidally compactified Type I theory with coupling constant $\lambda' = 1/\lambda$ and (in view of the Weyl transformation used to relate the low energy actions) compactification scale $r' = r/\lambda^{1/2}$. Thus, as $\lambda \to \infty$ for fixed r, λ' goes to zero, but r' also goes to zero, making the physical interpretation obscure. It is more helpful to make a T-duality transformation of the Type I theory to one with radius $r'' = 1/r'$. The T-duality transformation has a very unusual effect for Type I superstrings [11], mapping them to a system that is actually somewhat similar to a Type II orbifold; the relation of this unusual orbifold to the system considered in Section 4 merits further study. The T-duality transformation also changes the ten-dimensional string coupling constant to a new one λ'' which obeys

$$\frac{(r')^{10-d}}{(\lambda')^2} = \frac{(r'')^{10-d}}{(\lambda'')^2} \tag{5.3}$$

so that the d-dimensional effective Newton constant is invariant. Thus

$$\lambda'' = \lambda' \left(\frac{r''}{r'} \right)^{(10-d)/2} = \frac{\lambda^{(8-d)/2}}{r^{10-d}}. \tag{5.4}$$

So for $d = 9$, the strong coupling problem would be completely solved: as $\lambda \to \infty$ with fixed r, $\lambda'' \to 0$ (and $r'' \to \infty$, which gives further simplification). For $d = 8$, we have a story similar to what we have already found in $d = 5$ and 7 (and for Type IIA in $d = 10$): though λ'' is of order 1, the fact that $r'' \to \infty$ means that the coupling is weak at the compactification scale, so that one should have a weakly coupled description of the light degrees of freedom. But below $d = 8$, the transformation maps one strong coupling limit to another.

Of course, once we get down to seven dimensions, we have a conjecture about the heterotic string dynamics from the relation to Type II. Perhaps it is just as well that the speculative relation of the heterotic string to Type I does not give a simple answer below eight dimensions. If there were a dimension in which both approaches could be applied, then by comparing them we would get a relation between (say) a weakly coupled Type II string and a weakly coupled Type I string. Such a relation would very likely be false, so the fact that the speculative string–string duality in ten dimensions does not easily determine the strong coupling behavior below $d = 8$ could be taken as a further (weak) hint in its favor.

Acknowledgements

I would like to thank A. Borel, D. Morrison, R. Plesser and N. Seiberg for discussions.

References

[1] A. Sen, Strong–weak coupling duality in four-dimensional string theory, Int. J. Mod. Phys. A 9 (1994) 3707, hepth/9402002.
[2] J.H. Schwarz, Evidence for non-perturbative string symmetries, hep-th/9411178.
[3] C.M. Hull and P.K. Townsend, Unity of superstring dualities, Nucl. Phys. B 438 (1995) 109.
[4] C. Vafa, unpublished.
[5] W. Nahm, Supersymmetries and their representations, Nucl. Phys. B 135 (1978) 149.
[6] E. Cremmer, B. Julia and J. Scherk, Supergravity theory in 11 dimensions, Phys. Lett. B 76 (1978) 409.
[7] M. Duff, Strong/weak coupling duality from the dual string, hep-th/9501030, Nucl. Phys. B, to be published.
[8] K. Narain, New heterotic string theories in uncompactified dimensions < 10, Phys. Lett. B 169 (1986) 41;
K. Narain, M. Samadi and E. Witten, A note on the toroidal compactification of heterotic string theory, Nucl. Phys. B 279 (1987) 369.
[9] P. Ginsparg, On toroidal compactification of heterotic superstrings, Phys. Rev. D 35 (1987) 648.
[10] M. Dine, P. Huet and N. Seiberg, Large and small radius in string theory, Nucl. Phys. B 322 (1989) 301.
[11] J. Dai, R.G. Leigh and J. Polchinski, New connections between string theories, Mod. Phys. Lett. A 4 (1989) 2073.
[12] P. Townsend, The eleven-dimensional supermembrane revisited, hepth-9501068.
[13] M.F. Duff, Duality rotations in string theory, Nucl. Phys. B 335 (1990) 610;
M.F. Duff and J.X. Lu, Duality rotations in membrane theory, Nucl. Phys. B 347 (1990) 394.
[14] N. Seiberg, Electric–magnetic duality in supersymmetric non-abelian gauge theories, hepth/9411149.
[15] I. Bars, First massive level and anomalies in the supermembrane, Nucl. Phys. B 308 (1988) 462.
[16] J.L. Carr, S.J. Gates, Jr. and R.N. Oerter, $D = 10$, $N = 2a$ supergravity in superspace, Phys. Lett. B 189 (1987) 68.
[17] E. Witten and D. Olive, Supersymmetry algebras that include central charges, Phys. Lett. B 78 (1978) 97.
[18] G.W. Gibbons and C.M. Hull, A Bogolmol'ny bound for general relativity and solitons in $N = 2$ supergravity, Phys. Lett. B 109 (1982) 190.
[19] R. Kallosh, A. Linde, T. Ortin, A. Peet and A. Van Proeyen, Supersymmetry as a cosmic censor, Phys. Rev. D 46 (1992) 5278.
[20] S. Shenker, The strength of non-perturbative effects in string theory, *in* Proc. Cargèse Workshop on random surfaces, quantum gravity and strings (1990).
[21] N. Seiberg and E. Witten, Electric–magnetic duality, monopole condensation, and confinement in $N = 2$ supersymmetric Yang–Mills theory, Nucl. Phys. B 426 (1994) 19.

[22] M. Huq and M.A. Namazie, Kaluza–Klein supergravity in ten dimensions, Class. Quant. Grav. 2 (1985) 293.

[23] E. Cremmer and B. Julia, The SO(8) supergravity, Nucl. Phys. B 159 (1979) 141;
B. Julia, Group disintegrations, *in* Superspace and supergravity, ed. M. Rocek and S. Hawking (Cambridge Univ. Press, Cambridge, 1981) p. 331.

[24] A. Salam and E. Sezgin, eds., Supergravity in diverse dimensions (North-Holland, Amsterdam, 1989).

[25] A. Sen, Strong-weak coupling duality in three-dimensional string theory, hepth/9408083, Nucl. Phys. B 434 (1995) 179.

[26] A. Dabholkar and J.A. Harvey, Nonrenormalization of the superstring tension, Phys. Rev. Lett. 63 (1989) 719;
A. Dabholkar, G. Gibbons, J.A. Harvey and F. Ruiz Ruiz, Superstrings and solitons, Nucl. Phys. B 340 (1990) 33.

[27] A. Strominger, Heterotic solitons, Nucl. Phys. B 343 (1990) 167.

[28] C.G. Callan, Jr., J.A. Harvey and A. Strominger, World-sheet approach to heterotic instantons and solitons, Nucl. Phys. B 359 (1991) 611.

[29] J. Gauntlett and J.A. Harvey, *S*-duality and the spectrum of magnetic monopoles in heterotic string theory, hepth/9407111.

[30] N. Seiberg, Observations on the moduli space of superconformal field theories, Nucl. Phys. B 303 (1988) 286.

[31] P. Aspinwall and D. Morrison, String theory on K3 surfaces, DUK-TH-94-68, IASSNS-HEP-94/23.

[32] L. Romans, Self-duality for interacting fields: Covariant field equations for six-dimensional chiral supergravities, Nucl. Phys. B 276 (1986) 71.

[33] A. Giveon, M. Porrati and E. Rabinovici, Target space duality in string theory, Phys. Rep. 244 (1994) 77.

[34] A. Ceresole, R. D'Auria, S. Ferrara and A. Van Proeyen, Duality transformations in supersymmetric Yang–Mills theories coupled to supergravity, hepth-9502072, Nucl. Phys. B, to be published.

Particles, Strings and Cosmology (1996) ed J Bagger, G Domokos, A Falk and S Kovesi-Domokos, pp 271–85
Reprinted with permission by World Scientific Publishing

P-BRANE DEMOCRACY

P.K. TOWNSEND
DAMTP, University of Cambridge, Silver St.
Cambridge, England
E-mail: pkt10@amtp.cam.ac.uk

The ten or eleven dimensional origin of central charges in the N=4 or N=8 supersymmetry algebra in four dimensions is reviewed: while some have a standard Kaluza-Klein interpretation as momenta in compact dimensions, most arise from p-form charges in the higher-dimensional supersymmetry algebra that are carried by p-brane 'solitons'. Although $p = 1$ is singled out by superstring perturbation theory, U-duality of N=8 superstring compactifications implies a complete 'p-brane democracy' of the full non-perturbative theory. An 'optimally democratic' perturbation theory is defined to be one in which the perturbative spectrum includes all particles with zero magnetic charge. Whereas the heterotic string is optimally democratic in this sense, the type II superstrings are not, although the 11-dimensional supermembrane might be.

Soon after the advent of four-dimensional (D=4) supersymmetry, it was pointed out (Haag et al. 1975) that the N-extended supersymmetry algebra admits a central extension with $N(N-1)$ central charges: if Q^i_α, $(i = 1\dots, N)$, are the N Majorana-spinor supersymmetry charges and P_μ the 4-momentum, then

$$\{Q^i_\alpha, Q^j_\beta\} = \delta^{ij}(\gamma^\mu C)_{\alpha\beta}P_\mu + U^{ij}(C)_{\alpha\beta} + V^{ij}(C\gamma_5)_{\alpha\beta}\ , \tag{1}$$

where $U^{ij} = -U^{ji}$ and $V^{ij} = -V^{ji}$ are the central charges and C is the (antisymmetric) charge conjugation matrix. At first, the possibility of central charges was largely ignored. One reason for this is that the initial emphasis was naturally on $N = 1$ supersymmetry, for which there are no central charges. Another reason is that the emphasis was also on massless field theories; since the central charges appearing in (1) have dimension of mass they cannot be carried by any massless particle. Central charges acquired importance only when massive excitations of extended supersymmetric theories came under scrutiny. One way that massive excitations naturally arise is when a gauge group of an otherwise massless gauge theory is spontaneously broken by vacuum expectation values of scalar fields. Consider $N = 4$ super Yang-Mills (YM) theory with gauge group $SU(2)$ spontaneously broken to $U(1)$ by a non-vanishing vacuum expectation value of one of the six (Lie-algebra valued) scalar fields. This can be viewed as an $N = 4$ super-Maxwell theory coupled to a massive $N = 4$ complex vector supermultiplet. Clearly, this massive supermultiplet has maximum spin one, but all massive representations of the 'standard' $N = 4$ supersymmetry algebra contain fields of at least spin 2. The resolution of this puzzle is that the massive supermultiplet is a representation of the centrally-extended supersymmetry algebra, which has short multiplets of maximum spin one. The central charge is the $U(1)$ electric charge. Similar considerations apply to the magnetic monopole solutions of spontaneously broken gauge theories with N=4 supersymmetry (Witten and Olive 1978, Osborn 1979). Fermion zero modes in the presence of a magnetic monopole ensure that the states obtained by semi-classical quantization fall into supermultiplets. Since there cannot be bound states of spin greater than one in any field theory with maximum spin one (Weinberg and Witten 1980), these supermultiplets must have maximum spin one and must therefore carry a central charge, which is in fact the magnetic charge.

The appearance of central charges has usually been seen as a relic of additional compact dimensions. Consider $N = 2$ supersymmetry, for which there can be two central charges; to be specific, consider an N=2 super-YM theory with gauge group $SU(2)$ spontaneously broken to $U(1)$. The electric and magnetic charges associated with the unbroken $U(1)$ group can be interpreted as momenta in two extra dimensions, consistent with the natural interpretation of the $N = 2$ super-YM theory as a dimensionally-reduced six-dimensional super-YM theory. However, this cannot be the whole story because this interpretation of central charges fails when we consider $N > 2$. For example, the $N = 4$ super-YM theory can be obtained by dimensional reduction from ten dimensions. If we consider the momentum in each extra dimension above four as a possible central charge in the four-dimensional $N = 4$ supersymmetry algebra then we find a total of 6 central charges. But the total number of possible central charges is 12, not 6. One's first reaction to this discrepancy might be to suppose that it is due to central charges that are already present in the D=10 supersymmetry algebra, but the full N=1 D=10 super-Poincaré algebra does not admit central charges. It might therefore seem that no D=10 interpretation can be given to 6 of the 12 central charges in the D=4 N=4 supersymmetry algebra. What this argument overlooks is that central charges in D=4 might arise from charges that are *not* central in D=10. It is possible (in various spacetime dimensions) to include p-form charges that are central with respect to the supertranslation algebra but not with respect to the full super-Poincaré algebra (van Holten and Van Proeyen 1982). Together with the momenta in the extra dimensions, these p-form charges provide the higher-dimensional origin of all central charges in N-extended supersymmetry algebras (Abraham and Townsend 1991).

For the case under discussion, the 6 'surplus' central charges have their D=10 origin in a self-dual five-form central charge, Z^+, in the N=1 D=10 supertranslation algebra:

$$\{Q_\alpha, Q_\beta\} = (\mathcal{P}\Gamma^M C)_{\alpha\beta} P_M + (\mathcal{P}\Gamma^{MNPQR} C)_{\alpha\beta} Z^+_{MNPQR} , \tag{2}$$

where $\mathcal{P}$ is a chiral projector. Such p-form charges are excluded by the premises of the theorem of Haag, Lopuszanski and Sohnius, and so are not relevant to representations of supersymmetry by particle states. They are relevant to extended objects however; in the presence of a p-dimensional extended object, or p-brane, the supertranslation algebra aquires a p-form central charge (Azcárraga et al. 1989). We might therefore expect to find a fivebrane solution of D=10 super-YM theory corresponding to the five-form charge in (2) and there is indeed such a solution. It is found by interpreting the YM instanton of four dimensional Euclidean space as an 'extended soliton' in ten dimensional Minkowski space (Townsend 1988). By the inverse construction, which may be interpreted as 'wrapping' the fivebrane soliton around a 5-torus, one obtains an 'instantonic' soliton in five dimensions, i.e. a solution of the self-duality condition for YM fields on R^4. By solving these equations on $R^3 \times S^1$, which amounts to a further S^1 compactification to four spacetime dimensions, we obtain a BPS monopole of the four-dimensional YM/Higgs equations. The magnetic charge it carries is one of six possible types because there are six Higgs fields in the $N = 4$ super-YM multiplet. Which one gets an expectation value depends on which 5-torus is chosen for the first step in the construction, and this can be done in six ways. Thus, six central charges are momenta in the six extra dimensions but six more arise from the five-form charge in ten dimensions. From this interpretation, it is clear that we should expect to get 'surplus' central charges only for spacetime dimension $D \leq 5$ because only in these cases can a fiveform charge yield a scalar charge on dimensional reduction. As a check, consider the N=4 super YM theory in, say, D=7; the relevant supersymmetry algebra has an $SO(3)$ automorphism group, and the central charges in the relevant supersymmetry algebra belong to a **3** of $SO(3)$, i.e. there are three of them, just the number of extra dimensions. In D=5, the relevant supersymmetry algebra has a $USp(4)$ automorphism group and central charges belong to a $\mathbf{5} \oplus \mathbf{1}$ of $USp(4)$, i.e. there are five central charges for the five extra dimensions and one from dimensional reduction of the fiveform charge in ten dimensions. As a final point, observe that since the LHS of (2) is a 16×16 real symmetric matrix the maximal number of algebraically distinct charges that can appear on the RHS is 136, which is precisely the total number of components of P_M and Z^+_{MNPQR}.

Many of the observations made above concerning super YM theories can be generalized to supergravity theories. Certain N=2 and N=4 supergravity theories can be considered as compactifications of D=6 and D=10 supergravity on T^2 and T^6, respectively. The D=10 case is of particular interest because of its close connection with the heterotic string. Consider first D=10 supergravity coupled to a rank r semi-simple D=10 super-YM theory. The generic four dimensional massless field theory resulting from a compactification on T^6 is N=4 supergravity coupled to $(6+r)$ abelian N=4 vector supermultiplets, six of which contain the Kaluza-Klein (KK) gauge fields for $U(1)^6$, the isometry group of T^6. Since the graviton multiplet contains six vector fields, the total gauge symmetry group is $U(1)^{12+r}$ and the corresponding field strengths can be assigned to the irreducible vector representation of $SO(6, 6+r)$, which is a symmetry group of the four-dimensional effective action. The massive excitations in four dimensions discussed above for the pure super-YM theory are still present, since in addition to the massive vector mutiplets arising from the breaking of the rank r gauge group to $U(1)^r$, there are also gravitational analogues of the BPS monopoles (Harvey and Liu 1991, Gibbons et al. 1994, Gibbons and Townsend

1995), which can be viewed (Khuri 1992, Gauntlett et al. 1993) as 'compactifications' of Strominger's fivebrane solution (Strominger 1990) of the D=10 heterotic string. These excitations are sources for the $U(1)^r$ fields. However, there are now new massive excitations of KK origin; the electrically charged excitations arise from the harmonic expansion of the fields on T^6, while the magnetically charged ones are the states obtained by semi-classical quantization of KK monopoles. These serve as sources for the six KK gauge fields. This provides us with massive excitations carrying $2 \times (6+r)$ of the $2 \times (12+r)$ types of electric and magnetic charges. What about massive excitations carrying the remaining 6+6 electric and magnetic charges? There are none in the context of pure KK theory but the electrically charged states, at least, are present in string theory; they are the string winding modes around the six-torus. In D=10 the magnetic dual to a string is a fivebrane, so we should expect to find the corresponding magnetically charged states as compactifications on T^6 of a new D=10 fivebrane of essentially gravitational origin. Such a solution indeed exists. In terms of the string metric, the bosonic action of D=10 N=1 supergravity is

$$S = \int d^{10}x\, e^{-2\phi}[R + 4(\nabla\phi)^2 - \frac{1}{3}H^2] \tag{3}$$

where H is the three-form field strength of the two-form potential that couples to the string, and ϕ is the scalar 'dilaton' field. The field equations have a 'neutral' fivebrane solution (Duff and Lu 1991, Callan et al. 1991) for which the metric is

$$ds^2 = -dt^2 + d\mathbf{y}\cdot d\mathbf{y} + \left[1 + \frac{\mu_5}{\rho^2}\right]\left(d\rho^2 + \rho^2 d\Omega_3^2\right) , \tag{4}$$

where $\mathbf{y}$ are coordinates for $\mathbb{E}^5$, i.e. the fivebrane is aligned with the $\mathbf{y}$-axes, $d\Omega_3^2$ is the $SO(4)$-invariant metric on the three-sphere, and μ_5 is an arbitrary constant (at least in the classical theory; we shall return to this point below). This solution is geodesically complete because $\rho = 0$ is a null hypersurface at infinite affine parameter along any geodesic. Clearly, this solution of D=10 supergravity is also a solution of D=10 supergravity coupled to a D=10 super-YM theory because the YM fields vanish (hence the terminology 'neutral'). It may therefore be considered as an approximate solution of the heterotic string theory, but it is not an exact solution because the effective field theory of the heterotic string includes additional interaction terms involving the Lorentz Chern-Simons three-form. There is an exact solution, the 'symmetric' fivebrane, that takes account of these terms (Callan et al. 1991). Since the metric of the symmetric fivebrane solution is identical to that of the neutral fivebrane, it seems possible that, in the string theory context, the latter should be simply replaced by the former. In any case, the difference will not be of importance in this contribution. The main point for the present discussion is that there is an additional fivebrane in the supergravity context whose 'wrapping modes' are magnetically charged particles in D=4 that are the magnetic duals of the string winding modes.

We have now found massive excitations carrying all 28+28 electric and magnetic charges of the heterotic string (since $r = 22$ in this case). Because of the Dirac quantization condition, only 28 of the 56 possible types of charges can be carried by states in the perturbative spectrum, no matter how we choose the small parameter of perturbation theory. Heterotic string theory is 'optimal' in the sense that states carrying 28 different types of electric charge appear in string perturbation theory, whereas perturbative KK theory has states carrying only 22 electric charges. However, one can try to 'improve' KK theory by taking into account the fact that D=10 supergravity admits the extreme string solution (Dabholkhar et al. 1990) for which the metric is

$$ds^2 = \left[1 + \frac{\mu_1}{\rho^6}\right]^{-1} \left(- dt^2 + d\sigma^2\right) + \left(d\rho^2 + \rho^2 d\Omega_7^2\right) , \tag{5}$$

where σ is one of the space coordinates, i.e. the string is aligned with the σ-axis, $d\Omega_7^2$ is the $SO(8)$-invariant metric on the seven-sphere, and μ_1 is an arbitrary constant proportional to the string tension. Winding modes of this string 'soliton' carry the extra 6 electric charges that are missing from the KK theory. This suggests that we bring the extra 6 charges into perturbation theory by identifying this solitonic string with a fundamental string.

There is a suggestive analogy here with the Skyme model of baryons: in the limit of vanishing quark masses the pions are the massless fields of QCD and the non-linear sigma model their effective action, just as D=10 supergravity can be seen as the effective action for the massless modes of a string theory. The sigma-model action has Skyrmion solutions that carry a topological charge which can be identified as baryon number, and these solutions are *identified* with the baryons of QCD. A potential difficulty in the D=10 supergravity case is that the string solution (5) is not really solitonic because the singularity at $\rho = 0$ is a naked timelike singularity at finite affine parameter. This could of course be taken simply as a further indication that one should introduce the string as a fundamental one and relinquish any attempt to interpret the solution (5) as a soliton. On the other hand, there is a similar difficulty in the pion sigma model: there are no non-singular static solutions carrying baryon number unless an additional, higher-derivative, term is included in the action. Perhaps something similar occurs in D=10 supergravity.

Another potential difficulty in trying to interpret the string soliton (5) as the fundamental string is that the constant μ_1 in this solution is arbitrary. The resolution of this difficulty requires consideration of the quantum theory. First, we observe that the reason that (5) is called 'extreme' is that it saturates a Bogomolnyi-type lower bound on the string tension in terms of the charge $q_e = \oint e^{-2\phi} \star H$, where $\star$ is the Hodge dual of H and the integral is over the seven-sphere at 'transverse spatial infinity'. A similar result holds for the constant μ_5 of the fivebrane solution (4) but with the 'electric' charge q_e replaced by the 'magnetic' charge $q_m = \oint H$, where now the integral is over the three-sphere at transverse spatial infinity. Given these facts, the parameters μ_1 and μ_5 are proportional, with definite constants of proportionality whose precise values are not important here, to q_e and q_m respectively. Second, the existence of a non-singular magnetic dual fivebrane solution implies a quantization of q_e. Specifically, the product $q_e q_m$ is quantized as a consequence of a generalization of the Dirac quantization condition in dimension D to p-branes, and their duals of dimension $\tilde{p} = D - p - 4$ (Nepomechie 1985, Teitelboim 1986). Thus, not only is q_e quantized, but so too is q_m.

One way to establish the above mentioned Bogomolnyi bound on the constants μ_1, μ_5 is to make use of the supersymmetry algebra. In the fivebrane case one can relate the charge q_m to a contraction of the five-form charge in the D=10 algebra (2) with the five-vector formed from the outer-product of the five spatial translation Killing vectors of the fivebrane solution. The Bogomolnyi bound can then be deduced from the supersymmetry algebra by a procedure modeled on the derivation of Witten and Olive for magnetic monopoles in D=4. There might appear to be an asymmetry between strings and fivebranes in this respect because while (2) includes a five-form charge, carried by fivebranes, it does not include the corresponding one-form charge that we might expect to be carried by strings. To include such a charge we must modify the algebra (2) to

$$\{Q_\alpha, Q_\beta\} = (\mathcal{P}\Gamma^M C)_{\alpha\beta}(P_M + T_M) + (\mathcal{P}\Gamma^{MNPQR}C)_{\alpha\beta} Z^+_{MNPQR} \,. \qquad (6)$$

This algebra is isomorphic to the previous one, which is why the absence of the one-form charge T did not show up in the earlier counting exercise; classically, T can be absorbed into the definition of P. But suppose k is an everywhere non-singular spacelike Killing vector of spacetime, with closed orbits of length R; then the eigenvalues of the scalar operators $k \cdot P$ and $k \cdot T$ are multiples of R^{-1} and R, respectively, so in the quantum theory T cannot be absorbed into the definition of P. Moreover, it follows from the form of the Green-Schwarz action for the heterotic superstring that this one-form charge is indeed present in the algbra (Azcárraga et al. 1989, Townsend 1993).

The massive states of the heterotic string carry a total of 28 electric and 28 magnetic charges, as mentioned above, but only 12 linear combinations can appear as central charges in the N=4 D=4 supersymmetry algebra. This fact leads to some interesting consequences, e.g. symmetry enhancement at special vacua (Hull and Townsend 1995b). Here, however, I shall concentrate on theories which have N=8 supergravity as their effective D=4 field theory, e.g. type II superstrings compactified on a six-torus. In this case, the 56 electric or magnetic charges associated with the 28 abelian gauge fields of N=8 supergravity *all* appear as central charges in the N=8 supersymmetry algebra. From the standpoint of KK theory, only those massive states carrying the six electric KK charges appear in perturbation theory. Type II superstring theory improves on this by incorporating into perturbation theory the string winding modes, which carry six more electric charges, but this still leaves 16 electric charges unaccounted for. These 16 charges are those which would, if present in the spectrum, couple to the Ramond-Ramond (R-R) gauge fields of the D=4 type II string. They are absent in perturbation theory, however, because the R-R gauge fields couple to the string through their field strengths only. This has long been recognized as a problematic feature of type II superstrings, e.g. in the determination of the free type II string propagator (Mezincescu et al. 1989).

An alternative way to see that states carrying RR charges must be absent in perturbation theory is to note that the set of 56 electric plus magnetic central charges can be assigned to the irreducible **56** representation of the duality group $E_{7,7}$ (Cremmer and Julia 1978,1979), which becomes the U-duality group $E_7(\mathbf{Z})$ of the string theory (Hull and Townsend 1995a). Now $E_{7,7} \supset Sl(2;\mathbf{R}) \times SO(6,6)$, so the U-Duality group has as a subgroup the product of the S-Duality group $Sl(2;\mathbf{Z})$ and the type II T-Duality group $SO(6,6;\mathbf{Z})$. With respect to $Sl(2;\mathbf{R}) \times SO(6,6)$ the **56** of $E_{7,7}$ decomposes as

$$\mathbf{56} \to (\mathbf{2},\mathbf{12}) \oplus (\mathbf{1},\mathbf{32}) \ . \tag{7}$$

The analogous decomposition of the $(\mathbf{2},\mathbf{28})$ representation of the $S \times T$ duality group $Sl(2;\mathbf{R}) \times SO(6,22)$ of the generic T^6-compactified heterotic string is

$$(\mathbf{2},\mathbf{28}) \to (\mathbf{2},\mathbf{12}) \oplus 16 \times (\mathbf{2},\mathbf{1}) \ , \tag{8}$$

which makes it clear that the $(\mathbf{1},\mathbf{32})$ representation in (7) is that of the 16+16 electric and magnetic RR charges, which are therefore S-Duality inert and transform *irreducibly* under T-Duality. Since T-Duality is perturbative and magnetic monopoles cannot appear in perturbation theory, this means that both electric and magnetic RR charges must be non-perturbative. Moreover, while the complete absence from the spectrum of states carrying RR charges would be consistent with S and T duality, their presence is required by U-duality. In fact, these states are present (Hull and Townsend 1995a). They are p-brane 'wrapping modes' for $p > 1$, which explains their absence in perturbative string theory.

Thus, in contrast to the heterotic string, the type II string is *non-optimal*, in the sense that it does not incorporate into perturbation theory *all* electrically charged states. Just as string theory improves on KK theory in this respect, one wonders whether there is some theory beyond string theory that is optimal in the above sense. As a first step in this direction one can try to 'improve' string theory, as we tried to 'improve' KK theory, by incorporating p-brane solitons that preserve half the D=10 N=2 supersymmetry (no attempt will be made here to consider solitons that break more than half the supersymmetry). In order to preserve half the supersymmetry, a p-brane must carry a p-form central charge in the D=10 N=2 supertranslation algebra, so we shall begin by investigating the possibilities for such p-form charges. Consider first the N=2A D=10 supersymmetry algebra. Allowing for all algebraically inequivalent p-form charges permitted by symmetry, we have

$$\begin{aligned}\{Q_\alpha, Q_\beta\} &= (\Gamma^M C)_{\alpha\beta} P_M + (\Gamma_{11} C)_{\alpha\beta} Z + (\Gamma^M \Gamma_{11} C)_{\alpha\beta} Z_M + (\Gamma^{MN} C)_{\alpha\beta} Z_{MN} \\ &\quad + (\Gamma^{MNPQ} \Gamma_{11} C)_{\alpha\beta} Z_{MNPQ} + (\Gamma^{MNPQR} C)_{\alpha\beta} Z_{MNPQR} . \end{aligned} \tag{9}$$

The supersymmetry charges are 32 component non-chiral D=10 spinors, so the maximum number of components of charges on the RHS is 528, and this maximum is realized by the above algebra since

$$10 + 1 + 10 + 45 + 210 + 252 = 528 . \tag{10}$$

Note that in this case the Γ_{11} matrix distinguishes between the term involving the 10-momentum P and that involving the one-form charge carried by the type IIA superstring. This means that on compactification to D=4 we obtain an additional six electric central charges from this source, relative to the heterotic case. These are balanced by an additional six magnetic charges due to the fact that the five-form charge is no longer self-dual as it was in the heterotic case. Thus, there is now a total of 24 D=4 central charges carried by particles of KK, string, or fivebrane origin. These are the charged particles in the NS-NS sector of the superstring theory.

The remaining 32 D=4 central charges of the D=4 N=8 supersymmetry algebra have their D=10 origin in the zero-form, two-form and four-form charges of the D=10 algebra (9). One might suppose from this fact that these 32 charges would be carried by particles in the (non-perturbative) R-R sector whose D=10 origin is either a D=10 black hole, membrane or fourbrane solution of the IIA supergravity theory, but this is only partly correct. There are indeed R-R p-brane solutions of D=10 IIA supergravity for $p = 0, 2, 4$, in addition to the NS-NS p-brane solutions for $p = 1, 5$, but there is also a R-R p-brane solution for $p = 6$. (Horowitz and Strominger 1991). The IIA p-branes with $p = (0,6)\,;\,(1,5)\,;\,(2,4)$ are the (electric, magnetic) sources for the one-form, two-form and three-form gauge fields, respectively, of type IIA supergravity.

The reason for this mis-match can be traced to the fact that the four-form charge in (9) actually contributes 'twice' to the D=4 central charges because apart from the obvious 15 charges it also contributes an additional one central charge via the component $Z_{\mu\nu\rho\sigma}$, which is equivalent to a scalar in D=4. This scalar can alternatively be viewed as the obvious scalar charge in D=4 associated with a six-form charge in the D=10 algebra. However, a six-form charge is algebraically equivalent to a four-form charge, which explains why it is absent in (9) and why it is not needed to explain the 56 D=4 central charges. It is possible that the situation here for the six-form charge in the IIA algebra is analogous to the one-form charge in the heterotic case in that it may be necessary to include it as a separate charge in the quantum theory even though it is not algebraically independent of the other charges.

One objection that can be made to the association of p-brane solutions of a supergravity theory with p-form charges in the supersymmetry algebra is that the possibility just noted of replacing a four form by a six form in D=10 is of general applicability.

Thus, a p-form charge in a D-dimensional supersymmetry algebra could always be replaced by an algebraically equivalent $(D-p)$-form charge. Note that this is not simply a matter of exchanging one p-brane for its dual because the dual object is a $(D-p-4)$-brane, not a $(D-p)$-brane. As a result of this ambiguity, one cannot deduce from the algebra alone which p-brane solutions will occur as solutions of the supergravity theory; one needs additional information. Fortunately, this information is always available, and it is always the case that the supersymmetry algebra admits a p-form charge whenever the supergravity multiplet contains a $(p+1)$-form potential.

Let us now consider how this type of analysis fares when applied to the type IIB superstring. The IIB supersymmetry algebra has two Majorana-Weyl supercharges, Q^i_α, (i=1,2), of the same chirality and, allowing for p-form central charges, the supertranslation algebra is

$$\begin{aligned}\{Q^i_\alpha, Q^j_\beta\} &= \delta^{ij}(\mathcal{P}\Gamma^M C)_{\alpha\beta} P_M + (\mathcal{P}\Gamma^M C)_{\alpha\beta}\tilde{Z}^{ij}_M + \varepsilon^{ij}(\mathcal{P}\Gamma^{MNP}C)_{\alpha\beta} Z_{MNP} \\ &\quad + \delta^{ij}(\mathcal{P}\Gamma^{MNPQR}C)_{\alpha\beta}(Z^+)_{MNPQR} + (\mathcal{P}\Gamma^{MNPQR}C)_{\alpha\beta}(\tilde{Z}^+)^{ij}_{MNPQR}\,, \end{aligned} \tag{11}$$

where the tilde indicates the tracefree symmetric tensor of SO(2), equivalently a $U(1)$ doublet. The total number of components of all charges on the RHS of (11) is

$$10 + 2\times 10 + 120 + 126 + 2\times 126 = 528\,. \tag{12}$$

Moreover, *all* p-form charges are needed to provide a D=10 interpretation of the 56 central charges of the D=4 N=8 supersymmetry algebra. I emphasize this point because it is not what one might expect given that D=10 IIB supergravity admits p-brane solutions for $p = 1, 3, 5$, with there being two strings and two fivebranes because there are two two-form gauge potentials. Each of these p-brane solutions can be paired with a p-form charge in the supersymmetry algebra, but this leaves one self-dual five-form charge without an associated fivebrane solution. This is because there are *three* (self-dual) five-form charges, not two.

Perhaps even more surprising is that origin of this discrepancy lies in the NS-NS sector and not in the R-R sector. The reason for this has to do with the D=10 interpretation of the magnetic duals to the 6 KK charges, i.e. the magnetic charges carried by the KK monopoles. In type II string theory the KK charges are related by T-duality to the string winding modes. These have their origin in the D=10 NS-NS string for which the magnetic dual is a fivebrane. This fivebrane is associated with a five-form charge, so T-duality implies that the duals of the KK charges also have their D=10 origin in a fiveform charge. This is true for both the type IIA and the type IIB superstrings. In the type IIA case there was apparently only one five-form charge in the supesymmetry algebra but, in distinction to the heterotic case, it was not self dual. Thus, effectively there were two five-form charges: the self-dual one, in common with the heterotic string, and an additional anti-self-dual one. The additional one is *not* associated with a p-brane solution of the D=10 supergravity theory but, instead, is associated with the KK monopoles. From this perspective it is not surprising that there are three, rather than two, self-dual five-form charges in the type IIB supersymmetry algebra.

Having established the potential importance of p-brane solutions of the D=10 supergravity theories, the next step is to determine whether these solutions are non-singular. The NS-NS string is singular but it may be identified with the fundamental string and, as noted earlier, the NS-NS fivebrane solution (4) is geodesically complete, so we need concern ourselves only with the additional R-R p-branes. In the type IIB case these comprise a string, a threebrane and a fivebrane. While the threebrane is non-singular (Gibbons et al. 1995), the R-R string and fivebrane are singular (Townsend 1995a, Hull 1995), and the significance of this is unclear at present. In the type IIA case, the R-R p-branes comprise a zero-brane, i.e. extreme 'black hole', a two-brane, i.e. membrane, a four-brane and a six-brane. Again, all are singular but in this case there is a simple resolution of this difficulty, which we shall explain shortly.

At this point it should be clear that, for either the type IIA or type IIB superstring compactified on T^6, one can account for the existence of states in the spectrum carrying all 56 charges provided account is taken of the wrapping modes of the D=10 p-brane solitons associated with the p-form charges in the D=10 supersymmetry algebra. From the standpoint of perturbative string theory, $p = 1$ is a special value since string theory incorporates $p = 1$ wrapping modes, alias string winding modes, into perturbation theory. However, U-Duality of the non-perturbative D=4 string theory implies that the distinction between $p = 1$ and $p > 1$ is meaningless in the context of the full non-perturbative theory: i.e. *U-Duality implies a complete p-brane 'democracy'*, hence the title of this contribution. It is merely by convention that we continue to refer to this non-perturbative theory as 'string' theory.

Having just said, in effect, that 'all p-branes are equal', perhaps we can nevertheless allow ourselves the luxury of considering, following Orwell's dictum, that some are more equal than others. Specifically, it is convenient to divide the 56 central charges into the electric ones and the magnetic ones. I have emphasized above that the heterotic string is 'optimal' in that it incorporates all electric charges into perturbation theory; we might now say that it is 'as democratic' as a perturbative theory can be. Is there a similarly 'optimally democratic' perturbative theory underlying the type II string theories? The answer is a qualified 'yes', at least in the type IIA case, and it involves consideration of D=11 supergravity, to which we now turn our attention.

D=11 supergravity compactified on T^7 has in common with the type II superstrings compactified on T^6 that the effective D=4 field theory is N=8 supergravity (Cremmer and Julia 1978,1979). From the D=11 standpoint, seven of the 56 central charges can be interpreted as momenta in the extra dimensions, i.e. as KK electric charges, but this still leaves 49 unaccounted for. These remaining 49 charges must have a D=11 origin as p-form charges. Allowing for all possible p-form charges, the D=11 supersymmetry algebra is

$$\{Q_\alpha, Q_\beta\} = (\Gamma^M C)_{\alpha\beta} P_M + (\Gamma^{MN} C)_{\alpha\beta} Z_{MN} + (\Gamma^{MNPQR} C)_{\alpha\beta} Z_{MNPQR} \,. \tag{13}$$

That is, there is a two-form and a five-form charge. The total number of components of all charges on the RHS of (13) is

$$11 + 55 + 462 = 528 \,, \tag{14}$$

which is, algebraically, the maximum possible number. From this algebra one might guess that D=11 supergravity admits p-brane solutions that preserve half the supersymmetry for p=2 and p=5, and this guess is correct (Duff and Stelle 1991, Güven 1992). In the KK theory the only charged massive states are the KK modes carrying 7 of the 28 electric charges and the KK monopoles carrying the corresponding 7 magnetic charges. The remaining 21 electric charges are carried by the wrapping modes of the D=11 twobrane, i.e. membrane, while the corresponding 21 magnetic charges are carried by wrapping modes of the D=11 fivebrane. Note that all magnetic charges have their D=11 origin in the five-form charge; as for the type IIA string the five-form charge accounts not only for the fivebrane charges but also for the charges carried by KK monopoles.

The 11-metric for both the membrane and the fivebrane solution of D=11 supergravity can be written as

$$ds^2 = \left[1 + \frac{\mu_p}{\rho^{(8-p)}}\right]^{-\frac{2}{(p+1)}} (-dt^2 + d\mathbf{y}\cdot d\mathbf{y}) + \left[1 + \frac{\mu_p}{\rho^{(8-p)}}\right]^{\frac{2}{(8-p)}} \left(d\rho^2 + \rho^2 d\Omega^2_{(9-p)}\right) \,, \tag{15}$$

where $\mathbf{y}$ are coordinates of $\mathbf{E}^p$, so the p-brane is aligned with the $\mathbf{y}$ axes, and μ_p is a constant. In both cases there is a singularity at $\rho = 0$ and this was originally interpreted as due to a physical source. However, this singularity is merely a coordinate singularity, and the hypersurface $\rho = 0$ is an event horizon. Since the horizon can be reached and crossed in finite proper time, one might think that the appropriate generalization of the singularity theorems of General Relativity would imply the existence of a singularity behind the horizon. This is indeed the case for p=2, and the Carter-Penrose diagram in this case is rather similar to that of the extreme Reissner-Nordstrom (RN) solution of General Relativity, i.e. a timelike curvature singularity hidden behind an event horizon (Duff et al. 1994). For p=5, however, the the analytic continuation of the exterior metric through the horizon leads to an interior metric that is isometric to the exterior one. The maximally analytic extension of this exterior metric is therefore geodesically complete for p=5 (Gibbons et al. 1995). Thus, the fivebrane is genuinely solitonic while the membrane has a status similar to that of the extreme RN black hole in GR. This disparity suggests that we identify the membrane solution as the fields exterior to a *fundamental* supermembrane. Various reasons in favour of this idea can be found in the literature (Hull and Townsend

1995a, Townsend 1995a 1995b). Another one is that it could allow us to bring into perturbation theory the 21 electric charges that cannot be interpreted as momenta in the extra 7 dimensions. Thus, a fundamental supermembrane theory is 'optimal', in the sense that *all* electric charges appear in perturbation theory. This presupposes, of course, that some sense can be made of supermembrane perturbation theory. Until now, attempts in this direction have been based on the the worldvolume action for a D=11 supermembrane (Bergshoeff et al. 1987,1988), but this approach runs into the difficulty that the spectrum is most likely continuous (de Wit et al 1989), which would preclude an interpretation in terms of particles. We shall return to this point at the conclusion of this contribution.

This is a convenient point to summarize the p-brane solutions of the N=2 D=10 supergravity theories and of D=11 supergravity via the 'type II Branescan' of Table 1. Note that each p-brane has a dual of dimension $\tilde{p} = D - p - 4$, except the type IIB D=10 threebrane which is self-dual (Horowitz and Strominger 1991, Duff and Lu 1991b).

Table 1. **The TYPE II Branscan**

11			2			5	
10A	0	1	2		4	5	6
10B		1+1		3		5+5	

We are now in a position to return, as promised, to the resolution of the problem in type IIA superstring theory that the RR p-brane solutions are singular. Note first that the IIA supergravity can be obtained by dimensional reduction from D=11 supergravity, which was how it was first constructed (Gianni and Pernici 1984, Campbell and West 1984). The Green-Schwarz action for the type IIA superstring (Green and Schwarz 1984) can be obtained (Duff et al. 1987) by double dimensional reduction of the worldvolume action of the D=11 supermembrane. The D=10 extreme string solution is similarly related to the D=11 membrane solution (Duff and Stelle 1991) and this relation allows the singularity of the string solution to be reinterpreted as a mere coordinate singularity at the horizon in D=11 (Duff et al. 1994). As mentioned above, there remains a further singularity behind the horizon. It is not clear what the interpretation of this singularity should be. One can argue that 'clothed' singularities are not inconsistent with the soliton interpretation, as has been argued in the past for extreme RN black holes (Gibbons 1985), or one can argue that the D=11 membrane solution should be interpreted as a fundamental supermembrane.

It should be noted here that there are strong arguments against simply discarding the membrane solution: it is needed for U-duality of the D=4 type II superstring (Hull and Townsend 1995a) and for the symmetry enhancement in K_3-compactified D=11 supergravity (Hull and Townsend 1995b) needed for the proposed equivalence (Witten, 1995) to the T^3 compactified heterotic string.

Let us now turn to the other *p*-brane solutions of the type IIA theory. First, the fourbrane can be interpreted as a double dimensional reduction of the D=11 fivebrane, in which the singularity of the fourbrane becomes a coordinate singularity at the horizon of the fivebrane (Duff et al. 1994). Thus both the D=10A string and fourbrane in Table 1 are derived by double dimensional reduction from the D=11 membrane and fivebrane diagonally above. Second, the D=10 membrane and D=10 fivebrane can each be interpreted as a superposition of the corresponding D=11 solutions, so each of these D=10A solutions in Table 1 has a straightforward interpretation as the dimensional reduction of the D=11 solution directly above it. There is, however, an important distinction between the D=10A membrane and the D=10A fivebrane: in the membrane case it is *necessary* to pass to D=11 to remove the singularity, whereas this is optional for the fivebrane (as expected from the fact that this fivebrane solution must do triple purpose as both the type IIA fivebrane and the heterotic and NS-NS type IIB fivebrane). Third, the type IIA sixbrane solution can be interpreted as a direct analogue in D=10 of the Kaluza-Klein monopole in D=4. Just as the latter becomes non-singular in D=5, so the singular D=10 sixbrane becomes non-singular in D=11 (Townsend 1995a). At this point we may pause to note that all *magnetic* *p*-brane solutions have now been interpreted as *completely non-singular* solutions in D=11.

This leaves the electric D=10A 0-branes, alias extreme black holes. These carry the scalar central charge in the type IIA supersymmetry algebra (9). Since these black hole solutions are extreme they saturate a Bogomolnyi bound. Their mass is therefore a fixed multiple of their electric charge and, because of the existence of the magnetic six-brane dual, this electric charge is quantized. Hence their masses are quantized. Moreover, because they saturate a Bogomolnyi bound the corresponding ground state soliton supermultiplets are short ones of maximum spin 2. These are precisely the features exhibited by the tower of Kaluza-Klein states obtained by compactification of D=11 supergravity on S^1, and it is therefore natural to conjecture that the D=10 type IIA extreme black hole states should be *identified* as the KK states of D=11 supergravity (Townsend 1995a). Alternatively, or perhaps equivalently, one can think of the KK states of S^1 compactified D=11 supergravity as the *effective* description of the black hole states of the D=10 IIA superstring theory (Witten 1995). Another argument for the identification of the extreme black holes with KK states is that the former can be interpreted as parallel plane waves propagating at the speed of light in the compact direction (cf. Gibbons and Perry 1984), so the corresponding quanta can be interpreted as massless particles with momentum in the compact dimension, which is essentially a description of KK modes.

Thus, the type IIA superstring is really an eleven-dimensional theory. From the D=11 standpoint, the string coupling constant g is $g = R^{2/3}$ (Witten 1995), where R is the radius of the 11th dimension, so that weak coupling perturbation theory is a perturbation theory about $R = 0$. This explains why the critical dimension of the perturbative type IIA superstring theory is D=10. The strong coupling limit is associated with the decompactification limit $R \to \infty$ and D=11 supergravity can be interpreted as the effective field theory at strong coupling (Witten 1995). However, the type IIA superstring is really an 11-dimensional theory at *any* non-zero coupling, weak or strong, and the question arises as to whether there is an intrinsically 11-dimensional description of this theory that is not merely an effective one.

The only candidate for such a theory at present is the D=11 supermembrane but, as noted earlier, its quantization via its worldvolume action leads to difficulties. We can now explain why this should have been expected. While both the extreme string solution of D=10 supergravity, in the string metric, and the D=11 supermembrane solution of D=11 supergravity have a timelike singularity, consistent with their interpretation as fundamental extended objects, the two solutions differ in that the string singularity is naked whereas the membrane singularity is 'clothed'. The difference is significant. The fact that the string singularity is naked shows that the string is classically structureless. The worldsheet action is therefore an appropriate starting point for quantization. In contrast, the supermembrane has a finite core due to its horizon. Since the worldvolume action fails to take this classical structure into account it is not an appropriate starting point for quantization. An alternative approach (Townsend 1995a) that could circumvent this criticism would be to quantize the classical membrane solution of D=11 supergravity. This might run into difficulties caused by the singularity, however, in which case some hybrid approach would be required. Clearly, there is much to do before we can be sure whether a quantum 11-dimensional supermembrane makes physical sense.

References

Abraham E R C and Townsend P K, *Nucl. Phys.* **B351** (1991) 313.

de Azcárraga J A, Gauntlett J P, Izquierdo J M and Townsend P K, *Phys. Rev. Lett.* **63** (1989) 2443.

Bergshoeff E, Sezgin E and Townsend P K, *Phys. Lett.* **189B** (1987) 75.

Callan C, Harvey J A and Strominger A, *Nucl. Phys.* **B359** (1991); *ibid* **B367** (1991) 60.

Campbell I C G and West P, *Nucl. Phys.* **B243** (1984) 112.

Cremmer E and Julia B, *Phys. Lett.* **80B** (1978) 48; *Nucl. Phys.* **B159** (1979) 141.

Dabholkar A, Gibbons G W, Harvey J A and Ruiz-Ruiz F, *Nucl. Phys.* **B340** (1990) 33.

de Wit B, Lüscher M and Nicolai H, *Nucl. Phys.* **B320** (1989) 135.
Duff M J, Howe P S, Inami T and Stelle K S, *Phys. Lett.* **191B** (1987) 70.
Duff M J and Stelle K S, *Phys. Lett.* **253B** (1991) 113.
Duff M J and Lu J X, *Nucl. Phys.* **B354** (1991a) 141.
Duff M J and Lu J X, *Phys. Lett.* **273B** (1991b) 409.
Duff M J, Gibbons G W and Townsend P K, *Phys. Lett.* **332B** (1994) 321.
Gauntlett J P, Harvey J A, and Liu J T, *Nucl. Phys.* **B409** (1993) 363.
Gianni F and Pernici M, *Phys. Rev.* **D30** (1984) 325.
Gibbons G W and Perry M, *Nucl. Phys.* **B248** (1984) 629.
Gibbons G W, in *Supersymmetry, supergravity and related topics*, eds. del Aguila F, de Azcárraga J A and Ibañez L E (World Scientific 1985) pp. 123-181.
Gibbons G W, Kastor D, London L A J, Townsend P K and Traschen J, *Nucl. Phys.* **B416** (1994) 850.
Gibbons G W, Horowitz G T and Townsend P K, *Class. Quantum Grav.* **12** (1995) 297.
Gibbons G W and Townsend P K, *Antigravitating BPS monopoles and dyons*, *Phys. Lett.* **B**, *in press* (1995).
Green M B and Schwarz J H, *Phys. Lett.* **136B** (1984) 367.
Güven R, *Phys. Lett.* **276B** (1992) 49.
Haag R, Lopusanski J and Sohnius M, *Nucl. Phys.* **B88** (1975) 257.
Horowitz G T and Strominger A, *Nucl. Phys.* **B360** (1991) 197.
Hull C M, *String-string duality in ten dimensions*, hep-th/9506194 (1995).
Hull C M and Townsend P K, *Nucl. Phys.* **B438**, (1995a) 109.
Hull C M and Townsend P K, *Enhanced gauge symmetries in superstring theories*, *Nucl. Phys.* **B**, *in press* (1995b).
Harvey J A and Liu J T, *Phys. Lett.* **268B** (1991) 40.
Khuri R R, *Nucl. Phys.* **B387** (1992) 315.
Mezincescu L, Nepomechie R I and Townsend P K, *Nucl. Phys.* **B322** (1989) 127.
Nepomechie R I, *Phys. Rev.* **D31** (1985) 1921.
Osborn H, *Phys. Lett.* **83B** (1979) 321.
Strominger A, *Nucl. Phys.* **B343** (1990) 167.
Teitelboim C, *Phys. Lett.* **167B** 1986, 63; *ibid* 69.
Townsend P K, *Phys. Lett.* **202B** (1988) 53.
Townsend P K, in *Black Holes, Membranes, Wormholes and Superstrings* eds. S. Kalara and D. Nanopoulos (World Scientific 1993) pp. 85-96.
Townsend P K, *Phys. Lett.* **350B**, (1995a) 184.
Townsend P K, *Phys. Lett.* **354B**, (1995b) 247.
van Holten J W and Van Proeyen A, J. *Phys. A: Math. Gen.* **15** (1982) 3763.
Weinberg S and Witten E, *Phys. Lett.* **96B** (1980) 59.
Witten E and Olive D, *Phys. Lett.* **78B** (1978) 97.
Witten E, *Nucl. Phys.* **B443** (1995) 85.

Reprinted from Phys. Lett. B **367** (1996) 97–103

The power of M theory

John H. Schwarz

Rutgers University, Piscataway, NJ 08855-0849, USA
and California Institute of Technology, Pasadena, CA 91125, USA [1]

Received 23 October 1995
Editor: M. Dine

Abstract

A proposed duality between type IIB superstring theory on $\mathbb{R}^9 \times S^1$ and a conjectured 11D fundamental theory ("M theory") on $\mathbb{R}^9 \times T^2$ is investigated. Simple heuristic reasoning leads to a consistent picture relating the various p-branes and their tensions in each theory. Identifying the M theory on $\mathbb{R}^{10} \times S^1$ with type IIA superstring theory on $\mathbb{R}^{10}$, in a similar fashion, leads to various relations among the p-branes of the IIA theory.

1. Introduction

Recent results indicate that if one assumes the existence of a fundamental theory in eleven dimensions (let's call it the 'M theory' [2]), this provides a powerful heuristic basis for understanding various results in string theory. For example, type II superstrings can be understood as arising from a supermembrane in eleven dimensions [1] by wrapping one dimension of a toroidal supermembrane on a circle of the spatial geometry [2–5]. Similarly, when the spatial geometry contains a $K3$, one can obtain a heterotic string by wrapping a five-brane with the topology of $K3 \times S^1$ on the $K3$ [6,7]. This provides a very simple heuristic for understanding 'string-string duality' between type IIA and heterotic strings in six dimensions [8–10,6,11,12]. One simply considers the M theory on $\mathbb{R}^6 \times S^1 \times K3$. This obviously contains both type II strings and heterotic strings, arising by the two wrappings just described. Moreover, since the membrane and 5-brane are electric-magnetic duals in 11 dimensions, the two strings are dual in six dimensions, and so it is natural that the strong-coupling expansion of one corresponds to the weak-coupling expansion of the other. The remarkable thing about this kind of reasoning is that it works even though we don't understand how to formulate the M theory as a quantum theory. It is tempting to say that the success of the heuristic arguments that have been given previously, and those that will be given here, suggest that there really is a well-defined quantum M theory even when perturbative analysis is not applicable. The only thing that now appears to be special about strings is the possibility of defining a perturbation expansion. In other respects, all p-branes seem to be more or less equal [13,14].

Recently, I have analyzed heuristic relationships between Type II strings and the M theory [15]. The approach was to compare the 9D spectrum of the M theory on $\mathbb{R}^9 \times T^2$ with the IIB theory on $\mathbb{R}^9 \times S^1$. A nice correspondence was obtained between states arising from the supermembrane of the M theory and the strings of the IIB theory. The purpose of this paper

[1] Permanent address.
[2] This name was suggested by E. Witten.

0370-2693/96/$12.00 © 1996 Elsevier Science B.V. All rights reserved
SSDI 0370-2693(95)01429-2

is to extend the analysis to include higher p-branes of both theories, and to see what can be learned from imposing the natural identifications.

Let us begin by briefly recalling the results obtained in [15]. We compared the M theory compactified on a torus of area A_M in the canonical 11D metric $g^{(M)}$ with the IIB theory compactified on a circle of radius R_B (and circumference $L_B = 2\pi R_B$) in the canonical 10D IIB metric $g^{(B)}$. The canonical IIB metric is the convenient choice, because it is invariant under the $SL(2,\mathbb{R})$ group of IIB supergravity. By matching the 9D spectra of the two models (especially for BPS saturated states), the modular parameter τ of the torus was identified with the modulus $\lambda_0 = \chi_0 + ie^{-\phi_0}$, which is the vev of the complex scalar field of the IIB theory. This identification supports the conjectured nonperturbative $SL(2,\mathbb{Z})$ duality symmetry of the IIB theory. (This was also noted by Aspinwall [16].)

A second result was that the IIB theory has an infinite spectrum of strings, which forms an $SL(2,\mathbb{Z})$ multiplet. The strings, labelled by a pair of relatively prime integers (q_1, q_2), were constructed as solutions of the low-energy 10D IIB supergravity theory using results in Refs. [17,18]. They have an $SL(2,\mathbb{Z})$ covariant spectrum of tensions given by

$$T_{1q}^{(B)} = \Delta_q^{1/2} T_1^{(B)}, \tag{1}$$

where $T_1^{(B)}$ is a constant with dimensions of mass-squared, which defines the scale of the theory, and [3]

$$\Delta_q = e^{\phi_0}(q_1 - q_2\chi_0)^2 + e^{-\phi_0} q_2^2. \tag{2}$$

Note that strings with $q_2 \neq 0$, those carrying RR charge, have tensions that, for small string coupling $g_B = e^{\phi_0}$, scale as $g_B^{-1/2}$. The usual $(1,0)$ string, on the other hand, has $T \sim g_B^{1/2}$. In the string metric, these become g_B^{-1} and 1, respectively.

The mass spectrum of point particles (zero-branes) in nine dimensions obtained from the two different viewpoints were brought into agreement (for BPS saturated states, in particular) by identifying winding modes of the family of type IIB strings on the circle with Kaluza–Klein modes of the torus and by identifying Kaluza-Klein modes of the circle with wrappings of the supermembrane (2-brane) on the torus. [4] The 2-brane of the M theory has a tension (mass per unit area) in the 11D metric denoted $T_2^{(M)}$. If one introduces a parameter β to relate the two metrics ($g^{(B)} = \beta^2 g^{(M)}$), then one finds the following relations [15]

$$(T_1^{(B)} L_B^2)^{-1} = \frac{1}{(2\pi)^2} T_2^{(M)} A_M^{3/2}, \tag{3}$$

$$\beta^2 = A_M^{1/2} T_2^{(M)} / T_1^{(B)}. \tag{4}$$

Both sides of Eq. (3) are dimensionless numbers, which are metric independent, characterizing the size of the compact spaces. Note that, since $T_1^{(B)}$ and $T_2^{(M)}$ are fixed constants, Eq. (3) implies that $L_B \sim A_M^{-3/4}$.

Strings (1-branes) in nine dimensions were also matched. A toroidal 2-brane with one of its cycles wrapped on the spatial two-torus was identified with a type IIB string. When the wrapped cycle of the 2-brane is mapped to the (q_1, q_2) homology class of the spatial torus and taken to have minimal length $L_q = (A_M/\tau_2)^{1/2}|q_2\tau - q_1| = (A_M\Delta_q)^{1/2}$, this gives a spectrum of string tensions in the 11D metric $T_{1q}^{(M)} = L_q T_2^{(M)}$. Converting to the IIB metric by $T_{1q}^{(B)} = \beta^{-2} T_{1q}^{(M)}$ precisely reproduces the previous formula for $T_{1q}^{(B)}$ in Eq. (1), which therefore supports the proposed interpretation.

2. More consequences of M/IIB duality

Having matched 9D point particles (0-branes) and strings (1-branes) obtained from the IIB and M theory pictures, let us now explore what additional information can be obtained by also matching p-branes with $p = 2, 3, 4, 5$ in nine dimensions. [5] It should be emphasized that even though we use extremely simple classical reasoning, it ought to be precise (assuming the existence of an M theory), because we only

[3] Eq. (2) was given incorrectly in the original versions of my previous papers [15]. Also, $T_1^{(B)}$ and $T_2^{(M)}$ were called T and T_{11}, and A_M was called A_{11}. A more systematic notation is now desirable.

[4] The rule that gave sensible results was to allow the membrane to cover the torus any number of times (counting orientation), and to identify all the different ways of doing as equivalent. For other problems (such as Strominger's conifold transitions [19]) a different rule is required. As yet, a single principle that gives the correct rule for all such problems is not known. I am grateful to A. Strominger and A. Sen for correspondence concerning this issue.

[5] For useful background on p-branes see Refs. [20–22].

consider p-branes whose tensions are related to their charges by saturation of a BPS bound. This means that the relations that are obtained should not receive perturbative or non-perturbative quantum corrections. This assumes that the supersymmetry remains unbroken, which is certainly believed to be the case.

We begin with $p = 2$. In the M theory the 2-brane in 9D is the same one as in 11D. In the IIB description it is obtained by wrapping an S^1 factor in the topology of a self-dual 3-brane once around the spatial circle. Denoting the 3-brane tension by $T_3^{(B)}$, its wrapping gives a 2-brane with tension $L_B T_3^{(B)}$. Converting to the 11D metric and identifying the two 2-branes gives the relation

$$T_2^{(M)} = \beta^3 L_B T_3^{(B)}. \tag{5}$$

Using Eqs. (3) and (4) to eliminate L_B and β leaves the relation

$$T_3^{(B)} = \frac{1}{2\pi}\left(T_1^{(B)}\right)^2. \tag{6}$$

The remarkable thing about this result is that it is a relation that pertains entirely to the IIB theory, even though it was deduced from a comparison of the IIB theory and the M theory. It should also be noted that the tension $T_3^{(B)}$ is independent of the string coupling constant, which implies that in the string metric it scales as g_B^{-1}.

Next we consider 3-branes in nine dimensions. The only way they can arise in the M theory is from wrapping a 5-brane of suitable topology (once) on the spatial torus. In the IIB theory the only 3-brane is the one already present in ten dimensions. Identifying the tensions of these two 3-branes gives the relation

$$T_5^{(M)} A_M = \beta^4 T_3^{(B)}. \tag{7}$$

Eliminating β and substituting Eq. (6) gives

$$T_5^{(M)} = \frac{1}{2\pi}\left(T_2^{(M)}\right)^2. \tag{8}$$

This result pertains entirely to the M theory. Section 3 of Ref. [12] analyzed the implication of the Dirac quantization rule [23] for the charges of the 2-brane and 5-brane in the M theory. It was concluded that (in my notation) $\pi T_5^{(M)}/(T_2^{(M)})^2$ should be an integer. The present analysis says that it is 1/2. Indeed, I believe that Eq. (8) corresponds to the minimum product of electric and magnetic charges allowed by the quantization condition. It is amusing that simple classical reasoning leads to a non-trivial quantum result.

Next we compare 4-branes in nine dimensions. The IIB theory has an infinite $SL(2,\mathbb{Z})$ family of 5-branes. These are labeled by a pair of relatively prime integers (q_1, q_2), just as the IIB strings are. The reason is that they carry a pair of magnetic charges that are dual to the pair of electric charges carried by the strings. Let us denote the tensions of these 5-branes in the IIB metric by $T_{5q}^{(B)}$. Wrapping each of them once around the spatial circle gives a family of 4-branes in nine dimensions with tensions $L_B T_{5q}^{(B)}$. In the M theory we can obtain 4-branes in nine dimensions by considering 5-branes with an S^1 factor in their topology and mapping the S^1 to a (q_1, q_2) cycle of the spatial torus. Just as when we wrapped the 2-brane this way, we assume that the cycle is as short as possible, i.e., its length is L_q. Identifying the two families of 4-branes obtained in this way gives the relation

$$L_q T_5^{(M)} = \beta^5 L_B T_{5q}^{(B)}. \tag{9}$$

Substituting the relations [15]

$$L_q = \Delta_q^{1/2} T_1^{(B)} \beta^2 / T_2^{(M)} \tag{10}$$

and

$$L_B \beta^3 = 2\pi T_2^{(M)} / \left(T_1^{(B)}\right)^2 \tag{11}$$

and using Eq. (8) gives

$$T_{5q}^{(B)} = \frac{1}{(2\pi)^2}\Delta_q^{1/2}\left(T_1^{(B)}\right)^3. \tag{12}$$

This relation pertains entirely to the IIB theory. Since 5-brane charges are dual to 1-brane charges, they transform contragrediently under $SL(2,\mathbb{R})$. This means that *in this case* q_1 is a magnetic R-R charge and q_2 is a magnetic NS-NS charge. Thus 5-branes with pure R-R charge have a tension that scales as $g_B^{1/2}$ and ones with any NS-NS charge have tensions that scale as $g_B^{-1/2}$. Converting to the string metric, these give g_B^{-1} and g_B^{-2}, respectively. Of course, g_B^{-2} is the characteristic behavior of ordinary solitons, whereas g_B^{-1} is the remarkable intermediate behavior that is characteristic of all p-branes carrying R-R charge. It is gratifying that these expected properties emerge from matching M theory and IIB theory p-branes.

We have now related all 1-brane, 3-brane, and 5-brane tensions of the IIB theory in ten dimensions, so that they are determined by a single scale. We have also related the 2-brane and 5-brane tensions of the M theory in eleven dimensions, so they are also given by a single scale. The two sets of scales can only be related to one another after compactification, however, as the only meaningful comparison is provided by Eqs. (3) and (4).

All that remains to complete this part of the story, is to compare 5-branes in nine dimensions. Here something a little different happens. As is well-known, compactification on a space K with isometries (such as we are considering), so that the complete manifold is $K \times \mathbb{R}^d$, give rise to massless vectors in d dimensions. Electric charges that couple to these vectors correspond to Kaluza–Klein momenta and are carried by point-like 0-branes. The dual magnetic objects are $(d-4)$-branes. This mechanism therefore contributes "Kaluza–Klein 5-branes" in nine dimensions. However, which 5-branes are the Kaluza–Klein ones depends on whether we consider the M theory or the IIB theory. The original 5-brane of the M theory corresponds to the unique Kaluza–Klein 5-brane of the IIB theory, and the $SL(2,\mathbb{Z})$ family of 5-branes of the IIB theory corresponds to the Kaluza–Klein 5-branes of the M theory. The point is that there are three vector fields in nine dimensions which transform as a singlet and a doublet of the $SL(2,\mathbb{R})$ group. The singlet arises à la Kaluza–Klein in the IIB theory and from the three-form gauge field in the M theory. Similarly, the doublet arises from the doublet of two-form gauge fields in the IIB theory and à la Kaluza–Klein in the M theory.

We can now use the identifications described above to deduce the tensions of Kaluza–Klein 5-branes in nine dimensions. The KK 5-brane of the IIB theory is identified with the fundamental 5-brane of the M theory, which implies that its tension is $T_5^{(B)} = \beta^{-6} T_5^{(M)}$. Combining this with Eq. (11) gives

$$T_5^{(B)} = \frac{1}{(2\pi)^3} L_B^2 \left(T_1^{(B)}\right)^4. \tag{13}$$

Note that this diverges as $L_B \to \infty$, as is expected for a Kaluza–Klein magnetic p-brane. Similarly the $SL(2,\mathbb{Z})$ multiplet of KK 5-branes obtained from the M theory must have tensions that match the 5-branes of the 10D IIB theory. This implies that $T_{5q}^{(M)} = \beta^6 T_{5q}^{(B)}$. Substituting Eqs. (4) and (12) gives

$$T_{5q}^{(M)} = \frac{1}{(2\pi)^2} A_M^{3/2} \left(T_2^{(M)}\right)^3 \Delta_q^{1/2}. \tag{14}$$

This also diverges as $A_M \to \infty$, as is expected. As a final comment, we note that if all tensions are rescaled by a factor of 2π (in other words, equations are rewritten in terms of $\tilde{T} = T/2\pi$), then all the relations we have obtained in Eqs. (3)–(14) have a numerical coefficient of unity.

3. The IIA theory

The analysis given above is easily extended to the IIA theory in ten dimensions. The IIA theory is simply interpreted [4,5] as the M theory on $\mathbb{R}^{10} \times S^1$. Let $L = 2\pi r$ be the circumference of the circle in the 11D metric $g^{(M)}$. The string metric of the IIA theory is given by $g^{(A)} = \exp(2\phi_A/3) g^{(M)}$, where ϕ_A is the dilaton of the IIA theory. The IIA string coupling constant g_A is given by the vev of $\exp\phi_A$. These facts immediately allow us to deduce the tensions $T_p^{(A)}$ of IIA p-branes for $p = 1, 2, 4, 5$. The results are

$$T_1^{(A)} = g_A^{-2/3} L T_2^{(M)}, \tag{15}$$

$$T_2^{(A)} = g_A^{-1} T_2^{(M)}, \tag{16}$$

$$T_4^{(A)} = g_A^{-5/3} L T_5^{(M)}, \tag{17}$$

$$T_5^{(A)} = g_A^{-2} T_5^{(M)}. \tag{18}$$

Since $T_1^{(A)}$ and $T_2^{(M)}$ are constants, Eq. (15) gives the scaling rule $g_A \sim L^{3/2}$ [5,15]. Substituting Eqs. (15) and (8) into Eqs. (17) and (18) gives

$$T_4^{(A)} = \frac{1}{2\pi} g_A^{-1} T_1^{(A)} T_2^{(M)} = \frac{1}{2\pi} T_1^{(A)} T_2^{(A)}, \tag{19}$$

$$T_5^{(A)} = \frac{1}{2\pi} \left(T_2^{(A)}\right)^2. \tag{20}$$

Again we have found the expected scaling behaviors: g_A^{-1} for the 2-brane and 4-brane, which carry R-R charge, and g_A^{-2} for the NS-NS solitonic 5-brane. Combining Eqs. (19) and (20) gives

$$T_2^{(A)} T_4^{(A)} = T_1^{(A)} T_5^{(A)}. \tag{21}$$

This shows that the quantization condition for the corresponding charges is satisfied with the same (minimal) value in each case.

The IIA theory also contains an infinite spectrum of BPS saturated 0-branes (aka 'black holes') and a dual 6-brane, which are of Kaluza–Klein origin like those discussed earlier in nine dimensions. Since the Kaluza–Klein vector field is in the R-R sector, the tensions of these should be proportional to g_A^{-1}, as was demonstrated for the 0-branes in [5].

4. P-branes with $P \geq 7$

The IIB theory has a 7-brane, which carries magnetic χ charge. The way to understand this is that χ transforms under $SL(2,\mathbb{R})$ just like the axion in the 4D $N = 4$ theory. It has a Peccei-Quinn translational symmetry (broken to discrete shifts by quantum effects), which means that it is a 0-form gauge field. As a consequence, the theory can be recast in terms of a dual 8-form potential. Whether or not one does that, the classical supergravity equations have a 7-brane solution, which is covered by the general analysis of [20], though that paper only considered $p \leq 6$. Thus the 7-brane in ten dimensions is analogous to a string in four dimensions. Let us call the tension of the IIB 7-brane $T_7^{(B)}$.

The existence of the 7-brane in the 10D IIB theory suggests that after compactification on a circle, the resulting 9D theory has a 7-brane and a 6-brane. If so, these need to be understood in terms of the M theory. The 6-brane does not raise any new issues, since it is already present in the 10D IIA theory. It does, however, reinforce our confidence in the existence of the IIB 7-brane. A 9D 7-brane, on the other hand, certainly would require something new in the M theory. What could it be? To get a 7-brane after compactification on a torus requires either a 7-brane, an 8-brane, or a 9-brane in the 11D M theory. However, the cases of $p = 7$ and $p = 8$ can be ruled out immediately. They require the existence of a massless vector or scalar particle, respectively, in the 11D spectrum, and neither of these is present. The 9-brane, on the other hand, would couple to a 10-form potential with an 11-form field strength, which does not describe a propagating mode and therefore cannot be so easily excluded. Let us therefore consider the possibility that such a 9-brane with tension $T_9^{(M)}$ really exists and trace through its consequences in the same spirit as the preceding discussions.

First we match the 7-brane obtained by wrapping the hypothetical 9-brane of the M theory on the spatial torus to the 7-brane obtained from the IIB theory. This gives the relation

$$A_M T_9^{(M)} = \beta^8 T_7^{(B)}. \tag{22}$$

Substituting Eq. (4) gives

$$T_7^{(B)} = (A_M)^{-1} \frac{T_1^{(B)} T_9^{(M)}}{\left(T_2^{(M)}\right)^4}. \tag{23}$$

This formula is not consistent with our assumptions. A consistent picture would require $T_7^{(B)}$ to be independent of A_M or L_B, but we have found that $T_7^{(B)} \sim A_M^{-1} \sim L_B^{4/3}$. Also, the 8-brane and 9-brane of the IIA theory implied by a 9-brane in the M theory do not have the expected properties. I'm not certain what to make of all this, but it is tempting to conclude that there is no 9-brane in the M theory. Then, to avoid a paradox for 9D 7-branes, we must argue that they are not actually present. I suspect that the usual methods for obtaining BPS saturated p-branes in $d-1$ dimensions from periodic arrays of them in d dimensions break down for $p = d - 3$, because the fields are not sufficiently controlled at infinity, and therefore there is no 7-brane in nine dimensions. Another reason to be suspicious of a 9D 7-brane is that a $(d-2)$-brane in d dimensions is generically associated with a cosmological term, but straightforward compactification of the IIB theory on a circle does not give one.

In a recent paper [24], Polchinski has argued for the existence of a 9-brane in the 10D IIB theory and an 8-brane in the 10D IIA theory, both of which carry RR charges. (He also did a lot of other interesting things.) It ought to be possible to explore whether the existence of these objects is compatible with the reasoning of this paper, but it is unclear to me what the appropriate rules are for handling such objects.

5. Conclusion

We have shown that by assuming the existence of a quantum 'M theory' in eleven dimensions one can derive a number of non-trivial relations among various perturbative and non-perturbative structures of string theory. Specifically, we have investigated what can be learned from identifying M theory on $\mathbb{R}^9 \times T^2$ with type IIB superstring theory on $\mathbb{R}^9 \times S^1$ and matching (BPS saturated) p-branes in nine dimensions. Similarly, we identified the M theory on $\mathbb{R}^{10} \times S^1$ with type IIA superstring theory on $\mathbb{R}^{10}$ and matched p-branes in ten dimensions. Even though quantum M theory surely has no perturbative definition in 11D Minkowski space, these results make it more plausible that a non-perturbative quantum theory does exist. Of course, this viewpoint has been advocated by others – most notably Duff and Townsend – for many years.

Clearly, it would be interesting to explore other identifications like the ones described here. The natural candidate to consider next, which is expected to work in a relatively straightforward way, is a comparison of the M theory on $\mathbb{R}^7 \times K3$ with the heterotic string theory on $\mathbb{R}^7 \times T^3$. There is a rich variety of p-branes that need to be matched in seven dimensions. In particular, the M theory 5-brane wrapped on the $K3$ surface should be identified with the heterotic string itself.

The M theory on $\mathbb{R}^4 \times S^1 \times K$, where K is a Calabi–Yau space, should be equivalent to the type IIA superstring theory on $\mathbb{R}^4 \times K$. Kachru and Vafa have discussed examples for which there is a good candidate for a dual description based on the heterotic string theory on $\mathbb{R}^4 \times K3 \times T^2$ [25]. A new element, not encountered in the previous examples, is that while there is plausibly a connected moduli space of $N = 2$ models that is probed in this way, only part of it is accessed from the M theory viewpoint and a different (but overlapping) part from the heterotic string theory viewpoint. Perhaps this means that we still need to find a theory that is more fundamental than either the heterotic string theory or the putative M theory.

Acknowledgments

I wish to acknowledge discussions with R. Leigh, N. Seiberg, S. Shenker, L. Susskind, and E. Witten. I also wish to thank the Rutgers string theory group for its hospitality. This work was supported in part by the U.S. Dept. of Energy under Grant No. DE-FG03-92-ER40701.

References

[1] E. Bergshoeff, E. Sezgin and P.K. Townsend, Properties of the Eleven-Dimensional Supermembrane Theory, Phys. Lett. B 189 (1987) 75.

[2] M.J. Duff, P.S. Howe, T. Inami and K.S. Stelle, Superstrings in $D = 10$ from Supermembranes in $D = 11$, Phys. Lett. B 191 (1987) 70.

[3] M.J. Duff and K.S. Stelle, Multimembrane Solutions of $D = 11$ Supergravity, Phys. Lett. B 253 (1991) 113.

[4] P.K. Townsend, The Eleven-Dimensional Supermembrane Revisited, Phys. Lett. B 350 (1995), 184.

[5] E. Witten, String Theory Dynamics in Various Dimensions, Nucl. Phys. B 443 (1995) 85.

[6] J.A. Harvey and A. Strominger, The Heterotic String is a Soliton, hep-th/9504047.

[7] P.K. Townsend, String Supermembrane Duality in Seven Dimensions, Phys. Lett. B 354 (1995) 247.

[8] C. Hull and P. Townsend, Unity of Superstring Dualities, Nucl. Phys. B 438 (1995) 109.

[9] M.J. Duff, Strong/Weak Coupling Duality from the Dual String, Nucl. Phys. B 422 (1995) 47;
M. Duff and R. Khuri, Four-Dimensional String/String Duality, Nucl. Phys. B 411 (1994) 473.

[10] A. Sen, String String Duality Conjecture in Six Dimensions and Charged Sotitonic Strings, hep-th/9504027.

[11] C. Vafa and E. Witten, A One-Loop Test of String Duality, hep-th/9505053.

[12] M.J. Duff, J.T. Liu and R. Minasian, Eleven Dimensional Origin of String/String Duality: a One Loop Test, hep-th/9506126.

[13] P.K. Townsend, P-Brane Democracy, hep-th/9507048.

[14] K. Becker, M. Becker and A. Strominger, Fivebranes, Membranes, and Non-perturbative String Theory, hep-th/9507158.

[15] J.H. Schwarz, An $SL(2,\mathbb{Z})$ Multiplet of Type II Superstrings, hep-th/9508143; Superstring Dualities, hep-th/9509148.

[16] P.S. Aspinwall, Some Relationships Between Dualities in String Theory, hep-th/9508154.

[17] A. Dabholkar, G. Gibbons, J.A. Harvey and F. Ruiz Ruiz, Superstrings and Solitons, Nucl. Phys. B 340 (1990) 33;
A. Dabholkar and J.A. Harvey, Nonrenormalization of the Superstring Tension, Phys. Rev. Lett. 63 (1989) 478.

[18] C. Hull, String–String Duality in Ten Dimensions, hep-th/9506194.

[19] A. Strominger, Massless Black Holes and Conifolds in String Theory, Nucl. Phys. B 451 (1995) 96;
B. Greene, D. Morrison and A. Strominger, Black-Hole Condensation and the Unification of String Vacua, Nucl. Phys. B 451 (1995) 109.

[20] G.T. Horowitz and A. Strominger, Black Strings and p-Branes, Nucl. Phys. B 360 (1991) 197.

[21] P.K. Townsend, Three Lectures on Supersymmetry and Extended Objects, p. 317 in Salamanca 1992, Proceedings: Integrable Systems, Quantum Groups, and Quantum Field Theories.

[22] M.J. Duff, R.R. Khuri and J.X. Lu, String Solitons, Phys. Rep. 259 (1995) 213.

[23] R.I. Nepomechie, Magnetic Monopoles from Antisymmetric Tensor Gauge Fields, Phys. Rev. D 31 (1985) 1921;
C. Teitelboim, Monopoles of Higher Rank, Phys. Lett. B 167 (1986) 69.

[24] J. Polchinski, Dirichlet-Branes and Ramond-Ramond Charges, hep-th/9510017.

[25] S. Kachru and C. Vafa, Exact Results for $N = 2$ Compactifications of Heterotic Strings, Nucl. Phys. B 450 (1995) 69;
S. Ferrara, J. Harvey, A. Strominger and C. Vafa, Second Quantized Mirror Symmetry, hep-th/9505162;
A. Klemm, W. Lerche and P. Mayr, K3 Fibrations and Heterotic Type II String Duality, Phys. Lett. B 357 (1995) 313.

Reprinted from Nucl. Phys. B **460** (1996) 506–24

Heterotic and Type I string dynamics from eleven dimensions

Petr Hořava [a,1], Edward Witten [b,2]

[a] *Joseph Henry Laboratories, Princeton University, Jadwin Hall, Princeton, NJ 08544, USA*
[b] *School of Natural Sciences, Institute for Advanced Study, Olden Lane, Princeton, NJ 08540, USA*

Received 2 November 1995; accepted 29 November 1995

Abstract

We propose that the ten-dimensional $E_8 \times E_8$ heterotic string is related to an eleven-dimensional theory on the orbifold $\mathbb{R}^{10} \times S^1/\mathbb{Z}_2$ in the same way that the Type IIA string in ten dimensions is related to $\mathbb{R}^{10} \times S^1$. This in particular determines the strong coupling behavior of the ten-dimensional $E_8 \times E_8$ theory. It also leads to a plausible scenario whereby duality between $SO(32)$ heterotic and Type I superstrings follows from the classical symmetries of the eleven-dimensional world, just as the $SL(2, \mathbb{Z})$ duality of the ten-dimensional Type IIB theory follows from eleven-dimensional diffeomorphism invariance.

1. Introduction

In the last year, the strong coupling behavior of many supersymmetric string theories (or more exactly of what we now understand to be the one supersymmetric string theory in many of its simplest vacua) has been determined. For instance, the strong coupling behavior of most of the ten-dimensional theories and their toroidal compactifications seems to be under control [1]. A notable exception is the $E_8 \times E_8$ heterotic string theory in ten dimensions; no proposal has yet been made that would determine its low-energy excitations and interactions in the strong coupling regime. One purpose of this paper is to fill this gap.

[1] E-mail: horava@puhep1.princeton.edu.
Research supported in part by NSF grant PHY90-21984.
[2] E-mail: witten@sns.ias.edu.
Research supported in part by NSF grant PHY92-45317.

SSDI 0550-3213(95)00621-4

We also wish to further explore the relation of string theory to eleven dimensions. The strong coupling behavior of the Type IIA theory in ten dimensions has turned out [2,1] to involve eleven-dimensional supergravity on $\mathbb{R}^{10} \times S^1$, where the radius of the S^1 grows with the string coupling. An eleven-dimensional interpretation of string theory has had other applications, some of them explained in [3–5]. The most ambitious interpretation of these facts is to suppose that there really is a yet-unknown eleven-dimensional quantum theory that underlies many aspects of string theory, and we will formulate this paper as an exploration of that theory. (But our arguments, like some of the others that have been given, could be compatible with interpreting the eleven-dimensional world as a limiting description of the low-energy excitations for strong coupling, a view taken in [1].) As it has been proposed that the eleven-dimensional theory is a supermembrane theory but there are some reasons to doubt that interpretation,[3] we will non-committally call it the M-theory, leaving to the future the relation of M to membranes.

Our approach to learning more about the M-theory is to consider its behavior on a certain eleven-dimensional orbifold $\mathbb{R}^{10} \times S^1/\mathbb{Z}_2$. In the process, beyond making a proposal for how the $E_8 \times E_8$ heterotic string is related to the M-theory, we will make a proposal for relating the classical symmetries of the M-theory to the conjectured heterotic - Type I string duality in ten dimensions [1,7–9], much as the classical symmetries of the M-theory have been related to Type II duality symmetries [1,5,10]. These proposals suggest a common eleven-dimensional origin of all ten-dimensional string theories and their dualities.

2. The *M*-theory on an orbifold

The M-theory has for its low-energy limit eleven-dimensional supergravity. On an eleven-manifold, with signature $- + + \ldots +$, we introduce gamma matrices Γ^I, $I = 1, \ldots, 11$, obeying $\{\Gamma_I, \Gamma_J\} = 2\eta_{IJ}$ and (in an oriented orthonormal basis)

$$\Gamma^1\Gamma^2 \cdots \Gamma^{11} = 1. \tag{2.1}$$

We will assume that the M-theory has enough in common with what we know of string theory that it makes sense on a wide class of orbifolds – but possibly, like string theory, with extra massless modes arising at fixed points. We will consider the M-theory on the particular orbifold $\mathbb{R}^{10} \times S^1/\mathbb{Z}^2$, where $\mathbb{Z}_2$ acts on S^1 by $x^{11} \to -x^{11}$, reversing the orientation. Note that eleven-dimensional supergravity is invariant under orientation-reversal if accompanied by change in sign of the three-form $A^{(3)}$, so this makes sense at least for the massless modes coming from eleven dimensions.[4]

[3] To get the right spectrum of BPS saturated states after toroidal compactification, the eleven-dimensional theory should support stable macroscopic membranes of some sort, presumably described at long wavelengths by the supermembrane action [6,2]. We will indeed make this assumption later. But that the theory can be understood as a theory of fundamental membranes seems doubtful because (i) on the face of it, membranes cannot be quantized; (ii) there is no dilaton or coupling parameter that would justify a classical expansion in membranes.

[4] One might wonder whether there is a global anomaly that spoils parity conservation, as described on p. 309 of [11]. This does not occur, as there are no exotic twelve-spheres [12].

On $\mathbb{R}^{10} \times S^1$, the M-theory is invariant under supersymmetry generated by an arbitrary constant spinor ϵ. Dividing by $\mathbb{Z}_2$ kills half the supersymmetry; sign conventions can be chosen so that the unbroken supersymmetries are generated by constant spinors ϵ with $\Gamma^{11}\epsilon = \epsilon$. Together with (2.1), this condition means that

$$\Gamma^1\Gamma^2 \cdots \Gamma^{10}\epsilon = \epsilon, \tag{2.2}$$

so ϵ is chiral in the ten-dimensional sense.

The M-theory on $\mathbb{R}^{10} \times S^1/\mathbb{Z}_2$ thus reduces at low energies to a ten-dimensional Poincaré-invariant supergravity theory with one chiral supersymmetry. There are three string theories with that low-energy structure, namely the $E_8 \times E_8$ heterotic string and the two theories – Type I and heterotic – with $SO(32)$ gauge group. It is natural to wonder whether the M-theory on $\mathbb{R}^{10} \times S^1/\mathbb{Z}_2$ reduces, as the radius of the S^1 shrinks to zero, to one of these three string theories, just as the M-theory on $\mathbb{R}^{10} \times S^1$ reduces to the Type IIA superstring in the same limit. We will give three arguments that all show that if the M-theory on $\mathbb{R}^{10} \times S^1/\mathbb{Z}_2$ reduces for small radius to one of the three string theories, it must be the $E_8 \times E_8$ heterotic string. The arguments are based respectively on space-time gravitational anomalies, the strong coupling behavior, and world-volume gravitational anomalies.

(i) Gravitational anomalies. First we consider the gravitational anomalies of the M-theory on $\mathbb{R}^{10} \times S^1/\mathbb{Z}_2$; these should be computable without detailed knowledge of the M-theory because anomalies can be computed from only a knowledge of the low-energy structure. In raising the question, we understand a metric on $\mathbb{R}^{10} \times S^1/\mathbb{Z}_2$ to be a metric on $\mathbb{R}^{10} \times S^1$ that is invariant under $\mathbb{Z}_2$; a diffeomorphism of $\mathbb{R}^{10} \times S^1/\mathbb{Z}_2$ is a diffeomorphism of $\mathbb{R}^{10} \times S^1$ that commutes with $\mathbb{Z}_2$. The standard massless fermions of the M-theory are the gravitinos; by a gravitino mode on $\mathbb{R}^{10} \times S^1/\mathbb{Z}_2$ we mean a gravitino mode on $\mathbb{R}^{10} \times S^1$ that is invariant under $\mathbb{Z}_2$. With these specifications, it makes sense to ask whether the effective action obtained by integrating out the gravitinos on $\mathbb{R}^{10} \times S^1/\mathbb{Z}_2$ is anomaly-free, that is, whether it is invariant under diffeomorphisms.

First of all, on a smooth eleven-manifold, the effective action obtained by integrating out gravitinos is anomaly-free; purely gravitational anomalies are absent (except possibly for global anomalies in $8k$ or $8k+1$ dimensions) in any dimension not of the form $4k+2$ for some integer k. But the result on an orbifold is completely different. In the case we are considering, it is immediately apparent that the Rarita-Schwinger field has a gravitational anomaly. In fact, the eleven-dimensional Rarita-Schwinger field reduces in ten dimensions to a sum of infinitely many massive fields (anomaly-free) and the massless chiral ten-dimensional gravitino discussed above – which [11] gives an anomaly under ten-dimensional diffeomorphisms. Thus, at least under those diffeomorphisms of the eleven-dimensional orbifold that come from diffeomorphisms of $\mathbb{R}^{10}$ (times the trivial diffeomorphism of $S^1/\mathbb{Z}_2$), there is an anomaly.

To compute the form of this anomaly, it is not necessary to do anything essentially new; it is enough to know the standard Rarita-Schwinger anomaly on a ten-manifold, as well as the result (zero) on a smooth eleven-manifold. Thus, under a space-time diffeomorphism $\delta x^I = \epsilon v^I$ generated by a vector field v^I, the change of the effective action is on general grounds of the form

$$\delta\Gamma = i\epsilon \int_{\mathbb{R}^{10}\times S^1/\mathbb{Z}_2} d^{11}x\, \sqrt{g}\, v^I(x) W_I(x), \tag{2.3}$$

where g is the eleven-dimensional metric and $W_I(x)$ can be computed *locally* from the data at x. The existence of a local expression for W_I reflects the fact that the anomaly can be understood to result entirely from failure of the regulator to preserve the symmetries, and so can be computed from short distances.

Now, consider the possible form of $W_I(x)$ in our problem. If x is a *smooth* point in $\mathbb{R}^{10} \times S^1/\mathbb{Z}_2$ (not an orbifold fixed point), then lack of anomalies of the eleven-dimensional theory implies that $W_I(x) = 0$. W_I is therefore a sum of delta functions supported on the fixed hyperplanes $x^{11} = 0$ and $x^{11} = \pi$, which we will call H' and H'', so (2.3) actually takes the form

$$\delta\Gamma = i\epsilon \int_{H'} d^{10}x\, \sqrt{g'}\, v^I W_I' + i\epsilon \int_{H''} d^{10}x\, \sqrt{g''}\, v^I W_I'' \tag{2.4}$$

where now g' and g'' are the restrictions of g to H' and H'' and W', W'' are local functionals constructed from the data on those hyperplanes. Obviously, by symmetry, W'' is the same as W', but defined from the metric at H'' instead of H'. The form of W' and W'' can be determined as follows without any computation. Let the metric on $\mathbb{R}^{10} \times S^1/\mathbb{Z}^2$ be the product of an arbitrary metric on $\mathbb{R}^{10}$ and a standard metric on $S^1/\mathbb{Z}_2$, and take v to be the pullback of a vector field on $\mathbb{R}^{10}$. In this situation (as we are simply studying a massless chiral gravitino on $\mathbb{R}^{10}$ plus infinitely many massive fields), $\delta\Gamma$ must simply equal the standard ten-dimensional anomaly. The two contributions in (2.4) from $x^{11} = 0$ and $x^{11} = \pi$ must therefore each give *one-half* of the standard ten-dimensional answer. Though we considered a rather special configuration to arrive at this result, it was general enough to permit an arbitrary metric at $x^{11} = 0$ (or π) and hence to determine the functionals W', W'' completely.

Since the anomaly is not zero, the massless modes we know about cannot be the whole story for the M-theory on this orbifold. There will have to be additional massless modes that propagate only on the fixed planes; they will be analogous to the twisted sector modes of string theory orbifolds. The modes will have to be ten-dimensional vector multiplets because the vector multiplet is the only ten-dimensional supermultiplet with all spins ≤ 1. Let us determine what vector multiplets there may be.

First of all, part of the anomaly can be canceled by a generalized Green-Schwarz mechanism [13], with the fields B' and B'', defined as the components $A^{(3)}_{ij11}$ of the three-form on H' and H'', entering roughly as the usual B field does in the Green-Schwarz mechanism. There will be interactions $\int_{H'} B' \wedge Z_8'$ and $\int_{H''} B'' \wedge Z_8''$ at the fixed planes, with some eight-forms Z' and Z'', and in addition the gauge transformation law of $A^{(3)}$ will have terms proportional to delta functions supported on H' and H''. In this way – as in the more familiar ten-dimensional case – some of the anomalies can be canceled, but not all.

In fact, recall that the anomaly in ten-dimensional supergravity is constructed from a twelve-form Y_{12} that is a linear combination of $(\mathrm{tr}\,R^2)^3$, $\mathrm{tr}\,R^2 \cdot \mathrm{tr}\,R^4$, and $\mathrm{tr}\,R^6$ (with R the curvature two-form and tr the trace in the vector representation). The first two terms are "factorizable" and can potentially be canceled by a Green-Schwarz mechanism. The last term is "irreducible" and cannot be so canceled. The irreducible part of the anomaly must be canceled by additional massless modes – necessarily vector multiplets – from the "twisted sectors".

In ten dimensions, the story is familiar [13]. The irreducible part of the standard ten-dimensional anomaly can be canceled precisely by the addition of 496 vector multiplets, so that the possible gauge groups in ten-dimensional $N = 1$ supersymmetric string theory have dimension 496. We are in the same situation now except that the standard anomaly is divided equally between the two fixed hyperplanes. We must have therefore precisely 248 vector multiplets propagating on each of the two hyperplanes! 248 is, of course, the dimension of E_8. So if the M-theory on this orbifold is to be related to one of the three string theories, it must be the $E_8 \times E_8$ theory, with one E_8 propagating on each hyperplane. $SO(32)$ is not possible as gauge invariance would force us to put all the vector multiplets on one hyperplane or the other.

By placing one E_8 at each end, we cancel the irreducible part of the anomaly, but it may not be immediately apparent that the reducible part of the anomaly can be similarly canceled. To see that this is so, recall some facts about the standard ten-dimensional anomaly. With the gauge fields included, the anomaly is derived from a twelve-form $\widetilde{Y}_{12}$ that is a polynomial in $\mathrm{tr}\,F_1^2$ and $\mathrm{tr}\,F_2^2$ (F_1 and F_2 are the two E_8 curvatures; the symbol tr denotes $1/30$ of the trace in the adjoint representation) as well as $\mathrm{tr}\,R^2$, $\mathrm{tr}\,R^4$. ($\mathrm{tr}\,R^6$ is absent as that part has been canceled by adding vector multiplets.) $\widetilde{Y}_{12}$ has the properties

$$\frac{\partial^2 \widetilde{Y}_{12}}{\partial\,\mathrm{tr}\,F_1^2\ \partial\,\mathrm{tr}\,F_2^2} = 0\,,$$

$$\widetilde{Y}_{12} = \left(\mathrm{tr}\,F_1^2 + \mathrm{tr}\,F_2^2 - \mathrm{tr}\,R^2\right) \wedge \widetilde{Y}_8 \tag{2.5}$$

where the details of the polynomial $\widetilde{Y}_8$ will not be essential. The factorization in the second equation is the key to anomaly cancellation. The first equation reflects the fact that (as the massless fermions are in the adjoint representation) there is no massless fermion charged under each E_8, so that the anomaly has no "cross-terms" involving both E_8's.

Note that if we set $U_i = \mathrm{tr}\,F_i^2 - \frac{1}{2}\mathrm{tr}\,R^2$ for $i = 1, 2$, then the first equation in (2.5) implies that

$$\begin{aligned} &U_1 \wedge \left(\widetilde{Y}_8(U_1, U_2, \mathrm{tr}\,R^2, \mathrm{tr}\,R^4) - \widetilde{Y}_8(U_1, 0, \mathrm{tr}\,R^2, \mathrm{tr}\,R^4)\right) \\ &+ U_2 \wedge \left(\widetilde{Y}_8(U_1, U_2, \mathrm{tr}\,R^2, \mathrm{tr}\,R^4) - \widetilde{Y}_8(0, U_2, \mathrm{tr}\,R^2, \mathrm{tr}\,R^4)\right) = 0. \end{aligned} \tag{2.6}$$

Hence we can write

$$\begin{aligned}\widetilde{Y}_{12} = &(\mathrm{tr}\, F_1^2 - \tfrac{1}{2}\mathrm{tr}\, R^2) \wedge Z_8(\mathrm{tr}\, F_1^2, \mathrm{tr}\, R^2, \mathrm{tr}\, R^4)\\ &+(\mathrm{tr}\, F_2^2 - \tfrac{1}{2}\mathrm{tr}\, R^2) \wedge Z_8(\mathrm{tr}\, F_2^2, \mathrm{tr}\, R^2, \mathrm{tr}\, R^4).\end{aligned} \quad (2.7)$$

Here Z_8 is defined by $Z_8(\mathrm{tr}\, F_1^2, \mathrm{tr}\, R^2, \mathrm{tr}\, R^4) = \widetilde{Y}_8(U_1, 0, \mathrm{tr}\, R^2, \mathrm{tr}\, R^4)$. (2.7) is the desired formula showing how the anomalies can be canceled by a variant of the Green-Schwarz mechanism adapted to the eleven-dimensional problem. The first term, involving F_1 but not F_2, is the contribution from couplings of what was above called B', and the second term, involving F_2 but not F_1, is the contribution from B''.

(ii) Strong coupling behavior. If it is true that the M-theory on this orbifold is related to the $E_8 \times E_8$ superstring theory, then the relation between the radius R of the circle (in the eleven-dimensional metric) and the string coupling constant λ can be determined by comparing the predictions of the two theories for the low-energy effective action of the supergravity multiplet in ten dimensions. The analysis is precisely as in [1] and will not be repeated here. It gives the same relation

$$R = \lambda^{2/3} \quad (2.8)$$

that one finds between the M-theory on $\mathbb{R}^{10} \times S^1$ and Type IIA superstring theory.

In particular, then, for small R – where the supergravity cannot be a good description as R is small compared to the Planck length – the string theory is weakly coupled and can be a good description. On the other hand, we get a candidate for the strong coupling behavior of the $E_8 \times E_8$ heterotic string: it corresponds to supergravity on the $\mathbb{R}^{10} \times S^1/\mathbb{Z}_2$ orbifold, which is an effective description (of the low-energy interactions of the light modes) for large λ as then R is much bigger than the Planck length. If our proposal is correct, then what a low energy observer sees in the strongly coupled $E_8 \times E_8$ theory depends on where he or she is; a generic observer, far from one of the fixed hyperplanes, sees simply eleven-dimensional supergravity (or the M-theory), and does not distinguish the strongly coupled $E_8 \times E_8$ theory from a strongly coupled Type IIA theory, while an observer near one of the distinguished hyperplanes sees eleven-dimensional supergravity on a half-space, with an E_8 gauge multiplet propagating on the boundary.

We can also now see another reason that if the M-theory on $\mathbb{R}^{10} \times S^1/\mathbb{Z}_2$ is related to one of the three ten-dimensional string theories, it must be the $E_8 \times E_8$ heterotic string. Indeed, there is by now convincing evidence [1,7–9] that the strong coupling limit of the Type I superstring in ten dimensions is the weakly coupled $SO(32)$ heterotic string, and vice-versa, so we would not want to relate either of the two $SO(32)$ theories to eleven-dimensional supergravity. We must relate the orbifold to the $E_8 \times E_8$ theory whose strong coupling behavior has been previously unknown.

(iii) Extended membranes. As our third and last piece of evidence, we want to consider extended membrane states in the M-theory after further compactification to $\mathbb{R}^9 \times S^1 \times S^1/\mathbb{Z}_2$.

Our point of view is not that the M-theory "is" a theory of membranes but that it describes, among other things, membrane states. There is a crucial difference. For instance, any spontaneously broken unified gauge theory in four dimensions with an unbroken $U(1)$ describes, among other things, magnetic monopoles. That does not mean that the theory can be recovered by quantizing magnetic monopoles; that is presumably possible only in very special cases. Classical magnetic monopole solutions of gauge theory, because of their topological stability, can be quantized to give quantum states. But topologically unstable monopole-antimonopole configurations, while representing possibly interesting classical solutions, cannot ordinarily be quantized to understand photons and electrons. Likewise, we assume that when the topology is right, the M-theory has topologically stable membranes (presumably described if the length scale is large enough by the low-energy supermembrane action [2]) that can be quantized to give quantum eigenstates. Even when the topology is wrong – for instance on $\mathbb{R}^{11}$ where there is no two-cycle for the membrane to wrap around – macroscopic membrane solutions (with a scale much bigger than the Planck scale) will make sense, but we do not assume that they can be quantized to recover gravitons.

The most familiar example of a situation in which there are topologically stable membrane states is that of compactification of the M-theory on $\mathbb{R}^9 \times S^1 \times S^1$. With x^1 understood as the time and x^{10}, x^{11} as the two periodic variables, the classical membrane equations have a solution described by $x^2 = \ldots = x^9 = 0$. This solution is certainly topologically stable so (if the radii of the circles are big in Planck units) it can be reliably quantized to obtain quantum states. The solution is invariant under half of the supersymmetries, namely those obeying

$$\Gamma^1\Gamma^{10}\Gamma^{11}\epsilon = \epsilon, \tag{2.9}$$

so these will be BPS-saturated states. This latter fact gives the quantization of this particular membrane solution a robustness that enables one (even if the membrane in question can *not* for other purposes be usefully treated as elementary) to extrapolate to a regime in which one of the S^1's is small and one can compare to weakly coupled string theory.

Let us recall the result of this comparison, which goes under the name of double dimensional reduction of the supermembrane [14]. The membrane solution described above breaks the eleven-dimensional Lorentz group to $SO(1,2) \times SO(8)$. The massless modes on the membrane world-volume are the oscillations of $x^2, \ldots, x^9$, which transform as $(\mathbf{1}, \mathbf{8})$, and fermions that transform as $(\mathbf{2}, \mathbf{8}'')$. Here $\mathbf{2}$ is the spinor of $SO(1,2)$ and $\mathbf{8}$, $\mathbf{8}'$, and $\mathbf{8}''$ are the vector and the two spinors of $SO(8)$. To interpret this in string theory terms, one considers only the zero modes in the x^{11} direction, and decomposes the spinor of $SO(1,2)$ into positive and negative chirality modes of $SO(1,1)$; one thus obtains the world-sheet structure of the Type IIA superstring. So those membrane excitations that are low-lying when the second circle is small will match up properly with Type IIA states. Since many of these Type IIA states are BPS saturated and can be followed from weak to strong coupling, the membrane we started with was really needed to reproduce this part of the spectrum.

Now we move on to the $\mathbb{R}^{10} \times S^1/\mathbb{Z}_2$ orbifold. We assume that the classical membrane solution $x^2 = \ldots = x^9$ is still allowed on the orbifold; this amounts to assuming that the membranes of the M-theory can have boundaries that lie on the fixed hyperplanes. In the orbifold, unbroken supersymmetries (as we discussed at the outset of Section 2) correspond to spinors ϵ with $\Gamma^{11}\epsilon = \epsilon$; these transform as the **16** of $SO(1,9)$, or as $\mathbf{8}'_+ \oplus \mathbf{8}''_-$ of $SO(1,1) \times SO(8)$. The spinors unbroken in the field of the membrane solution also obey (2.9), or equivalently $\Gamma^1\Gamma^{10}\epsilon = \epsilon$. Thus, looking at the situation in string terms (for an observer who does not know about the eleventh dimension), the unbroken supersymmetries have positive chirality on the string world-sheet and transform as $\mathbf{8}'_+$ (where the + is the $SO(1,1)$ chirality) under $SO(1,1) \times SO(8)$. The massless world-sheet bosons, oscillations in $x^2, \ldots, x^9$, survive the orbifolding, but half of the fermions are projected out. The survivors transform as $\mathbf{8}''_-$; one can think of them [15] as Goldstone fermions for the $\mathbf{8}''_-$ supersymmetries that are broken by the classical membrane solution. The $-$ chirality means that they are right-moving.

So the massless modes we know about transform like the world-sheet modes of the heterotic string that carry space-time quantum numbers: left- and right-moving bosons transforming in the **8** and right-moving fermions in the $\mathbf{8}''$. Recovering much of the world-sheet structure of the heterotic string does not imply that the string theory (if any) related to the M-theory orbifold is a heterotic rather than Type I string; the Type I theory also describes among other things an object with the world-sheet structure of the heterotic string [9]. It is by considerations of anomalies on the membrane world-volume that we will reach an interesting conclusion.

The Dirac operator on the membrane three-volume is free of world-volume gravitational anomalies as long as the world-volume is a smooth manifold. (Recall that except possibly for discrete anomalies in $8k$ or $8k+1$ dimensions, gravitational anomalies occur only in dimensions $4k+2$.) In the present case, the world-volume is not a smooth manifold, but has orbifold singularities (possibly better thought of as boundaries) at $x^{11} = 0$ and at $x^{11} = \pi$. These singularities give rise to three-dimensional gravitational anomalies; this is obvious from the fact that the massless world-sheet fermions in the two-dimensional sense are the fermions $\mathbf{8}''_-$ of definite chirality. By analysis just like we gave in the eleven-dimensional case, the gravitational anomaly on the membrane world-volume vanishes at smooth points and is a sum of delta functions supported at $x^{11} = 0$ and $x^{11} = \pi$. As in the eleven-dimensional case, these delta functions each represent *one half* of the usual two-dimensional anomaly of the effective massless two-dimensional $\mathbf{8}''_-$ field.

As is perhaps obvious intuitively and we will argue below, the gravitational anomaly of the $\mathbf{8}''_-$ field is the usual gravitational anomaly of right-moving RNS fermions and superconformal ghosts in the heterotic string. So far we have only considered the modes that propagate in bulk on the membrane world-volume. If the membrane theory makes sense in the situation we are considering, the gravitational anomaly of the $\mathbf{8}''_-$ field must be canceled by additional world-volume "twisted sector" modes, supported at the orbifold fixed points $x^{11} = 0$ and $x^{11} = \pi$. If we are to recover one of the known string theories, these "twisted sector modes" should be left-moving current algebra modes with $c = 16$. (In any event there is practically no other way to maintain space-time supersymmetry.) Usually both $SO(32)$ and $E_8 \times E_8$ are possible, but in the present context the anomaly that must be cancelled is supported one half at $x^{11} = 0$ and one half at $x^{11} = \pi$, so the only possibility is to have $E_8 \times E_8$ with one E_8 supported at each end. This then is our third reason that if the M-theory on the given orbifold has a known string theory as its weak coupling limit, it must be the $E_8 \times E_8$ heterotic string.

It remains to discuss somewhat more carefully the gravitational anomalies that have just been exploited. Some care is needed here as there is an important distinction between objects that are quantized as elementary strings and objects that are only known as macroscopic strings embedded in space-time. (See, for instance, [16,17].) For elementary strings, one usually considers separately both right-moving and left-moving conformal anomalies. The sum of the two is the total conformal anomaly (which generalizes to the conformal anomaly in dimensions above two), while the difference is the world-sheet gravitational anomaly. For objects that are only known as macroscopic strings embedded in space-time, the total conformal anomaly is not a natural concept, since the string world-sheet has a natural metric (and not just a conformal class of metrics) coming from the embedding. But the gravitational anomaly, which was exploited above, still makes sense even in this situation, as the world-sheet is still not endowed with a natural coordinate system.

Let us justify the claim that the world-sheet gravitational anomaly of the $\mathbf{8}''_-$ fermions encountered above equals the usual gravitational anomaly from right-moving RNS fermions and superconformal ghosts. A detailed calculation is not necessary, as this can be established by the following simple means. First let us state the problem (as it appears after double dimensional reduction to the Green-Schwarz formulation of the heterotic string) in generality. The problem really involves, in general, a two-dimensional world-sheet Σ embedded in a ten-manifold M. The normal bundle N to the world-sheet is a vector bundle with structure group $SO(8)$. If S_- is the bundle of negative chirality spinors on Σ and N'' is the bundle associated to N in the $\mathbf{8}''$ representation of $SO(8)$, then the $\mathbf{8}''_-$ fermions that we want are sections of $L_- \otimes N''$. By making a triality transformation in $SO(8)$, we can replace the fermions with sections of $L_- \otimes N$ without changing the anomalies. Now using the fact that the tangent bundle of M is the sum of N and the tangent bundle of Σ - $TM = N \oplus T\Sigma$ - we can replace $L_- \otimes N$ by $L_- \otimes TM$ if we also subtract the contribution of fermions that take values in $L_- \otimes T\Sigma$. The $L_- \otimes TM$-valued fermions are the usual right-moving RNS fermions, and (as superconformal ghosts take values in $L_- \otimes T\Sigma$) subtracting the contribution of fermions valued in $L_- \otimes T\Sigma$ has the same effect as including the superconformal ghosts.

3. Heterotic - Type I duality from the M-theory

In this section, we will try to relate the eleven-dimensional picture to another interesting phenomenon, which is the conjectured duality between the heterotic and Type I $SO(32)$ superstrings.

So far we have presented arguments indicating that the $E_8 \times E_8$ heterotic string theory is related to the M-theory on $\mathbb{R}^{10} \times S^1/\mathbb{Z}_2$, just as the Type IIA theory is related to the

M-theory compactified on $\mathbb{R}^{10} \times S^1$.

We can follow this analogy one step further, and compactify the tenth dimension of the M-theory on S^1. Schwarz [5] and Aspinwall [10]explained how the $SL(2,\mathbb{Z})$ duality of the ten-dimensional Type IIB string theory follows from space-time diffeomorphism symmetry of the M-theory on $\mathbb{R}^9 \times T^2$. (For some earlier results in that direction see also [18].) Here we will argue that the $SO(32)$ heterotic - Type I duality similarly follows from classical symmetries of the M-theory on $\mathbb{R}^9 \times S^1 \times S^1/\mathbb{Z}_2$.

First we need several facts about T-duality of open-string models.

T-duality in Type I superstring theory. The Type I theory in ten dimensions can be interpreted as a generalized $\mathbb{Z}_2$ orbifold of the Type IIB theory [19–23]. The orbifold in question acts by reversing world-sheet parity, and acts trivially on the space-time. Projection of the Type IIB spectrum to $\mathbb{Z}_2$-invariant states makes the Type IIB strings unoriented; this creates an anomaly in the path integral over world-sheets with crosscaps, which must be compensated for by introducing boundaries. The open strings, which are usually introduced to cancel the anomaly, are naturally interpreted as the twisted states of the parameter-space orbifold.

More generally, one can combine the reversal of world-sheet orientation with a space-time symmetry, getting a variant of the Type I theory [20–22]. [5] A special case of this will be important here. Upon compactification to $\mathbb{R}^9 \times S^1$, Type IIB theory is T-dual to Type IIA theory. Analogously, the Type I theory – which is a $\mathbb{Z}_2$ orbifold of Type IIB theory – is T-dual to a certain $\mathbb{Z}_2$ orbifold of Type IIA theory. This orbifold is constructed by dividing the Type IIA theory by a $\mathbb{Z}_2$ that reverses the world-sheet orientation and acts on the circle by $x^{10} \to -x^{10}$. Note that the Type IIA theory is invariant under combined reversal of world-sheet and space-time orientations (but not under either one separately), so the combined operation is a symmetry. This theory has been called the Type I′ or Type IA theory. In this theory, the twisted states are open strings that have their endpoints at the fixed points $x^{10} = 0$ and $x^{10} = \pi$. To cancel anomalies, these open strings must carry Chan-Paton factors. If we want to treat the two fixed points symmetrically – as is natural in an orbifold – while canceling the anomalies, there must be $SO(16)$ Chan-Paton factors at each fixed point, so the gauge group is $SO(16) \times SO(16)$.

In fact, it has been shown [20–22] that the $SO(16) \times SO(16)$ theory just described is the T-dual of the vacuum of the standard Type I theory on $\mathbb{R}^9 \times S^1$ in which $SO(32)$ is broken to $SO(16) \times SO(16)$ by a Wilson line. This is roughly because T-duality exchanges the usual Neumann boundary conditions of open strings with Dirichlet boundary conditions, and gives a theory in which the open strings terminate at the fixed points.

Of course, the Type I theory on $\mathbb{R}^9 \times S^1$ has moduli corresponding to Wilson lines; by adjusting them one can change the unbroken gauge group or restore the full $SO(32)$. In the T-dual description, turning on these moduli causes the positions at which the open

[5] More generally still, one can divide by a group containing some elements that act only on space-time and some that also reverse the world-sheet orientation; the construction of the twisted states then has certain subtleties that were discussed in [24].

strings terminate to vary in a way that depends upon their Chan-Paton charges [25]. The vacuum with unbroken $SO(32)$ has all open strings terminating at the same fixed point.

Heterotic - Type I duality. We are now ready to try to relate the eleven-dimensional picture to the conjectured heterotic - Type I duality of ten-dimensional theories with gauge group $SO(32)$.

What suggests a connection is the following. Consider the Type I superstring on $\mathbb{R}^9 \times S^1$. Its T-dual is related, as we have discussed, to the Type IIA theory on an $\mathbb{R}^9 \times S^1/\mathbb{Z}_2$ orbifold. We can hope to identify the Type IIA theory on $\mathbb{R}^9 \times S^1/\mathbb{Z}_2$ with the M-theory on $\mathbb{R}^9 \times S^1/\mathbb{Z}_2 \times S^1$, since in general one hopes to associate Type IIA theory on any space X with M-theory on $X \times S^1$.

On the other hand, we have interpreted the $E_8 \times E_8$ heterotic string as M-theory on $\mathbb{R}^{10} \times S^1/\mathbb{Z}_2$, so the M-theory on $\mathbb{R}^9 \times S^1 \times S^1/\mathbb{Z}_2$ should be the $E_8 \times E_8$ theory on $\mathbb{R}^9 \times S^1$.

So we now have two ways to look at the M-theory on $X = \mathbb{R}^9 \times S^1/\mathbb{Z}_2 \times S^1$. (1) It is the Type IIA theory on $\mathbb{R}^9 \times S^1/\mathbb{Z}_2$ which is also the T-dual of the Type I theory on $\mathbb{R}^9 \times S^1$. (2) After exchanging the last two factors so as to write X as $\mathbb{R}^9 \times S^1 \times S^1/\mathbb{Z}_2$, the same theory should be the $E_8 \times E_8$ heterotic string on $\mathbb{R}^9 \times S^1$. So it looks like we can predict a relation between the Type I and heterotic string theories!

This cannot be right in the form stated, since the model in (1) has gauge group $SO(16) \times SO(16)$, while that in (2) has gauge group $E_8 \times E_8$. Without really understanding the M-theory, we cannot properly explain what to do, but pragmatically the most sensible course is to turn on a Wilson line in theory (2), breaking $E_8 \times E_8$ to $SO(16) \times SO(16)$.

At this point, it is possible that (1) and (2) are equivalent (under exchanging the last two factors in $\mathbb{R}^9 \times S^1/\mathbb{Z}_2 \times S^1$). The equivalence does not appear, at first sight, to be a known equivalence between string theories. We can relate it to a known equivalence by making a T-duality transformation on each side. In (1), a T-duality transformation will convert to the Type I theory on $\mathbb{R}^9 \times S^1$ (in its $SO(16) \times SO(16)$ vacuum). In (2), a T-duality transformation will convert to an $SO(32)$ heterotic string with $SO(32)$ spontaneously broken to $SO(16) \times SO(16)$.[6] At this point, theories (1) and (2) are Type I and heterotic $SO(32)$ theories (in their respective $SO(16) \times SO(16)$ vacua), so we can try to compare them using the conjectured heterotic - Type I duality. It turns out that this acts in the expected way, exchanging the last two factors in $X = \mathbb{R}^9 \times S^1/\mathbb{Z}_2 \times S^1$.

Since the logic may seem convoluted, let us recapitulate. On side (1), we start with the M theory on $\mathbb{R}^9 \times S^1/\mathbb{Z}_2 \times S^1$, and interpret is as the T-dual of the Type I theory on $\mathbb{R}^9 \times S^1$. On side (2), we start with the M-theory on $\mathbb{R}^9 \times S^1 \times S^1/\mathbb{Z}_2$, interpret it as the $E_8 \times E_8$ heterotic string on $\mathbb{R}^9 \times S^1$ and (after turning on a Wilson line) make a T-duality transformation to convert the gauge group to $SO(32)$. Then we compare

[6] $R \to 1/R$ symmetry, with R the radius of the circle in $\mathbb{R}^9 \times S^1$, maps the heterotic string vacuum with unbroken $SO(32)$ to itself, and maps the heterotic string vacuum with unbroken $E_8 \times E_8$ to itself, but maps a heterotic string vacuum with $E_8 \times E_8$ broken to $SO(16) \times SO(16)$ to a heterotic string vacuum with $SO(32)$ broken to $SO(16) \times SO(16)$. This follows from facts such as those explained in [26,27].

(1) and (2) using heterotic - Type I duality, which gives the same relation that was expected from the eleven-dimensional point of view. One is still left wishing that one understood better the meaning of T-duality in the M-theory.

We hope that this introduction will make the computation below easier to follow.

Side (1). We start with the M-theory on $\mathbb{R}^9 \times S^1/\mathbb{Z}_2 \times S^1$, with radii R_{10} and R_{11} for the two circles. We interpret this as the Type IA theory, which is the $\mathbb{Z}_2$ orbifold of the Type II theory on $\mathbb{R}^9 \times S^1$, a T-dual of the Type I theory in a vacuum with unbroken $SO(16) \times SO(16)$.

The relation between R_{10} and R_{11} and the Type IA parameters (the ten-dimensional string coupling λ_{IA} and radius R_{IA} of the S^1) can be computed by comparing the low-energy actions for the supergravity multiplet. The computation and result are as in [1]:[7]

$$R_{11} = \lambda_{IA}^{2/3}, \qquad R_{10} = \frac{R_{IA}}{\lambda_{IA}^{1/3}}. \tag{3.1}$$

Now we make a T-duality transformation to an ordinary Type I theory (with unbroken gauge group $SO(16) \times SO(16)$), by the standard formulas $R_I = 1/R_{IA}$, $\lambda_I = \lambda_{IA}/R_{IA}$. So

$$R_{11} = \frac{\lambda_I^{2/3}}{R_I^{2/3}}, \qquad R_{10} = \frac{\lambda_I^{1/3}}{R_I^{2/3}}. \tag{3.2}$$

Side (2). Now we start with the M-theory on $\mathbb{R}^9 \times S^1 \times S^1/\mathbb{Z}^2$, with the radii of the last two factors denoted as R'_{10} and R'_{11}. This is hopefully related to the $E_8 \times E_8$ heterotic string on $\mathbb{R}^9 \times S^1$, with the M-theory parameters being related to the heterotic string coupling λ_{E_8} and radius R_{E_8} by formulas

$$R'_{11} = \lambda_{E_8}^{2/3}, \qquad R'_{10} = \frac{R_{E_8}}{\lambda_{E_8}^{1/3}} \tag{3.3}$$

just like (3.1), and obtained in the same way. After turning on a Wilson line and making a T-duality transformation to an $SO(32)$ heterotic string, whose parameters λ_h, R_h are related to those of the $E_8 \times E_8$ theory by the standard T-duality relations $R_h = 1/R_{E_8}$, $\lambda_h = \lambda_{E_8}/R_{E_8}$, we get the analog of (3.2),

$$R'_{11} = \frac{\lambda_h^{2/3}}{R_h^{2/3}}, \qquad R'_{10} = \frac{1}{\lambda_h^{1/3} R_h^{2/3}}. \tag{3.4}$$

Comparison. Now we compare the two sides via the conjectured $SO(32)$ heterotic - Type I duality according to which these theories coincide with

$$\lambda_h = \frac{1}{\lambda_I}, \qquad R_h = \frac{R_I}{\lambda_I^{1/2}}. \tag{3.5}$$

[7] The second equation is equivalent to the Weyl rescaling $g_{10,M} = \lambda_{IIA}^{-2/3} g_{10,IIA}$ obtained in [1] between the ten-dimensional metrics as measured in the M-theory or Type IIA. Also, by λ_{IA} we refer to the *ten-dimensional* string coupling constant; similar convention is also used for all other string theories below.

A comparison of (3.2) and (3.4) now reveals that the relation of R_{10}, R_{11} to R'_{10}, R'_{11} is simply

$$R_{10} = R'_{11}\,, \qquad R_{11} = R'_{10}\,. \tag{3.6}$$

So – as promised – under this sequence of operations, the natural symmetry in eleven dimensions becomes standard heterotic - Type I duality.

4. Comparison to Type II dualities

We have seen a close analogy between the dualities that involve heterotic and Type I string theories and relate them to the M-theory, and the corresponding Type II dualities that relate the Type IIA theory to eleven dimensions and the Type IIB theory to itself. The reason for this analogy is of course that while the Type II dualities are all related to the compactification of the M-theory on $\mathbb{R}^9 \times \mathrm{T}^2$, the heterotic and Type I dualities are related to the compactification of the M-theory on a $\mathbb{Z}_2$ orbifold of $\mathbb{R}^9 \times \mathrm{T}^2$. It is the purpose of this section to make this analogy more explicit.

The moduli space of compactifications of the M-theory on a rectangular torus is shown in Fig. 1, following [10,5]. Let us first recall how one can see the $SL(2,\mathbb{Z})$ duality of the ten-dimensional Type IIB theory in the moduli space of the M-theory on $\mathbb{R}^9 \times \mathrm{T}^2$. The variables of the Type IIB theory on $\mathbb{R}^9 \times \mathrm{S}^1$ are related to the compactification radii of the M-theory on $\mathbb{R}^9 \times \mathrm{T}^2$ by

$$\lambda_{\mathrm{IIB}} = \frac{R_{11}}{R_{10}}\,, \qquad R_{\mathrm{IIB}} = \frac{1}{R_{10}R_{11}^{1/2}}\,. \tag{4.1}$$

The string coupling constant λ_{IIB} depends on the shape of the two-torus of the M-theory, but not on its area. As we send the radius R_{IIB} to infinity to make the Type IIB theory ten-dimensional at fixed λ_{IIB}, the radii R_{10} and R_{11} go to zero. So, the ten-dimensional Type IIB theory at arbitrary string coupling corresponds to the origin of the moduli space of the M-theory on $\mathbb{R}^9 \times \mathrm{T}^2$ as shown in Fig. 1. The $SL(2,\mathbb{Z})$ duality group of the ten-dimensional Type IIB theory can be identified with the modular group acting on the T^2 [5,10].

Similarly, the region of the moduli space where only one of the radii R_{10} and R_{11} is small corresponds to the weakly coupled Type IIA string theory. When both radii become large simultaneously, the Type IIA string theory becomes strongly coupled, and the low-energy physics of the theory is described by eleven-dimensional supergravity [1].

Now we can repeat the discussion for the heterotic and Type I theories. The moduli space of the M-theory compactified on $\mathbb{R}^9 \times \mathrm{S}^1/\mathbb{Z}_2 \times \mathrm{S}^1$ is sketched in Fig. 2. In the previous sections we discussed the relations between the Type IA theory on $\mathbb{R}^9 \times \mathrm{S}^1/\mathbb{Z}_2$, the Type I theory on $\mathbb{R}^9 \times \mathrm{S}^1$, and the M-theory on $\mathbb{R}^9 \times \mathrm{S}^1/\mathbb{Z}_2 \times \mathrm{S}^1$. This relation leads to the following expression for the Type I variables in terms of the variables of the M-theory:

$$\lambda_{\mathrm{I}} = \frac{R_{11}}{R_{10}}\,, \qquad R_{\mathrm{I}} = \frac{1}{R_{10}R_{11}^{1/2}}\,. \tag{4.2}$$

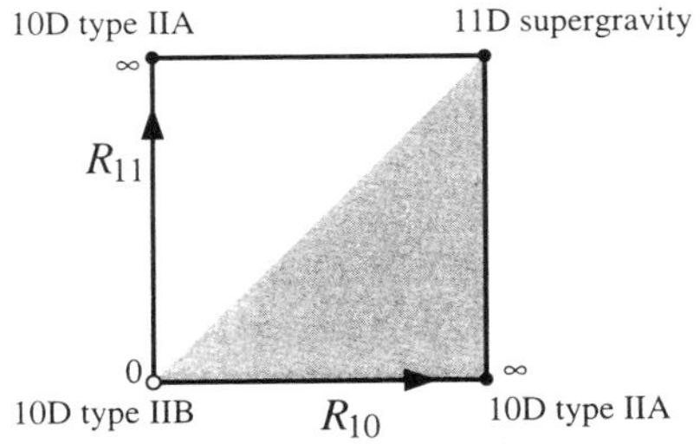

Fig. 1. A section of the moduli space of compactifications of the M-theory on $\mathbb{R}^9 \times T^2$. By virtue of the $\mathbb{Z}_2$ symmetry between the two compact dimensions, only the shaded half of the diagram is relevant.

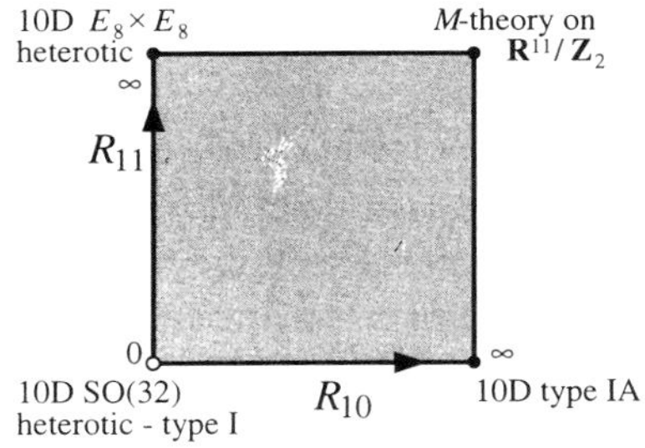

Fig. 2. A section of the moduli space of compactifications of the M-theory on $\mathbb{R}^9 \times S^1 \times S^1/\mathbb{Z}_2$. Here R_{10} is the radius of $S^1/\mathbb{Z}_2$.

The string coupling constant λ_I depends only on the shape of the two-torus of the M-theory. By the same reasoning as in the Type IIB case, the ten-dimensional Type I theory at arbitrary string coupling λ_I is represented by the origin of the moduli space, which should be – in both cases – more rigorously treated as a blow-up.

We have related the $SO(32)$ heterotic string is related to variables of the M-theory by

$$\lambda_h = \frac{R'_{11}}{R'_{10}}, \qquad R_h = \frac{1}{R'_{10} R'_{11}{}^{1/2}}. \tag{4.3}$$

Using heterotic - Type I duality, which simply exchanges the two radii,

$$R'_{10} = R_{11}, \qquad R'_{11} = R_{10}, \tag{4.4}$$

we can express this relation in terms of R_{10} and R_{11},

$$\lambda_h = \frac{R_{10}}{R_{11}}, \qquad R_h = \frac{1}{R_{11} R_{10}^{1/2}}. \tag{4.5}$$

Just like the ten-dimensional Type I theory, the ten-dimensional $SO(32)$ heterotic theory at all couplings corresponds to the origin of the moduli space. The heterotic - Type I duality maps one of these theories at strong coupling to the other theory at weak coupling, and vice-versa.

Now we would like to understand the regions of the moduli space where at least one of the radii R_{10}, R_{11} is large. Recall that R_{10} and R_{11} – as measured in the M-theory – are related to the string coupling constants by

$$R_{10} = \lambda_{E_8}^{2/3}, \qquad R_{11} = \lambda_{\mathrm{IA}}^{2/3}. \tag{4.6}$$

If one of the radii is large and the other one is small, the natural description of the physics at low energies is in terms of the weakly-coupled Type IA or $E_8 \times E_8$ heterotic string theory. Conversely, as we go to the limit where both R_{10} and R_{11} are large, both string theories are strongly coupled, and the low-energy physics is effectively described by the M-theory.

Comparison of the spectra. We can gain some more insight into the picture by looking at some physical states of the M-theory and interpreting them as states in different weakly-coupled string theories.

A particularly natural set of states in the M-theory on $\mathbb{R}^9 \times \mathrm{T}^2$ is given by the Kaluza-Klein (KK) states of the supergravity multiplet, that is the states carrying momentum in the tenth and eleventh dimension, along with the wrapping modes of the membrane. As measured in the M-theory, these states have masses

$$M^2 = \frac{\ell^2}{R_{10}^2} + \frac{m^2}{R_{11}^2} + n^2 R_{10}^2 R_{11}^2 \tag{4.7}$$

for certain values of m, n, ℓ.

We are of course interested in states of the M-theory on $\mathbb{R}^9 \times \mathrm{S}^1/\mathbb{Z}_2 \times \mathrm{S}^1$. In order to get the states that survive on the orbifold, we must project (4.7) to the $\mathbb{Z}_2$ invariant sector. Schematically, the orbifold group acts on the states with the quantum numbers of (4.7) as follows:

$$|\ell, m, n\rangle \to \pm|-\ell, m, n\rangle. \tag{4.8}$$

The action of the orbifold group on the KK modes follows directly from its action on the space-time coordinates. The action on the membrane wrapping modes indicates that the orbifold changes simultaneously the space-time orientation as well as the world-volume orientation of the membrane.

While m and n are conserved quantum numbers even in the orbifold, ℓ is not. Nevertheless, we include it in the discussion since ℓ is approximately conserved in some limits.

If our prediction about the relation of the M-theory on $\mathbb{R}^9 \times \mathrm{S}^1/\mathbb{Z}_2 \times \mathrm{S}^1$ to the heterotic and Type I string theories is correct, the stable states of the M-theory must have an interpretation in each of these string theories. The string masses of these states as measured by the Type I observer are

$$M_{\mathrm{I}}^2 = \ell^2 R_{\mathrm{I}}^2 + \frac{m^2 R_{\mathrm{I}}^2}{\lambda_{\mathrm{I}}^2} + \frac{n^2}{R_{\mathrm{I}}^2}. \tag{4.9}$$

The membrane wrapping modes can be identified with the KK modes of the Type I string, while the unstable states correspond to unstable winding modes of the elementary Type I string. The membrane KK modes along the eleventh dimension are non-perturbative states in the Type I theory. These states can be identified with winding modes of non-perturbative strings with tension $T_2 \propto \lambda_{\mathrm{I}}^{-1}$. We will see below that this is simply the solitonic heterotic string of the Type I theory [7–9].

Similarly, we can try to interpret the states in the T-dual, Type IA theory. A Type IA observer will measure the following masses of the states:

$$M_{\mathrm{IA}}^2 = \frac{\ell^2}{R_{\mathrm{IA}}^2} + \frac{m^2}{\lambda_{\mathrm{IA}}^2} + n^2 R_{\mathrm{IA}}^2. \tag{4.10}$$

As required by T-duality, the membrane wrapping modes correspond to the string winding modes. The unstable states correspond to the KK modes of the Type IA closed string; they are unstable because the tenth component of the momentum is not conserved in the Type IA theory. The stable KK states of the M-theory correspond to non-perturbative Type IA states. These Type IA states can be identified with the zero-branes (alias extremal black holes) of the Type IIA theory in ten dimensions. Notice that under the $\mathbb{Z}_2$ orbifold action, the quantum number that corresponds to the extremal black hole states is conserved, and the zero-brane states survive the orbifold projection.

In the $E_8 \times E_8$ heterotic theory, the masses of our states are given by

$$M_{E_8}^2 = \frac{m^2}{R_{E_8}^2} + n^2 R_{E_8}^2. \tag{4.11}$$

(We here omit ℓ, as the unstable states it labels have no clear interpretation for the weakly coupled heterotic string.) The stable KK modes along the eleventh dimension in the M-theory can be interpreted as the KK modes along the tenth dimension in the heterotic theory, while the membrane wrapping modes are the winding modes of the heterotic string.

In the $SO(32)$ heterotic string, the masses are

$$M_h^2 = m^2 R_h^2 + \frac{n^2}{R_h^2}. \tag{4.12}$$

Again, these are the usual momentum and winding states of the heterotic string. The formulas also make it clear that – as expected from the heterotic - Type I duality – the $m = 1$ non-perturbative Type I string state corresponds to the elementary heterotic string.

We already pointed out an analogy between the heterotic - Type I duality and the $SL(2, \mathbb{Z})$ duality of the Type IIB theory; now we actually see remnants of the $SL(2, \mathbb{Z})$ multiplet of Type IIB string states in the $SO(32)$ heterotic and Type I theories. This can be best demonstrated when we consider the weakly-coupled Type IIB theory, and look at the behavior of its spectrum under the $\mathbb{Z}_2$ orbifold group that leads to the Type I theory. The perturbative Type IIB string of the $SL(2, \mathbb{Z})$ multiplet is odd under the $\mathbb{Z}_2$ orbifold action, and so does not give rise to a stable string. But a linear combination of strings winding in opposite directions survives the projection and corresponds to the elementary Type I closed string, which is unstable but long-lived for weak coupling. The $SL(2,\mathbb{Z})$ Type IIB string multiplet also contains a non-perturbative state that is even under $\mathbb{Z}_2$, and we have just identified it with the elementary heterotic string. Upon orbifolding, the original $SL(2,\mathbb{Z})$ multiplet of Type IIB strings thus gives rise to both Type I and the heterotic string.

Twisted membrane states. The states we have discussed so far are analogous to untwisted states of string theory on orbifolds. The membrane world-volume is without boundary, but the membrane Hilbert space is projected onto $\mathbb{Z}_2$-invariant states; the $\mathbb{Z}_2$ simultaneously reverses the sign of x^{11} and the membrane orientation.

We must also add the twisted membrane states, which are analogous to open strings in parameter space orbifolds of Type II string theory reviewed above. Just like the open strings of the Type IA theory, the twisted membrane states have world-volumes with two boundary components, restricted to lie at one of the orbifold fixed points, $x^{10} = 0$ and $x^{10} = \pi$. Such a state might simply be localized near one of the fixed points (in which case the description as a membrane state might not really be valid), or it might wrap around $S^1/\mathbb{Z}_2 \times S^1$ a certain number of times (in which case the membrane description does make sense at least if the radii are large). The former states might be called twisted KK states, and should include the non-abelian gauge bosons discussed in Section 2. The latter states will be called twisted wrapping modes. The twisted states carry no momentum in the orbifold direction. Both the momentum $\widetilde{m}$ in the S^1 direction and the wrapping number $\widetilde{n}$ are conserved.

States of these twisted sectors have masses – as measured in the M-theory – given by

$$M^2 = \frac{\widetilde{m}^2}{R_{11}^2} + \widetilde{n}^2 R_{10}^2 R_{11}^2. \tag{4.13}$$

Just as in the untwisted sector, one has to project out the twisted states that are not invariant under the orbifold group action.

Again, the $\mathbb{Z}_2$-invariant twisted states should have a natural interpretation in the corresponding string theories. In the Type I theory, the twisted states of the M-theory have masses

$$M_{\rm I}^2 = \frac{\widetilde{m}^2 R_{\rm I}^2}{\lambda_{\rm I}^2} + \frac{\widetilde{n}^2}{R_{\rm I}^2}. \tag{4.14}$$

At generic points of our moduli space, the twisted states carry non-trivial representations of $SO(16) \times SO(16)$. The twisted wrapping modes of the membrane correspond to the KK modes of the open Type I string. The twisted KK modes of the M-theory are non-perturbative string states of the Type I theory, with masses $\propto R_{\rm I}/\lambda_{\rm I}$; they are charged under the gauge group, and should be identified with the charged heterotic soliton strings of the Type I theory.

In the Type IA theory, we obtain the following mass formula:

$$M_{\rm IA}^2 = \frac{\widetilde{m}^2}{\lambda_{\rm IA}^2} + \widetilde{n}^2 R_{\rm IA}^2. \tag{4.15}$$

While the twisted wrappping modes of the M-theory correspond to the perturbative winding modes of the Type IA open string, the twisted KK modes show up in the Type IA theory as additional non-perturbative black-hole states, charged under $SO(16) \times SO(16)$.

In the $E_8 \times E_8$ heterotic theory,

$$M_{E_8}^2 = \frac{\widetilde{m}^2}{R_{E_8}^2} + \widetilde{n}^2 R_{E_8}^2. \tag{4.16}$$

Both sectors are perturbative heterotic string states in non-trivial representations of $SO(16) \times SO(16)$. In the $SO(32)$ heterotic theory, the corresponding masses are

$$M_h^2 = \widetilde{m}^2 R_h^2 + \frac{\widetilde{n}^2}{R_h^2}, \tag{4.17}$$

which is of course in accord with T-duality.

One can go on and analyze spectra of other p-branes. Let us only notice here that the space-time orbifold singularities of the M-theory on $\mathbb{R}^9 \times S^1/\mathbb{Z}_2 \times S^1$ are intriguing M-theoretical analogs of Dirichlet-branes of string theory.

Acknowledgement

We would like to thank C.V. Johnson, J. Polchinski, J.H. Schwarz, and N. Seiberg for discussions.

References

[1] E. Witten, String theory dynamics in various dimensions, Nucl. Phys. B 443 (1995) 85.
[2] P.K. Townsend, The eleven-dimensional supermembrane revisited, Phys. Lett. B 350 (1995) 184.
[3] M.J. Duff, Electric/magnetic duality and its stringy origins, Texas A&M preprint hep-th/9509106; M.J. Duff, R.R. Khuri and J.X. Lu, String solitons, Phys. Rep. 259 (1995) 213.
[4] P.K. Townsend, p-Brane democracy, Cambridge preprint hep-th/9507048.
[5] J.H. Schwarz, An $SL(2, \mathbb{Z})$ multiplet of Type IIB superstrings, superstring dualities and the power of M theory, CALTECH preprints hep-th/9508143, 9509148 and 9510086.
[6] E. Bergshoeff, E. Sezgin and P.K. Townsend, Supermembranes and eleven-dimensional supergravity, Phys. Lett. B 189 (1987) 75; Properties of the eleven-dimensional supermembrane theory, Ann. Phys. 185 (1988) 330.
[7] A. Dabholkar, Ten-dimensional heterotic string as a soliton, Phys. Lett. B 357 (1995) 307.
[8] C.M. Hull, String-string duality in ten dimensions, Phys. Lett. B 357 (1995) 545.
[9] J. Polchinski and E. Witten, Evidence for heterotic - Type I string duality, Nucl. Phys. B 460 (1996) 525, next article in this issue.
[10] P.S. Aspinwall, Some relationships between dualities in string theory, Cornell preprint hep-th/9508154.
[11] L. Alvarez-Gaumé and E. Witten, Gravitational anomalies, Nucl. Phys. B 234 (1983) 269.
[12] M. Kervaire and J. Milnor, Groups of homotopy spheres I, Ann. Math. 77 (1963) 504.
[13] M.B. Green and J.H. Schwarz, Anomaly cancellations in supersymmetric $D = 10$ gauge theory and superstring theory, Phys. Lett. B 149 (1984) 117.
[14] M.J. Duff, P.S. Howe, T. Inami and K.S. Stelle, Superstrings in $D = 10$ from supermembranes in $D = 11$, Phys. Lett. B 191 (1987) 70.
[15] J. Hughes and J. Polchinski, Partially broken global supersymmetry and the superstring, Nucl. Phys. B 278 (1986) 147.

[16] M.J. Duff, J.T. Liu and R. Minasian, Eleven-dimensional origin of string/string duality: A one-loop test, Texas A&M preprint hep-th/9506126.

[17] J. Blum and J.A. Harvey, Anomaly inflow for gauge defects, Nucl. Phys. B 416 (1994) 119.

[18] E. Bergshoeff, C. Hull and T. Ortín, Duality in the Type-II superstring effective action, Nucl. Phys. B 451 (1995) 547;
E. Bergshoeff, B. Janssen and T. Ortín, Solution-generating transformations and the string effective action, hep-th/9506156.

[19] A. Sagnotti, Open strings and their symmetry groups, *in* Cargèse '87, Nonperturbative Quantum Field Theory, ed. G. Mack et al. (Pergamon, Oxford, 1988) p. 521.

[20] P. Hořava, Strings on world-sheet orbifolds, Nucl. Phys. B 327 (1989) 461.

[21] J. Dai, R.G. Leigh and J. Polchinski, New connections between string theories, Mod. Phys. Lett. 4 (1989) 2073.

[22] P. Hořava, Background duality of open string models, Phys. Lett. B 231 (1989) 351.

[23] J. Polchinski, Dirichlet-branes and Ramond-Ramond charges, ITP preprint hep-th/9510017.

[24] P. Hořava, Chern-Simons gauge theory on orbifolds: Open strings from three dimensions, hep-th/9404101.

[25] J. Polchinski, Combinatorics of boundaries in string theory, Phys. Rev. D 50 (1994) 6041.

[26] K.S. Narain, New heterotic string theories in uncompactified dimensions < 10, Phys. Lett. B 169 (1986) 41;
K.S. Narain, M.H. Sarmadi and E. Witten, A note on toroidal compactification of heterotic string theory, Nucl. Phys. B 279 (1987) 369.

[27] P. Ginsparg, On toroidal compactification of heterotic superstrings, Phys. Rev. D 35 (1987) 648.

Int. J. Mod. Phys. A **11** (1996) 5623–41
Reprinted with permission by World Scientific Publishing

M THEORY (THE THEORY FORMERLY KNOWN AS STRINGS)

M. J. DUFF*

Center for Theoretical Physics, Texas A&M University, College Station, Texas 77843, USA

Received 26 August 1996

Superunification underwent a major paradigm shift in 1984 when eleven-dimensional supergravity was knocked off its pedestal by ten-dimensional superstrings. This last year has witnessed a new shift of equal proportions: perturbative ten-dimensional superstrings have in their turn been superseded by a new nonperturbative theory called *M theory*, which describes supermembranes and superfivebranes, which subsumes all five consistent string theories and whose low energy limit is, ironically, eleven-dimensional supergravity. In particular, six-dimensional string/string duality follows from membrane/fivebrane duality by compactifying *M* theory on $S^1/Z_2 \times K3$ (heterotic/heterotic duality) or $S^1 \times K3$ (Type IIA/heterotic duality) or $S^1/Z_2 \times T^4$ (heterotic/Type IIA duality) or $S^1 \times T^4$ (Type IIA/Type IIA duality).

1. Ten to Eleven: It Is Not Too Late

The maximum space–time dimension in which one can formulate a consistent supersymmetric theory is eleven.[a] For this reason in the early 1980's many physicists looked to $D = 11$ supergravity,[2] in the hope that it might provide that superunification[3] they were all looking for. Then in 1984, superunification underwent a major paradigm shift: eleven-dimensional supergravity was knocked off its pedestal by ten-dimensional superstrings,[4] and eleven dimensions fell out of favor. This last year, however, has witnessed a new shift of equal proportions: perturbative ten-dimensional superstrings have in their turn been superseded by a new nonperturbative theory called *M theory*, which describes (amongst other things) supersymmetric extended objects with two spatial dimensions (*supermembranes*), and five spatial

*Research supported in part by NSF Grant PHY-9411543.
E-mail: duff@phys.tamu.edu
[a]The field-theoretic reason is based on the prejudice that there be no massless particles with spins greater than two.[1] However, as discussed in Sec. 5, $D = 11$ emerges naturally as the maximum dimension admitting super p-branes in Minkowski signature.

dimensions (*superfivebranes*), which subsumes all five consistent string theories and whose low energy limit is, ironically, eleven-dimensional supergravity.

The reason for this reversal of fortune of eleven dimensions is due, in large part, to the 1995 paper by Witten.[5] One of the biggest problems with $D = 10$ string theory[4] is that there are *five* consistent string theories: Type I SO(32), heterotic SO(32), heterotic $E_8 \times E_8$, Type IIA and Type IIB. As a candidate for a unique *theory of everything*, this is clearly an embarrassment of riches. Witten put forward a convincing case that this distinction is just an artifact of perturbation theory and that nonperturbatively these five theories are, in fact, just different corners of a deeper theory. Moreover, this deeper theory, subsequently dubbed *M theory*, has $D = 11$ supergravity as its low energy limit! Thus the five string theories and $D = 11$ supergravity represent six different special points[b] in the moduli space of M theory. The small parameters of perturbative string theory are provided by $\langle e^{\Phi} \rangle$, where Φ is the dilaton field, and $\langle e^{\sigma_i} \rangle$ where σ_i are the moduli fields which arise after compactification. What makes M theory at once intriguing and yet difficult to analyze is that in $D = 11$ there is neither dilaton nor moduli and hence the theory is intrinsically nonperturbative. Consequently, the ultimate meaning of M theory is still unclear, and Witten has suggested that in the meantime, M should stand for "magic," "mystery" or "membrane," according to taste.

The relation between the membrane and the fivebrane in $D = 11$ is analogous to the relation between electric and magnetic charges in $D = 4$. In fact, this is more than an analogy: electric/magnetic duality in $D = 4$ string theory[6,7] follows as a consequence of string/string duality in $D = 6$.[8] The main purpose of this paper is to show how $D = 6$ string/string duality[9–14,5] follows, in its turn, as a consequence of membrane/fivebrane duality in $D = 11$. In particular, heterotic/heterotic duality, Type IIA/heterotic duality, heterotic/Type IIA duality and Type IIA/Type IIA duality follow from membrane/fivebrane duality by compactifying M theory on $S^1/Z_2 \times K3$,[15] $S^1 \times K3$,[16] $S^1/Z_2 \times T^4$ and $S^1 \times T^4$, respectively.

First, however, I want to pose the question: "Should we have been surprised by the eleven-dimensional origin of string theory?"

2. Type II A&M Theory

The importance of eleven dimensions is no doubt surprising from the point of view of perturbative string theory; from the point of view of membrane theory, however, there were already tantalizing hints in this direction:

(i) *K3 compactification*

The four-dimensional compact manifold $K3$ plays a ubiquitous role in much of present day M theory. It was first introduced as a compactifying manifold in 1983[18]

[b]Some authors take the phrase *M theory* to refer merely to this sixth corner of the moduli space. With this definition, of course, M theory is no more fundamental than the other five corners. For us, *M theory* means the whole kit and caboodle.

when it was realized that the number of unbroken supersymmetries surviving compactification in a Kaluza–Klein theory depends on the *holonomy* group of the extra dimensions. By virtue of its SU(2) holonomy, $K3$ preserves precisely half of the supersymmetry. This means, in particular, that an $N = 2$ theory on $K3$ has the same number of supersymmetries as an $N = 1$ theory on T^4, a result which was subsequently to prove of vital importance for string/string duality. In 1986, it was pointed out[19] that $D = 11$ supergravity on $R^{10-n} \times K3 \times T^{n-3}$ [18] and the $D = 10$ heterotic string on $R^{10-n} \times T^n$ [20] not only have the same supersymmetry but also the same moduli spaces of vacua, namely

$$\mathcal{M} = \frac{\mathrm{SO}(16+n, n)}{\mathrm{SO}(16+n) \times \mathrm{SO}(n)} \,. \tag{2.1}$$

It took almost a decade for this "coincidence" to be explained but we now know that M theory $R^{10-n} \times K3 \times T^{n-3}$ is dual to the heterotic string on $R^{10-n} \times T^n$.

(ii) *Superstrings in $D = 10$ from supermembranes in $D = 11$*

Eleven dimensions received a big shot in the arm in 1987 when the $D = 11$ supermembrane was discovered.[21] The bosonic sector of its $d = 3$ world volume Green–Schwarz action is given by

$$S_3 = T_3 \int d^3\xi \left[-\frac{1}{2}\sqrt{-\gamma}\gamma^{ij}\partial_i X^M \partial_j X^N G_{MN}(X) + \frac{1}{2}\sqrt{-\gamma} \right.$$
$$\left. - \frac{1}{3!}\epsilon^{ijk}\partial_i X^M \partial_j X^N \partial_k X^P C_{MNP}(X) \right] , \tag{2.2}$$

where T_3 is the membrane tension, ξ^i $(i = 1, 2, 3)$ are the world volume coordinates, γ^{ij} is the world volume metric and $X^M(\xi)$ are the space–time coordinates ($M = 0, 1, \ldots, 10$). Kappa symmetry[21] then demands that the background metric G_{MN} and background three-form potential C_{MNP} obey the classical field equations of $D = 11$ supergravity,[2] whose bosonic action is

$$I_{11} = \frac{1}{2\kappa_{11}^2} \int d^{11}x \sqrt{-G} \left[R_G - \frac{1}{2 \cdot 4!} {K_4}^2 \right] - \frac{1}{12\kappa_{11}^2} \int C_3 \wedge K_4 \wedge K_4 \,, \tag{2.3}$$

where $K_4 = dC_3$ is the four-form field strength. In particular, K_4 obeys the field equation

$$d * K_4 = -\frac{1}{2}{K_4}^2 \tag{2.4}$$

and the Bianchi identity

$$dK_4 = 0 \,. \tag{2.5}$$

It was then pointed out[22] that in an $R^{10} \times S^1$ topology the weakly coupled $(d = 2, D = 10)$ Type IIA superstring follows by wrapping the $(d = 3, D = 11)$ supermembrane around the circle in the limit that its radius R shrinks to zero.

In particular, the Green–Schwarz action of the string follows in this way from the Green–Schwarz action of the membrane. It was necessary to take this $R \to 0$ limit in order to send to infinity the masses of the (at the time) unwanted Kaluza–Klein modes which had no place in weakly coupled Type IIA theory. The $D = 10$ dilaton, which governs the strength of the string coupling, is just a component of the $D = 11$ metric.

A critique of superstring orthodoxy *circa* 1987, and its failure to accommodate the eleven-dimensional supermembrane, may be found in Ref. 23.

(iii) *U-duality (when it was still non-U)*

Based on considerations of this $D = 11$ supermembrane, which on further compactification treats the dilaton and moduli fields on the same footing, it was conjectured[26] in 1990 that discrete subgroups of all the old noncompact global symmetries of compactified supergravity[24,25] (e.g. $SL(2, R)$, $O(6,6)$, E_7) should be promoted to duality symmetries of the supermembrane. Via the above wrapping around S^1, therefore, they should also be inherited by the Type IIA string.[26]

(iv) $D = 11$ *membrane/fivebrane duality*

In 1991, the supermembrane was recovered as an elementary solution of $D = 11$ supergravity which preserves half of the space–time supersymmetry.[27] Making the three/eight split $X^M = (x^\mu, y^m)$ where $\mu = 0, 1, 2$ and $m = 3, \ldots, 10$, the metric is given by

$$ds^2 = (1 + k_3/y^6)^{-2/3} dx^\mu dx_\mu + (1 + k_3/y^6)^{1/3}(dy^2 + y^2 d\Omega_7{}^2) \tag{2.6}$$

and the four-form field strength by

$$\tilde{K}_7 \equiv *K_4 = 6k_3\epsilon_7\,, \tag{2.7}$$

where the constant k_3 is given by

$$k_3 = \frac{2\kappa_{11}{}^2 T_3}{\Omega_7}\,. \tag{2.8}$$

Here ϵ_7 is the volume form on S^7 and Ω_7 is the volume. The mass per unit area of the membrane $\mathcal{M}_3$ is equal to its tension:

$$\mathcal{M}_3 = T_3\,. \tag{2.9}$$

This *elementary* solution is a singular solution of the supergravity equations coupled to a supermembrane source and carries a Noether "electric" charge

$$Q = \frac{1}{\sqrt{2}\kappa_{11}} \int_{S^7} (*K_4 + C_3 \wedge K_4) = \sqrt{2}\kappa_{11} T_3\,. \tag{2.10}$$

Hence the solution saturates the Bogomol'nyi bound $\sqrt{2}\kappa_{11}\mathcal{M}_3 \geq Q$. This is a consequence of the preservation of half the supersymmetries which is also intimately linked with the world volume kappa symmetry. The zero modes of this solution belong to a $(d = 3, n = 8)$ supermultiplet consisting of eight scalars and eight spinors (ϕ^I, χ^I), with $I = 1, \ldots, 8$, which correspond to the eight Goldstone bosons and their superpartners associated with breaking of the eight translations transverse to the membrane world volume.

In 1992, the superfivebrane was discovered as a soliton solution of $D = 11$ supergravity also preserving half the space–time supersymmetry.[28] Making the six/five split $X^M = (x^\mu, y^m)$ where $\mu = 0, 1, 2, 3, 4, 5$ and $m = 6, \ldots, 10$, the metric is given by

$$ds^2 = (1 + k_6/y^3)^{-1/3} dx^\mu dx_\mu + (1 + k_6/y^3)^{2/3}(dy^2 + y^2 d\Omega_4{}^2) \tag{2.11}$$

and the four-index field-strength by

$$K_4 = 3k_6 \epsilon_4 \,, \tag{2.12}$$

where the fivebrane tension $\tilde{T}_6$ is related to the constant k_6 by

$$k_6 = \frac{2\kappa_{11}{}^2 \tilde{T}_6}{3\Omega_4} \,. \tag{2.13}$$

Here ϵ_4 is the volume form on S^4 and Ω_4 is the volume. The mass per unit five-volume of the fivebrane $\mathcal{M}_6$ is equal to its tension:

$$\mathcal{M}_6 = \tilde{T}_6 \,. \tag{2.14}$$

This *solitonic* solution is a nonsingular solution of the source-free equations and carries a topological "magnetic" charge

$$P = \frac{1}{\sqrt{2}\kappa_{11}} \int_{S^4} K_4 = \sqrt{2}\kappa_{11} \tilde{T}_6 \,. \tag{2.15}$$

Hence the solution saturates the Bogomol'nyi bound $\sqrt{2}\kappa_{11}\mathcal{M}_6 \geq P$. Once again, this is a consequence of the preservation of half the supersymmetries. The covariant action for this $D = 11$ superfivebrane is still unknown (see Refs. 29 and 30 for recent progress) but consideration of the soliton zero modes[31,14,32] means that the gauged fixed action must be described by the same chiral antisymmetric tensor multiplet $(B^-{}_{\mu\nu}, \lambda^I, \phi^{[IJ]})$ as that of the Type IIA fivebrane.[33,34] Note that in addition to the five scalars corresponding to the five translational Goldstone bosons, there is also a two-form $B^-{}_{\mu\nu}$ whose three-form field strength is anti-self-dual and which describes three degrees of freedom.

The electric and magnetic charges obey a Dirac quantization rule[35,36]

$$QP = 2\pi n \,, \qquad n = \text{integer} \,. \tag{2.16}$$

Or, in terms of the tensions,[37,11]

$$2\kappa_{11}{}^2 T_3 \tilde{T}_6 = 2\pi n \,. \tag{2.17}$$

This naturally suggests a $D = 11$ membrane/fivebrane duality. Note that this reduces the three dimensionful parameters T_3, $\tilde{T}_6$ and κ_{11} to two. Moreover, it was recently shown[16] that they are not independent. To see this, we note from (2.2) that C_3 has period $2\pi/T_3$ so that K_4 is quantized according to

$$\int K_4 = \frac{2\pi n}{T_3}, \qquad n = \text{integer}. \tag{2.18}$$

Consistency of such C_3 periods with the space–time action, (2.3), gives the relation[c]

$$\frac{(2\pi)^2}{{\kappa_{11}}^2 T_3^3} \in 2Z. \tag{2.19}$$

From (2.17), this may also be written as

$$2\pi \frac{\tilde{T}_6}{{T_3}^2} \in Z. \tag{2.20}$$

Thus the tension of the singly charged fivebrane is given by

$$\tilde{T}_6 = \frac{1}{2\pi} {T_3}^2. \tag{2.21}$$

(v) *Hidden eleventh dimension*

We have seen how the $D = 10$ Type IIA string follows from $D = 11$. Is it possible to go the other way and discover an eleventh dimension hiding in $D = 10$? In 1993, it was recognized[40] that by dualizing a vector into a scalar on the gauge-fixed $d = 3$ world volume of the Type IIA supermembrane, one increases the number of world volume scalars (i.e. transverse dimensions) from 7 to 8 and hence obtains the corresponding world volume action of the $D = 11$ supermembrane. Thus the $D = 10$ Type IIA theory contains a hidden $D = 11$ Lorentz invariance! This device was subsequently used[41,42] to demonstrate the equivalence of the actions of the $D = 10$ Type IIA membrane and the Dirichlet twobrane.[43]

(vi) *U-duality*

Of the conjectured Cremmer–Julia symmetries referred to in (iii) above, the case for a target space $O(6, 6; Z)$ (*T-duality*) in perturbative string theory had already been made, of course.[44] Stronger evidence for an SL$(2, Z)$ (*S-duality*) in string theory was subsequently provided in Refs. 6 and 7 where it was pointed out that it corresponds to a *nonperturbative* electric/magnetic symmetry.

In 1994, stronger evidence for the combination of S and T into a discrete duality of Type II strings, such as $E_7(Z)$ in $D = 4$, was provided in Ref. 13, where

[c] This corrects a factor of two errors in Ref. 16 and brings us into agreement with a subsequent D-brane derivation[38] of (2.21). I am grateful to Shanta De Alwis[39] for pointing out the source of the error.

it was dubbed *U-duality*. Moreover, the BPS spectrum necessary for this U-duality was given an explanation in terms of the wrapping of either the $D = 11$ membrane or $D = 11$ fivebrane around the extra dimensions. This paper also conjectured a nonperturbative $SL(2, Z)$ of the Type IIB string in $D = 10$.

(vii) *Black holes*

In 1995, it was conjectured[32] that the $D = 10$ Type IIA superstring should be identified with the $D = 11$ supermembrane compactified on S^1, even for large R. The $D = 11$ Kaluza–Klein modes (which, as discussed in (ii) above, had no place in the *perturbative* Type IIA theory) were interpreted as charged extreme black holes of the Type IIA theory.

(viii) $D = 11$ *membrane/fivebrane duality and anomalies*

Membrane/fivebrane duality interchanges the roles of field equations and Bianchi identities. From (2.4), the fivebrane Bianchi identity reads

$$d\tilde{K}_7 = -\frac{1}{2}{K_4}^2 . \tag{2.22}$$

However, it was recognized in 1995 that such a Bianchi identity will in general require gravitational Chern–Simons corrections arising from a sigma-model anomaly on the fivebrane world volume[16]

$$d\tilde{K}_7 = -\frac{1}{2}K_4^2 + \frac{2\pi}{\tilde{T}_6}\tilde{X}_8 , \tag{2.23}$$

where the eight-form polynomial $\tilde{X}_8$, quartic in the gravitational curvature R, describes the Lorentz $d = 6$ world volume anomaly of the $D = 11$ fivebrane. Although the covariant fivebrane action is unknown, we know that the gauge-fixed theory is described by the chiral antisymmetric tensor multiplet $(B^-_{\mu\nu}, \lambda^I, \phi^{[IJ]})$, and it is a straightforward matter to read off the anomaly polynomial from the literature. For example see Ref. 45. We find

$$\tilde{X}_8 = \frac{1}{(2\pi)^4}\left[-\frac{1}{768}(\mathrm{tr}\ R^2)^2 + \frac{1}{192}\mathrm{tr}\ R^4\right] . \tag{2.24}$$

Thus membrane/fivebrane duality predicts a space–time correction to the $D = 11$ supermembrane action[16]

$$I_{11}(\mathrm{Lorentz}) = T_3\int C_3 \wedge \frac{1}{(2\pi)^4}\left[-\frac{1}{768}(\mathrm{tr}\ R^2)^2 + \frac{1}{192}\mathrm{tr}\ R^4\right] . \tag{2.25}$$

Such a correction was also derived in a somewhat different way in Ref. 17. This prediction is intrinsically M theoretic, with no counterpart in ordinary $D = 11$ supergravity. However, by simultaneous dimensional reduction[22] of ($d = 3$, $D = 11$) to ($d = 2$, $D = 10$) on S^1, it translates into a corresponding prediction for the Type IIA string:

$$I_{10}(\text{Lorentz}) = T_2 \int B_2 \wedge \frac{1}{(2\pi)^4}\left[-\frac{1}{768}(\text{tr } R^2)^2 + \frac{1}{192}\text{tr } R^4\right], \tag{2.26}$$

where B_2 is the string two-form and $T_2 = 1/2\pi\alpha'$ is the string tension.

As a consistency check we can compare this prediction with previous results found by explicit string one-loop calculations. These have been done in two ways: either by computing directly in $D = 10$ the Type IIA anomaly polynomial[46] following,[47] or by compactifying to $D = 2$ on an eight-manifold M and computing the B_2 one-point function.[48] We indeed find agreement. Thus using $D = 11$ membrane/fivebrane duality we have correctly reproduced the corrections to the B_2 field equations of the $D = 10$ Type IIA string (a mixture of tree-level and string one-loop effects) starting from the Chern–Simons corrections to the Bianchi identities of the $D = 11$ superfivebrane (a purely tree-level effect). It would be interesting to know, on the membrane side, what calculation in $D = 11$ M theory, when reduced on S^1, corresponds to this one-loop Type IIA string amplitude calculation in $D = 10$. Understanding this may well throw a good deal of light on the mystery of what M theory really is!

(ix) *Heterotic string from fivebrane wrapped around K3*

In 1995 it was shown that, when wrapped around $K3$ with its 19 self-dual and 3 anti-self-dual two-forms, the $d = 6$ world volume fields of the $D = 11$ fivebrane (or Type IIA fivebrane) $(B^{-}{}_{\mu\nu}, \lambda^I, \phi^{[IJ]})$ reduce to the $d = 2$ world sheet fields of the heterotic string in $D = 7$ (or $D = 6$).[49,50] The two-form yields $(19, 3)$ left and right moving bosons, the spinors yield $(0, 8)$ fermions and the scalars yield $(5, 5)$ which add up to the correct world sheet degrees of freedom of the heterotic string.[49,50]

A consistency check is provided[16] by the derivation of the Yang–Mills and Lorentz Chern–Simons corrections to the Bianchi identity of the heterotic string starting from the fivebrane Bianchi identity given in (viii). Making the seven/four split $X^M = (x^\mu, y^m)$ where $\mu = 0, \ldots, 6$ and $m = 7, 8, 9, 10$, the original set of $D = 11$ fields may be decomposed in a basis of harmonic p-forms on $K3$. In particular, we expand C_3 as

$$C_3(X) = C_3(x) + \frac{1}{2T_3}\sum C_1^I(x)\omega_2^I(y), \tag{2.27}$$

where ω_2^I, $I = 1, \ldots, 22$ are an integral basis of b_2 harmonic two-forms on $K3$. Following Ref. 12, let us define the dual string three-form $\tilde{H}_3$ by

$$T_2\tilde{H}_3 = \tilde{T}_6 \int_{K3} \tilde{K}_7 \,. \tag{2.28}$$

The dual string Lorentz anomaly polynomial, $\tilde{X}_4$, is given by

$$\tilde{X}_4 = \int_{K3} \tilde{X}_8 = \frac{1}{(2\pi)^2}\frac{1}{192}\text{tr } R^2 p_1(K3)\,, \tag{2.29}$$

where $p_1(K3)$ is the Pontryagin number of $K3$

$$p_1(K3) = -\frac{1}{8\pi^2}\int_{K3} \mathrm{tr}\ R_0^2 = -48\,. \tag{2.30}$$

We may now integrate (2.23) over $K3$, using (2.21) to find

$$d\tilde{H}_3 = -\frac{\alpha'}{4}[K_2^I K_2^J d_{IJ} + \mathrm{tr}\ R^2]\,, \tag{2.31}$$

where $K_2^I = dC_1^I$ and where d_{IJ} is the intersection matrix on $K3$, given by

$$d_{IJ} = \int_{K3} \omega_2^I \wedge \omega_2^J \tag{2.32}$$

which has $b_2^+ = 3$ positive and $b_2^- = 19$ negative eigenvalues. Thus we see that this form of the Bianchi identity corresponds to a $D = 7$ toroidal compactification of a heterotic string at a generic point on the Narain lattice.[20] Thus we have reproduced the $D = 7$ Bianchi identity of the heterotic string, starting from the $D = 11$ fivebrane.

For use in Sec. 3, we note that if we replace $K3$ by T^4 in the above derivation, the two-form now yields $(3,3)$ left and right moving bosons, the spinors now yield $(8,8)$ fermions and the scalars again yield $(5,5)$ which add up to the correct world sheet degrees of freedom of the Type IIA string. In this case, the Bianchi identity becomes $d\tilde{H}_3 = 0$ as it should be.

(x) $N = 1$ *in* $D = 4$

Also in 1995 it was noted[51–53,55,54,56,64] that $N = 1$ heterotic strings can be dual to $D = 11$ supergravity compactified on seven-dimensional spaces of G_2 holonomy which also yield $N = 1$ in $D = 4$.[57]

(xi) *Nonperturbative effects*

Also in 1995 it was shown[58] that membranes and fivebranes of the Type IIA theory, obtained by compactification on S^1, yield e^{-1/g_s} effects, where g_s is the string coupling.

(xii) SL$(2, Z)$

Also in 1995, strong evidence was provided for identifying the Type IIB string on $R^9 \times S^1$ with M theory on $R^9 \times T^2$.[38,64] In particular, the conjectured SL$(2, Z)$ of the Type IIB theory discussed in (vi) above is just the modular group of the M theory torus.[d]

[d]Two alternative explanations of this SL$(2, Z)$ had previously been given: (a) identifying it with the S-duality[16] of the $d = 4$ Born–Infeld world volume theory of the self-dual Type IIB superthreebrane,[65] and (b) using the four-dimensional heterotic/Type IIA/Type IIB triality[66] by noting that this SL$(2, Z)$, while nonperturbative for the Type IIB string, is perturbative for the heterotic string.

(xiii) $E_8 \times E_8$ *heterotic string*

Also in 1995 (that *annus mirabilis*!), strong evidence was provided for identifying the $E_8 \times E_8$ heterotic string[e] on R^{10} with M theory on $R^{10} \times S^1/Z_2$.[67]

This completes our summary of M theory before M theory was cool. The *phrase* M theory (though, as I hope to have shown, not the *physics* of M theory) first made its appearance in October 1995.[38,67] This was also the month that it was proposed[43] that the Type II p-branes carrying Ramond–Ramond charges can be given an exact conformal field theory description via open strings with Dirichlet boundary conditions, thus heralding the era of *D-branes*. Since then, evidence in favor of M theory and D-branes has been appearing daily on the internet, including applications to black holes,[62] length scales shorter than the string scale[59] and even phenomenology.[60,61] We refer the reader to the review by Schwarz[63] for these more recent developments in M theory, to the review by Polchinski[69] for developments in D-branes and to the paper by Aharony, Sonnenschein and Yankielowicz[70] for the connection between the two (since D-branes are intrinsically ten-dimensional and M theory is eleven-dimensional, this is not at all obvious). Here, we wish to focus on a specific application of M theory, namely the derivation of string/string dualities.

3. String/String Duality from M Theory

Let us consider M theory, with its fundamental membrane and solitonic fivebrane, on $R^6 \times M_1 \times \tilde{M}_4$ where M_1 is a one-dimensional compact space of radius R and $\tilde{M}_4$ is a four-dimensional compact space of volume V. We may obtain a fundamental string on R^6 by wrapping the membrane around M_1 and reducing on $\tilde{M}_4$. Let us denote fundamental string sigma-model metrics in $D = 10$ and $D = 6$ by G_{10} and G_6. Then from the corresponding Einstein Lagrangians

$$\sqrt{-G_{11}}R_{11} = R^{-3}\sqrt{-G_{10}}R_{10} = \frac{V}{R}\sqrt{-G_6}R_6\,, \tag{3.1}$$

we may read off the strength of the string couplings in $D = 10$[15]

$$\lambda_{10}{}^2 = R^3 \tag{3.2}$$

and $D = 6$

$$\lambda_6{}^2 = \frac{R}{V}\,. \tag{3.3}$$

Similarly we may obtain a solitonic string on R^6 by wrapping the fivebrane around $\tilde{M}_4$ and reducing on M_1. Let us denote the solitonic string sigma-model metrics in $D = 7$ and $D = 6$ by $\tilde{G}_7$ and $\tilde{G}_6$. Then from the corresponding Einstein Lagrangians

$$\sqrt{-G_{11}}R_{11} = V^{-3/2}\sqrt{-\tilde{G}_7}\tilde{R}_7 = \frac{R}{V}\sqrt{-\tilde{G}_6}\tilde{R}_6\,, \tag{3.4}$$

[e]It is ironic that, having hammered the final nail in the coffin of $D = 11$ supergravity by telling us that it can never yield a *chiral* theory when compactified on a manifold,[68] Witten pulls it out again by telling us that it does yield a chiral theory when compactified on something that is not a manifold!

we may read off the strength of the string couplings in $D = 7$[15]

$$\tilde{\lambda}_7^2 = V^{3/2} \tag{3.5}$$

and $D = 6$

$$\tilde{\lambda}_6^2 = \frac{V}{R}\,. \tag{3.6}$$

Thus we see that the fundamental and solitonic strings are related by a strong/weak coupling:

$$\tilde{\lambda}_6^2 = 1/{\lambda_6}^2\,. \tag{3.7}$$

We shall be interested in $M_1 = S^1$ (in which case from (ii) of Sec. 2 the fundamental string will be Type IIA) or $M_1 = S^1/Z_2$ (in which case from (xiii) of Sec. 2 the fundamental string will be heterotic $E_8 \times E_8$). Similarly, we will be interested in $\tilde{M}_4 = T^4$ (in which case from (ix) of Sec. 2 the solitonic string will be Type IIA) or $\tilde{M}_4 = K3$ (in which case from (ix) of Sec. 2 the solitonic string will be heterotic). Thus, there are four possible scenarios which are summarized in Table 1. (N_+, N_-) denotes the $D = 6$ space–time supersymmetries. In each case, the fundamental string will be weakly coupled as we shrink the size of the wrapping space M_1 and the dual string will be weakly coupled as we shrink the size of the wrapping space $\tilde{M}_4$.

Table 1. String/string dualities.

(N_+, N_-)	M_1	$\tilde{M}_4$	Fundamental string	Dual string
$(1,0)$	S^1/Z_2	$K3$	heterotic	heterotic
$(1,1)$	S^1	$K3$	Type IIA	heterotic
$(1,1)$	S^1/Z_2	T^4	heterotic	Type IIA
$(2,2)$	S^1	T^4	Type IIA	Type IIA

In fact, there is in general a topological obstruction to wrapping the fivebrane around $\tilde{M}_4$ provided by (2.18) because the fivebrane cannot wrap around a four-manifold that has $n \neq 0$.[f] This is because the anti-self-dual three-form field strength T on the world volume of the fivebrane obeys[41,17]

$$dT = K_4 \tag{3.8}$$

and the existence of a solution for T therefore requires that K_4 must be cohomologically trivial. For M theory on $R^6 \times S^1/Z_2 \times T^4$ this is no problem. For M theory

[f] Actually, as recently shown in Ref. 71, the object which must have integral periods is not $T_3K_4/2\pi$ but rather $T_3K_4/2\pi - p_1/4$ where p_1 is the first Pontryagin class. This will not affect our conclusions, however.

on $R^6 \times S^1/Z_2 \times K3$, with instanton number k in one E_8 and $24 - k$ in the other, however, the flux of K_4 over $K3$ is[15]

$$n = 12 - k\,. \tag{3.9}$$

Consequently, the M theoretic explanation of heterotic/heterotic duality requires $E_8 \times E_8$ with the symmetric embedding $k = 12$. This has some far-reaching implications. For example, the duality exchanges gauge fields that can be seen in perturbation theory with gauge fields of a nonperturbative origin.[15]

The dilaton $\tilde{\Phi}$, the string σ-model metric $\tilde{G}_{MN}$ and three-form field strength $\tilde{H}$ of the dual string are related to those of the fundamental string, Φ, G_{MN} and H by the replacements[11,12]

$$\Phi \to \tilde{\Phi} = -\Phi\,, \quad G_{MN} \to \tilde{G}_{MN} = e^{-\Phi} G_{MN}\,, \quad H \to \tilde{H} = e^{-\Phi} * H\,. \tag{3.10}$$

In the case of heterotic/Type IIA duality and Type IIA/heterotic duality, this operation takes us from one string to the other, but in the case of heterotic/heterotic duality and Type IIA/Type IIA duality, this operation is a discrete symmetry of the theory. This Type IIA/Type IIA duality is discussed in Ref. 78 and we recognize this symmetry as subgroup of the SO(5, 5; Z) U-duality[26,13,79] of the $D = 6$ Type IIA string.

Vacua with $(N_+, N_-) = (1, 0)$ in $D = 6$ have been the subject of much interest lately. In addition to DMW vacua[16] discussed above, obtained from M theory on $S^1/Z_2 \times K3$, there are also the GP vacua[72–74] obtained from the SO(32) theory on $K3$ and the MV vacua[85,86] obtained from F *theory*[75] on Calabi–Yau. Indeed, all three categories are related by duality.[80,85,83,97,84,76,86,81] In particular, the DMW heterotic strong/weak coupling duality gets mapped to a T-duality of the Type I version of the SO(32) theory, and the nonperturbative gauge symmetries of the DMW model arise from small Spin(32)/Z_2 instantons in the heterotic version of the SO(32) theory.[76] Because heterotic/heterotic duality interchanges world sheet and space–time loop expansions — or because it acts by duality on H — the duality exchanges the tree level Chern–Simons contributions to the Bianchi identity

$$dH = \alpha'(2\pi)^2 X_4\,, \qquad X_4 = \frac{1}{4(2\pi)^2}[\mathrm{tr}\ R^2 - \Sigma_\alpha v_\alpha\ \mathrm{tr}\ {F_\alpha}^2] \tag{3.11}$$

with the one-loop Green–Schwarz corrections to the field equations

$$d\tilde{H} = \alpha'(2\pi)^2 \tilde{X}_4\,, \qquad \tilde{X}_4 = \frac{1}{4(2\pi)^2}[\mathrm{tr}\ R^2 - \Sigma_\alpha \tilde{v}_\alpha\ \mathrm{tr}\ {F_\alpha}^2]\,. \tag{3.12}$$

Here F_α is the field strength of the αth component of the gauge group, tr denotes the trace in the fundamental representation, and v_α, $\tilde{v}_\alpha$ are constants. In fact, the Green–Schwarz anomaly cancellation mechanism in six dimensions requires that the anomaly eight-form I_8 factorize as a product of four-forms,

$$I_8 = X_4 \tilde{X}_4\,, \tag{3.13}$$

and a six-dimensional string–string duality with the general features summarized above would exchange the two factors.[12] Moreover, supersymmetry relates the coefficients v_α, $\tilde{v}_\alpha$ to the gauge field kinetic energy. In the Einstein metric $G^c{}_{MN} = e^{-\Phi/2}G_{MN}$, the exact dilaton dependence of the kinetic energy of the gauge field $F_{\alpha\,MN}$, is[96]

$$L_{\text{gauge}} = -\frac{(2\pi)^3}{8\alpha'}\sqrt{G^c}\Sigma_\alpha\big(v_\alpha e^{-\Phi/2} + \tilde{v}_\alpha e^{\Phi/2}\big)\text{tr } F_{\alpha\,MN}F_\alpha{}^{MN}\,. \tag{3.14}$$

So whenever one of the $\tilde{v}_\alpha$ is negative, there is a value of the dilaton for which the coupling constant of the corresponding gauge group diverges. This is believed to signal a phase transition associated with the appearance of tensionless strings.[88–90] This does not happen for the symmetric embedding discussed above since the perturbative gauge fields have $v_\alpha > 0$ and $\tilde{v}_\alpha = 0$ and the nonperturbative gauge fields have $v_\alpha = 0$ and $\tilde{v}_\alpha > 0$. Another kind of heterotic/heterotic duality may arise, however, in vacua where one may Higgs away that subset of gauge fields with negative $\tilde{v}_\alpha$, and be left with gauge fields with $v_\alpha = \tilde{v}_\alpha > 0$. This happens for the nonsymmetric embedding $k = 14$ and the appearance of nonperturbative gauge fields is not required.[80–82,85,86] Despite appearances, it is known from F theory that the $k = 12$ and $k = 14$ models are actually equivalent.[85,86]

Vacua with $(N_+, N_-) = (2, 0)$ arising from Type IIB on $K3$ also have an M theoretic description, in terms of compactification on T^5/Z_2.[87,17]

4. Four Dimensions

It is interesting to consider further toroidal compactification to four dimensions, replacing R^6 by $R^4 \times T^2$. Starting with a $K3$ vacuum in which the $E_8 \times E_8$ gauge symmetry is completely Higgsed, the toroidal compactification to four dimensions gives an $N = 2$ theory with the usual three vector multiplets S, T and U related to the four-dimensional heterotic string coupling constant and the area and shape of the T^2. When reduced to four dimensions, the six-dimensional string–string duality (3.10) becomes[8] an operation that exchanges S and T, so in the case of heterotic/heterotic duality we have a discrete $S - T$ interchange symmetry. This self-duality of heterotic string vacua does not rule out the the possibility that in $D = 4$ they are also dual to Type IIA strings compactified on Calabi–Yau manifolds. In fact, as discussed in Ref. 93 when the gauge group is completely Higgsed, obvious candidates are provided by Calabi–Yau manifolds with hodge numbers $h_{11} = 3$ and $h_{21} = 243$, since these have the same massless field content. Moreover, these manifolds do indeed exhibit the $S - T$ interchange symmetry.[92,91,94] Since the heterotic string on $T^2 \times K3$ also has R to $1/R$ symmetries that exchange T and U, one might expect a complete $S - T - U$ triality symmetry, as discussed in Ref. 66. In all known models, however, the $T - U$ interchange symmetry is spoiled by nonperturbative effects.[95,98]

An interesting aspect of the Calabi–Yau manifolds X appearing in the duality between heterotic strings on $K3 \times T^2$ and Type IIA strings on X, is that they can always be written in the form of a $K3$ fibration.[92] Once again, this ubiquity of $K3$ is presumably a consequence of the interpretation of the heterotic string as the $K3$ wrapping of a fivebrane. Consequently, if X admits *two* different $K3$ fibrations, this would provide an alternative explanation for heterotic dual pairs in four dimensions[83,85,86] and this is indeed the case for the Calabi–Yau manifolds discussed above.

5. Eleven to Twelve: Is It Still Too Early?

The M theoretic origin of the Type IIB string given in (xii) of Sec. 2 seems to require going down to nine dimensions and then back up to ten. An obvious question, therefore, is whether Type IIB admits a more direct higher-dimensional explanation, like Type IIA. Already in 1987 it was suggested[99] that the $(1,1)$-signature world sheet of the Type IIB string, moving in a $(9,1)$-signature space–time, may be descended from a supersymmetric extended object with a $(2,2)$-signature world volume, moving in a $(10,2)$-signature space–time. This idea becomes even more appealing if one imagines that the $\mathrm{SL}(2,Z)$ of the Type IIB theory[13] might correspond to the modular group of a T^2 compactification from $D=12$ to $D=10$ just as the $\mathrm{SL}(2,Z)$ of S-duality corresponds to the modular group of a T^2 compactification from $D=6$ to $D=4$.[8] In view of our claims that $D=11$ is the maximum space–time dimension admitting a consistent supersymmetric theory, however, this twelve-dimensional idea requires some explanation. So let us begin by recalling the $D=11$ argument.

As a p-brane moves through space–time, its trajectory is described by the functions $X^M(\xi)$ where X^M are the space–time coordinates $(M=0,1,\ldots,D-1)$ and ξ^i are the world volume coordinates $(i=0,1,\ldots,d-1)$. It is often convenient to make the so-called "static gauge choice" by making the $D=d+(D-d)$ split

$$X^M(\xi)=(X^\mu(\xi),Y^m(\xi))\,, \tag{5.1}$$

where $\mu=0,1,\ldots,d-1$ and $m=d,\ldots,D-1$, and then setting

$$X^\mu(\xi)=\xi^\mu\,. \tag{5.2}$$

Thus the only physical world volume degrees of freedom are given by the $(D-d)$ $Y^m(\xi)$, so the number of on-shell bosonic degrees of freedom is

$$N_B=D-d\,. \tag{5.3}$$

To describe the super p-brane we augment the D bosonic coordinates $X^M(\xi)$ with anticommuting fermionic coordinates $\theta^\alpha(\xi)$. Depending on D, this spinor could be Dirac, Weyl, Majorana or Majorana–Weyl. The fermionic κ-symmetry means that half of the spinor degrees of freedom are redundant and may be eliminated by a physical gauge choice. The net result is that the theory exhibits a *d-dimensional world volume supersymmetry*[100] where the number of fermionic generators is exactly half of the generators in the original space–time supersymmetry. This partial breaking of supersymmetry is a key idea. Let M be the number of real components of the minimal spinor and N the number of supersymmetries in D space–time dimensions and let m and n be the corresponding quantities in d world volume dimensions. Since κ-symmetry always halves the number of fermionic degrees of freedom and (for $d > 2$) going on-shell halves it again, the number of on-shell fermionic degrees of freedom is

$$N_F = \frac{1}{2}mn = \frac{1}{4}MN\,. \tag{5.4}$$

World volume supersymmetry demands $N_B = N_F$ and hence

$$D - d = \frac{1}{2}mn = \frac{1}{4}MN\,. \tag{5.5}$$

We note in particular that $D_{\max} = 11$ since $M = 32$ for $D = 11$ and we find the supermembrane with $d = 3$. For $D \geq 12$, $M \geq 64$ and hence (5.5) cannot be satisfied. Actually, the above argument is strictly valid only for p-branes whose world volume degrees of freedom are described by scalar supermultiplets. There are also p-branes with vector and/or antisymmetric tensor supermultiplets on the world volume,[33,34,40] but repeating the argument still yields $D_{\max} = 11$ where we find a superfivebrane with $d = 6$.[14]

The upper bound of $D = 11$ is thus a consequence of the jump from $M = 32$ to $M = 64$ in going from $D = 11$ to $D = 12$. However, this jump can be avoided if one is willing to pay the price of changing the signature to $(10, 2)$ where it is possible to define a spinor which is both Majorana and Weyl. A naive application of the above Bose–Fermi matching argument then yields $D_{\max} = 12$ where we find an extended object with $d = 4$ but with $(2, 2)$ signature.[99] The chiral nature of this object then naturally suggests a connection with the Type IIB string in $D = 10$, although the T^2 compactification would have to be of an unusual kind in order to preserve the chirality. Moreover, the chiral $(N_+, N_-) = (1, 0)$ supersymmetry algebra in $(10, 2)$ involves the anticommutator[101]

$$\{Q_\alpha, Q_\beta\} = \Gamma^{MN}{}_{\alpha\beta}P_{MN} + \Gamma^{MNPQRS}{}_{\alpha\beta}Z^{+}{}_{MNPQRS}\,. \tag{5.6}$$

The absence of translations casts doubt on the naive application of the Bose–Fermi matching argument, and the appearance of the self-dual six-form charge Z is suggestive of a sixbrane, rather than a threebrane.

Despite all the objections one might raise to a world with two time dimensions, and despite the above problems of interpretation, the idea of a $(2, 2)$ object moving in a $(10, 2)$ space–time has recently been revived in the context of F *theory*,[75] which involves Type IIB compactification where the axion and dilaton from the RR sector are allowed to vary on the internal manifold. Given a manifold M that has the structure of a fiber bundle whose fiber is T^2 and whose base is some manifold B, then

$$F \text{ on } M \equiv \text{Type IIB on } B\,. \tag{5.7}$$

The utility of F theory is beyond dispute and it has certainly enhanced our understanding of string dualities, but should the twelve-dimensions of F theory be taken seriously? And if so, should the F theory be regarded as more fundamental than M theory? Given that there seems to be no supersymmetric field theory with SO(10, 2) Lorentz invariance,[102] and given that the on-shell states carry only ten-dimensional momenta,[75] the more conservative interpretation is that the twelfth dimension is merely a mathematical artifact and that the F theory should simply be interpreted as a clever way of compactifying the IIB string.[103] Time will tell.

6. Conclusion

The overriding problem in superunification in the coming years will be to take the mystery out of M theory, while keeping the magic and the membranes.

References

1. W. Nahm, *Nucl. Phys.* **B135**, 409 (1978).
2. E. Cremmer, B. Julia and J. Scherk, *Phys. Lett.* **B76**, 409 (1978).
3. M. J. Duff, B. E. W. Nilsson and C. N. Pope, *Phys. Rep.* **130**, 1 (1986).
4. M. B. Green, J. H. Schwarz and E. Witten, *Superstring Theory* (Cambridge University Press, 1987).
5. E. Witten, *Nucl. Phys.* **B443**, 85 (1995).
6. A. Font, L. Ibanez, D. Lust and F. Quevedo, *Phys. Lett.* **B249**, 35 (1990).
7. S.-J. Rey, *Phys. Rev.* **D43**, 526 (1991).
8. M. J. Duff, *Nucl. Phys.* **B442**, 47 (1995).
9. M. J. Duff and J. X. Lu, *Nucl. Phys.* **B357**, 534 (1991).
10. M. J. Duff and R. R. Khuri, *Nucl. Phys.* **B411**, 473 (1994).
11. M. J. Duff and J. X. Lu, *Nucl. Phys.* **B416**, 301 (1994).
12. M. J. Duff and R. Minasian, *Nucl. Phys.* **B436**, 507 (1995).
13. C. M. Hull and P. K. Townsend, *Nucl. Phys.* **B438**, 109 (1995).
14. M. J. Duff, R. R. Khuri and J. X. Lu, *Phys. Rep.* **259**, 213 (1995).
15. M. J. Duff, R. Minasian and E. Witten, *Nucl. Phys.* **B465**, 413 (1996).
16. M. J. Duff, J. T. Liu and R. Minasian, *Nucl. Phys.* **B452**, 261 (1995).
17. E. Witten, *Nucl. Phys.* **B463**, 383 (1996).
18. M. J. Duff, B. E. W. Nilsson and C. N. Pope, *Phys. Lett.* **B129**, 39 (1983).
19. M. J. Duff and B. E. W. Nilsson, *Phys. Lett.* **B175**, 417 (1986).
20. K. S. Narain, *Phys. Lett.* **B169**, 41 (1986).
21. E. Bergshoeff, E. Sezgin and P. K. Townsend, *Phys. Lett.* **B189**, 75 (1987).
22. M. J. Duff, P. S. Howe, T. Inami and K. S. Stelle, *Phys. Lett.* **B191**, 70 (1987).
23. M. J. Duff, "Not the standard superstring review," in *Proc. Int. School of Subnuclear Physics — The Superworld II*, (Erice, 1987), ed. Zichichi (Plenum, 1990), CERN-TH. 4749/87.
24. E. Cremmer, S. Ferrara and J. Scherk, *Phys. Lett.* **B74**, 61 (1978).
25. E. Cremmer and B. Julia, *Nucl. Phys.* **B159**, 141 (1979).

26. M. J. Duff and J. X. Lu, *Nucl. Phys.* **B347**, 394 (1990).
27. M. J. Duff and K. S. Stelle, *Phys. Lett.* **B253**, 113 (1991).
28. R. Gueven, *Phys. Lett.* **B276**, 49 (1992).
29. E. Bergshoeff, M. de Roo and T. Ortin, "The eleven-dimensional five-brane," hep-th/9606118.
30. P. S. Howe and E. Sezgin, "Superbranes," hep-th/9607227.
31. G. W. Gibbons and P. K. Townsend, *Phys. Rev. Lett.* **71**, 3754 (1993).
32. P. K. Townsend, *Phys. Lett.* **B350**, 184 (1995).
33. C. G. Callan, J. A. Harvey and A. Strominger, *Nucl. Phys.* **B359**, 611 (1991).
34. C. G. Callan, J. A. Harvey and A. Strominger, *Nucl. Phys.* **B367**, 60 (1991).
35. R. I. Nepomechie, *Phys. Rev.* **D31**, 1921 (1985).
36. C. Teitelboim, *Phys. Lett.* **B167**, 63 (1986).
37. M. J. Duff and J. X. Lu, *Nucl. Phys.* **B354**, 141 (1991).
38. J. H. Schwarz, *Phys. Lett.* **B360**, 13 (1995).
39. S. P. De Alwis, "A note on brane tension and M-theory," hep-th/9607011.
40. M. J. Duff and J. X. Lu, *Nucl. Phys.* **B390**, 276 (1993).
41. P. Townsend, *Phys. Lett.* **B373**, 68 (1996).
42. C. Schmidhuber, *Nucl. Phys.* **B467**, 146 (1996).
43. J. Polchinski, *Phys. Rev. Lett.* **75**, 4724 (1995).
44. A. Giveon, M. Porrati and E. Rabinovici, *Phys. Rep.* **244**, 77 (1994).
45. L. Alvarez-Gaumé and E. Witten, *Nucl. Phys.* **B234**, 269 (1983).
46. James T. Liu, private communication.
47. A. N. Schellekens and N. P. Warner, *Nucl. Phys.* **B287**, 317 (1987).
48. C. Vafa and E. Witten, *Nucl. Phys.* **B447**, 261 (1995).
49. P. K. Townsend, *Phys. Lett.* **B354**, 247 (1995).
50. J. A. Harvey and A. Strominger, *Nucl. Phys.* **B449**, 535 (1995).
51. A. C. Cadavid, A. Ceresole, R. D'Auria and S. Ferrara, "11-dimensional supergravity compactified on Calabi–Yau threefolds," hep-th/9506144.
52. G. Papadopoulos and P. K. Townsend, *Phys. Lett.* **B357**, 300 (1995).
53. J. H. Schwarz and A. Sen, *Nucl. Phys.* **B454**, 427 (1995).
54. S. Chaudhuri and D. A. Lowe, *Nucl. Phys.* **B459**, 113 (1996).
55. J. A. Harvey, D. A. Lowe and A. Strominger, *Phys. Lett.* **B362**, 65 (1995).
56. B. S. Acharya, "$N = 1$ heterotic-supergravity duality and Joyce manifolds," hep-th/9508046.
57. M. A. Awada, M. J. Duff and C. N. Pope, *Phys. Rev. Lett.* **50**, 294 (1983).
58. K. Becker, M. Bec er and A. Strominger, *Nucl. Phys.* **B456**, 130 (1995).
59. M. R. Douglas, D. Kabat, P. Pouliot and S. H. Shenker, "D-branes and short distances in string theory," hep-th/9608024.
60. T. Banks and M. Dine, "Couplings and scales in strongly coupled heterotic string theory," hep-th/9605136.
61. P. Horava, "Gluino condensation in strongly coupled heterotic string theory," hep-th/9608019.
62. C. Vafa and A. Strominger, "Microscopic origin of the Bekenstein–Hawking entropy," hep-th/9601029.
63. J. H. Schwarz, "Lectures on superstring and M-theory dualities," hep-th/9607201.
64. P. S. Aspinwall, "Some relationships between dualities in string theory," hep-th/9508154.
65. M. J. Duff and J. X. Lu, *Phys. Lett.* **B273**, 409 (1991).
66. M. J. Duff, J. T. Liu and J. Rahmfeld, *Nucl. Phys.* **B459**, 125 (1996).
67. P. Horava and E. Witten, *Nucl. Phys.* **B460**, 506 (1996).

68. E. Witten, "Fermion quantum numbers in Kaluza–Klein theory," in *Proc. 1983 Shelter Island Conference on Quantum Field Theory and the Fundamental Problems in Physics* (1983), eds. Jackiw, Khuri and Weinberg (MIT Press, 1985).
69. J. Polchinski, "String duality," hep-th/9607050.
70. O. Aharony, J. Sonnenschein and S. Yankielowicz, "Interactions of strings and D-branes from M-theory," hep-th/9603009.
71. E. Witten, "On flux quantization in M-theory and the effective action," hep-th/9609122.
72. G. Pradisi and A. Sagnotti, *Phys. Lett.* **B216**, 59 (1989).
73. M. Bianchi and A. Sagnotti, *Nucl. Phys.* **B361**, 519 (1991).
74. E. G. Gimon and J. Polchinski, *Phys. Rev.* **D54**, 1667 (1996).
75. C. Vafa, *Nucl. Phys.* **B469**, 403 (1996).
76. M. Berkooz, R. G. Leigh, J. Polchinski, J. H. Schwarz, N. Seiberg and E. Witten, "Anomalies, dualities and topology of $D = 6N = 1$ superstring vacua," hep-th/9605184.
77. E. Witten, *Nucl. Phys.* **B460**, 541 (1996).
78. A. Sen and C. Vafa, *Nucl. Phys.* **B455**, 165 (1995).
79. R. Dijkgraaf, E. Verlinde and H. Verlinde, "BPS quantization of the five-brane," hep-th/9604055.
80. G. Aldazabal, A. Font, L. E. Ibanez and F. Quevado, *Phys. Lett.* **B380**, 42 (1996).
81. G. Aldazabal, A. Font, L. E. Ibanez and A. M. Uranga, "New branches of string compactifications and their F-theory duals," hep-th/9607121.
82. E. Gimon and C. V. Johnson, "Multiple realizations of $N = 1$ vacua in six dimensions," NSF-ITP-96-55, hep-th/9606176.
83. P. S. Aspinwall and M. Gross, "Heterotic–heterotic string duality and multiple $K3$ fibrations," hep-th/9602118.
84. S. Ferrara, R. Minasian and A. Sagnotti, "Low-energy analysis of M and F theories on Calabi–Yau manifolds, hep-th/9604097.
85. D. R. Morrison and C. Vafa, "Compactifications of F-theory on Calabi–Yau threefolds-I," hep-th/9602114.
86. D. R. Morrison and C. Vafa, "Compactifications of F-theory on Calabi–Yau threefolds-II," hep-th/9603161.
87. K. Dasgupta and S. Mukhi, *Nucl. Phys.* **B465**, 399 (1996).
88. O. J. Ganor and A. Hanany, "Small E_8 instantons and tensionless non-critical strings," hep-th/9603003.
89. N. Seiberg and E. Witten, "Comments on string dynamics in six dimensions," hep-th/9603003.
90. M. J. Duff, H. Lu and C. N. Pope, "Heterotic phase transitions and singularities of the gauge dyonic string, hep-th/9603037.
91. M. J. Duff, "Electric/magnetic duality and its stringy origins," hep-th/9509106.
92. A. Klemm, W. Lerche and P. Mayr, "$K3$-fibrations and heterotic-Type-II duality," hep-th/9506112.
93. S. Kachru and C. Vafa, "Exact results for $N = 2$ compactifications of heterotic strings," hep-th/9505105.
94. G. Lopez Cardoso, D. Lust and T. Mohaupt, "Non-perturbative monodromies in $N = 2$ heterotic string vacua," hep-th/9504090.
95. J. Louis, J. Sonnenschein, S. Theisen and S. Yankielowicz, "Non-perturbative properties of heterotic string vacua compactified on $K3 \times T^2$, hep-th/9606049.
96. A. Sagnotti, *Phys. Lett.* **B294**, 196 (1992).
97. E. Witten, "Phase transitions in M-theory and F-theory," hep-th/9603150.

98. G. Lopez Cardoso, G. Curio, D. Lust and T. Mohaupt, "Instanton numbers and exchange symmetries in $N = 2$ dual string pairs," hep-th/9603108.
99. M. J. Duff and M. Blencowe, *Nucl. Phys.* **B310**, 387 (1988).
100. A. Achucarro, J. M. Evans, P. K. Townsend and D. L. Wiltshire, *Phys. Lett.* **B198**, 441 (1997).
101. I. Bars, "Supersymmetry, P-brane duality and hidden dimensions," hep-th/9604139.
102. H. Nishino and E. Sezgin, "Supersymmetric Yang–Mills equations in $10 + 2$ dimensions," hep-th/9607185.
103. A. Sen, "F-theory and orientifolds," hep-th/9605150.

Reprinted from Phys. Rev. D **55** (1997) 5112–28

M theory as a matrix model: A conjecture

T. Banks,[1,*] W. Fischler,[2,†] S. H. Shenker,[1,‡] and L. Susskind[3,§]
[1]*Department of Physics and Astronomy, Rutgers University, Piscataway, New Jersey 08855-0849*
[2]*Theory Group, Department of Physics, University of Texas, Austin, Texas 78712*
[3]*Department of Physics, Stanford University, Standford, California 94305-4060*
(Received 13 December 1996)

We suggest and motivate a precise equivalence between uncompactified 11-dimensional *M* theory and the $N=\infty$ limit of the supersymmetric matrix quantum mechanics describing *D*0 branes. The evidence for the conjecture consists of several correspondences between the two theories. As a consequence of supersymmetry the simple matrix model is rich enough to describe the properties of the entire Fock space of massless well separated particles of the supergravity theory. In one particular kinematic situation the leading large distance interaction of these particles is exactly described by supergravity. The model appears to be a nonperturbative realization of the holographic principle. The membrane states required by *M* theory are contained as excitations of the matrix model. The membrane world volume is a noncommutative geometry embedded in a noncommutative spacetime. [S0556-2821(97)03308-0]

PACS number(s): 11.25.Sq, 11.30.Pb

I. INTRODUCTION

M theory [1] is the strongly coupled limit of type-IIA string theory. In the limit of infinite coupling, it becomes an 11-dimensional theory in a background-infinite flat space. In this paper *M* theory will always refer to this decompactified limit. We know very little about this theory except for the following two facts. At low energy and large distances, it is described by 11-dimensional supergravity. It is also known to possess membrane degrees of freedom with membrane tension $1/l_P^3$ where l_P is the 11-dimensional Planck length. It seems extremely unlikely that *M* theory is any kind of conventional quantum field theory. The degrees of freedom describing the short distance behavior are simply unknown. The purpose of this paper is to put forward a conjecture about these degrees of freedom and about the Hamiltonian governing them.

The conjecture grew out of a number of disparate facts about *M* theory, *D* branes [2], matrix descriptions of their dynamics [3], supermembranes [4,5,6], the holographic principle [7], and short distance phenomena in string theory [8,9]. Simply stated the conjecture is this. *M* theory, in the light cone frame, is exactly described by the large *N* limit of a particular supersymmetric matrix quantum mechanics. The system is the same one that has been used previously used to study the small distance behavior of *D*0 branes [9]. Townsend [10] was the first to point out that the supermatrix formulation of membrane theory suggested that membranes could be viewed as composites of *D*0 branes. Our work is a precise realization of his suggestion.

In what follows we will present our conjecture and some evidence for it. We begin by reviewing the description of string theory in the infinite momentum frame. We then present our conjecture for the full set of degrees of freedom of *M* theory and the Hamiltonian which governs them. Our strongest evidence for the conjecture is a demonstration that our model contains the excitations which are widely believed to exist in *M* theory, supergravitons and large metastable classical membranes. These are discussed in Secs. III and V. The way in which these excitations arise is somewhat miraculous, and we consider this to be the core evidence for our conjecture. In Sec. IV we present a calculation of supergraviton scattering in a very special kinematic region and argue that our model reproduces the expected result of low energy supergravity. The calculation depends on a supersymmetric nonrenormalization theorem whose validity we will discuss there. In Sec. VI we argue that our model may satisfy the holographic principle. This raises crucial issues about Lorentz invariance which are discussed there.

We emphasize that there are many unanswered questions about our proposed version of *M* theory. Nonetheless, these ideas seem of sufficient interest to warrant presenting them here. If our conjecture is correct, this would be the first nonperturbative formulation of a quantum theory which includes gravity.

II. INFINITE MOMENTUM FRAME AND THE HOLOGRAPHIC PRINCIPLE

The infinite momentum frame [11] is the old name for the misnamed light cone frame. Thus far this is the only frame in which it has proved possible to formulate string theory in Hamiltonian form. The description of *M* theory which we will give in this paper is also in the infinite momentum frame. We will begin by reviewing some of the features of the infinite momentum frame formulation of relativistic quantum mechanics. For a comprehensive review we refer the reader to [11]. We begin by choosing a particular spatial direction x^{11} called the longitudinal direction. The nine-dimensional space transverse to x^{11} is labeled x^i or $x^\perp$. Time

*Electronic address: banks@physics.rutgers.edu
†Electronic address: fischler@physics.utexas.edu
‡Electronic address: shenker@physics.rutgers.edu
§Electronic address: susskind@dormouse.stanford.edu

0556-2821/97/55(8)/5112(17)/$10.00

will be indicated by t. Now consider a system of particles with momenta (p_-^a, p_{11}^a) where a labels the particle. The system is boosted along the x^{11} axis until all longitudinal momenta are much larger than any scale in the problem. Further longitudinal boosting just rescales all longitudinal momenta in fixed proportion. Quantum field theory in such a limiting reference frame has a number of properties which will be relevant to us.

It is convenient to begin by assuming that the x^{11} direction is compact with a radius R. The compactification serves as an infrared cutoff. Accordingly, the longitudinal momentum of any system or subsystem of quanta is quantized in units of $1/R$. In the infinite momentum frame, all systems are composed of constituent quanta or partons. The partons all carry strictly positive values of longitudinal momentum. It is particularly important to understand what happens to quanta of negative or vanishing p_{11}. The answer is that as the infinite momentum limit is approached, the frequency of these quanta, relative to the Lorentz-time-dilated motion of the boosted system, becomes infinite and the zero and negative momentum quanta may be integrated out. The process of integrating out such fast modes may influence or even determine the Hamiltonian of the remaining modes. In fact, the situation is slightly more complicated in certain cases for the zero momentum degrees of freedom. In certain situations such as spontaneous symmetry breaking, these longitudinally homogeneous modes define backgrounds whose moduli may appear in the Hamiltonian of the other modes. In any case the zero and negative momentum modes do not appear as independent dynamical degrees of freedom.

Thus we may assume all systems have longitudinal momentum given by an integer multiple of $1/R$,

$$p_{11}=N/R, \tag{2.1}$$

with N strictly positive. At the end of a calculation we must let R and N/R tend to infinity to get to the uncompactified infinite momentum limit.

The main reason for the simplifying features of the infinite momentum frame is the existence of a transverse Galilean symmetry which leads to a naive nonrelativistic form for the equations. The role of nonrelativistic mass is played by the longitudinal momentum p_{11}. The Galilean transformations take the form

$$p_i \rightarrow p_i + p_{11} v_i. \tag{2.2}$$

As an example of the Galilean structure of the equations, the energy of a free massless particle is

$$E=\frac{p_\perp^2}{2p_{11}}. \tag{2.3}$$

For the 11-dimensional supersymmetric theory we will consider, the Galilean invariance is extended to the super-Galilean group which includes 32 real supergenerators. The supergenerators divide into two groups of 16, each transforming as spinors under the nine-dimensional transverse rotation group. We denote them by Q_α, and q_A, and they obey anticommutation relations

$$[Q_\alpha, Q_\beta]_+ = \delta_{\alpha\beta} H,$$
$$[q_A, q_B]_+ = \delta_{AB} P_{11},$$
$$[Q_\alpha, q_A] = \gamma^i_{A\alpha} P_i. \tag{2.4}$$

The Lorentz generators which do not preserve the infinite momentum frame mix up the two kinds of generators.

Let us now recall some of the features of string theory in the infinite momentum or light cone frame [12]. We will continue to call the longitudinal direction x^{11} even though in this case the theory has only ten space-time directions. The transverse space is of course eight dimensional. To describe a free string of longitudinal momentum p_{11}, a periodic parameter σ which runs from 0 to p_{11} is introduced. To regulate the world sheet theory, a cutoff $\delta\sigma=\epsilon$ is introduced. This divides the parameter space into $N=p_{11}/\epsilon$ segments, each carrying longitudinal momentum ϵ. We may think of each segment as a parton, but unlike the partons of quantum field theory, these objects always carry $p_{11}=\epsilon$. For a multiparticle system of total longitudinal momentum p_{11}(total), we introduce a total parameter space of overall length p_{11}(total), which we allow to be divided into separate pieces, each describing a string. The world sheet regulator is implemented by requiring each string to be composed of an integer number of partons of momentum ϵ. Interactions are described by splitting and joining processes in which the number of partons is strictly conserved. The regulated theory is thus seen to be a special case of Galilean quantum mechanics of N partons with interactions which bind them into long chains and allow particular kinds of rearrangements.

The introduction of a minimum unit of momentum ϵ can be given an interpretation as an infrared cutoff. In particular, we may assume that the x^{11} coordinate is periodic with length $R=\epsilon^{-1}$. Evidently, the physical limit $\epsilon \rightarrow 0$, $R\rightarrow\infty$ is a limit in which the number of partons N tends to infinity.

It is well known [7] that in this large N limit the partons become infinitely dense in the transverse space and that this leads to extremely strong interactions. This circumstance, together with the Bekenstein bound on entropy, has led to the *holographic* speculation that the transverse density of partons is strictly bounded to about one per transverse Planck area. In other words, the partons form a kind of incompressible fluid. This leads to the unusual consequence that the transverse area occupied by a system of longitudinal momentum p_{11} cannot be smaller than p_{11}/ϵ in Planck units.

The general arguments for the holographic behavior of systems followed from considerations involving the Bekenstein–'t Hooft bound on the entropy of a spatial region [13] and were not specific to string theory. If the arguments are correct, they should also apply to 11-dimensional theories which include gravitation. Thus we should expect that in M theory the radius of a particle such as the graviton will grow with p_{11} according to

$$\rho=\left(\frac{p_{11}}{\epsilon}\right)^{1/9} l_P=(p_{11}R)^{1/9} l_P, \tag{2.5}$$

where l_P is the 11-dimensional Planck length. In what follows we will see quantitative evidence for exactly this behavior.

At first sight the holographic growth of particles appears to contradict the boost invariance of particle interactions. Consider the situation of two low energy particles moving past one another with some large transverse separation, let us say of order a meter. Obviously these particles have negligible interactions. Now boost the system along the longitudinal direction until the size of each particle exceeds their separation. They now overlap as they pass each other. But longitudinal boost invariance requires that the scattering amplitude be still essentially zero. This would seem to require extremely special and unnatural cancellations. We will see below that one key to this behavior is the very special Bogomol'ni-Prasad-Sommerfield (BPS) property of the partons describing M theory. However, we are far from having a complete understanding of the longitudinal boost invariance of our system. Indeed, we view it as the key dynamical puzzle which must be unraveled in understanding the dynamics of M theory.

III. *M* THEORY AND *D*0 BRANES

M theory with a compactified longitudinal coordinate x^{11} is by definition a type-IIA string theory. The correspondences between the two theories include [1] the following.

(1) The compactification radius R is related to the string coupling constant by

$$R=g^{2/3}l_P=gl_s, \tag{3.1}$$

where l_s is the string length scale:

$$l_s=g^{-1/3}l_P. \tag{3.2}$$

(2) The Ramond-Ramond (RR) photon of IIA theory is the Kaluza-Klein (KK) photon which arises upon compactification of 11-dimensional supergravity.

(3) No perturbative string states carry RR charge. In other words, all perturbative string states carry vanishing momentum along the x^{11} direction. The only objects in the theory which do carry RR photon charge are the $D0$ branes of Polchinski. $D0$ branes are point particles which carry a single unit of RR charge. Equivalently, they carry longitudinal momentum

$$P_{11}^{D0\ \mathrm{branes}}=1/R. \tag{3.3}$$

The $D0$ branes carry the quantum numbers of the first massive KK modes of the basic 11-dimensional supergravity multiplet, including 44 gravitons, 84 components of a three-form, and 128 gravitinos. We will refer to these particles as supergravitons. As 11-dimensional objects, these are all massless. As a consequence, they are BPS saturated states in the ten-dimensional (10D) theory. Their 10D mass is $1/R$.

(4) Supergravitons carrying Kaluza-Klein momentum $p_{11}=N/R$ also exist, but are not described as elementary $D0$ branes. As shown in [3], their proper description is as bound composites of N $D0$ branes.

These properties make the $D0$ branes candidate partons for an infinite momentum limit description of M theory. We expect that if, as in quantum field theory, the degrees of freedom with vanishing and negative p_{11} decouple, then M theory in the infinite momentum frame should be a theory whose only degrees of freedom are $D0$ branes. Anti-$D0$ branes carry negative Kaluza-Klein momenta, and strings carry vanishing p_{11}. The decoupling of anti-$D0$ branes is particularly fortunate because brane-antibrane dynamics is something about which we know very little [14]. The BPS property of zero branes ameliorates the conflict between infinitely growing parton wave functions and low energy locality, which we noted at the end of the last section. We will see some partial evidence for this in a nontrivial scattering computation below. We will also discuss below the important point that a model containing only $D0$ branes actually contains large classical supermembrane excitations. Since the conventional story of the M-theoretic origin of strings depicts them as membranes wrapped around the compactified 11th dimension, we have some reason to believe that strings have not really been left out of the system.

All of these circumstances lead us to propose that M theory in the infinite momentum frame is a theory in which the only dynamical degrees of freedom or partons are $D0$ branes. Furthermore, it is clear in this case that all systems are built out of the composites of partons, each of which carries the minimal p_{11}. We note, however, that our system does have a set of degrees of freedom which go beyond the parton coordinates. Indeed, as first advocated in [3], the $D0$ brane coordinates of N partons have to be promoted to matrices. At distance scales larger than the 11-dimensional Planck scale, these degrees of freedom become very massive and largely decouple,[1] but their virtual effects are responsible for all parton interactions. These degrees of freedom are BPS states and are related to the parton coordinates by gauge transformations. Furthermore, when the partons are close together, they become low frequency modes. Thus they cannot be omitted in any discussion of the dynamics of $D0$ branes.

IV. *D*0 BRANE MECHANICS

If the infinite momentum limit of M theory is the theory of $D0$ branes, decoupled from the other string theory degrees of freedom, what is the precise form of the quantum mechanics of the system? Fortunately there is a very good candidate which has been extensively studied in another context in which $D0$ branes decouple from strings [9].

As emphasized at the end of the last section, open strings which connect $D0$ branes do not exactly decouple. In fact, the very short strings which connect the branes when they are practically on top of each other introduce a new kind of coordinate space in which the nine spatial coordinates of a system of N $D0$ branes become nine $N\times N$ matrices $X^i_{a,b}$ [3]. The matrices X are accompanied by 16 fermionic superpartners $\theta_{a,b}$, which transform as spinors under the SO(9) group of transverse rotations. The matrices may be thought of as the spatial components of ten-dimensional super Yang-Mills (SYM) fields after dimensional reduction to zero space directions. These Yang-Mills fields describe the open strings which are attached to the $D0$ branes. The Yang-Mills quantum mechanics has U(N) symmetry and is described (in

---

[1] Indeed, we will propose that this decoupling is precisely what defines the regime in which the classical notion of distance makes sense.

units with $l_s = 1$) by the Lagrangian

$$L=\frac{1}{2g}\{\mathrm{tr}\,\dot{X}^i\dot{X}^i+2\theta^T\dot{\theta}+\tfrac{1}{2}\,\mathrm{tr}[X^i,X^j]^2-2\theta^T\gamma_i[\theta,X^i]\}. \tag{4.1}$$

Here we have used conventions in which the fermionic variables are 16-component nine-dimensional spinors.

In [9] this Lagrangian was used to study the short distance properties of $D0$ branes in weakly coupled string theory. The 11D Planck length emerged as a natural dynamical length scale in that work, indicating that the system (4.1) describes some M-theoretic physics. In [9], Eq. (4.1) was studied as a low velocity effective theory appropriate to the heavy $D0$ branes of weakly coupled string theory. Here we propose Eq. (4.1) as the most general infinite momentum frame Lagrangian, with at most two derivatives, which is invariant under the gauge symmetry and the super-Galilean group[2] [15]. It would be consistent with our assumption that matrix $D0$ branes are the only degrees of freedom of M theory to write a Lagrangian with higher powers of first derivatives. We do not know if any such Lagrangians exist which preserve the full symmetry of the infinite momentum frame. What is at issue here is 11-dimensional Lorentz invariance. In typical infinite momentum frame field theories, the naive classical Lagrangian for the positive longitudinal momentum modes is renormalized by the decoupled infinite frequency modes. The criterion which determines the infinite momentum frame Lagrangian is invariance under longitudinal boosts and null plane rotating Lorentz transformations (the infamous angular conditions). Apart from simplicity, our main reason for suggesting the Lagrangian (4.1) is that we have found some partial evidence that the large N limit of the quantum theory it defines is indeed Lorentz invariant. A possible line of argument systematically leading to Eq. (4.1) is discussed in Sec. IX.

Following [9], let us rewrite the action in units in which the 11D Planck length is 1. Using Eqs. (3.1) and (3.2), the change of units is easily made and one finds

$$L=\mathrm{tr}\left\{\frac{1}{2R}D_tY^iD_tY^i-\tfrac{1}{4}R[Y^i,Y^j]^2-\theta^TD_t\theta \right.$$
$$\left.-R\theta^T\gamma_i[\theta,Y^i]\right\}, \tag{4.2}$$

where $Y=X/g^{1/3}$. We have also changed the units of time to 11D Planck units. We have restored the gauge field $(\partial_t\rightarrow D_t=\partial_t+iA)$ to this expression (previously we were in the $A=0$ gauge) in order to emphasize that the supersymmetric (SUSY) transformation laws (here ϵ and ϵ' are two independent 16-component anticommuting SUSY parameters)

$$\delta X^i=-2\epsilon^T\gamma^i\theta, \tag{4.3}$$

$$\delta\theta=\tfrac{1}{2}\{D_tX^i\gamma_i+\gamma_-+\tfrac{1}{2}[X^i,X^j]\gamma_{ij}\}\epsilon+\epsilon', \tag{4.4}$$

$$\delta A=-2\epsilon^T\theta \tag{4.5}$$

involve a gauge transformation. As a result, the SUSY algebra closes on the gauge generators and only takes on the form (2.4) when applied to gauge-invariant states.

The Hamiltonian has the form

$$H=R\,\mathrm{tr}\left\{\frac{\Pi_i\Pi_i}{2}-\frac{1}{4}[Y^i,Y^j]^2+\theta^T\gamma_i[\theta,Y^i]\right\}, \tag{4.6}$$

where Π is the canonical conjugate to Y. Note that in the limit $R\rightarrow\infty$, all finite energy states of this Hamiltonian have infinite energy. We will be interested only in states whose energy vanishes like $1/N$ in the large N limit, so that this factor becomes the inverse power of longitudinal momentum which we expect for the eigenstates of a longitudinal-boost-invariant system. Thus, in the correct infinite momentum frame limit, the only relevant asymptotic states of the Hamiltonian should be those whose energy is of order $1/N$. We will exhibit a class of such states below, the supergraviton scattering states. The difficult thing will be to prove that their S-matrix elements depend only on ratios of longitudinal momenta, so that they are longitudinally boost invariant.

To understand how this system represents ordinary particles, we note that when the Y's become large the commutator term in H becomes very costly in energy. Therefore for large distances the finite energy configurations lie on the flat directions along which the commutators vanish. In this system with 16 supercharges,[3] these classical zero energy states are in fact exact supersymmetric states of the system. In contrast to field theory, the continuous parameters which describe these states [the Higgs vacuum expectation values (VEV's) in the language of SYM theory] are not vacuum superselection parameters, but rather collective coordinates. We must compute their quantum wave functions rather than freeze them at classical values. They are, however, the slowest modes in the system, so that we can integrate out the other degrees of freedom to get an effective SUSY quantum mechanics of these modes alone. We will study some aspects of this effective dynamics below.

Along the flat directions the Y^i are simultaneously diagonalizable. The diagonal matrix elements are the coordinates of the $D0$ branes. When the Y are small, the cost in energy for a noncommuting configuration is not large. Thus for small distances there is no interpretation of the configuration space in terms of ordinary positions. Classical geometry and distance are only sensible concepts in this system in regions far out along one of the flat directions. We will refer to this as the long distance regime. In the short distance regime, we have a noncommutative geometry. Nevertheless, the full Hamiltonian (4.6) has the usual Galilean symmetry. To see this we define the center of mass of the system by

---

[2]The gauge invariance is in fact necessary to supertranslation invariance. The supergenerators close on gauge transformations and only satisfy the supertranslation algebra on the gauge-invariant subspace.

[3]The 16 supercharges which anticommute to the longitudinal momentum act only on the center of mass of the system and play no role in particle interactions.

$$Y(\text{c.m.}) = \frac{1}{N}\,\text{tr}Y. \tag{4.7}$$

A transverse translation is defined by adding a multiple of the identity to Y. This has no effect on the commutator term in L because the identity commutes with all Y. Similarly rotational invariance is manifest.

The center-of-mass momentum is given by

$$P(\text{c.m.}) = \text{tr}\Pi = \frac{N}{R}\,\dot{Y}(\text{c.m.}). \tag{4.8}$$

Using $p_{11} = N/R$ gives the usual connection between transverse velocity and transverse momentum:

$$\frac{1}{p_{11}}\,P(\text{c.m.}) = \dot{Y}(\text{c.m.}). \tag{4.9}$$

A Galilean boost is defined by adding a multiple of the identity to $\dot{Y}$. We leave it to the reader to show that this has no effect on the equations of motion. This establishes the Galilean invariance of H. The super-Galilean invariance is also completely unbroken. The alert reader may be somewhat unimpressed by some of these invariances, since they appear to be properties of the center-of-mass coordinate, which decouples from the rest of the dynamics. Their real significance will appear below when we show that our system possesses multiparticle asymptotic states, on which these generators act in the usual way as a sum of single-particle operators.

V. A CONJECTURE

Our conjecture is thus that M theory formulated in the infinite momentum frame is exactly equivalent to the $N \to \infty$ limit of the supersymmetric quantum mechanics described by the Hamiltonian (4.6). The calculation of any physical quantity in M theory can be reduced to a calculation in matrix quantum mechanics followed by an extrapolation to large N. In what follows we will offer evidence for this surprising conjecture.

Let us begin by examining the single-particle spectrum of the theory. For $N = 1$ the states with $p_\perp = 0$ are just those of a single $D0$ brane at rest. The states form a representation of the algebra of the 16 θ's with 2^8 components. These states have exactly the quantum numbers of the 256 states of the supergraviton. For nonzero $p_\perp$ the energy of the object is

$$E = \frac{R}{2}\,p_\perp^2 = \frac{p_\perp^2}{2p_{11}}. \tag{5.1}$$

For states with $N > 1$, we must study the U(N)-invariant Schrödinger equation arising from Eq. (4.6). H can easily be separated in terms of center-of-mass and relative motions. Define

$$Y = \frac{Y(\text{c.m.})}{N}\,I + Y_{\text{rel}}, \tag{5.2}$$

where I is the unit matrix and Y_{rel} is a traceless matrix in the adjoint of SU(N) representing relative motion. The Hamiltonian then has the form

$$H = H_{\text{c.m.}} + H_{\text{rel}}, \tag{5.3}$$

with

$$H_{\text{c.m.}} = \frac{P(\text{c.m.})^2}{2p_{11}}. \tag{5.4}$$

The Hamiltonian for the relative motion is the dimensional reduction of the supersymmetric 10D Yang-Mills Hamiltonian. Although a direct proof based on the Schrödinger equation has not yet been given, duality between IIA strings and M theory requires the relative Schrödinger equation to have normalizable threshold bound states with zero energy [3]. The bound system must have exactly the quantum numbers of the 256 states of the supergraviton. For these states the complete energy is given by Eq. (5.4). Furthermore, these states are BPS saturated. No other normalizable bound states can occur. Thus we find that the spectrum of stable single-particle excitations of Eq. (4.6) is exactly the supergraviton spectrum with the correct dispersion relation to describe massless 11-dimensional particles in the infinite momentum frame.

Next let us turn to the spectrum of widely separated particles. That a simple quantum mechanical Hamiltonian like Eq. (4.6) should be able to describe arbitrarily many well-separated particles is not at all evident and would certainly be impossible without the special properties implied by supersymmetry. Begin by considering commuting block diagonal matrices of the form

$$\begin{pmatrix} Y_1^i & 0 & 0 & \cdots \\ 0 & Y_2^i & 0 & \\ 0 & 0 & Y_e^i & \end{pmatrix}, \tag{5.5}$$

where Y_a^i are $N_a \times N_a$ matrices and $N_1 + N_2 + \cdots + N_n = N$. For the moment suppose all other elements of the Y's are constrained to vanish identically. In this case the Schrödinger equation obviously separates into n uncoupled Schrödinger equations for the individual block degrees of freedom. Each equation is identical to the original Schrödinger equation for the system of N_a $D0$ branes. Thus the spectrum of this truncated system includes collections of noninteracting supergravitons.

Now let us suppose the supergravitons are very distant from one another. In other words, for each pair the relative distance, defined by

$$R_{a,b} = \left| \frac{\text{tr}Y_a}{N_a} - \frac{\text{tr}Y_b}{N_b} \right|, \tag{5.6}$$

is asymptotically large. In this case the commutator terms in Eq. (4.6) cause the off-diagonal blocks in the Y's to have very large potential energy proportional to $R_{a,b}^2$. This effect can also be thought of as the Higgs effect giving mass to the broken generators ("W bosons") of U(N) when the symmetry is broken to U(N_1)×U(N_2)×U(N_3)$\cdots$. Thus one might naively expect the off-diagonal modes to leave the spectrum of very widely separated supergravitons unmodified. However, this is not correct in a generic situation. The off-diagonal modes behave like harmonic oscillators with frequency of order $R_{a,b}$, and their zero point energy will generally give rise to a potential energy of similar magnitude. This effect would certainly preclude an interpretation of the matrix model in terms of well-separated independent particles.

Supersymmetry is the ingredient which rescues us. In a well-known way, the fermionic partners of the off-diagonal bosonic modes exactly cancel the potential due to the bosons, leaving exactly flat directions. We know this from the nonrenormalization theorems for supersymmetric quantum mechanics with 16 supergenerators [15]. The effective Lagrangian for the collective coordinates along the flat directions must be supersymmetric, and the result of [15] guarantees that up to terms involving at most two derivatives the Lagrangian for these coordinates must be the dimensional reduction of $U(1)^n$ SYM theory, where n is the number of blocks (i.e., the number of supergravitons). This is just the Lagrangian for free motion of these particles. Furthermore, since we are doing quantum mechanics and the analogue of the Yang-Mills coupling is the dimensional quantity l_p^3, the coefficient of the quadratic term is uncorrected from its value in the original Lagrangian.

There are residual virtual effects at order p^4 from these heavy states which are the source of parton interactions. Note that the off-diagonal modes are manifestly nonlocal. The apparent locality of low energy physics in this model must emerge from a complex interplay between SUSY and the fact that the frequencies of the nonlocal degrees of freedom become large when particles are separated. We have only a limited understanding of this crucial issue, but in the next section we will provide some evidence that local physics is reproduced in the low energy, long distance limit.

The center of mass of a block of size $N^{(a)}$ is defined by Eq. (5.2). It is easy to see that the Hamiltonian for an asymptotic multiparticle state, when written in terms of center-of-mass transverse momenta, is just

$$H_{\mathrm{asymp}}=\sum_a \frac{R\mathbf{P}^{(a)2}}{N^{(a)}}=\sum_a \frac{\mathbf{P}^{(a)2}}{p_{11}^{(a)}}. \tag{5.7}$$

Note that the dispersion relation for the asymptotic particle states has the fully 11-dimensional Lorentz-invariant form. This is essentially due to the BPS nature of the asymptotic states. For large relative separations, the supersymmetric quantum state, corresponding to the supersymmetric classical flat direction in which the gauge symmetry is "broken" into n blocks, will be precisely the product of the threshold bound-state wave functions of each block subsystem. Each individual block is a BPS state. Its dispersion relation follows from the SUSY algebra and is relativistically invariant even when (e.g., for finite N) the full system is not.

We also note that the statistics of multisupergraviton states comes out correctly because of the residual block permutation gauge symmetry of the matrix model. When some subset of the blocks are in identical states, the original gauge symmetry instructs us to mod out by the permutation group, picking up minus signs depending on whether the states are constructed from an odd or even number of Grassmann variables. The spin statistics connection is the conventional one.

Thus the large N matrix model contains the Fock space of asymptotic states of 11-dimensional supergravity, and the free propagation of particles is described in a manner consistent with 11-dimensional Lorentz invariance. The field theory Fock space is, however, embedded in a system which, as we shall see, has no ultraviolet divergences. Particle statistics is embedded in a continuous gauge symmetry. We find the emergence of field theory as an approximation to an elegant finite structure one of the most attractive features of the matrix model approach to M theory.

VI. LONG RANGE SUPERGRAVITON INTERACTIONS

The first uncanceled interactions in the matrix model occur in the effective action at order $\dot{y}^4$ where $\dot{y}$ is the velocity of the supergravitons [9]. These interactions are calculated by thinking of the matrix model as SYM theory and computing Feynman diagrams. At one loop one finds an induced quartic term in the velocities which corresponds to an induced $F_{\mu\nu}^4$ term. The precise term for two $D0$ branes is given by

$$\frac{A[\dot{y}(1)-\dot{y}(2)]^4}{R^3 r^7}, \tag{6.1}$$

where r is the distance between the $D0$ branes and A is a coefficient of order 1, which can be extracted from the results of [9]. This is the longest range term which governs the interaction between the $D0$ branes as r tends to infinity. Thus the effective Lagrangian governing the low energy, long distance behavior of the pair is

$$L=\frac{\dot{y}(1)^2}{2R}+\frac{\dot{y}(2)^2}{2R}-A\frac{[\dot{y}(1)-\dot{y}(2)]^4}{R^3 r^7}. \tag{6.2}$$

The calculation is easily generalized to the case of two well-separated groups of N_1 and N_2 branes forming bound states. Keeping only the leading terms for large N (planar graphs), we find

$$L=\frac{N_1\dot{y}(1)^2}{2R}=\frac{N_2\dot{y}(2)^2}{2R}-AN_1N_2\frac{[\dot{y}(1)-\dot{y}(2)]^4}{R^3 r^7}. \tag{6.3}$$

To understand the significance of Eq. (6.3), it is first useful to translate it into an effective Hamiltonian. To leading order in inverse powers of r, we find

$$H_{\mathrm{eff}}=\frac{p_\perp(1)^2}{2p_{11}(1)}+\frac{p_\perp(2)^2}{2p_{11}(2)}+A\left[\frac{p_\perp(1)}{p_{11}(1)}-\frac{p_\perp(2)}{p_{11}(2)}\right]^4 \times\frac{p_{11}(1)p_{11}(2)}{r^7 R}. \tag{6.4}$$

From Eq. (6.4) we can compute a scattering amplitude in the Born approximation. Strictly speaking, the scattering amplitude is defined as a 10D amplitude in the compactified theory. However, it contains information about the 11D amplitude in the special kinematic situation where no longitudinal momentum is exchanged. The relation between the

10D amplitude and 11D amplitude at vanishing Δp_{11} is simple. They are essentially the same thing except for a factor of R, which is needed to relate the 10- and 11-dimensional phase space volumes. The relation between amplitudes is $A_{11}=RA_{10}$. Thus from Eq. (6.4) we find the 11D amplitude

$$A\left[\frac{p_{\perp}(1)}{p_{11}(1)}-\frac{p_{\perp}(2)}{p_{11}(2)}\right]^4\frac{p_{11}(1)p_{11}(2)}{r^7}. \tag{6.5}$$

The expression in Eq. (6.5) is noteworthy for several reasons. First of all, the factor r^{-7} is the 11D Green function (in space) after integration over x^{11}. In other words, it is the scalar Feynman propagator for vanishing longitudinal momentum transfer. Somehow the simple matrix Hamiltonian "knows" about massless propagation in 11D spacetime. The remaining momentum-dependent factors are exactly what is needed to make Eq. (6.5) identical to the single (super) graviton exchange diagram[4] in 11D. Even the coefficient is correct. This is closely related to a result reported in [9] where it was shown that the annulus diagram governing the scattering of two $D0$ branes has exactly the same form at very small and very large distances, which can be understood by noting that only BPS states contribute to this process on the annulus. This plus the usual relations between couplings and scales in type-IIA string theory and M theory guarantees that we obtain the correct normalization of 11-dimensional graviton scattering in supergravity (SUGRA). In the weak coupling limit, very long distance behavior is governed by single supergraviton exchange, while the ultrashort distances are governed by the matrix model. In [9] the exact equivalence between the leading interactions computed in these very different manners was recognized, but its meaning was not clear. Now we see that it is an important consistency criterion in order for the matrix model to describe the infinite momentum limit of M theory.

Let us next consider possible corrections to the effective action coming from higher loops. In particular, higher loops can potentially correct the quartic term in velocities. Since our interest lies in the large N limit, we may consider the leading (planar) corrections. Doing ordinary large N counting, one finds that the $\dot{y}^4$ term may be corrected by a factor which is a function of the ratio N/r^3. Such a renormalization by $f(N/r^3)$ could be dangerous. We can consider several cases which differ in the behavior of f as N/r^3 tends to infinity. In the first two, the function tends to zero or infinity. The meaning of this would be that the coupling to gravity is driven either to zero or infinity in the infinite momentum limit. Either behavior is intolerable. Another possibility is that the function f tends to a constant not equal to 1. In this case the gravitational coupling constant is renormalized by a constant factor. This is not supposed to occur in M theory. Indeed, supersymmetry is believed to protect the gravitational coupling from any corrections. The only other possibility is that $f\to 1$. The simplest way in which this can happen is if there are no corrections at all other than the one-loop term, which we have discussed.

We believe that there is a nonrenormalization theorem for this term which can be proven in the context of SUSY quantum mechanics with 16 generators. The closest thing we have been able to find in the literature is a nonrenormalization theorem for the $F_{\mu\nu}{}^4$ term in the action of ten-dimensional string theory[5] which has been proved by Tseytlin [16]. In the quantum-mechanical context, we believe that it is true and that the scattering of two supergravitons at large transverse distance and zero longitudinal momentum is exactly given in the matrix model by low energy 11D supergravity perturbation theory. Dine [17] has constructed the outlines of an argument which demonstrates the validity of the nonrenormalization theorem.

We have considered amplitudes in which vanishing longitudinal momentum is exchanged. Amplitudes with nonvanishing exchange of p_{11} are more complicated. They correspond to processes in 10D in which RR charge is exchanged. Such collisions involve rearrangements of the $D0$ branes in which the collision transfers $D0$ branes from one group to the other. We are studying such processes, but we have no definitive results as yet.

We have thus presented some evidence that the dynamics of the matrix model respects 11-dimensional Lorentz invariance. If this is correct, then the model reduces exactly to supergravity at low energies. It is clear, however, that it is much better behaved in the ultraviolet than a field theory. At short distances, as shown extensively in [9], restoration of the full matrix character of the variables cuts off all ultraviolet divergences. The correspondence limit by which M theory reduces to supergravity indicates that we are on the right track.

VII. SIZE OF A SUPERGRAVITON

As we have pointed out in Sec. II, the holographic principle requires the transverse size of a system to grow with the number of constituent partons. It is therefore of interest to estimate the size of the threshold bound state describing a supergraviton of longitudinal momentum N/R. According to the holographic principle, the radius should grow like $N^{1/9}$ in 11D Planck units. We will use a mean field approximation in which we study the wave function of one parton in the field of N others. We therefore consider the effective Lagrangian (6.3) for the case $N_1=1$, $N_2=N$. The action simplifies for $N\gg 1$ since in this case the N-particle system is much heavier than the single-particle system. Therefore we may set its velocity to zero. The Lagrangian becomes

$$L_1=\frac{\dot{y}^2}{2R}-N\frac{\dot{y}^4}{R^3y^7}. \tag{7.1}$$

---

[4]In [9] the amplitude was computed for $D0$ branes which have momenta orthogonal to their polarizations (this was not stated explicitly there, but was implicit in the choice of boundary state). The spin dependence of the amplitude is determined by the supersymmetric completion of the v^4/r^7 amplitude, which we have not computed. In principle, this gives another check of 11-dimensional Lorentz invariance. We suspect that the full answer follows by applying the explicit supersymmetries of the light cone gauge to the amplitude we have computed.

[5]We thank C. Bachas for pointing out this theorem to us.

where y refers to the relative coordinate between the two systems. We can remove all N and R dependence from the action $S=\int L_1 dt$ by scaling:

$$y \rightarrow N^{1/9} y,$$
$$t \rightarrow \frac{N^{2/9}}{R} t. \tag{7.2}$$

The characteristic length, time, and velocity ($v=\dot{y}$) scales are

$$y \sim N^{1/9},$$
$$t \sim \frac{N^{2/9}}{R}, \tag{7.3}$$
$$v \sim \frac{y}{t} \sim N^{-1/9} R.$$

That the size of the bound state wave function scales like $N^{1/9}$ is an indication of the incompressibility of the system when it achieves a density of order one degree of freedom per Planck area. This is in accordance with the holographic principle.

This mean field picture of the bound state, or any other description of it as a simple cluster, makes the problem of longitudinal boost invariance mentioned earlier very concrete. Suppose we consider the scattering of two bound states with N_1 and N_2 constituents, respectively, $N_1 \sim N_2 \sim N$. The mean field picture strongly suggests that scattering will show a characteristic feature at an impact parameter corresponding to the bound-state size $\sim N^{1/9}$. But this is not consistent with longitudinal boosts which take $N_1 \rightarrow \alpha N_1$, $N_2 \rightarrow \alpha N_2$. Boost invariance requires physics to depend only on the ratio N_1/N_2 or, said another way, only on the ratio of the bound-state sizes. This strongly suggests that a kind of scale invariance must be present in the dynamics that is clearly absent in the simple picture discussed above. In the string case the scale-invariant world sheet dynamics is crucial for longitudinal boost invariance.

The possibility that partons might form subclusters within the bound state was ignored in mean field discussion. A preliminary discussion of a hierarchical clustering model with many length scales is presented in Appendix A. Note also that wave functions of threshold bound states are power law behaved.

Understanding the dynamics of these bound states well enough to check longitudinal boost invariance reliably is an important subject for future research.

VIII. MEMBRANES

In order to be the strong coupling theory of IIA string theory, M theory must have membranes in its spectrum. Although in the decompactified limit there are no truly stable finite energy membranes, very-long-lived large classical membranes must exist. In this section we will show how these membranes are described in the matrix model, a result first found[6] in [5]. Townsend [10] first pointed out the connection between the matrix description of $D0$ brane dynamics and the matrix description of membranes, and speculated that a membrane might be regarded as a collective excitation of $D0$ branes. Our conjecture supplies a precise realization of Townsend's idea.

The formulation we will use to describe this connection is a version of the methods introduced in [5,18,19].

Begin with a pair of unitary operators U,V satisfying the relations

$$UV=e^{2\pi i/N}VU,$$
$$U^N=1,$$
$$V^N=1. \tag{8.1}$$

These operators can be represented on an N-dimensional Hilbert space as clock and shift operators. They form a basis for all operators in the space. Any matrix Z can be written in the form

$$Z=\sum_{n,m=1}^{N} Z_{nm} U^n V^m. \tag{8.2}$$

U and V may be thought of as exponentials of canonical variables p and q:

$$U=e^{ip},$$
$$V=e^{iq}, \tag{8.3}$$

where p,q satisfy the commutation relations

$$[q,p]=\frac{2\pi i}{N}. \tag{8.4}$$

From Eq. (8.2) we see that only periodic functions of p and q are allowed. Thus the space defined by these variables is a torus. In fact, there is an illuminating interpretation of these coordinates in terms of the quantum mechanics of particles on a torus in a strong background magnetic field. The coordinates of the particle are p,q. If the field is strong enough, the existence of a large gap makes it useful to truncate the space of states to the finite dimensional subspace of lowest Landau levels. On this subspace the commutation relation (8.4) is satisfied. The lowest Landau wave packets form minimum uncertainty packets which occupy an area $\sim 1/N$ on the torus. These wave packets are analogous to the "Planckian cells" which make up quantum phase space. The p,q space is sometimes called the noncommuting torus, the quantum torus, or the fuzzy torus. In fact, for large enough N we can choose other bases of N-dimensional Hilbert space which correspond to the lowest Landau levels of a charged particle propagating on an arbitrary Riemann surface wrapped by a constant magnetic field. For example, in [5], de Wit *et al.* construct the finite dimensional Hilbert space of

[6]We are grateful to M. Green for pointing out this paper to us when a preliminary version of this work was presented at the Santa Barbara Strings '96 conference.

lowest Landau levels on a sphere. This connection between finite matrix models and two-dimensional surfaces is the basis for the fact that the large N matrix model contains membranes. For finite N, the model consists of maps of quantum Riemann surfaces into a noncommuting transverse superspace; i.e., it is a model of a noncommuting membrane embedded in a noncommutative space.[7]

In the limit of large N, the quantum torus behaves more and more like classical phase space. The following correspondences connect the two.

(1) The quantum operators Z defined in Eq. (8.2) are replaced by their classical counterparts. Equation (8.2) becomes the classical Fourier decomposition of a function on phase space.

(2) The operation of taking the trace of an operator goes over to N times the integral over the torus:

$$\mathrm{tr}Z \rightarrow N \int Z(p,q)dp\, dq. \tag{8.5}$$

(3) The operation of commuting two operators is replaced by $1/N$ times the classical Poisson brackets:

$$[Z,W] \rightarrow \frac{1}{N}[\partial_q Z\, \partial_p W - \partial_q W\, \partial_p Z]. \tag{8.6}$$

We may now use the above correspondence to formally rewrite the matrix model Lagrangian. We begin by representing the matrices Y^i and θ as operator functions $Y^i(p,q)$ and $\theta(p,q)$. Now apply the correspondences to the two terms in Eq. (4.2). This gives

$$L = \frac{p_{11}}{2} \int dp\, dq\, \dot{Y}^i(p,q)^2 - \frac{1}{p_{11}} \int dp\, dq \times [\partial_q Y^i \partial_p Y^j - \partial_q Y^j \partial_p Y^i]^2 + \text{fermionic terms} \tag{8.7}$$

and a Hamiltonian

$$H = \frac{1}{2p_{11}} \int dp\, dq\, \Pi_{ii}(p,q)^2 + \frac{1}{p_{11}} \int dp\, dq \times [\partial_q Y^i \partial_p Y^j - \partial_q Y^j \partial_p Y^i]^2 + \text{fermionic terms}. \tag{8.8}$$

Equation (8.8) is exactly the standard Hamiltonian for the 11D supermembrane in the light cone frame. The construction shows us how to build configurations in the matrix model which represent large classical membranes. To do so we start with a classical embedding of a toroidal membrane described by periodic functions $Y^i(p,q)$. The Fourier expansion of these functions provides us with a set of coefficients Y^i_{mn}. Using Eq. (8.2), we then replace the classical Y's by operator functions of U,V. The resulting matrices represent the large classical membranes.

[7]Note that it is clear in this context that membrane topology is not conserved by the dynamics. Indeed, for fixed N a given matrix can be thought of as a configuration of many different membranes of different topology. It is only in the large N limit that stable topological structure may emerge in some situations.

If the matrix model membranes described above are to correspond to M-theory membranes, their tensions must agree. Testing this involves keeping track of the numerical factors of order 1 in the above discussion. We present this calculation in Appendix B where we show that the matrix model membrane tension exactly agrees with the M-theory membrane tension. This has also been verified by Berkooz and Douglas [20] using a different technique.

We do not expect static finite energy membranes to exist in the uncompactified limit. Nevertheless, let us consider the conditions for such a static solution. The matrix model equations of motion for static configurations is

$$[Y^i,[Y^j,Y^i]]=0. \tag{8.9}$$

It is interesting to consider a particular limiting case of an infinite membrane stretched out in the 8,9 plane for which a formal solution of Eq. (8.9) can be found. We first rescale p and q:

$$P=\sqrt{N}p,$$

$$Q=\sqrt{N}q,$$

$$[Q,P]=2\pi i. \tag{8.10}$$

In the $N\rightarrow\infty$ limit, the P,Q space becomes an infinite plane. Now consider the configuration

$$Y^8=R_8 P,$$

$$Y^9=R_9 Q, \tag{8.11}$$

with all other $Y^i=0$. Here R_i is the length of the corresponding direction, which should of course be taken to infinity. Since $[Y^8,Y^9]$ is a c number, Eq. (8.9) is satisfied. Thus we find the necessary macroscopic membranes require by M theory. This stretched membrane has the requisite "wrapping number" on the infinite plane. On a general manifold one might expect the matrix model version of the wrapping number of a membrane on a two-cycle to be

$$W=\frac{1}{N}\,\mathrm{Tr}\omega_{ij}(X^i(p,q))[X^i,X^j], \tag{8.12}$$

where ω is the two-form associated with the cycle. This expression approaches the classical winding number as we take the limit in which Poisson brackets replace commutators.

Another indication that we have found the right representation of the membrane comes from studying the supersymmetry transformation properties of our configuration.[8] The supermembrane should preserve half of the supersymmetries of the model. The SUSY transformation of the fermionic coordinates is

[8]This result was derived in collaboration with Seiberg, along with a number of other observations about supersymmetry in the matrix model, which will appear in a future publication.

$$\delta\theta=(P^i\gamma_i+[X^i,X^j]\gamma_{ij})\epsilon+\epsilon'. \quad (8.13)$$

For our static membrane configuration, $P^i=0$, and the commutator is proportional to the unit matrix; so we can choose ϵ' to make this variation vanish. The unbroken supergenerators are linear combinations of the IMF q_A and Q_α.

It is interesting to contemplate a kind of duality and complementarity between membranes and $D0$ branes. According to the standard light cone quantization of membranes, the longitudinal momentum p_{11} is uniformly distributed over the area of the p,q parameter space. This is analogous to the uniform distribution of p_{11} along the σ axis in string theory. As we have seen, the p,q space is a noncommuting space with a basic indivisible quantum of area. The longitudinal momentum of such a unit cell is $1/N$ of the total. In other words, the unit phase space cells that result from the noncommutative structure of p,q space are the $D0$ branes with which we began. The $D0$ branes and membranes are dual to one another. Each can be found in the theory of the other.

The two kinds of branes also have a kind of complementarity. As we have seen, the configurations of the matrix model which have classical interpretations in terms of $D0$ branes are those for which the Y's commute. On the other hand, the configurations of a membrane which have a classical interpretation are the extended membranes of large classical area. The area element is the Poisson bracket which in the matrix model is the commutator. Thus the very classical membranes are highly nonclassical configurations of $D0$ branes.

In the paper of de Wit, Luscher, and Nicolai [6], a pathology of membrane theory was reported. It was found that the spectrum of the membrane Hamiltonian is continuous. The reason for this is the existence of the unlifted flat directions along which the commutators vanish. Previously, it had been hoped that membranes would behave like strings and have discrete level structure and perhaps be the basis for a perturbation theory which would generalize string perturbation theory. In the present context this apparent pathology is exactly what we want. M theory has no small coupling analogous to the string splitting amplitude. The bifurcation of membranes when the geometry degenerates is expected to be an order-1 process. The matrix model, if it is to describe all of M theory, must inextricably contain this process. In fact, we have seen how important it is that supersymmetry maintains the flat directions. A model of a single noncommutative membrane actually contains an entire Fock space of particles in flat 11-dimensional space-time.

Another pathology of conventional membrane theories which we expect to be avoided in M theory is the nonrenormalizability of the membrane world volume field theory. For finite N, it is clear that ultraviolet divergences on the world volume are absent because the noncommutative nature of the space defines a smallest volume cell, just like a Planck cell in quantum-mechanical phase space (but we should emphasize here that this is a classical rather than quantum-mechanical effect in the matrix model). The formal continuum limit which gives the membrane Hamiltonian is clearly valid for describing the classical motion of a certain set of metastable semiclassical states of the matrix model. It should not be expected to capture the quantum mechanics of the full large N limit. In particular, it is clear that the asymptotic supergraviton states would look extremely singular and have no real meaning in a continuum membrane formalism. We are not claiming here to have a proof that the large N limit of the matrix quantum mechanics exists, but only that the issues involved in the existence of this limit are not connected to the renormalizability of the world volume field theory of the membrane.

There is one last point worth making about membranes. It involves evidence for 11D Lorentz invariance of the matrix model. We have considered in some detail the Galilean invariance of the infinite momentum frame and found that it is satisfied. But there is more to the Lorentz group. In particular, there are generators J^i which in the light cone formalism rotate the lightlike surface of initial conditions. The conditions for invariance under these transformations are the notorious angular conditions. We must also impose longitudinal boost invariance. The angular conditions are what makes Lorentz invariance so subtle in light cone string theory. It is clearly important to determine if the matrix model satisfies the angular conditions in the large N limit. In the full quantum theory the answer is not yet clear, but at the level of the classical equations of motion the answer is yes. The relevant calculations were done by de Wit, Marquard, and Nicolai [21]. The analysis is too complicated to repeat here, but we can describe the main points.

The equations for classical membranes can be given in covariant form in terms of a Nambu-Goto-type action. In the covariant form the generators of the full Lorentz group are straightforward to write down. In passing to the light cone frame, the expressions for the nontrivial generators become more complicated, but they are quite definite. In fact, they can be expressed in terms of the $Y(p,q)$ and their canonical conjugates $\Pi(p,q)$. Finally, using the correspondence between functions of p,q and matrices, we are led to matrix expressions for the generators. The expressions, of course, have factor-ordering ambiguities, but these, at least formally, vanish as $N\to\infty$. In fact, according to [21], the violation of the angular conditions goes to zero as $1/N^2$. Needless to say, a quantum version of this result would be very strong evidence for our conjecture.

We cannot refrain from pointing out that the quantum version of the arguments of [21] is apt to be highly nontrivial. In particular, the classical argument works for every dimension in which the classical supermembrane exists, while, by analogy with perturbative string theory, we only expect the quantum Lorentz group to be recovered in 11 dimensions. Further, the longitudinal boost operator of [21] is rather trivial and operates only on a set of zero mode coordinates, which we have not included in our matrix model. Instead, we expect the longitudinal boost generator to involve rescaling N in the large N limit and, thus, to relate the Hilbert spaces of different SUSY quantum-mechanical models. We have already remarked in the previous section that, as anticipated in [7], longitudinal boost invariance is the key problem in our model. We expect it to be related to a generalization of the conformal invariance of perturbative string theory.

IX. TOWARDS DERIVATION AND COMPACTIFICATION

In this section we would like to present a line of argument which may lead to a proof of the conjectured equivalence between the matrix model and M theory. It relies on a stringy extension of the conjectured nonrenormalization theorem combined with the possibility that all velocities in the large N cluster go to zero as $N\rightarrow\infty$.

Imagine that R stays fixed as $N\rightarrow\infty$. Optimistically, one might imagine that finite R errors are as small as in perturbative 11D supergravity, meaning that they are suppressed by powers of $k_{11}R$ or even $\exp(-k_{11}R)$ where k_{11} is the center-of-mass longitudinal momentum transfer. So for $k_{11} \sim 1/l_P$ we could imagine R fixed at a macroscopic scale and have very tiny errors. The mean field estimates discussed in Sec. VII give the velocity $v\sim N^{-1/9}R$, which, with R fixed, can be made arbitrarily small at large N. Although it is likely that the structure of the large N cluster is more complicated than the mean field description, it is possible that this general property of vanishing velocities at large N continues to hold. In particular, in Appendix A we present arguments that the velocities of the coordinates along some of the classical flat directions of the potential are small. We suspect that this can be generalized to all of the flat directions. If that is the case, then the only high velocities would be those associated with the "core" wave function of Appendix A. The current argument assumes that the amplitude for the core piece of the wave function vanishes in the large N limit.

Non-Abelian field strength is the correct generalization of velocity for membrane-type field configurations like those discussed in Sec. VIII. For classical configurations at least these field strengths are order $1/N$ and so are also small.

We have previously conjectured that the v^4 terms in the quantum-mechanical effective action are not renormalized beyond one loop. For computing the 11-dimensional supergravity amplitude, we needed this result in the matrix quantum mechanics, but is possible that this result holds in the full string perturbation theory. For example, the excited open string states can be represented as additional non-BPS fields in the quantum mechanics. These do not contribute to the one loop v^4 term because they are not BPS. Perhaps they do not contribute to higher loops for related reasons.

If these two properties hold, then the conjecture follows. The scattering of large N clusters of $D0$ branes can clearly be computed at small g (small R) using quantum mechanics. But these processes, by assumption, only involve low velocities independent of g and so only depend on the v^4 terms in the effective action which, by the stringy extension of the nonrenormalization theorem, would not receive g corrections. So the same quantum-mechanical answers would be valid at large g (large R).

This would prove the conjecture.

From this point of view, we have identified a subset of string theory processes (large-N $D0$ brane scattering) which are unchanged by stringy loop corrections and so are computable at strong coupling.

If this line of argument is correct, it gives us an unambiguous prescription for compactification. We take the quantum mechanics which describes 0-brane motion at weak string coupling in the compactified space and then follow it to strong coupling. This approach to compactification requires us to add extra degrees of freedom in the compactified theory. We will discuss an alternative approach in the next section.

For toroidal compactifications, it is clear at weak coupling that one needs to keep the strings which wrap any number of times around each circle. These unexcited wrapped string states are BPS states, and so they do contribute to the v^4 term and hence must be kept. In fact, these are the states which, in the annulus diagram, correct the power law in graviton scattering to its lower dimensional value and are crucial in implementing the various T dualities.

To be specific let us discuss the case of one coordinate X^9 compactified on a circle on radius R_9. Here we should keep the extra string winding states around X^9. An efficient way to keep track of them is to T dualize the X^9 circle. This converts $D0$ branes to $D1$ branes and winding modes to momentum modes. The collection of N $D1$ branes is described by a $(1+1)$-dimensional SU(N) super Yang-Mills quantum field theory with coupling $g^2_{\rm SYM}=R^2/(R_9 l_P^3)$ on a space of T-dual radius $R_{\rm SYM}=l_P^3/(RR_9)$. The dimensionless effective coupling of the super Yang-Mills theory is then $g^2_{\rm SYM}R^2_{\rm SYM}=(l_P/R_9)^3$, which is independent of R. For p-dimensional tori we get systems of Dp branes described by $(p+1)$-dimensional SU(N) super Yang-Mills theory. Related issues have also recently been discussed in [22].

For more general compactifications the rule would be to keep every BPS state which contributes to the v^4 term at large N. We are currently investigating such compactifications, including ones with less supersymmetry.

The line of argument presented in this section raises a number of questions. Is it permissible to hold R finite or to let it grow very slowly with N? Are there nonperturbative corrections to the v^4 term? Large N probably prohibits instanton corrections in the quantum mechanics, but perhaps not in the full string theory. This might be related to the effect of various wrapped branes in compactified theories.

Does the velocity stay low? A key problem here is that in the mean field theory cloud the 0 branes are moving very slowly. If two 0 branes encounter each other, their relative velocity is much less than the typical velocity in a bound pair $(v\sim R)$. It seems that the capture cross section to go into the pair bound state should be very large. Why is there no clumping into pairs? One factor which might come into play is the following. If the velocity is very low, the de Broglie wavelength of the particles might be comparable to the whole cluster (this is true in the mean field), and so there could be delicate phase correlations across the whole cluster—some kind of macroscopic quantum coherence. Whenever a pair is trying to form, another 0 brane might get between them and disrupt them. This extra coherent complexity might help explain the Lorentz invariance puzzle.

X. ANOTHER APPROACH TO COMPACTIFICATION

The conjecture which we have presented refers to an exact formulation of M theory in uncompactified 11-dimensional space-time. It is tempting to imagine that we can regain the compactified versions of the theory as particular collections of states in the large N limit of the matrix model. There is ample ground for suspicion that this may not be the case and that degrees of freedom that we have thrown away in the uncompactified theory may be required for compactification. Indeed, in IMF field theory the only general method for discussing theories with moduli spaces of vacua is implementable only when the vacua are visible in the classical approximation. Then we can shift the fields and do IMF quantization of the shifted theory. Different vacua correspond to different IMF Hamiltonians for the same degrees of freedom. The proposal of the previous section is somewhat in the spirit of IMF field theory. Different Hamiltonians and, indeed, different sets of degrees of freedom are required to describe each compactified vacuum.

We have begun a preliminary investigation of the alternative hypothesis, that different compactifications are already present in the model we have defined. This means that there must be collections of states which, in the large N limit, have S matrices which completely decouple from each other. Note that the large N limit is crucial to the possible existence of such superselection sectors. The finite N quantum mechanics cannot possibly have superselection rules. Thus the only way in which we could describe compactifications for finite N would be to add degrees of freedom or change the Hamiltonian. We caution the reader that the approach we will describe below is very preliminary and highly conjectural.

This approach to compactification is based on the idea that there is a sense in which our system defines a single "noncommuting membrane." Consider compactification of a membrane on a circle. Then there are membrane configurations in which the embedding coordinates do not transform as scalars under large diffeomorphisms of the membrane volume, but rather are shifted by large diffeomorphisms of the target space. These are winding states. A possible approach to identifying the subset of states appropriate to a particular compactification is to first find the winding states and then find all states which have nontrivial scattering from them in the large N limit. In fact, our limited study below seems to indicate that all relevant states, including compactified supergravitons, can be thought of as matrix model analogues of membrane winding states.

Let us consider compactification of the ninth transverse direction on a circle of radius $2\pi R_9$. A winding membrane is a configuration which satisfies

$$X^9(q,p+2\pi)=X^9(q,p)+2\pi R_9, \tag{10.1}$$

and the winding sector is defined by a path integral over configurations satisfying this boundary condition. A matrix analogue of this is

$$e^{-iNq}X^9e^{iNq}=X^9+2\pi R_9. \tag{10.2}$$

It is easy to see (by taking the trace) that this condition cannot be satisfied for finite N. However, if we take the large N limit in such a way that $q\rightarrow\sigma/N\otimes\mathbf{1}_{M\times M}$, with σ an angle variable, then this equation can be satisfied, with

$$X^9=\frac{2\pi R_9}{i}\frac{\partial}{\partial\sigma}\otimes\mathbf{1}_{M\times M}+x^9(\sigma), \tag{10.3}$$

where x^9 is an $M\times M$-matrix-valued function of the angle variable. The other transverse bosonic coordinates and all of the θ's are $M\times M$-matrix-valued functions of σ. These equations should be thought of as limits of finite matrices. Thus $2\pi R_9\mathcal{P}\equiv(2\pi R_9/i)\partial/\partial\sigma$ can be thought of as the limit of the finite matrices diag$(-2\pi PR_9\ \ldots\ 2\pi PR_9)$, with σ the obvious tridiagonal matrix in this representation. The total longitudinal momentum of such a configuration is $(2P+1)M/R$, and the ratio M/P is an effectively continuous parameter characterizing the states in the large N limit. We are not sure of the meaning of this parameter.

To get a feeling for the physical meaning of this proposal, we examine the extreme limits of large and small R_9. For large R_9 it is convenient to work in the basis where $\mathcal{P}$ is diagonal. If we take all of the coordinates X^i independent of σ, then our winding membrane approaches a periodic array of $(2P+1)$ collections of $D0$ branes, each with longitudinal momentum M/R. We can find a solution of the BPS condition by putting each collection into the M zero-brane threshold bound-state wave function. For large R_9 configurations of the X^i which depend on σ have very high frequency and can be integrated out. Thus, in this limit, the BPS state in this winding sector is approximately a periodic array of supergravitons. We identify this with the compactified supergraviton state. This state will have the right long range gravitational interactions (at scales larger than R_9) in the eight uncompactified dimensions. To obtain the correct decompactified limit, it would appear that we must rescale R, the radius of the longitudinal direction by $R\rightarrow(2P+1)R$, as we take P and M to infinity. With this rescaling, all trace of the parameter M/P seems to disappear in the decompactification limit.

For small R_9, our analysis is much less complete. However, string duality suggests an approximation to the system in which we keep only configurations with $M=1$ and $P\rightarrow\infty$. In this case the σ dependence of X^9 is pure gauge and the X^i all commute with each other. The matrix model Hamiltonian becomes the Hamiltonian of the Green-Schwarz type-IIA string:

$$H\rightarrow\int d\sigma(\dot{\mathbf{X}})^2+\left(\frac{\partial\mathbf{X}}{\partial\sigma}\right)^2+\theta^T\gamma_9\frac{\partial\theta}{\partial\sigma}. \tag{10.4}$$

As in previous sections, we will construct multiwound membrane states by making large block diagonal matrices, each block of which is the previous single-particle construction. Lest such structures appear overly baroque, we remind the reader that we are trying to make explicit constructions of the wave functions of a strongly interacting system with an infinite number of degrees of freedom. For large R_9 it is fairly clear that the correct asymptotic properties of multiparticle states will be guaranteed by the BPS condition (assuming that everything works as conjectured in the uncompactified theory).

If our ansatz is correct for small R_9, it should be possible to justify the neglect of fluctuations of the matrix variables away from the special forms we have taken into account, as well as to show that the correct string interactions (for multistring configurations defined by the sort of block diagonal construction we have used above) are obtained from the matrix model interactions. In this connection it is useful to note that in taking the limit from finite matrices, there is no meaning to the separation of configurations into winding sectors which we have defined in the formal large N limit. In particular, X^9 should be allowed to fluctuate. But we have seen that shifts of X^9 by functions of σ are pure gauge, so that all fluctuations around the configurations which we have kept give rise to higher derivative world sheet interactions. Since the $P\rightarrow\infty$ limit is the world sheet continuum limit, we should be able to argue that these terms are irrelevant operators in that limit. We have less understanding about how the sum over world sheet topologies comes out of our formalism, but it is tempting to think that it is in some way connected with the usual topological expansion of large N matrix models. In the Appendix we show that in 11 dimensions, dimensional analysis guarantees the dominance of planar graphs in certain calculations. Perhaps, in ten dimensions, the small dimensionless parameter, R_9/l_P must be scaled with a power of N in order to obtain the limit of the matrix model which gives IIA string theory.

These ideas can be extended to compactification on multidimensional tori. A wrapping configuration of a toroidal membrane can be characterized by describing the cycles on the target torus on which the a and b cycles of the membrane are mapped. This parametrization is redundant because of the SL(2,Z) modular invariance which exchanges the two membrane cycles. We propose that the analogue of these wrapping states, for a d-torus defined by modding out R^d by the shifts $X^i \rightarrow X^i + 2\pi R^i_a$, is defined by the conditions

$$e^{-i\sigma}X^i_{(\mathbf{m},\mathbf{n})}e^{i\sigma}=X^i_{(\mathbf{m},\mathbf{n})}+2\pi R^i_a n^a, \tag{10.5}$$

$$e^{2\pi i\mathcal{P}}X^i_{(\mathbf{m},\mathbf{n})}e^{-2\pi i\mathcal{P}}=X^i_{(\mathbf{m},\mathbf{n})}+2\pi R^i_a m^a, \tag{10.6}$$

where $\mathbf{n}$ and $\mathbf{m}$ are d-vectors of integers. The solutions to these conditions are

$$X^i_{(\mathbf{m},\mathbf{n})}=[2\rho R^i_a n^a\mathcal{P}+R^i_a m^a\sigma]\otimes \mathbf{1}_{M\times M}+x^i_{(\mathbf{m},\mathbf{n})}(\sigma), \tag{10.7}$$

where the x^i are periodic $M\times M$-matrix-valued functions of σ. The fermionic and noncompact coordinates are also matrix-valued functions of σ.

In order to discuss more complicated compactifications, we would have to introduce coordinates and find a group of large diffeomorphisms associated with one and two cycles around which membranes can wrap. Then we would search for embeddings of this group into the large N gauge group. Presumably, different coordinate systems would correspond to unitarily equivalent embeddings. We can even begin to get a glimpse of how ordinary Riemannian geometry would emerge from the matrix system. If we take a large manifold which breaks sufficient supersymmetries, the effective action for supergravitons propagating on such a manifold would be obtained, as before, by integrating over the off-diagonal matrices. Now, however, the nonrenormalization theorem would fail and the kinetic term for the gravitons would contain a metric. The obvious conjecture is that this is the usual Riemannian metric on the manifold in question. If this is the case, our prescription for compactification in the noncommutative geometry of the matrix model would reduce to ordinary geometry in the large radius limit.

A question which arises is whether the information about one and two cycles is sufficient to characterize different compactifications. We suspect that the answer to this is no. The moduli of the spaces that arise in string-theoretic compactifications are all associated with the homology of the space, but in general higher dimensional cycles (e.g., three cycles in Calabi-Yau threefolds) are necessary to a complete description of the moduli space. Perhaps in order to capture this information we will have to find the correct descriptions of five branes in the matrix model. If the theory really contains low energy SUGRA, then it will contain solitonic five branes, but it seems to us that the correct prescription is to define five branes as the D branes of membrane theory. We do not yet understand how to introduce this concept in the matrix model.

Finally, we would like to comment on the relation between the compactification schemes of this and the previous sections. For a single circle, if we take P to infinity and substitute the formula (10.3) into the matrix model Hamiltonian (as well as the prescription that all other coordinates and supercoordinates are functions of σ), then we find the Hamiltonian of (1+1)-dimensionally reduced 10 D SYM theory in $A_0=0$ gauge, with x^9 playing the role of the spatial component of the vector potential. Thus the prescription of the previous section appears to be a particular rule for how the large N limit should be taken in the winding configurations we have studied here. P is taken to infinity first, and then M is taken to infinity. The relation between the two approaches is reminiscent of the Eguchi-Kawai [23] reduction of large N gauge theory. It is clear once again that much of the physics of the matrix model is buried in the subtleties of the large N limit. For multidimensional tori, the relationship between the formalisms of this and the previous section is more obscure.

XI. CONCLUSIONS

Although the evidence we have given for the conjectured exact equivalence between the large N limit of supersymmetric matrix quantum mechanics and uncompactified 11-dimensional M theory is not definitive, it is quite substantial. The evidence includes the following.

(1) The matrix model has exact invariance under the super-Galilean group of the infinite momentum frame description of 11D Lorentz-invariant theories.

(2) Assuming the conventional duality between M theory and IIA string theory, the matrix model has normalizable marginally bound states for any value of N. These states have exactly the quantum numbers of the 11D supergraviton multiplet. Thus the spectrum of single-particle states is exactly that of M theory.

(3) As a consequence of supersymmetric nonrenormalization theorems, asymptotic states of any number of noninteracting supergravitons exist. These well-separated particles propagate in a Lorentz-invariant manner in 11 dimensions. They have the statistics properties of the supergravity Fock space.

(4) The matrix model exactly reproduces the correct long range interactions between supergravitons implied by 11D supergravity, for zero longitudinal momentum exchange. This one-loop result could easily be ruined by higher loop effects proportional to four powers of velocity. We believe that a highly nontrivial supersymmetry theorem protects us against all higher loop corrections of this kind.

(5) By examining the pieces of the bound-state wave function in which two clusters of particles are well separated from each other, a kind of mean field approximation, we find that the longest range part of the wave function grows with N exactly as required by the holographic principle. In par-

ticular, the transverse density never exceeds one parton per Planck area.

(6) The matrix model describes large classical membranes as required by M theory. The membrane world volume is a noncommutative space with a fundamental unit of area analogous to the Planck area in phase space. These basic quanta of area are the original $D0$ branes from which the matrix model was derived. The tension of this matrix model membrane is precisely the same as that of the M-theory membrane.

(7) At the classical level the matrix model realizes the full 11D Lorentz invariance in the large N limit.

Of course, many unanswered questions remain. Locality is extremely puzzling in this system. Longitudinal boost invariance, as we have stressed earlier, is very mysterious. Resolving this issue, perhaps by understanding the intricate dynamics it seems to require, will be crucial in deciding whether or not this conjecture is correct.

One way of understanding Lorentz invariance would be to search for a covariant version of the matrix model in which the idea of noncommutative geometry is extended to all of the membrane coordinates. An obvious idea is to consider functions of angular momentum operators and try to exploit the connection between spin networks and three-dimensional diffeomorphisms. Alternatively, one could systematically study quantum corrections to the angular conditions.

It is likely that more tests of the conjecture can be performed. In particular, it should be possible to examine the large distance behavior of amplitudes with nonvanishing longitudinal momentum transfer and to compare them with supergravity perturbation theory.

It will be important to to try to make precise the line of argument outlined in Sec. IX that may lead to a proof of the conjecture. The approaches to compactification discussed in Secs. IX and X should be explored further.

If the conjecture is correct, it would provide us with the first well-defined nonperturbative formulation of a quantum theory which includes gravitation. In principle, with a sufficiently large and fast computer any scattering amplitude could be computed in the finite N matrix model with arbitrary precision. Numerical extrapolation to infinite N is in principle, if not in practice, possible. The situation is much like that in QCD where the only known definition of the theory is in terms of a conjectured limit of lattice gauge theory. Although the practical utility of the lattice theory may be questioned, it is almost certain that an extrapolation to the continuum limit exists. The existence of the lattice gauge Hamiltonian formulation ensures that the theory is unitary and gauge invariant.

One can envision the matrix model formulation of M theory playing a similar role. It would, among other things, ensure that the rules of quantum mechanics are consistent with gravitation. Given that the classical long distance equations of 11D supergravity have black hole solutions, a Hamiltonian formulation of M theory would, at last, lay to rest the claim that black holes lead to a violation of quantum coherence

ACKNOWLEDGMENTS

A preliminary version of this work was presented in two talks by L. Susskind at the Strings 96 conference in Santa Barbara. We would like to thank the organizers and participants of the strings conference at Santa Barbara for providing us with a stimulating venue for the production of exciting physics. We would particularly like to thank M. Green for pointing out the relation to previous work on supermembranes, and B. de Wit and I. Bars for discussing some of their earlier work on this topic with us. We would also like to thank C. Bachas and M. Dine for conversations on nonrenormalization theorems, M. Dine for discussing his work with us, C. Thorn for discussions about light cone string theory, N. Seiberg for discussions about supersymmetry, and M. Berkooz, M. Douglas, and N. Seiberg for discussions about membrane tension. T.B., W.F., and S.H.S. would like to thank the Stanford Physics Department for its hospitality while some of this work was carried out. The work of T.B. and S.H.S. was supported in part by the U.S. Department of Energy under Grant No. DE-FG02-96ER40959 and that of W.F. was supported in part by the Robert A. Welch Foundation and by NSF Grant No. PHY-9511632. L.S. acknowledges the support of the NSF under Grant No. PHY-9219345.

APPENDIX A

In this appendix we will report on a preliminary investigation of the threshold bound-state wave function of N zero branes in the large N limit. In general, we may expect a finite probability for the N brane bound state to consist of p clusters of $N_1,\ldots,N_p$ branes separated by large distances along one of the flat directions of the potential. We will try to take such configurations into account by writing a recursion relation relating the N cluster to a k and $N-k$ cluster. This relation automatically incorporates multiple clusters since the pairs into which the original cluster is broken up will themselves contain configurations in which they are split up into further clusters. There may, however, be multiple cluster configurations which cannot be so easily identified as two such superclusters. We will ignore these for now, in order to get a first handle on the structure of the wave function.

The configuration of a pair of widely separated clusters has a single collective coordinate whose Lagrangian we have already written in our investigation of supergraviton scattering. The Lagrangian is

$$L=\frac{1}{2}\frac{k(N-k)}{N}v^2+\frac{k(N-k)}{r^7}v^4, \tag{A1}$$

where r is the distance between the clusters and v is their relative velocity. By scaling, we can write the solution of this quantum-mechanical problem as $\phi\{r(k[N-k])^{2/9}/N^{1/3}\}$, where ϕ is the threshold bound-state wave function of the Lagrangian

$$\frac{1}{2}v^2+\frac{v^4}{r^7}. \tag{A2}$$

This solution is valid when $r \gg l_P$.

We are now motivated to write the recursion relation

$$\Psi_N = \Psi_N^{(c)} + \frac{1}{2} P \sum_{k=1}^{N-1} A_{N,k} \Psi_k \Psi_{N-k} \times \phi\left(\frac{r(k[N-k])^{2/9}}{N^{1/3}}\right) e^{-r\,\mathrm{Tr} W_k^{\dagger} W_k}. \quad \text{(A3)}$$

Here we have chosen a gauge in order to make a block-diagonal splitting of our matrices. Ψ_j is the exact normalized threshold bound-state wave function for j zero branes. W_k are the off-diagonal $k \times N-k$ matrices which generate interactions between the two clusters. P is the gauge-invariant projection operator which rotates our gauge choice among all gauge-equivalent configurations. The $A_{N,k}$ are normalization factors, which in principle we would attempt to find by solving the Schrödinger equation. $\Psi_N^{(c)}$ is the "core" wave function, which describes configurations in which all of the zero branes are at a distance less than or equal to l_P from each other. We will describe some of its properties below. In this regime, the entire concept of distance breaks down, since the noncommuting parts of the coordinates are as large as the commuting ones.

The interesting thing which is made clear by this ansatz is that the threshold bound state contains a host of internal distance scales, which becomes a continuum as $N \to \infty$. This suggests a mechanism for obtaining scale-invariant behavior for large N, as we must if we are to recover longitudinal boost invariance. Note that the typical distance of cluster separation is largest as N goes to infinity when one of the clusters has only a finite number of partons. These are the configurations which give the $N^{1/9}$ behavior discussed in the text, which saturates the Bekenstein bound. By the uncertainty principle, these configurations have internal frequencies of the bound state $\sim N^{-2/9}$. Although these go to zero as N increases, they are still infinitely higher than the energies of supergraviton motions and interactions, which are of order $1/N$. As in perturbative string theory, we expect that this association of the large distance part of the wave function with modes of very high frequency will be crucial to a complete understanding of the apparent locality of low energy physics.

As we penetrate further in to the bound state, we encounter clusters of larger and larger numbers of branes. If we look for separated clusters carrying finite fractions of the total longitudinal momentum, the typical separation falls as N increases. Finally, we encounter the core $\Psi_N^{(c)}$, which we expect to dominate the ultimate short distance and high energy behavior of the theory in noncompact 11-dimensional space-time.

It is this core configuration to which the conventional methods of large N matrix models, which have so far made no appearance in our discussion, apply. Consider first gauge-invariant Green's functions of operators like $\mathrm{Tr} X_i^{2k}$, where i is one of the coordinate directions. We can construct a perturbation expansion of these Green's functions by conventional functional integral methods. When the time separations of operators are all short compared to the 11-dimensional Planck time, the terms in this expansion are well behaved. We can try to resum them into a large N series. The perturbative expansion parameter (the analogue of g_{YM}^2 if we think of the theory as dimensionally reduced Yang-Mills theory) is $R^3/l_P^6E^3$. Thus the planar Green's functions are functions of $R^3N/l_P^6E^3$.

The perturbative expansion, of course, diverges term by term as $E \to 0$. If we imagine that, as suggested by our discussion above, these Green's functions should be thought of as measuring properties of the core wave function of the system, there is no physical origin for such an infrared divergence. If, as in higher dimensions, the infrared cutoff is found already in the leading order of the $1/N$ expansion, then it must be of order $\omega_c \sim R l_P^{-2} N^{1/3}$. Note that this is much larger than any frequency encountered in our exploration of the parts of the wave function with clusters separated along a flat direction.

Now let us apply this result to the computation of the infrared-divergent expectation values of single gauge-invariant operators in the core of the bound-state wave function. The idea is to evaluate the graphical expansion of such an expression with an infrared cutoff and then insert the above estimate for the cutoff to obtain the correct large N scaling of the object. The combination of conventional large N scaling and dimensional analysis then implies that planar graphs dominate even though we are not taking the "gauge coupling" $R^3 l_P^{-6}$, to zero as we approach the large N limit. Dimensional analysis controls the otherwise unknown behavior of the higher order corrections in this limit. The results are

$$\left\langle \frac{1}{N} \mathrm{tr} X^{2k} \right\rangle \sim N^{2k/3}, \quad \text{(A4)}$$

$$\left\langle \frac{1}{N} \mathrm{tr}[X^i, X^j]^{2k} \right\rangle \sim N^{4k/3}, \quad \text{(A5)}$$

$$\left\langle \frac{1}{N} \mathrm{tr}(\bar{\theta}[\gamma_i X^i, \theta])^{2k} \right\rangle \sim N^{4k/3}, \quad \text{(A6)}$$

$$\left\langle \frac{1}{N} \mathrm{tr} \dot{X}^{2k} \right\rangle \sim N^{4k/3}. \quad \text{(A7)}$$

In the first of these expressions, X refers to any component of the transverse coordinates. In the second the commutator refers to any pair of the components. The final expression, whose lowest order perturbative formula has an ultraviolet divergence, is best derived by combining Eqs. (A5) and (A6) and the Schrödinger equation which says that the threshold bound state has zero binding energy. Note that these expressions are independent of R, the compactification radius of the 11th dimension. This follows from a cancellation between the R dependence of the infrared cutoff, that of the effective coupling and that of the scaling factor, which relates the variables X to conventionally normalized Yang-Mills fields.

The first of these equations says that the typical eigenvalue of any one coordinate matrix is of order $N^{1/3}$, much larger than the $N^{1/9}$ extension along the flat directions. The second tells us that this spectral weight lies mostly along the nonflat directions. In conjunction, the two equations can be read as a kind of "uncertainty principle of noncommutative geometry." The typical size of matrices is controlled by the size of their commutator. The final equation fits nicely with our estimate of the cutoff frequency. The typical velocity is such that the transit time of a typical distance[9] is the inverse of the cutoff frequency. It is clear that the high velocities encountered in the core of the wave function could invalidate our attempt to derive the matrix model by extrapolating from weakly coupled string theory. We must hope that the overall amplitude for this part of the wave function vanishes in the large N limit, relative to the parts in which zero branes are separated along flat directions.

It is important to realize that these estimates do not apply along the flat directions, but in the bulk of the N^2 dimensional configuration space. In these directions, it does not make sense to multiply together the "sizes" along different coordinate directions to make an area since the different coordinates do not commute. Thus there is no contradiction between the growth of the wave function in nonflat directions and our argument that the size of the bound state in conventional geometric terms saturates the Bekenstein bound.

APPENDIX B

In this appendix we compute the matrix model membrane tension and show that it exactly agrees with the M-theory membrane tension. The useful summary in [24] gives the tension of a Dp brane T_p in IIA string theory or M theory as

$$T_p=\frac{(2\pi)^{1/2}}{g_s}(2\pi)^{-p/2}\left(\frac{1}{2\pi\alpha'}\right)^{(1+p)/2}. \tag{B1}$$

where g_s is the fundamental string coupling and $1/2\pi\alpha'$ is the fundamental string tension.

The membrane tension T_2 is defined so that the mass $\mathcal{M}$ of a stretched membrane of area A is given by $\mathcal{M}=T_2A$. The mass squared for a light cone membrane with no transverse momentum described by the map $\mathcal{X}^i(\sigma_1,\sigma_2)$, $i=1,\dots,9$, can be written

$$\mathcal{M}^2=(2\pi)^4T_2^2\int_0^{2\pi}\frac{d\sigma_1}{2\pi}\int_0^{2\pi}\frac{d\sigma_2}{2\pi}\sum_{i<j}\{\mathcal{X}^i,\mathcal{X}^j\}^2, \tag{B2}$$

where the Poisson brackets of two functions $A(\sigma_1,\sigma_2),B(\sigma_1,\sigma_2)$ are defined by

$$\{A,B\}\equiv\frac{\partial A}{\partial\sigma_1}\frac{\partial B}{\partial\sigma_2}-\frac{\partial B}{\partial\sigma_1}\frac{\partial A}{\partial\sigma_2}. \tag{B3}$$

The coefficients in Eq. (B2) are set by demanding that $\mathcal{M}^2$ for the map $\mathcal{X}^8=(\sigma_1/2\pi)L$, $\mathcal{X}^9=(\sigma_2/2\pi)L$ is given by $\mathcal{M}^2=(T_2L^2)^2$.

To understand the relation to the matrix model, we write, as in Sec. VIII, the map $\mathcal{X}^i$ as a Fourier series:

$$\mathcal{X}^i(\sigma_1,\sigma_2)=\sum_{n_1,n_2}x^i_{n_1n_2}e^{i(n_1\sigma_1+n_2\sigma_2)}. \tag{B4}$$

Then the corresponding matrices X^i are given by

$$X^i=\sum_{n_1,n_2}x^i_{n_1n_2}U^{n_1}V^{n_2}, \tag{B5}$$

where the matrices U,V are elements of SU(N), have spectrum spec(U)=spec(V)=$\{1,\omega,\omega^2,\dots,\omega^{N-1}\}$, and obey $UV=\omega VU$, where $\omega=\exp(2\pi i/N)$. In a specific basis, $U=\mathrm{diag}(1,\omega,\omega^2,\dots,\omega^{N-1})$ and V is a cyclic forward shift.

The scale of Eq. (B5) is fixed since spec(U) and spec(V) go over as $N\to\infty$ to the unit circle $\exp(i\sigma)$. Note that $\mathcal{X}^i$ is real and so X^i is Hermitian.

The dynamics of the matrix model is governed by the Lagrangian

$$\frac{T_0}{2}\mathrm{Tr}\Big[\sum_i\dot{X}^i\dot{X}^i+C\sum_{i<j}[X^i,X^j]^2\Big]. \tag{B6}$$

The normalizations here are fixed by the requirement that Eq. (B6) describe $D0$-brane dynamics. The first term, for diagonal matrices describing $D0$-brane motion, is just the nonrelativistic kinetic energy $(m_0/2)\sum_{a=1}^N v_a^2$ since the 0-brane mass $m_0=T_0$. The rest energy of the system is just Nm_0, which in the M-theory interpretation is just p_{11}, and so

$$p_{11}=NT_0. \tag{B7}$$

The coefficient C is fixed by requiring that the small fluctuations around diagonal matrices describe harmonic oscillators whose frequencies are precisely the masses of the stretched strings connecting the 0 branes. This ensures that Eq. (B6) reproduces long range graviton interactions correctly. Expanding Eq. (B6) to quadratic order we find that $C=(1/2\pi\alpha')^2$.

The energy of a matrix membrane configuration with zero transverse momentum is given by the commutator term in Eq. (B6). We can evaluate this commutator in a semiclassical manner at large N as in Sec. VIII by introducing angular operators q with spectrum the interval $(0,2\pi)$ and $p=(2\pi/Ni)\partial/\partial q$ with the spectrum the discretized interval $(0,2\pi)$ so that $[p,q]=2\pi i/N$. The matrices U,V become $U=e^{ip}$, $V=e^{iq}$. By the Baker-Campbell-Hausdorff theorem, we see $UV=\omega VU$. The formal $\hbar$ in this algebra is given by $\hbar=2\pi/N$. Semiclassically, we have

$$[X,Y]\to i\hbar\{\mathcal{X},\mathcal{Y}\},$$

$$\mathrm{Tr}\to\int_0^{2\pi}\int_0^{2\pi}\frac{dp\,dq}{2\pi\hbar}. \tag{B8}$$

So we get

$$\mathrm{Tr}[X^i,X^j]^2\to-\frac{(2\pi)^2}{N}\int_0^{2\pi}\frac{d\sigma_1}{2\pi}\int_0^{2\pi}\frac{d\sigma_2}{2\pi}\{\mathcal{X}^i,\mathcal{X}^j\}^2. \tag{B9}$$

This commutator can also be evaluated for a given finite N matrix configuration explicitly with results that agree with Eq. (B9) as $N\to\infty$.

[9] In the space of eigenvalues, which in this noncommutative region is not to be confused with the classical geometrical distance between $D0$ branes.

Now we can perform the check. The value of the matrix model Hamiltonian on a configuration with no transverse momentum is

$$H=\frac{T_0}{2}\left(\frac{1}{2\pi\alpha'}\right)^2\frac{(2\pi)^2}{N}\int_0^{2\pi}\frac{d\sigma_1}{2\pi}\int_0^{2\pi}\frac{d\sigma_2}{2\pi}\sum_{i<j}\{\mathcal{X},\mathcal{X}\}^2. \tag{B10}$$

The conjecture interprets the matrix model Hamiltonian H as the infinite momentum frame energy $\sqrt{p_{11}^2+\mathcal{M}^2}-p_{11}\simeq\mathcal{M}^2/2p_{11}$. So the matrix membrane mass squared is $\mathcal{M}_{\rm mat}^2=2p_{11}H$. Using Eq. (B7), we find

$$\mathcal{M}_{\rm mat}^2=T_0^2\left(\frac{1}{2\pi\alpha'}\right)^2(2\pi)^2\int_0^{2\pi}\frac{d\sigma_1}{2\pi}\int_0^{2\pi}\frac{d\sigma_2}{2\pi}\sum_{i<j}\{\mathcal{X},\mathcal{X}\}^2. \tag{B11}$$

From Eq. (B2) we can now read off the matrix model membrane tension as

$$(T_2^{\rm mat})^2=T_0^2\left(\frac{1}{2\pi\alpha'}\right)^2\frac{1}{(2\pi)^2}. \tag{B12}$$

So we can write

$$\frac{T_2^2}{(T_2^{\rm mat})^2}=(2\pi)^2(2\pi\alpha')^2\left(\frac{T_2}{T_0}\right)^2=(2\pi)^2\left(\frac{1}{2\pi}\right)^2=1. \tag{B13}$$

So the M-theory and matrix model membrane tensions exactly agree.

---

[1] M. J. Duff, P. Howe, T. Inami, and K. S. Stelle, Phys. Lett. B **191**, 70 (1987); M. J. Duff and J. X. Lu, Nucl. Phys. **B347**, 394 (1990); M. J. Duff, R. Minasian, and James T. Liu, *ibid.* **B452**, 261 (1995); C. Hull and P. K. Townsend, *ibid.* **B438**, 109 (1995); P. K. Townsend, Phys. Lett. B **350**, 184 (1995); Phys. Lett. B **354**, 247 (1995); C. Hull and P. K. Townsend, Nucl. Phys. **B451**, 525 (1995); E. Witten, *ibid.* **B443**, 85 (1995).

[2] J. Polchinski, Phys. Rev. Lett. **75**, 4724 (1995); for a review see J. Polchinski, S. Chaudhuri, and C. V. Johnson, "Notes on *D*-Branes," ITP Report No. NSF-ITP-96-003, hep-th/9602052 (unpublished).

[3] E. Witten, Nucl. Phys. **B460**, 335 (1995).

[4] E. Bergshoeff, E. Sezgin, and P. K. Townsend, Phys. Lett. B **89**, 75 (1987); Ann. Phys. (N.Y.) **185**, 330 (1988); P. K. Townsend, in *Superstring '88*, Proceedings of the Trieste Spring School, edited by M. B. Green, M. T. Grisaru, R. Iengo, and A. Strominger (World Scientific, Singapore, 1989); M. J. Duff, Class. Quantum Grav. **5**, 189 (1988).

[5] B. de Wit, J. Hoppe, and H. Nicolai, Nucl. Phys. **B305** [FS23], 545 (1988).

[6] B. de Wit, M. Luscher, and H. Nicolai, Nucl. Phys. **B320**, 135 (1989).

[7] G. 't Hooft, in *Salamfestschrift*, Proceedings of the Conference, Trieste Italy, 1993, edited by A. Ali *et al.*, (World Scientific, Singapore, 1993), Report No. gr-qc/9310026 (unpublished); L. Susskind, J. Math. Phys. (N.Y.) **36**, 6377 (1995).

[8] S. H. Shenker, "Another Length Scale in String Theory?" Report No. hep-th/9509132 (unpublished).

[9] U. H. Danielsson, G. Ferretti, and B. Sundborg, Int. J. Mod. Phys. A **11**, 5463 (1996); D. Kabat, and P. Pouliot, Phys. Rev. Lett. **77**, 1004 (1996). M. R. Douglas, D. Kabat, P. Pouliot, and S. Shenker, Nucl. Phys. **B485**, 85 (1997).

[10] P. K. Townsend, Phys. Lett. B **373**, 68 (1996).

[11] S. Weinberg, Phys. Rev. **150**, 1313 (1966); J. Kogut and L. Susskind, Phys. Rep. **8**, 75 (1973).

[12] C. B. Thorn, in Proceedings of Sakharov Conference on Physics, Moscow, 1991, Report No. hep-th/9405069 (unpublished), pp. 447–454.

[13] J. D. Bekenstein, Phys. Rev. D **49**, 6606 (1994); G. 't Hooft [7].

[14] T. Banks and L. Susskind, "Brane-Antibrane Forces," Report No. hep-th/9511194 (unpublished).

[15] M. Baake, P. Reinicke, and V. Rittenberg, J. Math. Phys. (N.Y.) **26**, 1070 (1985); R. Flume, Ann. Phys. (N.Y.) **164**, 189 (1985); M. Claudson and M. B. Halpern, Nucl. Phys. **B250**, 689 (1985).

[16] A. A. Tseytlin, Phys. Lett. B **367**, 84 (1996); Nucl. Phys. **B467**, 383 (1996).

[17] M. Dine (in progress).

[18] D. Fairlie, P. Fletcher, and C. Zachos, J. Math. Phys. (N.Y.) **31**, 1088 (1990).

[19] J. Hoppe, Int. J. Mod. Phys. A **4**, 5235 (1989).

[20] M. Berkooz and M. R. Douglas, "Five-branes in *M*(atrix) Theory," Report No. hep-th/9610236 (unpublished).

[21] B. de Wit, V. Marquard, and H. Nicolai, Commun. Math. Phys. **128**, 39 (1990).

[22] W. Taylor, "*D*-brane field theory on compact spaces," Report No. hep-th/9611042 (unpublished).

[23] T. Eguchi, and H. Kawai, Phys. Rev. Lett. **48**, 1063 (1982).

[24] S. de Alwis, Phys. Lett. B **385**, 291 (1996).

Reprinted from Nucl. Phys. B **500** (1997) 3–42

Solutions of four-dimensional field theories via M-theory

Edward Witten [1]

School of Natural Sciences, Institute for Advanced Study, Olden Lane, Princeton, NJ 08540, USA

Received 28 April 1997; accepted 15 June 1997

Abstract

$\mathcal{N} = 2$ supersymmetric gauge theories in four dimensions are studied by formulating them as the quantum field theories derived from configurations of fourbranes, fivebranes, and sixbranes in Type IIA superstrings, and then reinterpreting those configurations in M-theory. This approach leads to explicit solutions for the Coulomb branch of a large family of four-dimensional $\mathcal{N} = 2$ field theories with zero or negative beta function. © 1997 Elsevier Science B.V.

1. Introduction

Many interesting results about field theory and string theory have been obtained by studying the quantum field theories that appear on the world-volume of string theory and M-theory branes. One particular construction that was considered recently in $2 + 1$ dimensions [1] and has been further explored in [2] and applied to $\mathcal{N} = 1$ models in four dimensions in [3] will be used in the present paper to understand the Coulomb branch of some $\mathcal{N} = 2$ models in four dimensions. The aim is to obtain for a wide class of four-dimensional gauge theories with $\mathcal{N} = 2$ supersymmetry the sort of description obtained in [4] for models with $SU(2)$ gauge group.

The construction in [1] involved branes of Type IIB superstring theory – to be more precise the Dirichlet threebranes and the solitonic and Dirichlet fivebranes. One considers, for example, NS fivebranes with threebranes suspended between them (Fig. 1). The fivebranes, being infinite in all six of their world-volume directions, are considered to be very heavy and are treated classically. The interest focuses on the quantum field theory on the world-volume of the threebranes. Being finite in one of their four dimensions,

[1] Research supported in part by NSF Grant PHY-9513835.

PII S0550-3213(97)00416-1

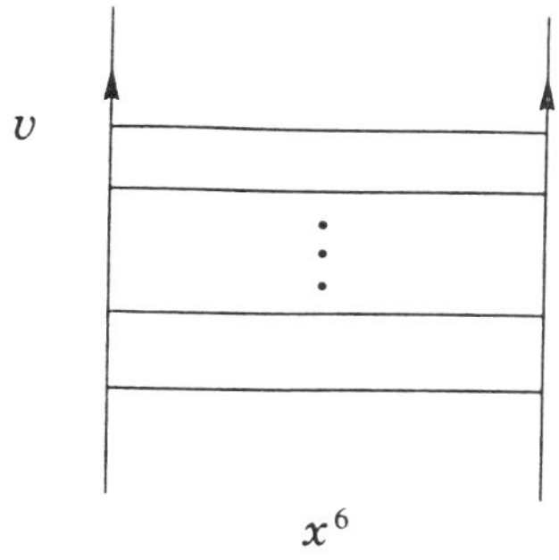

Fig. 1. Parallel fivebranes (vertical lines) with threebranes suspended between them (horizontal lines), as considered in Ref. [1].

the threebranes are macroscopically $2+1$ dimensional. The quantum field theory on this effective $2+1$-dimensional world has eight conserved supercharges, corresponding to $N=4$ supersymmetry in three dimensions or $\mathcal{N}=2$ in four dimensions. Many properties of such a model can be effectively determined using the description via branes.

To make a somewhat similar analysis of $3+1$-dimensional theories, one must replace the threebranes by fourbranes, suspended between fivebranes (and, as it turns out, also in the presence of sixbranes). Since the fourbrane is infinite in four dimensions (and finite in the fifth), the field theory on such a fourbrane is $3+1$-dimensional macroscopically.

Type IIB superstring theory has no fourbranes, so we will consider Type IIA instead. Type IIA superstring theory has Dirichlet fourbranes, solitonic fivebranes, and Dirichlet sixbranes. Because there is only one brane of each dimension, it will hopefully cause no confusion if we frequently drop the adjectives "Dirichlet" and "solitonic" and refer to the branes merely as fourbranes, fivebranes, and sixbranes.

One of the main techniques in [1] was to use $SL(2,\mathbf{Z})$ duality of Type IIB superstrings to predict a mirror symmetry of the $2+1$-dimensional models. For Type IIA there is no $SL(2,\mathbf{Z})$ self-duality. The strong coupling limit of Type IIA superstrings in ten dimensions is instead determined by an equivalence to eleven-dimensional M-theory; this equivalence will be used in the present paper to obtain solutions of four-dimensional field theories. As we will see, a number of facts about M-theory fit together neatly to make this possible.

In Section 2, we explain the basic techniques and solve models that are constructed from configurations of Type IIA fourbranes and fivebranes on $\mathbf{R}^{10}$. In Section 3, we incorporate sixbranes. In Section 4, we analyze models obtained by considering Type IIA fourbranes and fivebranes on $\mathbf{R}^9 \times \mathbf{S}^1$.[2] Many novel features will arise, including a geometric interpretation of the gauge theory beta function in Section 2 and a natural family of conformally invariant theories in Section 4. As we will see, each new step involves some essential new subtleties, though formally the brane diagrams are analogous (and related by T-duality) to those in [1].

[2] Compactification of such a brane system on a circle has been considered in [2] in the Type IIB context.

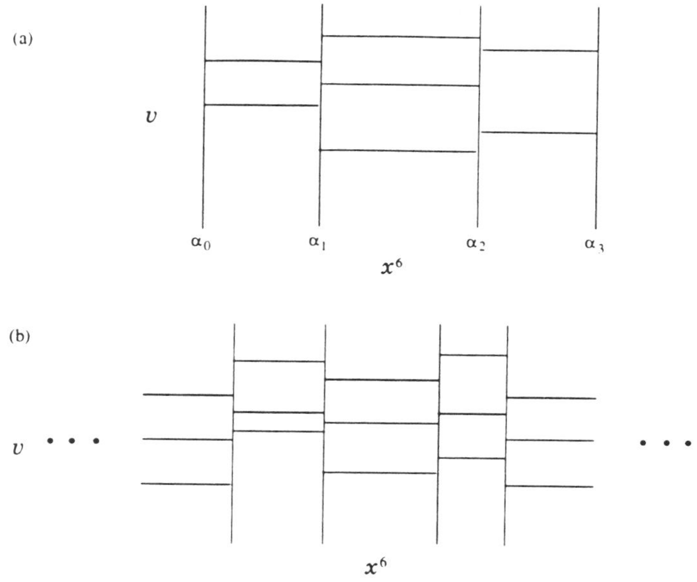

Fig. 2. (a) A chain of four fivebranes joined by fourbranes. (b) If one compactifies the x^6 direction to a circle, one can consider a periodic array of fivebranes and fourbranes. The left and right ends of this figure are to be identified.

2. Models with fourbranes and fivebranes

In this section we consider fourbranes suspended between fivebranes in Type IIA superstring theory on $\mathbf{R}^{10}$. Our fivebranes will be located at $x^7 = x^8 = x^9 = 0$ and – in the classical approximation – at some fixed values of x^6. The world-volume of the fivebrane is parametrized by the values of the remaining coordinates $x^0, x^1, \ldots, x^5$.

In addition, we introduce fourbranes whose world-volumes are parametrized by x^0, x^1, x^2, x^3, and x^6. However, our fourbranes will not be infinite in the x^6 direction. They will terminate on fivebranes. (Occasionally we will consider a semi-infinite fourbrane that terminates on a fivebrane at one end, and extends to $x^6 = \infty$ or $-\infty$ at the other end.) A typical picture is thus that of Fig. 2a. As in [1], we will examine this picture first from the fivebrane point of view and then from the fourbrane point of view.

It will be convenient to introduce a complex variable $v = x^4 + ix^5$. Classically, every fourbrane is located at a definite value of v. The same is therefore also true for its possible endpoints on a fivebrane.

2.1. Theory on fivebrane

A fact that was important in [1] is that on the world-volume of a Type IIB fivebrane there propagates a $U(1)$ gauge field. A system of k parallel but non-coincident fivebranes can be interpreted as a system with $U(k)$ gauge symmetry spontaneously broken to $U(1)^k$. Points at which Type IIB threebranes end on fivebranes carry magnetic charge in this spontaneously broken gauge theory.

Even though one draws similar brane pictures in the Type IIA case, the interpretation is rather different. Type IIA fivebranes do not carry gauge fields, but rather self-dual antisymmetric tensors. When parallel fivebranes become coincident, one gets not enhanced gauge symmetry but a strange critical point with tensionless strings [5], concerning which too little is known for it to be useful in the present paper.

However, the endpoints of a fourbrane on a fivebrane do behave as charges in an appropriate sense. A fivebrane on which fourbranes end does not really have a definite value of x^6 as the classical brane picture suggests. The fourbrane ending on a fivebrane creates a "dimple" in the fivebrane. What one would like to call the x^6 value of the fivebrane is really the x^6 value measured at $v = \infty$, far from the disturbances created by the fourbranes.

To see whether this makes sense, note that x^6 is determined as a function of v by minimizing the total fivebrane world-volume. For large v the equation for x^6 reduces to a Laplace equation,

$$\nabla^2 x^6 = 0. \tag{2.1}$$

Here ∇^2 is the Laplacian on the fivebrane world-volume. x^6 is a function only of the directions normal to the fourbrane ends, that is only of v and $\bar{v}$. Since the Green's function of the Laplacian in two dimensions is a logarithm, the large v behavior of x^6 is determined by (2.1) to be

$$x^6 = k \ln|v| + \text{constant} \tag{2.2}$$

for some k. Thus, in general, there is no well-defined large v limit of x^6. This contrasts with the situation considered in [1] where (because of considering threebranes instead of fourbranes) x^6 obeys a *three*-dimensional Laplace equation, whose solution approaches a constant at infinity. The limiting value $x^6(\infty)$ is then the "x^6 value of the fivebrane" which appears in the classical brane diagram and was used in [1] to parametrize the configurations.

Going back to the Type IIA case, for a fivebrane with a single fourbrane ending on it from, say, the left, k in (2.2) is an absolute constant that depends only on the fourbrane and fivebrane tensions (and hence the Type IIA string coupling constant). However, a fourbrane ending on a fivebrane on its right pulls in the opposite direction and contributes to k with the opposite sign from a fivebrane ending on the left. If a_i, $i = 1, \ldots, q_L$ and b_j, $j = 1, \ldots, q_R$ are the v values of fourbranes that end on a given fivebrane on its left and on its right, respectively, then the asymptotic form of x^6 is

$$x^6 = k \sum_{i=1}^{q_L} \ln|v - a_i| - k \sum_{j=1}^{q_R} \ln|v - b_j| + \text{constant}. \tag{2.3}$$

We see that x^6 has a well-defined limiting value for $v \to \infty$ if and only if $q_L = q_R$, that is if there are equal forces on the fivebrane from both left and right.

For any finite chain of fivebranes with fourbranes ending on them, as in Fig. 2a, it is impossible to obey this condition, assuming that there are no semi-infinite fourbranes that go off to $x^6 = \infty$ or $x^6 = -\infty$. At least the fivebranes at the ends of the chain are subject to unbalanced forces. The "balanced" case, a chain of fivebranes each connected by the same number of fourbranes, as in Fig. 2b, is most natural if one compactifies the x^6 direction to a circle, so that all fourbranes are finite in extent. It is very special and will be the subject of Section 4.

Another important question is affected by a related infrared divergence. For this, we consider the motion of fourbranes. When the fourbranes move, the disturbances they produce on the fivebranes move also, producing a contribution to the fourbrane kinetic energy. We consider a situation in which the a_i and b_j vary as a function of the first four coordinates x^μ, $\mu = 0, \ldots, 3$ (which are the "space-time" coordinates of the effective four-dimensional field theories studied in this paper). The fivebrane kinetic energy has a term $\int d^4x\, d^2v \sum_{\mu=0}^{3} \partial_\mu x^6 \partial^\mu x^6$. With x^6 as in (2.3), this becomes

$$k^2 \int d^4x\, d^2v \left| \mathrm{Re} \left(\sum_i \partial_\mu a_i \left(\frac{1}{v - a_i} \right) - \sum_j \partial_\mu b_j \left(\frac{1}{v - b_j} \right) \right) \right|^2 . \tag{2.4}$$

The v integral converges if and only if

$$\partial_\mu \left(\sum_i a_i - \sum_j b_j \right) = 0, \tag{2.5}$$

so that

$$\sum_i a_i - \sum_j b_j = q_\alpha, \tag{2.6}$$

where q_α is a constant characteristic of the αth fivebrane. While the q_α are constants that we will eventually interpret in terms of "bare masses," the remaining a's and b's are free to vary; they are indeed "order parameters" which depend on the choice of quantum vacuum of the four-dimensional field theory.

The above discussion of the large v behavior of x^6 and its kinetic energy is actually only half of the story. From the point of view of the four-dimensional $\mathcal{N} = 2$ supersymmetry of our brane configurations, x^6 is the real part of a complex field that is in a vector multiplet. The imaginary part of this superfield is a scalar field that propagates on the fivebrane. If Type IIA superstring theory on $\mathbf{R}^{10}$ is reinterpreted as M-theory on $\mathbf{R}^{10} \times \mathbf{S}^1$, the scalar in question is the position of the fivebrane in the eleventh dimension. We have labeled the ten dimensions of Type IIA as $x^0, x^1, \ldots, x^9$, so we will call the eleventh dimension x^{10}. In generalizing (2.3) to include x^{10}, we will use M-theory units (which differ by a Weyl rescaling from Type IIA units used in (2.3)). Also, we understand x^{10} to be a periodic variable with period $2\pi R$.

With this understood, the generalization of (2.3) to include x^{10} is

$$x^6 + ix^{10} = R \sum_{i=1}^{q_L} \ln(v - a_i) - R \sum_{j=1}^{q_R} \ln(v - b_j) + \text{constant}. \tag{2.7}$$

The fact that x^6+ix^{10} varies holomorphically with v is required by supersymmetry. The imaginary part of this equation states that x^{10} jumps by $\pm 2\pi R$ when one circles around one of the a_i or b_j in the complex v plane. In other words, the endpoints of fourbranes on a fivebrane behave as vortices in the fivebrane effective theory (an overall constant in (2.7) was fixed by requiring that the vortex number is one). This is analogous, and related by T-duality, to the fact that the endpoint of a threebrane on a fivebrane looks like a magnetic monopole, with magnetic charge one, in the fivebrane theory; this fact was extensively used in [1]. The interpretation of brane boundaries as charges on other branes was originally described in [5].

In terms of $s=(x^6+ix^{10})/R$, the last formula reads

$$s=\sum_{i=1}^{q_L}\ln(v-a_i)-\sum_{j=1}^{q_R}\ln(v-b_j)+\text{constant}. \tag{2.8}$$

2.2. *Four-dimensional interpretation*

Now we want to discuss what the physics on this configuration of branes looks like to a four-dimensional observer.

We consider a situation, shown in Fig. 2a in a special case, with $n+1$ fivebranes, labeled by $\alpha=0,\ldots,n$. Also, for $\alpha=1,\ldots,n$, we include k_α fourbranes between the $(\alpha-1)$th and αth fivebranes.

It might seem that the gauge group would be $\prod_{\alpha=1}^n U(k_\alpha)$, with each $U(k_\alpha)$ factor coming from the corresponding set of k_α parallel fourbranes. However, (2.6) means precisely that the $U(1)$ factors are "frozen out." To be more precise, in (2.6), $\sum_i a_i$ is the scalar part of the $U(1)$ vector multiplet in one factor $U(k_\alpha)$, and $\sum_j b_j$ is the scalar part of the $U(1)$ multiplet in the "next" gauge group factor $U(k_{\alpha+1})$. Eq. (2.6) means that the difference $\sum_i a_i-\sum_j b_j$ is "frozen," and therefore, by supersymmetry, an entire $U(1)$ vector supermultiplet is actually missing from the spectrum. Since such freezing occurs at each point in the chain, including the endpoints (the fivebranes with fourbranes ending only on one side), the $U(1)$'s are all frozen out and the gauge group is actually $\prod_{\alpha=1}^n SU(k_\alpha)$.

What is the hypermultiplet spectrum in this theory? By reasoning exactly as in [1], massless hypermultiplets arise (in the classical approximation of the brane diagram) precisely when fourbranes end on a fivebrane from opposite sides at the same point in space-time. Such a hypermultiplet is charged precisely under the gauge group factors coming from fourbranes that adjoin the given fivebrane. So the hypermultiplets transform, in an obvious notation, as $(\mathbf{k}_1,\overline{\mathbf{k}}_2)\oplus(\mathbf{k}_2,\overline{\mathbf{k}}_3)\oplus\ldots\oplus(\mathbf{k}_{n-1},\overline{\mathbf{k}}_n)$. The constants q_α in (2.6) determine the bare masses m_α of the $(\mathbf{k}_\alpha,\overline{\mathbf{k}}_{\alpha+1})$ hypermultiplets, so in fact arbitrary bare masses are possible. The bare masses are actually

$$m_\alpha=\frac{1}{k_\alpha}\sum_i a_{i,\alpha}-\frac{1}{k_{\alpha+1}}\sum_j a_{j,\alpha+1}, \tag{2.9}$$

where $a_{i,\alpha}$, $i = 1, \ldots, k_\alpha$ are the positions in the v plane of the fourbranes between the $(\alpha - 1)$th and αth fivebrane. In other words, m_α is the difference between the average position in the v plane of the fourbranes to the left and right of the αth fivebrane. m_α is not simply a multiple of q_α, but the q_α for $\alpha = 1, \ldots, n$ determine the m_α.

Now, we come to the question of what is the coupling constant of the $SU(k_\alpha)$ gauge group. Naively, if x^6_α is the x^6 value of the αth fivebrane, then the gauge coupling g_α of $SU(k_\alpha)$ should be given by

$$\frac{1}{g_\alpha^2} = \frac{x^6_\alpha - x^6_{\alpha-1}}{\lambda}, \tag{2.10}$$

where λ is the string coupling constant.

We have here a problem, though. What precisely is meant by the objects x^6_α? As we have seen above, these must be understood as functions of v which in general diverge for $v \to \infty$. Therefore, we must interpret g_α as a function of v:

$$\frac{1}{g_\alpha^2(v)} = \frac{x^6_\alpha(v) - x^6_{\alpha-1}(v)}{\lambda}. \tag{2.11}$$

We interpret v as setting a mass scale, and $g_\alpha(v)$ as the effective coupling of the $SU(k_\alpha)$ theory at mass $|v|$. Then $1/g_\alpha^2(v)$ generally, according to (2.3), diverges logarithmically for $v \to \infty$. But that is familiar in four-dimensional gauge theories: the inverse gauge coupling of an asymptotically free theory diverges logarithmically at high energies. We thus interpret this divergence as reflecting the one loop beta function of the four-dimensional theory.

It is natural to include x^{10} along with x^6, and thereby to get a formula for the effective theta angle θ_α of the $SU(k_\alpha)$ gauge theory, which is determined by the separation in the x^{10} direction between the $(\alpha - 1)$th and αth fivebranes. Set

$$\tau_\alpha = \frac{\theta_\alpha}{2\pi} + \frac{4\pi i}{g_\alpha^2}. \tag{2.12}$$

Then in terms of $s = (x^6 + ix^{10})/R$ (with distances now measured in M-theory units) we have

$$-i\tau_\alpha(v) = s_\alpha(v) - s_{\alpha-1}(v). \tag{2.13}$$

(A multiplicative constant on the right-hand side has been set to one by requiring that under $x^{10}_\alpha \to x^{10}_\alpha + 2\pi R$, the theta angle changes by $\pm 2\pi$.) But according to (2.8), at large v one has $s_\alpha(v) = (k_\alpha - k_{\alpha+1}) \ln v$, so

$$-i\tau_\alpha(v) \cong (2k_\alpha - k_{\alpha-1} - k_{\alpha+1}) \ln v. \tag{2.14}$$

The standard asymptotic freedom formula is $-i\tau = b_0 \ln v$, where $-b_0$ is the coefficient of the one-loop beta function. So (2.14) amounts to the statement that the one-loop beta function for the $SU(k_\alpha)$ factor of the gauge group is

$$b_{0,\alpha} = -2k_\alpha + k_{\alpha-1} + k_{\alpha+1}. \tag{2.15}$$

This is in agreement with a standard field theory computation for this model. In fact, for $\mathcal{N} = 2$ supersymmetric QCD with gauge group $SU(N_c)$ and N_f flavors, one usually has $b_0 = -(2N_c - N_f)$. In the case at hand, $N_c = k_\alpha$, and the $(\mathbf{k}_{\alpha-1}, \overline{\mathbf{k}}_\alpha) \oplus (\mathbf{k}_\alpha, \overline{\mathbf{k}}_{\alpha+1})$ hypermultiplets make the same contribution to the $SU(k_\alpha)$ beta function as $k_{\alpha-1} + k_{\alpha+1}$ flavors, so the effective value of N_f is $k_{\alpha-1} + k_{\alpha+1}$.

2.3. *Interpretation via M-theory*

By now we have identified a certain class of models that can be constructed with fivebranes and fourbranes only. The remaining question is of course how to analyze these models. For this we will use M-theory.

First of all, the reason that one may effectively go to M-theory is that according to (2.10), a rescaling of the Type IIA string coupling constant, if accompanied by a rescaling of the separations of the fivebranes in the x^6 direction, does not affect the field theory coupling constant and so is irrelevant. One might be concerned that (2.10) is just a classical formula. But in fact, we have identified in the brane diagram all marginal and relevant operators (the coupling constants and hypermultiplet bare masses) of the low energy $\mathcal{N} = 2$ field theory, so any additional parameters (such as the string coupling constant) really are irrelevant. Therefore we may go to the regime of large λ.

What will make this useful is really the following. A fourbrane ending on a fivebrane has no known explicit conformal field theory description. The end of the fourbrane is a kind of singularity that is hard to understand in detail. That is part of the limitation of describing this system via Type IIA superstrings. But in M-theory everything we need can be explicitly understood using only the low energy limit of the theory. The Type IIA fivebrane on $\mathbf{R}^{10}$ is simply an M-theory fivebrane on $\mathbf{R}^{10} \times \mathbf{S}^1$, whose world-volume, roughly, is located at a point in $\mathbf{S}^1$ and spans a six-manifold in $\mathbf{R}^{10}$. A Type IIA fourbrane is an M-theory fivebrane that is wrapped over the $\mathbf{S}^1$ (so that, roughly, its world-volume projects to a five-manifold in $\mathbf{R}^{10}$). Thus, *the fourbrane and fivebrane come from the same basic object in M-theory*. The Type IIA singularity where the fourbrane appears to end on a fivebrane is, as we will see, completely eliminated by going to M-theory.

The Type IIA configuration of parallel fivebranes joined by fourbranes can actually be reinterpreted in M-theory as a configuration of a single fivebrane with a more complicated world history. The fivebrane world-volume be described as follows. (1) It sweeps out arbitrary values of the first four coordinates $x^0, x^1, \ldots, x^3$. It is located at $x^7 = x^8 = x^9 = 0$. (2) In the remaining four coordinates x^4, x^5, x^6, and x^{10} – which parametrize a four-manifold $Q \cong \mathbf{R}^3 \times \mathbf{S}^1$ – the fivebrane world-volume spans a two-dimensional surface Σ. (3) If one forgets x^{10} and projects to a Type IIA description in terms of branes on $\mathbf{R}^{10}$, then one gets back, in the limit of small R, the classical configuration of fourbranes and fivebranes that we started with. (4) Finally, $\mathcal{N} = 2$ supersymmetry means that if we give Q the complex structure in which $v = x^4 + ix^5$ and $s = x^6 + ix^{10}$ are holomorphic, then Σ is a complex Riemann surface in Q. This makes $\mathbf{R}^4 \times \Sigma$ a supersymmetric cycle in the sense of [6] and so ensures space-time supersymmetry.

In the approximation of the Type IIA brane diagrams, Σ has different components that are described locally by saying that s is constant (the fivebranes) or that v is constant (the fourbranes). But the singularity that appears in the Type IIA limit where the different components meet can perfectly well be absent upon going to M-theory; and that will be so generically, as we will see. Thus, for generic values of the parameters, Σ will be a smooth complex Riemann surface in Q.

This smoothness is finally the reason that going to M-theory leads to a solution of the problem. For large λ, all distances characteristic of the Riemann surface Σ are large and it will turn out that there are generically no singularities. So obtaining and analyzing the solution will require only a knowledge of the low energy long wavelength approximation to M-theory and its fivebranes.

Low energy effective action

We will now work out the low energy four-dimensional physics that will result from such an M-theory configuration. The discussion is analogous to, but more elementary than, a situation considered in [7] where an $\mathcal{N} = 2$ theory in four dimensions was related to a brane of the general form $\mathbf{R}^4 \times \Sigma$.

Vector multiplets will appear in four dimensions because on the world-volume of an M-theory fivebrane there is a chiral antisymmetric tensor field β, that is, a two-form β whose three-form field strength T is self-dual. Consider in general a fivebrane whose world-volume is $\mathbf{R}^4 \times \overline{\Sigma}$, where $\overline{\Sigma}$ is a compact Riemann surface of genus g. According to [8], in the effective four-dimensional description, the zero-modes of the antisymmetric tensor give g abelian gauge fields on $\mathbf{R}^4$. The coupling constants and theta parameters of the g abelian gauge fields are described by a rank g abelian variety which is simply the Jacobian $J(\overline{\Sigma})$.

These conclusions are reached as follows. Let

$$T = F \wedge \Lambda + *F \wedge *\Lambda, \tag{2.16}$$

where F is a two-form on $\mathbf{R}^4$, Λ is a one-form on $\overline{\Sigma}$, and $*$ is the Hodge star. This T is self-dual, and the equation of motion $dT = 0$ gives Maxwell's equations $dF = d * F = 0$ along with the equations $d\Lambda = d * \Lambda = 0$ for Λ. So Λ is a harmonic one-form, and every choice of a harmonic one-form Λ gives a way of embedding solutions of Maxwell's equations on $\mathbf{R}^4$ as solutions of the equations for the self-dual three-form T. If $\overline{\Sigma}$ has genus g, then the space of self-dual (or anti-self-dual) Λ's is g dimensional, giving g positive helicity photon states (and g of negative helicity) on $\mathbf{R}^4$. The low energy theory thus has gauge group $U(1)^g$. The terms quadratic in F in the effective action for the gauge fields are obtained by inserting (2.16) in the fivebrane kinetic energy $\int_{\mathbf{R}^4 \times \overline{\Sigma}} |T|^2$; the Jacobian of $\overline{\Sigma}$ enters by determining the integrals of wedge products of Λ's and $*\Lambda$'s.

In our problem of $n+1$ parallel Type IIA fivebranes joined by fourbranes, the M-theory fivebrane is $\mathbf{R}^4 \times \Sigma$, where Σ is not compact. So the above discussion does not immediately apply. However, Σ could be compactified by adding $n+1$ points. Indeed, for a single fivebrane, Σ would be a copy of $\mathbf{C}$ (the v plane), which is $\mathbf{CP}^1$ with a point deleted. So if there are no fourbranes, we have just $n+1$ disjoint copies of $\mathbf{CP}^1$ with a point omitted from each. Including a fourbrane means cutting holes out of adjoining fivebranes and connecting them with a tube. This produces (if all k_α are positive) a connected Riemann surface Σ which can be compactified by adding $n+1$ points. Note that the deleted points are "at infinity"; the metric on Σ that is obtained from its embedding in Q is complete and looks "near each puncture" like the flat complex plane with the puncture being the point "at infinity."

The reason that non-compactness potentially modifies the discussion of the low energy effective action is that in (2.16), one must ask for Λ to be square-integrable, in the metric on Σ which comes from its embedding in Q, as well as harmonic. Since the punctures are "at infinity," square-integrability implies that Λ has vanishing periods on a contour that surrounds any puncture. (A harmonic one-form Λ' that has a non-vanishing period on such a contour would look near $v=\infty$ like $\Lambda' = dv/v$, leading to $\int \Lambda' \wedge *\Lambda' = \int dv \wedge d\bar{v}/|v|^2 = \infty$.) Hence Λ extends over the compactification $\overline{\Sigma}$ of Σ. Since moreover the equation for a one-form on $\overline{\Sigma}$ to be self-dual is conformally invariant and depends only on the complex structure of $\overline{\Sigma}$, the square-integrable harmonic one-forms on Σ are the *same* as the harmonic one-forms on $\overline{\Sigma}$. So finally, in our problem, the low energy effective action of the vector fields is determined by the Jacobian of $\overline{\Sigma}$.

It is thus of some interest to determine the genus of $\overline{\Sigma}$. We construct $\overline{\Sigma}$ beginning with $n+1$ disjoint copies of $\mathbf{CP}^1$, of total Euler characteristic $2(n+1)$. Then we glue in a total of $\sum_{\alpha=1}^{n} k_\alpha$ tubes between adjacent $\mathbf{CP}^1$'s. Each time such a tube is glued in, the Euler characteristic is reduced by two, so the final value is $2(n+1) - 2\sum_{\alpha=1}^{n} k_\alpha$. This equals $2-2g$, where g is the genus of $\overline{\Sigma}$. So we get $g = \sum_{\alpha=1}^{n}(k_\alpha - 1)$. This is the expected dimension of the Coulomb branch for the gauge group $\prod_{\alpha=1}^{n} SU(k_\alpha)$. In particular, this confirms that the $U(1)$'s are "missing"; for the gauge group to be $\prod_{\alpha=1}^{n} U(k_\alpha)$, the genus would have to be $\sum_{\alpha=1}^{n} k_\alpha$.

So far we have emphasized the effective action for the four-dimensional gauge fields. Of course, the rest of the effective action is determined from this via supersymmetry. For instance, the scalars in the low energy effective action simply determine the embedding of $\mathbf{R}^4 \times \Sigma$ in space-time, or more succinctly the embedding of Σ in Q; and their kinetic energy is obtained by evaluating the kinetic energy for motion of the fivebrane in space-time.

The integrable system

In general, the low energy effective action for an $\mathcal{N}=2$ system in four dimensions is determined by an integrable Hamiltonian system in the complex sense. The expectation values of the scalar fields in the vector multiplets are the commuting Hamiltonians; the orbits generated by the commuting Hamiltonian flows are the complex tori which determine the kinetic energy of the massless four-dimensional vectors. This structure was noticed in special cases [9–11] and deduced from the generalities of low energy supersymmetric effective field theory [12].

A construction of many complex integrable systems is as follows. Let X be a two-dimensional complex symplectic manifold. Let Σ be a complex curve in X. Let $\mathcal{W}$ be the deformation space of pairs $(\Sigma', \mathcal{L}')$, where Σ' is a curve in X to which Σ can be deformed and $\mathcal{L}'$ is a line bundle on Σ' of specified degree. Then $\mathcal{W}$ is an integrable system; it has a complex symplectic structure which is such that any functions that depend only on the choice of Σ' (and not of $\mathcal{L}'$) are Poisson-commuting. The Hamiltonian flows generated by these Poisson-commuting functions are the linear motions on the space of $\mathcal{L}'$'s, that is, on the Jacobian of Σ'.

This integrable system was described in [13], as a generalization of a gauge theory construction by Hitchin [14]; a prototype for the case of non-compact Σ is the extension of Hitchin's construction to Riemann surfaces with punctures in [15]. The same integrable system has appeared in the description of certain BPS states for Type IIA superstrings on K3 [16].

In general, fix a hyper-Kähler metric on the complex symplectic manifold X (of complex dimension two) and consider M-theory on $\mathbf{R}^7 \times X$. Consider a fivebrane of the form $\mathbf{R}^4 \times \Sigma$, where $\mathbf{R}^4$ is a fixed linear subspace of $\mathbf{R}^7$ (obtained by setting three linear combinations of the seven coordinates to constants) and Σ is a complex curve in X. Then the effective $\mathcal{N} = 2$ theory on $\mathbf{R}^4$ is controlled by the integrable system described in the last paragraph, with the given X and Σ. This follows from the fact that the scalar fields in the four-dimensional theory parametrize the choice of a curve Σ' to which Σ can be deformed (preserving its behavior at infinity) while the Jacobian of Σ' determines the couplings of the vector fields.

The case of immediate interest is the case that $X = Q$ and Σ is related to the brane diagram with which we started the present section. The merit of this case (relative to an arbitrary pair (X, Σ)) is that because of the Type IIA interpretation, we know a gauge theory whose solution is given by this special case of the integrable model. Some generalizations that involve different choices of X are in Sections 3 and 4.

BPS states

The spectrum of massive BPS states in models constructed this way can be analyzed roughly as in Ref. [7], by using the fact that M-theory twobranes can end on fivebranes [5,17]. BPS states can be obtained by considering suitable twobranes in $\mathbf{R}^7 \times X$. To ensure the BPS property, the twobrane world volume should be a product $\mathbf{R} \times D$, where $\mathbf{R}$ is a straight line in $\mathbf{R}^4 \subset \mathbf{R}^7$ (representing "the world line of the massive particle in space-time") and $D \subset X$ is a complex Riemann surface with a non-empty boundary C that lies on Σ. By adjusting D to minimize the area of D (keeping fixed the holomogy class of $C \subset \Sigma$), one gets a twobrane world-volume whose quantization gives a BPS state.

2.4. Solution of the models

We now come to the real payoff, which is the solution of the models.

The models are to be described in terms of an equation $F(s,v) = 0$, defining a complex curve in Q.

Since s is not single-valued, we introduce

$$t = \exp(-s) = \exp(-(x^6 + ix^{10})/R) \tag{2.17}$$

and look for an equation $F(t,v) = 0$.

Now if $F(t,v)$ is regarded as a function of t for fixed v, then the roots of F are the positions of the fivebranes (at the given value of v). The degree of F as a polynomial in t is therefore the number of fivebranes. To begin with, we will consider a model with only two fivebranes. F will therefore be quadratic in t.

Classically, if one regards $F(t,v)$ as a function of v for fixed t, with a value of t that is "in between" the two fivebranes, then the roots of $F(t,v)$ are the values of v at which there are fourbranes. We will set the number of fourbranes suspended between the two fivebranes equal to k, so $F(t,v)$ should be of degree k in v. (If t is "outside" the classical position of the fivebranes, the polynomial $F(t,v)$ still vanishes for k values of v; these roots will occur at large v and are related to the "bending" of the fivebranes for large v.)

So such a model will be governed by a curve of the form

$$A(v)t^2 + B(v)t + C(v) = 0, \tag{2.18}$$

with A, B, and C being polynomials in v of degree k. We set $F = At^2 + Bt + C$.

At a zero of $C(v)$, one of the roots of (2.18) (regarded as an equation for t) goes to $t = 0$. According to (2.17), $t = 0$ is $x^6 = \infty$. Having a root of the equation which goes to $x^6 = \infty$ at a fixed limiting value of v (where $C(v)$ vanishes) means that there is a semi-infinite fourbrane to the "right" of all of the fivebranes.

Likewise, at a zero of $A(v)$, a root of F goes to $t = +\infty$, that is to say, to $x^6 = -\infty$. This corresponds to a semi-infinite fourbrane on the "left."

Since there are k fourbranes between the two fivebranes, these theories will be $SU(k)$ gauge theories. As in [1], a semi-infinite fourbrane, because of its infinite extent in x^6, has an infinite kinetic energy (relative to the fourbranes that extend a finite distance in x^6) and can be considered to be frozen in place at a definite value of v. The effect of a semi-infinite fourbrane is to add one hypermultiplet in the fundamental representation of $SU(k)$.

We first explore the "pure gauge theory" without hypermultiplets. For this we want no zeroes of A or C, so A and C must be constants and the curve becomes after a rescaling of t

$$t^2 + B(v)t + 1 = 0. \tag{2.19}$$

In terms of $\widetilde{t} = t + B/2$, this reads

$$\tilde{t}^2 = \frac{B(v)^2}{4} - 1. \tag{2.20}$$

By rescaling and shifting v, one can put B in the form

$$B(v) = v^k + u_2 v^{k-2} + u_3 v^{k-3} + \ldots + u_k. \tag{2.21}$$

(2.20) is our first success; it is a standard form of the curve that governs the $SU(k)$ theory without hypermultiplets [18,19].

We chose $F(t, v)$ to be of degree k in v so that, for a value of t that corresponds to being "between" the fivebranes, there would be k roots for v. Clearly, however, the equation $F(t, v) = 0$ has k roots for v for *any* non-zero t (we recall that $t = 0$ is "at infinity" in the original variables). What is the interpretation of these roots for very large or very small v, to the left or right of the fivebranes? For t very large, the roots for v are approximately at

$$t \cong c \cdot v^k, \tag{2.22}$$

and for t very small they are approximately at

$$t \cong c' \cdot v^{-k}; \tag{2.23}$$

here c, c' are constants. The formulas $t \cong v^{\pm k}$ are actually special cases of (2.8); they represent the "bending" of the fivebranes as a result of being pulled on by fourbranes. The formulas (2.22) and (2.23) show that for $x^6 \to \pm\infty$, the roots of F, as a function of v for fixed t, are at very large $|v|$. These roots do not correspond, intuitively, to positions of fourbranes but are points "near infinity" on the bent fivebranes.

We can straightforwardly incorporate hypermultiplet flavors in this discussion. For this, we merely incorporate zeroes of A or C. For example, to include N_f flavors we can take $A = 1$ and $C(v) = f\prod_{j=1}^{N_f}(v - m_j)$ where the m_j, being the zeroes of C, are the positions of the semi-infinite fourbranes or in other words the hypermultiplet bare masses, and f is a complex constant. Eq. (2.20) becomes

$$\tilde{t}^2 = \frac{B(v)^2}{4} - f\prod_{j=1}^{N_f}(v - m_j). \tag{2.24}$$

We set now

$$B(v) = e(v^n + u_2 v^{n-2} + u_3 v^{n-3} + \ldots + u_n) \tag{2.25}$$

with e and the u_i being constants. We have shifted v by a constant to remove the v^{n-1} term. This is again equivalent to the standard solution [20,21] of the $SU(k)$ theory with N_f flavors. As long as $N_f \neq 2k$, one can rescale $\tilde{t}$ and v to set $e = f = 1$. Of course, shifting v by a constant to eliminate the v^{k-1} term in B will shift the m_j by a constant. This is again a familiar part of the solution of the models.

Of special interest is the case $N_f = 2k$ where the beta function vanishes. In this case, by rescaling $\tilde{t}$ and v, it is possible to remove only one combination of e and f.

The remaining combination is a modulus of the theory, a coupling constant. This is as expected: four-dimensional quantum Yang–Mills theory has a dimensionless coupling constant when and only when the beta function vanishes.

The coupling constant for $N_f = 2k$ is coded into the behavior of the fivebrane for $z, t \to \infty$. This behavior, indeed, is a "constant of the motion" for finite energy disturbances of the fivebrane configuration and hence can be interpreted as a coupling constant in the four-dimensional quantum field theory. The behavior at infinity for $N_f = 2k$ is

$$t \cong \lambda_\pm v^k, \tag{2.26}$$

where $\lambda_\pm$ are the two roots of the quadratic equation

$$y^2 + ey + f = 0. \tag{2.27}$$

This follows from the fact that the asymptotic behavior of the equation is

$$t^2 + e(v^k + \ldots)t + f(v^{2k} + \ldots) = 0. \tag{2.28}$$

y can be identified as t/v^k. The fact that the two fivebranes are parallel at infinity – on both branches $t \cong v^k$ for $v \to \infty$ – means that the distance between them has a limit at infinity, which determines the gauge coupling constant.

A rescaling of t or v rescales $\lambda_\pm$ by a common factor, leaving fixed the function $w = -4\lambda_+\lambda_-/(\lambda_+ - \lambda_-)^2$ which is also invariant under exchange of the λ's. This function can be constructed as a product of cross ratios of the four distinguished points $0, \infty, \lambda_+$ and λ_- on the y plane. Let $\mathcal{M}_{0,4;2}$ be the moduli space of the following objects: a smooth Riemann surface of genus zero with four distinct marked points, two of which (0 and ∞) are distinguished and ordered, while the others (λ_+ and λ_-) are unordered. The choice of a value of w (not equal to zero or infinity) is the choice of a point in $\mathcal{M}_{0,4;2}$. The point $w = 1$ is a $\mathbf{Z}_2$ orbifold point in $\mathcal{M}_{0,4;2}$.

In the conventional description of this theory, one introduces a coupling parameter τ appropriate near one component of "infinity" in $\mathcal{M}_{0,4;2}$ – near $w \to 0$ (which corresponds for instance to $\lambda_+ \to 0$ at fixed λ_-), where the $SU(k)$ gauge theory is weakly coupled.[3] In [20], the solution (2.24) is expressed in terms of τ. Near $w = 0$ one has $w = e^{2\pi i\tau}$; the inverse function $\tau(w)$ is many-valued. The fact that the theory depends only on w and not on τ is from the standpoint of weak coupling interpreted as the statement that the theory is invariant under a discrete group of duality transformations. This group is $\Gamma = \pi_1(\mathcal{M}_{0,4;2})$. It can be shown that Γ is isomorphic to the index three subgroup of $SL(2, \mathbf{Z})$ consisting of integral unimodular matrices

$$\begin{pmatrix} a & b \\ c & d \end{pmatrix} \tag{2.29}$$

with b even; this group is usually called $\Gamma_0(2)$.

[3] The other possible degeneration is $w \to \infty$ ($\lambda_+ \to \lambda_-$). This is in M-theory the limit of coincident fivebranes, and a weakly coupled description in four dimensions is not obvious.

The case of a positive beta function

What happens when the $SU(k)$ gauge theory has positive beta function, that is for $N_f > 2k$? The fivebrane configuration (2.24) still describes something, but what? The first main point to note is that for $N_f > 2k$, the two fivebranes are parallel near infinity; both branches of (2.24) behave for large v as $\tilde{t} \sim v^{N_f/2}$. I interpret this to mean that the four-dimensional theory induced from the branes is conformally invariant at short distances and flows in the infrared to the $SU(k)$ theory with N_f flavors.

What conformally invariant theory is this? A key is that for $N_f \geqslant 2k+2$, there are additional terms that can be added to (2.24) without changing the asymptotic behavior at infinity (and cannot be absorbed in redefining $\tilde{t}$ and v). Such terms really should be included because they represent different vacua of the same quantum system.

There are two rather different cases to consider. If $N_f = 2k'$ is even, the general curve with the given behavior at infinity is

$$\tilde{t}^2 = \frac{1}{4}e'(v^{k'} + \ldots)^2 - f\prod_{i=1}^{2k'}(v - m_i). \tag{2.30}$$

This describes the $SU(k')$ theory with $2k'$ flavors, a theory that is conformally invariant in the ultraviolet and which by suitably adjusting the parameters can reduce in an appropriate limit (taking e' to zero while rescaling v and some of the other parameters) to the solution (2.24) for the $SU(k)$ theory with $N_f > 2k$ flavors. The $SU(k')$ theory with $2k'$ flavors has of course a conventional Lagrangian description, valid when the coupling is weak.

The other case is $N_f = 2k' + 1$, with $k' \geqslant k$. The most general curve with the same asymptotic behavior as (2.24) is then

$$\tilde{t}^2 = \frac{1}{4}e'(v^{k'} + \ldots)^2 - f\prod_{i=1}^{2k'+1}(v - m_i). \tag{2.31}$$

There is no notion of weak coupling here; the asymptotic behavior of the fivebranes is $\tilde{t} = \lambda_{\pm}v^{n'+1/2}$ with $\lambda_- = -\lambda_+$ so that w has the fixed value 1. (We recall that this is the $\mathbf{Z}_2$ orbifold point on $\mathcal{M}_{0,4;2}$.) (2.31) describes a strongly coupled fixed point with no obvious Lagrangian description and no dimensionless "coupling constant," roughly along the lines of the fixed point analyzed in [22]. By specializing some parameters, it can flow in the infrared to the $SU(k)$ theory with $2k'+1$ flavors for any $k < k'$.

Also, the $SU(k'+1)$ theory with $2k'+2$ flavors can flow in the infrared to the fixed point just described. This is done starting with (2.24) by taking one mass to infinity while shifting and adjusting the other variables in an appropriate fashion.

In the rest of this paper, we concentrate on models of zero or negative beta function. Along just the above lines, models of positive beta function can be derived from conventional fixed points like the one underlying (2.30) or unconventional ones like the one underlying (2.31); the conventional and unconventional fixed points are linked by renormalization group flows.

2.5. Generalization

Now we will consider a more general model, with a chain of $n+1$ fivebranes labeled from 0 to n, the $(\alpha-1)$th and αth fivebranes, for $\alpha = 1, \ldots, n$, being connected by k_α fourbranes. We assume that there are no semi-infinite fourbranes at either end.[4]

The gauge group is thus $\prod_{\alpha=1}^{n} SU(k_\alpha)$, and the coefficient in the one-loop beta function of $SU(k_\alpha)$ is

$$b_{0,\alpha} = -2k_\alpha + k_{\alpha+1} + k_{\alpha-1}. \tag{2.32}$$

(We understand here $k_0 = k_{n+1} = 0$.) We will assume $b_{0,\alpha} \leqslant 0$ for all α. Otherwise, as in the example just treated, the model is not really a $\prod_\alpha SU(k_\alpha)$ gauge theory at short distances but should be interpreted in terms of a different ultraviolet fixed point. Note that $\sum_{\alpha=1}^{n} b_{0,\alpha} < 0$ (in fact $\sum_{\alpha=1}^{n} b_{0,\alpha} = -k_1 - k_n$), so it is impossible in a model of this type for all beta functions to vanish. (Models with vanishing beta function can be obtained by including semi-infinite fourbranes at the ends of the chain, as above, or by other generalizations made in sections three and four.)

If the position $t_\alpha(v)$ of the αth fivebrane, for $\alpha = 0, \ldots, n$, behaves for large v as

$$t_\alpha(v) \sim h_\alpha v^{a_\alpha} \tag{2.33}$$

with $a_0 \geqslant a_1 \geqslant a_2 \ldots \geqslant a_n$ and constants h_α, then from our analysis of the relation of the beta function to "bending" of fivebranes, we have

$$a_\alpha - a_{\alpha-1} = -b_{0,\alpha}, \quad \text{for } \alpha = 1, \ldots, n. \tag{2.34}$$

The fivebrane world-volume will be described by a curve $F(v,t) = 0$, for some polynomial F. F will be of degree $n+1$ in t so that for each v there are $n+1$ roots $t_\alpha(v)$ (already introduced in (2.33)), representing the v-dependent positions of the fivebranes. F thus has the general form

$$F(t,v) = t^{n+1} + f_1(v)t^n + f_2(v)t^{n-1} + \ldots + f_n(v)t + 1. \tag{2.35}$$

As in the special case considered in Section 2.4, the coefficients of t^{n+1} and t^0 are non-zero constants to ensure the absence of semi-infinite fourbranes; those coefficients have been set to 1 by scaling F and t. Alternatively, we can factor F in terms of its roots:

$$F = \prod_{\alpha=0}^{n} (t - t_\alpha(v)). \tag{2.36}$$

The fact that the t^0 term in (2.35) is independent of v implies that

[4] By analogy with the $SU(k)$ theory with N_f hypermultiplets treated in the last subsection, semi-infinite fourbranes would be incorporated by taking the coefficients of t^{n+1} and t^0 in the polynomial $P(t,v)$ introduced below to be polynomials in v of positive degree. This gives solutions of models that are actually special cases of models that will be treated in Section 3.

$$\sum_{\alpha=0}^{n} a_\alpha = 0, \tag{2.37}$$

and this, together with the n equations (2.34), determines the a_α for $\alpha = 0, \ldots, n$. The solution is in fact

$$a_\alpha = k_{\alpha+1} - k_\alpha \tag{2.38}$$

with again $k_0 = k_{n+1} = 0$.

If the degree of a polynomial $f(v)$ is denoted by $[f]$, then the factorization (2.36) and asymptotic behavior (2.33) imply that

$$[f_1] = a_0, \qquad [f_2] = a_0 + a_1, \qquad [f_3] = a_0 + a_1 + a_2, \qquad \ldots \tag{2.39}$$

Together with (2.38), this implies simply

$$[f_\alpha] = k_\alpha, \quad \text{for } \alpha = 1, \ldots, n. \tag{2.40}$$

If we rename f_α as $p_{k_\alpha}(v)$, where the subscript now equals the degree of a polynomial, then the polynomial $F(t, v)$ takes the form

$$F(t, v) = t^{n+1} + p_{k_1}(v)t^n + p_{k_2}(v)t^{n-1} + \ldots + p_{k_n}(v)t + 1. \tag{2.41}$$

The curve $F(t, v) = 0$ thus describes the solution of the model with gauge group $\prod_{\alpha=1}^{n} SU(k_\alpha)$ and hypermultiplets in the representation $\sum_{\alpha=1}^{n-1}(\mathbf{k}_\alpha, \overline{\mathbf{k}}_{\alpha+1})$. The fact that the coefficient of t^i, for $1 \leqslant i \leqslant n$, is a polynomial of degree k_i in v has a clear intuitive interpretation: the zeroes of $p_{k_\alpha}(v)$ are the positions of the k_α fourbranes that stretch between the αth and $(\alpha + 1)$th fivebrane.

The polynomial p_{k_α} has the form

$$p_{k_\alpha}(v) = c_{\alpha,0}v^{k_\alpha} + c_{\alpha,1}v^{k_\alpha - 1} + c_{\alpha,2}v^{k_\alpha - 2} + \ldots + c_{\alpha,k_\alpha}. \tag{2.42}$$

The leading coefficients $c_{\alpha,0}$ determine the asymptotic positions of the fivebranes for $v \to \infty$, or more precisely the constants h_α in (2.33). In fact by comparing the factorization $F(t, v) = \prod_\alpha (t - t_\alpha(v)) = \prod_\alpha (t - h_\alpha t^{v_\alpha} + O(t^{v_\alpha - 1}))$ to the series (2.41) one can express the h_α in terms of the $c_{0,\alpha}$.

The h_α determine the constant terms in the asymptotic freedom formula

$$-i\tau_\alpha = -b_{0,\alpha} \ln v + \text{constant} \tag{2.43}$$

for the large v behavior of the inverses of the effective gauge couplings. Thus, the $c_{\alpha,0}$'s should be identified with the gauge coupling constants. Of course, one combination of the $c_{\alpha,0}$'s can be eliminated by rescaling the v's; this can be interpreted as a renormalization group transformation via which (as the beta function coefficients $b_{0,\alpha}$ are not all zero) one coupling constant can be eliminated.

In particular, the $c_{\alpha,0}$ are constants that parametrize the choice of a quantum system, not order parameters that determine the choice of a vacuum in a fixed quantum system.

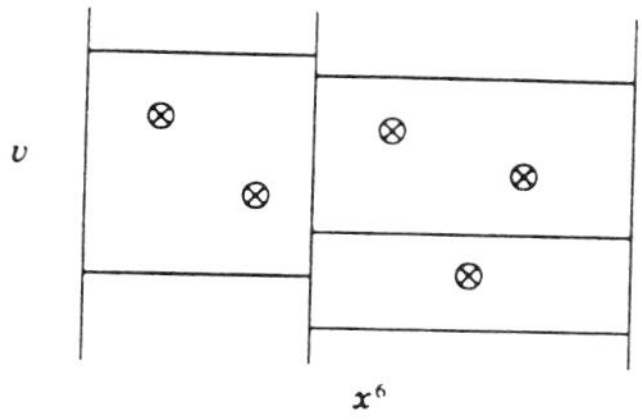

Fig. 3. A system of fourbranes, fivebranes, and sixbranes. The v direction runs vertically and the x^6 direction runs horizontally. Fivebranes and fourbranes are depicted as vertical and horizontal lines, and sixbranes are depicted by the symbol ⊗. This is meant to indicate that the sixbranes are "perpendicular" to the figure and occupy definite values of v and x^6.

The $c_{\alpha,1}$ are likewise constants, according to (2.5); they determine the hypermultiplet bare masses. (One of the $c_{\alpha,1}$ can be removed by adding a constant to v; in fact there are n $c_{\alpha,1}$'s and only $n-1$ hypermultiplet bare masses.) The $c_{\alpha,s}$ for $s = 2, \ldots, k_\alpha$ are the order parameters on the Coulomb branch of the $SU(k_\alpha)$ factor of the gauge group.

3. Models with sixbranes

3.1. Preliminaries

The goal in the present section is to incorporate sixbranes in the models of the previous section. The sixbranes will enter just like the D fivebranes in [1] and for some purposes can be analyzed quite similarly.

Thus we consider again the familiar chain of $n+1$ fivebranes, labeled from 0 to n, with k_α fourbranes stretched between the $(\alpha-1)$th and αth fivebranes, for $\alpha = 1, \ldots, n$. But now we place d_α sixbranes between the $(\alpha-1)$th and αth fivebranes, for $\alpha = 1, \ldots, n$. A special case is sketched in Fig. 3. In the coordinates introduced at the beginning of section two, each sixbrane is located at definite values of x^4, x^5, and x^6 and has a world-volume that is parametrized by arbitrary values of $x^0, x^1, \ldots, x^3$ and x^7, x^8, and x^9.

Given what was said in section two and in [1], the interpretation of the resulting model as a four-dimensional gauge theory is clear. The gauge group is $\prod_{\alpha=1}^{n} SU(k_\alpha)$. The hypermultiplets consist of the $(\mathbf{k}_\alpha, \overline{\mathbf{k}}_{\alpha+1})$ hypermultiplets that were present without the sixbranes, plus additional hypermultiplets that become massless whenever a fourbrane meets a sixbrane. As in [1], these additional hypermultiplets transform in d_α copies of the fundamental representation of $SU(k_\alpha)$, for each α. The bare masses of these hypermultiplets are determined by the positions of the sixbranes in $v = x^4 + ix^5$. As in [1], the positions of the sixbranes in x^6 decouple from many aspects of the low energy four-dimensional physics.

One difference from section two is that (even without semi-infinite fourbranes) there are many models with vanishing beta function. In fact, for each choice of k_α such that the models considered in section two had all beta functions zero or negative, there is upon inclusion of sixbranes a unique choice of the d_α for which the beta functions all vanish, namely

$$d_\alpha = 2k_\alpha - k_{\alpha+1} - k_{\alpha-1} \tag{3.1}$$

(where we understand that $k_0 = k_{n+1} = 0$). By solving all these models, we will get a much larger class of solved $\mathcal{N} = 2$ models with zero beta function than has existed hitherto. For each such model, one expects to find a non-perturbative duality group generalizing the duality group $SL(2, \mathbf{Z})$ of four-dimensional $\mathcal{N} = 4$ super Yang–Mills theory. From the solutions we will get, the duality groups turn out to be as follows. Let $\mathcal{M}_{0,n+3:2}$ be the moduli space of objects of the following kind: a smooth Riemann surface of genus zero with $n + 3$ marked points, two of which are distinguished and ordered while the other $n + 1$ are unordered. Then the duality group of a model with $n + 1$ fivebranes is the fundamental group $\pi_1(\mathcal{M}_{0,n+3:2})$. One can think roughly of the genus-zero Riemann surface in the definition of $\mathcal{M}_{0,n+3:2}$ as being parametrized by the variable t of section two, with the marked points being 0, ∞, and the positions of the $n + 1$ fivebranes.

In contrast to section two, we would gain nothing essentially new by incorporating semi-infinite fourbranes at the two ends of the chain. This gives hypermultiplets in the fundamental representation of the groups $SU(k_1)$ and $SU(k_n)$ that are supported at the ends of the chain; we will anyway generate an arbitrary number of such hypermultiplets via sixbranes. Another generalization that would give nothing essentially new would be to include fourbranes that connect fivebranes to sixbranes. Using a mechanism considered in [1], one can by moving the sixbranes in the x^6 direction reduce to the case that all fourbranes end on fivebranes. One could also add sixbranes to the left or to the right of all fivebranes. In fact, we will see how this generalization can be incorporated in the formulas. In the absence of fourbranes ending on them, sixbranes that are to the left or right of everything else simply decouple from the low energy four-dimensional physics.

Another generalization is to consider fourbranes that end on sixbranes at both ends. As in [1], such a fourbrane supports a four-dimensional hypermultiplet, not a vector multiplet, and configurations containing such fourbranes must be included to describe Higgs branches (and mixed Coulomb–Higgs branches) of these theories. We will briefly discuss the Higgs branches in Section 3.5.

3.2. *Interpretation in M-theory*

Since our basic technique is to interpret Type IIA brane configurations in M-theory, we need to know how to interpret the Type IIA sixbrane in M-theory. This was first done in [23].

Consider M-theory on $\mathbf{R}^{10} \times \mathbf{S}^1$. This is equivalent to Type IIA on $\mathbf{R}^{10}$, with the $U(1)$ gauge symmetry of Type IIA being associated in M-theory with the rotations of the $\mathbf{S}^1$. States that have momentum in the $\mathbf{S}^1$ direction are electrically charged with respect to this $U(1)$ gauge field and are interpreted in Type IIA as Dirichlet zerobranes. The sixbrane is the electric-magnetic dual of the zerobrane, so it is magnetically charged with respect to this same $U(1)$.

The basic object that is magnetically charged with respect to this $U(1)$ is the "Kaluza–Klein monopole" or Taub-NUT space. This is derived from a hyper-Kähler solution of the four-dimensional Einstein equations. The metric is asymptotically flat, and the space-time looks near infinity like a non-trivial $\mathbf{S}^1$ bundle over $\mathbf{R}^3$. The Kaluza–Klein magnetic charge is given by the twisting of the $\mathbf{S}^1$ bundle, which is incorporated in the formula given below by the appearance of the Dirac monopole potential.

Using conventions of [24] adapted to the notation of the present paper, if we define a three-vector $\vec{\mathbf{r}} = (x^4, x^5, x^6)R$, and set $r = |\vec{\mathbf{r}}|$ and $\tau = x^{10}/R$, then the Taub-NUT metric is

$$ds^2 = \frac{1}{4}\left(\frac{1}{r} + \frac{1}{R^2}\right) d\vec{\mathbf{r}}^2 + \frac{1}{4}\left(\frac{1}{r} + \frac{1}{R^2}\right)^{-1} (d\tau + \vec{\omega}\cdot d\vec{\mathbf{r}})^2. \tag{3.2}$$

Here $\vec{\omega}$ is the Dirac monopole potential (which one can identify locally as a one-form obeying $\vec{\nabla} \times \vec{\omega} = \vec{\nabla}(1/r)$).

To construct a sixbrane on $\mathbf{R}^{10} \times \mathbf{S}^1$, we simply take the product of the metric (3.2) with a flat metric on $\mathbf{R}^7$ (the coordinates on $\mathbf{R}^7$ being $x^0, \ldots, x^3$ and $x^7, \ldots, x^9$). We will be interested in the case of many parallel sixbranes, which is described by the multi-Taub-NUT metric [25]:

$$ds^2 = \frac{V}{4} d\vec{\mathbf{r}}^2 + \frac{V^{-1}}{4}(d\tau + \vec{\omega}\cdot d\vec{r})^2, \tag{3.3}$$

where now

$$V = 1 + \sum_{a=1}^{d} \frac{1}{|\mathbf{r} - \mathbf{x}_a|} \tag{3.4}$$

and $\vec{\nabla} \times \omega = \vec{\nabla} V$. This describes a configuration of d parallel sixbranes, whose positions are the $\mathbf{x}_a$.

The reason that by going to eleven dimensions we will get some simplification in the study of sixbranes is that, in contrast to the ten-dimensional low energy field theory in which the sixbrane core is singular, in M-theory the sixbrane configuration is described by the multi-Taub-NUT metric (3.4), which is complete and smooth (as long as the $\mathbf{x}_a$ are distinct). This elimination of the sixbrane singularity was in fact emphasized in [23]. In going from M-theory to Type IIA, one reduces from eleven to ten dimensions by dividing by the action of the vector field $\partial/\partial\tau$. This produces singularities at d points at which $\partial/\partial\tau$ vanishes; those d points are interpreted in Type IIA as positions of sixbranes. In general in physics, appearance of singularities in a long wavelength description means that to understand the behavior of a system one needs more information. The fact that the sixbrane singularity is eliminated in going to M-theory means that, if the radius R of the

x^{10} circle is big,[5] the M-theory can be treated via low energy supergravity. This is just analogous to what happened in Section 2; the singularity of Type IIA fourbranes ending on fivebranes was eliminated upon going to M-theory, as a result of which low energy supergravity was an adequate approximation. The net effect is that unlike either long wavelength ten-dimensional field theory or conformal field theory, the long wavelength eleven-dimensional field theory is an adequate approximation for the problem.

In this paper we will really not use the hyper-Kähler metric of the multi-Taub-NUT space, but only the structure (or more exactly one of the structures) as a complex manifold. If as before we set $v = x^4 + ix^5$, then in one of its complex structures the multi-Taub-NUT space can be described by the equation

$$yz = \prod_{a=1}^{d}(v - e_a) \tag{3.5}$$

in a space $\mathbf{C}^3$ with three complex coordinates y, z, and v. Here e_a are the positions of the sixbranes projected to the complex v plane. Note that (3.5) admits the $\mathbf{C}^*$ action

$$y \to \lambda y, \qquad z \to \lambda^{-1} z, \tag{3.6}$$

which is the complexification of the $U(1)$ symmetry of (3.3) that is generated by $\partial/\partial\tau$. For the special case that there are no fivebranes, this $\mathbf{C}^*$ corresponds to the transformation $t \to \lambda t$ where $t = \exp\left(-(x^6 + ix^{10})/R\right)$. Hence very roughly, for large y with fixed or small z, y corresponds to t and for large z with fixed or small y, z corresponds to t^{-1}. (As there is a symmetry exchanging y and z, their roles could be reversed in these assertions.)

In Section 3.6, we will use the approach of [24] to show that the multi-Taub-NUT space is equivalent as a complex manifold to (3.5). The formulas in Section 3.6 can also be used to make the asymptotic identification of y and z with t and t^{-1} more precise. For now, we note the following facts, which may orient the reader. When all e_a are coincident at, say, $v = 0$, (3.5) reduces to the A_{n-1} singularity $yz = v^n$. A system of parallel and coincident sixbranes in Type IIA generates a $U(n)$ gauge symmetry; the A_{n-1} singularity is the mechanism by which such enhanced gauge symmetry appears in the M-theory description. In general, (3.5) describes the unfolding of the A_{n-1} singularity.

The complex structure (3.5) does not uniquely fix the hyper-Kähler metric, not even the behavior of the metric at infinity. The *same* complex manifold (3.5) admits a family of "asymptotically locally Euclidean" (ALE) metrics, which look at infinity like $\mathbf{C}^2/\mathbf{Z}_n$. (They are given by the same formula (3.3), but with a somewhat different choice of V.) The metrics (3.3) are not ALE but are "asymptotically locally flat" (ALF).

Even if one asks for ALF behavior at infinity, the hyper-Kähler metric involves parameters that do not appear in (3.5). The hyper-Kähler metric (3.3) depends on the

[5] We recall that we can assume this radius to be big since it corresponds to an "irrelevant" parameter in the field theory.

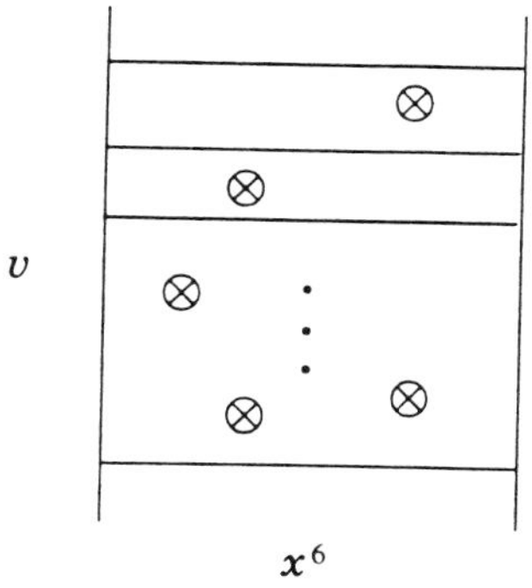

Fig. 4. A specific configuration of k fourbranes (horizontal lines), two fivebranes (vertical lines) and d sixbranes (depicted by the symbol $\otimes$) that gives a representation of $\mathcal{N} = 2$ supersymmetric QCD in four dimensions, with gauge group $SU(k)$ and d hypermultiplet flavors in the fundamental representation.

positions $\vec{\mathbf{x}}_a$ of the sixbranes, while in (3.5) one sees only the projections e_a of those positions to the v plane. From the point of view of the complex structure that is exhibited in (3.5), the x^6 component of the sixbrane positions is coded in the Kähler class of the metric (3.3).

In studying the Coulomb branch of $\mathcal{N} = 2$ models, we will really need only the complex structure (3.5); the x^6 positions of sixbranes will be irrelevant. This is analogous to the fact that in studying the Coulomb branch of $N = 4$ models in three dimensions by methods of [1], the x^6 positions of Dirichlet fivebranes are irrelevant. As that example suggests, the x^6 positions are relevant for understanding the Higgs branches of these models.

In one respect, the description (3.5) of the complex structure is misleading. Whenever $e_a = e_b$ for some a and b, the complex manifold (3.5) gets a singularity. The hyper-Kähler metric, however, becomes singular only if two sixbranes have equal positions in x^6 and not only in v. When two sixbranes have the same position in v but not in x^6, the singular complex manifold (3.5) must be replaced by a smooth one that is obtained by blowing up the singularities, replacing each A_k singularity by a configuration of k curves of genus zero. This subtlety will be important when, and only when, we briefly examine the Higgs branches of these models.

3.3. $\mathcal{N} = 2$ *supersymmetric QCD revisited*

Now we want to solve for the Coulomb branch of a model that is constructed in terms of Type IIA via a configuration of fourbranes, fivebranes, and sixbranes. The only change from Section 2 is that to incorporate sixbranes we must replace $Q = \mathbf{R}^3 \times \mathbf{S}^1$, in which the M-theory fivebrane propagated in Section 2, by the multi-Taub-NUT space $\widetilde{Q}$ that was just introduced. We write the defining equation of $\widetilde{Q}$ as

$$yz = P(v). \tag{3.7}$$

with $P(v) = \prod_{a=1}^{d}(v - e_a)$. Type IIA fourbranes and fivebranes are described as before by a complex curve Σ in $\widetilde{Q}$. Σ will be described by an equation $F(y, v) = 0$. Note that we can assume that F is independent of z, because z could be eliminated via $z = P(v)/y$.

For our first attempt to understand the combined system of fourbranes, fivebranes, and sixbranes, we consider the example in Fig. 4 of two parallel fivebranes connected by k fourbranes, with d sixbranes between them. We assume that there are no semi-infinite fourbranes extending to the left or right of the figure. This configuration should correspond to $\mathcal{N} = 2$ supersymmetric QCD, that is to an $SU(k)$ gauge theory with d hypermultiplets in the fundamental representation of $SU(k)$.

As in Section 2, the fact that there are two fivebranes means that the equation $F(y, v) = 0$, regarded as an equation in y for fixed v, has generically two roots. Thus, F is quadratic in y and has the general form

$$A(v)y^2 + B(v)y + C(v) = 0. \tag{3.8}$$

By clearing denominators and dividing by common factors, we can assume that A, B, and C are relatively prime polynomials.

Now we must interpret the statement that there are no semi-infinite fourbranes. This means, as in Section 2, that it is impossible for y or z (which correspond roughly to t and t^{-1} in the notation of Section 2) to go to infinity at a finite value of v. The requirement that y never diverges at finite v means that – if A, B, and C are understood to have no common factors – $A(v)$ is a constant, which we can take to equal 1. So the defining equation of Σ reduces to

$$y^2 + B(v)y + C(v) = 0. \tag{3.9}$$

Now let us express this in terms of $z = P(v)/y$. We get

$$C(v)z^2 + B(v)P(v)z + P(v)^2 = 0. \tag{3.10}$$

z will diverge at zeroes of C unless both BP and P^2 are divisible by C. Such divergence would represent the existence of a semi-infinite fourbrane.

In particular, the absence of semi-infinite fourbranes implies that P^2 is divisible by C. So any zero of C is a zero of P, that is, it is one of the e_a. Moreover, in the generic case that the e_a are distinct, each e_a can appear as a root of C with multiplicity at most two. Thus, we can label the e_a in such a way that e_a is a root of C with multiplicity 2 for $a \leqslant i_0$, of multiplicity 1 for $i_0 < a \leqslant i_1$, and of multiplicity 0 for $a > i_1$. We then have

$$C = f\prod_{a=1}^{i_0}(v - e_a)^2 \prod_{b=i_0+1}^{i_1}(v - e_b) \tag{3.11}$$

with some non-zero complex constant f. The requirement that BP should be divisible by C now implies that the e_a of $a \leqslant i_0$ are roots of B, so

$$B(v) = \widetilde{B}(v) \prod_{a \leqslant i_0} (v - e_a) \tag{3.12}$$

for some polynomial $\widetilde{B}$.

Eq. (3.8) now reduces to

$$y^2 + \widetilde{B}(v) \prod_{a \leqslant i_0} (v - e_a) y + f \prod_{a \leqslant i_0} (v - e_a)^2 \prod_{b=i_0+1}^{i_1} (v - e_b) = 0. \tag{3.13}$$

In terms of $\widetilde{y} = y / \prod_{a \leqslant i_0} (v - e_a)$, this is

$$\widetilde{y}^2 + \widetilde{B}(v) \widetilde{y} + f \prod_{a=i_0+1}^{i_1} (v - e_a) = 0. \tag{3.14}$$

If $\widetilde{B}(v)$ is a polynomial of degree k, this is (for $i_1 - i_0 \leqslant 2k$; otherwise as at the end of Section 2 one encounters a new ultraviolet fixed point) the familiar solution of the $SU(k)$ gauge theory with $i_1 - i_0$ flavors in the fundamental representation, written in the same form in which it appeared in Section 2. The e_a with $a \leqslant i_0$ or $a > i_1$ have decoupled from the gauge theory.

This suggests the following interpretation: the sixbranes with $a \leqslant i_0$ are to the left of all fivebranes, the sixbranes with $i_0 + 1 \leqslant a \leqslant i_1$ are between the two fivebranes, and the sixbranes with $a > i_1$ are to the right of all fivebranes. If so then (in the absence of fourbranes ending on the sixbranes) the sixbranes with $a \leqslant i_0$ or $a > i_1$ would be decoupled from the four-dimensional gauge theory, and the number of hypermultiplet copies of the fundamental representation of $SU(k)$ would be $i_1 - i_0$, as we have just seen. We will now justify that interpretation.

Interpretation of i_0 and i_1

The manifold $\widetilde{Q}$ defined by $yz = P(v)$ maps to the complex v plane, by forgetting y and z. Let Q_v be the fiber of this map for a given value of v. For generic v, the fiber is a copy of $\mathbf{C}^*$. Indeed, whenever $P(v) \neq 0$, the fiber Q_v, defined by

$$yz = P(v), \tag{3.15}$$

is a copy of $\mathbf{C}^*$ (the complex y plane with $y = 0$ deleted). This copy of $\mathbf{C}^*$ is actually an orbit of the $\mathbf{C}^*$ action (3.6) on $\widetilde{Q}$.

We recall from Section 3.2 that if z or y is large with the other fixed, then the asymptotic relation between z, y, and $t = \exp\left(-(x^6 + ix^{10})/R\right)$ is $y \cong t$ or $z \cong t^{-1}$. $t \to 0$ means large x^6, which we call "being on the right"; $t \to \infty$ means $x^6 \to -\infty$, which we call "being on the left." Thus z much larger than y or vice versa corresponds to being on the right or on the left in x^6.

The surface Σ is defined by an equation $F(y, v) = 0$ where F is quadratic in y; it intersects each Q_v in two points. (Q_v is not complete, but we have chosen F so that no root goes to $y = \infty$ or $z = \infty$ for v such that $P(v) \neq 0$.) These are the two points with fivebranes, for the given value of v.

Now consider the special fibers with $F(v) = 0$. This means that for some a, v is equal to e_a, the position in the v plane of the ath sixbrane. The fiber F_v is for such v defined by

$$yz = 0, \tag{3.16}$$

and is a union of two components C_v and C'_v with, respectively, $z = 0$ and $y = 0$. The total number of intersection points of Σ with F_v is still 2, but some intersections lie on C_v and some lie on C'_v. Without passing through any singularity, we can go to the case that the intersections on C_v are at large y and those on C'_v are at large z. Hence, fivebranes that correspond to intersections with C_v are to the left of the ath sixbrane (y is much bigger than z so they are at a smaller value of x^6) and fivebranes that correspond to intersections with C'_v are to the right of the ath sixbrane (they are at a larger value of x^6).

The intersection points on C_v are the zeroes of (3.13) which as $v \to e_a$ do not go to $y = 0$. The intersection points on C'_v are likewise the zeroes of that polynomial that *do* vanish as $v \to e_a$. The number of such intersections with C'_v is two if $a \leqslant i_0$, one if $i_0 + 1 \leqslant a \leqslant i_1$, and zero otherwise. This confirms that the number of sixbranes to the left of both fivebranes is i_0, the number which are to the left of one and to the right of the other is $i_1 - i_0$, and the number which are to the right of both is i_1.

3.4. Generalization

We will now use similar methods to solve for the Coulomb branch of a more general model with $n + 1$ fivebranes, joined in a similar way by fourbranes and with sixbranes between them.

The curve Σ will now be defined by the vanishing of a polynomial $F(y, v)$ that is of degree $n + 1$ in y:

$$y^{n+1} + A_1(v)y^n + A_2(v)y^{n-1} + \ldots + A_{n+1}(v) = 0. \tag{3.17}$$

The $A_\alpha(v)$ are polynomials in v. We assume that there are no semi-infinite fourbranes and therefore have set the coefficient of y^{n+1} to 1. Substituting $y = P(v)/z$, we get

$$A_{n+1}z^{n+1} + A_nPz^n + A_{n-1}P^2z^{n-1} + \ldots + P^{n+1} = 0. \tag{3.18}$$

Hence absence of semi-infinite fourbranes implies that $A_\alpha P^{n+1-\alpha}$ is divisible by A_{n+1} for all α with $0 \leqslant \alpha \leqslant n$. (In this assertion we understand $A_0 = 1$.) In particular, P^{n+1} is divisible by A_{n+1}.

It follows that all zeroes of A_{n+1} are zeroes of P, and occur (if the e_a are distinct) with multiplicity at most $n + 1$. As in the example considered before, zeroes of P that occur as zeroes of A_{n+1} with multiplicity 0 or $n + 1$ make no essential contribution (they correspond to sixbranes that are to the left or the right of everything else and can be omitted). So we will assume that all zeroes of P occur as zeroes of A_{n+1}

with some multiplicity between 1 and n. There are therefore integers $i_0, i_1, \ldots, i_n$ with $i_0 = 0 \leqslant i_1 \leqslant i_2 \leqslant \ldots \leqslant i_{n-1} \leqslant i_n = n$ such that if for $1 \leqslant s \leqslant n$

$$J_s = \prod_{a=i_{s-1}+1}^{i_s} (v - e_a) \tag{3.19}$$

then

$$A_{n+1} = f \prod_{s=1}^{n} J_s^{n+1-s} \tag{3.20}$$

with f a constant. By an argument along the lines given at the end of Section 3.3, we can interpret i_α as the number of sixbranes to the left of the αth fivebrane. So $d_\alpha = i_\alpha - i_{\alpha-1}$ is the number of sixbranes between the $(\alpha - 1)$th and αth fivebranes. The number of hypermultiplets in the fundamental representation of the αth factor of the gauge group will hence be d_α.

The requirement that $A_\alpha P^{n+1-\alpha}$ is divisible by A_{n+1} is then equivalent to the statement

$$A_\alpha = g_\alpha(v) \prod_{s=1}^{\alpha-1} J_s^{\alpha-s} \tag{3.21}$$

with some polynomial $g_\alpha(v)$. We interpret $g_\alpha(v)$ as containing the order parameters for the αth factor of the gauge group. So if $g_\alpha(v)$ is of degree k_α, then the gauge group is

$$G = \prod_{\alpha=1}^{n} SU(k_\alpha). \tag{3.22}$$

The hypermultiplet spectrum consists of the usual $(\mathbf{k}_\alpha, \overline{\mathbf{k}}_{\alpha+1})$ representations plus d_α copies of the fundamental representation of $SU(k_\alpha)$.

The curve describing the solution of this theory should thus be

$$\begin{aligned} &y^{n+1} + g_1(v)y^n + g_2(v)J_1(v)y^{n-1} + g_3(v)J_1(v)^2 J_2(v)y^{n-2} \\ &\quad + \ldots + g_\alpha(v)\prod_{s=1}^{\alpha-1} J_s^{\alpha-s} \cdot y^{n+1-\alpha} + \ldots + f\prod_{s=1}^{n} J_s^{n+1-s} = 0. \end{aligned} \tag{3.23}$$

This of course reduces in the absence of sixbranes to the solution found in (2.41); it likewise gives back the standard solution of $\mathcal{N} = 2$ supersymmetric QCD when there are precisely two fivebranes. As a further check, let us examine the condition on the d_α and the k_α under which the beta function vanishes. Note that the coefficient of y^n is of degree v^{k_1}. All fivebranes will be parallel at large v, and the beta function will vanish, if the coefficient of y^{n+1-m} is of order v^{mk_1} for $m = 1, \ldots, n+1$. Those conditions can be evaluated to give $k_2 + d_1 = 2k_1$, $k_3 + k_1 + d_2 = 2k_2$, and so on – the standard conditions for vanishing beta function of the gauge theory.

In this case of vanishing beta function, let the polynomials $g_\alpha(v)$ be of the form $g_\alpha(v) = h_\alpha v^{k_\alpha} + O(v^{k_\alpha - 1})$. Then the asymptotic behavior of the roots of (3.23) (re-

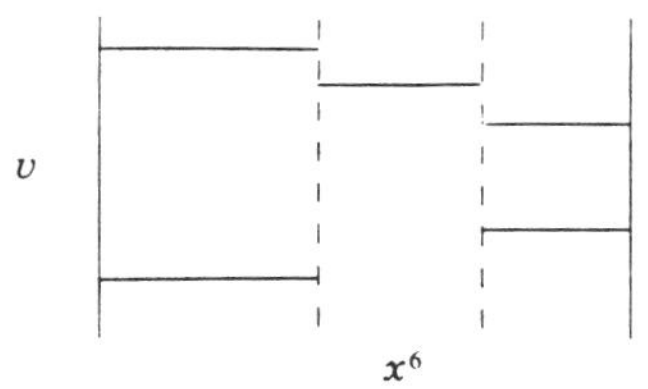

Fig. 5. A configuration representing a mixed Coulomb–Higgs branch. Here as before fivebranes are shown as vertical solid lines and fourbranes as horizontal solid lines. But in contrast to Figs. 3 and 4, sixbranes are depicted (as in Ref. [1]) as vertical dashed lines. This makes it easier to visualize the hypermultiplet moduli of fourbranes that end on parallel sixbranes. Such a modulus appears whenever there is a fourbrane suspended between two sixbranes as in this example.

garded as an equation for y) is $y \sim \lambda_i v^{k_1}$, where the λ_i are the roots of the polynomial equation

$$x^{n+1} + h_1 x^n + h_2 x^{n-1} + \ldots + h_n x + f = 0. \tag{3.24}$$

On the x plane, there are $n+3$ distinguished points, namely 0, ∞, and the λ_i. The λ_i are of course defined only up to permutation and (as one could rescale y and x) up to multiplication by a common complex scalar. A choice of the λ_i, modulo those equivalences, determines the asymptotic distances between fivebranes and hence the bare gauge coupling constants. The same choice also determines a point in the moduli space $\mathcal{M}_{0,n+3:2}$ that was introduced in Section 3.1. In any description by a Lagrangian field theory with coupling parameters τ_i, the fundamental group $\pi_1(\mathcal{M}_{0,n+3:2})$ would be interpreted as the group of discrete duality symmetries.

3.5. Higgs branches

In this subsection, we will sketch how the transition to a Higgs branch (or a mixed Higgs–Coulomb branch) can be described from the present point of view.

We recall that the transition to a Higgs branch is a process in which the genus of Σ drops by one (or more) and a transition is made to a new branch of vacua in which there are massless hypermultiplets. In terms of Type IIA brane diagrams, massless hypermultiplets result (as in [1]) from fourbranes suspended between fivebranes, a configuration shown in Fig. 5.

For a transition to a Higgs branch to occur, it is necessary for two hypermultiplet bare masses to become equal. From the present point of view, this means that the positions of two sixbranes in v become equal. It is *not* necessary for the two sixbranes to have equal positions in x^6. In fact, the semiclassical brane diagram of Fig. 5 cannot be drawn if the x^6 values of the sixbranes are equal.

The hypermultiplet bare masses are the roots of $P(v) = \prod_a (v - e_a)$. We therefore want to consider the case that two e_a are coincident at, say, the origin. The other e_a will play no material role, and we may as well take the case of only two sixbranes. So we take $P(v) = v^2$. The equation $yz = P(v)$ is in this case

$$xy = v^2, \tag{3.25}$$

and describes a manifold Q_0 which has a singularity at the point P with coordinates $x = y = z = 0$.

We recall, however, from the discussion in Section 3.2 that in case two sixbranes coincide in v but not in x^6, such a singularity should be blown up. Thus, the multi-Taub-NUT manifold $\widetilde{Q}$ does not coincide with Q_0, but is a smooth surface obtained by blowing up the singularity in Q_0. In the blow-up, P is replaced by a smooth curve C of genus zero.

Now we consider a curve Σ in $\widetilde{Q}$ (or Q_0) representing a point on the Coulomb branch of one of the models considered in this section. Let g be the generic genus of Σ. Nothing essential will be lost if we consider the case of supersymmetric QCD – two fivebranes; gauge group $SU(n)$. So Σ is defined by a curve of the form

$$y^2 + By + fv^2 = 0. \tag{3.26}$$

Nothing of interest will happen unless Σ passes through the singular point $y = z = v = 0$. That is so if and only if B vanishes at $v = 0$ (if B is non-zero at $v = 0$ then either y is non-zero for $v \to 0$, or $y \sim v^2$ for $v \to 0$ and z is non-vanishing at $v = 0$), so generically $B = bv + O(v^2)$ with a non-zero constant b.

So near P, Σ looks like

$$y^2 + bvy + fv^2 = 0. \tag{3.27}$$

This curve has a singularity at $y = v = 0$. In fact, the quadratic polynomial $y^2 + bvy + fv^2$ has a factorization as $(y + \gamma v)(y + \gamma' v)$. Generically, the two factors correspond, near P, to two branches of Σ that meet "transversely" at P, giving the singularity. The genus of Σ drops by one when this singularity appears. So Σ now has genus $g - 1$.

We actually want to consider the case in which the two sixbranes are not coincident in x^6, so we must consider the curve defined by (3.26) not in the singular manifold Q_0 but in its smooth resolution $\widetilde{Q}$. This curve has two components. One is a *smooth* curve Σ' of genus $g - 1$ and the other is a copy of the genus-zero curve C in $\widetilde{Q}$ that is obtained by the blowup of P. Σ' is smooth (generically) because after the blowup the two branches $y + \gamma v = 0$ and $y + \gamma' v = 0$ of Σ no longer meet. A copy of C is present because the polynomial $y^2 + By + v^2$ vanishes on P and hence (when pulled back to $\widetilde{Q}$) on C.

At this point, by adding a constant to B, we could deform the two-component curve $\Sigma' + C$ (which is singular where Σ' and C meet) back to a smooth irreducible curve of genus g that does not pass through P or C. Instead, we want to make the transition to the Higgs branch.

We recall that in the present paper, the curve Σ is really an ingredient in the description of a fivebrane in eleven dimensions. The fivebrane propagates in $\mathbf{R}^7 \times \widetilde{Q}$. $\mathbf{R}^7$ has coordinates $x^0, x^1, \ldots, x^3$ and x^7, x^8, x^9. The fivebrane world-volume is of the form $\mathbf{R}^4 \times \Sigma$, where Σ is a curve in $\widetilde{Q}$ and $\mathbf{R}^4$ is a subspace of $\mathbf{R}^7$ defined by (for instance) $x^7 = x^8 = x^9 = 0$.

The transition to the Higgs branch can be described as follows. When Σ degenerates to a curve that is a union of two branches Σ' and C, the fivebrane degenerates to two branches $\mathbf{R}^4 \times \Sigma'$ and $\mathbf{R}^4 \times C$. At this point, it is possible for the two branches to move independently in $\mathbf{R}^7$. $\mathbf{R}^4 \times C$ can move to $\widetilde{\mathbf{R}}^4 \times C$, where $\widetilde{\mathbf{R}}^4$ is a *different* copy of $\mathbf{R}^4$ embedded in $\mathbf{R}^7$. For unbroken supersymmetry, $\widetilde{\mathbf{R}}^4$ should be parallel to $\mathbf{R}^4$, so it is defined in $\mathbf{R}^7$ by $(x^7, x^8, x^9) = \vec{w}$ for some constant $\vec{w}$.

The four-dimensional field theory derived from a fivebrane on $\widetilde{\mathbf{R}}^4 \times C$ has no massless vector multiplets, as C has genus zero. It has one massless hypermultiplet, whose components are $\vec{w}$ and $\int_C \beta$, where β is the chiral two-form on the fivebrane world-volume.

A motion of $\mathbf{R}^4 \times \Sigma'$ in the x^7, x^8, x^9 directions, analogous to the above, is *not* natural because Σ' is non-compact and such a motion would entail infinite action per unit volume on $\mathbf{R}^4$. The allowed motions of $\mathbf{R}^4 \times \Sigma'$ are the motions of Σ' in $\widetilde{Q}$ that determine the order parameters on the Coulomb branch and that we have been studying throughout this paper. The four-dimensional field theory derived from a fivebrane on $\mathbf{R}^4 \times \Sigma'$ has $g - 1$ massless vector multiplets, because Σ' is a curve of genus $g - 1$, and one hypermultiplet. The combined system of fivebranes on $\mathbf{R}^4 \times \Sigma'$ and on $\widetilde{\mathbf{R}}^4 \times C$ has $g - 1$ massless vector multiplets and one hypermultiplet.

There is no way to deform Σ' to a curve of genus g. It is only $\Sigma' + C$ that can be so deformed. So once C has moved to $\vec{w} = 0$, there is no way to regain the gth massless vector multiplet except by first moving C back to $\vec{w} = 0$. The transition to the Higgs branch has been made.

3.6. *Metric and complex structure*

Finally, using the techniques of [24], we will briefly describe how to exhibit the complex structure (3.5) of the ALF manifold (3.3). In that paper, the formula (3.3) for the ALF hyper-Kähler metric is obtained in the following way.

Let $\mathbf{H}$ be a copy of $\mathbf{R}^4$ with the flat hyper-Kähler metric. Let $\mathcal{M} = \mathbf{H}^d \times \mathbf{H}$, with coordinates q_a, $a = 1, \ldots, d$, and w. Consider the action on $\mathcal{M}$ of an abelian group G, locally isomorphic to $\mathbf{R}^m$, for which the hyper-Kähler moment map is

$$\mu_a = \frac{1}{2}\mathbf{r}_a + \mathbf{y}, \tag{3.28}$$

where $\mathbf{r} = q_a i \bar{q}_a$ and $\mathbf{y} = (w - \overline{w})/2$. Notation is as explained in [24]. G is a product of d factors; the ath factor, for $a = 1, \ldots, d$, acts on q_a by a one-parameter group of rotations that preserve the hyper-Kähler metric, on w by translations, and trivially on the other variables. The manifold defined as $\mu^{-1}(\mathbf{e})/G$, with an arbitrary constant $\mathbf{e}$, carries a natural hyper-Kähler metric, which is shown in [24] to coincide with (3.3). The choice of $\mathbf{e}$ determines the positions $\vec{\mathbf{x}}_a$ of the sixbranes in (3.3).

To exhibit the structure of this hyper-Kähler manifold as a complex manifold, one may proceed as follows. In any one of its complex structures, $\mathbf{H}$ can be identified as

$\mathbf{C}^2$. One can pick coordinates so that each q_a consists of a pair of complex variables y_a, z_a, and w consists of a pair v, v', such that the action of G is

$$y_a \longrightarrow e^{i\theta_a} y_a, \qquad z_a \longrightarrow e^{-i\theta_a} z_a,$$
$$v \longrightarrow v, \qquad v' \longrightarrow v' - \sum_{a=1}^{d} \theta_a, \tag{3.29}$$

where the θ_a are real parameters.

Once a complex structure is picked, the moment map μ breaks up as a complex moment map $\mu_{\mathbf{C}}$ and a real moment map $\mu_{\mathbf{R}}$. A convenient way to exhibit the complex structure of the ALF manifold is the following. Instead of setting μ to a constant value and dividing by G, one can set $\mu_{\mathbf{C}}$ to a constant value and divide by $G_{\mathbf{C}}$, the complexification of G (whose action is given by the formulas (3.29) with the θ_a now complex-valued).[6] The advantage of this procedure is that the complex structure is manifest.

The components of $\mu_{\mathbf{C}}$ are

$$\mu_{\mathbf{C},a} = y_a z_a - v. \tag{3.30}$$

Setting the $\mu_{\mathbf{C},a}$ to constants, which we will call $-e_a$, means therefore taking

$$y_a z_a = v - e_a. \tag{3.31}$$

Dividing by $G_{\mathbf{C}}$ is accomplished most simply by working with the $G_{\mathbf{C}}$-invariant functions of y_a, z_a, v, and v'. In other words, the $G_{\mathbf{C}}$-invariants can be regarded as functions on the quotient $\widetilde{Q} = \mu_{\mathbf{C}}^{-1}(-e_a)/G_{\mathbf{C}}$.

The basic invariants are $y = e^{iv'} \prod_{a=1}^{d} y_a$, $z = e^{-iv'} \prod_{a=1}^{d} z_a$, and v. The relation that they obey is, in view of (3.31),

$$yz = \prod_{a=1}^{d} (v - e_a), \tag{3.32}$$

which is the formula by which we have defined the complex manifold $\widetilde{Q}$. This exhibits the complex structure of the ALF manifold, for generic sixbrane positions e_a.

[6] The quotient should be taken in the sense of geometric invariant theory. This leads to the fact, exploited in Section 3.5, that when two sixbranes coincide in v but not in x^6, the ALF manifold (3.3) is equivalent as a complex manifold not to $yz = \prod_a (v - e_a)$ but to the smooth resolution $\widetilde{Q}$ of that singular surface. We will treat the invariant theory in a simplified way which misses the precise behavior for $e_a = e_b$. The calculation we do presently with invariants really proves not that the ALF manifold is isomorphic to $\widetilde{Q}$, but only that it has a holomorphic and generically one-to-one map to $\widetilde{Q}$. When $\widetilde{Q}$ is smooth (as it is for generic e_a), the additional fact that the ALF manifold is hyper-Kähler implies that it must coincide with $\widetilde{Q}$.

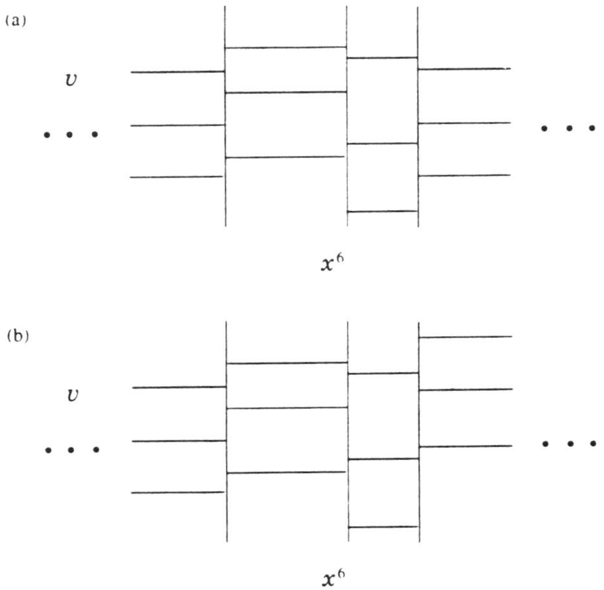

Fig. 6. (a) A periodic chain of fourbranes and fivebranes wrapped around a circle in the x^6 direction. The two ends of the chain are to be identified. (b) A generalization of the configuration in (a), in which one has periodicity in x^6 only modulo a translation in v. This generalization is needed to incorporate arbitrary hypermultiplet bare masses.

4. Elliptic models

4.1. Description of the models

In this section we compactify the x^6 direction to a circle, of radius L, and consider a chain of n fivebranes arranged around this circle, as in Fig. 6.[7] Let k_α be the number of fourbranes stretching between the $(\alpha - 1)$th and αth fivebrane, and let d_α be the number of sixbranes localized at points between the $(\alpha - 1)$th and αth fivebrane. The beta function of the $SU(k_\alpha)$ factor in the gauge group is then

$$b_{0,\alpha} = -2k_\alpha + k_{\alpha-1} + k_{\alpha+1} + d_\alpha. \tag{4.1}$$

Since $\sum_\alpha b_{0,\alpha} = \sum_\alpha d_\alpha$, and the d_α are all non-negative, the only case in which all beta functions are zero or negative is that case that all $b_{0,\alpha} = d_\alpha = 0$. Then writing $0 = \sum_\alpha k_\alpha(-2k_\alpha + k_{\alpha-1} + k_{\alpha+1}) = -\sum_\alpha (k_\alpha - k_{\alpha-1})^2$, we see that this occurs if and only if all k_α are equal to a fixed integer k. The present section will be devoted to analyzing this case.

The gauge group is $G = U(1) \times SU(k)^n$. Only the occurrence of a $U(1)$ factor requires special comment. The condition (2.6) "freezes out" the difference between the

[7] In the context of three-dimensional models with $\mathcal{N} = 4$ supersymmetry, configurations of fivebranes arranged around a circle were studied in [2].

$U(1)$ factors in the gauge group supported on alternate sides of any given fivebrane. In Sections 2 and 3, we considered a finite chain of fivebranes with $U(1)$'s potentially supported only in the "interior" of the chain, and this condition sufficed to eliminate all $U(1)$'s. In the present case of n fivebranes arranged around a circle with fourbranes connecting each neighboring pair, (2.6) eliminates $n-1$ of the $U(1)$'s, leaving a single (diagonal) $U(1)$ factor in the gauge group.

Hypermultiplets arise from fourbranes that meet a single fivebrane at the same point in space from opposite sides. If the symbol $\mathbf{k}_\alpha$ represents the fundamental representation of the αth $SU(k)$ factor in G, then the hypermultiplets transform as $\oplus_{\alpha=1}^{n}\mathbf{k}_\alpha\otimes\overline{\mathbf{k}}_{\alpha+1}$. Note that all of these hypermultiplets are neutral under the $U(1)$, so that all beta functions vanish including that of the $U(1)$. The $U(1)$, while present, is thus completely decoupled in the model. The curve Σ that we will eventually construct will have the property that its Jacobian determines the coupling constant of the $U(1)$ factor as well as the structure of the $SU(k)^n$ Coulomb branch.

A special case that merits some special discussion is the case $n=1$. In that case the gauge group consists just of a single $SU(k)$ (times the decoupled $U(1)$) and the $\mathbf{k}\otimes\overline{\mathbf{k}}$ hypermultiplet consists of a copy of the adjoint representation of $SU(k)$ plus a neutral singlet. This in fact corresponds to the $\mathcal{N}=4$ theory with gauge group $U(k)$; however, we will study it eventually in the presence of a hypermultiplet bare mass that breaks $\mathcal{N}=4$ to $\mathcal{N}=2$. Precisely this model has been solved in [12], and we will recover the description in that paper.

Hypermultiplet bare masses

Before turning to M-theory, we will analyze, in terms of Type IIA, the hypermultiplet bare masses.

Let $a_{i,\alpha}$, $i=1,\ldots,k$ be the v values of the fourbranes between the $(\alpha-1)$th and αth fivebranes. According to (2.9), the bare mass m_α of the $\mathbf{k}_\alpha\otimes\overline{\mathbf{k}}_{\alpha+1}$ hypermultiplet is

$$m_\alpha=\frac{1}{k}\left(\sum_i a_{i,\alpha}-\sum_j a_{j,\alpha+1}\right). \qquad (4.2)$$

This formula seems to imply that the m_α are not all independent, but are restricted by $\sum_\alpha m_\alpha=0$. However, that restriction can be avoided if one chooses correctly the space-time in which the branes propagate.

So far, we have described the positions of the fourbranes and fivebranes in terms of x^6 and $v=x^4+ix^5$. Since we are now compactifying the x^6 direction to a circle, this part of the space-time is so far $T=\mathbf{S}^1\times\mathbf{C}$, where $\mathbf{S}^1$ is the circle parametrized by x^6 and $\mathbf{C}$ is the v plane.

We can however replace $\mathbf{S}^1\times\mathbf{C}$ by a certain $\mathbf{C}$ bundle over $\mathbf{S}^1$. In other words, we begin with x^6 and v regarded as coordinates on $\mathbf{R}^3=\mathbf{R}\times\mathbf{C}$, and instead of dividing simply by $x^6\to x^6+2\pi L$ for some L, we divide by the combined operation

$$x^6\to x^6+2\pi L,\qquad v\to v+m, \qquad (4.3)$$

for an arbitrary complex constant m. Starting with the flat metric on $\mathbf{R}^3$, this gives a $\mathbf{C}$ bundle over $\mathbf{S}^1$ with a flat metric; we call this space T_m. Now when one goes all the way around the x^6 circle, one comes back with a shifted value of v, as suggested in Fig. 6b. The result is that the formula $\sum_\alpha m_\alpha = 0$ which one would get on $\mathbf{R} \times \mathbf{C}$ is replaced on T_m by

$$\sum_\alpha m_\alpha = m. \tag{4.4}$$

Thus arbitrary hypermultiplet bare masses are possible, with a judicious choice of the space-time.

4.2. *Interpretation in M-theory*

Now we want to study these models via M-theory.

Going to M-theory means first of all including another circle, parametrized by a variable x^{10} with $x^{10} \cong x^{10} + 2\pi R$. Now because in the present section we are compactifying also the x^6 direction to a circle, we have really two circles. The metric structure, however, need not be a simple product $\mathbf{S}^1 \times \mathbf{S}^1$. Dividing $x^6 \to x^6 + 2\pi L$ can be accompanied by a shift of x^{10}, the combined operation being

$$x^6 \to x^6 + 2\pi L, \qquad x^{10} \to x^{10} + \theta R \tag{4.5}$$

with some angle θ. We also still divide by $x^{10} \to x^{10} + 2\pi R$, as in uncompactified Type IIA. In the familiar complex structure in which $s = x^6 + ix^{10}$ is holomorphic, the quotient of the s plane by these equivalences is a complex Riemann surface E of genus-one which – by varying L and θ for fixed R (that is fixed ten-dimensional Type IIA string coupling constant) – can have an arbitrary complex structure. E also has a flat metric with an area that (if we let R vary) is arbitrary; this, however, will be less important, since we are mainly studying properties that are controlled by the holomorphic data.

The interpretation of this generalization for our problem of gauge theory on branes is as follows. The αth fivebrane has, in the M-theory description, a position x^{10}_α in the x^{10} direction, as well as a position x^6_α in the x^6 direction. The theta angle θ_α of the αth $SU(k)$ factor in the gauge group is

$$\theta_\alpha = \frac{x^{10}_\alpha - x^{10}_{\alpha-1}}{R}. \tag{4.6}$$

If metrically $x^6 - x^{10}$ space were a product $\mathbf{S}^1 \times \mathbf{S}^1$ (or in other words if $\theta = 0$ in (4.5)) then (4.6) would imply that $\sum_\alpha \theta_\alpha = 0$. Instead, via (4.5), we arrange that when one goes around a circle in the x^6 direction, one comes back with a shifted valued of x^{10}; as a result one has

$$\sum_\alpha \theta_\alpha = \theta. \tag{4.7}$$

In a Type IIA description, one would not see the x^{10} coordinate. The fact that x^{10} shifts by θ under $x^6 \to x^6 + 2\pi L$ would be expressed by saying that the holonomy around the x^6 circle of the Ramond–Ramond $U(1)$ gauge field of Type IIA is $e^{i\theta}$. The x^{10} positions of a fivebrane would be coded in the value of a certain scalar field that propagates on the fivebrane.

Duality group

In general, E is a (smooth) genus-one Riemann surface with an arbitrary complex structure, and the fivebranes are at n arbitrary points $p_1, \ldots, p_n$ on E. By varying in an arbitrary fashion the complex structure of E and the choice of the p_α, the bare couplings and theta angles of $G' = \prod_{\alpha=1}^{k} SU(k)$ can be varied in arbitrarily. (The coupling and theta angle of the $U(1)$ factor in the full gauge group $G = U(1) \times G'$ is then determined in terms of those.) The duality group of these models can thus be described as follows. Let $\mathcal{M}_{1,n}$ be the moduli space of smooth Riemann surfaces of genus one with n distinct, unordered marked points. The duality group is then $\pi_1(\mathcal{M}_{1,n})$. For $n = 1$, $\pi_1(\mathcal{M}_{1,1})$ is the same as $SL(2, \mathbf{Z})$, and this becomes the usual duality group of $\mathcal{N} = 4$ super Yang–Mills theory. For $n > 1$, $\pi_1(\mathcal{M}_{1,n})$ is a sort of hybrid of $SL(2, \mathbf{Z})$ and the duality group found in Section 3.

Incorporation of v

We now want to consider also the position of the fivebranes in $v = x^4 + ix^5$. An important special case is that in which the fivebranes propagate in $X = E \times \mathbf{C}$, where $\mathbf{C}$ is the complex v plane. However, from the discussion of (4.3), it is clear that in general we should consider not a product $E \times \mathbf{C}$ but a $\mathbf{C}$ bundle over E. In general, we start with $\mathbf{R} \times \mathbf{S}^1 \times \mathbf{C}$ (with respective coordinates x^6, x^{10}, and v) and divide by the combined symmetry

$$\begin{aligned} x^6 &\to x^6 + 2\pi L\,, \\ x^{10} &\to x^{10} + \theta\,, \\ v &\to v + m\,. \end{aligned} \tag{4.8}$$

The quotient is a complex manifold that we will call X_m; it can be regarded as a $\mathbf{C}$ bundle over E. From the discussion at the Type IIA level, it is clear that the parameter m must be identified with the sum of the hypermultiplet bare masses.

The complex manifold X_m will actually not enter as an abstract complex manifold;[8] the map $X_m \to E$ (by forgetting $\mathbf{C}$) will be an important part of the structure. As a $\mathbf{C}$ bundle over E, X_m is an "affine bundle"; this means that the fibers are all copies of $\mathbf{C}$ but there is no way to globally define an "origin" in $\mathbf{C}$, in a fashion that varies holomorphically. Such affine bundles over E, with the associated complex line bundle (in which one ignores shifts of the fibers) being trivial, are classified by the sheaf cohomology group $H^1(E, \mathcal{O}_E)$, which is one dimensional; the one complex parameter that enters is what we have called m. If X_m is viewed just as a complex manifold with

[8] As such it is isomorphic to $\mathbf{C}^* \times \mathbf{C}^*$.

map to E, m could be set to 1 (given that it is non-zero) by rescaling v, but we prefer not to do that since the fivebrane effective action is not invariant under rescaling of v.

The complex manifold X_m appeared in [12], where the $SU(k)$ theory with massive adjoint hypermultiplet – in other words, the $n = 1$ case of the series of models considered here – was described in terms of an appropriate curve in X_m, rather as we will do below. Actually, in what follows we will consider curves in X_m that "go to infinity" at certain points, corresponding to the positions of fivebranes. In [12], a "twist" of X_m was made to keep the curve from going to infinity.

4.3. Solution of the models

What remains is to describe the solution of the models. First we consider the special case that the sum of the hypermultiplet bare masses is zero,

$$\sum_{\alpha} m_\alpha = 0, \tag{4.9}$$

so that the model will be described by a curve Σ in $X = E \times \mathbf{C}$. There are n fivebranes at points $p_1, p_2, \ldots, p_n$ in E; and to use a classical Type IIA language (which we will presently reformulate in a way more suitable in M-theory) each pair of adjacent fivebranes is connected by k fourbranes.

First of all, the elliptic curve E can be described by a Weierstrass equation, $zy^2 = 4x^3 - g_2xz^2 - g_3z^3$ in homogeneous coordinates x, y, z; g_2 and g_3 are complex constants. Usually we work in coordinates with $z = 1$ and write simply

$$y^2 = 4x^3 - g_2x - g_3. \tag{4.10}$$

E admits an everywhere non-zero holomorphic differential

$$\omega = \frac{dx}{y}. \tag{4.11}$$

To incorporate the classical idea that there are k fourbranes between each pair of fivebranes, we proceed as follows. X maps to E by forgetting $\mathbf{C}$; under this map, the curve $\Sigma \subset X$ maps to E. Via the map $\Sigma \to E$, Σ can be interpreted as a k-fold cover of E, the k branches being the positions of the fourbranes in $\mathbf{C}$. In other words, Σ is defined by an equation $F(x, y, v) = 0$, where F is of degree k in v:

$$F(x, y, v) = v^k - f_1(x, y)v^{k-1} + f_2(x, y)v^{n-2} \mp \ldots + (-1)^k f_k(x, y). \tag{4.12}$$

The functions $f_i(x, y)$ are meromorphic functions on E (and hence are rational functions of x and y) obeying certain additional conditions that will be described.

The idea here is that for generic x and y, the equation $F(x, y, v)$ has k roots for v, which are the positions of the fourbranes in the v plane. Call those roots $v_i(x, y)$. Unless the f_i are all constants, there will be points on E at which some of the f_i have poles. At such a point, at least one of the $v_i(x, y)$ diverges.

We would like to interpret the poles in terms of positions of fivebranes. Let us first explain why such an interpretation exists. An M-theory fivebrane located at $v = v_0$ would be interpreted in Type IIA as a fourbrane at $v = v_0$. A Type IIA fivebrane located at some point $p \in E$ also corresponds to a fivebrane in Type IIA. The equation for such a fivebrane is, say, $s = s_0$ where s is a local coordinate on E near p and $s = s_0$ at p. The combined Type IIA fourbrane–fivebrane system can be described in M-theory by a fivebrane with the world-volume

$$(v - v_0)(s - s_0) = 0. \tag{4.13}$$

The space of solutions of this equation has two branches, $v = v_0$ and $s = s_0$; these are interpreted in Type IIA as the fourbrane and fivebrane, respectively. There is a singularity where the two branches meet. Now without changing the asymptotic behavior of the curve described in (4.13) – in fact, while changing only the microscopic details – one could add a constant to the equation, getting

$$(v - v_0)(s - s_0) = \epsilon. \tag{4.14}$$

The singularity has disappeared; what in Type IIA is a fourbrane and a fivebrane appears in this description as a single, smooth, irreducible object. On the other hand, if we solve (4.14) for v we get

$$v = v_0 + \frac{\epsilon}{s - s_0}. \tag{4.15}$$

We see that a fivebrane corresponds to a simple (first order) pole in v.

Poles of the f_i will lead to singularities of the v_i. It is now possible to determine what kind of singularities we should allow in the f_i. At a point p_α at which a fivebrane is located, one of the v_i should have a simple pole, analogous to that in (4.15), and the others should be regular. The v_i will behave in this way if and only if the f_i have simple poles at p_α. So *the functions* $f_1, \ldots, f_k$ *have simple poles at the points* $p_1, \ldots, p_n$ *and no other singularities.*

This then almost completes the description of the solution of the models: they are described by curves $F(x, y, v) = 0$ in $E \times \mathbf{C}$, where F is as in (4.12) and the allowed functions f_i are characterized by the property just stated. What remains is to determine which parameters in the f_i are hypermultiplet bare masses and which ones are order parameters describing the choice of a quantum vacuum.

First let us count all parameters. By the Riemann–Roch theorem, the space of meromorphic functions on E with simple poles allowed at $p_1, \ldots, p_n$ is n dimensional. As we have k such functions, there are kn parameters in all. Of these, $n - 1$ should be hypermultiplet bare masses (because of (4.9) there are only $n - 1$ hypermultiplet bare masses), leaving $n(k - 1) + 1$ order parameters. The gauge group $G = U(1) \times SU(k)^n$ has rank $n(k - 1) + 1$, so $n(k - 1) + 1$ is the dimension of the Coulomb branch, and hence is the correct number of order parameters. It remains then to determine which $n - 1$ parameters are the hypermultiplet bare masses.

Let us note the following interpretation of the function f_1: in view of the factorization $F(x, y, v) = \prod_{i=1}^{k}(v - v_i(x, y))$, one has

$$f_1(x, y) = \sum_{i=1}^{k} v_i(x, y). \tag{4.16}$$

The generic behavior is that near any one of the p_σ, all of the v_i except one remain finite, and the remaining one, say $v_1(x, y)$, has a simple pole. So according to (4.16) the singular behavior of v_1 is the same as the singular behavior of f_1. In other words, the singular part of f_1 determines the behavior of Σ near infinity. Since hypermultiplet bare masses are always coded in the behavior of the curve Σ at infinity – as we saw in (2.5), that is why the bare masses are constant – the hypermultiplet bare masses must be coded in the singular part of f_1.

The singular part of f_1 depends only on $n - 1$ complex parameters. In fact, f_1 itself depends on n complex parameters, but as one is free to add a constant to f_1 without affecting its singular behavior, the singular part of f_1 depends on $n - 1$ parameters. Thus, fixing the hypermultiplet bare masses completely fixes the singular part of f_1. The additive constant in f_1 and the parameters in f_j, $j > 1$ are the order parameters specifying a choice of quantum vacuum. Actually, the additive constant in f_1 is the order parameter on the Coulomb branch of the $U(1)$ factor in the gauge group; this constant can be shifted by adding a constant to v and so does not affect the Jacobian of Σ, in agreement with the fact that the $U(1)$ is decoupled. The order parameters of the $SU(k)^n$ theory are the $n(k-1)$ coefficients in $f_2, f_3, \ldots, f_n$.

To be more complete, one would like to know which functions of the singular part of f_1 are the hypermultiplet bare masses m_α. One approach to this question is to think about the integrable system that controls the structure of the Coulomb branch. We recall from Section 2.3 that a point in the phase space of this integrable system is given by the choice of a curve $\Sigma \subset E \times \mathbf{C}$ with fixed behavior at infinity together with the choice of a line bundle on the compactification of Σ. As in Section 17 of the second paper in [4], the cohomology class of the complex symplectic form on the phase space should vary linearly with the masses. How to implement this condition for integrable systems of the kind considered here is explained in Section 2 of [12]. The result is as follows: the hypermultiplet bare masses are the residues of the differential form $\beta = f_1(x, y)\omega$. Since the sum of the residues of a meromorphic differential form vanishes, this claim is in accord with (4.9).

4.4. *Extension to arbitrary masses*

What remains is to eliminate the restriction (4.9) and solve the models with arbitrary hypermultiplet bare masses. For this, as we have discussed in Section 4.2, it is necessary to consider curves Σ not in $X = E \times \mathbf{C}$, but in an affine bundle over E that we have called X_m.

X_m differs from the trivial product bundle $X = E \times \mathbf{C} \to E$ by twisting by an element of $H^1(E, \mathcal{O}_E)$. That cohomology group vanishes if a point is deleted from E. We can pick that point to be the point p_∞ with $x = y = \infty$ in the Weierstrass model (4.10). To preserve the symmetry among the points p_σ at which there are fivebranes, we take p_∞ to be distinct from all of the p_σ. Because X_m coincides with X away from the fiber over p_∞, we can describe the curve Σ away from p_∞ by the same equation as before, $F(x, y, v) = 0$ with

$$F(x, y, v) = v^k - f_1(x, y)v^{k-1} + f_2(x, y)v^{k-2} \mp \ldots + (-1)^k f_k(x, y). \qquad (4.17)$$

Away from $x = y = \infty$, the functions $f_i(x, y)$ are subject to the same conditions as before – no singularities except simple poles at the points p_σ.

Previously, we required that the roots $v_i(x, y)$ were finite at $x = y = \infty$ (since there are no fivebranes there) and hence that the f_i were finite at $x = y = \infty$. For describing a curve on X_m, that is not the right condition. The trivialization of the affine bundle X_m over E minus the point at infinity breaks down at $x = y = \infty$. A good coordinate near infinity is not v but

$$\widetilde{v} = v + \left(\frac{m}{2k}\right)\frac{y}{x}. \qquad (4.18)$$

(Instead of y/x one could use any other function with a simple pole at $x = y = \infty$. For the moment one should think of the $m/2k$ on the right-hand side (4.18) as an arbitrary constant.) It is not v but $\widetilde{v}$ that should be finite at $x = y = \infty$.

Thus the restrictions on the f_i that are needed to solve the model with arbitrary hypermultiplet bare masses can be stated as follows:

(1) The functions $f_i(x, y)$ are meromorphic functions on E with no singularities except simple poles at the p_σ, $\sigma = 1, \ldots, n$, and poles (of order i) at $x = y = \infty$.

(2) The singular part of the function $F(v, x, y)$ near $x = y = \infty$ disappears if this function is expressed in terms of $\widetilde{v}$ instead of v.

The hypermultiplet bare masses m_α are the residues of the differential form $\beta = f_1\omega$ at the points p_σ. Since the sum of the residues of β will vanish, β has a pole at $x = y = \infty$ with residue $-\sum_\alpha m_\alpha$. We can now relate this expression to the parameter m in (4.18). Since condition (2) above implies that the singular behavior of f_1 is $f_1 = -my/2x + \ldots$, and since the differential form $(dx/y)(y/2x)$ has a pole at infinity with residue 1, the residue of β is in fact $-m$, so we get

$$m = \sum_\alpha m_\alpha. \qquad (4.19)$$

This relation between the coefficient m by which X_m is twisted and the hypermultiplet bare masses m_α was anticipated in (4.4).

Just as in the case $m = 0$ that we considered first, the order parameters on the Coulomb branch are the parameters not fixed by specifying the singular part of f_1.

In [12], the solution of this model for the special case $n = 1$ was expressed in an equivalent but slightly different way. Since – to adapt the discussion to the present language – there was only one fivebrane, the fivebrane was placed at p_∞ without any loss of symmetry. In place of conditions (1) and (2), the requirements on the f_i were the following:

(1′) The functions $f_i(x,y)$ are meromorphic functions on E with no singularities except a pole of order at most i at $x = y = \infty$.

(2′) After the change of variables (4.18), the singularity of the function $F(x,y,v)$ at $x = y = \infty$ is only a simple pole.

These conditions were used as the starting point for fairly detailed calculations of the properties of the model.

For the general case of n fivebranes, if we choose one of the fivebrane locations, say p_1, to equal p_∞, then (1) and (2) can be replaced by the following conditions:

(1″) The functions $f_i(x,y)$ are meromorphic functions on E whose possible singularities are simple poles at $p_2,\ldots,p_n$ and a pole of order i at $x = y = \infty$.

(2″) After the change of variables (4.18), the singularity of the function $F(x,y,t)$ at $x = y = \infty$ is only a simple pole.

These conditions are equivalent to (1) and (2), up to a translation on E that moves p_1 to infinity and a change of variables $v \to v + a(x,y)$ for some function a.

References

[1] A. Hanany and E. Witten, Type IIB Superstrings, BPS Monopoles, and Three-Dimensional Gauge Dynamics, hep-th/9611230, Nucl. Phys. B 492 (1997) 152.

[2] Jan de Boer, Kentaro Hori, Hirosi Ooguri, Yaron Oz and Zheng Yin, Mirror Symmetry in Three-Dimensional Gauge Theories, $SL(2,Z)$ and D-Brane Moduli Spaces, hep-th/9612131, Nucl. Phys. B 493 (1997) 148;
Jan de Boer, Kentaro Hori, Yaron Oz and Zheng Yin, Branes and Mirror Symmetry in $N = 2$ Supersymmetric Gauge Theories in Three Dimensions, hep-th/9702154, Nucl. Phys. B (1997), to be published.

[3] S. Elitzur, A. Giveon and D. Kutasov, Branes and $N = 1$ Duality in String Theory, hep-th/9702014, Phys. Lett. B 400 (1997) 269.

[4] N. Seiberg and E. Witten, Electric-Magnetic Duality, Monopole Condensation, and Confinement in $N = 2$ Supersymmetric Yang–Mills Theory, Nucl. Phys. B 426 (1994) 19; Monopoles, Duality, and Chiral Symmetry Breaking in $N = 2$ Supersymmetric QCD, Nucl. Phys. B 341 (1994) 484.

[5] A. Strominger, Open p-Branes, hep-th/9512059, Phys. Lett. B 383 (1996) 44.

[6] K. Becker, M. Becker and A. Strominger, Fivebranes, Membranes, and Non-perturbative String Theory, Nucl. Phys. B 456 (1995) 130.

[7] A. Klemm, W. Lerche, P. Mayr, C. Vafa and N. Warner, Self-Dual Strings and $N = 2$ Supersymmetric Field Theory, hep-th/9604034, Nucl. Phys. B 477 (1996) 746.

[8] E. Verlinde, Global Aspects of Electric-Magnetic Duality, hep-th/9506011, Nucl. Phys. B 455 (1995) 211.

[9] A. Gorsky, I. Krichever, A. Marshakov, A. Mironov and A. Morozov, Integrability and Seiberg–Witten Exact Solution, hep-th/9505035, Nucl. Phys. B 459 (1996) 97.

[10] E. Martinec and N. Warner, Integrable Systems and Supersymmetric Gauge Theory, hep-th/9509161, Nucl. Phys. B 459 (1996) 97.

[11] T. Nakatsu and K. Takasaki, Whitham–Toda Hierarchy and $N = 2$ Supersymmetric Yang–Mills Theory, hep-th/9509162, Mod. Phys. Lett. A 11 (1996) 157.

[12] R. Donagi and E. Witten, Supersymmetric Yang–Mills Theory and Integrable Systems, hep-th/9510101, Nucl. Phys. B 460 (1996) 299.

[13] R. Donagi, L. Ein and R. Lazarsfeld, A Non-Linear Deformation of the Hitchin Dynamical System, alg-geom/9504017.

[14] N. Hitchin, The Self-Duality Equations on a Riemann Surface, Proc. London Math. Soc. 55 (1987) 59; Stable Bundles and Integrable Systems, Duke Math. J. 54 (1987) 91.

[15] E. Markman, Spectral Curves and Integrable Systems, Comp. Math. 93 (1994) 255.

[16] M. Bershadsky, V. Sadov and C. Vafa, D-Branes and Topological Field Theories, hep-th/9511222, Nucl. Phys. B 463 (1996) 420.

[17] P. Townsend, *D*-Branes from M-Branes, hep-th/9512062, Phys. Lett. B 373 (1996) 68.

[18] P. Argyres and A. Faraggi, The Vacuum Structure and Spectrum of $N = 2$ Supersymmetric $SU(N)$ Gauge Theory, hep-th/9411057, Phys. Rev. Lett. 74 (1995) 3931.

[19] S. Klemm, W. Lerche, S. Theisen and S. Yankielowicz, Simple Singularities and $N = 2$ Supersymmetric Yang–Mills Theory, hep-th/9411048, Phys. Lett. B 344 (1995) 169.

[20] P.C. Argyres, M.R. Plesser and A.D. Shapere, The Coulomb Phase of $N = 2$ Supersymmetric QCD, hep-th/9505100, Phys. Rev. Lett. 75 (1995) 1699.

[21] A. Hanany and Y. Oz, On the Quantum Moduli Space of Vacua of $N = 2$ Supersymmetric $SU(N_C)$ Gauge Theories, hep-th/9505075, Nucl. Phys. B 452 (1995) 283.

[22] P. Argyres and M. Douglas, New Phenomena in $SU(3)$ Supersymmetric Gauge Theory, hep-th/9505062, Nucl. Phys. B 448 (1995) 93.

[23] P. Townsend, The Eleven-Dimensional Supermembrane Revisited, Phys. Lett. B 350 (1995) 184.

[24] G.W. Gibbons and P. Rychenkova, HyperKähler Quotient Construction of BPS Monopole Moduli Spaces, hep-th/9608085.

[25] S. Hawking, Gravitational Instantons, Phys. Lett. A 60 (1977) 81.

Advances in Theoretical and Mathematical Physics **2** (1998) 231–52
Reprinted with permission by International Press, Cambridge, MA, USA

The Large N Limit of Superconformal field theories and supergravity

Juan Maldacena[1]

Lyman Laboratory of Physics
Harvard University
Cambridge, MA 02138, U.S.A.

Abstract

We show that the large N limit of certain conformal field theories in various dimensions include in their Hilbert space a sector describing supergravity on the product of Anti-deSitter spacetimes, spheres and other compact manifolds. This is shown by taking some branes in the full M/string theory and then taking a low energy limit where the field theory on the brane decouples from the bulk. We observe that, in this limit, we can still trust the near horizon geometry for large N. The enhanced supersymmetries of the near horizon geometry correspond to the extra supersymmetry generators present in the superconformal group (as opposed to just the super-Poincare group). The 't Hooft limit of 3+1 $\mathcal{N} = 4$ super-Yang-Mills at the conformal point is shown to contain strings: they are IIB strings. We conjecture that compactifications of M/string theory on various Anti-deSitter spacetimes is dual to various conformal field theories. This leads to a new proposal for a definition of M-theory which could be extended to include five non-compact dimensions.

1 General Idea

In the last few years it has been extremely fruitful to derive quantum field theories by taking various limits of string or M-theory. In some cases this

[1] malda@pauli.harvard.edu

is done by considering the theory at geometric singularities and in others by considering a configuration containing branes and then taking a limit where the dynamics on the brane decouples from the bulk. In this paper we consider theories that are obtained by decoupling theories on branes from gravity. We focus on conformal invariant field theories but a similar analysis could be done for non-conformal field theories. The cases considered include N parallel D3 branes in IIB string theory and various others. We take the limit where the field theory on the brane decouples from the bulk. At the same time we look at the near horizon geometry and we argue that the supergravity solution can be trusted as long as N is large. N is kept fixed as we take the limit. The approach is similar to that used in [1] to study the NS fivebrane theory [2] at finite temperature. The supergravity solution typically reduces to $p+2$ dimensional Anti-deSitter space (AdS_{p+2}) times spheres (for D3 branes we have $AdS_5 \times S^5$). The curvature of the sphere and the AdS space in Planck units is a (positive) power of $1/N$. Therefore the solutions can be trusted as long as N is large. Finite temperature configurations in the decoupled field theory correspond to black hole configurations in AdS spacetimes. These black holes will Hawking radiate into the AdS spacetime. We conclude that excitations of the AdS spacetime are included in the Hilbert space of the corresponding conformal field theories. A theory in AdS spacetime is not completely well defined since there is a horizon and it is also necessary to give some boundary conditions at infinity. However, local properties and local processes can be calculated in supergravity when N is large if the proper energies involved are much bigger than the energy scale set by the cosmological constant (and smaller than the Planck scale). We will conjecture that the full quantum M/string-theory on AdS space, plus suitable boundary conditions is dual to the corresponding brane theory. We are not going to specify the boundary conditions in AdS, we leave this interesting problem for the future. The $AdS \times$(spheres) description will become useful for large N, where we can isolate some local processes from the question of boundary conditions. The supersymmetries of both theories agree, both are given by the superconformal group. The superconformal group has twice the amount of supersymmetries of the corresponding super-Poincare group [3,4]. This enhancement of supersymmetry near the horizon of extremal black holes was observed in [5,6] precisely by showing that the near throat geometry reduces to $AdS \times$(spheres). AdS spaces (and branes in them) were extensively considered in the literature [7–13], including the connection with the superconformal group.

In section 2 we study $\mathcal{N} = 4$ d=4 $U(N)$ super-Yang-Mills as a first example, we discuss several issues which are present in all other cases. In section 3 we analyze the theories describing M-theory five-branes and M-theory two-branes. In section 4 we consider theories with lower supersymmetry which are related to a black string in six dimensions made with D1 and D5 branes. In section 5 we study theories with even less supersymmetry involving black strings in five dimensions and finally we mention the theories related to extremal Reissner-Nordström black holes in four spacetime dimensions (these last cases will be more speculative and contain some unresolved puzzles). Finally in section 6 we make some comments on the relation to matrix theory.

2 D3 Branes or $\mathcal{N} = 4$ $U(N)$ Super-Yang-Mills in $d = 3 + 1$

We start with type IIB string theory with string coupling g, which will remain fixed. Consider N parallel D3 branes separated by some distances which we denote by r. For low energies the theory on the D3 brane decouples from the bulk. It is more convenient to take the energies fixed and take

$$\alpha' \to 0 \,, \qquad U \equiv \frac{r}{\alpha'} = \text{fixed} \,. \tag{2.1}$$

The second condition is saying that we keep the mass of the stretched strings fixed. As we take the decoupling limit we bring the branes together but the the Higgs expectation values corresponding to this separation remains fixed. The resulting theory on the brane is four dimensional $\mathcal{N} = 4$ $U(N)$ SYM. Let us consider the theory at the superconformal point, where $r = 0$. The conformal group is SO(2,4). We also have an $SO(6) \sim SU(4)$ R-symmetry that rotates the six scalar fields into each other[2]. The superconformal group includes twice the number of supersymmetries of the super-Poincare group: the commutator of special conformal transformations with Poincare supersymmetry generators gives the new supersymmetry generators. The precise superconformal algebra was computed in [3]. All this is valid for any N.

Now we consider the supergravity solution carrying D3 brane charge [14]

$$\begin{aligned} ds^2 &= f^{-1/2} dx_{\|}^2 + f^{1/2}(dr^2 + r^2 d\Omega_5^2) \,, \\ f &= 1 + \frac{4\pi g N \alpha'^2}{r^4} \,, \end{aligned} \tag{2.2}$$

where $x_{\|}$ denotes the four coordinates along the worldvolume of the three-brane and $d\Omega_5^2$ is the metric on the unit five-sphere[3]. The self dual five-form field strength is nonzero and has a flux on the five-sphere. Now we define the new variable $U \equiv \frac{r}{\alpha'}$ and we rewrite the metric in terms of U. Then we

[2] The representation includes objects in the spinor representations, so we should be talking about SU(4), we will not make this, or similar distinctions in what follows.

[3] We choose conventions where $g \to 1/g$ under S-duality.

take the $\alpha' \to 0$ limit. Notice that U remains fixed. In this limit we can neglect the 1 in the harmonic function (2.2). The metric becomes

$$ds^2 = \alpha' \left[\frac{U^2}{\sqrt{4\pi g N}} dx_{||}^2 + \sqrt{4\pi g N} \frac{dU^2}{U^2} + \sqrt{4\pi g N} d\Omega_5^2 \right] . \tag{2.3}$$

This metric describes five dimensional Anti-deSitter (AdS_5) times a five-sphere[4]. We see that there is an overall α' factor. The metric remains constant in α' units. The radius of the five-sphere is $R_{sph}^2/\alpha' = \sqrt{4\pi g N}$, and is the same as the "radius" of AdS_5 (as defined in the appendix). In ten dimensional Planck units they are both proportional to $N^{1/4}$. The radius is quantized because the flux of the 5-form field strength on the 5 sphere is quantized. We can trust the supergravity solution when

$$gN \gg 1 . \tag{2.4}$$

When N is large we have approximately ten dimensional flat space in the neighborhood of any point[5]. Note that in the large N limit the flux of the 5 form field strength per unit Planck (or string) 5-volume becomes small.

Now consider a near extremal black D3 brane solution in the decoupling limit (2.1). We keep the energy density on the brane worldvolume theory (μ) fixed. We find the metric

$$\begin{aligned} ds^2 &= \alpha' \left\{ \frac{U^2}{\sqrt{4\pi g N}} \left[-(1 - U_0^4/U^4) dt^2 + dx_i^2 \right] \right. \\ &\left. + \sqrt{4\pi g N} \frac{dU^2}{U^2(1 - U_0^4/U^4)} + \sqrt{4\pi g N} d\Omega_5^2 \right\} . \\ U_0^4 &= \frac{2^7}{3} \pi^4 g^2 \mu \end{aligned} \tag{2.5}$$

We see that U_0 remains finite when we take the $\alpha' \to 0$ limit. The situation is similar to that encountered in [1]. Naively the whole metric is becoming of zero size since we have a power of α' in front of the metric, and we might incorrectly conclude that we should only consider the zero modes of all fields. However, energies that are finite from the point of view of the gauge theory, lead to proper energies (measured with respect to proper time) that remain finite is in α' units (or Planck units, since g is fixed). More concretely, an excitation that has energy ω (fixed in the limit) from the point of view of the gauge theory, will have proper energy $E_{proper} = \frac{1}{\sqrt{\alpha'}} \frac{\omega (gN4\pi)^{1/4}}{U}$. This also means that the corresponding proper wavelengths remain fixed. In

[4] See the appendix for a brief description of AdS spacetimes.

[5] In writing (2.4) we assumed that $g \leq 1$, if $g > 1$ then the condition is $N/g \gg 1$. In other words we need large N, not large g.

other words, the spacetime action on this background has the form $S \sim \frac{1}{\alpha'^4}\int d^{10}x\sqrt{G}R + \cdots$, so we can cancel the factor of α' in the metric and the Newton constant, leaving a theory with a finite Planck length in the limit. Therefore we should consider fields that propagate on the AdS background. Since the Hawking temperature is finite, there is a flux of energy from the black hole to the AdS spacetime. Since $\mathcal{N} = 4$ d=4 $U(N)$ SYM is a unitary theory we conclude that, for large N, *it includes in its Hilbert space the states of type IIB supergravity on* $(AdS_5 \times S_5)_N$, where subscript indicates the fact that the "radii" in Planck units are proportional to $N^{1/4}$. In particular the theory contains gravitons propagating on $(AdS_5 \times S_5)_N$. When we consider supergravity on $AdS_5 \times S_5$, we are faced with global issues like the presence of a horizon and the boundary conditions at infinity. It is interesting to note that the solution is nonsingular [15]. The gauge theory should provide us with a specific choice of boundary conditions. It would be interesting to determine them.

We have started with a quantum theory and we have seen that it includes gravity so it is natural to think that this correspondence goes beyond the supergravity approximation. We are led to the conjecture that *Type IIB string theory on* $(AdS_5 \times S^5)_N$ *plus some appropriate boundary conditions (and possibly also some boundary degrees of freedom) is dual to N =4 d=3+1 U(N) super-Yang-Mills.* The SYM coupling is given by the (complex) IIB string coupling, more precisely $\frac{1}{g_{YM}^2} + i\frac{\theta}{8\pi^2} = \frac{1}{2\pi}(\frac{1}{g} + i\frac{\chi}{2\pi})$ where χ is the value of the RR scalar.

The supersymmetry group of $AdS_5 \times S^5$, is known to be the same as the superconformal group in 3+1 spacetime dimensions [3], so the supersymmetries of both theories are the same. This is a new form of "duality": a large N field theory is related to a string theory on some background, notice that the correspondence is non-perturbative in g and the $SL(2, Z)$ symmetry of type IIB would follow as a consequence of the $SL(2, Z)$ symmetry of SYM[6]. It is also a strong-weak coupling correspondence in the following sense. When the effective coupling gN becomes large we cannot trust perturbative calculations in the Yang-Mills theory but we can trust calculations in supergravity on $(AdS_5 \times S^5)_N$. This is suggesting that the $\mathcal{N} = 4$ Yang-Mills master field is the anti-deSitter supergravity solution (similar ideas were suggested in [17]). Since N measures the size of the geometry in Planck units, we see that quantum effects in $AdS_5 \times S^5$ have the interpretation of $1/N$ effects in the gauge theory. So Hawking radiation is a $1/N$ effect. It would be interesting to understand more precisely what the horizon means from the gauge theory point of view. IIB supergravity on $AdS_5 \times S^5$ was studied in [7, 9].

The above conjecture becomes nontrivial for large N and gives a way to answer some large N questions in the SYM theory. For example, suppose

[6]This is similar in spirit to [16] but here N is not interpreted as momentum.

that we break $U(N) \to U(N-1) \times U(1)$ by Higgsing. This corresponds to putting a three brane at some point on the 5-sphere and some value of U, with world volume directions along the original four dimensions ($x_{||}$). We could now ask what the low energy effective action for the light U(1) fields is. For large N (2.4) it is the action of a D3 brane in $AdS_5 \times S^5$. More concretely, the bosonic part of the action becomes the Born-Infeld action on the AdS background

$$
\begin{aligned}
S &= -\frac{1}{(2\pi)^3 g} \int d^4x\, h^{-1} \\
&\quad \cdot \left[\sqrt{-\operatorname{Det}(\eta_{\alpha\beta} + h\partial_\alpha U \partial_\beta U + U^2 h g_{ij} \partial_\alpha \theta^i \partial_\beta \theta^j + 2\pi\sqrt{h} F_{\alpha\beta})} - 1\right] \\
h &= \frac{4\pi g N}{U^4} ,
\end{aligned}
\tag{2.6}
$$

with $\alpha, \beta = 0,1,2,3$, $i,j = 1,..,5$; and g_{ij} is the metric of the unit five-sphere. As any low energy action, (2.6) is valid when the energies are low compared to the mass of the massive states that we are integrating out. In this case the mass of the massive states is proportional to U (with no factors of N). The low energy condition translates into $\partial U/U \ll U$ and $\partial \theta^i << U$, etc.. So the nonlinear terms in the action (2.6) will be important only when gN is large. It seems that the form of this action is completely determined by superconformal invariance, by using the broken and unbroken supersymmetries, in the same sense that the Born Infeld action in flat space is given by the full Poincare supersymmetry [18]. It would be very interesting to check this explicitly. We will show this for a particular term in the action. We set $\theta^i =$ const and $F = 0$, so that we only have U left. Then we will show that the action is completely determined by broken conformal invariance. This can be seen as follows. Using Lorentz invariance and scaling symmetry (dimensional analysis) one can show that the action must have the form

$$
S = \int d^{p+1}x U^{p+1} f(\partial_\alpha U \partial^\alpha U / U^4) , \tag{2.7}
$$

where f is an arbitrary function. Now we consider infinitesimal special conformal transformations

$$
\begin{aligned}
\delta x^\alpha &= \epsilon^\beta x_\beta x^\alpha - \epsilon^\alpha (x^2 + \frac{\tilde{R}^4}{U^2})/2 , \\
\delta U &\equiv U'(x') - U(x) = -\epsilon^\alpha x_\alpha U ,
\end{aligned}
\tag{2.8}
$$

where ϵ^α is an infinitesimal parameter. For the moment $\tilde{R}$ is an arbitrary constant. We will later identify it with the "radius" of AdS, it will turn out that $\tilde{R}^4 \sim gN$. In the limit of small $\tilde{R}$ we recover the more familiar form of the conformal transformations (U is a weight one field). Usually conformal transformations do not involve the variable U in the transformations of x. For constant U the extra term in (2.8) is a translation in x, but we will take U to be a slowly varying function of x and we will determine $\tilde{R}$ from other facts that we know. Demanding that (2.7) is invariant under (2.8) we find that the function f in (2.7) obeys the equation

$$f(z) + \text{const} = 2\left(z + \frac{1}{\tilde{R}^4}\right) f'(z) \tag{2.9}$$

which is solved by $f = b[\sqrt{1 + \tilde{R}^4 z} - a]$. Now we can determine the constants $a, b, \tilde{R}$ from supersymmetry. We need to use three facts. The first is that there is no force (no vacuum energy) for a constant U. This implies $a = 1$. The second is that the ∂U^2 term (F^2 term) in the $U(1)$ action is not renormalized. The third is that the only contribution to the $(\partial U)^4$ term (an F^4 term) comes from a one loop diagram [19]. This determines all the coefficients to be those expected from (2.6) including the fact that $\tilde{R}^4 = 4\pi gN$. It seems very plausible that using all 32 supersymmetries we could fix the action (2.6) completely. This would be saying that (2.6) is a consequence of the symmetries and thus not a prediction[7]. However we can make very nontrivial predictions (though we were not able to check them). For example, if we take g to be small (but N large) we can predict that the Yang-Mills theory contains strings. More precisely, in the limit $g \to 0$, $gN = \text{fixed} \gg 1$ ('t Hooft limit) we find free strings in the spectrum, they are IIB strings moving in $(AdS_5 \times S^5)_{gN}$.[8] The sense in which these strings are present is rather subtle since there is no energy scale in the Yang-Mills to set their tension. In fact one should translate the mass of a string state from the *AdS* description to the Yang-Mills description. This translation will involve the position U at which the string is sitting. This sets the scale for its mass. As an example, consider again the D-brane probe (Higgsed configuration) which we described above. From the type IIB description we expect open strings ending on the D3 brane probe. From the point of view of the gauge theory these open strings have energies $E = \frac{U}{(4\pi gN)^{1/4}}\sqrt{N_{open}}$ where N_{open} is the integer charaterizing the massive open string level. In this example we

[7] Notice that the action (2.6) includes a term proportional to v^6 similar to that calculated in [20]. Conformal symmetry explains the agreement that they would have found if they had done the calculation for 3+1 SYM as opposed to 0+1.

[8] In fact, Polyakov [21] recently proposed that the string theory describing bosonic Yang-Mills has a new dimension corresponding to the Liouville mode φ, and that the metric at $\varphi = 0$ is zero due to a "zig-zag" symmetry. In our case we see that the physical distances along the directions of the brane contract to zero as $U \to 0$. The details are different, since we are considering the $\mathcal{N} = 4$ theory.

see that α' disappears when we translate the energies and is replaced by U, which is the energy scale of the Higgs field that is breaking the symmetry.

Now we turn to the question of the physical interpretation of U. U has dimensions of mass. It seems natural to interpret motion in U as moving in energy scales, going to the IR for small U and to the UV for large U. For example, consider a D3 brane sitting at some position U. Due to the conformal symmetry, all physics at energy scales ω in this theory is the same as physics at energies $\omega' = \lambda\omega$, with the brane sitting at $U' = \lambda U$.

Now let us turn to another question. We could separate a group of D3 branes from the point were they were all sitting originally. Fortunately, for the extremal case we can find a supergravity solution describing this system. All we have to do is the replacement

$$\frac{N}{U^4} \to \frac{N-M}{U^4} + \frac{M}{|\vec{U}-\vec{W}|^4}, \tag{2.10}$$

where $\vec{W} = \vec{r}/\alpha'$ is the separation. It is a vector because we have to specify a point on S^5 also. The resulting metric is

$$ds^2 = \alpha' \left[U^2 \frac{1}{\sqrt{4\pi g}\left(N - M + \frac{MU^4}{|\vec{U}-W|^4}\right)^{1/2}} dx_{||}^2 \right. \\ \left. + \sqrt{4\pi g}\frac{1}{U^2}\left(N - M + \frac{MU^4}{|\vec{U}-W|^4}\right)^{1/2} d\vec{U}^2 \right] . \tag{2.11}$$

For large $U \gg |W|$ we find basically the solution for $(AdS_5 \times S^5)_N$ which is interpreted as saying that for large energies we do not see the fact that the conformal symmetry was broken, while for small $U \ll |W|$ we find just $(AdS_5 \times S^5)_{N-M}$, which is associated to the CFT of the unbroken $U(N-M)$ piece. Furthermore, if we consider the region $|\vec{U}-\vec{W}| \ll |\vec{W}|$ we find $(AdS_5 \times S^5)_M$, which is described by the CFT of the $U(M)$ piece.

We could in principle separate all the branes from each other. For large values of U we would still have $(AdS_5 \times S^5)_N$, but for small values of U we would not be able to trust the supergravity solution, but we naively get N copies of $(AdS_5 \times S^5)_1$ which should correspond to the $U(1)^N$.

Now we discuss the issue of compactification. We want to consider the YM theory compactified on a torus of radii R_i, $x_i \sim x_i + 2\pi R_i$, which stay fixed as we take the $\alpha' \to 0$ limit. Compactifying the theory breaks conformal invariance and leaves only the Poincare supersymmetries. However one can still find the supergravity solutions and follow the above procedure, going near the horizon, etc. The *AdS* piece will contain some identifications. So we will be able to trust the supergravity solution as long as the physical length of these compact circles stays big in α' units. This implies that we can trust the supergravity solution as long as we stay far from the horizon (at $U = 0$)

$$U \gg \frac{(gN)^{1/4}}{R_i} , \tag{2.12}$$

for all i. This is a larger bound than the naive expectation $(1/R_i)$. If we were considering near extremal black holes we would require that U_0 in (2.5) satisfies (2.12), which is, of course, the same condition on the temperature gotten in [22].

The relation of the three-brane supergravity solution and the Yang-Mills theory has been studied in [23–26]. All the calculations have been done for near extremal D3 branes fall into the category described above. In particular the absorption cross section of the dilaton and the graviton have been shown to agree with what one would calculate in the YM theory [24,25]. It has been shown in [26] that some of these agreements are due to non-renormalization theorems for $\mathcal{N} = 4$ YM. The black hole entropy was compared to the *perturbative* YM calculation and it agrees up to a numerical factor [23]. This is not in disagreement with the correspondence we were suggesting, It is expected that large gN effects change this numerical factor, this problem remains unsolved.

Finally notice that the group $SO(2,4) \times SO(6)$ suggests a twelve dimensional realization in a theory with two times [27].

3 Other Cases with 16 → 32 Supersymmetries, M5 and M2 Brane Theories

Basically all that we have said for the D3 brane carries over for the other conformal field theories describing coincident M-theory fivebranes and M-theory twobranes. We describe below the limits that should be taken in each of the two cases. Similar remarks can be made about the entropies [28], and the determination of the probe actions using superconformal invariance. Eleven dimensional supergravity on the corresponding AdS spaces was studied in [8, 10, 11, 15].

3.1 M5 Brane

The decoupling limit is obtained by taking the 11 dimensional Planck length to zero, $l_p \to 0$, keeping the worldvolume energies fixed and taking the separations $U^2 \equiv r/l_p^3 =$ fixed [29]. This last condition ensures that the membranes stretching between fivebranes give rise to strings with finite tension.

The metric is[9]

$$ds^2 = f^{-1/3} dx_{||}^2 + f^{2/3}(dr^2 + r^2 d\Omega_4^2) \; ,$$
$$f = 1 + \frac{\pi N l_p^3}{r^3} \; , \tag{3.1}$$

We also have a flux of the four-form field strength on the four-sphere (which is quantized). Again, in the limit we obtain

$$ds^2 = l_p^2 [\frac{U^2}{(\pi N)^{1/3}} dx_{||}^2 + 4(\pi N)^{2/3} \frac{dU^2}{U^2} + (\pi N)^{2/3} d\Omega_4^2] \; , \tag{3.2}$$

where now the "radii" of the sphere and the AdS_7 space are $R_{sph} = R_{AdS}/2 = l_p(\pi N)^{1/3}$. Again, the "radii" are fixed in Planck units as we take $l_p \to 0$, and supergravity can be applied if $N \gg 1$.

Reasoning as above we conclude that this theory contains seven dimensional Anti-deSitter times a four-sphere, which for large N looks locally like eleven dimensional Minkowski space.

This gives us a method to calculate properties of the large N limit of the six dimensional (0,2) conformal field theory [30]. The superconformal group again coincides with the algebra of the supersymmetries preserved by $AdS_7 \times S^4$. The bosonic symmetries are $SO(2,6) \times SO(5)$ [4]. We can do brane probe calculations, thermodynamic calculations [28], etc.

The conjecture is now that *the (0,2) conformal field theory is dual to M-theory on* $(AdS_7 \times S^4)_N$, the subindex indicates the dependence of the "radius" with N.

3.2 M2 Brane

We now take the limit $l_p \to 0$ keeping $U^{1/2} \equiv r/l_p^{3/2} =$ fixed. This combination has to remain fixed because the scalar field describing the motion of the twobrane has scaling dimension 1/2. Alternatively we could have derived this conformal field theory by taking first the field theory limit of D2 branes in string theory as in [31–33], and then taking the strong coupling limit of that theory to get to the conformal point as in [34–36]. The fact that the theories obtained in this fashion are the same can be seen as follows. The D2 brane gauge theory can be obtained as the limit $\alpha' \to 0$, keeping $g_{YM}^2 \sim g/\alpha' =$ fixed. This is the same as the limit of M-theory two branes in the limit $l_p \to 0$ with $R_{11}/l_p^{3/2} \sim g_{YM} =$ fixed. This is a theory where

[9]In our conventions the relation of the Planck length to the 11 dimensional Newton constant is $G_N^{11} = 16\pi^7 l_p^9$.

one of the Higgs fields is compact. Taking $R^{11} \to \infty$ we see that we get the theory of coincident M2 branes, in which the SO(8) R-symmetry has an obvious origin.

The metric is

$$ds^2 = f^{-2/3} dx_{||}^2 + f^{1/3}(dr^2 + r^2 d\Omega_7^2) ,$$
$$f = 1 + \frac{2^5 \pi^2 N l_p^6}{r^6} , \tag{3.3}$$

and there is a nonzero flux of the dual of the four-form field strength on the seven-sphere. In the decoupling limit we obtain $AdS_4 \times S^7$, and the supersymmetries work out correctly. The bosonic generators are given by $SO(2,3) \times SO(8)$. In this case the "radii" of the sphere and AdS_4 are $R_{sph} = 2R_{AdS} = l_p(2^5\pi^2 N)^{1/6}$.

The entropy of the near extremal solution agrees with the expectation from dimensional analysis for a conformal theory in 2+1 dimensions [28], but the N dependence or the numerical coefficients are not understood.

Actually for the case of the two brane the conformal symmetry was used to determine the v^4 term in the probe action [37], we are further saying that conformal invariance determines it to all orders in the velocity of the probe. Furthermore the duality we have proposed with M-theory on $AdS_4 \times S^7$ determines the precise numerical coefficient.

When M-theory is involved the dimensionalities of the groups are suggestive of a thirteen dimensional realization [38].

4 Theories with 8 → 16 Supersymmetries, the D1+D5 System

Now we consider IIB string theory compactified on M^4 (where $M^4 = T^4$ or $K3$) to six spacetime dimensions. As a first example let us start with a D-fivebrane with four dimensions wrapping on M^4 giving a string in six dimensions. Consider a system with Q_5 fivebranes and Q_1 D-strings, where the D-string is parallel to the string in six dimensions arising from the fivebrane. This system is described at low energies by a 1+1 dimensional (4,4) superconformal field theory. So we take the limit

$$\alpha' \to 0 , \qquad \frac{r}{\alpha'} = \text{fixed} , \qquad v \equiv \frac{V_4}{(2\pi)^4 \alpha'^2} = \text{fixed} , \qquad g_6 = \frac{g}{\sqrt{v}} = \text{fixed} \tag{4.1}$$

where V_4 is the volume of M^4. All other moduli of M^4 remain fixed. This is just a low energy limit, we keep all dimensionless moduli fixed. As a six dimensional theory, IIB on M^4 contains strings. They transform under the

U-duality group and they carry charges given by a vector q^I. In general we can consider a configuration where $q^2 = \eta_{IJ}q^I q^J \neq 0$ (the metric is the U-duality group invariant), and then take the limit (4.1).

This theory has a branch which we will call the Higgs branch and one which we call the Coulomb branch. On the Higgs branch the 1+1 dimensional vector multiplets have zero expectation value and the Coulomb branch is the other one. Notice that the expectation values of the vector multiplets in the Coulomb branch remain fixed as we take the limit $\alpha' \to 0$.

The Higgs branch is a SCFT with (4,4) supersymmetry. This is the theory considered in [39]. The above limit includes also a piece of the Coulomb branch, since we can separate the branes by a distance such that the mass of stretched strings remains finite.

Now we consider the supergravity solution corresponding to D1+D5 branes [40]

$$\begin{aligned} ds^2 &= f_1^{-1/2} f_5^{-1/2} dx_{\|}^2 + f_1^{1/2} f_5^{1/2} (dr^2 + r^2 d\Omega_3^2) , \\ f_1 &= \left(1 + \frac{g\alpha' Q_1}{v r^2}\right) , \qquad f_5 = \left(1 + \frac{g\alpha' Q_5}{r^2}\right) , \end{aligned} \tag{4.2}$$

where $dx_{\|}^2 = -dt^2 + dx^2$ and x is the coordinate along the D-string. Some of the moduli of M^4 vary over the solution and attain a fixed value at the horizon which depends only on the charges and some others are constant throughout the solution. The three-form RR-field strength is also nonzero.

In the decoupling limit (4.1) we can neglect the 1's in f_i in (4.2) and the metric becomes

$$ds^2 = \alpha' \left[\frac{U^2}{g_6\sqrt{Q_1 Q_5}} dx_{\|}^2 + g_6\sqrt{Q_1 Q_5}\frac{dU^2}{U^2} + g_6\sqrt{Q_1 Q_5} d\Omega_3^2 \right] . \tag{4.3}$$

The compact manifold $M^4(Q)$ that results in the limit is determined as follows. Some of its moduli are at their fixed point value which depends only on the charges and not on the asymptotic value of those moduli at infinity (the notation $M^4(Q)$ indicates the charge dependence of the moduli) [41][10]. The other moduli, that were constant in the black hole solution, have their original values. For example, the volume of M^4 has the fixed point value $v_{fixed} = Q_1/Q_5$, while the six dimensional string coupling g_6 has the original value. Notice that there is an overall factor of α' in (4.3) which can be removed by canceling it with the factor of α' in the Newton constant as explained above. We can trust the supergravity solution if Q_1, Q_5 are large, $g_6 Q_i \gg 1$. Notice that we are talking about a six dimensional supergravity

[10]The fixed values of the moduli are determined by the condition that they minimize the tension of the corresponding string (carrying charges q^I) in six dimensions [41]. This is parallel to the discussion in four dimensions [42].

solution since the volume of M^4 is a constant in Planck units (we keep the Q_1/Q_5 ratio fixed). The metric (4.3) describes three dimensional AdS_3 times a 3-sphere. The supersymmetries work out correctly, starting from the 8 Poincare supersymmetries we enhance then to 16 supersymmetries. The bosonic component is $SO(2,2) \times SO(4)$. In conformal field theory language $SO(2,2)$ is just the $SL(2,R) \times SL(2,R)$ part of the conformal group and SO(4) $\sim SU(2)_L \times SU(2)_R$ are the R-symmetries of the CFT [43].

So the conjecture is that *the 1+1 dimensional CFT describing the Higgs branch of the D1+D5 system on* M^4 *is dual to type IIB string theory on* $(AdS_3 \times S^3)_{Q_1Q_5} \times M^4(Q)$. The subscript indicates that the radius of the three sphere is $R^2_{sph} = \alpha' g_6 \sqrt{Q_1Q_5}$. The compact fourmanifold $M^4(Q)$ is at some particular point in moduli space determined as follows. The various moduli of M^4 are divided as tensors and hypers according to the (4,4) supersymmetry on the brane. Each hypermultiplet contains four moduli and each tensor contains a modulus and an anti-self-dual B-field. (There are five tensors of this type for T^4 and 21 for $K3$). The scalars in the tensors have fixed point values at the horizon of the black hole, and those values are the ones entering in the definition of $M^4(Q)$ (Q indicates the dependence on the charges). The hypers will have the same expectation value everywhere. It is necessary for this conjecture to work that the 1+1 dimensional (4,4) theory is indendent of the tensor moduli appearing in its original definition as a limit of the brane configurations, since $M^4(Q)$ does not depend on those moduli. A non renomalization theorem like [44,45] would explain this. We also need that the Higgs branch decouples from the Coulomb branch as in [46,47].

Finite temperature configurations in the 1+1 conformal field theory can be considered. They correspond to near extremal black holes in AdS_3. The metric is the same as that of the BTZ 2+1 dimensional black hole [48], except that the angle of the BTZ description is not periodic. This angle corresponds to the spatial direction x of the 1+1 dimensional CFT and it becomes periodic if we compactify the theory[11] [49–51] [12]. All calculations done for the 1D+5D system [39, 53, 54] are evidence for this conjecture. In all these cases [54] the nontrivial part of the greybody factors comes from the AdS part of the spacetime. Indeed, it was noticed in [55] that the greybody factors for the BTZ black hole were the same as the ones for the five-dimensional black hole in the dilute gas approximation. The

[11] I thank G. Horowitz for many discussions on this correspondence and for pointing out ref. [49] to me. Some of the remarks the remarks below arose in conversations with him.

[12] The ideas in [49–51] could be used to show the relation between the AdS region and black holes in M-theory on a light like circle. However the statement in [49–51] that the $AdS_3 \times S^3$ spacetime is U-dual to the full black hole solution (which is asymptotic to Minkowski space) should be taken with caution because in those cases the spacetime has identifications on circles that are becoming null. This changes dramatically the physics. For examples of these changes see [32, 52].

dilute gas condition $r_0, r_n \ll r_1 r_5$ [53] is automatically satisfied in the limit (4.1) for finite temperature configurations (and finite chemical potential for the momentum along $\hat{x}$). It was also noticed that the equations have an $SL(2,R) \times SL(2,R)$ symmetry [56], these are the isometries of AdS_3, and part of the conformal symmetry of the 1+1 dimensional field theory. It would be interesting to understand what is the gravitational counterpart of the full conformal symmetry group in 1+1 dimensions.

5 Theories with 4 → 8 Supersymmetries

The theories of this type will be related to black strings in five dimensions and Reissner-Nordström black holes in four dimensions. This part will be more sketchy, since there are several details of the conformal field theories involved which I do not completely understand, most notably the dependence on the various moduli of the compactification.

5.1 Black String in Five Dimensions

One can think about this case as arising from M-theory on M^6 where M^6 is a CY manifold, $K3 \times T^2$ or T^6. We wrap fivebranes on a four-cycle $P_4 = p^A \alpha_A$ in M^6 with nonzero triple self intersection number, see [57]. We are left with a one dimensional object in five spacetime dimensions. Now we take the following limit

$$l_p \to 0 \qquad (2\pi)^6 v \equiv V_6/l_p^6 = \text{fixed} \qquad U^2 \equiv r/l_p^3 = \text{fixed} \ , \tag{5.1}$$

where l_p is the eleven dimensional Planck length. In this limit the theory will reduce to a conformal field theory in two dimensions. It is a (0,4) CFT and it was discussed in some detail in a region of the moduli space in [57]. More generally we should think that the five dimensional theory has some strings characterized by charges p^A, forming a multiplet of the U-duality group and we are taking a configuration where the triple self intersection number p^3 is nonzero (in the case $M^6 = T^6$, $p^3 \equiv D \equiv D_{ABC} p^A p^B p^C$ is the cubic E_6 invariant).

We now take the corresponding limit of the black hole solution. We will just present the near horizon geometry, obtained after taking the limit. Near the horizon all the vector moduli are at their fixed point values [58]. So the near horizon geometry can be calculated by considering the solution with constant moduli. We get

$$ds^2 = l_p^2 \left[\frac{U^2 v^{1/3}}{D^{1/3}} (-dt^2 + dx^2) + \frac{D^{2/3}}{v^{2/3}} \left(4\frac{dU^2}{U^2} + d\Omega_2^2 \right) \right] \ . \tag{5.2}$$

In this limit M^6 has its vector moduli equal to their fixed point values which depend only on the charge while its hyper moduli are what they were at infinity. The overall size of M^6 in Planck units is a hypermultiplet, so it remains constant as we take the limit (5.1). We get a product of three dimensional AdS_3 spacetime with a two-sphere, $AdS_3 \times S^2$. Defining the five dimensional Planck length by $l_{5p}^3 = l_p^3/v$ we find that the "radii" of the two sphere and the AdS_3 are $R_{sph} = R_{AdS}/2 = l_{5p}D^{1/3}$. In this case the superconformal group contains as a bosonic subgroup $SO(2,2) \times SO(3)$. So the R-symmetries are just $SU(2)_R$, associated to the 4 rightmoving supersymmetries.

In this case we conjecture that this (0,4) conformal field theory is dual, for large p^A, to M-theory on $AdS_3 \times S^2 \times M_p^6$. The hypermultiplet moduli of M_p^6 are the same as the ones entering the definition of the (0,4) theory. The vector moduli depend only on the charges and their values are those that the black string has at the horizon. A necessary condition for this conjecture to work is that the (0,4) theory should be independent of the original values of the vector moduli (at least for large p). It is not clear to me whether this is true.

Using this conjecture we would get for large N a compactification of M theory which has five extended dimensions.

5.2 Extremal 3+1 Dimensional Reissner-Nordström

This section is more sketchy and contains an unresolved puzzle, so the reader will not miss much if he skips it.

We start with IIB string theory compactified on M^6, where M^6 is a Calabi-Yau manifold or $K3 \times T^2$ or T^6. We consider a configuration of $D3$ branes that leads to a black hole with nonzero horizon area. Consider the limit

$$\alpha' \to 0 \qquad (2\pi)^6 v \equiv \frac{V_6}{\alpha'^3} = \text{fixed} \qquad U \equiv \frac{r}{\alpha'} = \text{fixed}\,. \tag{5.3}$$

The string coupling is arbitrary. In this limit the system reduces to quantum mechanics on the moduli space of the three-brane configuration.

Taking the limit (5.3) of the supergravity solution we find

$$ds^2 = \alpha' \left[\frac{U^2}{g_4^2 N^2} dt^2 + g_4^2 N^2 \frac{dU^2}{U^2} + g_4^2 N^2 d\Omega_2^2 \right] \tag{5.4}$$

where N is proportional to the number of D3 branes. We find a two dimensional AdS_2 space times a two-sphere, both with the same radius $R = l_{4p} N$, where $l_{4p}^2 = g^2\alpha'/v$. The bosonic symmetries of $AdS_2 \times S^2$ are $SO(2,1) \times SO(3)$. This superconformal symmetry seems related to the symmetries of the *chiral* conformal field theory that was proposed in [59] to describe the Reissner-Nordström black holes. Here we find a puzzle, since in the limit (5.3) we got a quantum mechanical system and not a 1+1 dimensional conformal field theory. In the limit (5.3) the energy gap (mentioned in [59, 60]) becomes very large[13]. So it looks like taking a large N limit at the same time will be crucial in this case. These problems might be related to the large ground state entropy of the system.

If this is understood it might lead to a proposal for a non perturbative definition of M/string theory (as a large N limit) when there are four non-compact dimensions.

It is interesting to consider the motion of probes on the AdS_2 background. This corresponds to going into the "Coulomb" branch of the quantum mechanics. Dimensional analysis says that the action has the form (2.7) with $p = 0$. Expanding f to first order we find $S \sim \int dt \frac{\dot{U}^2}{U^3} \sim \int dt v^2/r^3$, which is the dependence on r that we expect from supergravity when we are close to the horizon. A similar analysis for Reissner-Nordström black holes in five dimensions would give a term proportional to $1/r^4$ [17]. It will be interesting to check the coefficient (note that this is the *only* term allowed by the symmetries, as opposed to [17]).

6 Discussion, Relation to Matrix Theory

By deriving various field theories from string theory and considering their large N limit we have shown that they contain in their Hilbert space excitations describing supergravity on various spacetimes. We further conjectured that the field theories are dual to the full quantum M/string theory on various spacetimes. In principle, we can use this duality to give a definition of M/string theory on flat spacetime as (a region of) the large N limit of the field theories. Notice that this is a non-perturbative proposal for defining such theories, since the corresponding field theories can, *in principle*, be defined non-perturbatively. We are only scratching the surface and there are many things to be worked out. In [61] it has been proposed that the large N limit of D0 brane quantum mechanics would describe eleven dimensional M-theory. The large N limits discussed above, also provide a definition of M-theory. An obvious difference with the matrix model of [61] is that here N is not interpreted as the momentum along a compact direction. In our case, N is related to the curvature and the size of the space where the theory is defined. In both cases, in the large N limit we expect to get flat, non-compact spaces. The matrix model [61] gives us a prescription to build asymptotic states, we have not shown here how to construct graviton states,

[13] I thank A. Strominger for pointing this out to me.

and this is a very interesting problem. On the other hand, with the present proposal it is more clear that we recover supergravity in the large N limit.

This approach leads to proposals involving five (and maybe in some future four) non-compact dimensions. The five dimensional proposal involves considering the 1+1 dimensional field theory associated to a black string in five dimensions. These theories need to be studied in much more detail than we have done here.

It seems that this correspondence between the large N limit of field theories and supergravity can be extended to non-conformal field theories. An example was considered in [1], where the theory of NS fivebranes was studied in the $g \to 0$ limit. A natural interpretation for the throat region is that it is a region in the Hilbert space of a six dimensional "string" theory[14]. And the fact that contains gravity in the large N limit is just a common feature of the large N limit of various field theories. The large N master field seems to be the anti-deSitter supergravity solutions [17].

When we study non extremal black holes in *AdS* spacetimes we are no longer restricted to low energies, as we were in the discussion in higher dimensions [44, 54]. The restriction came from matching the *AdS* region to the Minkowski region. So the five dimensional results [53, 54] can be used to describe arbitrary non-extremal black holes in three dimensional Anti-deSitter spacetimes. This might lead us to understand better where the degrees of freedom of black holes really are, as well as the meaning of the region behind the horizon. The question of the boundary conditions is very interesting and the conformal field theories should provide us with some definite boundary conditions and will probably explain us how to interpret physically spacetimes with horizons. It would be interesting to find the connection with the description of 2+1 dimensional black holes proposed by Carlip [63].

In [8, 13] super-singleton representations of *AdS* were studied and it was proposed that they would describe the dynamics of a brane "at the end of the world". It was also found that in maximally supersymmetric cases it reduces to a free field [8]. It is tempting therefore to identify the singleton with the center of mass degree of freedom of the branes [6, 13]. A recent paper suggested that super-singletons would describe all the dynamics of *AdS* [51]. The claim of the present paper is that all the dynamics of *AdS* reduces to previously known conformal field theories.

It seems natural to study conformal field theories in Euclidean space and relate them to deSitter spacetimes.

Also it would be nice if these results could be extended to four-dimensional gauge theories with less supersymmetry.

[14] This possibility was also raised by [62], though it is a bit disturbing to find a constant energy flux to the UV (that is how we are interpreting the radial dimension).

Acknowledgments

I thank specially G. Horowitz and A. Strominger for many discussions. I also thank R. Gopakumar, R. Kallosh, A. Polyakov, C. Vafa and E. Witten for discussions at various stages in this project. My apologies to everybody I did not cite in the previous version of this paper. I thank the authors of [64] for pointing out a sign error.

This work was supported in part by DOE grant DE-FG02-96ER40559.

Appendix

$D = p + 2$-dimensional anti-deSitter spacetimes can be obtained by taking the hyperboloid

$$-X_{-1}^2 - X_0^2 + X_1^2 + \cdots + X_p^2 + X_{p+1}^2 = -R^2 , \qquad \text{(A.1)}$$

embedded in a flat D+1 dimensional spacetime with the metric $\eta = \text{Diag}(-1, -1, 1, \cdots, 1)$. We will call R the "radius" of AdS spacetime. The symmetry group $SO(2, D-1) = SO(2, p+1)$ is obvious in this description. In order to make contact with the previously presented form of the metric let us define the coordinates

$$\begin{aligned} U &= (X_{-1} + X_{p+1}) \\ x_\alpha &= \frac{X_\alpha R}{U} \qquad \alpha = 0, 1, \cdots, p \\ V &= (X_{-1} - X_{p+1}) = \frac{x^2 U}{R^2} + \frac{R^2}{U} . \end{aligned} \qquad \text{(A.2)}$$

The induced metric on the hyperboloid (A.1) becomes

$$ds^2 = \frac{U^2}{R^2} dx^2 + R^2 \frac{dU^2}{U^2} . \qquad \text{(A.3)}$$

This is the form of the metric used in the text. We could also define $\tilde{U} = U/R^2$ so that metric (A.3) has an overall factor of R^2, making it clear that R is the overall scale of the metric. The region outside the horizon corresponds to $U > 0$, which is only a part of (A.1). It would be interesting to understand what the other regions in the AdS spacetime correspond to. For further discussion see [65].

References

[1] J. Maldacena and A. Strominger, hep-th/9710014.

[2] N. Seiberg, *Phys. Lett.* **B408** (1997) 98.

[3] R. Haag, J. Lopuszanski, and M. Sohnius, *Nucl. Phys.* **B88** (1975) 257.

[4] W. Nahm, *Nucl. Phys.* **B135** (1978) 149.

[5] G. Gibbons, *Nucl. Phys.* **B207** (1982) 337; R. Kallosh and A. Peet, *Phys. Rev.* **D46** (1992) 5223; S. Ferrara, G. Gibbons, R. Kallosh, *Nucl. Phys.* **B500** (1997) 75.

[6] G. Gibbons and P. Townsend, *Phys. Rev. Lett.* **71** (1993) 5223.

[7] A. Salam and E. Sezgin, *"Supergravities in Diverse Dimensions"*, Vol. **1** and **2**, North-Holland 1989. Look for "gauged" supergravities.

[8] C. Frondal, *Phys. Rev.* **D26** (1988) 82; D. Freedman and H. Nicolai, *Nucl. Phys.* **B237** (1984) 342; K. Pilch, P. van Nieuwenhuizen and P. Townsend, *Nucl. Phys.* **B242** (1984) 377; M. Günaydin, P. van Nieuwenhuizen and N. Warner, *Nucl. Phys.* **B255** (1985) 63; M. Günaydin and N. Warner, *Nucl. Phys.* **B272** (1986) 99; M. Günaydin, N. Nilsson, G. Sierra and P. Townsend, *Phys. Lett.* **B176** (1986) 45; E. Bergshoeff, A. Salam, E. Sezgin and Y. Tanii, *Phys. Lett.* **B205** (1988) 237; *Nucl. Phys.* **D305** (1988) 496; E. Bergshoeff, M. Duff, C. Pope and E. Sezgin, *Phys. Lett.* **B224** (1989) 71.

[9] M. Günaydin and N. Marcus, *Class. Quant. Grav.* **2** (1985) L11; *Class. Quant. Grav.* **2** (1985) L19; H. Kim, L. Romans and P. van Nieuwenhuizen, *Phys. Lett.* **B143** (1984) 103; M. Günaydin, L. Romans and N. Warner, *Phys. Lett.* **B154** (1985) 268; *Phys. Lett.* **B164** (1985) 309; *Nucl. Phys.* **B272** (1986) 598.

[10] M. Blencowe and M. Duff, *Phys. Lett.* **B203** (1988) 229; *Nucl. Phys.* **B310** (1988) 387.

[11] H. Nicolai, E. Sezgin and Y. Tanii, *Nucl. Phys.* **B305** (1988) 483.

[12] M. Duff, G. Gibbons and P. Townsend, hep-th/9405124.

[13] P. Claus, R. Kallosh and A. van Proeyen, hep-th/9711161.

[14] G. Horowitz and A. Strominger, *Nucl. Phys.* **B360** (1991) 197.

[15] G. Gibbons, G. Horowitz and P. Townsend, hep-th/9410073.

[16] L. Susskind, hep-th/9611164; O. Ganor, S. Ramgoolam and W. Taylor IV, *Nucl. Phys.* **B492** (1997) 191.

[17] M. Douglas, J. Polchinski and A. Strominger, hep-th/9703031.

[18] M. Aganagic, C. Popescu and J. Schwarz, *Nucl. Phys.* **B495** (1997) 99, hep-th/9612080.

[19] M. Dine and N. Seiberg, *Phys. Lett.* **B409** (1997) 239.

[20] K. Becker, M. Becker, J. Polchinski and A. Tseytlin, *Phys. Rev.* **D56** (1997) 3174.

[21] A. Polyakov, hep-th/9711002.

[22] T. Banks, W. Fishler, I. Klebanov and L. Susskind, hep-th/9709091.

[23] S. Gubser, I. Klebanov and A. Peet, *Phys. Rev.* **D54** (1996) 3915; A. Strominger, unpublised.

[24] I. Klebanov, *Nucl. Phys.* **B496** (1997) 231.

[25] S. Gubser, I. Klebanov and A. Tseytlin, *Nucl. Phys.* **B499** (1997) 217.

[26] S. Gubser and I. Klebanov, hep-th/9708005.

[27] C. Vafa, *Nucl. Phys.* **B469** (1996) 403.

[28] I. Klebanov and A. Tseytlin, *Nucl. Phys.* **B475** (1996) 164.

[29] A. Srominger, *Phys. Lett.* **B383** (1996) 44.

[30] E. Witten, hep-th/9507121; N. Seiberg and E. Witten, *Nucl. Phys.* **B471** (1996) 121.

[31] J. Maldacena, *Proceedings of Strings'97*, hep-th/9709099.

[32] N. Seiberg, *Phys. Rev. Lett.* **79** (1997) 3577.

[33] A. Sen, hep-th/9709220.

[34] N. Seiberg, hep-th/9705117.

[35] S. Sethi and L. Susskind, *Phys. Lett.* **B400** (1997) 265.

[36] T. Banks and N. Seiberg, *Nucl. Phys.* **B497** (1997) 41.

[37] T. Banks, W. Fishler, N. Seiberg and L. Susskind, *Phys. Lett.* **B408** (1997) 111.

[38] I. Bars, *Phys. Rev.* **D55** (1997) 2373.

[39] A. Strominger and C. Vafa, *Phys. Lett.* **B379** (1996) 99.

[40] G. Horowitz, J. Maldacena and A. Strominger, *Phys. Lett.* **B383** (1996) 151.

[41] L. Andrianopoli, R. D'Auria and S. Ferrara, *Int. J. Mod. Phys* **A12** (1997) 3759.

[42] S. Ferrara, R. Kallosh and A. Strominger, *Phys. Rev.* **D52** (1995) 5412; S. Ferrara and R. Kallosh, *Phys. Rev.* **D54** (1996) 1514; *Phys. Rev.* **D54** (1996) 1525.

[43] J. C. Breckenridge, R. C. Myers, A. W. Peet and C. Vafa, *Phys. Lett.* **B391** (1997) 93.

[44] J. Maldacena, *Phys. Rev.* **D55** (1997) 7645.

[45] D. Diaconescu and N. Seiberg, *J. High Energy Phys.* **7** (1997) 001.

[46] O. Aharony, M. Berkooz, S. Kachru, N. Seiberg and E. Silverstein, hep-th/9707079.

[47] E. Witten, hep-th/9707093.

[48] Bañados, Teitelboim and Zanelli, *Phys. Rev. Lett.* **69** (1992) 1849.

[49] S. Hyun, hep-th/9704005.

[50] H. Boonstra, B. Peeters and K. Skenderis, hep-th/9706192

[51] K. Sfetsos and K. Skenderis, hep-th/9711138.

[52] S. Hellerman and J. Polchinski, hep-th/9711037.

[53] G. Horowitz and A. Strominger, *Phys. Rev. Lett.* **77** (1996) 2368, hep-th/9602051.

[54] A. Dhar, G. Mandal and S. Wadia *Phys. Lett.* **B388** (1996) 51; S. Das and S. Mathur, *Nucl. Phys.* **B478** (1996) 561; *Nucl. Phys.* **B482** (1996) 153; J. Maldacena and A. Strominger, *Phys. Rev.* **D55** (1997) 861; S. Gubser, I. Klebanov, *Nucl. Phys.* **B482** (1996) 173; C. Callan, Jr., S. Gubser, I. Klebanov and A. Tseytlin, *Nucl. Phys.* **B489** (1997) 65; I. Klebanov and M. Krasnitz, *Phys. Rev.* **D55** (1997) 3250; I. Klebanov, A. Rajaraman and A. Tseytlin, *Nucl. Phys.* **B503** (1997) 157; S. Gubser, hep-th/9706100; H. Lee, Y. Myung and J. Kim, hep-th/9708099; K. Hosomichi, hep-th/9711072.

[55] D. Birmingham, I. Sachs and S. Sen, hep-th/9707188.

[56] M. Cvetic and F. Larsen, *Phys. Rev.* **D56** (1997) 4994; hep-th/9706071.

[57] J. Maldacena, A. Strominger and E. Witten, hep-th/9711053.

[58] A. Chamseddine, S. Ferrara, G. Gibbons and R Kallosh, *Phys. Rev.* **D55** (1997) 3647.

[59] J. Maldacena and A. Strominger, *Phys. Rev.* **D56** (1997) 4975.

[60] J. Maldacena and L. Susskind, *Nucl. Phys.* **B475** (1996) 679.

[61] T. Banks, W. Fischler, S. Shenker and L. Susskind, hep-th/9610043.

[62] O. Aharony, S. Kachru, N. Seiberg and E. Silverstein, private communication.

[63] S. Carlip, *Phys. Rev.* **D51** (1995) 632.

[64] R. Kallosh, J. Kumar and A. Rajaraman, hep-th/9712073.

[65] S. Hawking and J. Ellis, *"The large scale structure of spacetime"*, Cambrige Univ. Press 1973, and references therein.